Sources and Studies
in the History of Mathematics
and Physical Sciences

**Sources and Studies in the
History of Mathematics and Physical Sciences**

K. Andersen
Brook Taylor's Work on Linear Perspective

K. Andersen
The Geometry of An Art

H.J.M. Bos
Redefining Geometrical Exactness: Descartes' Transformation of the Early Modern Concept of Construction

J. Cannon/S. Dostrowsky
The Evolution of Dynamics: Vibration Theory From 1687 to 1742

B. Chandler/W. Magnus
The History of Combinatorial Group Theory

A.I. Dale
History of Inverse Probability: From Thomas Bayes to Karl Pearson, Second Edition

A.I. Dale
Pierre-Simon de Laplace, Philosophical Essay on Probabilities, Translated from the fifth French edition of 1825, with Notes by the Translator

A.I. Dale
Most Honourable Remembrance: The Life and Work of Thomas Bayes

P.J. Federico
Descartes On Polyhedra: A Study of the *De Solidorum Elementa*

B.R. Goldstein
The Astronomy of Levi Ben Gerson (1288–1344)

H.H. Goldstine
A History of Numerical Analysis from the 16th Through the 19th Century

H.H. Goldstine
A History of the Calculus of Variations From the 17th Through the 19th Century

G. Graßhoff
The History of Ptolemy's Star Catalogue

A. Hermann/K. von Meyenn/V.F. Weisskopf (Eds.)
Wolfgang Pauli: Scientific Correspondence I: 1919–1929

A. Hermann/K. von Meyenn/V.F. Weisskopf (Eds.)
Wolfgang Pauli: Scientific Correspondence II: 1930–1939

C.C. Heyde/E. Seneta, I.J.
Bienaymé: Statistical Theory Anticipated

J.P. Hogendijk
Ibn Al-Haytham's *Completion of the Conics*

J. Høyrup
Length, Widths, Surfaces: A Portrait of Old Babylonian Algebra and Its Kin

A. Jones
Pappus of Alexandria, Book 7 of the *Collection*

Continued after References

Jöran Friberg

A Remarkable Collection of Babylonian Mathematical Texts

Manuscripts in the Schøyen Collection
Cuneiform Texts I

Jöran Friberg
Department of Mathematical Sciences
Chalmers University of Technology
Gothenburg, Sweden

Sources and Series Editor:
Jesper Lützen
Institute for Mathematical Sciences
University of Copenhagen
DK-2100 Copenhagen
Denmark

Mathematics Subject Classification (2000): 01A05 01A17

Library of Congress Control Number: 2006932408

ISBN-13: 978-0-387-34543-7   e-ISBN-13: 978-0-387-48977-3

Printed on acid-free paper.

9 8 7 6 5 4 3 2 1

springer.com

Acknowledgements

In an unparalleled way, Martin Schøyen has made the numerous items of various parts of his great collection of ancient manuscripts available to scholars for evaluation and publication. As for the sub-collection edited in this volume, I am full of admiration for the way in which Martin Schøyen has managed to bring together from the antiquities market, in the 1980's-1990's, clay tablets representing nearly all aspects of the whole corpus of mathematical cuneiform texts. Without his untiring efforts, for which I am very grateful, these exceedingly interesting documents would have been scattered in small collections all over the world, inaccessible to scholars and unknown to the general public.

In an ideal world, unprovenanced texts coming from the antiquities market would not exist. In the real world, where such texts do exist, some people excitedly claim that serious scholars should have nothing to do with them, for various contrived reasons. However, one must keep in mind that great parts of most collections of non-European cultural heritage, even those of the greatest museums in Europe and the United States, are unprovenanced texts from the antiquities market. Thus, for instance, the classical works about mathematical cuneiform texts, *Mathematische Keilschrift-Texte* by Neugebauer, 1935-1937, and *Mathematical Cuneiform Texts* by Neugebauer and Sachs, 1945, would never have appeared if their authors had refused to work with texts without known provenance. Now, as then, the task of a serious scholar must be to attempt to make all kinds of texts publicly known and understood, and to try to trace the origin of the texts if it is not known.

I want to express my profound gratitude to my friend Professor Farouk N. Al-Rawi, who has assisted me in various ways, in particular by helping me to read particularly difficult mathematical cuneiform texts in the Schøyen Collection, and by making beautiful hand copies for me of several of them. Without his help, the work with this volume could never have been completed. (However, since I have not always followed the advice he gave me, I alone am responsible for all mistakes in transliterations and translations of the texts.)

Other friends and colleagues have assisted me in further ways in my work, for which I want to thank them here. Professor Jens Braarvig, in particular, has been very supportive and encouraging and has shown me how to make excellent digital photos of all the texts discussed in this volume. Also Professor Robert K. Englund has helped and encouraged me in several ways. So has Elizabeth G. Sørensen, editor of the catalog of the Schøyen collection. Dr. Renee Kovacs has helped me to read some of the colophons, and Professor Peter Damerow has kindly supplied me with a photo of tokens from a pre-literate spherical envelope in the Schøyen Collection. Finally, with extreme generosity Professor David Owen has allowed me to publish in this volume, in cooperation with Dr. Rudi Mayr, two extraordinary Early Dynastic/Old Akkadian metro-mathematical texts from the collections of the Department of Near Eastern Studies at Cornell University.

Last, but not least, I want to thank my wife, Dr. Ingegerd Friberg, for her constant support and encouragement and for her help with all kinds of details regarding my manuscript, but above all for having put up with my nearly total absorption in my work for the past three years.

This work has been supported by Stiftelsen Längmanska Kulturfonden and by The Royal Society of Arts and Sciences in Gothenburg.

Hässleholm, Sweden, June 13, 2004   Jöran Friberg
(final revision December 31, 2006)

Introduction

The sub-collection of mathematical cuneiform texts in the Schøyen Collection makes a substantial addition to the known corpus of such texts. It contains 121 texts, not counting 151 multiplication tables and 53 small weight stones. According to the catalog at the end of the Index of Subjects below, where those 121 mathematical texts are ordered by content, nearly all known kinds, and some new kinds, of mathematical cuneiform texts are represented in the collection. Therefore it has been possible to organize the present work as a broad general account of Mesopotamian mathematics, illustrated mainly by texts from the Schøyen Collection, but occasionally also by previously published texts. The general disposition of the book is borrowed from my own concise but comprehensive survey of Mesopotamian mathematics in the article on "Mathematics" in *Reallexikon der Assyriologie*, vol. 7 (1990).

My ambition has been to make the account easily accessible to all kinds of readers, yet still as detailed and exhaustive as possible. For that purpose, there is, for instance, an introductory **Chapter 0** on "how to get a better understanding of mathematical cuneiform texts". The chapter begins with a discussion of the danger of unintentional anachronisms in translations of pre-Greek mathematical texts, and continues with a presentation of the kind of "conform" transliterations, translations, and interpretations, true to the original, that will be used throughout the book in discussions of individual texts. There is also a rather detailed discussion of sexagesimal numbers in Babylonian "relative" place-value notation, with explicit examples showing how to count with such numbers.

Chapter 1 contains a discussion of several small square clay tablets inscribed with large cuneiform number signs, apparently some beginners' first exposure to mathematics in the form of multiplication exercises, squaring exercises, and even a division exercise. More advanced beginners' exercises are devoted to operations with "many-place regular sexagesimal numbers", in particular to simple applications of the powerful "trailing part algorithm", an ingenious factorization method for the computation of reciprocals or square roots of regular sexagesimal numbers.

Chapter 2 starts with examples of Old Babylonian standard tables of squares, square sides (square roots), cube sides (cube roots), and reciprocals, as well as two examples of non-standard tables of "quasi-cube sides". A paragraph about multiplication tables in the Schøyen collection mentions, among other things, a single multiplication table with a new head number, smaller than any previously known head number, and a combined multiplication table with a sub-table for a previously unknown head number. The chapter ends with the first complete explanation of the whole set of documented head numbers for Old Babylonian multiplication tables.

Chapter 3 is devoted to a discussion of Old Babylonian metrological tables of standard type for capacity numbers (system *C*), weight numbers (system *M*), area numbers (system *A*), and (length numbers (systems *Ln* and *Lc*). **Chapter 4** contains a related discussion of "weight stones", in particular a Kassite talent weight with a long inscription in Sumerian. In **Chapter 5,** there is another related discussion of area numbers in two "field plan texts" from the Neo-Sumerian Ur III period.

In **Chapter 6** is presented the oldest mathematical text in the Schøyen collection. It is a clay tablet from the Early Dynastic III period (c. 2600-2350 BC) with a curious table of areas of large rectangles on the obverse and an even more curious geometric progression of area numbers on the reverse.

Some Old Babylonian texts with interesting "practical" and almost exclusively numerical mathematical exercises are discussed in **Chapter 7**. The first example is a quite small clay tablet with a cleverly devised division exercise, closely related to a known text from Ur where the division problem is expressed in terms of shepherds and their herds of sheep or goats. Next comes a paragraph with several examples of "combined market rate exercises", on round or square hand tablets, with solutions to problems where several commodities at given different market rates (inverted prices) are bought in equal quantities for a given amount of silver. A particularly important text is a small square clay tablet, inscribed on the obverse with a series of calculations of "carrying numbers" for bricks or mud, computed as the products of a fixed "walking number" and diverse "loading numbers". The carrying numbers are a kind of work norms for men transporting full loads of bricks or mud over varying distances. On the reverse of the same small clay tablet are inscribed the numerical parameters for a related "combined work norm exercise". The chapter ends with a discussion of two clay tablets with inheritance problems, one of them a quite massive round hand tablet with an interesting computation (with round-off) of a geometric progression and its sum.

Chapter 8 is devoted to the subject of hand tablets with geometric exercises. It contains, in particular, a paragraph about texts with "figures within figures", unfortunately only in the form of hand tablets with drawings and a few related numbers. The first of these texts shows how the space between two "concentric" (and parallel) equilateral triangles can be divided into a chain of three trapezoids. Four other hand tablets with drawings show an equilateral triangle inscribed in a circle, a square with diagonals inscribed in a circle, and a circle in the middle of a square or a regular hexagon.

There is also in this chapter a paragraph about Old Babylonian labyrinths and mazes. Two clay tablets are inscribed with careful drawings of a pair of complicated labyrinths of completely new types, one square, the other rectangular. It is shown how the two labyrinths can be constructed step by step by use of certain intricate algorithms, and these algorithms are compared with the well known algorithm for the construction of the familiar "Greek" labyrinth.

Two clay tablets with the beginning and the end of the famous Sumerian King List are the topic of **Chapter 9**. One contains a list of the reigns of the kings of the Ur III and Isin dynasties, the other an "antediluvian" king list with the reigns of eight kings in five Mesopotamian cities that allegedly held the kingship before the flood. Although not directly mathematical, these two texts are included here because they contain interesting numbers, such as the numbers given for the reigns of the antediluvian kings, numbers so large that it is clear that they are non-historical and unrealistic.

Three Old Babylonian quite extensive "problem texts" are discussed in **Chapter 10**. The vocabulary used in the three texts show that they are all from Uruk, a southern site in Mesopotamia, and therefore not later than Samsuiluna 11 (1739 BC), the year when the southern cities were abandoned. The first of the three texts is a mathematical "recombination text" with mixed exercises (a collection of excerpts from various earlier large theme texts). One of the exercises is a parallel to a previously known text from a northern, peripheral site, with a "metric algebra" problem for a rectangle with a given diagonal and a given area. There is also a series of five "igi-igi.bi problems" of a new type, closely associated with the famous table text Plimpton 322. Another, somewhat larger, mathematical recombination text begins with a series of four problems, also of a new type, concerned with walls in which a breach has to be repaired, or through which a hole has been drilled. The problems lead to systems of equations, apparently solved by use of metric algebra in a quite sophisticated way. Then follows another example of an igi-igi.bi problem, and two more, badly preserved problems. The text ends with a summary of the topics in the text. There is no summary of a similar kind in any previously known Old Babylonian mathematical problem text. A third Old Babylonian problem text contains two exercises with vaguely stated, but intricately constructed problems of a new type for a ramp built by four teams of "soldiers".

Three more problem texts are discussed in **Chapter 11**. One of them is only a small fragment of a mathematical recombination text, luckily with another well preserved summary of the same type as the one mentioned above. One of the preserved problems in this text is a surprising application of the Old Babylonian "diagonal rule" in three dimensions, with the computation of the "inner diagonal" of a gate. The solution to the problem is a multiple of the "diagonal quartet" 13, 12, 4, 3. The second problem text in Chapter 11 is a substantial fragment of a large mathematical recombination text. On the obverse of this text there is a series of metric algebra problems for the sides of one or several squares, while on the reverse there is a series of metric algebra problems for the sides of a rectangle. The third text, finally, contains three closely related problems for 20 equilateral triangles, where the number 20 is mysteriously explained as $(6 - 1) \cdot 4$. The individual equilateral triangles are referred to as "gaming-piece figures", and some object constructed by use of the 20 equilateral triangles is called a "horn figure". Although there is no drawing showing what a horn figure looks like, the only conceivable interpretation seems to be that the horn figure is an icosahedron, a regular polyhedron bounded by 20 equilateral triangles. This text is inscribed on a clay tablet of a very unusual format. The only other known mathematical cuneiform text on a clay tablet of a similar format is also the only previously known Kassite (and therefore post-Old-Babylonian) mathematical cuneiform text. The shared unusual format as well as other details seem to suggest that also the text with the 20 equilateral triangles is a Late Kassite mathematical text, hence from the 14th to 13th century BC.

It has been fascinating in many ways to write the present book. In particular, it has been interesting to be able to follow the Old Babylonian scribe school students' progress through the successive stages of their mathematical education, as revealed for instance by the size of the cuneiform signs they write, from the overly large number signs used in elementary arithmetical exercises, through the signs of normal size used in various arithmetical and metrological table texts, all the way to the quite minute signs used in some of the problem texts. This variation in size of the signs, by the way, is why in this book all cuneiform texts are reproduced in 2/3 of their actual size, with the exception of most of the problem texts and some other important texts, which are reproduced in the scale 1 : 1.

The detailed discussion in Chapters 1-11 of the mathematical cuneiform texts in the Schøyen Collection is rounded off with a series of ten appendices in which related matters are discussed or included for easy reference. **Appendix 1**, for instance, contains a discussion of the use of "subtractive numbers" in mathematical cuneiform texts. There, the use of special notations for '19' $(20 - 1)$ in Old Babylonian multiplication tables is compared with the use of subtractive numbers in a Neo-Sumerian table of reciprocals and in an Early Dynastic table of areas of small squares. **Appendix 2** consists entirely of a reconstruction of the complete Old Babylonian combined multiplication table.

In **Appendix 3** are shown two examples of an Old Babylonian combined arithmetical algorithm text, where the "trailing part algorithm" for the computation of reciprocals of regular sexagesimal numbers is combined in an ingenious way with the "doubling-and-halving algorithm".

Appendix 4 contains a survey of cuneiform systems of notations for numbers and measures. The corresponding "factor diagrams" are presented for the familiar Old Babylonian systems of numbers or measures, as well as for various systems of numbers and measures appearing in cuneiform texts from the third millennium BC, or in proto-literate texts from the end of the fourth millennium. There is even a discussion of tentative factor diagrams for some pre-literate systems of numbers and measures, hypothetically documented by sets of tokens enclosed in so called "spherical envelopes", for instance at least one in the Schøyen Collection.

Appendix 5 contains a detailed survey of the contents of Old Babylonian metrological tables for the main systems $C, M, A, Ln,$ and $Lc,$ and ends with a discussion of a family of apparently related subscripts mentioning the gods Nisaba and Haia and appearing as subscripts on several important table texts inscribed on large clay tablets, cylinders, and hexagonal prisms.

The discussion in **Appendix 6** of a number of mathematical cuneiform texts from the third millennium BC forms an interesting background to the discussion of Old Babylonian mathematical texts in the rest of the book. It is particularly instructive to see how mathematical problems of various kinds could be solved, and how table texts could be constructed, before the invention of sexagesimal numbers in place value notation. (Contrary to what has been claimed by some scholars, there is no trace of counting with place value numbers in mathematical texts before the Neo-Sumerian Ur III period.)

The discussion in Appendix 6 is continued in **Appendix 7** with a discussion of the large Early Dynastic combined metro-mathematical table text CUNES 50-08-001 with its separate sub-tables for the areas of squares with sides measured in multiples of the ninda or of its various fractions. This text is remarkable not least because it clearly demonstrates that there may have been an Early Dynastic origin (dating to the first half of the third millennium BC) for several of the features that are characteristic for Old Babylonian mathematics, nearly a millennium later. In Appendix 7 is also discussed another remarkable text, the lexical or metrological text CUNES 47-12-176, a decreasing list of Early Dynastic/Old Akkadian weight measures.

Appendix 8 contains an updated discussion of the famous table text Plimpton 322, based on my own discussion of the text in *HM* 8 (1981) and on the new evidence provided by the igi-igi.bi problems in two of the texts discussed in Chapter 10. It is shown how the data for Plimpton 322 must have been constructed with departure from a series of cleverly and systematically constructed pairs of reciprocal sexagesimal numbers, called igi and igi.bi. It is also revealed what the captions for the columns of the table may really mean, and that the trailing part algorithm is behind the computations of the data for the table.

In **Appendix 9** is presented for the first time an interpretation of a fragment of a Late Babylonian explicit multiplication algorithm (actually the only one of its kind known) for the computation of the square of the 13-place regular sexagesimal number $3^{46} = 4\ 04\ 17\ 40\ 45\ 13\ 17\ 45\ 52\ 14\ 42\ 12\ 09$.

Appendix 10, finally, contains a series of color photos of texts discussed in the present volume.

The addition of the many mathematical cuneiform texts from the Schøyen Collection to the previously known corpus of such texts has enriched and consolidated the corpus in several ways. Perhaps the most important lesson to be drawn from this enlargement of the corpus is the following: It is still far from clear what the true extent and depth of Mesopotamian mathematics really was. Indeed, every new mathematical cuneiform problem text tends to contain a surprise. Spectacular examples from the present volume are the discoveries that Old Babylonian mathematicians were familiar with the icosahedron and with a three-dimensional version of the diagonal equation (the "Pythagorean equation"), and that the purpose of the columns of numbers on Plimpton 322 really, as suspected, was to serve as a systematic source of data for a series of igi-igi.bi problems.

Note: In many places in this book, the notation (?) is inserted with the intention of pointing out that a proposed numerical value or specific statement is *probably but not certainly correct*. The use of such a notation is regrettably unavoidable, in view of the fact much of the source material is either obscure, or damaged, or both.

Statement of Provenance of Near Eastern Pictographic and Cuneiform Tablets in the Schøyen Collection

The large holdings of pictographic and cuneiform tablets in the Schøyen Collection derive from a great variety of collections and sources. Collected in the late 1980's and 1990's it would not have been possible to collect such a great number of tablets and of such major textual importance, if the undertaking had not been based on the endeavour of some of the greatest collectors in earlier times. These collections are:

1. Institute of Antiquity and Christianity, Claremont Graduate School. Claremont, California (1970-1994)
2. Erlenmeyer Collection and Foundation, Basel (ca. 1935-1988)
3. Cumberland Clark Collection, Bournemouth, UK (1920'ies-1941)
4. Lord Amherst of Hackney, UK (1894-1909)
5. Crouse Collection, Hong Kong and New England (1920'ies-1980'ies)
6. Dring Collection, Surrey, UK (1911-1990)
7. Lindgren Collection, San Francisco, California (1965-1985)
8. Rosenthal Collection, San Francisco, California (1953-1988)
9. Kevorkian Collection, New York (ca. 1930-1959) and Fund (1960-1977)
10. Kohanim Collection, Tehran, Paris and London (1959-1985)
11. Simmonds Collection, UK (1944-1987)
12. Schaeffer Collection, College de France, Zürich (1950'ies)
13. Henderson Collection, Boston, Massachusetts (1930'ies-1950'ies)
14. Pottesman Collection, London (1904-1978)
15. Geuthner Collection, France (1960'ies–1980'ies)
16. Harding Smith Collection, UK (1893-1922)

These 16 collections are the source of almost all the tablets. Some tablets were acquired through Christie's and Sotheby's, where the names of the former owners in some cases were not revealed.

Some of the older of these collections are also the source of some of the later collections. A large number of the tablets in the Crouse collection f. inst., came from Cumberland Clark, Kohanim, Amherst, and Simmonds collections, as well as others. The Claremont tablets came from the Schaeffer collection, and the Dring tablets came from the Harding Smith collection. The sources of the oldest collections like Amherst, Harding Smith and Cumberland Clark were antiquity sellers who acquired the materials in the Near East in the 1890's – 1930's. F. inst. in the summers of 1893 and 1894 some 30 000 tablets from Tello (Lagash) came on the market. Most of these were bought by museums, but several hundreds were acquired by Amherst and other collectors.

Besides Lagash, the original archaeological context of the materials in The Schøyen Collection are libraries and archives of numerous temples, palaces, schools, houses, and administrative centres in Sumer, Elam, Babylonia, Assyria and various city states in present Syria, Turkey, Iraq, and Iran. Details of this context will not be known until all texts have been read, published and compared with other published museum collections.

Oslo, December 14, 2004   Martin Schøyen

MANUSCRIPTS IN THE SCHØYEN COLLECTION
CUNEIFORM TEXTS

Vol. 1. J. Friberg, *A Remarkable Collection of Babylonian Mathematical Texts*. Sources and Studies in the History of Mathematics and Physical Sciences. Springer: New York (2007).

Vol. 2. B. Alster, *Sumerian Proverbs in the Schøyen Collection*. Occasional Publications of the Department of Near Eastern Studies and the Program of Jewish Studies, Cornell University. Bethesda: CDL Press (in press)

Other volumes in preparation

Abbreviations

AB	*Assyriologische Bibliothek* (Leipzig)
AfO	*Archiv für Orientforschung* (Wien)
AmSUH	*Abhandlungen aus dem mathematischen Seminar der Universität Hamburg*
AMM	*American Mathematical Monthly* (Buffalo)
AnOr	*Analecta Orientalia* (Rome)
AoF	*Altorientalische Forschungen* (Berlin)
AOAT	*Alter Orient und Altes Testament* (Kevelaer/Neukirchen-Vluyn)
AOS	*American Oriental Series* (New Haven)
AS	*Assyriological Studies* (Chicago)
ASJ	*Acta Sumerologica* (Hiroshima)
BagM	*Baghdader Mitteilungen* (Berlin)
BE	*The Babylonian Expedition of the University of Pennsylvania* (Philadelphia)
BIN	*Babylonian Inscriptions in the Collection of J. B. Nies* (New Haven)
BRM	*Babylonian Records in the Library of J. Pierpoint Morgan* (New Haven)
CDLJ	*Cuneiform Digital Library Journal*
CTMMA	*Cuneiform Texts from the Metropolitan Museum of Art* (New York)
CT	*Cuneiform Texts from Babylonian Tablets in the British Museum* (London)
GMS	*Grazer Morgenländische Studien* (Graz)
HM	*Historia Mathematica* (Orlando, Fla.)
HSc	*History of Science* (London)
HSJ	*Historia Scientiarum* (Tokyo)
JCS	*Journal of Cuneiform Studies* (New Haven)
JESHO	*Journal of the Economic and Social History of the Orient* (Leiden)
JNES	*Journal of Near Eastern Studies* (Chicago)
MAD	*Materials for the Assyrian Dictionary* (Chicago)
MDP	*Mémoires de la Délégation (Archéologique) en Perse/Iran* (Paris)
MEE	*Materiali epigrafici di Ebla* (Naples)
MSL	*Materials for the Sumerian Lexicon* (Rome)
MSVO	*Materialien zu den Frühen Schriftzeugnissen des Vorderen Orients* (Berlin)
OECT	*Oxford Editions of Cuneiform Texts* (Oxford)
OIP	*Oriental Institute Publications* (Chicago)
OLZ	*Orientalistische Literatur-Zeitung* (Berlin)
OrNS	*Orientalia, Nova Series* (Rome)
OrSP	*Orientalia, Series Prior* (Rome)
PBS	*University of Pennsylvania, Publications of the Babylonian Section* (Philadelphia)
PCHM	*Proceedings of the Cultural History of Mathematics*
PIHANS	*Publications de l'Institut historique-archéologique néerlandais de Stamboul* (Istanbul)
RA	*Revue d'assyriologie et d'archéologie orientale* (Paris)
RAI	*Proceedings of the Rencontre assyriologique internationale*
RlA	*Reallexikon der Assyriologie und vorderasiatischen Archäologie* (Berlin/New York)
RHM	*Revue d'histoire des mathématiques* (Paris)
SBM	*Studies in Babylonian Mathematics* (Japan)

SpTU	*Spätbabylonische Texte aus Uruk* (Mainz am Rhein)
TAPS	*Transactions of the American Philosophical Society* (Philadelphia)
TMH	*Texte und Materialien der Frau Professor Hilprecht Collection, Jena*
UE	*Ur Excavations* (Oxford/Philadelphia)
VDI	*Vestnik drevnej istorii* (Moscow)
VO	*Vicino Oriente* (Rome)
WVDOG	*Wissenschaftliche Veröffentlichungen der Deutschen Orient-Gesellschaft* (Leipzig/Berlin)
YOS	*Yale Oriental Series* (New Haven)
ZA	*Zeitschrift für Assyriologie und vorderasiatische Archäologie* (Leipzig/Berlin)

Table of Contents

Remark. The successive chapters of this book have been ordered into what initially seemed to be a logical succession of increasingly sophisticated topics. After the work with the book had been finished, much too late to be considered, C. Proust's dissertation *TMN* (2004) became available, with its thorough statistical analysis of nearly the whole corpus of known mathematical cuneiform texts from Nippur. According to Proust, at the elementary level of the education in the Old Babylonian scribe schools at Nippur, metrological and mathematical table texts were studied in the following order: a) *metrological lists* for systems $C, M. A, L$, b) *metrological tables* for systems $C, M. A, Ln, Lc$, c) *tables of reciprocals*, *multiplication tables*, *tables of squares*, d) *tables of square sides and cube sides*. All other mathematical cuneiform texts are relegated by Proust to the category of *exercises at an advanced level*.

It is not clear to what extent the results of Proust's analysis are applicable to the mathematical cuneiform texts in the Schøyen Collection. The great majority of the mathematical texts used by Proust for her statistical analysis are (fragments of) clay tablets of the so called type IIa/IIb, and very few belong to what she calls the advanced level. In contrast to this, of the Old Babylonian mathematical tablets in the Schøyen Collection very few, if any, are of type IIa/IIb, and all the texts discussed in Chapters 1(?), 7, 8, 10, and 11 of this book belong to what Proust calls the advanced level.

0
How to Get a Better Understanding of Mathematical Cuneiform Texts

0.1. On Avoiding Anachronisms in Translations of Mathematical Terms

The terminology used in modern elementary mathematics has a mixed origin, which may be characterized as Greek/Latin/early modern. The terminology used in Babylonian mathematical cuneiform texts, on the other hand, is pre-Greek. Therefore, the use of Greek and Latin words in a discussion of a mathematical cuneiform text is in itself an anachronism. If one wants to convey a proper understanding of the essence of Mesopotamian mathematics, one should try to avoid anachronisms by using *literal translations* of technical terms in cuneiform mathematical texts, whenever possible.

A clear example is the seemingly self-evident term *triangle*. It is derived from the Latin word *triangulum* which, like its Greek predecessor τρίγωνον, means 'three-cornered' or 'three-angled'. However, the idea of an angle between two sides in a rectilinear plane figure can be traced no further back than to classical Greek geometry. More specifically, in classical Greek geometry, a triangle was referred to as 'the triangle ABC' where A, B, C are the three corners of the triangle. In Old Babylonian geometry, on the other hand, a triangle was always specified in terms of the lengths of two or three of its sides, never in terms of its angles or the positions of its corners. The term used for 'triangle' was invariably the Sumerian word sag.kak, possibly with the literal meaning 'peg-head'. (According to the dictionaries, the corresponding Akkadian (i. e. Babylonian) word is *santakkum* 'wedge', but there is not necessarily an actual connection between the two words.) Other examples of modern mathematical terms, in the Greek tradition referring to angles or corners, are *rectangle* 'right-angled', *diagonal* 'across corners', and *pentagon* 'five-cornered', *hexagon* 'six-cornered', *heptagon* 'seven-cornered', for the latter of which the Old Babylonian names were sag.5 '5-front', sag.6 '6-front', sag.7 '7-front'.

Since the concept of angles was unknown (or, at least, never explicitly mentioned) in Old Babylonian mathematics, the concept *right-angled* was also unknown. It is likely that what we call a right-angled triangle was thought of as one of the two triangles into which a rectangle can be divided by an oblique transversal of maximal length. Perhaps a better word for such a triangle is simply a "right triangle". The term *ṣiliptum* used for such a transversal in a rectangle was derived from a verb meaning 'to cross over', while the rectangle itself was called simply uš sag, where uš 'length' (š = sh) is the term used for the long side of a rectangle, and sag 'head, front' the term used for the short side.

The modern terms 'square' and 'quadratic' are both derived from the Latin word *quadratum* 'fourish', probably derived from the Greek word τετράγωνον 'four-angled'. The terms normally used in Old Babylonian mathematical texts are the Sumerian íb.si₈ or the Akkadian *mitḫartum*, both derived from verbs with the meaning 'to be equal'. Another term, documented for the first time in MS 5112 (Sec. 11.2 below) is téš.a.sì, possibly with the literal meaning 'given together' or 'given equal'. A well known peculiarity of the Old Babylonian mathematical terminology is that all these words can denote both a square figure and a side of such a figure, depending on the context. As a reminder of this fact, and for lack of better alternatives, "equalside" is suggested here as a fairly literal translation of íb.si₈/*mitḫartum* and "sameside" as a translation of téš.a.sì.

It is clearly important that (more or less) *literal translations should be used for Babylonian or Sumerian mathematical terms*. One direct consequence of this principle is that different translations should be used for different words in the original texts, even when they correspond to the same word in modern mathematical terminology. It was this simple idea that led Jens Høyrup to the surprising observation that Old Babylonian mathematicians distinguished between different kinds of addition (in particular, *joining together* two equally important entities, as opposed to *adding* one entity to another, more important entity). Old Babylonian mathematicians also distinguished between different kinds of multiplication, *etc*. This observation, in its turn, led Høyrup to the important discovery that geometric models play a much greater role than earlier realized in Old Babylonian problems for squares and rectangles. The idea has been further developed by Høyrup in a long series of publications since 1982. See, for instance, Høyrup, *AoF* 17 (1990), *HSc* 34 (1996), and *LWS* (2002). See also the related discussion of "metric algebra" problems in the theme text MS 5112 in Sec. 11.2 below.

0.2.   Conform Transliterations, Translations, and Interpretations

The idea of "conform" transliterations and translations of mathematical cuneiform texts was developed by the present author in the early 1980's as a reaction to the extremely reader-unfriendly way in which Babylonian mathematics was presented by Otto Neugebauer in the classical works *MKT = Mathematische Keilschrift-Texte* I-III (1935-37), by François Thureau-Dangin in *TMB = Textes mathématiques babyloniens* (1938), and by Otto Neugebauer together with Abraham Sachs in *MCT = Mathematical Cuneiform Texts* (1945). In *TMB*, for instance, all Sumerian words in the mathematical cuneiform texts were interpreted as "sumerograms" and replaced by the Akkadian (Babylonian) words that they were assumed to represent. For mathematically oriented readers trying to compare transliterations of mathematical cuneiform texts *sign by sign* with hand copies of the original texts, this made the task much more difficult than it needed to be. In all three of the classical works, transliterations of mathematical cuneiform texts were followed by translations into *standard* German, French, or English. For readers trying to compare the transliterations *word by word* with the translations, this, too, made the task much more difficult than it needed to be.

The following example of a *conform translations*, word by word (and sentence by sentence), is taken from MS 5112 § 2 b (Sec. 11.2 d below) :

1	a.šà 2 téš.a.sì gar.gar-*ma* 21 40	The fields of 2 samesides (I) heaped, then 21 40.
	téš.a.sì gar.gar-*ma* 50 /	The samesides (I) heaped, then 50.
2	téš.a.sì.meš en.nam	The samesides (are) what?

A translation of the same passage into standard English would be like this:

I added the areas of 2 squares: 21 40. I added the sides of the squares: 50. What are the sides of the squares?

Here is another example, taken from MS 5112 § 9 (Sec. 11.2 l):

1	uš sag gu₇.gu₇-*ma* 50 a.šà	The length (and) the front (I made) eat (each other), then 50, the field.
	i-na uš 30 ninda zi-*ma*	From the length 30 ninda (I) tore out, then
2	*a-na* sag / 5 ninda daḫ-*ma* 50 a.šà	to the front 5 ninda (I) added, then 50, the field.
	uš sag en.nam	The length (and) the front (are) what?

In this case, a translation to standard English would be like this:

I multiplied the length and the width: area 50. I subtracted 30 ninda from the length, and I added 5 ninda to the width: area 50. What are the length and the width?

There are several differences between conform translations and translations into standard English. The most obvious difference is that in the conform translations the word order is (essentially) the same as in the transliterated texts. This should make it possible even for readers who know little Sumerian or Akkadian to connect most of the words in the transliterations with their translations.

Another difference is that in the conform translations an effort has been made to let the translations of individual words be as close as possible to the normal (non-mathematical) meanings of the words in the original texts. Note, in particular, that in the first of the examples above the term used for an addition (of the areas of two squares) is gar.gar '(make a) heap', while in the second example a different term is used for an addition (of an extra 5 ninda to the front), namely daḫ 'add'. Note also the peculiar use in the first example of the term téš.a.sì to denote both two squares (of which the areas are added together) and the sides of those squares (which are also added together).

The un-English word order and the strange-looking words in a conform translation should be no great obstacles to the average reader. After a little while the inverted word order will be familiar and the unusual words will be well known.[1] The trouble caused by the conform translation may even give the reader a pleasurable feeling of getting closer to the original language of the mathematical cuneiform text!

A *conform transliteration* is one where the cuneiform signs in the original text have been transliterated directly, without any attempt to replace sumerograms by the Akkadian words that they can be assumed to represent. In transliterations in this work, Sumerian words or sumerograms are written with plain letters with extra spacing, while Akkadian words are written with Italic letters. In order to emphasize the difference, dots are used the separate from each other the components of composite Sumerian words, while dashes are used to separate the syllables of Akkadian words. (The common practice of letting capital letters indicate uncertain readings of parts of Sumerian words is ignored here.)

Sometimes it is desirable to find a form for a conform transliteration which is even closer to the form of the original text. This is the case, in particular, when the text is damaged so that it becomes important to make clear where the damaged parts of the text are located and how great the chance is of being able to make a credible reconstruction of lost or damaged words. It is also often of interest to have a transliteration which clearly shows how the original text was structured, how the layout of the text on the clay tablet was organized, *etc.* In all such cases, what is needed is *a conform transliteration within an outline of the clay tablet*. See, for instance, the examples of such conform transliterations, side by side with hand copies of simple clay tablets with arithmetical exercises in Fig. 1.1.1 below, or the examples of conform transliterations within the outlines of large clay tablets with mathematical problem texts in Figs. 10.1.1, 10.2.2, 10.3.1, *etc.*

A final and crucial step in the complete presentation of a new mathematical cuneiform text is the interpretation of the text. In this work it will be attempted to give what may be called "conform interpretations", disclosing as clearly as possible the intentions of the writers of the texts, and explaining as faithfully as possible the methods they used to construct and solve the stated problems in the larger collections of exercises. Although normal language and normal mathematical terminology will be used to a great extent in these interpretations, also here an effort will be made to avoid possible anachronisms. In particular, no use will be made of modern abstract notations such as x, y, and z, or D_1, D_2, R_1, R_2, *etc.* Instead, easily remembered acrophonic abbreviations will be used, such as u and s for uš 'length' and sag 'front', or h, A, V for heights, areas, and volumes. Squares and square roots will never be expressed by use of exponents and square root signs. Instead the square of a number a will be written sq. a, and square roots, or rather *square sides*, will be written as sqs. b. Solutions to intricate mathematical problems will never be presented in the form of complicated mathematical formulas. Instead, *the successive steps of complicated solution procedures will be presented one by one* in "quasi-modern notations" that are only mildly anachronistic.

0.3. Babylonian Sexagesimal Numbers

0.3 a. Sexagesimal Numbers

---

1. E. Robson has less confidence in her own or her readers' ability to get past such minor obstacles. In her book *MMTC* (1999), 5, she writes: "Any of these translations have their own merits and demerits: it is almost impossible to find a satisfactory translation for any Old Babylonian mathematical word in modern English, as the concepts behind them are so different from ours. Friberg and Høyrup have each tried to overcome this difficulty in recent years by inventing a new mathematical vocabulary. The exercise, although well intentioned, has been to a great deal self-defeating, as one then has to translate a jargon-filled and visually unattractive 'conform translation' into standard English to understand it."

Every given *integer* (whole number) a can be written in the form

$$a = a_{n-1} \cdot 60^{n-1} + \cdots + a_2 \cdot 60^2 + a_1 \cdot 60^1 + a_0,$$

where n is a sufficiently large whole number, and where

all the n "double digits" or "sexagesimal places" $a_{n-1}, ..., a_2, a_1, a_0$ are whole numbers between 1 and 59.

In *sexagesimal place value notation*, an integer of this form is expressed more compactly as

$$a = a_{n-1} \dots a_2\, a_1\, a_0.$$

Take, for instance, the large decimal number 1,000,000 (a million). It can be *converted into a sexagesimal number* (= a number in sexagesimal place value notation) in the following way:

1,000,000/60 = 16,666.66..., **16,666**.66.../60 = 277.77833..., **277**.77833.../60 = 4.62963... .

Consequently, $n = 4$ and the first sexagesimal place is 4. Now, subtract 4 and multiply by 60:

$$0.62963... \cdot 60 = \mathbf{37}.7778333... .$$

Therefore, the second sexagesimal place is 37. Subtract 37 and multiply by 60:

$$0.7778333... \cdot 60 = \mathbf{46}.66... .$$

Thus, the third sexagesimal place is 46. Now, subtract 46 and multiply by 60:

$$0.66... \cdot 60 = \mathbf{40}.$$

This shows that the fourth and final sexagesimal place is 4. Therefore, the computation shows that

$$1,000,000 = 4\ 37\ 46\ 40.$$

Conversely, the sexagesimal number 4 37 46 40 can be *converted into a decimal number* (= a number in decimal place value notation) most easily by use of the following systematic procedure:

$$
\begin{aligned}
4\ 37 = \quad & \mathbf{4 \cdot 60 + 37} = 240 + 37 = \mathbf{277}, \\
4\ 37\ 46 = \quad & \mathbf{277 \cdot 60 + 46} = 16,620 + 46 = \mathbf{16,666}, \\
4\ 37\ 46\ 40 = \quad & \mathbf{16,666 \cdot 60 + 40} = 999,960 + 40 = \mathbf{1,000,000}.
\end{aligned}
$$

The methods used in this couple of examples can be used generally to convert given decimal numbers into sexagesimal numbers or, conversely, given sexagesimal numbers into decimal numbers.

Sexagesimal place value notation can be used also for *fractions* or for numbers with both an integral and a fractional part. A "sexagesimal semi-colon" is then inserted to separate the integral part from the fractional part of the number. It is not true, however, that *any* fraction can be written as a (finite) sexagesimal number. (Neither can *any* fraction be written as a (finite) decimal number.) *The only numbers that can be written as sexagesimal numbers with a (finite) fractional part are numbers that are equal to a whole number divided by some power of 60.*

Important examples are

$$
\begin{aligned}
1/3 &= 20/60 = ;20 \\
1/2 &= 30/60 = ;30 \\
2/3 &= 40/60 = ;40.
\end{aligned}
$$

Other examples are

$$
\begin{aligned}
1,000,000/60 &= \quad 4\ 37\ 46;40 \\
1,000,000/60^2 &= \quad 4\ 37;46\ 40 \\
1,000,000/60^3 &= \quad 4;37\ 46\ 40 \\
1,000,000/60^4 &= \quad ;04\ 37\ 46\ 40 \\
1,000,000/60^5 &= \quad ;00\ 04\ 37\ 46\ 40.
\end{aligned}
$$

Two useful *decimal-sexagesimal* and *sexagesimal-decimal* conversion tables are:

60	1 00	;00 01	1/3,600	
70	1 10	;01	1/60	*Continued*

80	1 20	1 00	60
90	1 30	2 00	120
100	1 40	3 00	180
200	3 20	4 00	240
300	5 00	5 00	300
400	6 40	6 00	360
500	8 20	7 00	420
600	10 00	8 00	480
700	11 40	9 00	540
800	13 20	10 00	600
900	15 00	1 00 00	3,600
1,000	16 40	1 00 00 00	216,000

It is important to keep in mind that the predominantly Semitic population in Mesopotamia in the 2nd and 1st millennia BC normally counted with decimal numbers, using their own *Semitic decimal number words*. Only educated scribes had learned how to count with the originally Sumerian sexagesimal numbers in place value notation. Therefore, it is no wonder that there existed special "conversion tables" for the conversion of decimal numbers into sexagesimal numbers. An Old Babylonian conversion table is MS 3970, shown in Fig. 2.7.1 below. A Late Babylonian example is BM 36841, shown in Fig. 2.7.2. It is also no wonder that there existed multiplication tables for both 1 40 = 100 and 16 40 = 1,000. See, for instance, the reverse of the combined multiplication table MS 3974 (Fig. 2.6.13).

0.3 b.   Sexagesimal Numbers in Relative Place Value Notation

The Old Babylonians (or, rather, their Sumerian predecessors in the Ur III period) invented place value notation for sexagesimal numbers, but they did not invent *final zeros* to distinguish, for instance, 1 from 1 00 = 60. Neither did they invent a *separator* like the sexagesimal semi-colon to separate the integral part of a sexagesimal number from its fractional part. They also did not invent *initial zeros* to distinguish, for instance, ;01 and ;00 01 from 1. As a result, sexagesimal numbers in Old (and Late) Babylonian mathematical texts have only "relative" (or "floating") values. All Babylonian sexagesimal numbers are written as if they were integers, but the *intended* value of a sexagesimal number in a Babylonian cuneiform text can be its "nominal" value multiplied by any positive or negative power of 60. Thus, *only the context* can decide what the intended "absolute value" is of a "relative sexagesimal number" in such a text. This can seem to have been an awkward handicap for Old Babylonian calculators, but the truth is that it can be very convenient to be able to count with sexagesimal numbers without having to bother about the absolute value of the numbers or about the precise positions of the separating semicolons and final or initial zeros.

It is important to remember this peculiarity of Babylonian sexagesimal numbers. Therefore, it is advisable to follow the practice in this book, where in all *transliterations* and most *translations* of mathematical texts no use is made of separating semi-colons or of final and initial zeros in sexagesimal numbers. As a compromise, and only for the sake of clarity, *internal zeros* are inserted where needed, to indicate missing zeros or ones in the middle of a sexagesimal number. In the *commentaries*, on the other hand, full use is made of both separating semi-colons and final or initial zeros.

The Babylonians themselves overcame the ambiguity of their relative sexagesimal numbers in several ways. Thus, for instance, a missing internal double zero in a sexagesimal number could be indicated by a gap or by a special sign. An example is offered by the number 1 19 00 44 26 40 in MS 3049 § 5, line 9 (see Fig. 11.1.4 in Ch. 11 below), which is written as '1 19   44 26 40'. (Actually, the scribe at first wrote 44 where the gap should be, but then corrected himself by erasing 44 and writing it again a step to the right.)

Typically, in an Old Babylonian mathematical problem text, the *question* was stated in terms of "traditional" length numbers, area numbers, capacity numbers, *etc.*, for which unambiguous special notations were used. In the ensuing *solution procedure*, the given traditional numbers were converted into sexagesimal numbers and all computations were carried out using those numbers. The *answer*, finally, was often again given in the form of traditional numbers.

A clear example of this practice is offered by the exercise MS 3052 § 1 c (Sec. 10.2.a below), where in the question the length of a wall is given as 5 uš (5 'lengths' = 5 · 60 ninda), the volume of the wall as 2_{iku} (2 'dykes' = 2 · 100 square ninda), and the length of a hole drilled through the wall as 1 kùš 7 1/2 šu.si ('1 cubit 7 1/2 fingers'). When these data reappear in the solution procedure, they are expressed as 5 (meaning 5 00 ninda), 3 20 (meaning 3 20 square ninda), and 6 15 (meaning ;06 15 ninda), respectively. (The ninda was the basic Old Babylonian unit of length measure, equal to 12 cubits of 30 fingers each.) The end result of the computation is the relative sexagesimal number 8, rephrased in the answer as 1/2 ninda 2 cubits (= 8 cubits).

In another exercise, MS 3049 § 5 (Sec. 11.1 d), the height of a gate is given in the question as '5 cubits, 25, and 10 fingers, 1 40', but reappears in the solution procedure as 26 40 (= ;25 + ;01 40).

A more confusing Old Babylonian way of specifying the intended size of a relative sexagesimal number is exemplified by BM 96957+ § 5 b (also in Sec. 11.1 d below), where the height of a gate is given in the question as 40 kùš '40, cubits' (meaning not 40 cubits but ;40 ninda = 8 cubits).

0.4.   Counting with Sexagesimal Numbers in Relative Place Value Notation

0.4 a.   Addition of Sexagesimal Numbers

It is not known how complicated additions of sexagesimal numbers were carried out by Old Babylonian school boys or their teachers. No exercise tablets with additions have been found, and when sexagesimal numbers are added as one of the steps of the solution procedure in an Old Babylonian mathematical problem text, the result is always given directly. There was simply not space enough on a clay tablet for non-essential text, such as the details of an addition algorithm. Thus, additions that could not be done in the head were carried out, presumably, either on some kind of counting board or on a separate clay tablet that was erased after each computation.

A simple example is the addition of the numbers of men in four troops of soldiers in the mathematical exercise MS 2792 # 2 (see Fig. 10.3.7). The numbers to be added are 4 30, 8 (00), 6 40, and 7 30, and the text states, quite laconically (in line 7): 'Add the soldiers, 26 40'. It is clear that the addition must have been done (essentially) in the following way:

$$\underline{1}$$

4 30	a) 30 + 40 + 30 = (100 =) 1 40, where
8	40 is used as the last sexagesimal place in the sum, while
6 40	1 is carried to the preceding column of sexagesimal places.
+ 7 30	b) 1 + 4 + 8 + 6 + 7 = 26, the preceding sexagesimal place in the sum.
26 40	

The example shows that the addition algorithm is similar to the familiar addition algorithm for *decimal* numbers. The only differences are that in sexagesimal addition sexagesimal *places* or *double digits* are used instead of decimal *single digits*, and that in sexagesimal addition *sixties* in the sum of a column of sexagesimal double-digits are carried to the preceding column of double-digits, while in decimal addition *tens* in the sum of a column of decimal digits are carried to the preceding column.

A more interesting example is the addition of three squares of sexagesimal numbers in MS 3049 § 5, an application of the Old Babylonian "diagonal rule" in three dimensions (see Fig. 11.1.5 below). The three squares are given in relative numbers as 26 40, 8 53 20, and 6 40, and their squares are, respectively, 11 51 06 40, 1 19 (00) 44 26 40, and 44 26 40. In the text, the addition of the three squares is announced simply as follows (in line 12): 'Add them, 13 54 34 14 26 40 you will see'.

The author of the text must have had some idea about the *absolute* values of the three numbers and their squares. Otherwise he would not have known how to add the squares correctly. As a matter of fact, the given numbers have to be interpreted as ;26 40 (= 12/27 = 4/9), ;08 53 20 (= 4/27), and ;06 40 (= 3/27 = 1/9). The corresponding interpretations of their squares as absolute sexagesimal numbers are ;11 51 06 40, ;01 19 00 44

26 40, and ;00 44 26 40.

Below is shown how the addition of the three squares can be set up in relative sexagesimal numbers (the display to the left), and in absolute sexagesimal numbers (the display to the right):

	1 2			1 2
	11 51 06 40			;11 51 06 40
	1 19 00 44 26 40			;01 19 00 44 26 40
+	44 26 40		+	;00 44 26 40
	13 54 34 04 26 40			;13 54 34 04 26 40

It is clear that it required considerable skill to get the addition right using only relative sexagesimal numbers!

0.4 b. Subtraction of Sexagesimal Numbers

In MS 2792 # 1, the object considered is a leaning ramp built by four troops of soldiers. The side of a ramp has the form of a trapezoid with three transversals parallel to the base and the top of the trapezoid (see Fig. 10.3.3). In the computation of the lengths of the three transversals, three subtractions have to be done (in lines 12, 15, 17). No details are given. Below is shown how the three subtractions can have been set up:

			60		~~60~~ 60
7 49 26 40			5̶ 36 06 40		3 4̶5̶
− 2 13 20			− 1 51 06 40		− 1 28 53 20
5 36 06 40			3 45		2 16 06 40

The first example is simple. The sexagesimal places or double-digits in the lower sexagesimal number are subtracted from the places in the upper number, one at a time, starting from the right.

The second example is more complicated. In the third place from the right, 51 cannot be subtracted from 36, so 1 is *borrowed* from the fourth place, being worth 60 in the third place.

In the third example, there are two such borrowings.

Thus, the indicated sexagesimal subtraction algorithm is similar to the usual decimal subtraction algorithm, with the exception that decimal digits are replaced by sexagesimal double-digits and that *sixties* are borrowed each time instead of *tens*.

0.4 c. Multiplication of Sexagesimal Numbers

Three Old Babylonian clay tablets with elementary multiplication exercises are known, **MS 2728, 2729**, and **3944** (see Fig. 1.1.1). As could be expected, all the multiplication exercises consist of only a question and an answer, with a total lack of detailed computations.

There are several methods that an Old Babylonian school boy conceivably may have used to find the products of the numbers given in the mentioned multiplication exercises. Take, for instance, the first exercise in MS 2729, to find the product of 35 and 30. The most direct way of finding the product would have been through addition of "partial products":

$$35 \cdot 30 = (30 + 5) \cdot 30 = 30 \cdot 30 + 5 \cdot 30 = (900 + 150 =) \ 15 \ (00) + 2 \ 30 = 17 \ 30.$$

A somewhat more ingenious way would be to think of 30 as the fraction 1/2 = ;30. Then, in *relative* place value notation,

$$35 \cdot 30 = 35/2 = 17 \ 1/2 = 17 \ 30.$$

In a similar way, the answers to the second multiplication exercise on MS 2729 may have been obtained as follows:

$$40 \cdot 35 = 2/3 \cdot 35 = 2/3 \cdot 30 + 2/3 \cdot 5 = 20 + 3 \ 20 = 23 \ 20.$$

The remaining multiplication exercises in MS 2728 and 2729 are equally simple. Somewhat more complicated are the multiplication exercises in MS 3944. Take, for instance, the first exercise in MS 3944, to find the product of 25 and 17 30. It can be used to illustrate three available techniques for finding the product of two sexagesimal numbers.

The method that may appear most natural to someone familiar only with *decimal* arithmetics is to convert the given sexagesimal numbers to decimal numbers, then multiply the decimal numbers in the usual way, and finally convert the result back to a sexagesimal number, as follows:

$$17\ 30 = 17 \cdot 60 + 30 = 1{,}020 + 30 = 1{,}050,$$
$$25 \cdot 17\ 30 = 25 \cdot 1{,}050 = 26{,}250,$$
$$26{,}250 = 7\ 17\ 30\ (\text{see Sec. 0.3 a above}).$$

The most obvious method for an Old Babylonian school boy probably was to rely on the multiplication table for 25, which he may have known by heart (see Appendix 2). Then:

$$25 \cdot 17\ 30 = 25 \cdot 17 + 25 \cdot 30 = 7\ 05 + 12\ 30 = 7\ 17\ 30\ (\text{in } \textit{relative}\ \text{place value notation}).$$

Another method that might have appealed to an Old Babylonian school boy is an application of the following "factorization method", requiring only the multiplication table for 5, a small number:

$$25 \cdot 17\ 30 = 5 \cdot 5 \cdot 17\ 30,\ \ 5 \cdot 17\ 30 = 1\ 25 + 2\ 30 = 1\ 27\ 30,\ \ 5 \cdot 1\ 27\ 30 = 5 + 2\ 15 + 2\ 30 = 7\ 17\ 30.$$

It has not been known before how Babylonian teachers and students of mathematics carried out multiplications of "many-place" sexagesimal numbers, too long to be multiplied in the head. There have been speculations about the use in Mesopotamia of some kind of abacus or counting board, but no archaeological remains of such devices have ever been found. Now, however, the present author has been able to show that some fragments of texts from the Late Babylonian/Seleucid period in Mesopotamia (in the second half of the first millennium BC) can be explained as what remains of explicit written multiplication algorithms for many-place sexagesimal numbers, organized in very much the same way as our own decimal multiplication algorithms. See Sec. A9 b in Appendix 9 below.

0.4 d.   *Division of Sexagesimal Numbers*

In most cases, divisions in Old Babylonian mathematical texts were transformed into multiplications by use of the following simple device: A given sexagesimal number is "regular" if it is possible to find another sexagesimal number igi $a = a'$, the "reciprocal" of a, such that $a \cdot a' = $ '1' (= some power of 60). It is well known that this happens if, and only if, a is a product of powers of 2, 3, and 5 (2, 3, and 5 are the prime factors of the sexagesimal base 60). Now, if the divisor is a regular sexagesimal number a, then division by a can be replaced by a multiplication by igi a, the reciprocal of a.

Here are some examples of the application of this method in various exercises in MS 5112 (Sec. 11.2 below) and in MS 3052 (Sec. 10.2 below):

30/16	=	30 · igi 16	=	30 · 3 45	= 1 52 30	(MS 5112 § 12, lines 8-9)
20/1 20	=	20 · igi 1 20	=	20 · 45	= 15	(MS 5112 § 13, line 10)
2 30/3 20	=	2 30 · igi 3 20	=	2 30 · 18	= 45	(MS 3052 § 1 a, line 7)
26 15/22 30	=	26 15 · igi 22 30	=	26 15 · 2 40	= 1 10	(MS 3052 § 1 b, line 4)

The correctness of the last of the examples above may not be completely obvious, in particular since the reciprocal of 22 30 is not listed in the Old Babylonian standard table of reciprocals (Sec. 2.5 below). On the other hand, it is easy to find the reciprocal of 22 30, for instance as follows: Clearly 22 30 is equal to 1/ 2 · 45. Therefore, the reciprocal of 22 30 is equal to 2 times the reciprocal of 45, so that igi 22 30 = 2 · 1 20 = 2 40.

It is also possible to compute 26 15/22 30 by use of a detour over decimal arithmetic. Indeed, since 26 15 = 1,200 + 360 + 15 = 1,575 and 22 30 = 1,200 + 120 + 30 = 1,350, it follows that 26 15/22 30 = 1,575/1,350 = 1.166... = 1 1/6 = 1;10.

It is, of course, not always possible to transform a division problem into an equivalent multiplication problem. This happens when the divisor is either a "many-place" regular sexagesimal number, or a non-regular sexagesimal number. An example of the first kind can be found in MS 3871 (Fig. 1.3.1), the only known example of an Old Babylonian division exercise. No details of the division algorithm are given in MS 3871, but it is likely that the division there of 4 37 46 40 by 11 34 26 40 was achieved by use of a quite straightforward factorization method. (See the discussion in Sec. 1.3 below.)

Interesting examples of divisions by non-regular sexagesimal numbers can be found in some of the "combined market rate exercises" in Sec. 7.2 below, for instance MS 2299 (Fig. 7.2.1, bottom). There, four commodities have a "combined unit price" of ;57 shekel, while the silver available is given as 16 53 20, probably meaning 16;53 20 shekels. To find out how many units of each kind can be bought for this amount of silver, one has to divide 16 53 20 by 57. Since 57 = 3 · 19, it is a "semiregular" sexagesimal number, that is, the product of a regular and a non-regular number. Similarly,

$$16\ 53\ 20 = 20 \cdot 50\ 40 = 20 \cdot 40 \cdot 1\ 16 = 20 \cdot 40 \cdot 4 \cdot 19.$$

Therefore, 16 53 20/57 = 20 · 40 · 4 / 3 = 17 46 40.

Another example of a division of a semiregular sexagesimal number by a non-regular sexagesimal number can be found in the combined market rate exercise MS 2268/19 (Fig. 7.2.2, top). There, the combined price for 3 1/2 = 3;30 units of each of five different commodities is 2;49 shekels, while the silver available is 1;00 05 20 shekels. To find out how many units of each kind can be bought for this amount of silver, one has to divide 1 00 05 20 by 2 49, and then multiply by 3 30. Here 2 49 is non-regular, since it is equal to the square of 13. The answer given in the text is that the wanted number is 1 14 40, probably meaning 1;14 40 units of each kind. This number can be explained as follows:

$$(1\ 00\ 05\ 20/2\ 49) \cdot 3\ 30 = 21\ 20 \cdot 3\ 30 = 1\ 14\ 40 \quad \text{(in relative sexagesimal numbers)}.$$

Since 1 00 05 20 is semiregular, the division of 1 00 05 20 by 2 49 can have been achieved as follows:

$$1\ 00\ 05\ 20 = 20 \cdot 3\ 00\ 16 = 20 \cdot 16 \cdot 11\ 16 = 20 \cdot 16 \cdot 4 \cdot 2\ 49 = 21\ 20 \cdot 2\ 49.$$

A more difficult task is to divide a sexagesimal number by another sexagesimal number, when neither number is semi-regular. How this could be done is shown in the discussion of MS 2317 in § 7.1. There, it is suggested that the division of the "funny number" 1 01 01 01 by the non-regular number 13 was achieved by use of a certain elegant division algorithm, known from division problems in mathematical cuneiform texts from the third millennium BC. (See Appendix 6, Secs. A6 e-g.) Essentially, that method is based on the computation of a succession of increasingly more accurate *approximate reciprocals* to the non-regular divisor.

0.4 e.   Computing Square Sides (Square Roots) of Sexagesimal Numbers

Computations of square roots (or rather *square sides*) are common in Old Babylonian mathematical problem texts, in solution procedures involving quadratic equations. Here are, for instance, examples from MS 5112 (Sec. 11.2 below), a recombination text with the theme 'equations for squares or rectangles':

sqs. 2 01 = 11	(MS 5112 § 1)	sqs. 5 03 45 = 2 15	(MS 5112 § 9)
sqs. 10 25 = 25	(MS 5112 § 2 c)	sqs. 16 01 = 31	(MS 5112 § 12)
sqs. 42 15 = 6 30	(MS 5112 § 5)	sqs. 56 15 = 7 30	(MS 5112 § 13)
sqs. 1 46 40 = 1 20	(MS 5112 § 8)		

In MS 5112, the questions and answers are typically of the form

16 01, what is the square side? 31 each (way) is the square side   (MS 5112 § 12, lines 7-8)

Note that the reference to 'each (way)', probably meaning 'in each direction', is a clear indication that the computation was interpreted as the computation of *all sides* of a square with the given area. However, no details of the computation method are given. In the first two and in the last but one of the examples listed above, the square sides can have been found by use of a *table of square sides* (see Sec. 2.2 below). In the remaining cases, a table of square sides would be of no help. Instead, it is likely that the square sides were computed by use of what may be called the "Old Babylonian square side rule", incorrectly known as "Heron's rule" (see Friberg, *BagM* 28 (1997) § 8). This is a rule for the computation of relatively good approximations to a square side. When the given area is an exact square, the computation will yield an exact answer, as in the following example:

$$\text{sqs. } 5\ 03\ 45 = \text{sqs. (sq. } 2\ (00) + 1\ 03\ 45) = \text{appr. } 2\ (00) + 1\ 03\ 45/(2 \cdot 2\ (00)) = \text{appr. } 2\ (00) + 15 = 2\ 15$$
$$\text{sq. } 2\ 15 = \text{sq. } (2\ (00) + 15) = 4\ (00\ 00) + 1\ (00\ 00) + 3\ 45 = 5\ 03\ 45.$$

The method works also in more complicated examples, as in MS 3049 § 5 (Sec. 11.1 d below). There the "inside diagonal" of a gate is computed by use of the three-dimensional diagonal rule. The last step of the solution procedure is the computation of the square side of 13 54 34 04 26 40. The answer is given directly as 28 53 20. No information is given about how this result was obtained. However, *if* the square side was obtained by repeated use of the square side rule, then the successive steps of the computation were (essentially) the following:

1. sq. 29 = 14 01,   13 54 = 14 01 – 7   (13 54 being the first couple of sexagesimal places in 13 54 | 34 04 | 26 40).
2. sqs. (13 54) = sqs. (sq. 29 – 7) = appr. 29 – 7 / (2 · 29) = appr. 29 – 7 / (2 · 30) = 28;53.
3. sq. 28 53 = 13 54 14 49,   13 54 | 34 04 = 13 54 14 49 + 19 15.
4. sqs. (13 54 | 34 04) = sqs. (sq. 28 53 + 19 15) = appr. 28 53 + 19 15 / (2 · 28 53)   = appr. 28 53 + 20 / (2 · 30) = 28 53;20.
5. sq. 28 53 20 = 13 54 34 04 26 40, the given number.

Another possibility is that the square side of 13 54 34 04 26 40 was computed by use of a variant of the Old Babylonian "trailing part algorithm". (See the detailed discussion in Sec. 1.5 a below.)

Other interesting examples of computations of square sides of semiregular sexagesimal numbers can be found in Appendix 8 below, in connection with the discussion of the famous table text Plimpton 322. Thus, for instance, in Sec. A8 a it is shown that

$$\text{sqs. } 1\ 56\ 56\ 58\ 14\ 50\ 06\ 15 = 3\ 13 \cdot 5^5 \cdot 30 = 1\ 23\ 46\ 02\ 30, \text{ and that}$$
$$\text{sqs. } 56\ 56\ 58\ 14\ 50\ 06\ 15 = 56\ 07 \cdot 5^3 \cdot 30 = 58\ 27\ 17\ 30.$$

0.4 f. Number Signs for Sexagesimal Numbers With and Without Place Value Notation

When writing sexagesimal numbers *in place value notation*, the only numbers signs needed are signs for the 'ones' from 1 to 9, and for the 'tens' from 10 to 50. The normal Babylonian forms of the cuneiform signs for the ones and tens are displayed in Fig. 0.4.1 below.

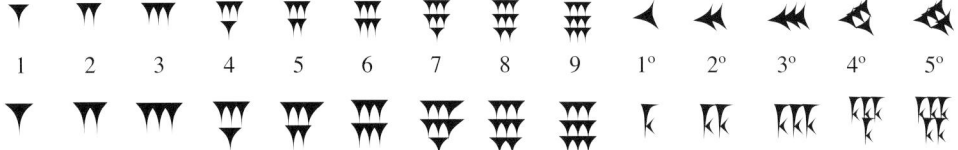

Figure 0.4.1. Babylonian cuneiform number signs in place value notation.

Note that since there are special cuneiform signs for the tens, it is appropriate to call those signs for instance 1°, 2°, 3°, 4°, and 5°, as here, rather than 10, 20, 30, 40, and 50.

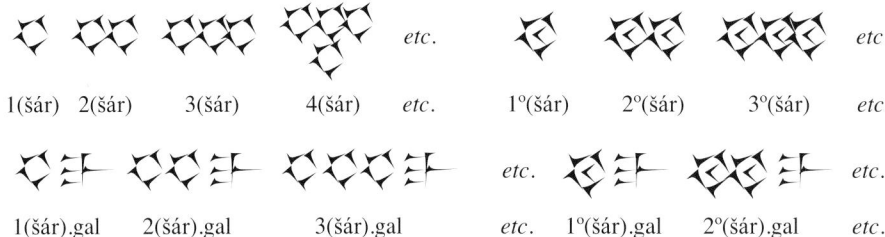

Figure 0.4.2. Babylonian cuneiform number signs in "sign value notation".

Occasionally sexagesimal numbers are written *without the use of place value notation* in Babylonian cuneiform texts, by use of notations inherited from the Sumerians. The Babylonian forms of the sexagesimal numbers signs in such "sign value notation" are displayed in Fig. 0.4.2 above. In the transliterations of the number signs, the following Sumerian words are used:

géš = 60,   šár = 60 · 60,   šár.gal = 60 · 60 · 60 'the great šár'.

See also the much more complete discussion of cuneiform systems of notations for numbers and measures in Appendix 4 below!

For completeness, also the forms of the Babylonian cuneiform signs for the frequently occurring "basic fractions" are shown in Fig. 0.4.3 below. (Cf. Fig. A4.2 in Appendix 4.)

3' (1/3)   2' (1/2)   3" (2/3)   6" (5/6)

Figure 0.4.3. Babylonian cuneiform signs for the "basic fractions".

1
Old Babylonian Arithmetical Hand Tablets

The elementary multiplication, squaring, and division exercises in Secs. 1.1-1.3 are written *with large cuneiform signs* on square fist-sized "hand tablets". They are clearly beginners' exercises.

1.1. Old Babylonian Multiplication Exercises

Four small tablets in the Schøyen Collection are inscribed with simple multiplication exercises. Three of the tablets in question are displayed in Fig. 1.1.1 below, the fourth in Fig. 1.1.3. The texts will be presented individually below. *No previously published parallel texts are known!*

1.1 a. MS 2728 and 2729. Two Linked Triples of Consecutive Multiplication Exercises

On **MS 2729**, there are three computations:

1.	[35]	·	30	=	17 30	(1,050)
2.	40	·	3[5]	=	23 20	(1,400)
3.	4[5]	·	[40]	=	30 (00)	(1,800)

In a modern context, these three computations would be understood as simple *arithmetical* multiplication exercises. In the context of Babylonian mathematics, on the other hand, it is more likely that they should be understood as *geometric* multiplication exercises, more precisely *examples of computations of areas of rectangles*. If the numbers to the left are interpreted as the 'lengths' (long sides) and 'fronts' (short sides) of three rectangles, then the numbers to the right are the corresponding 'fields' (areas). The silently understood unit of length must be the ninda (about 6 meters), and the corresponding unit of area the square ninda.

The way the data were chosen is obvious. The author of the text started in exercise # 1 with two nearly equal numbers (rectangle sides), both multiples of 5 (ninda). Then he made the short side (the front) in # 2 equal to the long side (the length) in # 1, and the short side in # 3 equal to the long side in # 2. Thus, the three exercises are *chained together* in a systematic way. See the first of the three diagrams displayed in Fig. 1.1.2 below.

Another text of the same kind is **MS 2728**. The three successive computations in that text are:

1.	50	·	45	=	37 30	(2,250)
2.	55	·	50	=	45 50	(2,750)
3.	1 (00)	·	55	=	55 (00)	(3,300)

The three computations are consecutive in the same way as the computations on MS 2729, with the short side in exercise # 2 equal to the long side in exercise # 1, and the short side in exercise # 3 equal to the long side in exercise # 2. Thus, in MS 2729 just as in MS 2728 the three exercises can be interpreted as computations of the areas of three rectangles, *chained together pairwise*. See the second diagram in Fig. 1.1.2. In addition, since the long side of a rectangle in MS 2729 is equal to the short side of a rectangle in MS 2728, *the three rectangles in the former text can be linked to the three rectangles in the latter text* as shown in the third diagram in Fig. 1.1.2.

obv.

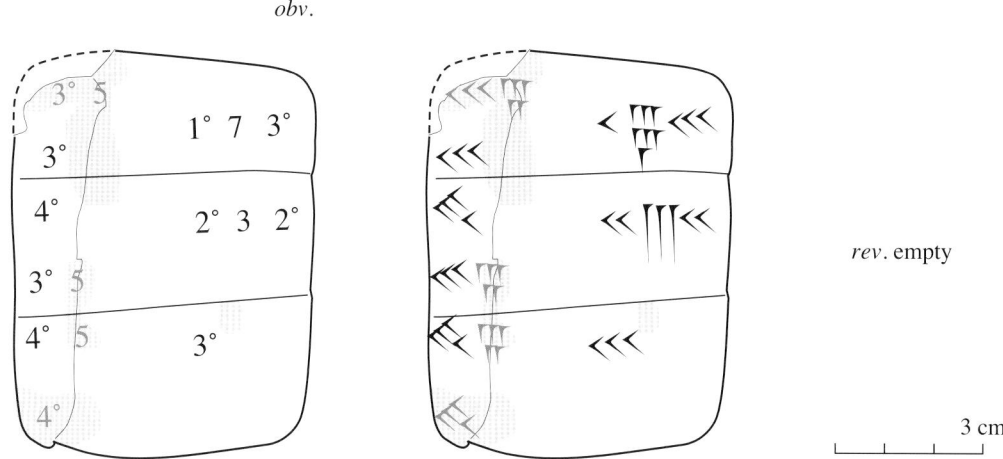

MS 2729. Three consecutive multiplication exercises.

obv.

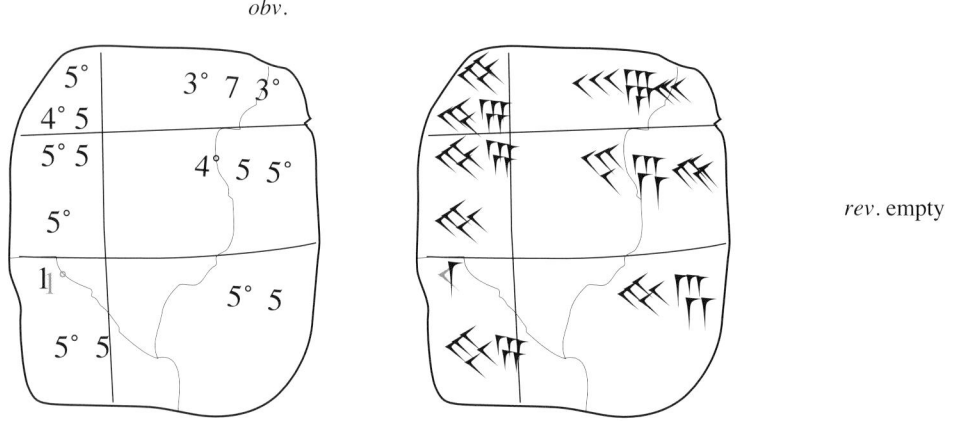

MS 2728. Three similarly consecutive multiplication exercises.

obv.

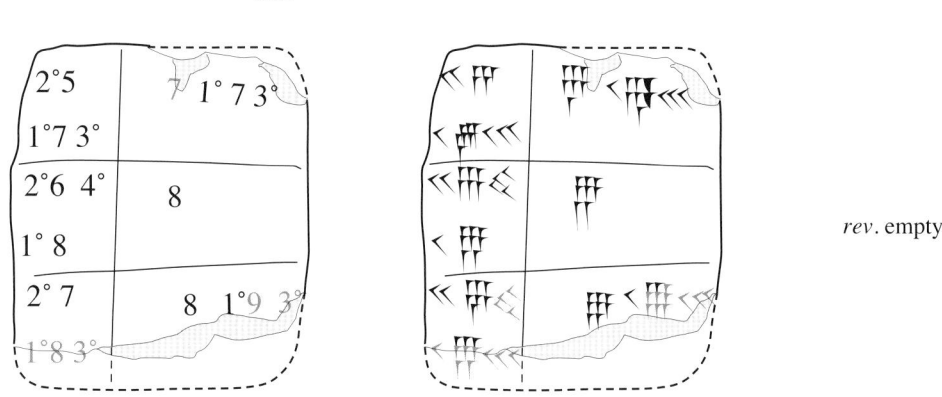

MS 3944. Three multiplication exercises with gradually increasing data.

Fig. 1.1.1. Three hand tablets with intricately constructed multiplication exercises.

Whether this geometric interpretation is correct or not, it is clear that the three exercises on MS 2728 constitute a "continuation" of the three exercises on MS 2729. Thus, this pair of texts is another example of the way in which *Old Babylonian teachers of mathematics used to hand out series of "consecutive" assignments to their students, identical except for the choice of parameters in the problems*. The archetypal example is consecutive single or multiple multiplication tables (Sec. 2.6 below.)

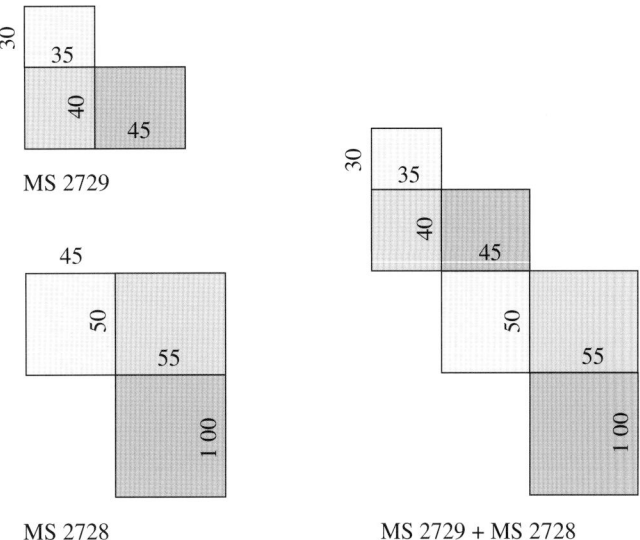

Fig. 1.1.2. Geometric interpretation of the six consecutive multiplication exercises in MS 2728 and MS 2729.

1.1 b.   MS 3944. Another Triple of Consecutive Multiplication Exercises

MS 3944 (Fig. 1.1.1, bottom) is a third example of a small clay tablet inscribed with three simple multiplication exercises (or area calculations). The three exercises may be understood as

$$
\begin{array}{llll}
1. & 25 & \cdot\ 17;30 & = & [7]\ 17;30 \\
2. & 26;40 & \cdot\ 18 & = & 8\ (00) \\
3. & 27 & \cdot\ [18;30]? & = & 8\ 1[9\ 30]?
\end{array}
$$

These exercises are not obviously direct continuations of each other, although, in what appears to be a half systematic way, the three first long sides (the lengths) 25, 26;40, 27 form a slowly increasing sequence of numbers, just as the three short sides (the fronts) 17 30, 18, [18;30]. The three lengths are "regular sexagesimal numbers". (See Sec. 0.4 d for the meaning of this term.)

The suggestions concerning the absolute values of the three given numbers (for instance, that the second length is 26;40 rather than 26 40) are motivated by the observation that in Old Babylonian mathematical cuneiform texts *the sides of a field normally amount to one or several tens, or at most a few sixties, of the basic length unit, the* ninda.

As in so many other cases in Old Babylonian mathematical texts, data that appear to be arbitrarily chosen on closer inspection turn out to be carefully constructed. In the present case,

the lengths 25, 26;40, 27 = 1/3 times 1 15, 1 20, 1 21, where 1 15, 1 20, 1 21 are *consecutive regular integers.*

(It is easy to check that 1 20 is the only integer between 1 15 and 1 21 that is also a regular sexagesimal number.) Therefore, in MS 3944 *the three given lengths are consecutive regular integers (1 15, 1 20, 1 21) times 1/3, while the three fronts are consecutive integers (35, 36, 37) times 1/2!*

1.1 c.   MS 3955. Four Multiplication Exercises with Funny Numbers

MS 3955, the fourth example of a text with multiplication exercises (Fig. 1.1.3 below), is particularly interesting. It is a clay tablet with originally eight multiplication exercises, of which six are fairly well preserved, four on the obverse and two on the reverse.

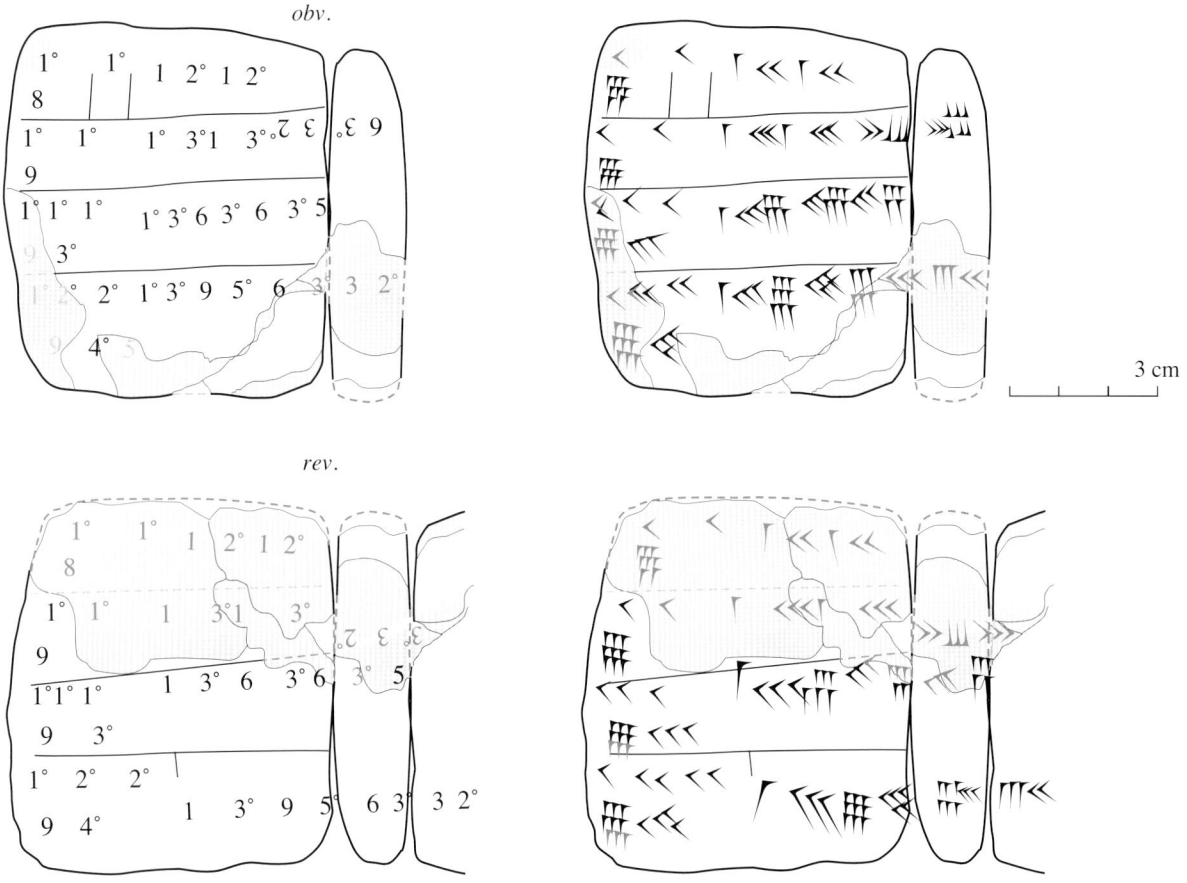

Fig. 1.1.3. MS 3955 *obv*., four multiplication exercises with funny numbers, the teacher's model.
The same four exercises are repeated on the reverse, the student's copy.

The writing on the reverse of this tablet is somewhat careless, and parts of both obverse and reverse are lost. Nevertheless (as will be demonstrated below), the text of all eight exercises can be reconstructed, in the following form:

1.	[10] 10	·	8	=	1 21 20
2.	10 10	·	9	=	1 31 30
3.	[10] 10 10	·	[9] 30	=	1 36 36 35
4.	[10 2]0 20	·	[9] 40	=	1 39 56 [33 20]
5.	[10 10]	·	[8]	=	[1 21 20]
6.	10 [10]	·	9	=	[1 31 30]
7.	10 10 10	·	9 30	=	1 36 36 [3]5
8.	10 20 20	·	9 40	=	1 39 56 33 20

The three first exercises in this text are examples of what happens when a one-place sexagesimal number like 8, 9, or a two-place sexagesimal number like 9 30 is multiplied by a "funny" sexagesimal number like 10 10 or 10 10 10. The most likely way of setting up the computations in the first two examples, exercises # 1 and # 2, is shown below:

10 10 · 8 =	1 20		10 10 · 9 =	1 30
	+ 1 20			+ 1 30
	1 21 20			1 31 30

Once the student has understood the principle, he can carry out computations like these in his head.

The most likely setup of the computation in exercise # 3 takes the idea one step further:

$$10\ 10\ 10 \cdot 9\ 30 = \quad 1\ 35 \qquad\qquad (10 \cdot 9\ 30 = 1\ 30(00) + 5(00) = 1\ 35\ (00))$$

$$\begin{array}{r} 1\ 35 \\ 1\ 35 \\ +\quad\ \ 1\ 35 \\ \hline 1\ 36\ 36\ 35 \end{array}$$

Exercise # 4 is similar:

$$10\ 20\ 20 \cdot 9\ 40 = 10 \cdot 9\ 40 \cdot 1\ 02\ 02 = 1\ 36\ 40 \cdot 1\ 02\ 02 = \quad 1\ 36\ 40$$

$$\begin{array}{r} 3\ 13\ 20 \\ +\quad\ \ 3\ 13\ 20 \\ \hline 1\ 39\ 56\ 33\ 20 \end{array}$$

The funny numbers in MS 3955 can be compared with similar funny numbers in **Ist. Ni 2739**, a small fragment of a combined multiplication table from Nippur (Neugebauer, *MKT 1* (1935), 79), which may have originally contained multiplication tables for all "head numbers" from 8 to [1 15] and [1 12]?, followed by a table of square sides. The last, atypical entries of this table of square sides are

1	.e 1	[íb.si$_8$]	sqs. 1 = 1
1 02 01	.e 1 01	[íb.si$_8$]	sqs. 1 02 01 = 1 01
1 02 03 02 01	.e 1 01 [01]	[íb.si$_8$]	sqs. 1 02 03 02 01 = 1 01 01
1 02 03 04 03 02 01	.e [1 01 01 01]	[íb.si$_8$]	sqs. 1 02 03 04 03 02 01 = 1 01 01 01.

For verification, the squares of the "funny numbers" 1 01, 1 01 01, and 1 01 01 01 can be computed by use of the same method as the one apparently used in MS 3955, ## 1-4.

On the reverse of MS 3955 only the last two exercises, ## 3-4, are well preserved. They turn out to be identical with the last two exercises on the obverse. Also the beginning of exercise # 2 on the reverse is identical with the beginning of the corresponding exercise on the obverse. The obvious conclusion is that *the reverse is a copy of the obverse*. The copy is not perfect, however. The clumsily written number signs of the sexagesimal numbers in the two preserved lines of the reverse occupy more space than the corresponding neatly written numbers on the obverse. One of them even spills over onto the right edge of the tablet and continues from there onto the obverse! This observation confirms the suspicion that the text on the obverse is a model written by a teacher, while the text on the reverse is a student's imperfect copy.

Similar Old Babylonian "teacher-student texts" have been published before, both mathematical and non-mathematical. A new example is MS 3925 (Fig. 3.2.4), a round clay tablet inscribed on the obverse and the reverse with the same four lines from a metrological table for large weight measures.

1.1 d.   *The Proto-Literate Field Expansion Procedure*

The observation that in Old Babylonian mathematical cuneiform texts the sides of a field are normally counted in tens (or a few sixties) of the basic length unit, the nin d a, has already been mentioned once in this section. In view of this rule, it can safely be assumed that the silently understood absolute sizes of the numbers in ## 3-4 on MS 3955 were

$$\begin{array}{lll} 3. & 10;10\ 10\ \text{n.} \cdot 9;30\ \text{n.} = 1\ 36;36\ 35 & \text{sq. n.} \\ 4. & 10;20\ 20\ \text{n.} \cdot 9;40\ \text{n.} = 1\ 39;56\ 33\ 20 & \text{sq. n.} \end{array}$$

It can hardly be a coincidence that the resulting area number in exercise # 4 is very close to 1 40 (100) square nin d a, the size of the Babylonian minor area unit 1 iku. The key to what is going on here is provided by a number of proto-cuneiform texts from the end of the 4th millennium BC. (See Friberg *AfO* 44/45 (1997/98).) In those texts, apparently, a certain "proto-literate field expansion procedure" was used in order *to find the sides of a rectangle with (approximately) a given area and a given ratio between the sides*. There exist quite a few indications that the same field expansion procedure was still operative in the Old Babylonian period. Against this background, the construction of the side numbers appearing in exercise # 4 on MS 3955 can be explained as follows. (Cf. the extensive discussion in Sec. 8.1 b below of the numbers associated with a drawing of a trapezoid on the round clay tablet MS 2107. See also the discussion in Appendix 6 below of the Old Akkadian exercises *DPA* 34 (Fig. A6.1) and A 786 (Fig. A6.2).)

In exercise # 3, the author of the text had shown that a nearly square rectangle with the sides 10;10 10 n i n d a and 9;30 n i n d a has the area 1 36;36 35 square n i n d a. Suppose that he now wanted to extend the sides of this "initial" rectangle by simple fractions of their lengths so that the resulting new nearly square rectangle would have an area as close as possible to the round area number 1 40 square n i n d a = 1 i k u. He would start by computing the "deficit" (1 40 – 1 36;36 35) sq. n., approximately equal to 3;20 sq. n. Thus, the approximate deficit was equal to about 1/30 of the initial area 1 36;36 35 sq. n. Extending one of the sides by 1/60 of its length, the author of the text could increase the initial area by 1/60 of its size., and increasing also the other side by 1/60 of its length he could increase the area by another 1/60 of its size. Thus, increasing both sides by 1/60 of their lengths, he could increase the initial area by $2 \cdot 1/60 = 1/30$ of its size, which is just what he wanted to do.

Now, 1/60 of 9;30 n. is approximately 1/60 of 10 n., which is ;10 n. Therefore the extended shorter side of the initial near-square would be

$$9;30 \text{ n.} + 1/60 \text{ of } 9;30 \text{ n.} = \text{(approximately)} \ 9;30 \text{ n.} + ;10 \text{ n.} = 9;40 \text{ n.}$$

Similarly, 1/60 of 10;10 10 n. is approximately equal to ;10 10 n. Therefore the extended longer side of the initial rectangle would be

$$10;10 \text{ 10 n.} + 1/60 \text{ of } 10;10 \text{ 10 n.} = \text{(approximately)} \ 10;10 \text{ 10 n.} + ;10 \text{ 10 n.} = 10;20 \text{ 20 n.}$$

Since fractions of the square n i n d a were often not mentioned in Old Babylonian texts, the area of the expanded rectangle would be close enough to the wanted area. Indeed,

$$10;20 \text{ 20 n.} \cdot 9;40 \text{ n.} = 1 \text{ 39};56 \text{ 33 20 sq. n.} = \text{(approximately)} \ 1 \text{ 40 sq. n.} = 1 \text{ i k u.}$$

As for the lengths of the sides expressed in conventional Old Babylonian units of length, one has to know that 1 ninda was equal to 12 cubits, and that 1 cubit was equal to 30 fingers. Consequently, 1 cubit = ;05 n., and 1 finger = ;00 10 n. This means that the result of exercise # 4 on MS 3955 can be expressed in the following way, in the appropriate units of length and area measure:

A rectangular field with the sides 10 n i n d a 4 cubits 2 fingers and 9 1/2 n i n d a 2 cubits has an area close to 1 i k u.

More exactly, since an area-shekel was 1/60 of a square n i n d a and an area-barley-corn 1/180 of an area-shekel, the area of a rectangular field with the indicated sides would be equal to precisely

$$10 \text{ n i n d a 4 cubits 2 fingers} \cdot 9 \text{ 1/2 n i n d a 2 cubits} = 1 \text{ 39}; 56 \text{ 33 20 sq. n i n d a} = 1 \text{ 40 sq. n i n d a} – ;03 \text{ 26 40 sq. n i n d a}$$
$$= 1 \text{ i k u} – 3 \text{ 1/3 area-shekels 20 area-barley-corns.}$$

1.2.   Old Babylonian Squaring Exercises

MS 2831 (Fig. 1.2.1) is a small rectangular tablet with a brief *non-standard table of squares* on the obverse. The five entries are all arranged as follows:

$$\begin{array}{c} a \\ a \end{array} \quad \text{(sq. } a\text{)}$$

just like the entries in the four multiplication exercises discussed in Sec. 1.1 above. This fact is ignored in the following simplified transliteration of the text. In the transliteration, semi-colons are inserted, where needed, to indicate the probably *intended absolute values* of the numbers:

1.	10	·	10	= 1 40	(100)
2.	11	·	11	= 2 01	(121)
3.	11;40	·	11;40	= 2 16;06 40	(136 1/9)
4.	12	·	12	= 2 24	(144)
5.	12;30	·	12;30	= 2 36;15	(156 1/4)

obv.

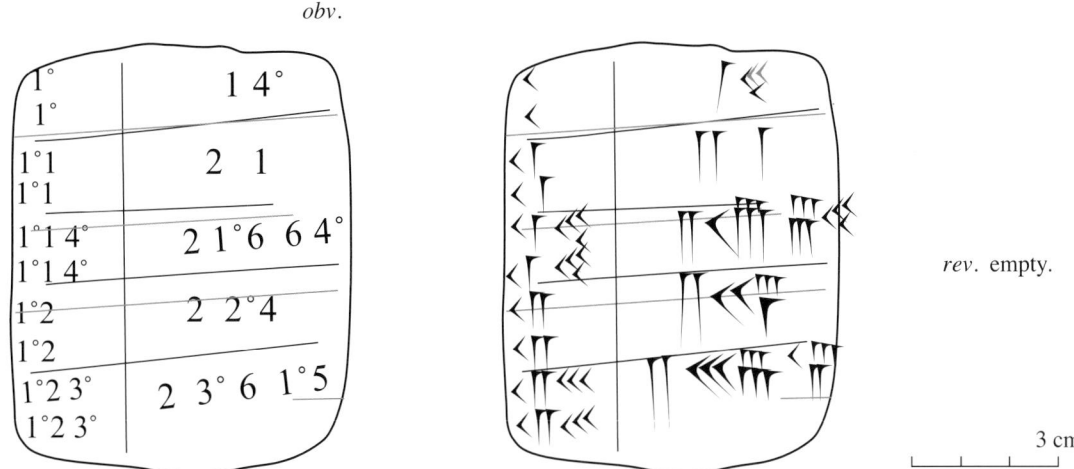

MS 2831. Computation of the squares of 5 numbers, 10 and 10 + 1/*n* · 10, with *n* = 10, 6, 5, 4.

obv.

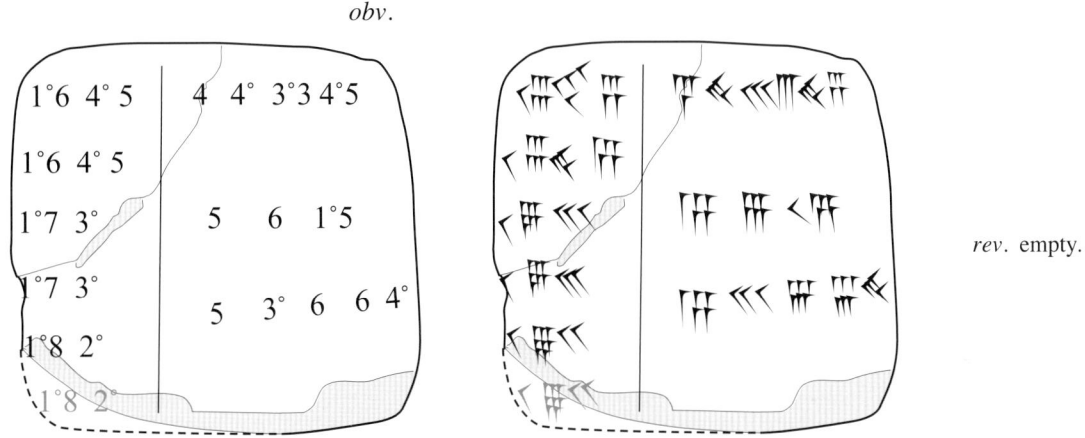

MS 3045. Computation of the squares of 3 numbers, 16;40 + 1/*n* · 16;40, with *n* = 200, 20, 10.

Fig. 1.2.1. Two non-standard tables of squares.

The numbers 10, 11, 11;40, 12, 12;30 of which the squares are computed in this text at first sight appear to be randomly chosen numbers between 10 and 12 30, ordered by size. However, a small text like MS 2831, with a brief series of (results of) related computations, can often be explained as the answers handed in by a student to an assignment given to him by his teacher. (Note that the given numbers in the left column are written with small and well formed cuneiform signs, while the computed numbers in the right column are written with large and awkwardly formed signs. It is clear that, in this particular text, the given numbers were written by the teacher, and the computed numbers by the student.)

It is to be expected that the teacher chose the given numbers in such an assignment with some care. (Cf. the explanations of the data in the multiplication exercises discussed in Sec. 1.1 above.) In the case of the present text, MS 2831, the five given numbers can be explained as the "round" number 10, followed by four "almost round numbers" of the form '10 plus a simple fraction of 10'. Indeed,

10			=	10		
11	=	10 + 1	=	10 +	1/10	of	10
11;40	=	10 + 1;40	=	10 +	1/6	of	10
12	=	10 + 2	=	10 +	1/5	of	10
12;30	=	10 + 2;30	=	10 +	1/4	of	10

Why the teacher chose to ask his student to compute the squares of such a series of almost round numbers is not directly obvious. Anyway, in handing out this assignment, the teacher's purpose was probably to give the student training in the application of the simple *binomial rule*

$$\text{sq.}\,(a+b) = \text{sq.}\,a + 2 \cdot a \cdot b + \text{sq.}\,b.$$

Cf. the discussion of a number of Old Akkadian "square-side-and-area exercises" in Appendix 6 below. See, in particular, the diagrams in Figs. A6.4 and A6.6.

A straightforward *geometric derivation of this rule* can be obtained through reference to a square of side $a + b$ divided by two transversal lines into a square of side a, a square of side b, and two rectangles with the sides a, b. Thus, the student may have been expected to compute the square of 11 as follows (if he did not already know the answer):

$$\text{sq.}\,11 = \text{sq.}\,(10+1) = \text{sq.}\,10 + 2 \cdot 10 \cdot 1 + \text{sq.}\,1 = 1\,40 + 20 + 1 = 2\,01.$$

In the next example, he could continue in various ways, computing for instance like this:

$$\text{sq.}\,11;\!40 = \text{sq.}\,(11 + ;\!40) = \text{sq.}\,11 + 2 \cdot 11 \cdot ;\!40 + \text{sq.}\,;\!40 =\quad 2\,01$$
$$14;\!40$$
$$\underline{+\quad ;\!26\,40}$$
$$2\,16;\!06\,40$$

Similarly, in the fourth example, he could proceed as follows:

$$\text{sq.}\,12;\!30 = \text{sq.}\,(12 + ;\!30) = \text{sq.}\,12 + 2 \cdot 12 \cdot ;\!30 + \text{sq.}\,;\!30 =\quad 2\,24$$
$$12$$
$$\underline{+\quad ;\!15}$$
$$2\,36;\!15$$

MS 3045 (Fig. 1.2.1, bottom) is a square tablet with another brief *non-standard table of squares* on the obverse. The entries are arranged as in MS2831, although this fact is ignored in the following simplified transliteration of the text, with semi-colons inserted in the supposedly correct places:

1.	16;45 ·	16;45	=	4 40;33 45
2.	17;30 ·	17;30	=	5 06;15
3.	18;20 ·	18;20	=	5 36;06 40

In line with the proposed explanation of the data on MS 2831, the given numbers (square sides) in the left column of MS 3045 can be explained as three "almost round numbers", all of the form '16;40 plus a simple (decimal!) fraction of 16;40':

16;45	=	16;40 +	;05	=	16;40 +	1/200 of 16;40
17;30	=	16;40 +	;50	=	16;40 +	1/20 of 16;40
18;20	=	16;40 +	1;40	=	16;40 +	1/10 of 16;40

If the student knew the square of 16;40, he could compute as follows:

$$\text{sq.}\,16;\!45 = \text{sq.}\,(16;\!40 + ;\!05) = \text{sq.}\,16;\!40 + 2 \cdot 16;\!40 \cdot ;\!05 + \text{sq.}\,;\!05 =\ 4\,37;\!46\,40$$
$$2;\!46\,40$$
$$\underline{+\quad ;\!00\,25}$$
$$4\,40;\!33\,45$$

And so on, in the other examples.

Remark: Since 16 40 = 1,000, the three square sides can also, perhaps, be explained as follows:

16 45 =	16 40 +	5	=	16 40 +	1/200	of 16 40	= 1,005
17 30 =	16 40 +	50	=	16 40 +	1/20	of 16 40	= 1,050
18 20 =	16 40 +	1 40	=	16 40 +	1/10	of 16 40	= 1,100

MS 3946 (Fig. 1.2.2 below) is a squarish hand tablet with a brief non-standard table of squares on obverse and reverse. The entries on MS 3946 are arranged as in MS 2831, although this fact is ignored, as usual, in the following transliteration of the text (written now *without inserted semicolons and final zeros!*):

1. 1 45 · 1 45 = 2 30 33 45 (error!)   (sq. 1;45 = 3;03 45 = appr. 3)
2. 1 54 · 1 54 = 3 36 36   (sq. 1;52 30 = appr. 3;30;   1;54 may be a bad approximation)
3.   2 · 2   = 4   (sq. 2 = 4)
4. 4 35 · 4 35 = 21 25 (?)   (one would expect here sq. 2;07 30 = appr. 4;30)

On the reverse, the computed value of the square of 4 35 is badly preserved. It appears to be written here *without the use of an internal zero*. The correct value of the square of 4 35 is 21 00 25.

The error in line 1 is interesting. The square of 1 45 should have been computed as follows:

$$\text{sq. } 1\ 45 = \text{sq. } (1 + 45) = \text{sq. } 1 + 2 \cdot 1 \cdot 45 + \text{sq. } 45 = \begin{array}{r} 1 \\ 1\ 30 \\ +\ 33\ 45 \\ \hline 3\ 03\ 45 \end{array}$$

(The modern formulation of this result, with anachronistically inserted zeros, is that

$$\text{sq. } 1\ 45 = \text{sq. } (1\ 00 + 45) = 1\ 00\ 00 + 1\ 30\ 00 + 33\ 45 = 3\ 03\ 45.)$$

The incorrect result in the text corresponds to the following careless addition (a "telescoping error"):

$$\text{sq. } 1\ 45 = \text{sq. } (1 + 45) = \text{sq. } 1 + 2 \cdot 1 \cdot 45 + \text{sq. } 45 = \begin{array}{r} 1 \\ 1\ 30 \\ +\ \ \ \ 33\ 45 \\ \hline 2\ 30\ 33\ 45 \end{array}$$

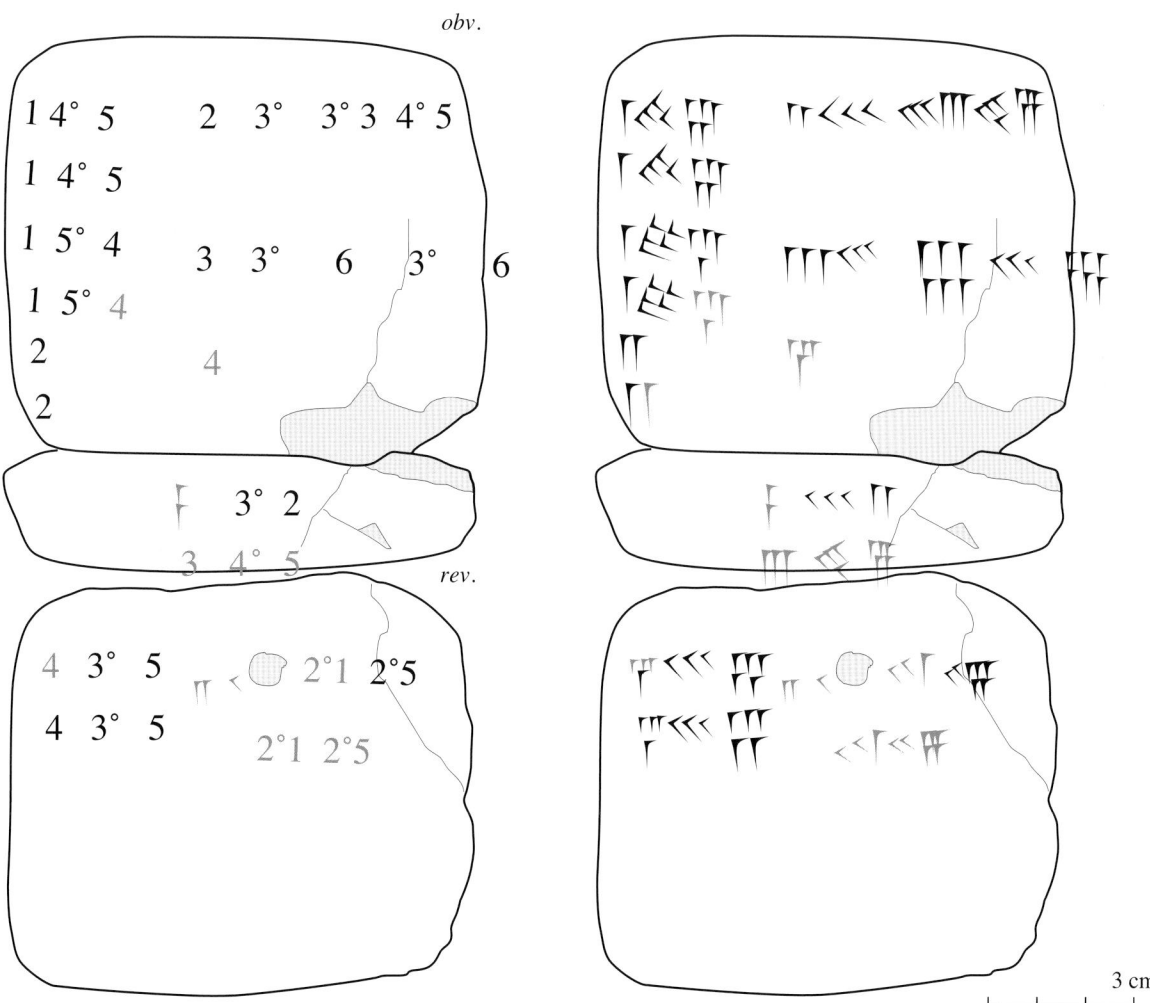

Fig. 1.2.2. MS 3946, four squaring exercises with a puzzling choice of data and an interesting error.

The choice of data in this text is puzzling, in particular the inclusion of the trivial computation sq. 2 = 4. There is no obvious explanation in style with the proposed explanations of the choice of data on MS 2831 and MS 3045 (Fig. 1.2.1).

The text on the lower edge seems to be scribbled notations without any obvious relation to the computations on the obverse and the reverse.

Two previously published parallels to the squaring exercises in MS 2831, 3045, 3922, 3946 and 3947 are **YBC 7294** and **YBC 10801** (Neugebauer and Sachs, *MCT* (1945), 35; photos in Nemet-Nejat, *UOS* (2002), 276-277). In YBC 7294 is recorded only the square of 2 30, and in YBC 10801 the square of 4 35. Note that 4 35 = 275 = 25 · 11 = 4 10 + 1/10 of 4 10, where 4 10 = 250.

1.3.   An Old Babylonian Division Exercise

MS 3871 (Fig. 1.3.1 below) is a clay tablet of the same squarish format as the hand tablets with the multiplication and squaring exercises in Secs. 1.1-1.2. It is inscribed with three lines of sexagesimal numbers:

1.   4 37 46 40
2.   11 34 26 40
3.   2 30.

The three numbers can easily be factorized by use of the trailing part algorithm. (See Sec. 1.4 a below.) In modern terms, the result is that

$$\begin{aligned}4\ 37\ 46\ 40 &= 2^6 \cdot 5^6 \\ 11\ 34\ 26\ 40 &= 2^5 \cdot 5^7 \\ 2\ 30 &= 2 \cdot 3 \cdot 5^2\end{aligned}$$

Thus, the three recorded numbers are all regular sexagesimal numbers, and

$$4\ 37\ 46\ 40 \cdot 2\ 30 = 2^7 \cdot 3 \cdot 5^8 = 2^2 \cdot 3 \cdot 5 \cdot 11\ 34\ 26\ 40 = 60 \cdot 11\ 34\ 26\ 40.$$

This analysis of the recorded numbers suggests that the text is a "division exercise" of a new kind. *There is no previously published parallel text.*

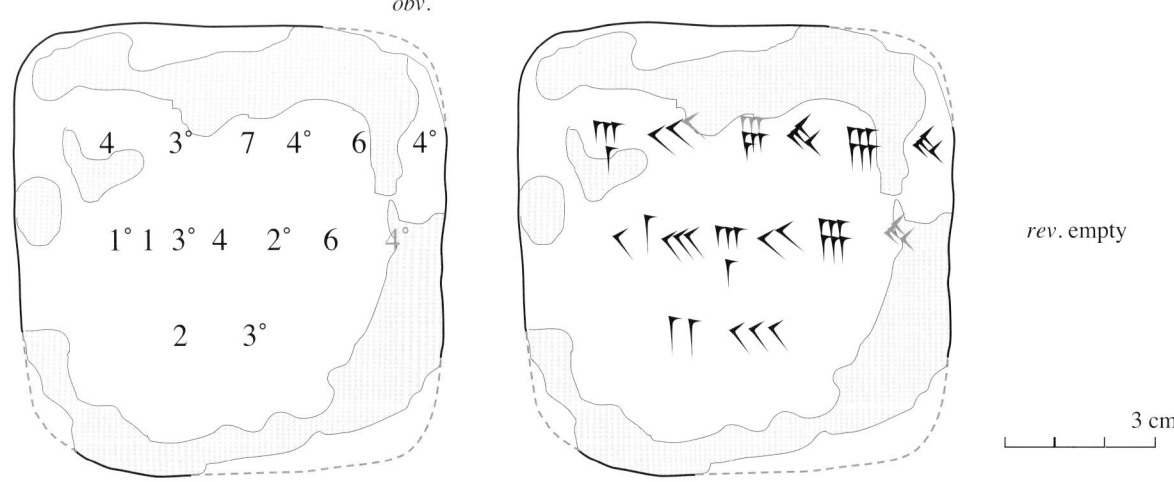

Fig. 1.3.1. MS 3871. A division exercise: What times 4 37 46 40 equals 11 34 26 40? Answer: 2 30.

In Old Babylonian mathematical problem texts, it often happens that the need arises to divide one given sexagesimal number (*a*) by another (*b*). If the reciprocal igi *b* is known, then *a/b* is routinely computed as *a* · igi *b*. (See Sec. 0.4 d above.) Otherwise, the situation is expressed with a couple of standard phrases of the following form:

igi *b ú-ul ip-pa-aṭ-ṭa-ar* (nu.du₈)   the opposite of *b* cannot be resolved
mi-nam a.na *b lu-uš-ku-un* (ḫé.gar) *ša a i-na-di-nam* (in.sì)   what as much as *b* shall I set that will give me *a*?

(It cannot be excluded that the correct translation of the second phrase is instead 'what to (*a-na*) *b* shall I set that will give me *a*?'.) The answer to this question is always given directly; there is no known text where the question is followed by an explicit computation.

In the case of MS 3871, it is likely that the student was given the assignment to find out what number he should multiply 4 37 46 40 with in order to get 11 34 26 40. The recorded answer, 2 30, may have been computed by the student or it may have been provided by the teacher. In either case, it was probably the student's task to supply the details of the computation. It is possible that he proceeded as follows, using the "trailing part algorithm" (see Sec. 1.4 a below):

4 37 46 **40**	9	11 34 26 40
41 40	36	1 44 10
25	2 24	1 02 30
1		2 30

A division algorithm like this amounts to finding a factorization of the reciprocal of *b* = 4 37 46 40 and multiplying *a* = 11 34 26 40 by the computed factors of the reciprocal, one at a time. The algorithm works every time that *b* is a regular sexagesimal number.

It is interesting to investigate how the division algorithm may have been modified in the case when *b* is only a *semi-regular* sexagesimal number, that is, the product of a regular and a non-regular sexagesimal number. A particularly interesting example is the single problem text **VAT 7532** (Høyrup, *LWS* (2002), 209). This is an example of a "broken reed problem" (Friberg, *RlA 7* (1990), Sec. 5.4b; cf. **MS 3971** § 1 in Sec. 10.1 a below), where the sides of a trapezoid are measured by use of a reed of unknown length *r*, from which a sixth of its original length has been broken off, so that the remaining length is $r* = (1 – 1/6) \cdot r = ;50 \cdot r$. The problem is reduced to a quadratic equation which, in quasi-modern terms, can be expressed as follows:

6 14 24 sq. *r** – 12 00 ninda · *r** = 1 00 00 sq. ninda,
where 6 14 24 = 2 36 · 2 24 is the a.šà lul 'false area',
and where 1 00 00 sq. ninda is twice the 'true area'.

The quadratic equation is solved by use of a routine application of metric algebra (cf. Friberg, *op. cit.* Sec. 5.7c). In one of the last steps of the solution algorithm, the problem has been reduced to a linear equation:

6 14 24 · *r** = 2 36 (00) ninda.

The division of 2 36 (00) by 6 14 24 is not explicitly given, but was possibly achieved as follows:

6 14 **24**	2 30	2 36
15 **36**	1 40	6 30
26	30	10 50
13		5 25

Thus, the removal of all regular factors in 6 14 24 shows that the equation 6 14 24 · *r** = 2 36 (00) ninda can be simplified to 13 (00) · *r** = 5 25 ninda. In the multiplication table for the head number 25 one of the entries says that 13 · 25 = 5 25. It follows, as in the text of VAT 7532, that

$r* = ;25$ ninda, and $r = (1 + 1/5) \cdot r* = ;30$ ninda = 1 'reed'.

Note: In Sec. 7.1 a below is considered a curious division exercise (**MS 2317**) involving a given number (1 01 01 01) with *only* non-regular sexagesimal numbers as factors.

1.4.   Old Babylonian Operations with Many-Place Regular Sexagesimal Numbers

A "many-place" sexagesimal number is a sexagesimal number with more than a few sexagesimal places (double digits). Old Babylonian assignments involving many-place (regular) sexagesimal numbers are often, for obvious reasons, inscribed on clay tablets that are *more wide than tall*.

A given sexagesimal number is "regular" if it is possible to find another sexagesimal number a', the "reciprocal" of a, such that $a \cdot a' = '1'$ (= some power of 60). (See Sec. 0.4 d.)

1.4 a. Factorization by Use of the Trailing Part Algorithm

MS 2242 (Fig. 1.4.1, top) is a small clay tablet inscribed with a sequence of six sexagesimal numbers of decreasing length. The initial entry is an 8-place number, the final entry only 2-place. All the six numbers have their last two places equal to 03 45. The six numbers are:

1.   46 20 54 51 30 14 03 45
2.   12 21 34 37 44 03 45
3.   3 17 45 14 03 45
4.   52 44 03 45
5.   14 03 45
6.   3 45

It is likely that this text is a student's answer to an assignment given to him by his teacher:

Factorize 46 20 54 51 30 14 03 45, a given many-place regular sexagesimal number.

The method the student was supposed to use was almost certainly the Old Babylonian "trailing part algorithm". (See Friberg, *RlA 7* (1990), Sec. 5.3b, and Sachs, *JCS* 1 (1947).). For an illustration of how the method works, consider the initially given number 46 20 54 51 30 14 03 45. Its 2-place "trailing part" is 03 45. Since 15 is a factor in 1 00 = 60, the base of the sexagesimal system, it is clear that 3 45, its square, is a factor in 1 00 00 = 60 · 60. Therefore, 3 45 is a factor in every sexagesimal number with the trailing part 03 45, in particular in the given number

46 20 54 51 30 14 03 45 = 46 20 54 51 30 14 · 1 00 00 + 3 45.

Each factor 3 45 can be removed through multiplication by 16, which is the reciprocal of 3 45 in the sense that 16 · 3 45 = 1 (00 00). In the present case, for instance, the number in line 2 on MS 2242 was probably computed as follows (*in relative numbers* without final zeros):

46 20 54 51 30 14 03 45 · 16 = 46 20 54 51 30 14 · 16 + 1 = 12 21 34 37 44 03 45.

This new number, too, has the trailing part 03 45. Therefore, the procedure can be repeated. Indeed, it can be repeated without change as long as every calculated number has the trailing part 03 45.

In the example recorded on MS 2242, the procedure was repeated 5 times. Each time, the new, reduced number had the trailing part 03 45. Hence, the successive steps of the algorithm were:

46 20 54 51 30 14 03 45
12 21 34 37 44 03 45   (46 20 54 51 30 14 · 16 + 1)
3 17 45 14 03 45   (12 21 34 37 44 · 16 + 1)
52 44 03 45   (3 17 45 14 · 16 + 1)
14 03 45   (52 44 · 16 + 1)
3 45   (14 · 16 + 1)

Since one factor 3 45 was removed every time the procedure was repeated, it follows that the given number contained 3 45 as a factor 6 times. In other words, *the application of the trailing part algorithm in MS 2242 resulted in an effective factorization of the given number*, showing that

46 20 54 51 30 14 03 45 = the 6th power of 3 45 (= the 12th power of 15).

This means, by the way, that the six numbers recorded on MS 2242 are *the first six powers of 3 45, in descending order*

obv.

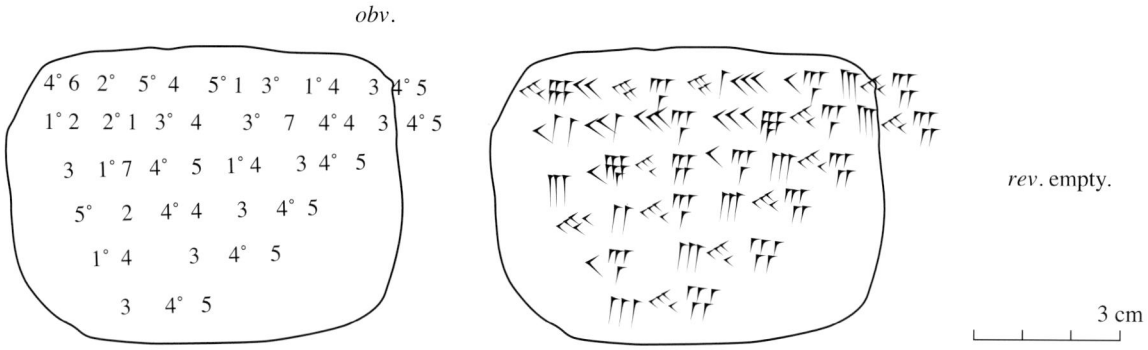

rev. empty.

MS 2242. Factorization of 46 20 54 51 30 14 03 45 (the 6th power of 3 45) by use of the trailing part algorithm.

obv.

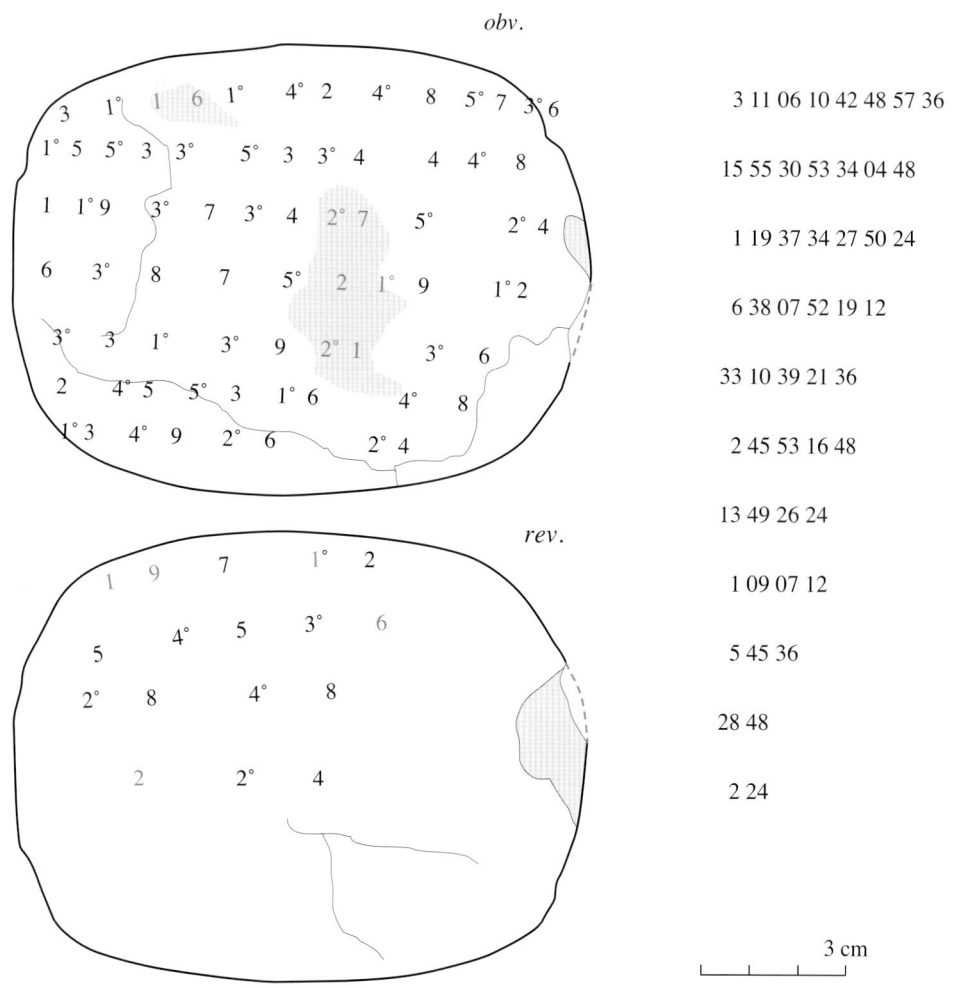

	3 11 06 10 42 48 57 36
	15 55 30 53 34 04 48
	1 19 37 34 27 50 24
	6 38 07 52 19 12
	33 10 39 21 36
	2 45 53 16 48
	13 49 26 24
	1 09 07 12
	5 45 36
	28 48
	2 24

MS 3037. Factorization of 3 11 06 10 42 48 57 36 (the 12th power of 12) by use of the trailing part algorithm.

Fig. 1.4.1. Two examples of factorizations of many-place regular sexagesimal numbers.

No previously published Old Babylonian mathematical text is a direct parallel to MS 2242.[1] On the other hand, there exists a *Late Babylonian text*, about 1500 years younger, which starts with a given regular sexagesimal number, from which factors are removed one at a time by use of the trailing part algorithm. The text is von Weiher, *SpTU 5, 316* (1998), a round hand tablet with 10 lines of numbers. With some minor errors corrected, the text is the following (see Fig. A9.1 in Appendix 9 below):

1 02 12 28 48	$1\ 02\ 12\ 28\ 48 \cdot 5\ =\ 5\ 11\ 02\ 24$
5 11 02 24	$5\ 11\ 02\ 24 \cdot 5\ =\ 25\ 55\ 12$
25 55 12	$25\ 55\ 12 \cdot 5\ =\ 2\ 09\ 36$
2 09 36	$2\ 09\ 36 \cdot 5\ =\ 10\ 48$
10 48	$10\ 48 \cdot 5\ =\ 54$
54	$54 \cdot 5\ =\ 4\ 30$ (a badly chosen multiplier)
4 30	$4\ 30 \cdot 2\ =\ 9$ (here the mistake is corrected)
9	$9 \cdot 20\ =\ 3$
3	$3 \cdot 20\ =\ 1$
1	

The algorithm demonstrates that, in *relative* sexagesimal numbers,

the reciprocal of the given number is $\quad 5 \cdot 5 \cdot 5 \cdot 5 \cdot 5 \cdot 10! \cdot 20 \cdot 20$,
equal to the 5th power of 2 times the 8th power of 5.

Hence, again in *relative* sexagesimal numbers,

the given number itself is $\quad 12 \cdot 12 \cdot 12 \cdot 12 \cdot 12 \cdot 6 \cdot 3 \cdot 3$,
equal to the 11th power of 2 times the 8th power of 3.

MS 3037 (Fig. 1.4.1, bottom) is an oval medium sized clay table inscribed with a sequence of eleven sexagesimal numbers of decreasing length. As in MS 2242, the initial entry is an 8-place number, the final entry only 2-place. The eleven numbers are:

1.   3 11 06 10 42 48 57 36
2.   15 55 30 53 34 04 48
3.   1 19 37 34 27 50 24
4.   6 38 07 52 19 12
5.   33 10 39 21 36
6.   2 45 53 16 48
7.   13 49 26 24
8.   1 09 07 12
9.   5 45 36
10.   28 48
11.   2 24

The last sexagesimal places of these numbers cycle through the values 36, 48, 24, 12, all of which contain 12 as a factor. Since 12 is also a factor in 60, it is clear that 12 is a common factor in all the numbers. Each factor 12 can be removed through multiplication with 5, the reciprocal of 12 in the sense that $5 \cdot 12 = 1(00)$. Hence, the first few steps of the trailing part algorithm in this example are

---

1. However, *increasing* geometric progressions corresponding to the first 10 powers of 3 45, a so called "table of powers", are recorded on four Old Babylonian tablets from Kish, **Ist. O 3816, 3826, 3862, 4583** (Neugebauer, *MKT 1*, 77-78). On an Old Babylonian tablet from Larsa (**IM 73355**; Arnaud (1994), *BBVOT 3*, pl. 35; see Fig. A5.7 in Appendix 5) are recorded, in increasing order, the first 10 powers of 3 45, followed by the first 10 powers of 16, the reciprocal of 3 45. Cf. the Old Babylonian text **BM 22706** in Nissen/Damerow/Englund, *ABK* (1993), 150. On the obverse of that tablet are recorded the first 10 powers of 1 40, while on the reverse are recorded the first 10 powers of 5.

<table>
<tr><td>3 11 06 10 42 48 57 36</td></tr>
<tr><td>15 55 30 53 34 04 48</td></tr>
<tr><td>1 19 37 34 27 50 24</td></tr>
<tr><td>6 38 07 52 19 12</td></tr>
<tr><td>.........</td></tr>
</table>

$$3\ 11\ 06\ 10\ 42\ 48\ 57 \cdot 5 + 3 = 15\ 55\ 30\ 53\ 34\ 04\ 48$$
$$15\ 55\ 30\ 53\ 34\ 04 \cdot 5 + 4 = 1\ 19\ 37\ 34\ 27\ 50\ 24$$
$$1\ 19\ 37\ 34\ 27\ 50 \cdot 5 + 2 = 6\ 38\ 07\ 52\ 19\ 12$$
$$6\ 38\ 07\ 52\ 19 \cdot 5 + 1 = 33\ 10\ 39\ 21\ 36$$

1.4 b.   Reciprocals of Many-Place Regular Sexagesimal Numbers

MS 2730 (Fig. 1.4.2, top) is a rectangular clay tablet, inscribed with cuneiform number signs in two lines of text near the upper edge:

1. 4 51 16 16 03 45
2. 12 21 34 37 44 03 45

The 7-place sexagesimal number in the second line of MS 2730 happens to be identical with the number in the second line of the text MS 2242 (Fig. 1.4.1, top). Therefore, it is equal to the 5th power of 3 45, a regular sexagesimal number. Its reciprocal is then the 5th power of 16, which can be shown to be the number 4 51 16 16. This observation is a clue to the correct explanation of the first line of text on MS 2730, which must now be interpreted as *the regular sexagesimal number 4 51 16 16 followed by the number 3 45, the reciprocal of 16, the one-place trailing part of 4 51 16 16.*

The text can be interpreted as an assignment, where a student was asked to compute the reciprocal of 4 51 16 16, the 5th power of 16 (or the 20th power of 2), by use of the trailing part algorithm. One possibility is that the assignment was given in the form of the first and the last line of the algorithm. Another is that the assignment was just to find the reciprocal of the given number 4 51 16 16, and that the student gave his answer on this tablet in the form of the first step of the algorithm together with the computed reciprocal of the given number. In either case, *the full form of the algorithm table* that the student was supposed to produce is shown below:

<table>
<tr><td>4 51 16 16</td><td>3 45</td></tr>
<tr><td>18 12 16</td><td>3 45</td></tr>
<tr><td>1 08 16</td><td>3 45</td></tr>
<tr><td>4 16</td><td>3 45</td></tr>
<tr><td>16</td><td>3 45</td></tr>
<tr><td></td><td>14 03 45</td></tr>
<tr><td></td><td>52 44 03 45</td></tr>
<tr><td></td><td>3 17 45 14 03 45</td></tr>
<tr><td></td><td>12 21 34 37 44 03 45</td></tr>
</table>

$$4\ 51\ 16 \cdot 3\ 45 + 1 = 18\ 12\ 16$$
$$18\ 12 \cdot 3\ 45 + 1 = 1\ 08\ 16$$
$$1\ 08 \cdot 3\ 45 + 1 = 4\ 16$$
$$4 \cdot 3\ 45 + 1 = 16$$
$$\text{rec. } 16 = 3\ 45$$

the 2nd power of 3 45
the 3rd power of 3 45
the 4th power of 3 45
the 5th power of 3 45

In the first five lines of this algorithm table, five factors 16 in the given number 4 51 16 16 are removed, one at a time, through multiplication by 3 45, the reciprocal of 16. The result is a *decreasing* table of powers of 16, to the left, and 5 copies of the reciprocals 3 45, to the right. In the five last lines of the table, these reciprocals are multiplied together, one at a time. The result is an *increasing* table of powers of 3 45. The number recorded in the last line is then the 5th power of 3 45, which is the reciprocal of the given number.

The general rule applied here, well known by Old Babylonian mathematicians, is that

> The reciprocal of a regular sexagesimal number is equal to the product of the reciprocals of the factors of the number.

Another general rule, also well known by Old Babylonian mathematicians, is that

> The reciprocal of the reciprocal of a given regular sexagesimal number is equal to the given number.

As a consequence of this second rule, after the computation of the reciprocal of a given many-place regular sexagesimal number, the correctness of the computation can be verified by computing the reciprocal of the computed reciprocal, which should be equal to the given number.

obv.

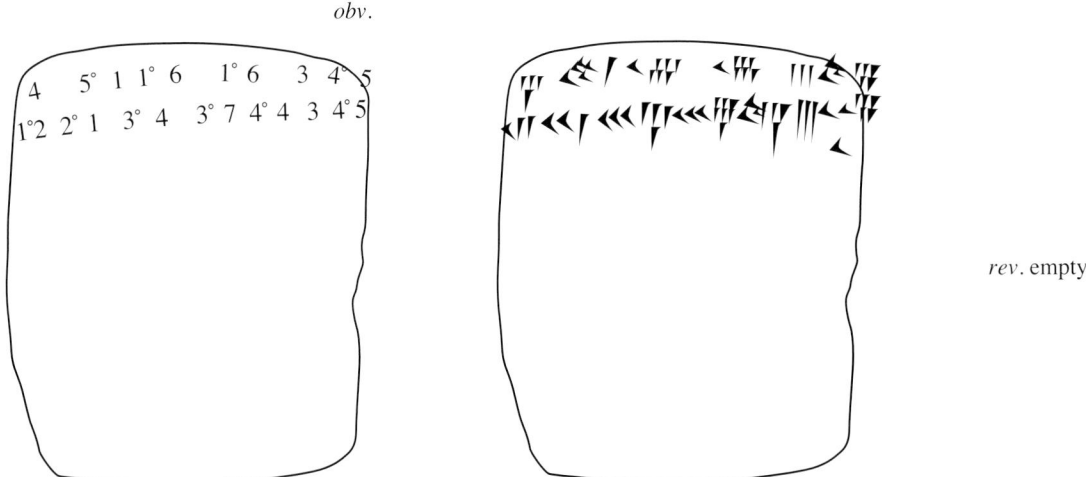

MS 2730. An assignment: Use of the trailing part algorithm to find the reciprocal of 4 51 16 16 (the 20th power of 2).

obv.

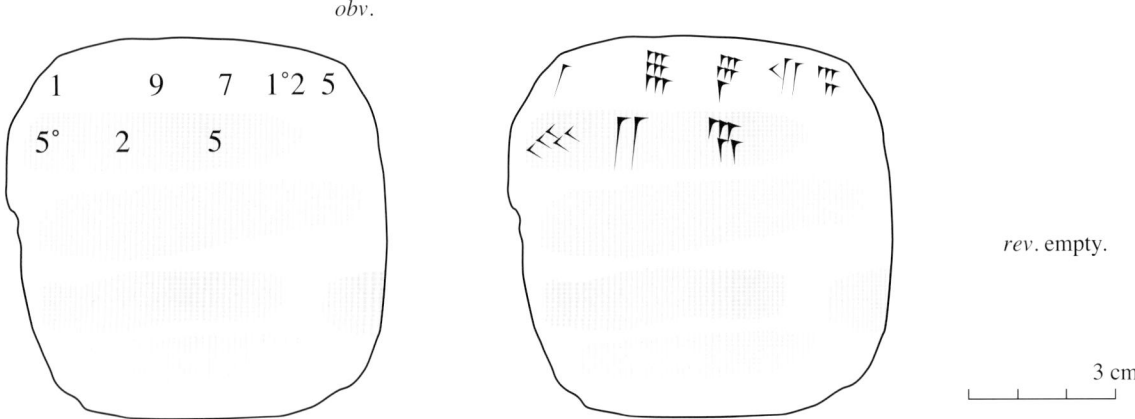

rev. empty.

3 cm

MS 2732. An assignment: Use of the trailing part algorithm to find the reciprocal of 1 08 07 12 (the 5th power of 12).

Fig. 1.4.2. Assignments: To compute the reciprocals of given many-place regular sexagesimal numbers.

Several *explicit* examples of Old Babylonian algorithm tables of this kind have been published before. See, in particular, the discussion of the reverse of the round hand tablet **UET 6/2, 295** in Friberg, *RA* 94 (2000) § 2 c, which is followed by a survey of other known examples. In *UET 6/2, 295*, it is shown that the reciprocal of 2 05 (the third power of 5) is equal to 28 48 (the third power of 12). Then, by reversal of the process, it is shown that the reciprocal of 28 48 is 2 05:

2 05	12	rec. 5 =	12	2 · 12 + 1 = 25
25	2 24	rec. 25 =	2 24	2 24 · 12 = 28 48
28 48	1 15	rec. 48 =	1 15	28 · 1 15 + 1 = 36
36	1 40	rec. 36 =	1 40	1 40 · 1 15 = 2 05
2 05				

In the Late Babylonian period, in the latter half of the 1st millennium BC, the reciprocals of many-place regular sexagesimal numbers were still computed by use of essentially the same method. This fact is demonstrated by **W 23021**, a large fragment of a round clay tablet from Uruk. (See Friberg, *BagM* 30 (1999)). On that tablet were recorded the computations of the reciprocals of eight given numbers by use of the trailing part algorithm. In the table below, the given numbers are the ones to the left, while their computed reciprocals are the ones to the right. An explanation for the choice of given numbers is given in Friberg, *op. cit*. (Note that these numbers and their reciprocals appear in the columns for igi *n* and *n* in the Late Babylonian "First Tablet of many-place reciprocals" in Appendix 9.)

52 40 29 37 46 40	1 08 20 37 30
51 50 24	1 09 26 40
51 26 25 11 06 40	1 09 59 02 24
51 12	1 10 18 45
50 34 04 26 40	1 11 11 29 03 45
50	1 12
49 22 57 46 40	1 12 54
49 09 07 12	1 13 14 31 52 30

MS 2732 (Fig. 1.4.2, bottom) is a crudely fashioned square clay tablet with rounded corners. There are traces on it of several lines of erased numbers, but it is newly inscribed with cuneiform number signs in two lines of text near the upper edge:

1.   1 09 07 12 5   2.   52 05

Just like the two lines of numbers on MS 2730, the two lines of numbers on MS 2732 can be explained as the first and last lines of an algorithm table associated with the trailing part algorithm:

1 09 07 12	5	rec. 12 = 51 09 07 · 5 + 1 = 5 45 36
5 45 36	1 40	rec. 36 = 1 40 5 45 · 1 40 + 1 = 9 36
9 36	1 40	rec. 36 = 1 40 1 40 · 9 + 1 = 16
16	3 45	rec. 16 = 3 45 3 45 · 1 40 = 6 15
	6 15	6 15 · 1 40 = 10 25
	10 25	10 25 · 5 = 52 05
	52 05	

In the first 4 lines of this algorithm table, the four factors 12, 36, 36, and 16 are removed, one at a time, from the given number 1 09 07 12, through multiplication by the reciprocals of those factors, that is 5, 1 40, 1 40, and 3 45. This shows, incidentally, that

$$1\ 09\ 07\ 12 = 12 \cdot 36 \cdot 36 \cdot 16 = \text{the 5th power of 12.}$$

In the 4 last lines of the table, the reciprocals 5, 1 40, 1 40, and 3 45 are multiplied together, in inverse order. The result, recorded in the last line of the table, is the reciprocal of the given number. Hence, this reciprocal can be factorized as follows (in *relative* sexagesimal numbers):

$$52\ 05 = 3\ 45 \cdot 1\ 40 \cdot 1\ 40 \cdot 5 = 15 \cdot 15 \cdot 4 \cdot 25 \cdot 4 \cdot 25 \cdot 5 = \text{the 5th power of 5.}$$

MS 2793 (Fig. 1.4.3, top) is a nearly square clay tablet, like MS 2730 and MS 2732 inscribed with cuneiform number signs in two lines of text near the upper edge:

1.   41 25 30 48 32   2.   1 26 54 12 51 34 11 22

Just like the two lines of numbers on MS 2730 and on MS 2732, the two lines of numbers on MS 2732 can be explained as the first and last lines of an algorithm table, the computation of the reciprocal of 41 25 30 48 32 by use of the trailing part algorithm. Note that, on the clay tablet, the last line is incomplete. The last 3 places of the number 1 26 54 12 51 34 11 22 01 52 30 are missing, although one would have expected to find them inscribed across the right edge and the reverse. A likely explanation is that the two lines on this tablet were copies of two lines on another tablet, the teacher's model, where the last 3 places of the second number where written on the right edge, so that they were inadvertently missed by the student making the copy!

The correct form of the whole algorithm table would be

41 25 30 48 32	30	rec. 2 = 30	41 25 30 48 · 30 + 1 = 20 42 45 24 16
20 42 45 24 16	3 45	rec. 16 = 3 45	20 42 45 24 · 3 45 + 1 = 1 17 40 20 16
1 17 40 20 16	3 45	rec. 16 = 3 45	1 17 40 20 · 3 45 + 1 = 4 51 16 16
4 51 16 16	3 45	rec. 16 = 3 45	4 51 16 · 3 45 + 1 = 18 12 16
18 12 16	3 45	rec. 16 = 3 45	18 12 · 3 45 + 1 = 1 08 16
1 08 16	3 45	rec. 16 = 3 45	1 08 · 3 45 + 1 = 4 16
4 16	3 45	rec. 16 = 3 45	4 · 3 45 + 1 = 16
16	3 45	rec. 16 = 3 45	
	14 03 45		3 45 · 3 45
	52 44 03 45		14 03 45 · 3 45
	3 17 45 14 03 45		52 44 03 45 · 3 45
	12 21 34 37 44 03 45		3 17 45 14 03 45 · 3 45
	46 20 54 51 30 14 03 45		12 21 34 37 44 03 45 · 3 45
	2 53 48 25 43 08 22 44 03 45		46 20 54 51 30 14 03 45 · 3 45
	1 26 54 12 51 34 11 22 01 52 30		2 53 48 25 43 08 22 44 03 45 · 30

Note that in the first line of the algorithm table, one might have expected to see the reciprocal of 32, rather than the reciprocal of 2, figuring in the right column. The reason why the reciprocal of 32 is not used here is that it is a 3-place sexagesimal number, 1 52 30, for which no Old Babylonian single multiplication table existed! (More about this below, in Sec. 2.6 f.)

The algorithm table above makes it clear that the given number on MS 2793 can be factorized as

$$41\ 25\ 30\ 48\ 32 = 2 \cdot 16 \cdot 16 \cdot 16 \cdot 16 \cdot 16 \cdot 16 \cdot 16 = \text{the 29th power of 2.}$$

MS 2894 (Fig. 1.4.3, middle) is a finger-shaped clay tablet, just large enough to hold three lines of number signs. An 8-place(?) sexagesimal number is inscribed in line 1, and a 9-place number in lines 2-3. The beginning of the first number is lost:

$$[1\ 2]6^?\ 13\ 24\ 45\ 11\ 06\ 40\ /\ \text{igi}^?\ 41\ 42\ 49\ 22\ 21\ /\ 12\ 39\ 22\ 30$$

It looks as if the first number might be a regular sexagesimal number, since its 2-place trailing part 06 40 is a regular number. There even exist regular sexagesimal numbers sharing the 4-place trailing part 45 11 06 40 with the first number, for instance $19\ 45\ 11\ 06\ 40 = 2^{14} \cdot 5^6$. On the other hand, it is easy to show that there are *no* regular sexagesimal numbers sharing the 6-place trailing part 13 24 45 11 06 40 with the first number. Indeed, if one tries to use the trailing part algorithm to remove regular factors from any number with this 6-place trailing part, one gets the following result:

$$\begin{array}{ll}
(\ldots)\ 13\ 24\ 45\ 11\ 06\ 40 & 9 \\
(\ldots)\ 00\ 42\ 46\ 40 & 9 \\
(\ldots)\ 06\ 25 & 12 \\
(\ldots)\ 17 &
\end{array}$$

A sexagesimal number with the last place 17 cannot be a regular sexagesimal number. Since any number of the form (...) 13 24 45 11 06 40 is equal to 6 40 · 6 40 · 5 · (...) 17, a number with the 6-place trailing part 13 24 45 11 06 40 cannot be a regular sexagesimal number either.

The second number recorded on MS 2894 is easily recognized as a regular sexagesimal number. Indeed, compare with sub-table # 1 of the fragment UM 29.13.21 (Neugebauer and Sachs, *MCT* (1945), 13-14; Fig. A3.2), a slightly imperfect copy of what seems to have been the standard example of the Old Babylonian "doubling and halving algorithm" (Friberg, *RlA 7* (1990) Sec. 5.3b; see also Appendix 3 below).This standard example is a list of 30 reciprocal pairs of regular sexagesimal numbers, beginning with the pair (2 05, 28 48) and ending with the pair

$$(1\ 26\ 18\ 09\ 11\ 06\ 40,\ 41\ 42\ 49\ 22\ 21\ 12\ 39\ 22\ 30) = (2\ 05 \cdot 2^{29},\ 28\ 48 \cdot 2^{-29}).$$

Actually, the last entry in § 1 of UM 29.13.21 is

$$1\ 25\ 20\ \text{KUD}\ 58\ 09\ 11\ 06\ 40\ /\ \text{igi.bi}\ 41\ 42\ 49\ 22\ 21\ 12\ /\ 39\ 22\ 30.$$

obv.

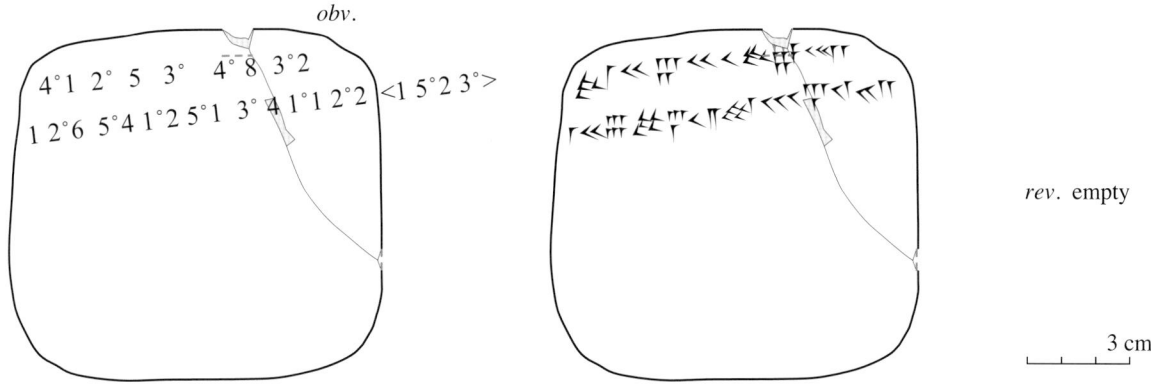

MS 2793. The regular sexagesimal number 41 25 30 48 32 (the 29th power of 2) and its reciprocal (incomplete).

obv.

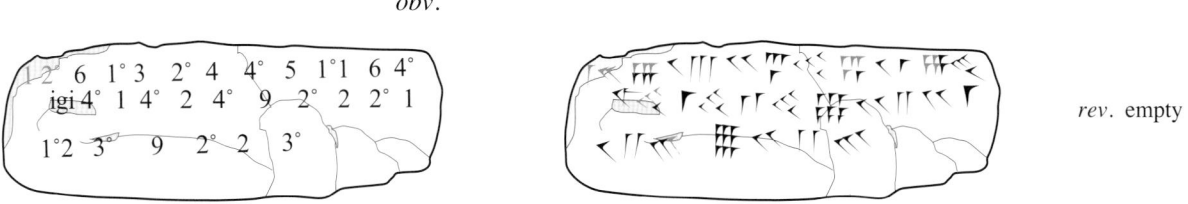

MS 2894. The regular sexagesimal number 2 05 doubled 29 times and its reciprocal, 41 42 49 22 21 12 39 22 30. The number in the first line is incorrect, [1 2]6 13 24 45 11 06 40 instead of 1 26 18 09 11 06 40.

obv.   *rev.* empty

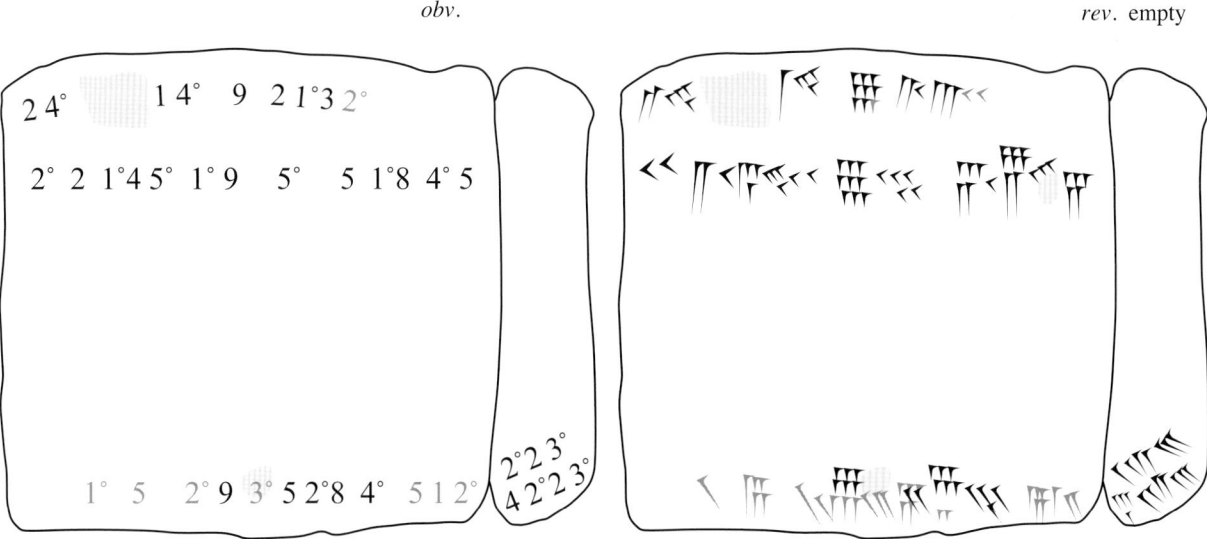

MS 2699. The regular sexagesimal number 2 41 49 02 13 20 (2 05 doubled 24 times) and its reciprocal. The meaning of the partly erased number (or numbers?) near the lower edge is not clear.

Fig. 1.4.3. Three examples of many-place regular sexagesimal numbers with their reciprocals.

According to Sachs (*JCS 1* (1947)), the sign KUD in the first number of this entry is a separation sign, incorrectly splitting the number 1 26 18 09 11 06 40 in two parts. Such telescoping errors are typical for careless counting with sexagesimal numbers in Babylonian place value notation. Cf. the example in MS 3946 above, Fig. 1.2.2 (2 30 33 45 instead of 3 03 45). If the two parts 1 25 20 and 58 09 11 06 40 are joined in the right way, the result is 1 26 18 09 11 06 40, as shown below:

$$\begin{array}{r} 1\ 25\ 20 \\ +\quad 58\ 09\ 11\ 06\ 40 \\ \hline 1\ 26\ 18\ 09\ 11\ 06\ 40 \end{array}$$

The same type of error seems to have been committed in the text of MS 2894. Consider the first number in this text, [1 2]6 13 24 45 11 06 40, which appears to be an error for what ought to be 1 26 18 09 11 06 40. The incorrect number can be divided into two parts, 1 26 13 24 and 45 11 06 40, which can then be joined again in the following way:

$$\begin{array}{r} 1\ 26\ 13\ 24 \\ +\quad 45\ 11\ 06\ 40 \\ \hline 1\ 26\ 14\ 09\ 11\ 06\ 40 \end{array}$$

This is the expected sexagesimal number, except for 14 instead of 18 in the third position. Therefore, the explanation for the strange form of first number in the "reciprocal pair" recorded on MS 2894 seems to be that the person who calculated the number made two mistakes, one a *telescoping error*, the other a simple *counting error*, 13 instead of 17 in the third position of the first number.

It is likely that the teacher who handed out this incorrect assignment had copied the numbers from the end of a table text similar to UM 29.13.21 # 1.

An interesting example of a telescoping error *in reverse* can be found in the text **VAT 5457** (Neugebauer and Sachs, *MCT* (1945), 16; Fig. 1.4.4 below), inscribed only with two lines of numbers on the obverse:

Fig. 1.4.4. VAT 5457. The regular sexagesimal number 9 06 08 and its reciprocal.

The recorded numbers can easily be misinterpreted as the pair

$$9\ 06\ 08\ 07\ 30, \quad 6\ 35\ 58\ 07\ 32,$$

but must actually be understood as

$$\begin{array}{ll} 9\ 06\ 08 & 7\ 30 \\ 6\ 35\ 58\ 07\ 30 & 2. \end{array}$$

This is clearly an assignment, where the student is asked to find first the reciprocal of 9 06 08 by use of the trailing part algorithm and then the reciprocal of the computed reciprocal by a repeated application of the trailing part algorithm. Here 9 06 08 = the 15th power of 2 (= 2^{15}). Its reciprocal is given in the text as 6 35 **58** 07 30, which is a mistake, since the correct value is 6 35 **30 28** 07 30. The error can be explained as follows: The result of an application of the trailing part algorithm would be that the given number and its reciprocal have the factorizations

$$9\ 06\ 08 = 8 \cdot 16 \cdot 16 \cdot 16, \text{ and}$$
$$\text{rec.}\ 9\ 06\ 08 = 3\ 45 \cdot 3\ 45 \cdot 3\ 45 \cdot 7\ 30.$$

The last step of the computation of the reciprocal would be to compute the product of 52 44 03 45 (the third power of 3 45) and 7 30. Apparently the author of VAT 5457 computed the product incorrectly, in the following way, by use of a multiplication table with the head number 7 30:

$$52\ 44\ 03\ 45 \cdot 7\ 30 = \begin{array}{r} 6\ 30 \\ 5\ 30 \\ 22\ 30 \\ + \quad 5\ 37\ 30 \\ \hline 6\ 35\ 58\ 07\ 30 \end{array}$$

The correct value, however, is computed as follows:

$$52\ 44\ 03\ 45 \cdot 7\ 30 = \begin{array}{r} 6\ 30 \\ 5\ 30 \\ 22\ 30 \\ + \quad 5\ 37\ 30 \\ \hline 6\ 35\ 30\ 28\ 07\ 30 \end{array}$$

(It is almost impossible to avoid "positional errors" of this kind without the use of final zeros. Note, in particular, that the correct values of the products $52 \cdot 7\ 30$ and $44 \cdot 7\ 30$ are 6 30 00 and 5 30 00.)

MS 2699 (Fig. 1.4.3, bottom) is a square tablet inscribed with two lines of numbers near the upper edge of the obverse and one or two additional lines of numbers near the lower edge. The upper pair of numbers is easily explained, since it is the same as the 25th entry in the doubling and halving algorithm table UM 29.13.21:

$$2\ 41\ 49\ 02\ 13\ 20,\ 22\ 14\ 50\ 19\ 55\ 18\ 45 = 2\ 05 \cdot 2^{24},\ 28\ 48 \cdot 2^{-24}.$$

There is a big gap (an erasure?), between the digits 40 and 1 in the second place of the first number.

The number signs near the lower edge are hard to read. Some of the signs are weakly impressed, others are apparently written on top of each other. The numbers near the lower edge are followed by several perfectly legible number signs on the lower part of the right edge.

What seems to be clear is that the last of the number signs near the lower edge, and the numbers signs on the right edge near the lower edge, together form a 3-place trailing part of some many-place regular sexagesimal number, that trailing part being … 24 22 30. The whole many-place number is at most 8-place.

Now, a very useful computer-generated table of *all* at most 11-place regular sexagesimal numbers and their reciprocals (often much more than 11-place) was published by Gingerich in *TAPS* 55 (1965). In that table, there is no number beginning 15 29 … and ending with … 24 22 30. Therefore, if the number near the lower edge of MS 2699 was meant to be a regular sexagesimal number, it must contain an error. Interestingly, however, Gingerich's table can be used (with considerable effort) to make a list of possible candidates for reasonably short regular sexagesimal numbers with the 3-place trailing part 24 22 30. As it turns out, all such numbers are the reciprocals of either 5 doubled a number of times or the 9-th power of 5 doubled a number of times. Particularly short (that is not exceedingly many-place) examples of such numbers are

5 16 24 22 30	(5-place)	the reciprocal of	$11\ 22\ 40 = 5 \cdot 2^{13}$
19 46 31 24 22 30	(6-place)	the reciprocal of	$3\ 02\ 02\ 40 = 5 \cdot 2^{17}$
1 14 05 27 46 24 22 30	(8-place)	the reciprocal of	$48\ 32\ 42\ 40 = 5 \cdot 2^{21}$

and

2 33 46 24 22 30	(6-place)	the reciprocal of	$23\ 24\ 39\ 50\ 07\ 24\ 26\ 40 = 5^9 \cdot 2^{25}$
9 26 39 01 24 22 30	(7-place)	the reciprocal of	$6\ 14\ 34\ 37\ 21\ 58\ 31\ 06\ 40 = 5^9 \cdot 2^{29}$
36 02 26 20 16 24 22 30	(8-place)	the reciprocal of	$1\ 39\ 53\ 13\ 57\ 51\ 36\ 17\ 46\ 40 = 5^9 \cdot 2^{33}$

Unfortunately, none of these numbers matches the number inscribed near the lower edge of MS 2699, not even if minor errors are taken into account.

MS 3264 (Fig. 1.4.5 below) is a provisional designation for a tablet in the Schøyen Collection, known only from a couple of photos. The tablet itself seems to be lost, and the photos do not show its catalog number. Anyway, two occurrences of the term igi.bi, Sumerian for 'its reciprocal' makes it clear that *two pairs of reciprocal numbers* are recorded on the tablet. The first pair of numbers is badly preserved with several of the digits missing and others hard to read.

Note that although the tablet is more wide than tall, obviously in order to accommodate many-place numbers, the second line (the first reciprocal) continues onto the reverse of the tablet. Even more remarkable is that the fourth line (the second reciprocal) continues onto the reverse and across the right edge of the reverse, with the last few places recorded on a line below! More about this later.

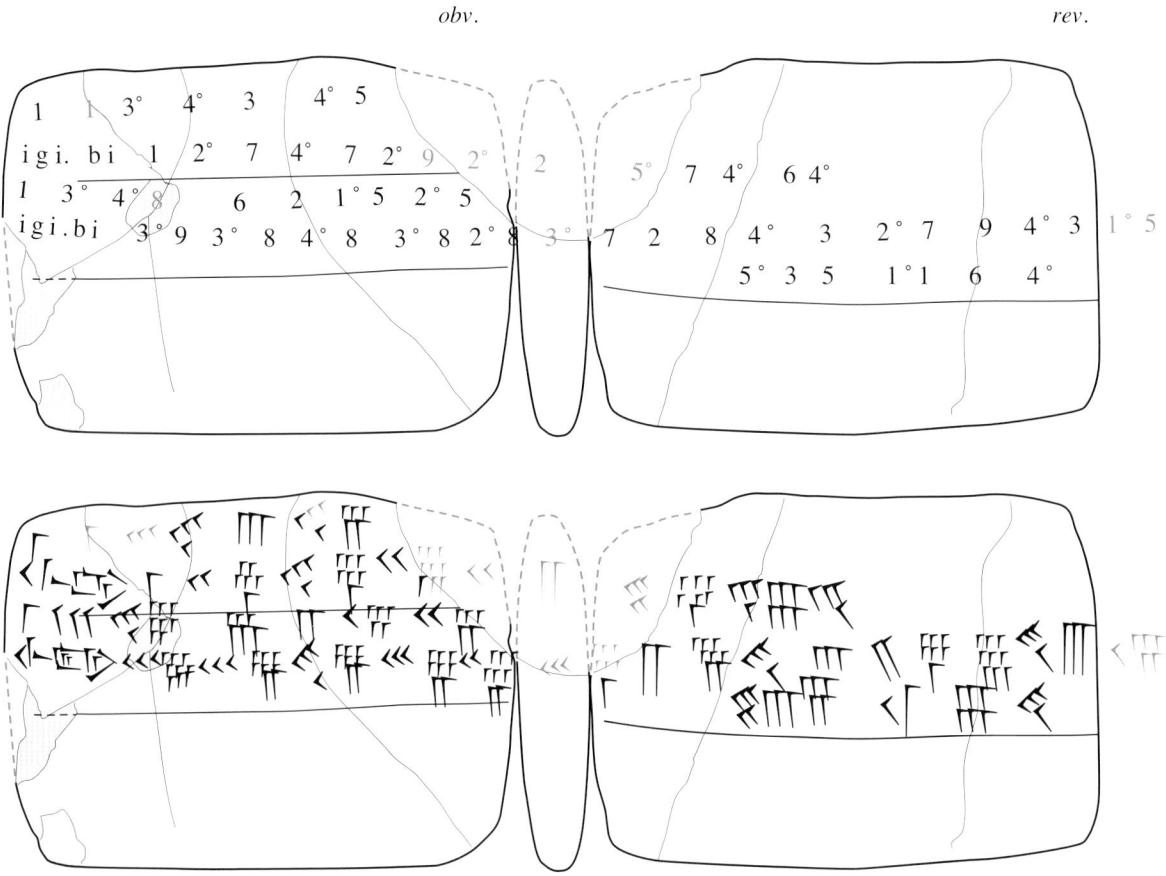

The reciprocal of the 7-place sexagesimal number in line 3 is an 18-place number beginning in line 4 of the obverse with 5 sexagesimal places and continuing with 13 additional places, written on the edge and in two lines on the reverse. An interesting mistake was made in the computation of the first reciprocal.

Fig. 1.4.5. MS 3264. Two many-place regular sexagesimal numbers and their reciprocals.

The two pairs of numbers on MS 3264 can be transliterated as

1.   1 [01] 30 43 45
     igi.bi 1 27 47 2[9 22 5]7 46 40
2.   1 30 4[8] 06 02 15 25
     igi.bi 39 38 48 38 28 [3]7 02 08 43 27 09 43 [15] / 53 05 11 06 40.

A quick look in Gingerich's table of 11-digit regular numbers and their reciprocals (*op. cit.*) is enough in order to find out that the number in line 1 must be a mistake for

1 01 30 **33** 45 = the square of 5, tripled 12 times ($3^{12} \cdot 5^2$).

The number below it, preceded by igi.bi, must be

1 27 47 2[9 22 5]7 46 40 = the 9th power of 5, doubled 21 times ($2^{21} \cdot 5^9$).

Strangely enough, this number is not the reciprocal of 1 01 30 33 45. The teacher or student who computed this number must have made a fatal mistake. What that mistake was can be demonstrated in the following way: The product of a number and its reciprocal should always be equal to some power of the sexagesimal base 60. However in this case, the product of the two given numbers is instead

$$3^{12} \cdot 5^2 \cdot 2^{21} \cdot 5^9 = 2^{21} \cdot 3^{12} \cdot 5^{11} = 3/2 \cdot (4 \cdot 3 \cdot 5)^{11} = 3/2 \cdot 60^{11}.$$

This means that whoever computed the reciprocal of 1 01 30 33 45 got a result that was 3/2 times greater than the correct result. It is obvious how this could happen, since the reciprocal, computed by use of the trailing part algorithm, should have been obtained as the twelfth power of 20 times the second power of 12. At some point in the course of the calculation, an intermediate result must by mistake have been multiplied by 30 instead of by 20!

For an explanation of the second pair of reciprocals on MS 3264, Gingerich's table of 11-digit regular numbers and their reciprocals again turns out to be useful. According to that table, the first number of the pair must be a mistake for

1 30 48 06 02 15 **20 15** $(= 3^{25} \cdot 5)$,   with the reciprocal

39 38 48 38 28 37 02 08 43 **37** 09 43 15 53 05 11 06 40 $(= 2^{50} \cdot 5^{24})$.

The bold digits indicate errors, The first number, as recorded, ends with 25 instead of with 20 15, and in the second number, a correct 37 is replaced by an incorrect 27.

Here are some details of the computation of the reciprocal of the second given number on MS 3264, showing to some extent how laborious the student's computations were:

1 30 48 06 02 15 20 15	4	1 30 48 06 02 15 · 4 + 1 = 6 03 12 24 09 01 21
6 03 12 24 09 01 21	44 26 40	6 03 12 24 09 · 44 26 40 + 1 = 4 29 02 31 13 21
4 29 02 31 13 21	44 26 40	4 29 02 31 12 · 44 26 40 + 1 = 3 19 17 25 21
3 19 17 25 21	44 26 40	3 19 17 24 · 44 26 40 + 1 = 2 27 37 21
2 27 37 21	44 26 40	2 27 36 · 44 26 40 + 1 = 1 49 21
1 49 21	44 26 40	1 48 · 44 26 40 + 1 = 1 21
1 21	44 26 40	
	32 55 18 31 06 40	
	24 23 11 29 42 42 57 46 40	the 2nd power of 44 26 40
	18 03 50 44 13 51 49 27 54 04 26 40	the 3rd power of 44 26 40
	13 22 50 54 59 09 29 58 26 43 17 31 51 06 40	the 4th power of 44 26 40
	9 54 42 09 37 09 15 32 10 54 17 25 48 58 16 17 46 40	the 5th power of 44 26 40
	39 38 48 38 28 37 02 08 43 37 09 43 15 53 05 11 06 40	the 6th power of 44 26 40
		the 6th power of 44 26 40, times 4

(Here 44 26 40 is the reciprocal of 1 21. It is one of the known head numbers for Old Babylonian single multiplication tables, so it would not have been too difficult to accomplish the needed multiplications by 44 26 40. Alternatively, since 44 26 40 is the fourth power of 20, each multiplication by 44 26 40 can be replaced by four multiplications by 20.) It is a wonder that such massive computations could be carried out with very few errors! It is, by the way, hard to imagine an abacus or a counting table big enough to allow computations with up to 18-place (that is 36-digit) numbers. The only remaining alternative seems to be that the computations were done with cuneiform numbers on clay tablets used as temporary "scratch pads".

An interesting aspect of MS 3264 is the curious layout of the text. On the obverse, the two pairs of numbers are surrounded by the upper edge of the tablet and two lines parallel to the edge. In both pairs, the reciprocals are too long to be contained between these boundaries. The first reciprocal, 1 27 47 29 22 57 46 40, is an 8-place sexagesimal number, with 5 sexagesimal places on the obverse and the right edge, continued on the reverse with the remaining 3 places. The second reciprocal, 39 38 48 38 28 37 02 08 43 27 09 43 15 53 05 11 06 40, is an 18-place number, with 6 places on the obverse and the edge, continued with 12 additional places, written in two lines on the reverse.

This curious layout suggests that a student was given the assignment to find the reciprocals of the regular sexagesimal numbers 1 01 30 33 45 = $2^{12} \cdot 5^2$ and 1 30 48 06 02 15 20 15 = $5 \cdot 3^{25}$. Not anticipating that the reciprocals of the given numbers would have many more places than the given numbers, he started by copying the two given numbers onto his tablet and drawing the two lines, allocating as much place to the reciprocals as to the given numbers, that is just one line of text on the obverse for each one of them. Then, he computed the reciprocals, probably using the trailing part algorithm to find factorizations of the given numbers and corresponding factorizations of the reciprocals. After the computations were done, he started to record the reciprocals in the allocated spaces but rapidly found that there was not space enough, so that he had to continue onto the reverse.

One difficulty with this interpretation is that there are errors in both of the given numbers recorded on MS 3264. These errors would have made it impossible for the student to compute the reciprocals, because the trailing part algorithm would not have worked. One way to get around this difficulty is to assume that MS 3264 is a faulty copy of another student's clay tablet.

YBC 4704 (Neugebauer and Sachs, *MCT*, 16; Fig. 1.4.6 below) is a parallel to MS 2894 (Fig. 1.4.3) and MS 3264. It is a small rectangular clay tablet inscribed with three regular sexagesimal numbers and their many-place reciprocals. The three regular numbers are:

1.	3 41 26 01 30	$= 10 \cdot 3^{14}$	5-place
2.	33 12 54 13 30	$= 10 \cdot 3^{16}$	5-place
3.	6 03 12 24 09 01 21	$= 3^{24}$	7-place.

Their reciprocals are

1.	16 15 27 39 48 28 38 31 06 40	10-place
2.	1 48 23 04 25 23 10 56 47 24 26 40	12-place
3.	9 54 42 09 37 09 **15 32 10 54 17** 25 48 58 16 17 46 40	18-place.

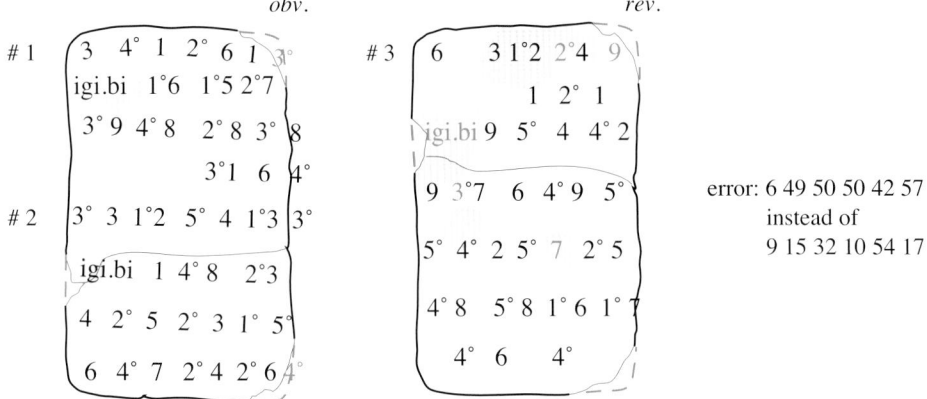

Fig. 1.4.6. YBC 4704. Three regular sexagesimal numbers and their many-place reciprocals.

The student who wrote this text made it wide enough to accommodate 5-place sexagesimal numbers, never suspecting that he would encounter longer numbers. Consequently, he had to split, for instance, the 7-place third number in two, writing it on two consecutive lines, and its 18-place reciprocal in five parts, written on five consecutive lines. Note that, as remarked in *MCT*, 16, fn. 76b, in the last two lines on the obverse, the number 56 is split in two parts, with 50 on one line and 6 on the next. This is an excellent demonstration of why it is convenient to use notations like $1°, 2°, 3°, 4°, 5°$ instead of $10, 20, 30, 40, 50$ in conform transliterations.

MS 3920 is inscribed with four regular sexagesimal numbers in various places on the obverse. There are also traces of an erased number. The four numbers can be interpreted as follows:

1.	5 55 33 20	$= 2^{11} \cdot 5^4$
2.	1 33 45	$= 3^2 \cdot 5^4$
3.	46 01 40	$= 2^{34} \cdot 5^3$
4.	5 16 24 22 30	$= 2 \cdot 3^7 \cdot 5^6.$

The first of these numbers, 5 55 33 20, seems to be followed by the number 18 (only partly visible due to damage to the tablet). Since 18 is the reciprocal of the trailing part 3 20, this may be another example of an assignment in the form of the first line of an application of the trailing part algorithm.

Since there is no obvious connection between the four numbers recorded on MS 3920, MS 3290 is a quite uninteresting text.

A thorough discussion of many-place pairs of reciprocal numbers in *Late Babylonian* cuneiform texts can be found in Friberg, *BagM* 30 (1999), *BagM* 31 (2000), and *CTMMA 2* (2005).

1.5. Old Babylonian Squares and Squares of Squares of Many-Place Sexagesimal Numbers

1.5 a. Squares of Many-Place Regular or Semiregular Sexagesimal Numbers

MS 2318/2 (Fig. 1.5.1, top) is a small clay tablet, with only one line of text, what may be mistaken for a 7-place sexagesimal number. However, a closer analysis reveals that what is recorded is *an initial number 6, followed by a six-place sexagesimal number*:

6 1 59 34 27 12 36

The sexagesimal number ends in 36, a regular sexagesimal number equal to the square of 6. This circumstance suggests that the text may be *an assignment to compute the square side of the given number by use of a variant of the trailing part algorithm.*

The idea behind the algorithm is the following: Since the given number 1 59 34 27 12 36 has the last place 36, there is a good chance that the whole number is a multiple of 36. That is, indeed, the case, since (in modern notation)

$$1\ 59\ 34\ 27\ 12\ 36 = 1\ 59\ 34\ 27 \cdot 1\ 00\ 00 + 12 \cdot 1\ 00 + 36$$
$$= (1\ 59\ 34\ 27 \cdot 1\ 40 + 20 + 1) \cdot 36.$$

This factor 36 has the reciprocal 1 40, since

$$36 \cdot 1\ 40 = (36 \cdot 100 = 3600 =)\ 1\ (00\ 00).$$

Therefore, in Babylonian place value notation without zeros, *the square factor 36 can be removed from the given number through multiplication of the number by 140, the reciprocal of 36. After the removal of this first square factor, the procedure can be repeated, until all the square factors of the given number have been found. The square side of the given number is then equal to the product of the square sides of all removed square factors.*

An excellent explicit example of how the algorithm works is given by the inscription on the reverse of the round hand tablet ***UET 6/2 222*** (Friberg, *RA* 94 (2000), § 2 d):

		1 03 45		sq. 1 03 45 = 1 07 44 03 45
		1 03 45		
sqs. 3 45 = 15	15	1 07 44 03 45	16	1 07 44 · 16 + 1 = 18 03 45
sqs. 3 45 = 15	15	18 03 45	16	18 · 16 + 1 = 4 49
sqs. 4 49 = 17	17	4 49		
		3 45		15 · 15 = 3 45
		1 03 45		3 45 · 17 = 1 03 45

If a similar layout had been used for the computation of the square side of the number recorded on MS 2318/2, the result would have been a compact algorithm table of the following form:

obv.

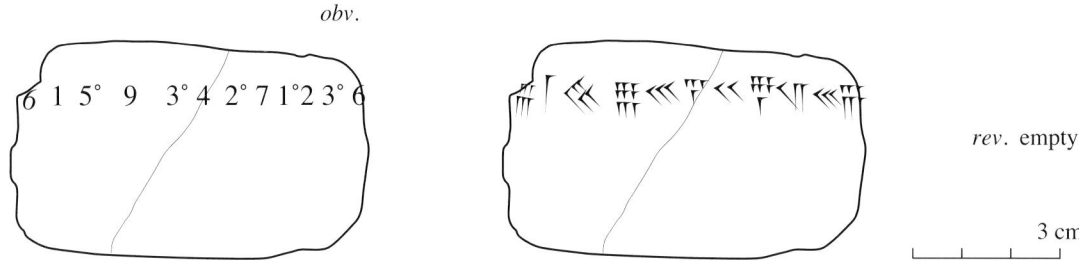

rev. empty.

3 cm

MS 2318/2. The 6-place sexagesimal number 1 59 34 27 12 36, equal to the square of the regular
sexagesimal number 10 56 06 (2 times the 9th power of 3).

obv.

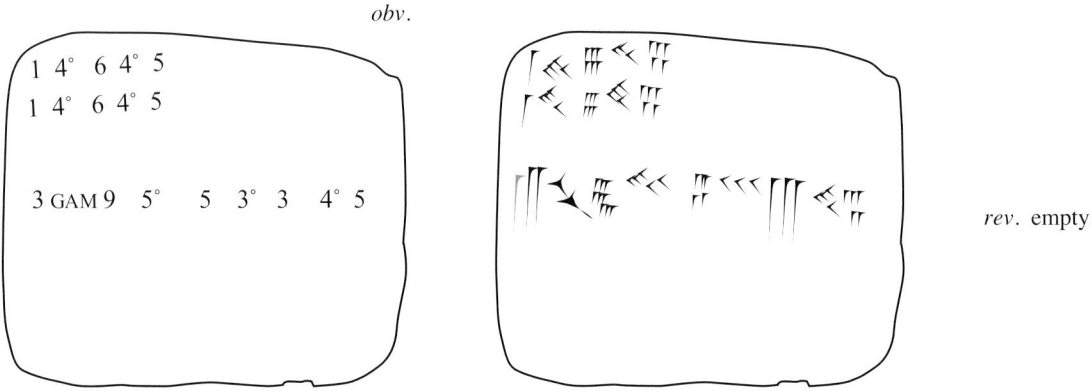

rev. empty.

MS 2731. Computation of the square of a number with a funny number (1 01) as a factor.

obv.

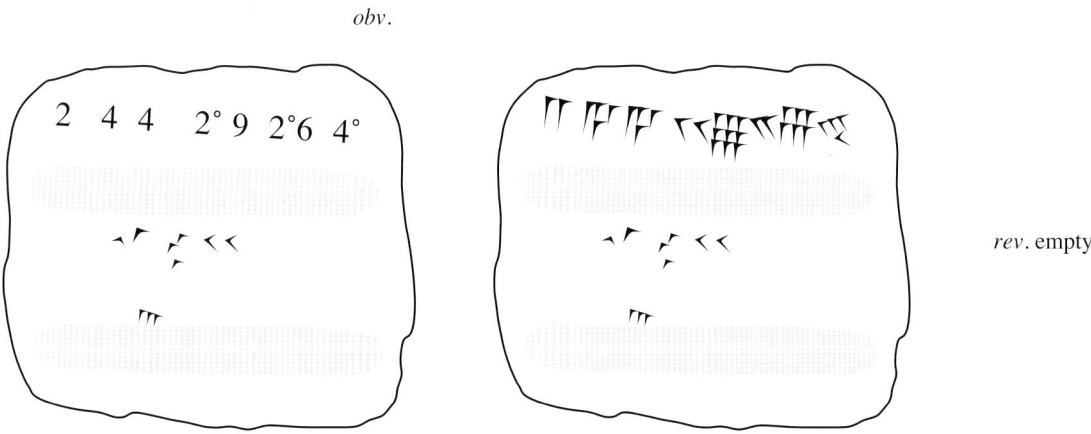

rev. empty

MS 3917. The 6-place sexagesimal number 2 04 04 29 26 40, equal to the square of the semi-regular
sexagesimal number 11 08 20 (1 40 times 6 41).

Fig. 1.5.1. Many-place sexagesimal numbers equal to squares of regular or semiregular sexagesimal numbers.

sqs. 36 = 6	6	**1 59 34 27 12 36**	1 40		1 59 34 27 12 · 1 40 + 1 = 3 19 17 25 21
sqs. 1 21 = 9	9	3 19 17 25 21	44 26 40		3 19 17 24 · 44 26 40 + 1 = 2 27 37 21
sqs. 1 21 = 9	9	2 27 37 21	44 26 40		2 27 36 · 44 26 40 + 1 = 1 49 21
sqs. 1 21 = 9	9	1 49 21	44 26 40		1 48 · 44 26 40 + 1 = 1 21
sqs. 1 21 = 9	9	1 21	44 26 40		
		1 21			9 · 9 = 1 21
		12 09			1 21 · 9 = 12 09
		1 49 21			12 09 · 9 = 1 49 21
		10 56 06			1 49 21 · 6 = 10 56 06

The number in the last line of this algorithm table is the square side of the given number, equal to 6 times the fourth power of 9, or 2 times the 9th power of 3. Note the obvious similarity with the computation of reciprocals of regular sexagesimal numbers by use of the trailing part algorithm. Note also, in the first line of the algorithm, the position of the number 6 immediately to the left of the given number, exactly as it is written in the single line of numbers on MS 2318/2!

The algorithm for the computation of a square side on the reverse of *UET 6/2 222*, a round hand tablet from Ur, has an interesting parallel on **3N-T 611 = A 30279**, a square hand tablet from Nippur. (The conform transliteration below is based on the hand copy in Robson, *UOS* (2002), Fig. 18):

obv.:only random signs

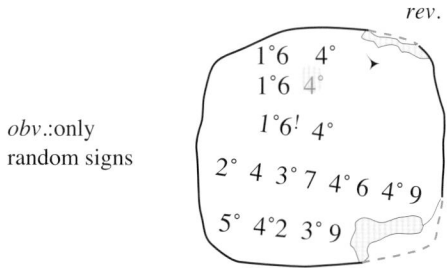

Fig. 1.5.2. A 30279, a square hand tablet with an algorithm for the computation of a square side.

Robson correctly identified 4 37 46 40 as the square of 16 40. (It is not clear why there is an extra third copy of 16 40 recorded on the clay tablet above 4 37 46 40.) She then mistakenly saw the final 9 in line 4 as an indication that the purpose of the text was to compute the reciprocal of 4 37 46 40, but correctly noted that 42 39 in line 5 can be explained as an incorrect sum of 41 39 and 1 (41;39 + 1 = 42;39, instead of 41 39 + 1 = 41 40). She ended her commentary by saying, in fn. 77, that "I do not have any explanation for the 20 and 50 written to the left of the calculation; presumably they relate to intermediate steps in the calculation". Actually, however, the correct interpretation of the text is the following:

		16 40		sq. 16 40 = 4 37 46 40
		<<16 [40]>>		
		16 40		
sqs. 6 40 = 20	20	**4 37 46 40**	9	9 · 4 37 40 + 1 = 41 39 + 1 = 42 39 (should be 41 40)
sqs. 41 40 = 50	50	42 39		(50 · 20 = 16 40)

Trying to compute the square side of 4 37 46 40, the square of 16 40, the student who wrote the text first computed the square side 20 of the trailing part 6 40 and noted it to the left in line 4. He eliminated the factor 6 40 in 4 37 46 40 through multiplication by its reciprocal 9, noted to the right in the same line, and got the (incorrect) result 41 39 + 1 = 42 39, instead of 41 40. Then he wrote 50 (the square side of 41 40) to the left in line 5, obviously because he knew that the wanted square side 16 40 (= 1,000) is the product of 50 and 20. In other words, he *cheated* and did not actually try to compute the square side of his incorrect number 42 39, because he thought that he already knew what it should be. (Thus, Robson was mistaken when she stated (*op. cit.*, 357) that among the mathematical texts from "House F" at Nippur there was not, as among the round hand tablets from Ur with mathematics on their reverses, any example of her observed rule that "fifth, and most speculatively, the students knew the results they were aiming for, and were not above fudging their calculations to fit".)

MS 2731 (Fig. 1.5.1, middle) is a rectangular clay tablet with a single squaring exercise:

$$1\ 46\ 45$$
$$3\ 09\ 55\ 33\ 45$$
$$1\ 46\ 45$$

In the text, the 5-place number 3 09 55 33 45 is written as

3' GAM 9 55 33 45,   where the cuneiform sign GAM (an oblique double wedge) is used as a notation for 'zero'.

It is not clear why the scribe felt obliged to insert a notation for zero in front of the 9. Explicit notations for zero are very rare in Old Babylonian mathematical texts. (See Friberg, *RlA 7* (1960), Sec. 3.2.)

The number being squared in MS 2731 has the following interesting factorization:

$$1\ 46\ 45 = 1\ 45 \cdot 1\ 01 = 7 \cdot 15 \cdot 1\ 01.$$

Here, 1 01 is a "funny number" similar to the numbers 10 10, and 10 10 10 in computations ## 1- 3 of the multiplication exercise text MS 3955 (Fig. 1.1.3), and to the funny number 1 01 01 01 in the division problem MS 2317, discussed in Sec. 7.1 below. The other factor, 1 45, can be explained in two different ways. It may be an *almost round number* of the type appearing in the squaring exercises on MS 2831 (Sec. 1.2 above), in which case it should be understood as

$$1\ 45 = 7 \cdot 15 = (1 + 1/6) \cdot 6 \cdot 15 = (1 + 1/6) \cdot 1\ 30.$$

On the other hand, it is just as likely that 1 45 was chosen as a factor in the given number because 1;45 (= 7/4) was an Old Babylonian approximation to sqs. 3. Indeed, sq. 1;45 = 3;03 45 (= 49/16).

A reasonable conjecture is that the square of 1 46 45 in MS 2731 was computed as follows:

$$\text{sq. } 1\ 46\ 45 = \text{sq. } (1\ 45 \cdot 1\ 01) = \text{sq. } 1\ 45 \cdot \text{sq. } 1\ 01 = 3\ 03\ 45 \cdot 1\ 02\ 01 = \quad \begin{array}{r} 3\ 03\ 45 \\ 6\ 07\ 30 \\ +\quad 3\ 03\ 45 \\ \hline 3\ 09\ 55\ 33\ 45 \end{array}$$

It is likely that the squaring of the given number 1 46 45 in MS 2731, just as the squaring of the given number in *UET 6/2 222, rev.*, was only *the first half of an assignment, the second half of the assignment being to compute the square side of the computed square, which ought to be equal to the initially given number*. Compare with the way in which the computation of a reciprocal of a given regular number in texts like *UET 6/2 295* (Sec. 1.4 above) is followed by the computation of the reciprocal of the computed reciprocal, which turns out to be equal to the initially given number.

A sexagesimal number with both regular and non-regular factors is what has been called above a *semi-regular* sexagesimal number. How Old Babylonian mathematicians computed square sides of squares of semi-regular sexagesimal numbers is demonstrated by *UET 6/2 222, rev.*, already mentioned above as a model for how square sides could be computed by use of the trailing part algorithm. In the example on *UET 6/2 222*, the given number is the square of the semiregular number 1 03 45, equal to the product of the regular number 16 and the non-regular number 17. Another well known example is the round hand tablet **Ist. Si 428** (Friberg, *RlA 7* (1990) Sec. 5.3 a), where the trailing part algorithm, applied to the given "funny number" $a = 2\ 02\ 02\ 02\ 04\ 05\ 04$, leads to the factorization

$$a = 4 \cdot 16 \cdot 4 \cdot 28\ 36\ 06\ 06\ 49,\ \text{where } 28\ 36\ 06\ 06\ 49 \text{ is a 5-place non-regular sexagesimal number.}$$

The final result given in Ist. Si 428 is that

$$\text{sqs. } 2\ 02\ 02\ 02\ 04\ 05\ 04 = 2 \cdot 4 \cdot 2 \cdot 5\ 20\ 53 = 1\ 25\ 34\ 08,$$

but there is no indication of how the square side of the 5-place non-regular number 28 36 36 06 49 was computed. (A detailed discussion of known or tentatively reconstructed Babylonian methods for approximation of square sides of is offered in Friberg, *BagM* 28 (1997), 316-334. The simplest of those methods is the one traditionally ascribed to the Greek mathematician Heron of Alexandria.)

In the case of the assignment on MS 2731, the student's computation of the square side of the semiregular number 2 09 55 33 20 would almost certainly have been based on a quite brief application of the trailing part algorithm:

sqs. 3 45 = 15	15	**3 09 55 33 45**	16
sqs. 50 38 49 = 7 07	7 07	50 38 49	
		1 46 45	

3 09 55 30 · 16 + 1 = 50 38 49

7 07 · 15 = 1 46 45

It remains to explain how the student could compute the square side of the 3-place non-regular number 50 38 49. His trick would be to use the geometrically obvious "additive square side rule" (Friberg, *op. cit*, 317), in the following way. The given sexagesimal number 50 38 49 can be interpreted *in absolute numbers* as, for instance, 50;38 49. Then, clearly, 7 is a *good initial approximation* to the square side, since sq. 7 = 49. The exact deficit is 1;38 49. A corresponding *improved second approximation* is given by the additive square side rule in the form

sqs. 50;38 49 = (approximately) 7 + 1;38/(2 · 7) = 7 + ;07 = 7;07.

It is easy to check that 7;07 is *the exact answer*, not only a good approximation.

MS 3917 (Fig. 1.5.1, bottom) is a rectangular clay tablet inscribed with a single 6-place sexagesimal number close to the upper edge. The tablet had clearly been used for other purposes before. There are traces of half erased numbers and two whole lines of text appear to have been erased with the tip of a finger. The inscribed number is 2 04 04 29 26 40. Since this number has the trailing part 6 40, and since 6 40 (400) is the square of 20, it can be suspected that the number is the square of a regular, or at least semiregular, sexagesimal number, and that the text is another example of an assignment to compute a square side. Indeed, proceeding as in the case of MS 2731 above, one rapidly finds the square side in the following way:

sqs. 6 40 = 20	20	**2 04 04 29 26 40**	9
sqs. 25 = 5	5	18 36 40 25	2 24
sqs. 44 40 01 = 6 41	6 41	44 40 01	
		33 25	
		11 08 20	

2 04 04 29 20· 9 + 1 = 18 36 40 25

18 36 40 · 2 24 + 1 = 44 40 01

6 41 · 5 = 33 25

33 25 · 20 = 11 08 20

The square side of 44 40 01 could be computed as follows: A first application of the additive square side rule shows that: sqs. 44 40 01 = (appr.) 6 + 8/(2 · 6) = 6;40. Since 2;10 01 ends with a '1', it is now easy to guess that the exact square side is 6;41 rather than 6;40. The correctness of this result is easy to check by computing the square of 6;41.

1.5 b.   Non-Square Semiregular Sexagesimal Numbers

MS 3956 (Fig. 1.5.3 below) is a small rectangular clay tablet, inscribed with two 4-place sexagesimal numbers on the obverse, two 3-place numbers on the reverse, and the 1-place number 7 on the right edge.

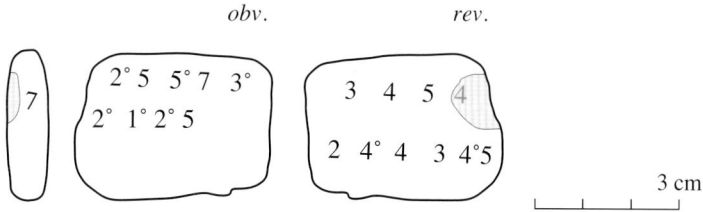

Fig. 1.5.3. MS 3956. Four non-square, non-regular numbers, each with a square regular factor.

These four numbers on the obverse and the reverse can be factorized as follows:

1.	25 57 30 = sq. 5 ·	1 02 18,	where	1 02 18 =	7 · 8 54,
2.	20 10 25 = sq. 5 ·	48 25,	where	48 25 =	7 · 6 55
3.	3 04 05 04 = sq. 4 ·	11 30 19,	where	11 30 19 =	7 · 1 38 37
4.	2 44 03 45 = sq. 15 ·	1 45,	where	1 45 =	7 · 15.

The surprising conclusion is that each of the four numbers on MS 3956 is a product of two factors, one that is the square of a regular sexagesimal number, and another that is a non-square, non-regular multiple of 7. The purpose of such a text is not at all clear. It may have been an assignment to compute the approximate square sides of the four numbers, in each case by first removing the square regular factor and then applying the additive square side rule. The curious presence of 7 as a factor in all the given numbers may possibly have the following explanation:

There are many indications that *when an Old Babylonian teacher handed out assignments to a number of students, he often gave them assignments identical except for the choice of consecutive values for certain numerical data*. In this way, the assignments were at the same time equal and individual. In the case of MS 3956, the teacher may have given all his students a set of numbers like this to work on (in whichever way was required):

1.	$a \cdot$ sq.5 ·	8 54
2.	$a \cdot$ sq.5 ·	6 55
3.	$a \cdot$ sq.4 ·	1 38 37
4.	$a \cdot$ sq. 15 ·	15

Only the value of the factor a was different for different students. For one of them, the chosen value was 7, and this value was noted on the left edge of his clay tablet!

1.5 c.   Many-Place Squares of Squares of Regular or Semiregular Numbers

obv.

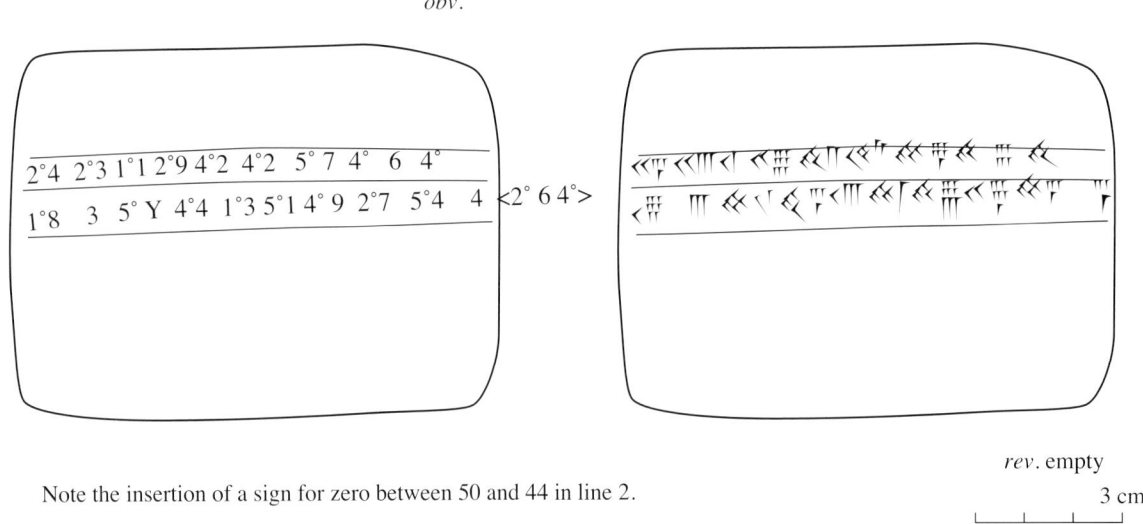

Note the insertion of a sign for zero between 50 and 44 in line 2.

rev. empty

3 cm

Fig. 1.5.4. MS 2205. The squares of the squares of the 3rd and 4th powers of 20.

MS 2205 (Fig. 1.5.4 above) is a rectangular clay tablet, inscribed near the middle of the tablet with two many-place sexagesimal numbers. The two numbers can be analyzed as

   1. 24 23 11 29 42 42 57 46 40 = the 12th power of 20
   2. 18 03 50 44 13 51 49 27 54 04 <26 40> = the 16th power of 20 (with the last two sexagesimal places missing).

A mathematically equivalent, but more interesting, way of characterizing the two numbers is as

   1. 24 23 11 29 42 42 57 46 40 = the 4th power of 2 13 20 (the 3rd power of 20)
   2. 18 03 50 44 13 51 49 27 54 04 <26 40> = the 4th power of 44 26 40 (the 4th power of 20)

Note that both 2 13 20 and 44 26 40 are present in the second column of the Old Babylonian standard table of reciprocals, 2 13 20 as the reciprocal of 27, and 44 26 40 as the reciprocal of 1 21.

The characterization of the two numbers as fourth powers, or, less anachronistically "squares of squares" is particularly interesting in view of the fact that there are two known fragments of a Seleucid table of squares of squares of many-place regular sexagesimal numbers. The fragments in question are **BM 55557** (Britton, *JCS* 43-45 (1991-93), 71-87) and **BM 32584** (Appendix 9 below).

The two numbers on MS 2205 may be the *answers* to an assignment, to compute the squares of the squares of 2 13 20 and 44 26 40. On the other hand, it is just as likely that they were *given numbers* in an assignment to compute either the reciprocals or the square sides (or even the square sides of the square sides) of those numbers, presumably by use of the trailing part algorithm.

If the text really was an assignment, with a student being asked to compute the square sides of the two given numbers, then it is strange that the trailing part, 26 40, of the second given number is missing. This would have made it impossible to take even the first step of the trailing part algorithm in order to compute the square side. Omitting the trailing part of the number may, of course, have been a stupid mistake committed by a careless student copying a number presented to him by his teacher.

A much more interesting alternative explanation of the missing trailing part is that the teacher *intentionally* made it impossible for the student to compute the square side by use of the trailing part algorithm. That would have forced the student to make use of some other method, in order to compute at least an *approximation* to the square side of the given number.

Note that in the conform transliteration of MS 2205 in Fig. 1.5.4, the second number recorded on the tablet is transliterated as

18 3 5Y 44 13 51 49 27 54 4 <26 40>, where Y denotes a sign resembling the cuneiform writing of 11.

The sign Y, which resembles a cuneiform '11' may be a partly erased number sign, but it is also possible that it is a sign for a missing digit, in other words a zero. An Old Babylonian zero of this form has not ben documented before. (Cf. the zero written as GAM in MS 2731 (Fig. 1.5.1 above).)

MS 2351 (Fig. 1.5.5) is a clay tablet of an unusual, finger-like format. It is inscribed with the very long, actually 15-place, sexagesimal number

13 22 50 54 59 09 29 58 26 43 17 31 51 06 40.

The number is so long that it was *written in two lines on the obverse and continued onto the reverse*. Its trailing part is 06 40, where 6 40 is the square of 20. This is a circumstance suggesting (as in the case of MS 2318/2 above, Fig. 1.5.1) that the given number may be the square of a regular sexagesimal number, and that the text was an assignment to compute the square side of the given number. It is not difficult, although tedious, to proceed with the actual calculation, which shows that

13 22 50 54 59 09 29 58 26 43 17 31 51 06 40 = sq. 3 39 28 43 27 24 26 40.

Repeating the process, one can show that, in its turn,

3 39 28 43 27 24 26 40 = sq. 14 48 53 20.

Collecting the results, one finds that

13 22 50 54 59 09 29 58 26 43 17 31 51 06 40 = sq. sq. 14 48 53 20,

and it is easy to show that

14 48 53 20 is the 5th power of 20 or, in other words, the reciprocal of 4 03, the 5th power of 3.

It is interesting to note that this means that MS 2351 can be interpreted as a continuation of MS 2205, since the two texts together exhibit the squares of the squares of the 3rd, 4th, and 5th powers of 20! As for the factorization of the given number itself, it follows that

13 22 50 54 59 09 29 58 26 43 17 31 51 06 40 = sq. sq. 14 48 53 20 is the 20th power of 20.

obv.   *rev.*

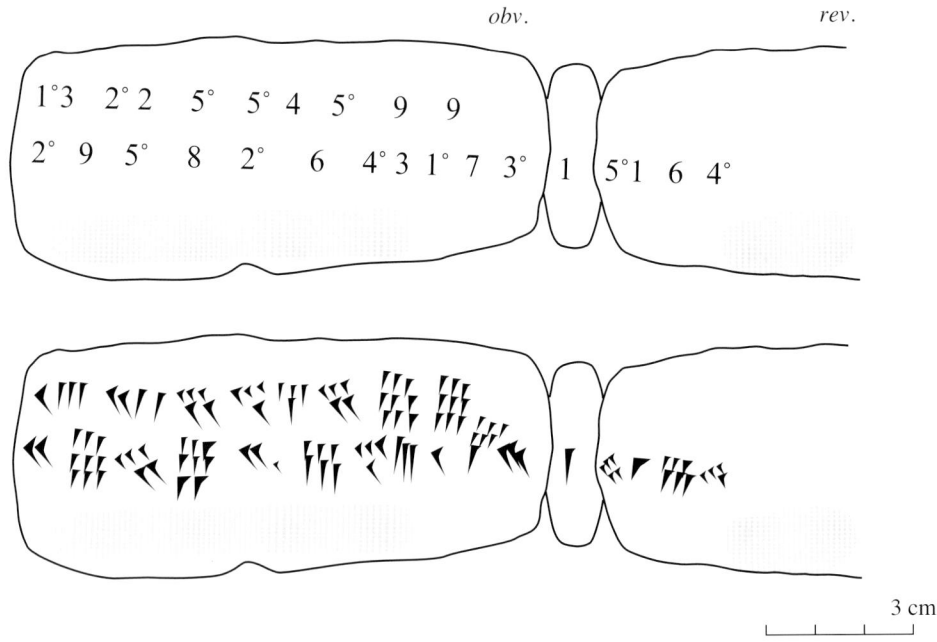

1°3   2°2   5°   5°4   5°   9   9

2° 9 5°   8   2°   6   4°3 1°7 3°   1   5°1 6 4°

3 cm

13 22 50 54 59 09 / 29 58 26 43 17 31 51 06 40, a 15-place regular sexagesimal number.

Fig. 1.5.5. MS 2351. The square of the square of the 5th power of 20.

2
Old Babylonian Arithmetical Table Texts

Old Babylonian clay tablets inscribed with arithmetical or metrological table texts are typically much taller than wide, often rectangular with the sides more or less in the ratio 3 : 2.

2.1. Old Babylonian Tables of Squares

2.1 a. Standard Tables of Squares

(Cf. Neugebauer, *MKT 1* (1935), 69-70, 74, 85; Neugebauer and Sachs *MCT* (1945), 33.)

A complete "unabridged" Old Babylonian table of squares contains entries for *the squares of all integers from 1 to 59 and 1 (00)*. For examples, see *MKT 1* § 4b: 5-13. The entries in all lines (except possibly the first line of the table) are normally of one of the following two types,

$$\text{type a:} \quad n \qquad \text{a.rá} \quad n \qquad \text{sq. } n$$
$$\text{type b:} \quad n \qquad n \qquad \text{sq. } n$$

(This is a simplification of the set of types of Old Babylonian tables of squares that was proposed by Neugebauer in *MKT 1*, 74.)

A complete unabridged table of squares with entries of type a looks like this:

1	**a.rá**	**1**	**1**		31	a.rá	31	16 01
2	**a.rá**	**2**	**4**		32	a.rá	32	17 04
3	**a.rá**	**3**	**9**		33	a.rá	33	18 09
4	**a.rá**	**4**	**16**		34	a.rá	34	19 16
5	**a.rá**	**5**	**25**		35	a.rá	35	20 25
6	**a.rá**	**6**	**36**		36	a.rá	36	21 36
7	**a.rá**	**7**	**49**		37	a.rá	37	22 49
8	**a.rá**	**8**	**1 04**		38	a.rá	38	24 04
9	**a.rá**	**9**	**1 21**		39	a.rá	39	25 21
10	**a.rá**	**10**	**1 40**		**40**	**a.rá**	**40**	**26 40**
11	**a.rá**	**11**	**2 01**		41	a.rá	41	28 01
12	**a.rá**	**12**	**2 24**		42	a.rá	42	29 24
13	**a.rá**	**13**	**2 49**		43	a.rá	43	30 49
14	**a.rá**	**14**	**3 16**		44	a.rá	44	32 16
15	**a.rá**	**15**	**3 45**		45	a.rá	45	33 45
16	**a.rá**	**16**	**4 16**		46	a.rá	46	35 16
17	**a.rá**	**17**	**4 49**		47	a.rá	47	36 49
18	**a.rá**	**18**	**5 24**		48	a.rá	48	38 24
19	**a.rá**	**19**	**6 01**		49	a.rá	49	40 01
20	**a.rá**	**20**	**6 40**		**50**	**a.rá**	**50**	**41 40**
21	a.rá	21	7 21		51	a.rá	51	43 21
22	a.rá	22	8 04		52	a.rá	52	45 04
23	a.rá	23	8 49		53	a.rá	53	46 49

Continued

24	a.rá	24	9 36		54	a.rá	54	48 36
25	a.rá	25	10 25		55	a.rá	55	50 25
26	a.rá	26	11 16		56	a.rá	56	52 16
27	a.rá	27	12 09		57	a.rá	57	54 09
28	a.rá	28	13 04		58	a.rá	58	56 04
29	a.rá	29	14 01		59	a.rá	59	58 01
30	**a.rá**	**30**	**15**		1	a.rá	1	1

Two Old Babylonian tables of squares of standard type are present in the Schøyen Collection. One of them, **MS 2794** (Fig. 2.1.1, top), is a complete "abridged" table of squares of *type a*. A complete "abridged" Old Babylonian table of squares contains entries only for *the squares of the 23 integers 1, 2, …,19, 20, 30, 40, 50*, in the same way as Old Babylonian single multiplication tables contain entries for the head number multiplied by these integers. For examples, see *MKT 1* § 4b: 1-4. In the unabridged table above, bold style indicates the entries in an abridged table of squares.

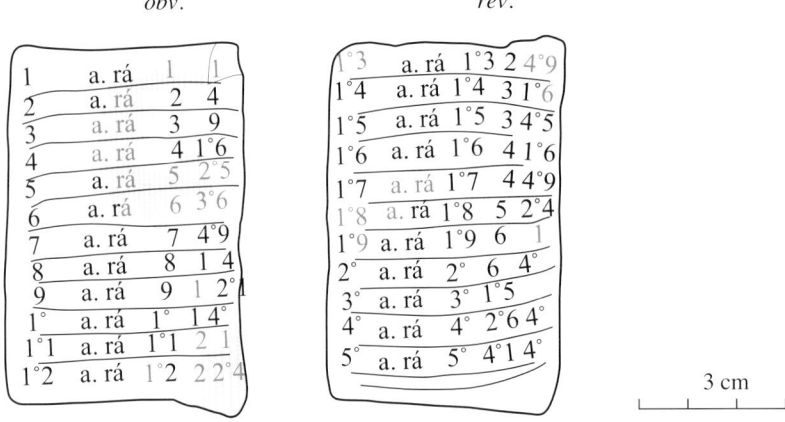

MS 2794. An OB abridged table of squares, for integers from 1 to 20, and 30, 40, 50 (23 lines). Type a.

MS 2706. An excerpt from an OB table of squares, for integers from 1 to 14 (14 lines). Type a.

Fig. 2.1.1. Two Old Babylonian tables of squares, one complete but abridged, the other incomplete.

MS 2706 (Fig. 2.1.1, bottom) is an *excerpt* (a partial copy of the text) from a table of squares with entries of *type a*. For some reason, its 14 lines do not even cover the whole obverse of the clay tablet. The work with the tablet seems to have been interrupted before it was finished.

2.1 b. Special Tables of Squares

Note. There is no big difference between "clay tablets with squaring exercises" (Sec. 1.2 above) and "special tables of squares" (the present paragraph). Loosely speaking, the only difference is that a clay tablet with squaring exercises looks very much like a clay tablet with multiplication exercises (Sec. 1.1 above), while a special table of squares looks like a standard table of squares (Sec. 2.1 a above).

MS 3937 (Fig. 2.1.2, top) is a special table of squares. More precisely, it is a table of *squares of multiples of 5*, from 10 = 5 · 2 to 2 05 = 5 · 25. The entries are all of *type b*.

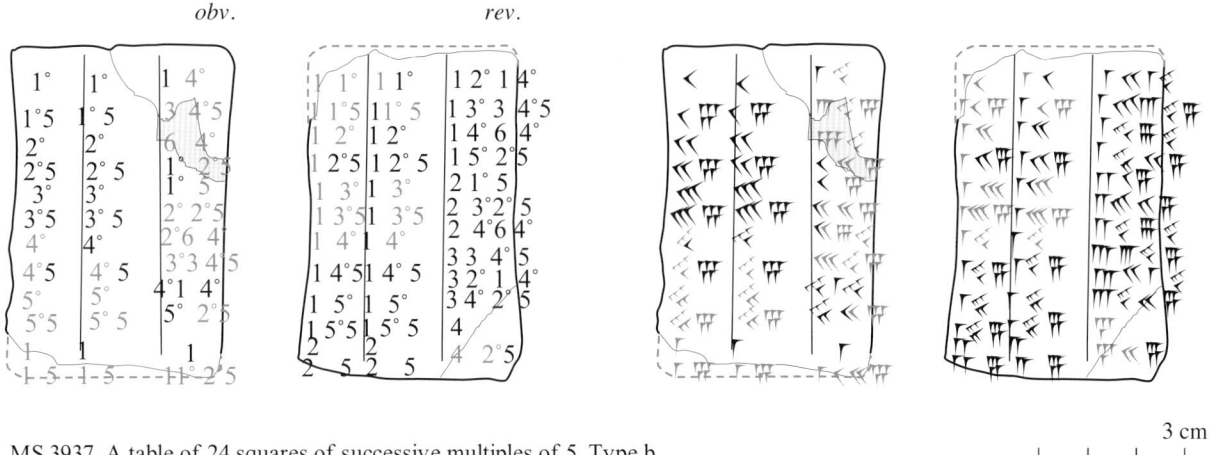

MS 3937. A table of 24 squares of successive multiples of 5. Type b.
Two errors in the text are easy to explain.

3 cm

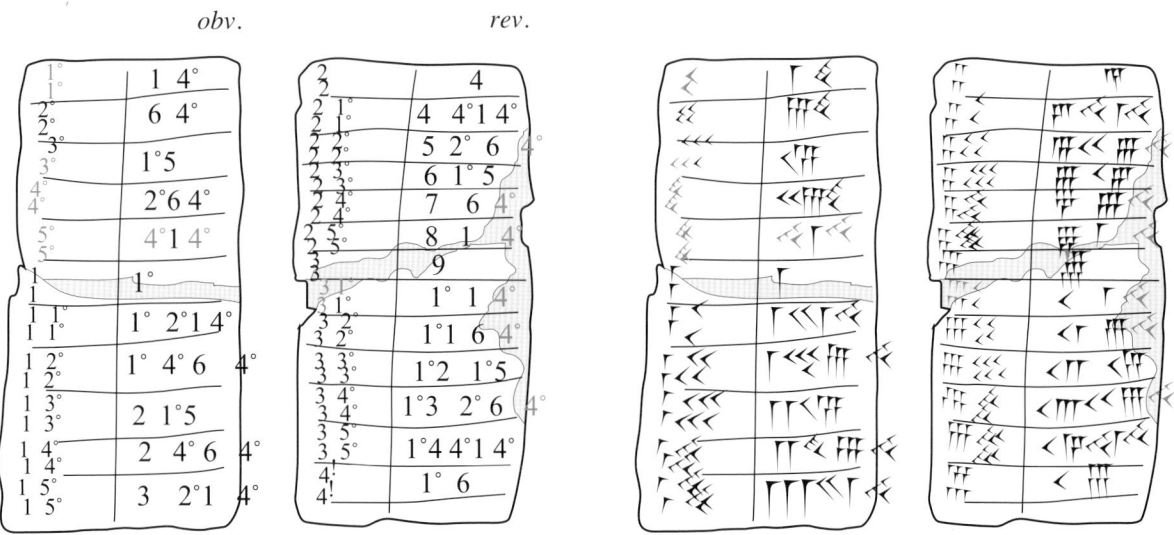

MS 3906. A table of 24 squares of successive multiples of 10. Type b$_V$.

Fig. 2.1.2. Two examples of Old Babylonian special tables of squares.

In the transliteration below of the table on MS 3937, reconstructed values of number signs that are lost or poorly visible on the clay tablet are in *italics*.

	obv.				*rev.*		
10	10	1 40		*1 10*	*1 10*	1 21 40	
15	15	*3 45*		*1 15*	*1 15*	1 33 45	
20	20	*6 40*		*1 20*	1 20	1 46 40	
25	25	10 25		*1 25*	1 25	1 50 25	should be 2 00 25
30	30	15		*1 30*	1 30	2 15	
35	35	20 25		*1 35*	1 35	2 30 25	
40	40	26 40		*1 40*	1 40	2 46 40	
45	45	33 45		1 45	1 45	3 03 45	
50	50	41 40		1 50	1 50	3 21 40	
55	55	50 25		1 55	1 55	3 40 25	
1	1	1		2	2	4	
1 05	1 05	1 10 25		2 05	2 05	4 25	should be 4 20 25

It is not unlikely that some of the entries of the table of squares on MS 3937 were computed by use of a standard table of squares (Sec. 2.1 a), in conjunction with the observation that

$$\text{sq.}\,(5 \cdot n) = 25 \cdot \text{sq.}\,n.$$

Thus, for instance, since $35 = 5 \cdot 7$, it follows that the square of 35 is equal to $25 \cdot 49 = 20\ 25$.

The idea to compute the square of a multiple of 5 in this way was old already in the Old Babylonian period. The Old Sumerian mathematical text ***TSS*** **188**, from the ancient Mesopotamian city Shuruppak, ca. 2600 BC, is a "metro-mathematical" squaring exercise, the computation of the square of $5 \cdot (10 \cdot 60 \text{ ninda})$. (See Fig. 6.2 below.) Except for a trivial error (corrected here), and for the use of *non-positional* sexagesimal numbers, the computation appears to have proceeded as follows:

$$
\begin{aligned}
5 \cdot (10 \cdot 60\ \text{ninda}) \cdot 5 \cdot (10 \cdot 60\ \text{ninda}) \ &= 25 \cdot \text{sq.}\,(10 \cdot 60\ \text{ninda}) \\
&= 25 \cdot 3\ 20\ \text{bùr} \quad (1\ \text{bùr} = 30\ 00\ \text{sq. ninda}) \\
&= 2\ 1/2 \cdot 10 \cdot 3\ 20\ \text{bùr} = 2\ 1/2 \cdot 33\ 20\ \text{bùr} = 1\ 23\ 20\ \text{bùr}.
\end{aligned}
$$

The two errors in the text of MS 3937 are easy to explain. The error in the square of 2 05 amounts (in *conform* transliteration) to writing 4 2° 5 instead of 4 2° 2° 5. Thus, the error is a simple "notational error", writing one instead of two signs for 20. The error in the square of 1 25 can possibly be explained in the following way. The correct way to compute the square of 1 25 would be like this:

$$
\text{sq.}\,1\ 25 = \text{sq.}\,(5 \cdot 17) = 25 \cdot \text{sq.}\,17 = 25 \cdot 4\ 49 = \quad
\begin{array}{r}
\underline{1} \\
1\ 40 \\
16\ 40 \\
+\ \ \underline{3\ 45} \\
2\ 00\ 25
\end{array}
$$

Thus, the mistake made would be a simple incorrect addition, $1 + 40 + 16 + 3 = 50$, instead of $1\ 00$.

MS 3922 is an excerpt from a table of squares of multiples of 5 like MS 3937. It contains only three entries, giving the squares of 35, 40, and 45.

MS 3947 is still another square hand tablet with a similar non-standard table of squares. The numbers squared are 4 10, 4 20, 4 30, and 4 40, four consecutive multiples of 10.

MS 3906 (Fig. 2.1.2, bottom) is a table of squares of multiples of 10, from $10 = 10 \cdot 1$ to $4\ (00) = 10 \cdot 24$. In the transliteration below of the table of squares on MS 3906, indications have been included of how the entries may have been computed by use of either the binomial rule

$$\text{sq.}\,(a + b) = \text{sq.}\,a + 2 \cdot a \cdot b + \text{sq.}\,b,$$

or the observation that

$$\text{sq.}\,(10 \cdot n) = \text{sq.}\,10 \cdot \text{sq.}\,n = 1\ 40 \cdot \text{sq.}\,n.$$

MS 3906

10 · 10 = 1 40			
20 · 20 = 6 40			
30 · 30 = 15			
40 · 40 = 26 40			
50 · 50 = 41 40			
1 · 1 = 1			
1 10 · 1 10 = 1 21 40	1 + 20 + 1 40	or	1 40 · 49
1 20 · 1 20 = 1 46 40	1 + 40 + 6 40		1 40 · 1 04
1 30 · 1 30 = 2 15	1 + 1 + 15		1 40 · 1 21
1 40 · 1 40 = 2 46 40	1 + 1 20 + 26 40		1 40 · 1 40
1 50 · 1 50 = 3 21 40	1 + 1 40 + 41 40		1 40 · 2 01
2 · 2 = 4			
2 10 · 2 10 = 4 41 40	4 + 40 + 1 40		1 40 · 2 49
2 20 · 2 20 = 5 26 40	4 + 1 20 + 6 40		1 40 · 3 16
2 30 · 2 30 = 6 15	4 + 2 + 15		1 40 · 3 45
2 40 · 2 40 = 7 06 [40]	4 + 2 40 + 26 40		1 40 · 4 16
2 50 · 2 50 = 8 [01 40]	4 + 3 20 + 41 40		1 40 · 4 49
3 · 3 = 9			
3 10 · 3 10 = 10 01 40	9 + 1 + 1 40		1 40 · 6 01
3 20 · 3 20 = 11 06 40	9 + 2 + 6 40		1 40 · 6 40
3 30 · 3 30 = 12 15	9 + 3 + 15		1 40 · 7 21
3 40 · 3 40 = 13 26 40	9 + 4 + 26 40		1 40 · 8 04
3 50 · 3 50 = 14 41 40	9 + 5 + 41 40		1 40 · 8 49
4 · 4 = 15			

Strictly speaking, the entries in MS 3906 are not of type b, that is arranged like this:

type b: *n*   *n*   sq. *n*

Instead, they are arranged in a *variant of type b*, in the following way,

type b$_v$: $\begin{matrix} n \\ n \end{matrix}$ sq. *n*

just like the entries in the multiplication and squaring exercises discussed in Secs. 1.1-1.2 above.

A parallel text: The round hand tablet **Böhl 1328** (Neugebauer, *MKT 3* (1937), 51) is a table of squares of multiples of 5, from 1 (00) = 5 · 12 to 1 15 = 5 · 15. No photo or hand copy of that text is available, but the transcription offered in *MKT 3* shows that the entries are of type b$_v$.

2.2. Old Babylonian Tables of Square Sides

(Cf. *MKT 1*, 70-71, 75; *MCT*, 33-34.)

MS 3963, 2185, and 3864 (Fig. 2.2.1 below) are three examples of Old Babylonian tables of square sides. The tables of square sides on the three tablets are not complete, they are only *excerpts* from a complete table of square sides. A *complete* table of square sides lists the square sides of the squares of *all* integers from 1 to 1 (00), while MS 3963 and MS 2185 list only the square sides of squares of integers from 21 to 40, and MS 3864 only those of squares of integers from 40 to 50.

Normally, Old Babylonian tables of square sides are of "type a", in the sense that each line in such a table is of the form

sq. *n* .e   *n*   íb.si$_8$ sq. *n* makes *n* equalsided   (where *n* is an integer)

Here, the *ad hoc* translation to the right is an attempt to describe what is meant by the Sumerian phrase to the left. The ergative postfix .e means that the square sq. *n* makes something to *n*. The verb si$_8$ has the basic meaning 'to be equal'. The whole phrase should probably be understood in a *geometric* sense, as saying that

a square field with the area sq. *n* has the side *n*.

A complete Old Babylonian table of square sides looks like this:

1	.e	1	íb.si₈	(or ba.si₈)		16 01	.e	31	íb.si₈	(or ba.si₈)
4	.e	2	íb.si₈			17 04	.e	32	íb.si₈	
9	.e	3	íb.si₈			18 09	.e	33	íb.si₈	
16	.e	4	íb.si₈			19 16	.e	34	íb.si₈	
25	.e	5	íb.si₈			20 25	.e	35	íb.si₈	
36	.e	6	íb.si₈			21 36	.e	36	íb.si₈	
49	.e	7	íb.si₈			22 49	.e	37	íb.si₈	
1 04	.e	8	íb.si₈			24 04	.e	38	íb.si₈	
1 21	.e	9	íb.si₈			25 21	.e	39	íb.si₈	
1 40	.e	10	íb.si₈			26 40	.e	40	íb.si₈	
2 01	.e	11	íb.si₈			28 01	.e	41	íb.si₈	
2 24	.e	12	íb.si₈			29 24	.e	42	íb.si₈	
2 49	.e	13	íb.si₈			30 49	.e	43	íb.si₈	
3 16	.e	14	íb.si₈			32 16	.e	44	íb.si₈	
3 45	.e	15	íb.si₈			33 45	.e	45	íb.si₈	
4 16	.e	16	íb.si₈			35 16	.e	46	íb.si₈	
4 49	.e	17	íb.si₈			36 49	.e	47	íb.si₈	
5 24	.e	18	íb.si₈			38 24	.e	48	íb.si₈	
6 01	.e	19	íb.si₈			40 01	.e	49	íb.si₈	
6 40	.e	20	íb.si₈			41 40	.e	50	íb.si₈	
7 21	.e	21	íb.si₈			43 21	.e	51	íb.si₈	
8 04	.e	22	íb.si₈			45 04	.e	52	íb.si₈	
8 49	.e	23	íb.si₈			46 49	.e	53	íb.si₈	
9 36	.e	24	íb.si₈			48 36	.e	54	íb.si₈	
10 25	.e	25	íb.si₈			50 25	.e	55	íb.si₈	
11 16	.e	26	íb.si₈			52 16	.e	56	íb.si₈	
12 09	.e	27	íb.si₈			54 09	.e	57	íb.si₈	
13 04	.e	28	íb.si₈			56 04	.e	58	íb.si₈	
14 01	.e	29	íb.si₈			58 01	.e	59	íb.si₈	
15	.e	30	íb.si₈			1	.e	1	íb.si₈	

MS 3963 and MS 3864 are both of type a. However, MS 2185 is of a new type, which may be called "type a'", in which each line of the table is of the form

$$\text{sq. } n \text{ .e } n \text{ ba.si}_8 \qquad \text{sq. } n \text{ makes } n \text{ equalsided} \quad (\text{where } n \text{ is an integer}).$$

There is no real difference between type a and type a', since íb. and ba. are two Sumerian grammatical prefixes with more or less the same function, but with no direct counterparts in English.

The Sumerian term for a (geometric) 'square' is íb.si₈. It is a term characterizing a square as a (rectangular) *field with equal sides*, hence, for lack of a better alternative, the conform translation 'equalside'. The Sumerian term for 'side of a square' is the same(!) as the term for a square. Therefore, 'equalside' can stand for both a square and the side of a square. Moreover, in some circumstances íb.si₈ has to be regarded as a verb, in other circumstances as a noun. This double ambivalence of the term íb.si₈ makes it even harder to find a good translation of the mentioned key phrase in Old Babylonian tables of square sides. (For a more thorough discussion of the matter, see Høyrup, *LWS* (2002), 25-27.)

In addition to being of the unusual type a', the table of square sides MS 2185 has a couple of other special features. Of minor importance is that it ends with a brief subscript or colophon, an invocation of the god Nisaba, the Sumerian god of grain, writing, and wisdom, and as such the patron of scribes. More interesting is that it makes use of what may be called "variant number signs" 7_v, 8_v, 9_v for the numbers 7, 8, 9. The standard Old Babylonian 7 is written 3-3-1 (3 wedges on top of 3 wedges on top of 1 wedge), while the variant 7_v is written 4-3 (4 wedges

obv.   rev.

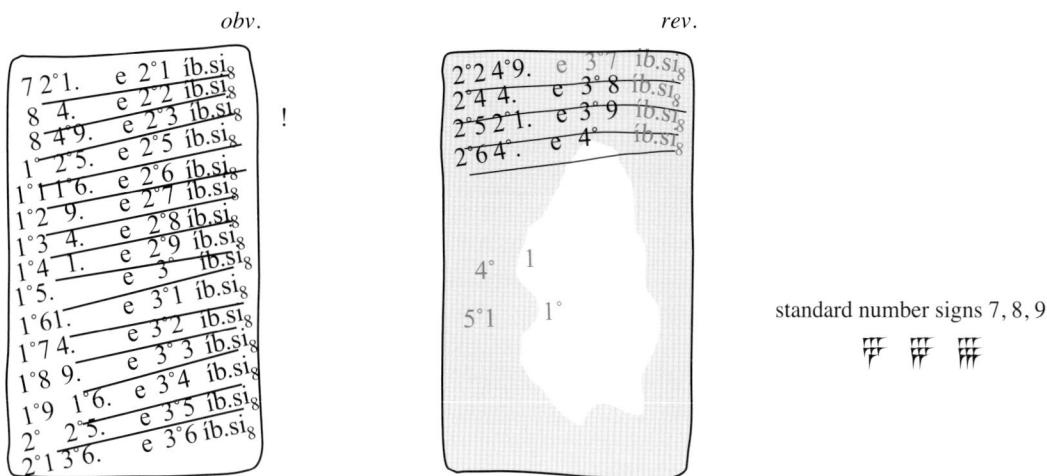

standard number signs 7, 8, 9:

MS 3963. Square sides of squares of integers, from 21 to 40 (20 lines). Type a, with íb.si₈.
The line '9 36.e 24 ib.sig' is missing.

obv.   rev.

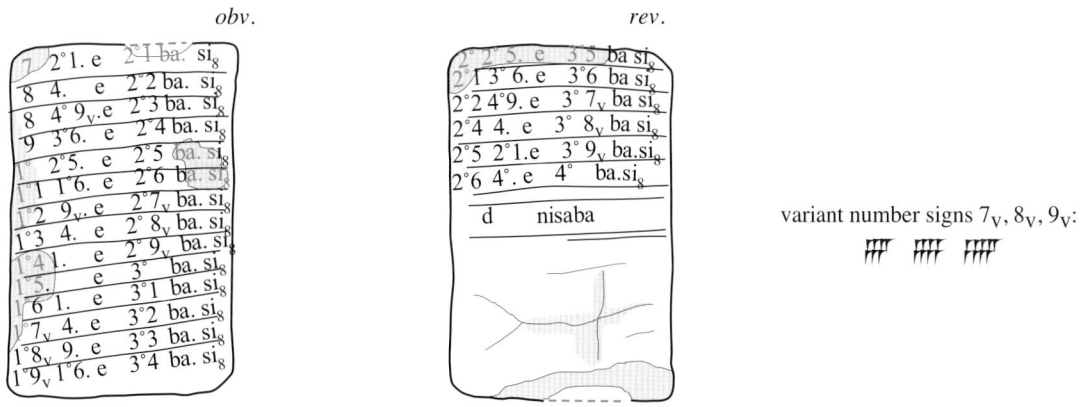

variant number signs 7ᵥ, 8ᵥ, 9ᵥ:

MS 2185. Square sides of squares of integers, from 21 to 40 (20 lines). Type a' (new), with ba.si₈.
Variant number signs for 7, 8, 9.

obv.   rev.

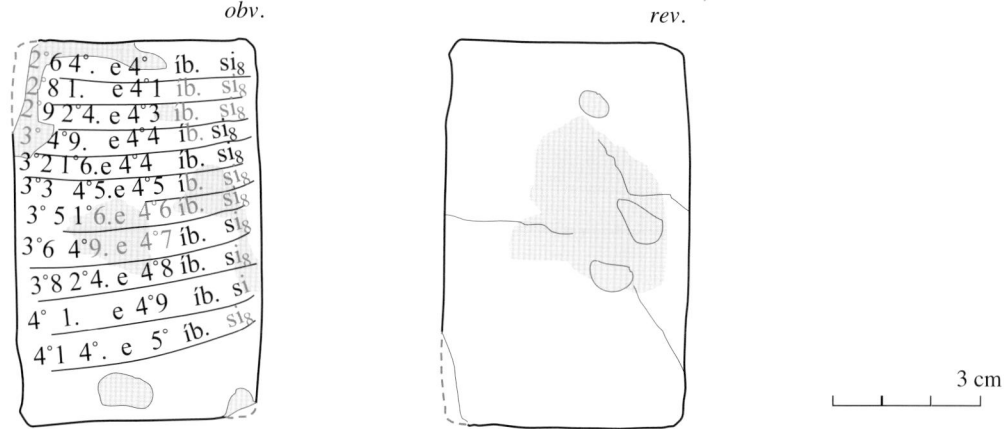

3 cm

MS 3864. Square sides of squares of integers, from 40 to 50 (11 lines). Type a, with íb.si₈.

Fig. 2.2.1. Three excerpts of the Old Babylonian complete table of square sides.

on top of 3 wedges). Similarly, the standard Old Babylonian 8 and 9 are written 3-3-2 and 3-3-3, respectively, while the variants 8_v and 9_v are written 4-4 and 5-4. (See the illustrations in Fig. 2.2.1 above.) Using the variant number signs is an archaic feature in an Old Babylonian text, since the variant signs were the ones used in the Neo-Sumerian Ur III period which preceded the Old Babylonian period.

A rapid survey shows that variant number signs can be found, often alternating with standard number signs, exclusively in Old Babylonian mathematical cuneiform texts from *southern* Mesopotamian cities such as Uruk, Larsa, and Ur. Such mathematical texts can be dated to before the fall of the southern cities in the year Sam-suiluna 11, 1739 BC. In terms of the Goetze/Høyrup/Friberg classification of unprovenanced mathematical cuneiform texts, variant number signs are used only in texts belonging to groups 1, 2, and 3 or to the groups of series texts Sa and Sb (see § 7 and Appendix 1 in Friberg, *RA* 94 (2000)).

Although it may seem to be the case that the use of variant number signs is characteristic for relatively *early* Old Babylonian mathematical texts, from a wider perspective the situation is quite complex and confusing. See the discussion in Oelsner, *ChV* (2001), where the use of variant number signs in mathematical cuneiform texts of unknown dates is compared with the use of such number signs in securely dated non-mathematical (legal and administrative) texts.

2.3.   Old Babylonian Tables of Cube Sides

A complete Old Babylonian table of cube sides looks like this:

1.	e	1	ba.si		8 16 31.	e	31	ba.si
8.	e	2	ba.si		9 06 08.	e	32	ba.si
27.	e	3	ba.si		9 58 57.	e	33	ba.si
1 04.	e	4	ba.si		10 55 04.	e	34	ba.si
2 05.	e	5	ba.si		11 54 35.	e	35	ba.si
3 36.	e	6	ba.si		12 57 36.	e	36	ba.si
5 43.	e	7	ba.si		14 04 13.	e	37	ba.si
8 32.	e	8	ba.si		15 14 32.	e	38	ba.si
12 09.	e	9	ba.si		16 28 39.	e	39	ba.si
16 40.	e	10	ba.si		17 46 40.	e	40	ba.si
22 11.	e	11	ba.si		19 08 41.	e	41	ba.si
28 48.	e	12	ba.si		20 34 48.	e	42	ba.si
36 37.	e	13	ba.si		22 05 07.	e	43	ba.si
45 44.	e	14	ba.si		23 39 44.	e	44	ba.si
56 15.	e	15	ba.si		25 18 45.	e	45	ba.si
1 08 16.	e	16	ba.si		27 02 16.	e	46	ba.si
1 21 53.	e	17	ba.si		28 50 23.	e	47	ba.si
1 37 12.	e	18	ba.si		30 43 12.	e	48	ba.si
1 54 19.	e	19	ba.si		32 40 49.	e	49	ba.si
2 13 20.	e	20	ba.si		34 43 20.	e	50	ba.si
2 34 21.	e	21	ba.si		36 50 51.	e	51	ba.si
2 57 28.	e	22	ba.si		39 03 28.	e	52	ba.si
3 22 47.	e	23	ba.si		41 21 17.	e	53	ba.si
3 50 24.	e	24	ba.si		43 44 24.	e	54	ba.si
4 20 25.	e	25	ba.si		46 12 55.	e	55	ba.si
4 52 56.	e	26	ba.si		48 46 56.	e	56	ba.si
5 28 03.	e	27	ba.si		51 26 33.	e	57	ba.si
6 05 52.	e	28	ba.si		54 11 52.	e	58	ba.si
6 46 29.	e	29	ba.si		57 02 59.	e	59	ba.si
7 30.	e	30	ba.si		1.	e	1	ba.si

MS 3863, and 3913 (Fig. 2.3.1 below) are two examples of Old Babylonian tables of cube sides. A *complete* table of cube sides lists the cube sides of the cubes of *all* integers from 1 to 1 (00) = 60. MS 3863 is an *excerpt* from a complete table of cube sides, with entries for cube sides from 41 to 1 (00) = 60, the last third of a complete table of cube sides. MS 3913 is an *abridged* table of cube sides, with entries only for cube sides from 1 to 20, and for 30, 40 and 50. Three other examples in the Schøyen Collection of such abridged tables of cube sides are **MS 3981, MS 3972,** and **MS 3985**.

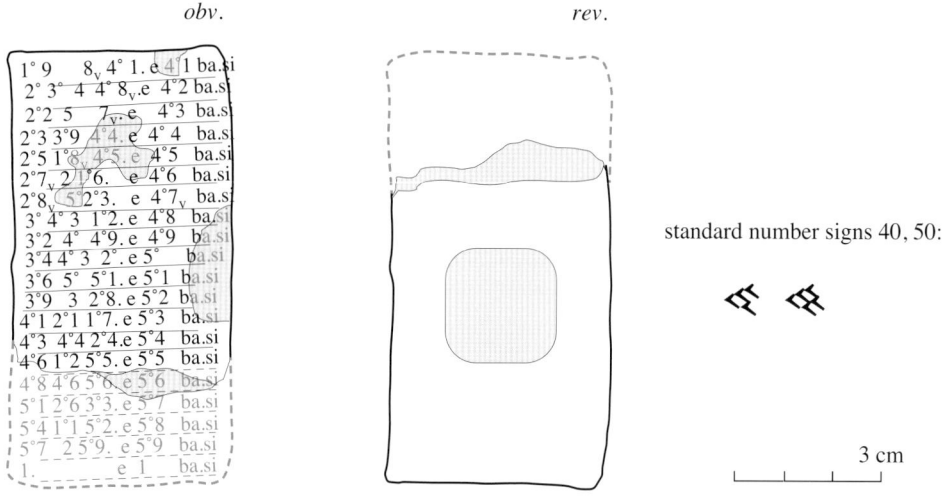

MS 3863. Cube sides from 41 to [1(00)] (20 lines). Type a. Variant number signs only for 7 and 8.

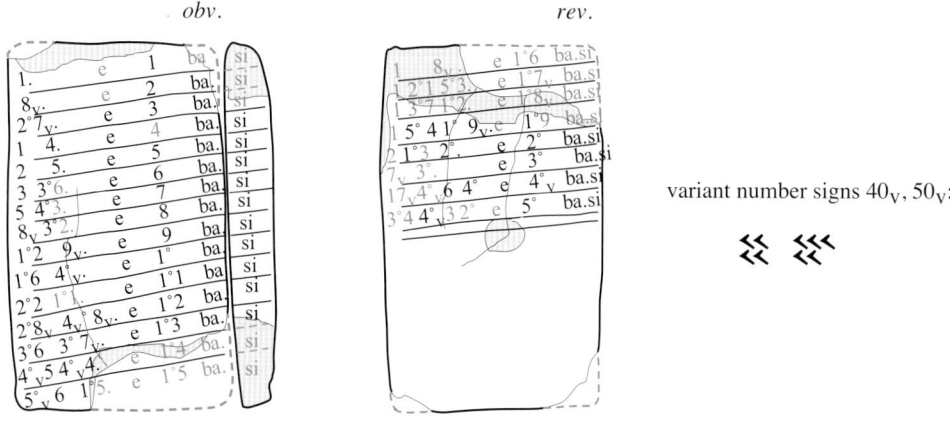

MS 3913.   Cube sides from 1 to 20, and 30, 40, 50 (23 lines). Type a.
Variant number signsfor 7, 8, 9, 40, and 50.

Fig. 2.3.1. Two excerpts of Old Babylonian tables of cube sides, type a.

MS 3863 makes use of variant number signs only for 7 and 8, not for 9. On the reverse of MS 3863 there is a large but shallow square impression, of unknown significance, stamped into the clay.

MS 3913 is unusual in that it makes use of variant number signs not only for 7, 8, and 9, but also for 40, and 50. In their standard forms, 40 and 50 are written as 3 oblique wedges above 1 or 2 oblique wedges, respectively, in *slanting* rows. The variant form for 40 is 2 oblique wedges over 2, in *horizontal* rows, and the variant form for 50 is 3 oblique wedges over 2, again in horizontal rows. See the illustrations in Fig. 2.3.1 to the right, top and bottom.

Normally, tables of cube sides are of "type a", in the sense that each line in such a table is of the form

cu. n. e n ba.si   cu. n makes n likesided   (where n is an integer)

Here, the cuneiform sign si is a homophone to the sign si_8. Both signs stand for a Sumerian verb with the meaning 'to be equal'. It is not clear why, in most cases, the sign si_8 is used in connection with squares and square sides, while the sign si is used in connection with cube sides. A possible explanation is that tables of cube sides was a relatively late invention, and that the way of writing the term /si/ had changed with time. What is known is that the sign si_8 (or sá) was used to describe the equalsidedness of squares already in Early Dynastic/Old Sumerian tables of squares (cf. the discussion in Ch. 6 below), but that *no pre-Babylonian tables of cube sides have yet been found*.

On the other hand, *no pre-Babylonian tables of square sides have been found either*. So, to complete the attempted explanation, it must be added that while it is natural to consider square courtyards, square bricks, and so on, there are no corresponding situations where it is natural to consider cube-shaped objects. For this reason, it is not really surprising that there are *no known examples of Sumerian or Babylonian tables of cubes*. It is tempting to draw the conclusion that, on one hand, the existence of Sumerian and Babylonian tables of squares can be explained as a consequence of the fact that squares are commonly occurring artefacts, while, on the other hand, the existence of Old Babylonian tables of square sides and cube sides can be explained only as a consequence of Old Babylonian mathematicians' preoccupation with quadratic and cubic equations! (Cf. Sec. 2.4 below with its discussion of tables of quasi-cube sides.)

Six tables of cube sides in the Schøyen Collection are of type a. There are, in addition to MS 3863, and 3913, mentioned above, the following three:

MS 2987, an excerpt from a table of cube sides, from 21 to 40 (15 + 5 lines, with a subscript in 2 lines)

MS 3931, another excerpt from 21 to 40 (16 + 4 lines)

MS 3962, a brief excerpt from a table of cube sides, from 50 to 1 (00) (11 lines)

MS 3972, a complete abridged table of cube sides, like MS 3913 (23 lines)

Three tables of cube sides in the Schøyen Collection are of other types than type a. One of them is **MS 3973/1** (Fig. 2.3.2, top), a table of cube sides from 31 to 1 (00) = 60, which is of "type b" in the sense that each line in the table is of the *purely numerical* form

cu. n n (where n is an integer)

There is an error in this text, in line 3 on the reverse, where the cube of 54 is given as 43 13 43 instead of 43 44 24. There is no obvious explanation of this error.

MS 3966 (Fig. 2.3.2, middle) is a small square tablet, inscribed on the obverse with only five lines from a table of cube sides, with n from 13 to 17. Its type is a somewhat strange variant of the purely numerical type b, in that the first line of numbers in the brief table is followed on the right edge by the word íb.si-*tam* = (possibly) *mithartam* (*mithartum* is the Akkadian counterpart to the Sumerian íb.si 'likesided', *mithartam* is the accusative of *mithartum*, and -*tam* would then be a phonetic complement to íb.si, presumably showing that íb.si was understood as a logogram for *mithartum*). The accusative shows that the line should be translated as '36 37 (makes) 13 likesided'.

Note: In an attempt to make a distinction between cube sides and square sides, the term 'likesided' was used above in connection with cube sides, but 'equalsided' in connection with square sides. After all, while the Akkadian counterpart of the Sumerian terms íb.si_8/ba.si_8 is known to have been *mithartum*, it is not impossible that the Akkadian counterpart of ba.si/íb.si may have been some other Akkadian word.

MS 2996 (Fig. 2.3.2, bottom) is a table of cube sides from 46 to 1 (00) = 60. It is of a new type, which may be called "type c" with entries of the form

cu. n n.àm   ba.si   cu. n, n it is, likesided

(It is difficult to know what a more precise translation of the Sumerian phrase should be.) This type is related to, but not identical with, what Neugebauer called "type 2 a" (*MKT 1*, 75).

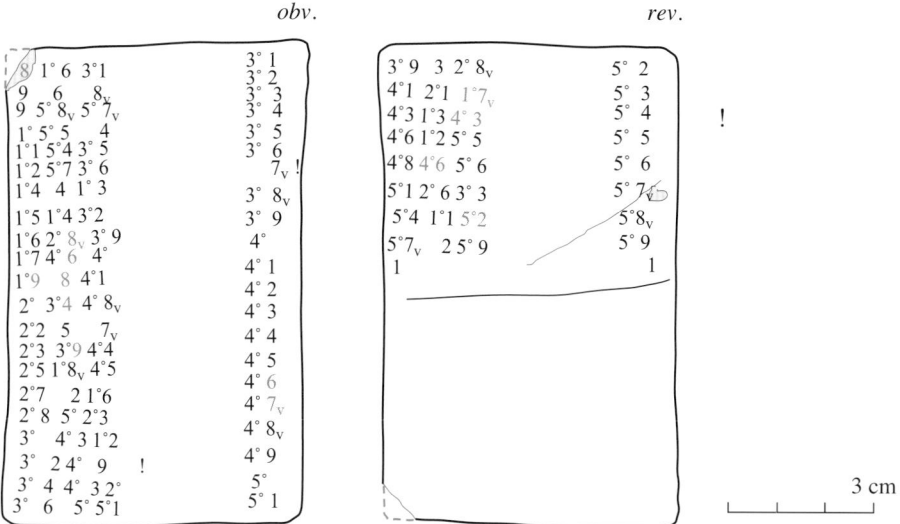

MS 3973/1. Cube sides from 31 to 1 (00) (30 lines). Type b (new). Variant number signs for 7 and 8.
The entry in line 3 on the reverse is incorrect, with 43 13 53 instead of 43 44 24.

MS 3966. Cube sides from 13 to 17 (5 lines). Type b' (new).

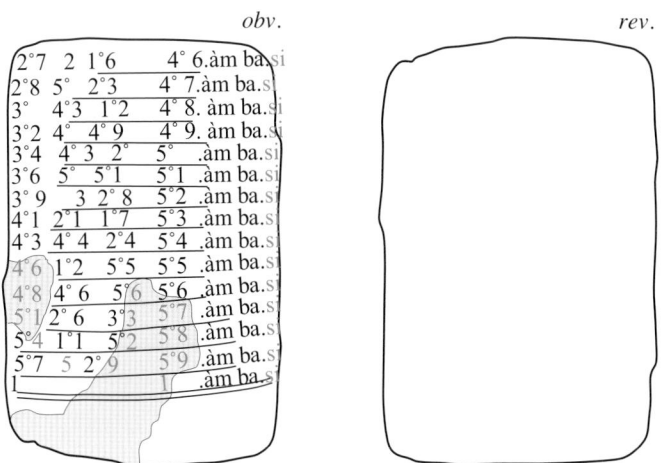

MS 2996. Cube sides from 46 to 1 (00) (15 lines). Type c (new).

Fig. 2.3.2. Three excerpts of Old Babylonian tables of cube sides, of new types.

2.4. Old Babylonian Tables of Quasi-Cube Sides

2.4 a.   MS 3899. A Table of n · n · (n + 1) Sides

MS 3899. A table of $n \cdot n \cdot (n + 1)$ sides from 1 to 12.

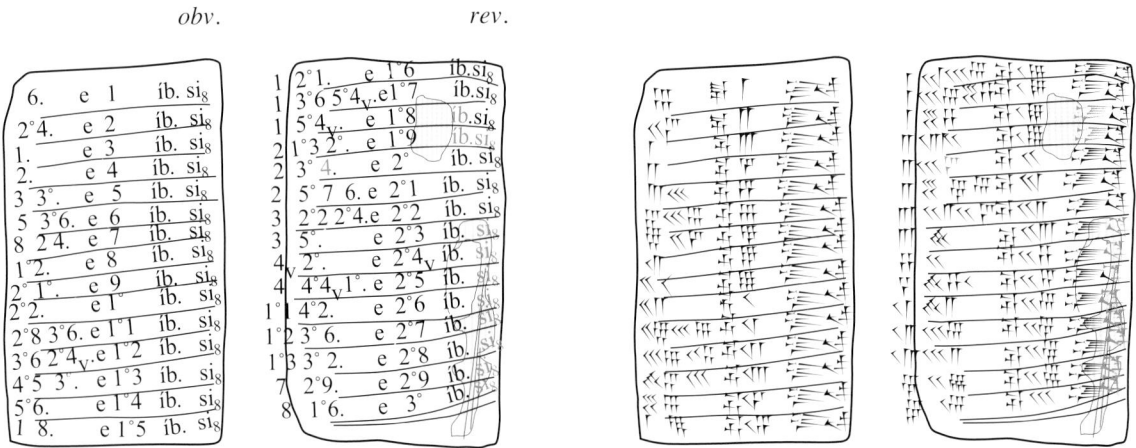

MS 3048. A table of $n \cdot (n + 1) \cdot (n + 2)$ sides from 1 to 30.

Fig. 2.4.1. Two Old Babylonian tables of quasi-cube sides, of two different kinds.

MS 3899 (Fig. 2.4.1, top) is a small, squarish tablet, inscribed with a table text on the obverse and the lower edge. Much of the text on the obverse is damaged, but at least the lines close to and on the lower edge are clearly legible. The general layout of the text is that of a table of cube sides, with entries of the form

$$m \text{ .e } \quad n \quad \text{ba.si} \qquad\qquad m \text{ makes } n \text{ likesided}$$

In the 12 lines of text, the 'likesided' n proceeds from 1 to 12, and it is easy to check that whenever the value of m is preserved, it satisfies the equation

$$m = n \cdot n \cdot (n + 1).$$

Hence, it may be motivated to call MS 3899 a table of "quasi-cube sides", or, more precisely, a table of "$n \cdot n \cdot (n + 1)$ sides". (Cf. Friberg, *RlA 7* (1990) Sec. 5.2 f.)

A reconstruction of the whole text of the table is given below.

MS 3899

[2]	.e	1	ba.[si]	$1 \cdot 1 \cdot 2$	=	2
[12]	.e	[2]	ba.[si]	$2 \cdot 2 \cdot 3$	=	12
[3]6	.e	[3]	ba.si	$3 \cdot 3 \cdot 4$	=	36
[1] 20	[.e	4	ba.si]	$4 \cdot 4 \cdot 5$	=	1 20
[2 30	.e	5	ba.si]	$5 \cdot 5 \cdot 6$	=	2 30
[4 12	.e	6	ba.si]	$6 \cdot 6 \cdot 7$	=	4 12
[6 32	.e	7	ba.si]	$7 \cdot 7 \cdot 8$	=	6 32
[9 36	.e	8]	ba.si	$8 \cdot 8 \cdot 9$	=	9 36
[1]3 30	.e	[9]	ba.si	$9 \cdot 9 \cdot 10$	=	13 30
[18] 20	.e	10	ba.si	$10 \cdot 10 \cdot 11$	=	18 20
[2]4 1[2]	.e	11	[ba.si]	$11 \cdot 11 \cdot 12$	=	24 12
31 12	.e	12	ba.si	$12 \cdot 12 \cdot 13$	=	31 12

VAT 8492 § 3. Another Table of $n \cdot n \cdot (n + 1)$ Sides, for n from [1] to [1 00]

The large table text **VAT 8492** is mentioned repeatedly by Neugebauer in *MKT 1*, unfortunately without any reference to a photo or hand copy of the tablet. The first paragraph of the text (col. *i* 1 - col. *ii* 15; *MKT 1*, 70) is a table of square sides, with each line of the form

m .e n íb.si$_8$ m makes n equalsided   (for n from [1] to 1 00)

The second paragraph (col. *ii* 16 - col. *iii* 30; *MKT 1*, 73) is a table of cube sides,

m .e n ba.si m makes n likesided   (for n from 1 to [1 00])

The third, and last paragraph (col. *iii* 31 - col. *iv* 45; *MKT 1*, 76) is a table of $n \cdot n \cdot (n + 1)$ sides, again with each line of the form

m .e n ba.si m makes n likesided   (for n from [1] to [1 00])

The transliteration of the quasi-cube table VAT 8492 offered in *MKT* is incomplete. A more complete transliteration is offered below:

VAT 8492 § 3

………	…	…	… …	… … … … …		… … … … …
[1 42 36	.e	1]8$_v$	ba.si	$18 \cdot 18 \cdot 19$	=	1 42 36
[2 00 20	.e	1]9	ba.si	$19 \cdot 19 \cdot 20$	=	2 00 20
[2 20	.e]	20	ba.si	$20 \cdot 20 \cdot 21$	=	2 20 00
[2 41 42	.e]	21	ba.si	$21 \cdot 21 \cdot 22$	=	2 41 42
[3 05 32	.e]	22	ba.si	$22 \cdot 22 \cdot 23$	=	3 05 32
[3 31 36	.e]	23	ba.si	$23 \cdot 23 \cdot 24$	=	3 31 36
[4	.e]	24	ba.si	$24 \cdot 24 \cdot 25$	=	4 00 00
[4 30 50]	.e	25	ba.si	$25 \cdot 25 \cdot 26$	=	4 30 50
5 04 [12]	.e	26	ba.si	$26 \cdot 26 \cdot 27$	=	5 04 12
5 40 12	.e	27$_v$	[ba.si]	$27 \cdot 27 \cdot 28$	=	5 40 12
6 18$_v$ [5]6	.e	28$_v$	[ba.si]	$28 \cdot 28 \cdot 29$	=	6 18 56
7$_v$ 30.	e	29	[ba.si]	$29 \cdot 29 \cdot 30$	=	7 00 30!
7$_v$ 45	.e	30	[ba.si]	$30 \cdot 30 \cdot 31$	=	7 45 00
8$_v$ [32 32.	e]	3[1	ba.si]	$31 \cdot 31 \cdot 32$	=	8 32 32
9 23 [12.	e]	3[2	ba.si]	$32 \cdot 32 \cdot 33$	=	9 32 12
10 17$_v$ [06.	e]	3[3	ba.si]	$33 \cdot 33 \cdot 34$	=	10 17 06
11 14 20	[.e	34	ba.si]	$34 \cdot 34 \cdot 35$	=	11 14 20
12 15.	[e	35	ba.si]	$35 \cdot 35 \cdot 36$	=	12 15 00
13 19 1[2	.e	36	ba.si]	$36 \cdot 36 \cdot 37$	=	13 19 12
14 27$_v$ [02	.e	37$_v$	ba.si]	$37 \cdot 37 \cdot 38$	=	14 27 02
15 38$_v$ 3[6	.e	38$_v$	ba.si]	$38 \cdot 38 \cdot 39$	=	15 38 36
16 [5]4	[.e	39	ba.si]	$39 \cdot 39 \cdot 40$	=	16 54 00

Continued

18 13 20	[.e	40	ba.si]		40 · 40 · 41	=	18 13 20
19 3[6 42	.e	41	ba.si]		41 · 41 · 42	=	19 36 42
21 [04 12	.e	42	ba.si]		42 ·42 · 43	=	21 04 12
22 [35 36	.e	43	ba.si]		43 · 43 · 44	=	22 35 36
24 [12	.e	44	ba.si]		44 · 44 · 45	=	24 12 00
25 [52 30	.e	45	ba.si]		45 · 45 · 46	=	25 52 30
27 [37 32	.e	46	ba.si]		46 · 46 · 47	=	27 37 32
2[9] 27 [12	.e	47	ba.si]		47 · 47 · 48	=	29 27 12
31 [2]1 [36	.e	48	ba.si]		48 · 48 · 49	=	31 21 36
3[3 20 50	.e	49	ba.si]		49 · 49 · 50	=	33 20 50
35 [25	.e	50	ba.si]		50 · 50 · 51	=	35 25 00
… … …	…	…	… …		… … … …		… … …

Below, in Fig. 2.4.2, is presented a new conform transliteration of the quasi-cube table on VAT 8492:

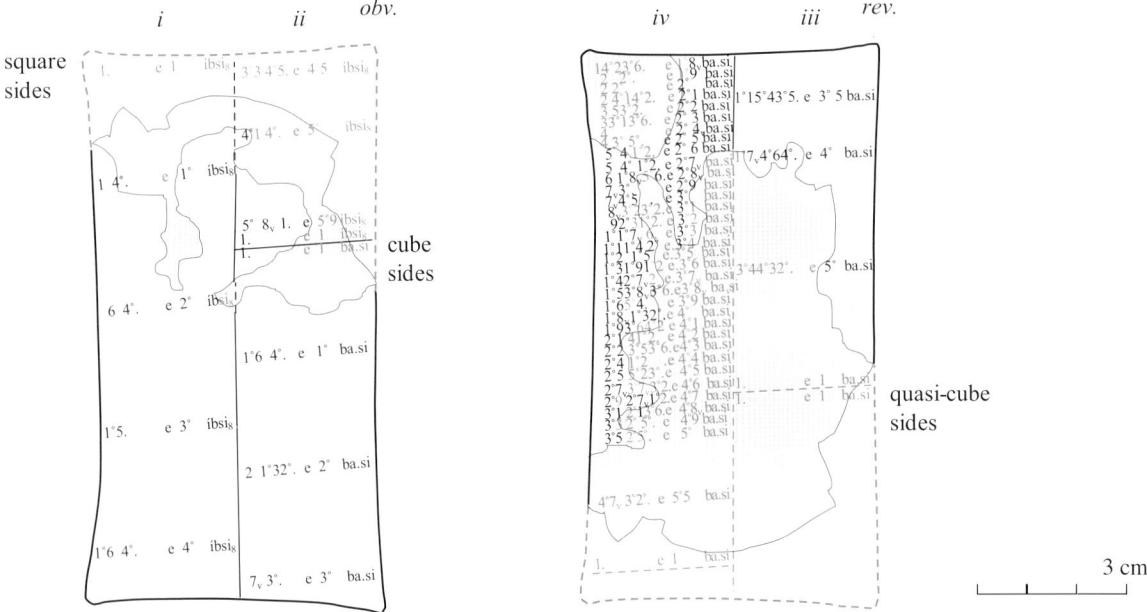

Fig. 2.4.2. VAT 8492. A combined table for square sides, cube sides (in outlines only), and quasi-cube sides.

2.4 b.   For What Could a Table of n · n · (n + 1) Sides Possibly be Used?

The answer to this question is well known (see Neugebauer, *MKT 1* (1935), 210, and Høyrup, *LWS* (2002), 149-154). The fact that the product $n \cdot n \cdot (n + 1)$ is equal to *the sum of the cube of n and the square of n* is behind the utilization of a table of $n \cdot n \cdot (n + 1)$ sides in the large Old Babylonian mathematical theme text **BM 85200 + VAT 6599 ## 5 and 23**. The theme of that text is quadratic or cubic equations for the sides of an exca-vated box-like room (a "cellar"). In addition to ## 5 and 23, other exercises in the text dealing with cubic equa-tions are ## 6-7 ## 20-22, and (probably) the now lost ## 1-4. The solution methods used in ## 6-7 and 20-22 are, almost certainly, trivial factorizations. Anyway, what makes the series of cubic equations in BM 85200 + VAT 6599 particularly interesting is that no other known cuneiform texts deal with cubic equations, other than extractions of cube sides in particularly easy cases. Here follows a transliteration and translation of one of the exercises making use of the table of $n \cdot n \cdot (n + 1)$ sides:

BM 85200 + VAT 6599 # 5. (Høyrup, *LWS*, 138.) **A cubic equation for the sides of an excavated room.**

… … … … … … …	… … … … … … …
1' [uš sa]g en.nam /	Length and front are what?
2' [··· ··· ··· ··· 3 *ta-mar*]	··· ··· ··· ··· 3 you see.
2' 3 *ḫe-pé* 1 30 *ta-mar* /	1/2 of 3 break, 1 30 you see.
3' [igi 1 30 du₈.a]	*The opposite of 1 30 resolve,*
40 *ta-mar* bal sag	40 you see, the ratio of the front.
igi 12 bal gam	The opposite of 12, the ratio of the depth,
4' du₈.a / [5 *ta-mar*]	resolve, *5 you see.*
[*a-na* 1] *i-ši* 5 *ta-mar*	*To 1* raise, 5 you see.
a-na 40 *i-ši* 3 20 *ta-mar* /	to 40 raise, 3 20 you see.
5' [3 20] *a-na* 5 *i-ši* 16 40 *ta-mar*	*3 20* to 5 raise, 16 40 you see.
igi 16 40 du₈.a 3 36 *ta-mar*	The opposite of 16 40 resolve, 3 36 you see.
6' 3 36 / *a-na* 1 10 *i-ši* 4 12 *ta-mar*	3 36 to 1 10 raise, 4 12 you see,
6 íb.si₈	6 equalsided.
6 *a-na* 5 *i-ši* 30 *ta-<mar>*	6 to 5 raise, 30 you <see>.
7' 6 *a-na* 3 20 *i-<ši>* / 20 sag	6 to 3 20 r<aise>, 20, the front.
6 *a-na* 1 *i-ši* 6 *ta-mar* gam	6 to 1 raise, 6 you see, the depth.
8' *ki-a-am* / *ne-pé-šum*	Such is the procedure.

(The form of the sign 2' for '1/2' is shown in Fig. A4.2 in Appendix 4.) The question asked in the lost first few lines of this exercise may have been formulated in something like the following way:

> An excavated room. As much as two times the length is three times the front. As much as the length is the depth. The mud I have torn out. The ground and the mud together, 1 10. Length and front are what?

In modern symbolic notation, the question can be reformulated as follows:

> Let u be the long side (the length), s the short side (the front), and d the depth of a rectangular excavated room.
> Let V be the volume (the mud), and let A be the base area (the ground) of the room.
> Assume that $s \cdot 3 = u \cdot 2$, that $d = u$, and that $A + V = 1;10$ (šar). Find u, s (and d).

A certain complication of the situation is caused by a well known peculiarity of Sumerian/Old Babylonian metrology, namely that the *horizontal* dimensions of the room, u and s, are measured in multiples of a ninda (= 6 meters), and multiplied by 'mutual eating (or holding)' (×), while the *vertical* dimension d is measured in multiples of a cubit (= 1/2 meter, since 1 ninda = 12 cubits), and is 'raised' to the base area (·). (See Høyrup, *LWS*, 20-21.) The base area is measured in sq. ninda (area šar), and the volume in sq. ninda times cubits (volume šar).

A second complication arises because, strictly speaking, the base area A, measured in area šar, cannot freely be added to the volume V, measured in volume šar. As first pointed out by Høyrup, Old Babylonian mathematicians apparently interpreted the expression "the ground and the mud together" as meaning the sum of two volumes, the first equal to the base area *multiplied by a unit of depth* (1 cubit), the second equal to the base area multiplied by the actual depth. In other words,

$$\text{"}A + V\text{"} = A \cdot 1 \text{ cubit} + A \cdot d = A \cdot (d + 1 \text{ cubit}).$$

Therefore, the question presumably stated in the first lines of # 5 can be further reformulated and made more precise as follows:

> Let $u = a \cdot 1$ ninda, $s = a \cdot 2/3$ ninda, and $d = a \cdot 12$ cubits.
> Find a so that "$A + V$" = 1;10 volume šar.
> Since "$A + V$" = $a \cdot 1 \times a \cdot 2/3 \cdot (a \cdot 12 + 1)$ volume šar, this means that
> a must be a solution to the cubic equation $a \cdot a \cdot 2/3 \cdot (a \cdot 12 + 1) = 1;10$.

In the absence of modern symbolic notations, the cuneiform text mentions

;40, the 'ratio of the front' (bal sag), an expression corresponding to the modern equation $s = a \cdot 2/3$,

and

12, the 'ratio of the depth' (bal gam), an expression corresponding to the modern equation $d = a \cdot 12$.

The equation $a \cdot a \cdot 2/3 \cdot (a \cdot 12 + 1) = 1;10$ cannot be solved directly by use of the $n \cdot n \cdot (n + 1)$ table. A further transformation is needed, in modern terms the introduction of $n = a \cdot 12$ as a new unknown, so that $a = n \cdot ;05$. Since $d = a \cdot 12$ cubits, this transformation can be interpreted as *the use of the depth rather than the length as a new primary unknown*. Hence, the following final reformulation of the stated question in # 5:

> Let $d = n$ cubits. Then $u = n \cdot ;05 \cdot 1$ ninda $= n \cdot ;05$ ninda, and $s = n \cdot ;40 \cdot ;05$ ninda $= n \cdot ;03\ 20$ ninda.
> Consequently, "$A + V$" $= n \cdot ;05 \times n \cdot ;03\ 20 \cdot (n + 1)$ volume šar $= n \cdot n \cdot (n + 1) \cdot ;00\ 16\ 40$ volume šar, and
> n must be a solution to the cubic equation $n \cdot n \cdot (n + 1) \cdot ;00\ 16\ 40 = 1;10$.
> After multiplication by $3\ 36$, the reciprocal of $;00\ 16\ 40$, the equation is reduced to $n \cdot n \cdot (n + 1) = 4\ 12$.

It is easy to check that, indeed,

> $1 \cdot ;05 = ;05$, $;40 \cdot ;05 = ;03\ 20$, $;05 \cdot ;03\ 20 = ;00\ 16\ 40$, $;00\ 16\ 40 \cdot 3\ 36 = 1$,   and $3\ 36 \cdot 1;10 = 4\ 12$.

These computations are explicitly mentioned in lines 3' - 6' of # 5. The unknown n was probably thought of as a correction factor for 1, the new bal gam 'ratio of the depth'. Its value could be obtained from the line of an $n \cdot n \cdot (n + 1)$ table saying

> $4\ 12.e$ 6 ib.si$_8$ $4\ 12$ makes 6 equalsided

With $n = 6$, it follows, as stated in lines 6' - 7' of # 5, that

> $u = 6 \cdot ;05$ (ninda) $= ;30$ (ninda), $s = 6 \cdot ;03\ 20$ (ninda) $= ;20$ (ninda),   and $d = 6 \cdot 1$ (cubit) $= 6$ (cubits).

BM 85200 + VAT 6599 # 23 is a second exercise making use of the $n \cdot n \cdot (n + 1)$ table. It is shown below in transliteration and translation:

BM 85200 + VAT 6599 # 23. (Høyrup, *LWS* (2002), 145.) **Another cubic equation.**

1	túl.sag	An excavated room (a cellar).
	ma-la uš-tam-ḫir	As much as I made equalsided,
	ù 1 kùš diri gam-*ma*	and 1 cubit beyond, the depth, then
	1 45 saḫar.ḫi.a [ba].zi/	1 45 of mud I tore *out*.
2	za.e	You:
	5 diri *a-na* 1 bal *i-ši* 5 *ta-mar*	The 5 beyond to 1, the ratio, raise, 5 you see.
	a-na 12 *i-š*[*i* 1] *ta-mar* /	To 12 rai*se*, *1* you see.
3	5 *šu-tam-*<ḫir> 25 *ta-mar*	5 make equalsi<ded>, 25 you see.
	25 *a-na* 1 *i-ši* 25 *ta-mar*	25 to 1 raise, 25 you see.
4	igi [25 du$_8$.a] / 2 24 *ta-mar*	The opposite of *25 resolve*, 2 24 you see.
	2 24 *a-na* 1 45 *i-ši* 4 12 *ta-mar* /	2 24 to 1 45 raise, 4 12 you see.
5	*i-na* íb.si$_8$ 1 daḫ.ḫa	From the "equalside 1 added",
	6 *ša*$^?$ íb.s[i$_8$]	6 is that which is made equals*ided.*
	6 *a-na* 5 *i-*[*ši* 30] *ta-*<mar> *im-*<ta-ḫar> /	6 to 5 rai*se*, *30* you see, it is made equalsided.
6	6 gam	7$^!$ the depth.
	ne-pé-š[*um*]	The procedu*re*.

The question stated in line 1 of this exercise can be reformulated in modern symbolic notation as

> Let u be the long side (the length), s the short side (the front), d the depth, and V the volume of a rectangular room.
> Assume that $s = u$, that $d = u + 1$ cubit, and that $V = 1;45$ (šar). (Find u, s and d.)

The situation is simpler in this case than it was in the case of # 5. The question can immediately be made more precise, still in modern terms, as

> Let $u = s = a \cdot 1$ ninda, and let $d = (a \cdot 12 + 1)$ cubits.
> Then, $V = a \cdot 1 \times a \cdot 1 \cdot (a \cdot 12 + 1)$ volume šar,
> and a must be a solution to the cubic equation $a \cdot a \cdot (a \cdot 12 + 1) = 1;45$.

The equation $a \cdot a \cdot (a \cdot 12 + 1) = 1;45$ cannot be directly solved by use of the $n \cdot n \cdot (n + 1)$ table. As in the case of # 5, a further transformation is needed, in modern terms the introduction of $n = a \cdot 12$ as a new unknown. Then a can be expressed as a multiple of n, $a = n \cdot ;05$. Hence, the following final reformulation of the stated question in # 23:

Let $d = (n + 1)$ cubits, and let $u = s = n \cdot {;}05 \cdot 1$ ninda $= n \cdot {;}05$ ninda.

Then, $V = n \cdot {;}05 \times n \cdot {;}05 \cdot (n + 1)$ volume šar $= n \cdot n \cdot (n + 1) \cdot {;}00\ 25$ volume šar, and

n must be a solution to the cubic equation $n \cdot n \cdot (n + 1) \cdot {;}00\ 25 = 1{;}45$.

After multiplication by 2 24, the reciprocal of ;00 25, the equation is reduced to $n \cdot n \cdot (n + 1) = 4\ 12$.

It is easy to check that, indeed,

$$1 \cdot {;}05 = {;}05, \quad 12 \cdot {;}05 = 1, \quad {;}05 \cdot {;}05 = {;}00\ 25, \quad {;}00\ 25 \cdot 2\ 24 = 1, \quad \text{and} \quad 2\ 24 \cdot 1{;}45 = 4\ 12.$$

These computations are explicitly mentioned in lines 2-4 of # 23. The value of n can now, just as in # 5, be obtained from the line of an $n \cdot n \cdot (n + 1)$ table saying

4 12.e   6 íb.si$_8$ 4 12 makes 6 equalsided.

With $n = 6$, it follows that

$$u = s = 6 \times {;}05 \ (\text{ninda}) = {;}30 \ (\text{ninda}), \text{ and } d = 6 + 1 \ (\text{cubits}) = 7 \ (\text{cubits}).$$

The answer provided in lines 5-6 of # 23 amounts to saying that the horizontal sides of the excavated room are equal to $6 \cdot {;}05 = 30$, and that the depth is 6, a mistake for 7. Maybe the author of the text forgot that he had assumed the depth to be $n + 1$ (cubits), not n (cubits) as in # 5.

2.4 c.   *VAT 8521. A Problem Text with References to a Table of $n \cdot n \cdot (n − 1)$ Sides*

The large table text VAT 8492 mentioned above contains tables for n from 1 to 1 00 for square sides, cube sides, and $n \cdot n \cdot (n + 1)$ sides, showing that the table of $n \cdot n \cdot (n + 1)$ sides was regarded as a natural generalization of the more common table of cube sides. Problems ## 5 and 23 of the mathematical theme text BM 85200 + VAT 6599 demonstrated that the table of $n \cdot n \cdot (n + 1)$ sides was so familiar to Old Babylonian mathematicians that it was possible to include a reference to that kind of table in an Old Babylonian mathematical exercise.

In a similar way, the curious mathematical theme text **VAT 8521** (Neugebauer, *MKT 1*, 352) demonstrates that a table of "$n \cdot n \cdot (n − 1)$ sides" was regarded as another natural generalization of the common table of cube sides, and that it was possible to include a reference to that kind of table, too, in an Old Babylonian mathematical exercise. VAT 8521 is a theme text with four similarly worded exercises, all ostensibly concerned with interest problems. The interest rate is the usual 12 shekels per mina (that is, 20%), but the questions are quite nonsensical, asking what the initial capital will be if one year's interest is required to be a square (## 1 and 3), a cube (# 2) or a number of the form $n \cdot n \cdot (n − 1)$ (# 4). The answers provided are the following:

1: if the interest is, for instance, 1 40 (the square of 10), then the initial capital is 8 20

2: if the interest is, for instance, 7 30 (00) (the cube of 30), then the initial capital is 37 30 (00)

3: if the interest is, for instance, 36 (the square of 6), then the initial capital is 3 (00)

4: if the interest is, for instance, 18 (= $n \cdot n \cdot (n − 1)$ with $n = 3$), then the initial capital is 1 30

This text is historically very interesting, because it is the only known Old Babylonian precursor of a problem type that reappears nearly 2000 years later in Diophantus' *Arithmetica*.

Below is presented the full text of problem # 4, in transliteration and translation:

VAT 8521 # 4. An artificial interest problem.

1	*a-na* 1 ma.na [1]2 gín *i-di-in-ma* /	To 1 mina *12* shekels he gave, then
2	máš ba.si.1.lá *li-id-di-kum* /	interest (as) "likeside-1-minus" may he give to you.
3	1 ma.na [gar.r]a 12 ṣ[í]-*ip-tam* gar.ra /	1 mina *set,* 12, the interest, set,
4	*ù* 18 a.r[á *š*]*a* ba.si 1.lá	and 18 (as) the fac*tor* that "likeside-1-minus"
	i-[*na-di-nu-kum* gar.ra-*ma*] /	*will give to you set, then*
5	1 ma.na *a-na* 12 máš íl 12	1 mina to 12, the interest, raise, 12.
	igi 12 p[u-ṭ]*ur-ma* 5 /	The opposite of 12 *reso*lve, then 5.
6	5 *a-na* 18 a.rá *ša ša-ak-nu* íl 1 30 /	5 to 18, the factor that was set, raise, 1 30.
7	1 30 sag kù.babbar /	1 30 is the initial silver.

8	*šum-ma* 1 30 sag kù.babbar	If 1 30 is the initial silver, (and)
9	*a-na* 1 ma.na 12 *lu-ud-di-im-ma* /	for 1 mina 12 he may give, then
10	18 máš.bi 1 *li-id-di-nam* /	18 its interest 1(?) may he give to me.
11	1 ma.na *a-na* 12 *ṣí-ip-tim* íl 12 /	1 mina to 12, the interest, raise, 12.
12	12 *a-na* 1 30 sag kù.babbar íl 18 máš.bi /	12 to 1 30, the initial silver, raise, 18 its interest.
13	ba.si 18.e$^?$ 1.lá en.nam 3	18 (as) "likeside-1-minus" is what? 3.

In lines 1-2 of this problem text, the interest is requested to be in the form of ba.si.1.lá, a term meaning something like 'cube-minus-1'. In line 4 it is suggested that 18 should be chosen as the value of the cube-minus-1. In lines 5-7, the initial capital is then found to be 1 30. The verification follows in lines 8-12, and in line 13 it is noted that 18 is the cube-minus-1 corresponding to (the side) 3. (It is difficult to understand the syntax of the text in line 13, but the meaning is clear.)

2.4 d.   MS 3048. A Table of $n \cdot (n + 1) \cdot (n + 2)$ Sides

MS 3048 (Fig. 2.4.1 above) is an Old Babylonian table of quasi-cube sides of a completely new kind, a table of "$n \cdot (n + 1) \cdot (n + 2)$ sides". The integer n proceeds from 1 to 15 on the obverse of the clay tablet and from 16 to 30 on the reverse, as shown in the transliteration below. Errors are marked with bold style.

6	.e	1	íb.si$_8$	$1 \cdot 2 \cdot 3$	=	6			
24	.e	2	íb.si$_8$	$2 \cdot 3 \cdot 4$	=	24			
1	.e	3	íb.si$_8$	$3 \cdot 4 \cdot 5$	=	1 00			
2	.e	4	íb.si$_8$	$4 \cdot 5 \cdot 6$	=	2 00			
3 30	.e	5	íb.si$_8$	$5 \cdot 6 \cdot 7$	=	3 30			
5 36	.e	6	íb.si$_8$	$6 \cdot 7 \cdot 8$	=	5 36			
8 24	.e	7	íb.si$_8$	$7 \cdot 8 \cdot 9$	=	8 24			
12	.e	8	íb.si$_8$	$8 \cdot 9 \cdot 10$	=	12 00			
20 10	.e	9	íb.si$_8$	$9 \cdot 10 \cdot 11$	=	16 30 !	$11 \cdot 10 \cdot 11$	=	20 10
22	.e	10	íb.si$_8$	$10 \cdot 11 \cdot 12$	=	22 00			
28 36	.e	11	íb.si$_8$	$11 \cdot 12 \cdot 13$	=	28 36			
36 24	.e	12	íb.si$_8$	$12 \cdot 13 \cdot 14$	=	36 24			
45 30	.e	13	íb.si$_8$	$13 \cdot 14 \cdot 15$	=	45 30			
56	.e	14	íb.si$_8$	$14 \cdot 15 \cdot 16$	=	56 00			
1 08	.e	15	íb.si$_8$	$15 \cdot 16 \cdot 17$	=	1 08 00			
1 21 36	.e	16	íb.si$_8$	$16 \cdot 17 \cdot 18$	=	1 21 36			
1 36 54	.e	17	íb.si$_8$	$17 \cdot 18 \cdot 19$	=	1 36 54			
1 54	.e	18	íb.si$_8$	$18 \cdot 19 \cdot 20$	=	1 54 00			
2 13 **20**	.e	19	íb.si$_8$	$19 \cdot 20 \cdot 21$	=	2 13 00 !	$20 \cdot 6$ **40**	=	2 13 **20**
2 34	.e	20	íb.si$_8$	$20 \cdot 21 \cdot 22$	=	2 34 00			
2 57 06	.e	21	íb.si$_8$	$21 \cdot 22 \cdot 23$	=	2 57 06			
3 22 24$^!$	.e	22	íb.si$_8$	$22 \cdot 23 \cdot 24$	=	3 22 24			
3 50$^!$	.e	23	íb.si$_8$	$23 \cdot 24 \cdot 25$	=	3 50 00			
4 20	.e	24	íb.si$_8$	$24 \cdot 25 \cdot 26$	=	4 20 00			
4 44 10	.e	25	íb.si$_8$	$25 \cdot 26 \cdot 27$	=	4 52 30 !	$25 \cdot 11$ **22**	=	4 **44 10**
11 42	.e	26	íb.si$_8$	$26 \cdot 27 \cdot 28$	=	5 27 36 !	$26 \cdot 27$	=	11 42
12 36	.e	27	íb.si$_8$	$27 \cdot 28 \cdot 29$	=	6 05 24 !	$27 \cdot 28$	=	12 36
13 32	.e	28	íb.si$_8$	$28 \cdot 29 \cdot 30$	=	6 36 00 !	$28 \cdot 29$	=	13 32
7 29	.e	29	íb.si$_8$	$29 \cdot 30 \cdot 31$	=	7 29 30 !	$30 \cdot 14$ **58**	=	7 29 **00**
8 16	.e	30	íb.si$_8$	$30 \cdot 31 \cdot 32$	=	8 16 00			

It is somewhat surprising that the "sides" in this table are called íb.si$_8$ 'equalsides', the term normally used in Old Babylonian tables of square sides, while ba.si 'likesides' is the term normally used in Old Babylonian tables of cube sides. Equally surprising is the use of the term íb.si$_8$ in the cubic equation problems BM 85200 + VAT 6599 # 5 and # 23, mentioned above. On the other hand, ba.si is the term used in the tables of $n \cdot n \cdot$

(n + 1) sides MS 3899 and VAT 8492 # 3, as well as in the artificial interest problem VAT 8521 # 4.

There is an error in line 9 of MS 3048, where 20 10 occurs instead of the correct value 16 30. The error can be explained as a trivial mistake, the computation of the product of 11, 10, and 11, instead of 9, 10, and 11.

The error in line 19 of the table, 2 13 20 instead of 2 13 (00), can probably be explained as follows. The author of the table, feeling smart, may have intended to compute $19 \cdot 20 \cdot 21$ as

$$19 \cdot 20 \cdot 21 = 20 \cdot 19 \cdot 21 = 20 \cdot (\text{sq. } 20 - 1) = 20 \cdot (6\ 40 - 1) = 2\ 13\ 20 - 20 = 2\ 13.$$

For some reason, he ended up counting with $20 \cdot \text{sq. } 20$ instead of with $20 \cdot (\text{sq. } 20 - 1)$.

The error in line 29 of the table can be explained in a similar way. The author of the table, again feeling smart, may have intended to compute $29 \cdot 30 \cdot 31$ as

$$30 \cdot 29 \cdot 31 = 30 \cdot 14\ 59, \text{ arguing that } 29 \cdot 31 = (30 - 1) \cdot (30 + 1) = \text{sq. } 30 - \text{sq. } 1 = 15\ (00) - 1 = 14\ 59.$$

For some reason, he ended up counting with $30 \cdot 14\ 58$ instead of with $30 \cdot 14\ 59$.

The error in line 25 of the table is even simpler to explain, since $26 \cdot 27 = 11\ 42$. The author of the table simply made the mistake of multiplying 25 with 11 22 instead of with 11 42.

The curious series of related errors in lines 26-28 seem to suggest that the author of the table had recourse to a table of "$n \cdot (n + 1)$ sides" (*a table of quasi-square sides*). He intended to facilitate his computations by computing, for instance, $26 \cdot 27 \cdot 28$ as $(26 \cdot 27) \cdot 28$, with the value of $26 \cdot 27$ fetched from the table of $n \cdot (n + 1)$ sides. Being bored, after too many computations, he did not pay attention to what he was doing and ended up writing three lines of the table of $n \cdot (n + 1)$ sides in the place of three lines of the table of $n \cdot (n + 1) \cdot (n + 2)$ sides!

An alternative, although related, explanation is that the table of $n \cdot (n + 1) \cdot (n + 2)$ sides on MS 3048 was copied from a larger table with values of $n \cdot (n + 1)$ and values of $n \cdot (n + 1) \cdot (n + 2)$ side by side, as in the following tentative reconstruction:

n	$n \cdot (n + 1)$	$n \cdot (n + 1) \cdot (n + 2)$
...	... ...	... ...
26	11 42	5 27 36
27	12 36	6 05 24
28	13 32	6 36 00
...	... ...	... ...

In such a case, the errors in lines 26-29 would have resulted from values inadvertently being copied from three lines of col. 2 instead of from the corresponding three lines of col. 3.

2.4 e.   *For What Could a Table of $n \cdot (n + 1) \cdot (n + 2)$ Sides Possibly be Used?*

As shown above, cubic equations for the sides of an excavated room were solved by use of a table of $n \cdot n \cdot (n + 1)$ sides in ## 5 and 23 of the large Old Babylonian mathematical theme text BM 85200 + VAT 6599. In # 5, the front (the short side) of the room is 2/3 of the length (the long side), and the depth is equal to length, In addition, the sum of the volume and the bottom area of the room is given. Actually, what is called "the sum of the volume and the area" is the volume plus *the bottom area multiplied by 1 cubit*. Therefore, that sum is the same thing as the bottom area multiplied by *the depth plus 1 cubit*. In # 23, the front of the room is equal to the length, the depth is equal to *the length plus 1 cubit*, and the volume is given. In either case, after suitable reformulations, the given problem could be reduced to finding a solution to the cubic equation $n \cdot n \cdot (n + 1) = 4\ 12$. The table of $n \cdot n \cdot (n + 1)$ sides immediately gave the answer, that $n = 6$. And so on.

Is it possible that the table of $n \cdot (n + 1) \cdot (n + 2)$ sides was used in a similar way, as a tool for solving a certain class of cubic equations? The answer is yes, and it is not difficult to think of (constructed) examples. Consider, for instance, a problem of the following type (Fig. 2.4.3):

An excavated room. The length exceeds the front by 1 cubit. As much as the length is the depth. The mud I have torn out. The ground and the mud together, 1;27 30. Length and front are what?

In modern symbolic notations, the question can be reformulated as follows:

Let u be the long side (the length), s the short side (the front), and d the depth of a rectangular room.
Let V be the volume (the mud), and let A be the base area (the ground) of the room.
Assume that $s = u - 1$ cubit, that $d = u$, and that $V + A = 1;27\ 30$ (šar). Find u, s and d.

Proceeding as in BM 85200 + VAT 6599 # 5, one can solve this problem in the following way:

Let $d = n$ cubits. Then $u = n \cdot\ ;05$ ninda, and $s = (n - 1) \cdot\ ;05$ ninda. Consequently,
"$A + V$" $= n \cdot\ ;05 \times (n - 1) \cdot\ ;05 \cdot (n + 1)$ volume šar $= (n - 1) \cdot n \cdot (n + 1) \cdot\ ;00\ 25$ volume šar.
Hence, n must be a solution to the cubic equation $(n - 1) \cdot n \cdot (n + 1) \cdot\ ;00\ 25 = 1;27\ 30$.
After multiplication by $2\ 24$, the reciprocal of $;00\ 25$, the equation is reduced to $(n - 1) \cdot n \cdot (n + 1) = 3\ 30$.
This equation has the solution $n = 6$. Hence, $u = 6$ cubits $= ;30$ ninda, $s = 5$ cubits $= ;25$ ninda, and $d = 6$ cubits.

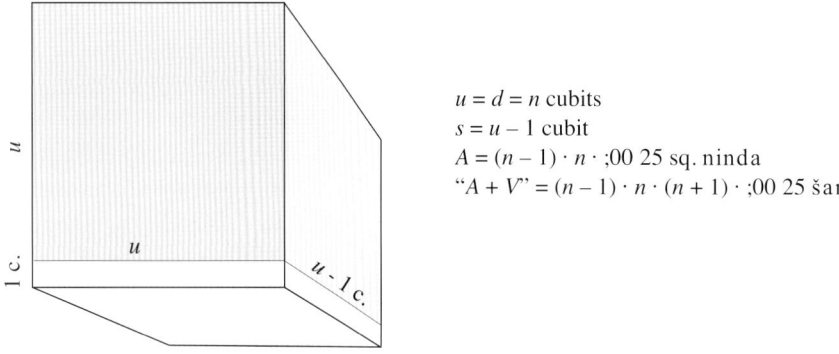

$u = d = n$ cubits
$s = u - 1$ cubit
$A = (n - 1) \cdot n \cdot\ ;00\ 25$ sq. ninda
"$A + V$" $= (n - 1) \cdot n \cdot (n + 1) \cdot\ ;00\ 25$ šar

Fig. 2.4.3. Explanation of $(n - 1) \cdot n \cdot (n + 1)$ as the "area plus volume" of an excavated room.

This explanation of the reason for the existence of an Old Babylonian table of $n \cdot (n + 1) \cdot (n + 2)$ sides is not supported by the existence of any known Old Babylonian problem texts in which cubic equations are solved by use of such a table. Therefore, the explanation may or may not be correct, and it is comforting to know that an alternative explanation exists for the existence of an Old Babylonian table of $n \cdot (n + 1) \cdot (n + 2)$ sides (quasi-cube sides). Recall that the errors in lines 26-28 of the table on MS 3048 suggest that the author of the table may have had recourse to a table of $n \cdot (n + 1)$ sides (quasi-square sides). Now, it is well known among the mathematically educated in our own time that the sum of the first n integers is equal to $n \cdot (n + 1)/2$ and that, similarly, the sum of the first n squares of integers is equal to $n \cdot (n + 1) \cdot (n + 2)/6$. The familiar arithmetical proof of the first of these propositions is very simple. The routine proof of the second proposition is somewhat more complicated, relying on the method of induction.

It is quite unlikely that general *arithmetical* identities of the kind mentioned above could be formulated and proved by Old Babylonian mathematicians, at least in the modern way. On the other hand, it is not unlikely that the same feat could be achieved by Old Babylonian mathematicians by means of *geometric* methods. Remember that Old Babylonian mathematicians apparently visualized and solved problems involving quadratic equations by use of geometric representations and "cut-and-glue" operations with geometric figures. The method has been called "naive geometry" by Høyrup (*LWS* (2002), Ch. IV) and "metric algebra" by Friberg (*BagM* 28 (1997), Ch. 1).

Consider, for instance, the sum of the first 10 integers. A suitable geometric representation of that sum is the total area of 10 rectangles, from 1 to 10 units long, but all 1 unit wide, together forming a "step triangle" with 10 steps. See the grey region in Fig. 2.4.4 below.

As shown by the figure, a rectangle of length (long side) $n + 1$ and front (short side) n with, for instance, $n = 10$, can be divided into two such step triangles, one a slightly offset mirror image of the other. This is a simple geometric proof of the identity

$$1 + 2 + \ldots + n = 1/2 \cdot n \cdot (n + 1), \text{ for all positive integers } n.$$

A word of caution: The Old Babylonian mathematicians had no way of indicating a "general" integer n. What they did instead, typically, was to choose some *arbitrary* integer and let it stand for *any* integer. Thus, an Old Babylonian mathematician would express an identity like the one above in something like the following way:

$$1 + 2 + \ldots + 10 = 1/2 \cdot 10 \cdot (10 + 1),$$

with the silent understanding that the 10 could be replaced by any other (positive) integer.

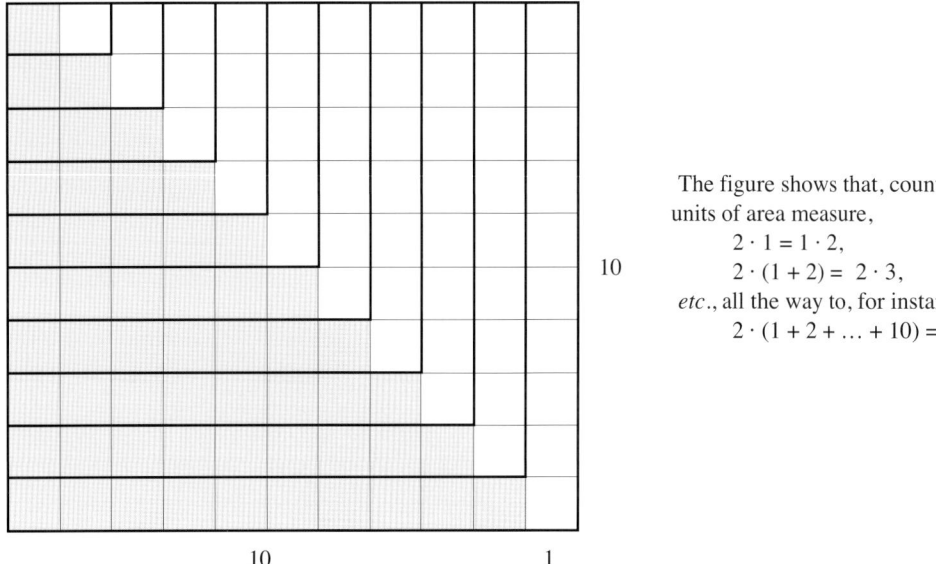

10

The figure shows that, counting with units of area measure,
$$2 \cdot 1 = 1 \cdot 2,$$
$$2 \cdot (1 + 2) = 2 \cdot 3,$$
etc., all the way to, for instance
$$2 \cdot (1 + 2 + \ldots + 10) = 10 \cdot 11.$$

10   1

Fig. 2.4.4. Geometric interpretation of $n \cdot (n + 1)$ as 2 times the sum $1 + 2 + \ldots + n$.

See Friberg, *RlA 7* (1990) Sec. 5.7 h, for a survey of summation rules for arithmetical or geometric progressions in Old and Late Babylonian mathematical texts. To that survey can now be added the Old Babylonian mathematical exercise BM 85 196 # 13. According to Robson, *MMTC* (1999), 80, the question in that exercise must be understood as follows: *How many days will a man spend carrying home one sheaf at a time if 6 sixties of sheaves are placed in a line before him, 5 ninda (30 meters) apart?* It is silently understood that the man's daily work norm is walking with a load 45 00 ninda (16 kilometers). (See Sec. 7.3 a below.) Therefore, to bring home all the sheaves, the man has to walk

$$(1 + 2 + \ldots + 6\,00) \cdot 5 \text{ ninda} = 1/2 \cdot 6\,00 \cdot 6\,01 \cdot 5 \text{ ninda} = 15\,00 \cdot 6\,01 \text{ ninda}.$$

To accomplish this, the number of days he has to spend carrying sheaves is

$$15\,00 \cdot 6\,01/45\,00 = 6\,01/3 = 2\,00\ 1/3 \quad \text{(4 months of 30 days plus 1/3 day)}.$$

The construction in Fig. 2.4.4 above provides a geometric interpretation *in two dimensions* of numbers of the type $n \cdot (n + 1)$. The construction may have been known to Old Babylonian mathematicians. In a similar way, the construction in Fig. 2.4.5 below provides a geometric interpretation *in three dimensions* of numbers of the type $n \cdot (n + 1) \cdot (n + 2)$. This construction, too, may have been known to Old Babylonian mathematicians, even if there is no direct support for this conjecture.

In the two-dimensional case illustrated by Fig. 2.4.4, it is immediately clear that *a rectangle of sides n and n + 1 can be divided into* two equal step triangles *with n steps*, and that this circumstance can be used for a geometric interpretation of the identity $1 + 2 + \ldots + n = 1/2 \cdot n \cdot (n + 1)$. In the three-dimensional case illustrated by Fig. 2.4.5 below, the situation is more complicated, because the interior of a three-dimensional object is not visible. Anyway, the situation in three dimensions is not very different from the one in two dimensions. What is suggested by the construction in Fig. 2.4.4 is that *a rectangular block of front n, length n + 1, and depth n + 2 can be divided into three equal step pyramids, each* with n steps. This circumstance can be used for a geometric interpretation of the identity

$$1 \cdot 2 + 2 \cdot 3 + \dots + n \cdot (n + 1) = 1/3 \cdot n \cdot (n + 1) \cdot (n + 2).$$

Indeed, what this identity says is that the volume of each one of the three equal step pyramids is one third of the volume of the whole rectangular block.

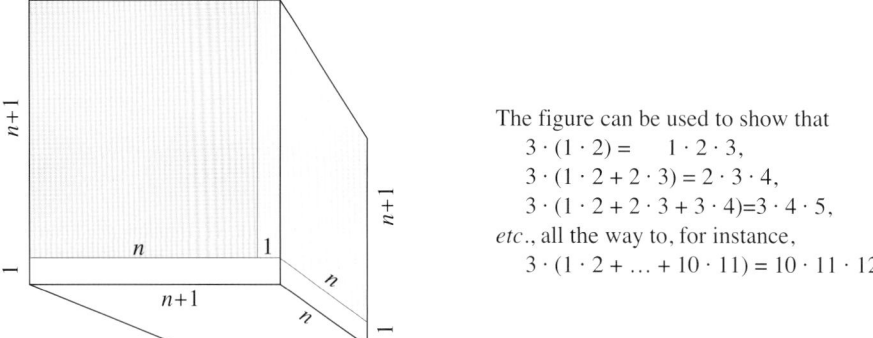

The figure can be used to show that
$$3 \cdot (1 \cdot 2) = \quad 1 \cdot 2 \cdot 3,$$
$$3 \cdot (1 \cdot 2 + 2 \cdot 3) = 2 \cdot 3 \cdot 4,$$
$$3 \cdot (1 \cdot 2 + 2 \cdot 3 + 3 \cdot 4) = 3 \cdot 4 \cdot 5,$$
etc., all the way to, for instance,
$$3 \cdot (1 \cdot 2 + \dots + 10 \cdot 11) = 10 \cdot 11 \cdot 12.$$

Fig. 2.4.5. Geometric interpretation of $n \cdot (n + 1) \cdot (n + 2)$ as 3 times the sum $1 \cdot 2 + 2 \cdot 3 + \dots + n \cdot (n + 1)$.

The idea that Old Babylonian mathematicians may have known that a rectangular block of front n, length $n + 1$, and depth $n + 2$ can be divided into three step pyramids of equal volumes is supported by the fact that Old Babylonian mathematicians knew, in certain specific cases, that the volume of a pyramid is equal to one third of the product of its base area and its height. (See the discussion of the large Old Babylonian mathematical recombination text **BM 96954 + BM 102366 + SÉ 93** in Friberg, *PCHM* 6 (1996), Friberg, *UL* (2005), Fig. 4.8,4, and in Robson, *MMTC* (1999), Appendix 3.)

A related text is exercise # 1 in the Egyptian demotic mathematical papyrus **P.BM 10520** (Parker *DMP* (1972)), where the iterated sum of the integers from 1 to 10, called "1 filled up twice to 10", is computed as 55 · 12/3, that is, as $\{n \cdot (n + 1)/2\} \cdot (n + 2)/3$. There is no indication of how the Egyptians had found this summation rule, or for what purpose it was used by them.

2.4 f.   A Note on Excerpts from Old Babylonian Arithmetical Table Texts

The maximal extent of an Old Babylonian table of squares, square sides, or cube sides seems to have been 60 lines (from 1 to 1(00)). (Compare with the surveys in *MKT 1*, 70-71 and 73.) Since n goes from [1] to [1 00] in the table of $n \cdot n \cdot (n + 1)$ sides in VAT 8492, that table, too, must have been a table of maximal extent. On the other hand, the small table of $n \cdot n \cdot (n + 1)$ sides on the hand tablet MS 3899, with only 12 lines (Fig. 2.4.1, top), is a brief excerpt from a table of maximal extent. This fact, in itself, is interesting, because it suggests that the table of $n \cdot n \cdot (n + 1)$ sides, although apparently not as common as ordinary tables of cube sides, must have been one of the mathematical "standard texts" which school boys in the Old Babylonian edubba had to study and make excerpts from.

Similarly, the table of $n \cdot (n + 1) \cdot (n + 2)$ sides MS 3048 with its 30 lines (Fig. 2.4.1, bottom), apparently is one half of a table of maximal extent. Hence, this kind of table, too, seems to have been common enough to have been an Old Babylonian mathematical standard text, even if not as common as the ordinary table of cube sides.

Here follows a survey of all tables of squares, square sides, cube sides, and quasi-cube sides in the Schøyen Collection, showing how large a part each one is of a table of maximal extent:

MS 2794	t. of squares	Fig. 2.1.1	lines 1-20, 30, 40, 50	complete abridged
MS 2706	"	"	lines 1-14	1st of 4 parts
MS 3963	t. of square sides	Fig. 2.2.1	lines 21-40	2nd of 3 parts

MS 2185	"	"	lines 21-40	2nd of 3 parts
MS 3864	"	"	lines 40-50	4th of 6 parts
MS 3913	t. of cube sides	Fig. 2.3.1	lines 1-20, 30, 40, 50	complete abridged
MS 3863	"	"	lines 41-60	3rd of 3 parts
MS 3973/1	"	Fig. 2.3.2	lines 31-60	2nd of 2 parts
MS 3966	"	"	lines 13-17	?
MS 2996	"	"	lines 46-60	4th of 4 parts
MS 2987	"		lines 21-40	2nd of 3 parts
MS 3931	"		lines 21-40	2nd of 3 parts
MS 3912	"		lines 50-60	5th of 6 parts
MS 3899	t. of quasi-cube sides	Fig. 2.4.1	lines 1-12	1st of 5 parts
MS 3048		"	lines 1-30	1st of 2 parts

The common pattern is obvious. With the exception of the very small clay tablet MS 3966, all the others are either complete abridged tables or one of 2, 3, 4, 5, or 6 equal (or almost equal) parts of a table of maximal extent. What this means is probably that excerpts from arithmetical table texts were not meant to be tools to assist in computations, but were instead *different writing exercises of equal size excerpted from complete tables, given out by the teacher to the individual members in classes of from 1 to 6 students!*

2.5. The Old Babylonian Standard Table of Reciprocals

It may be helpful to recall here the meaning of (modern) notations such as "regular sexagesimal number", "reciprocal of a regular sexagesimal number", *etc*. In modern discussions of Babylonian mathematics, *n* is called a regular sexagesimal number when *some* power of 60 can be divided by *n* exactly, without a remainder. However, in Babylonian *relative* place value notation *all* powers of 60 were written in the same way, as the single digit '1'. Therefore, a simpler definition is that

A regular sexagesimal number is a number *n* for which exact division of '1' by *n* is possible.

A special Babylonian name for regular sexagesimal numbers is not known, and may never have existed. What we call a regular sexagesimal number was thought of by Old Babylonian mathematicians as a sexagesimal number *n* for which an 'opposite number' igi *n* exists. This opposite number was required to be the *reciprocal of n* in the (generalized) sense that *n* · igi *n* = '1'. In modern terms,

A given sexagesimal number *n* is called regular when it is possible to find another sexagesimal number igi *n* satisfying the requirement that *n* · igi *n* = some power of 60.

Precisely because the sexagesimal base 60 itself is equal to 2 · 2 · 3 · 5, it follows that a sufficiently high power of 60 can be divided exactly by any given integer which is a product of an arbitrary number of factors 2, 3, and 5, but not by numbers containing other prime factors. Therefore,

A given sexagesimal number is regular when it is equal to (some positive or negative power of 60 times)*an integer containing no other prime factors than 2, 3, or 5.*

The well known *Old Babylonian* "standard table of reciprocals" enumerated all pairs of "1-place" regular sexagesimal numbers and their reciprocals, from igi 2 = 30 to igi 54 = 1 06 40, and(!) igi 1 = 1. At the end of the table, there were two additional lines, igi 1 04 = 56 15 and igi 1 21 = 44 26 40. Three further lines, igi 1 12 = 50, igi 1 15 = 48, and igi 1 20 = 45, were actually redundant and were often omitted, probably because of the redundancy. The reason why the "2-place" regular numbers 1 04 and 1 21 were added to the list of regular 1-place numbers in the standard table of reciprocals is probably that 1 04 (= 64) = 2^5 and 1 21 (= 81) = 3^4. Thus, both numbers are high powers of 2 and 3, respectively, yet close to '1'. Note, that, for instance

$$1\ 21 = 3^4, \text{hence igi } 1\ 21 = 60^4/3^4 = 20^4 = 44\ 26\ 40.$$

(The atypical OB table of reciprocals BM 106444 (Robson, *AfO* 50 (2003/04)) ends with igi 2 05 (5^3) = 28 48!)

Old Babylonian standard tables of reciprocals of the most common type, "type a", normally begin with *an extra line for the reciprocal of 2/3*, often in the following form

1.da 3".bi 40.àm   1, its 2/3, 40 it is

(For explicit examples, see the surveys of the beginnings of 28 tables of reciprocals in *MKT 1*, 10-12, and of 13 tables of reciprocals in *MCT*, 12.) The proposed translation of this initial phrase is only tentative. The exact meaning of the phrase is difficult to establish, because .da and .bi can both be grammatical suffixes, but they can also both be phonetic complements. In the latter case, /da/ may the last syllable of the Sumerian word gešda = 'sixty', and /bi/ may be the last syllable of the Sumerian word šanabi 'two-thirds', written with the sign 3". The form of this sign is shown in Fig. 0.3 in Sec. 0.4.6.

The second line of Old Babylonian standard tables of reciprocals of type a is normally of the form

šu.ri.a.bi (or šu.ri.bi) 30.àm   its half, 30 it is

All the other lines of an Old Babylonian standard table of reciprocals of type a are of the form

igi.*n*.gál.bi *n*'   the opposite of *n* exists of it (and is) *n*'

A standard table of reciprocals (of type a) looks like this (28 or 31 entries):

1.da 3".bi 40.àm		igi. 27. gál.bi 2 13 20	
šu.ri.a.bi 30.àm		igi. 30. gál.bi 2	
igi. 3. gál.bi 20		igi. 32. gál.bi 1 52 30	
igi. 4. gál.bi 15		igi. 36. gál.bi 1 40	
igi. 5. gál.bi 12		igi. 40. gál.bi 1 30	
igi. 6. gál.bi 10		igi. 45. gál.bi 1 20	
igi. 8. gál.bi 7 30		igi. 48. gál.bi 1 15	
igi. 9. gál.bi 6 40		igi. 50. gál.bi 1 12	
igi.10. gál.bi 6		igi. 54. gál.bi 1 06 40	
igi.12. gál.bi 5		igi. 1. gál.bi 1	
igi.15. gál.bi 4		igi.1 04. gál.bi 56 15	
igi.16. gál.bi 3 45		(igi.1 12.gál.bi 50)	
igi.18. gál.bi 3 20		(igi.1 15.gál.bi 48)	
igi.20. gál.bi 3		(igi.1 20.gál.bi 45)	
igi.24. gál.bi 2 30		igi.1 21. gál.bi 44 26 40	
igi.25. gál.bi 2 24			

The problematic formulation igi.*n*.gál.bi *n*' can be compared with the simpler phrases used in Old Babylonian mathematical assignments to compute the reciprocals of given numbers. Examples:

n igi.bi *n*'	*n*, its opposite (is) *n*'	MS 3264	Fig. 1.4.5
n igi *n*'	*n*, opposite *n*'	MS 2894	Fig. 1.4.3
n *n*'	*n*, *n*'	MS 2793	Fig. 1.4.3

The Old Babylonian standard table of reciprocals can also be compared with three tables of reciprocals, **HS 201**, **Ist. T 7375**, and **Ist. Ni 374**, all assumed to be from the Neo-Sumerian Ur III period, which immediately preceded the Old Babylonian period. (Oelsner, *ChV* (2001); Appendix 1 below, Fig. A1.2; Proust, *TMN* (2004), vol. 2, pl. 1; Friberg, *CDLJ* 2005:2, Fig. 6). The common layout of the three Ur III tables differs in several ways from the layout of the Old Babylonian standard table of reciprocals of type a. Thus, the initial entry on Ist. T 7375 is lost, but the initial entry on HS 201 is

1.da igi.2.gál.bi 30   1 (· 60), the opposite of 2 exists of it (and is) 30

This entry corresponds to line 2 in the Old Babylonian standard table of reciprocals. There is no entry corresponding to line 1 in the standard table (the one for the reciprocal of 2/3). All the other entries in HS 201, Ist. T. 7375, and Ist. Ni 374 are of one or the other of the following two forms

n igi *n*'	*n*, opposite *n*'	(when *n* is a regular sexagesimal number)
n igi nu	*n*, opposite none	(when *n* is a non-regular sexagesimal number!)

It is also remarkable that the table on HS 201 proceeds from *n* = 2 to *n* = 32, the table on Ist. T 7375 from *n* = [2] to *n* = 1 (00), and the table on Ist. Ni 374 from *n* = [2] to *n* = 1 40! A further striking dissimilarity is that while Old Babylonian tables of reciprocals normally are written on single column tablets, the Ur III tables are written in two columns on each side of the clay tablet. This is probably because the Ur III tables contain many more entries than the Old Babylonian standard table.

MS 3874 (Fig. 2.5.1) is a standard table of reciprocals of type a. It ends with a subscript in 2 lines. The first line of the subscript states that this is an im.gíd.da written by the student *ši-ip*-suen. The term im.gíd.da 'long clay', or rather 'long tablet', is often seen in subscripts of this type and seems to allude to the fact that table texts often are written on clay tablets of a longish format. The second line of the subscript begins with the sign u_4 'day'. Although the remaining part of this line is damaged, it is clear that what was written here was a date of the standard type 'day *n* of month so and so'.

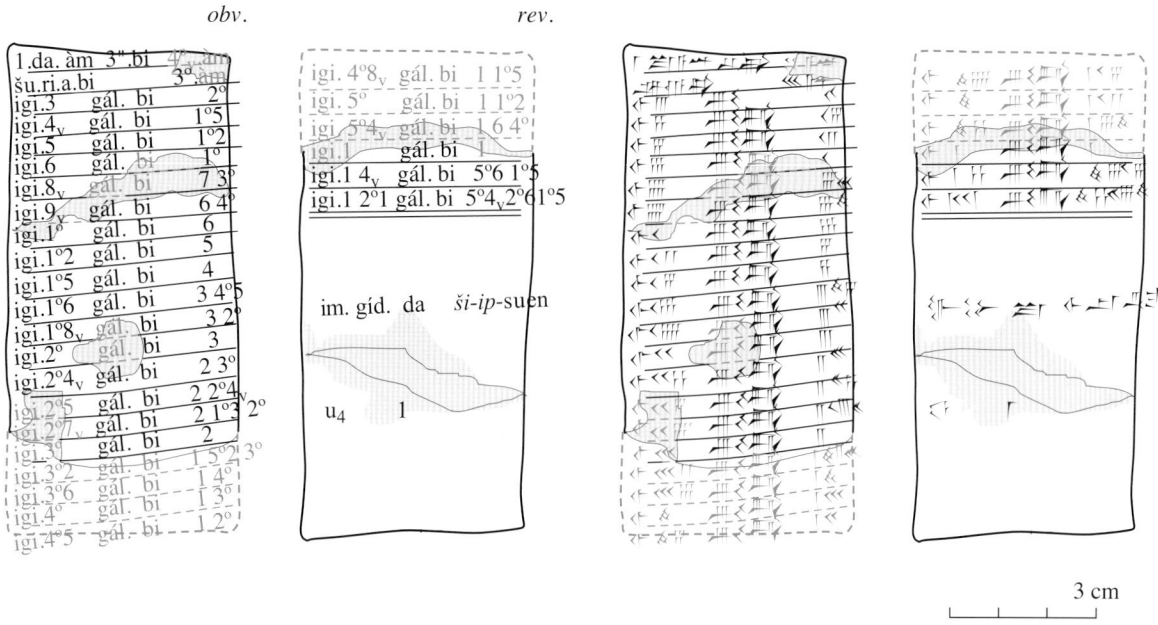

obv.   *rev.*

3 cm

MS 3874. A standard table of reciprocals of type a (28 lines), with a colophon (scribe's name and date). Variant number signs for 4, 7, 8, 9, and 40, and the early OB form of the sign for .bi.

Fig. 2.5.1. An example of the Old Babylonian standard table of reciprocals, probably early Old Babylonian.

MS 3874 is particularly interesting because the text makes use of variant number signs for 6, 7, and 8, and even for 4 and 40. Also some of the non-numerical signs in the text, in particular the sign for the suffix .bi has an "archaic" form. All this suggests that MS 3874 is an "early" Old Babylonian table text from some southern city in Mesopotamia. (Cf. Friberg, *RA* 94 (2000) § 7c.)

MS 3869/5 and **MS 3890** (Fig. 2.5.2) are two other Old Babylonian standard tables of reciprocals of type a. They both use standard Old Babylonian forms for the number signs and for non-numerical signs, such as .bi. They are therefore probably younger than MS 3874 and not from any of the southern Mesopotamian cities.

MS 3869/5 has lost its lower part. The last preserved line on this tablet is igi.36.gál 1 40. Presumably the whole standard table of reciprocals was inscribed on either just the obverse, or on the obverse and the uppermost part of the reverse. Anyway, what remains of the reverse is empty.

MS 3890 contains the full standard table of reciprocals, except for the three optional pairs between igi.1 04.gál 56 15 and igi.1 21.gál 44 26 40. The first line of the text is

1.da.àm 40.bi 40.

Compare with MS 3874 and MS 3869/5, where the corresponding first lines of text are

1.da.àm 3".bi 40.[àm]   and   1.da 3".bi 40

A dissatisfied and irate teacher(?) has mutilated both obverse and reverse with several long scratches. The student's error, whatever it was, may have disappeared behind the scratches.

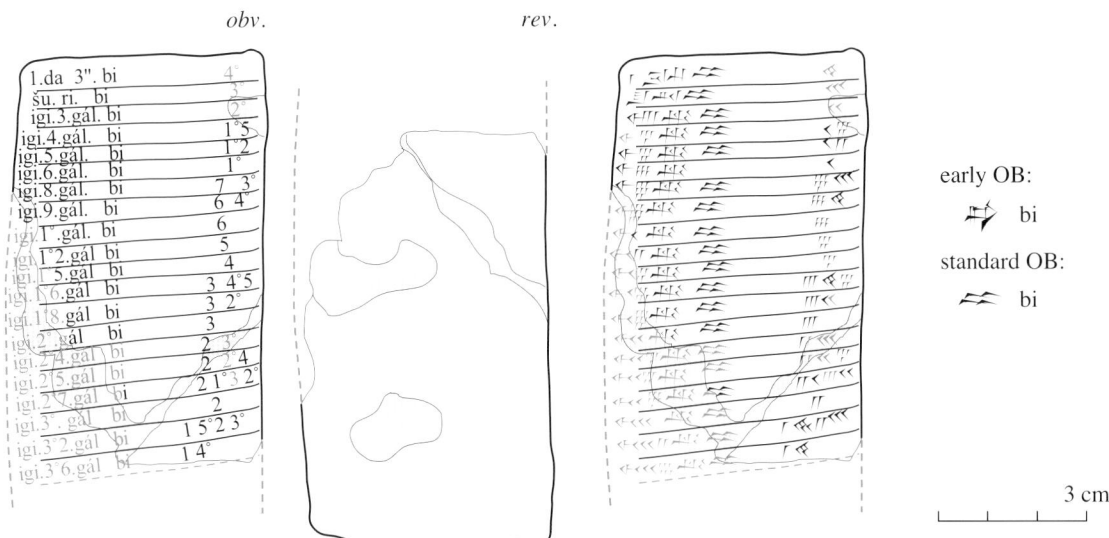

MS 3869/5. A standard table of reciprocals of type a. The lower edge of the clay tablet is lost.

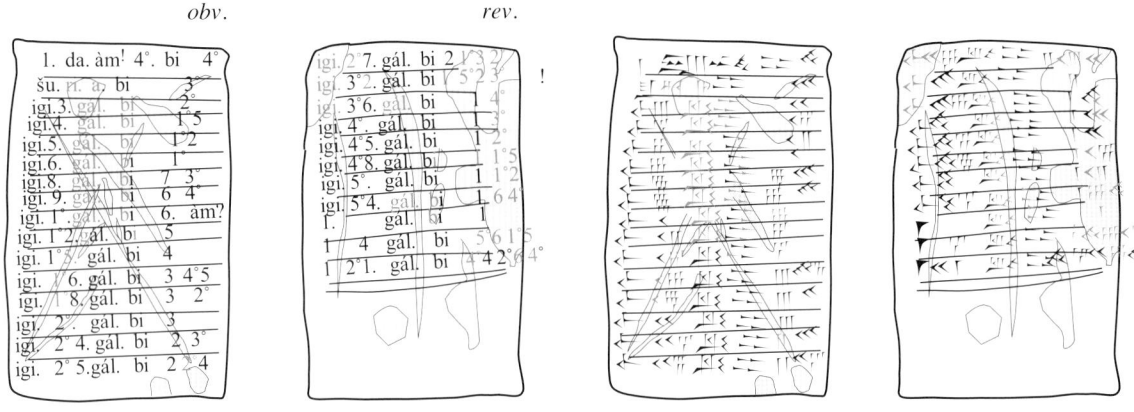

MS 3890. A standard table of reciprocals of type a (27 lines), mutilated in antiquity by an irate teacher.

Fig. 2.5.2. Two examples of the Old Babylonian standard table of reciprocals of type a.

Possibly, the teacher had noticed that the student had forgotten the sign igi at the beginning of the last three lines of the table, and also the whole line igi.30.gál.bi 2, all on the reverse.

In addition to the three examples mentioned above, all of type a, there is also a fourth example of an Old Babylonian standard table of reciprocals in the Schøyen Collection. That fourth example, is **MS 2877, *obv.*** (Fig. 2.6.7, bottom), a table of reciprocals of an atypical, purely numerical format, which may be called "type b*" because of its similarity with the purely numerical format called type b* in the case of single multiplication tables. (The text on the reverse of MS 2877 is a multiplication table for the "head number" 50.) Two pairs of reciprocals are missing in this table, probably by mistake, the pairs 27 — 2 13 20 and 50 — 1 12. In addition, the initial lines with the sexagesimal representations of the fractions 2/3 and 1/2 are absent from this table of reciprocals.

2.6. Old Babylonian Multiplication Tables

2.6 a.   Single Multiplication Tables

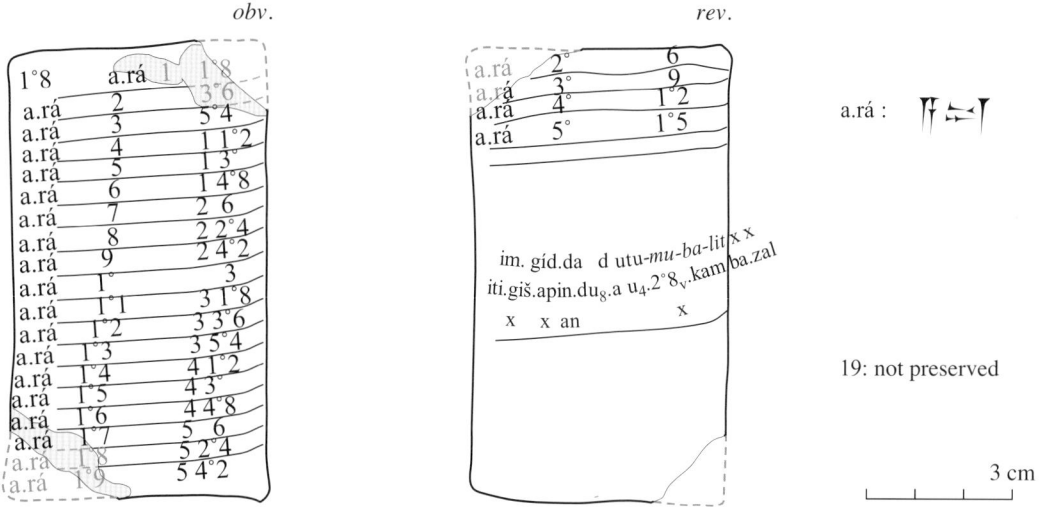

MS 2708. A single multiplication table (18 ×) of type a, the most common type (23 lines).
Colophon: "Long tablet inscribed by Shamash-Muballit, day 28, month 8, is completed."

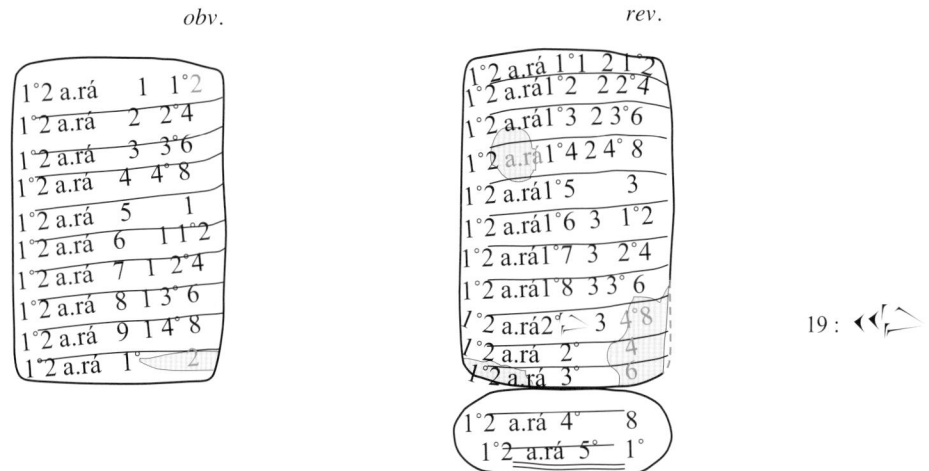

MS 2184/3. A single multiplication table (12 ×) of type a'.

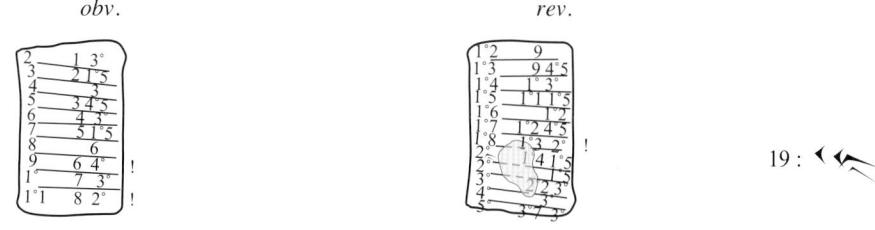

MS 3044/3. A single multiplication table (45 ×) of the purely numerical type b* (new).
Errors in line 9: 6 40 for 6 45, line 11: 8 20 for 8 15, and line 18: 13 20 for 13 30.

Fig. 2.6.1. Three examples, of three different types of Old Babylonian single multiplication tables.

An Old Babylonian "single multiplication table" is a clay tablet with a single table where a given number, the "head number" p, is multiplied first by the integers from 1 to 19, then by 20, 30, 40, and 50. This is the same arrangement as in, for instance, the complete abridged table of squares MS 2794 (Fig. 2.1.1), or the complete abridged table of cube sides MS 3913 (Fig. 2.3.1). Single multiplication tables constitute the most common category of mathematical cuneiform texts. Previously published single multiplication tables include, in particular, 85 texts in *MKT* (1935-37) (see *MKT 1*, 32-34, 36-43; *MKT 2*, 36-37, and *MKT 3*, 50), and 77 texts in *MCT* (1945), 19-24.

There are as many as 148 recognizable single multiplication tables in the Schøyen Collection. Ten representative examples are presented with full details in the present paragraph. The primary aim of the presentation is to illustrate different types of single multiplication tables that can be found among the tablets in the Schøyen Collection. Those types are:

	first line:			subsequent lines:		
type a:		p a.rá 1	p		a.rá n	$p \cdot n$
type a':		p a.rá 1	p		p a.rá n	$p \cdot n$
type a":		p a.šà 1	p		p a.šà n	$p \cdot n$
type b:		1	p		n	$p \cdot n$
type b*:		——			n	$p \cdot n$
type c:		1 a.rá	p		n a.rá	$p \cdot n$

These definitions and designations differ to some extent from the ones proposed in *MCT*, 20. However, types a and a' above are the same as types A and A' in *MCT*. Type a is, by far, the most common of all the types. Type b* is purely numerical, with an omitted first line. It is represented by two examples in the Schøyen Collection but does not appear among the texts in *MKT* and *MCT*, where instead types B, B', B", and C have various first lines, while all subsequent lines are of the same form as all the lines in tables of type b*. Type b is identical with type C in *MCT*.

Type a" is a new type, represented by two texts in the Schøyen Collection. In tables of type a", the standard term a.rá, a Sumerian word actually meaning 'step', 'going', *etc.*, is replaced by the term a.šà, a Sumerian word with the meaning 'field', 'area'. It is not clear if this is a *bona fide* new type, or a mistake (provided that the two tables of type a" were written by the same hapless student). Type c, too, may be not a new type, but a mistake. It appears only on a single tablet with a beginner's unfinished multiplication table.

From a mathematical point of view, the meaning of the phrase p a.rá n $p \cdot n$ is clear. It is obviously 'p times n (equals) $p \cdot n$'. From a linguistic point of view, the situation is less clear. What does 'p step n $p \cdot n$' really mean? It is possible that the phrase originally meant something like 'n steps, each of length p, equals $p \cdot n$'.

MS 2708 (Fig. 2.6.1, top) is a single multiplication table of type a with the head number 18. Of the 23 lines of the table, the first 19 are written on the obverse of the tablet, the remaining 4 on the reverse. That leaves plenty of room for a subscript in 3 lines:

im.gíd.da ^{d}utu-*mu-ba-liṭ* x / iti.^{giš}apin.du$_8$.a u$_4$.28.kam ba.zal / x x x x

'Long tablet inscribed by Shamash-Muballit, the month of plowing (month 8), day 28 is completed. x x x x'

The month name iti.^{giš}apin.du$_8$.a (Akk. *Araḫsamnu*) is the Sumerian name for the eighth month of the Babylonian lunar year.

MS 2184/3 (Fig. 2.6.1, middle) is written on a smaller clay tablet than MS 2708. Consequently, the 23 lines of its multiplication table cover almost the entire surface of the tablet, with 10 lines on the obverse, 11 on the reverse, and the remaining 2 lines on the edge below the reverse. The multiplication table is of type a', with the head number 12. That the table is of type a' means that the number 12 is repeated at the beginning of each line of the table. Thus, the table begins with the lines

12 a.rá 1   12 / 12 a.rá 2   24   *etc.*, all the way to   12 a.rá 50   10

The number 19 is written in this text as 20 followed by a cuneiform sign that is hard to identify. This way of writing 19 seems to be more of a rule than an exception in single multiplication tables. Indeed, 10 examples of single multiplication tables are displayed in Figs. 2.6.1 and 2.6.3-5 in this paragraph. In all cases where the number 19 is well enough preserved so that its form can be determined, it is written as 20 followed by a cuneiform sign that is hard to identify. The unidentifiable sign is *different* in all those cases!

The result of a partial survey of the occurrence of this phenomenon in single multiplication texts from the Schøyen Collection is exhibited in Fig. 2.6.2 below, a table of 12 different forms of the sign for 19. The table shows that 19 written as 20 followed by a squiggle occurs in single multiplication tables of all types, and in conjunction with both standard and early Old Babylonian forms of the number signs for 7, 8, *etc.*

MS 2184/2			type a'	
MS 2184/3	Fig. 2.6.1		type a'	
MS 2184/7			type a	
MS 2286/2			type a'	
MS 2804/1			type a	 20.lá 1ₕ
MS 2875	Fig. 2.6.3		type a	
MS 2895		early OB	type a'	
MS 3044/3	Fig. 2.6.1		type b*	
MS 3849	Fig. 2.6.3		type a	
MS 3866	Fig. 2.6.4		type a	
MS 3909/3	Fig. 2.6.5		type b*	
MS 3967/1	Fig. 2.6.4		type a	
MS 3967/2		early OB	type a	 20.lá 1ₕ

Fig. 2.6.2. Signs for '19' in selected OB multiplication tables from the Schøyen Collection.

The origin of this curious way of writing 19 is well known. In Sumerian and Old Akkadian texts from the third millennium BC, both administrative and mathematical, it was a common practice to write numbers in a "subtractive form". If a given number or measure a was slightly less than a relatively large round number or measure b, then it could be written as the large number *minus* a small number. Two explanations for the practice suggest themselves. One is that the large and round number b was the *expected* outcome of a measurement, a the *actual* outcome, and writing a as $b - d$ was a way of showing how far the actual outcome was from the expected one. The other explanation is that, in certain kinds of operations with the numbers, notably multiplications, it was easier to count with $b - d$ than with a.

An example is offered by the Old Sumerian table of square areas *OIP 14*, 70 (Edzard, *AOAT 1* (1969); Fig. A1.4 in Appendix 1), where, for instance the square of (10 – 1) cubits is written in the form 2' šar 4 gín – igi.4. (A cubit is 1/12 of a ninda and a šar is a square ninda. Therefore the computation behind this entry can be explained (anachronistically) in the following way, in terms of sexagesimal numbers in place value notation: sq. (9 cubits) = sq. (;45 ninda) = ;33 45 sq. ninda = (;34 – ;00 15) sq. ninda = 1/2 šar 4 gín – 1/4 gín.)

Another example is offered by Ist. T 7375, an Ur III table of reciprocals, where, for instance, 37, 38, 39 are written in the form '40 – 3', '40 – 2', '40 – 1'. (See Figs. A1.2-3 in Appendix 1 below.)

Among the examples in Fig. 2.6.2, there are two that support this explanation of the origin of curious number signs for 19. Indeed, the signs for 19 in MS 2804/1 and MS 3967/2 can be read as

$$20. \text{lá } 1_h \qquad\qquad \text{'20 less 1' that is '20 – 1'.}$$

Here, 1_h is a *horizontal* sign for '1'. (It is not clear why it is horizontal.) The cuneiform sign lá, possibly the image of some kind of weighing apparatus, is a Sumerian logogram meaning

$$\text{lá } = \check{s}aq\bar{a}lu \text{ 'to weigh', 'to pay'} \qquad \text{or} \qquad \text{lá } = mat\hat{u} \text{ 'to be less'.}$$

Two difficult questions remain and will perhaps never be answered. One is why the clear form of 20.lá 1_h exhibited in MS 2804/1 and MS 3967/2 is more or less severely corrupted in so many other Old Babylonian single multiplication tables, and also, as will be shown below, in Old Babylonian "double" and "combined" multiplication tables. It is definitely *not* as if the curious way of writing 19 is a simplification of the straightforward way of writing 19 as a 10 followed by a 9! The other question is why this curious way of writing 19 is not extended to an equally curious way of writing 9, 29, *etc.*, and why it is does not make its appearance in any other genres of Old Babylonian arithmetical table texts or, generally, in any other kinds of Old Babylonian mathematical texts.

MS 3044/3 (Fig. 2.6.1, bottom) is an unusually small clay tablet with a single multiplication table for the head number 45. Because the tablet is so small, the script is tiny, and the multiplication table is of type b*, omitting the word a.rá 'times' in all lines of the table, and also omitting the first line '45 (times) 1 (equals) 45'.

There are three numerical errors in MS 3044/3. Two of them may be copying errors by the student: 6 40 instead of 6 45 in line 9 and 13 20 instead of 13 30 in line 18. The remaining error, 8 20 instead of 8 15 in line 11 looks like a mistake in the computation of 11 · 45.

MS 2875 (Fig. 2.6.3, top) is a small clay tablet with a single multiplication table, head number 30. It is of a curious new type, here called type a", which resembles type a' in that the head number is repeated at the beginning of each line in the table. It differs from type a' in a curious way in that the usual term a.rá 'step, times' is replaced by the term a.šà 'field, area'. Phrases such as

$$30 \text{ a.šà } 12 \text{ } 6 \qquad\qquad 30 \text{ field } 12 \text{ (equals) } 6 \text{ (00)}$$

seem to make little sense. The use of a.šà instead of a.rá can possibly be explained as a mistake made by an inattentive student, who misheard the teacher's dictation of the text. If this somewhat unlikely explanation is correct, then another (or the same?) student must have repeated this mistake when writing the single multiplication table for the head number 7 on the small clay tablet **MS 2829** (Fig. 2.6.3, bottom). In addition, this second student forgot one whole line of text, and he made a simple error (a copying error?) in line 19, when he wrote 2 instead of 2 06.

MS 3866 (Fig. 2.6.4, top) is a clay tablet with a single multiplication table of type a. The text is interesting for two reasons. One is that the head number for the multiplication table is 1 12, smaller than the head number for any previously published Old Babylonian multiplication table. See Secs. 2.6 e-f below. The text is also interesting because the multiplication table ends with a line giving *the square of the head number*:

$$1 \text{ } 12 \text{ a.rá } 1 \text{ } 12 \quad 1 \text{ } 26 \text{ } 24 \qquad\qquad \text{'1 12 steps 1 12 (is) 1 26 24'.}$$

MS 3967/1 (Fig. 2.6.4, bottom) is a fragment of a clay tablet with a single multiplication table of type a, head number 2. The table ends by giving *the square of the head number and the square side of that square*:

$$2 \text{ a.rá } 2 \quad 4 \text{ } 4 \text{ / .e } 2.\text{àm íb.si}_8 \qquad\qquad \text{'2 steps 2 (is) 4, 4 makes 2 equalsided'.}$$

After the table follows a brief subscript indicating *day and time of day*:

$$u_4.9.\text{kam ba.zal} \qquad\qquad \text{'the 9th day is completed'.}$$

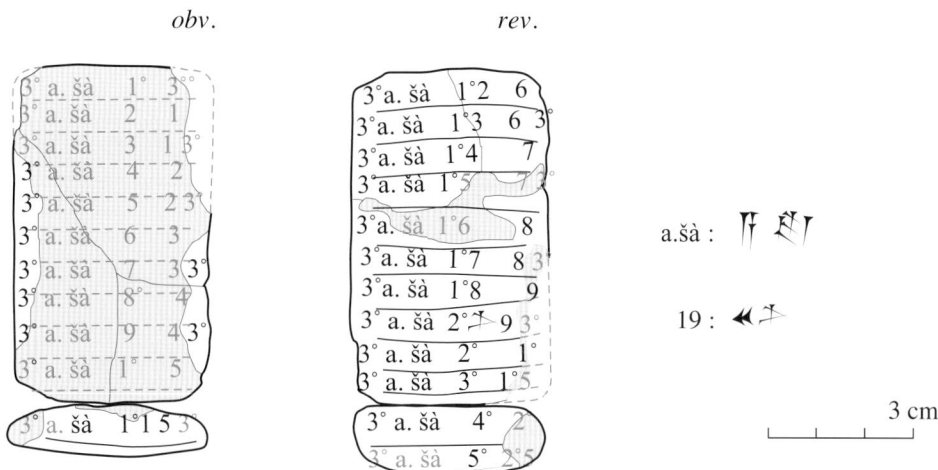

MS 2875. A multiplication table (30 ×) of type a″ (new), with a.šà instead of a.rá. Variant number sign for 19.

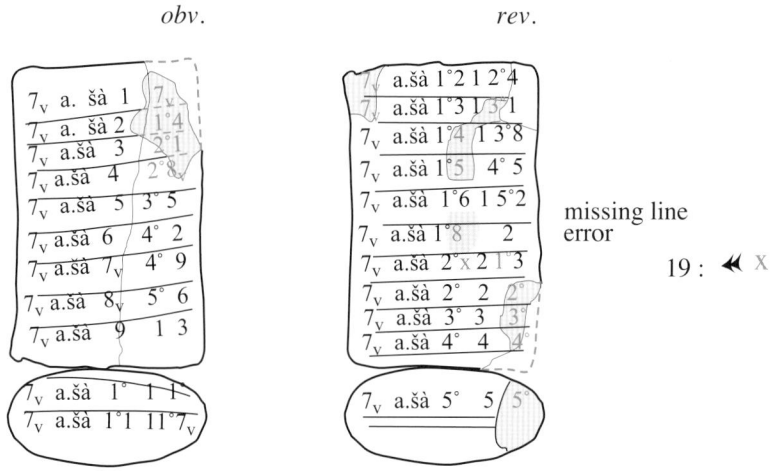

MS 2829. A multiplication table (7 ×) similar to MS 2875 above. Variant number signs for 7, 8, and 19.
Missing line: 7 a.šà 17 1 59. Error in line 18: 2 instead of 2 06.

Fig. 2.6.3. Two curiously atypical Old Babylonian single multiplication tables.

Of the 77 single multiplication tables published in *MCT* and the 85 single multiplication tables published in *MKT* (together 162 tables), 21 (about 27%) end by giving the square of the head number. Of those 21, only 2 (2.5% of all the multiplication tables in *MCT* and *MKT*) give also the square side of the square of the head number. (Cf. *MKT 1*, 65.) The corresponding numbers for the Schøyen Collection are 152 single multiplication tables, of which 2 (1.3%) give the square of the head number and 1 (.65%) also the square side of that square.

MS 3873/1 (Fig. 2.6.5, top) is a single multiplication table of type a, head number 24. The text ends with a subscript indicating *writer, month, and day*:

<im.>gíd.da ᵈen.zu.[x] / iti.ab.è 24.kam   'long (tablet), PN, month 10, (day) 24th'.

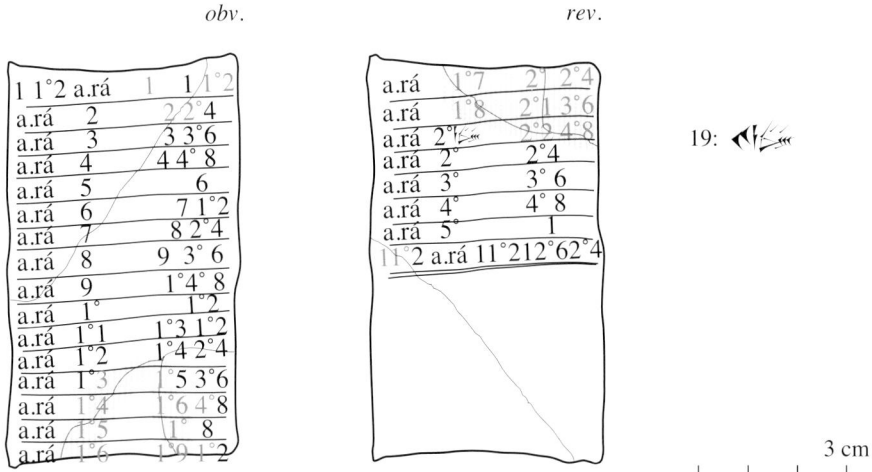

MS 3866. A single multiplication table (1 12 ×) of type a.
The smallest known head number! Ends with the square of the head number.

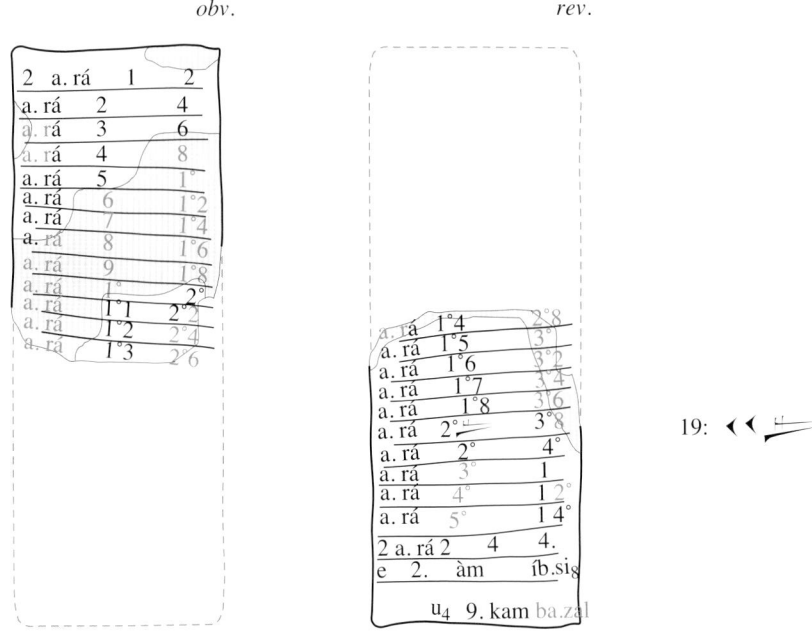

MS 3967/1. A single multiplication table (2 ×) of type a.
Ends with the square of 2, and the square root of 4. Colophon: 'Day 9 is completed'

Fig. 2.6.4. Two Old Babylonian single multiplication tables ending with the square of the principal number.

MS 3909/3, finally (Fig. 2.6.5, bottom) is a single multiplication table of the unusual type b*, head number 40. On the tablet, the multiplication table is followed by a subscript indicating *writer, month,* [*day*], *and year*:

1	im.gíd.da ^den.zu.[x] /	long tablet, PN
2	iti.še.kin.ku₅ u₄.[x.kam] /	month 12, day *xth,*
3	mu *sa-am-su-[i-lu-na* x x]	the year (when) Samsu*iluna x x*

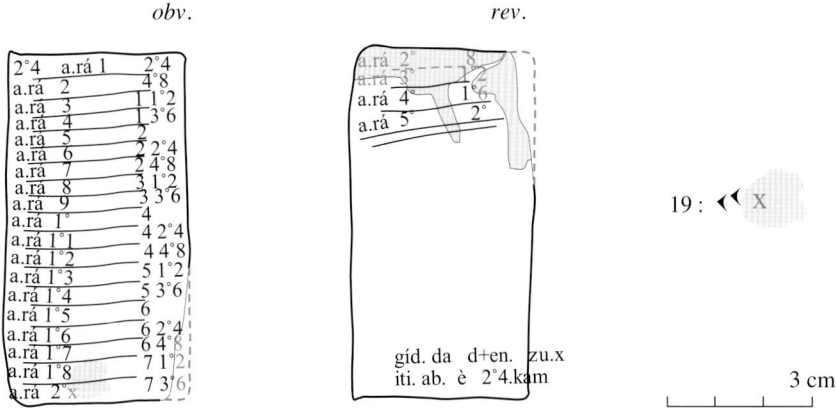

obv.   *rev.*

MS 3873/1. A single multiplication table (24 ×) of type a.
Colophon indicating writer, month and day.

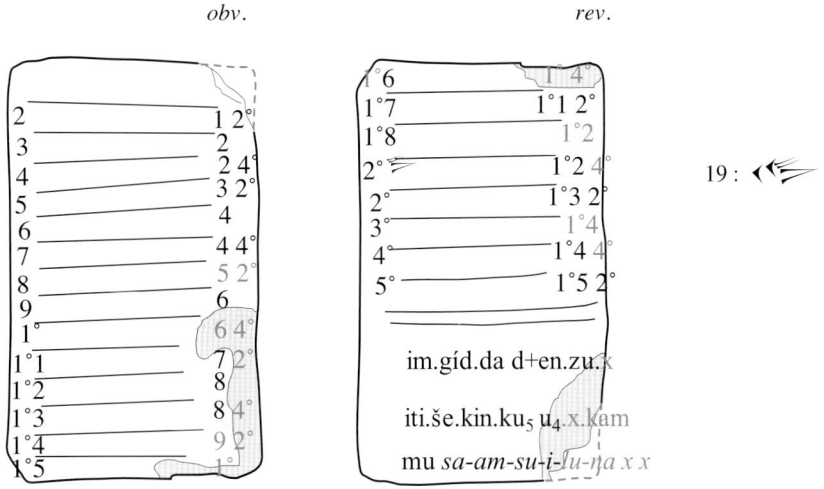

obv.   *rev.*

MS 3909/3. A single multiplication table (40 ×) of type b*.
Colophon indicating writer, month, [day], and year.

Fig. 2.6.5. Two Old Babylonian single multiplication tables with subscripts of standard type.

2.6 b. Double Multiplication Tables

(Cf. *MKT 1*, 44-59; *MCT*, 25-33.)

Nine examples of "double multiplication tables" are present in the Schøyen Collection. Copies of three of them, plus a double table of a related kind, will be shown below.

MS 2719 (Fig. 2.6.6, top) is a first example of an Old Babylonian double multiplication table. It is inscribed with two single multiplication tables of type a, one with the head number 18 (on the obverse of the clay tablet), the other with the head number 16 40 (on the reverse of the tablet). Variant number signs for 4 are used on both obverse and reverse. A variant number sign for 19 is used on the reverse. The corresponding number on the obverse is not preserved.

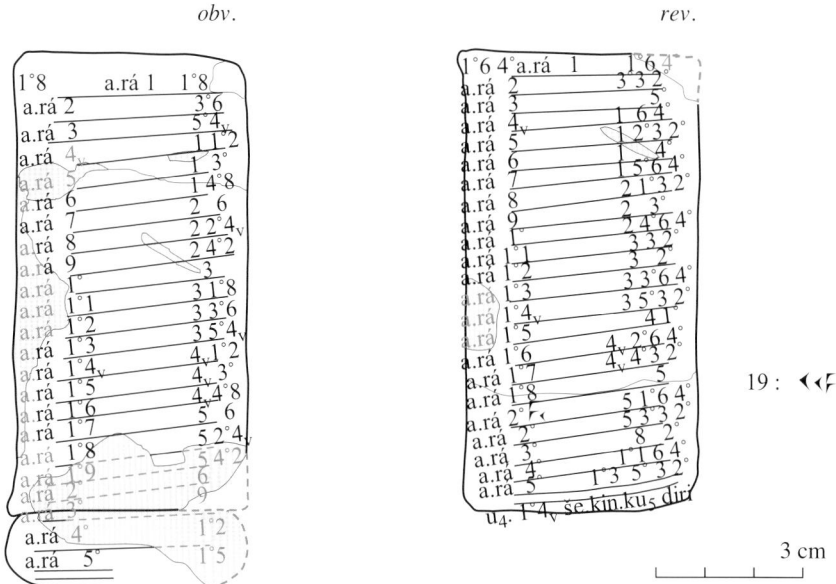

MS 2719. A double multiplication table (18 × and 16 40 ×) of type a.
Variant number signs for 4 and 19.
The table ends with a colophon: day 14th, month 12 II.

MS 3964. A double multiplication table (7 30 × and 7 12 ×) of type a.
Variant number signs for 7, 8, and 19.

Fig. 2.6.6. Two double excerpts from an Old Babylonian combined multiplication table of type b*.

The text on the reverse ends with a subscript, indicating *day and month*:

u₄ 14ᵥ še.kin.ku₅.diri   'day 14, <month> 12 II'.

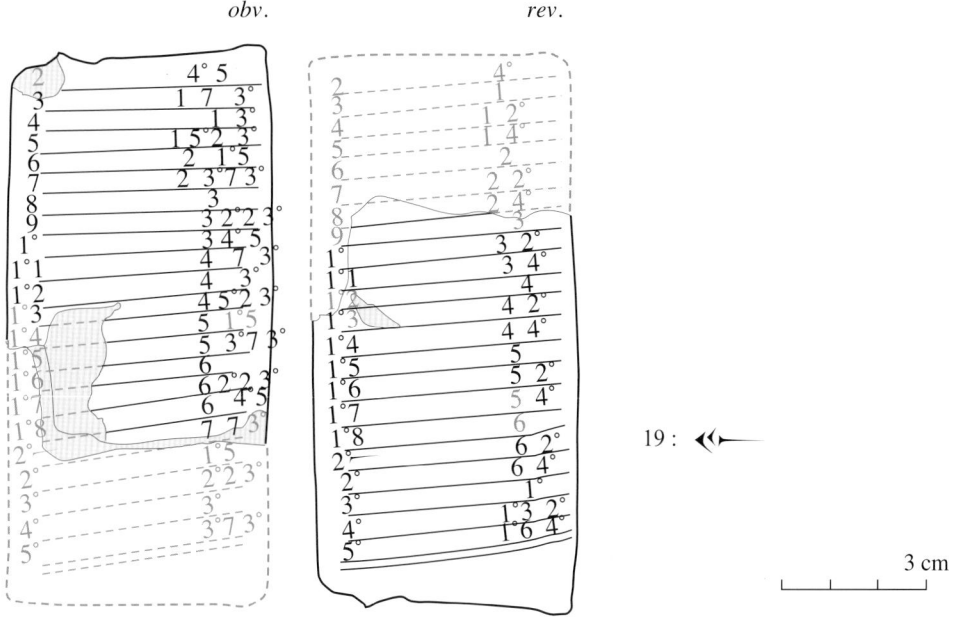

MS 2878. A double multiplication table: (22 30 × and 20 ×) of type b*. Variant number sign for 19.

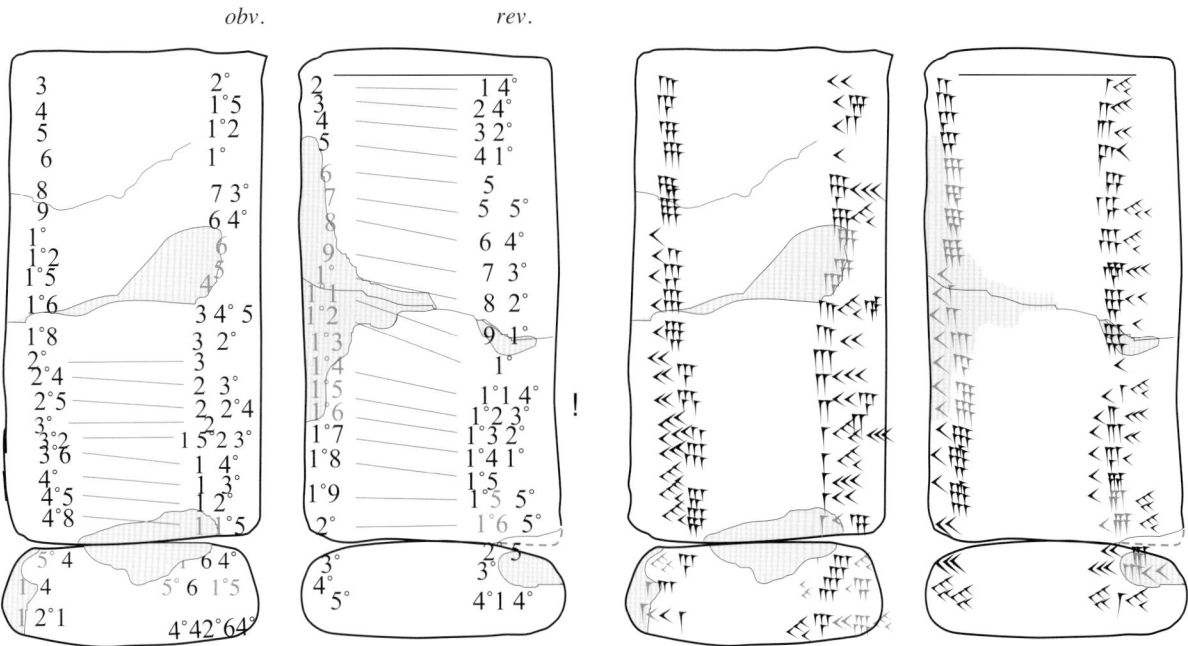

MS 2877. The first two sub-tables of a combined multiplication table: (reciprocals and 50 ×, both of type b*).
        Standard number signs are used for all numbers, even 19.

Fig. 2.6.7. Two more double excerpts from an Old Babylonian combined multiplication table.

MS 3964 (Fig. 2.6.6, bottom) is another example of an Old Babylonian double multiplication table of type a. A table on the obverse with the head number 7 30 is immediately followed by a table with the head number 7 12. The second table begins near the end of the obverse and continues onto the reverse. Variant number signs are used on both obverse and reverse for 7 and 8, but not for 4. A variant number sign for 19 is used on the obverse. The corresponding number on the reverse is not preserved.

MS 2878 (Fig. 2.6.7, top) is an example of an Old Babylonian double multiplication table of the very unusual, purely numerical type b*. A single multiplication table on the obverse has the head number 22 30, and a second table on the reverse has the head number 20. A variant number sign for 19 is used on the reverse. The corresponding number on the obverse is not preserved.

MS 2877 (Fig. 2.6.7, bottom) is not, properly speaking, a double multiplication table, yet it is a text of a similar kind. It is inscribed on the obverse with a table of reciprocals of an atypical, purely numerical format, which may be called "type b*" because of its similarity with the purely numerical format called type b* in the case of single multiplication tables. Two pairs of reciprocals are missing in this table, probably by mistake, the pairs 27 — 2 13 20 and 50 — 1 12. In addition, the initial couple of lines are absent from this table of reciprocals, the ones displaying the sexagesimal representations of the fractions 2/3 and 1/2.

On the reverse, MS 2877 is inscribed with a single multiplication table, head number 50. This multiplication table, too, is of the very unusual, purely numerical type b*, just like the two single multiplication tables on MS 2878, and like the table of reciprocals on the obverse. Another unusual feature is that there are no variant number signs on MS 2877. Even the sign for 19 is written in the standard form, as a ten followed by a nine, not as twenty followed by a squiggle.

The layout on the reverse of MS 2877 is badly organized. (See the help lines inserted into the conform transliteration of the reverse in Fig. 2.6.7.) Numbers belonging together are not consistently facing each other, probably for the reason that the student who wrote the text copied first the entire left column of numbers from another text, then the right column, instead of copying from the other text line by line. The resulting confusion has made the writer forget to insert the number '10 50' (13 times 50) in its proper place in the right column.

2.6 c. *Multiple Multiplication Tables*

(Cf. *MKT 1*, 35, 44-59; *MCT*, 24-33.)

Three examples of "multiple multiplication tables" are present in the Schøyen Collection. They are all shown below. A multiple multiplication table contains more than two but not a full range of single multiplication tables. (This is a somewhat arbitrary, but convenient distinction.)

MS 3891 (Fig. 2.6.8 below) is a fairly well preserved example of a multiple multiplication table. It is inscribed with three single multiplication tables, in the following referred to as "sub-tables", with the head numbers

7, 6 40, and 6.

The first sub-table is inscribed on the greater part of the obverse. The second sub-table was inscribed around the lower edge of the obverse (now lost) and on the upper two-thirds of the reverse. The last sub-table is almost completely preserved. It is inscribed close to and on the lower edge of the reverse, and on the left edge of the clay tablet.

MS 3891 resembles the double table MS 2877 in that all sub-tables are of type b*, but whereas the standard sign is used for 19 in MS 2877, a variant number sign is used for 19 in MS 3891.

MS 3939 is a fragment of a multiple multiplication table. The middle half of the tablet is preserved, although it is not in a very good shape. The writing is miniature. Yet it is possible to see that there were originally four sub-tables on MS 3939, single multiplication tables with the head numbers

8, 7 30, 7 12, and 7.

A suggested reconstruction of the text is shown in Fig. 2.6.8.

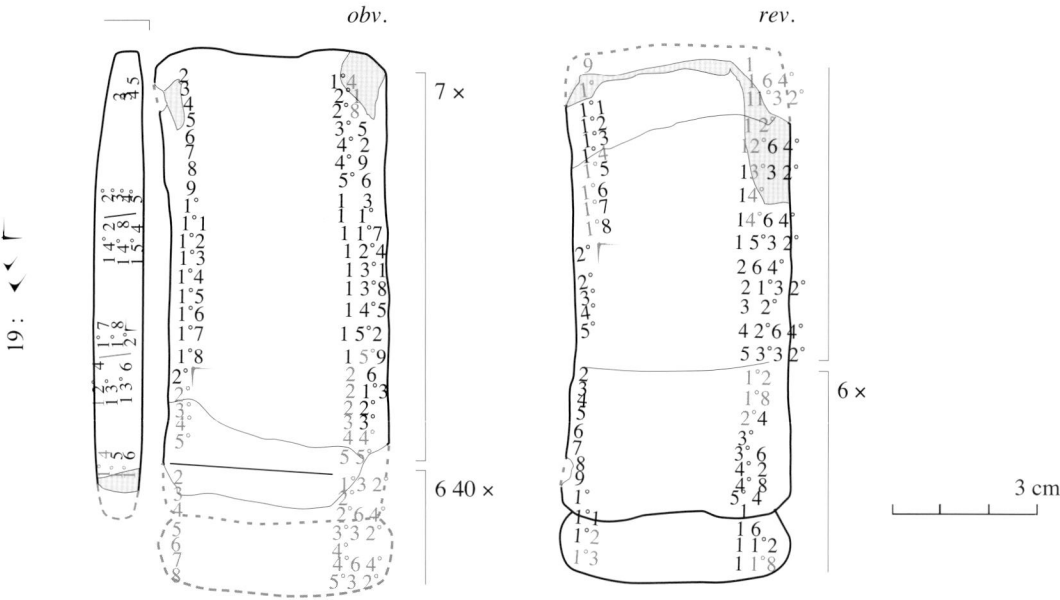

MS 3891. A multiple multiplication table (7, 6 40, 6 ×) of type b*. Variant number sign for 19.

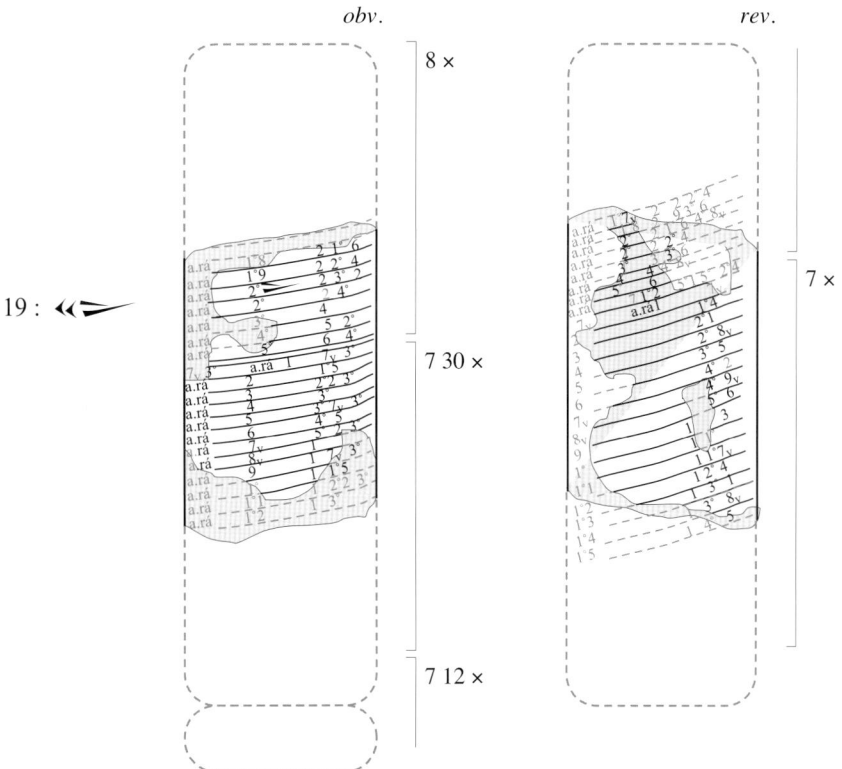

MS 3939. A multiple multiplication table (8, 7 30, 7 12, 7 ×) of mixed types a and b.
Variant number signs are used for 7, 8, and 19.

Fig. 2.6.8. Two examples of Old Babylonian multiple multiplication tables.

The first three sub-tables are of the common type a, but for some reason, probably because the writer got tired, the simpler type b is used for the fourth sub-table, which (apparently) begins with the standard line '7_v a.rá 1 7_v' but then continues with purely numerical entries of the kind 'n $n \cdot 7$'

MS 3870 (Fig. 2.6.9) is a fragment very much like MS 3939, but the writing on MS 3870 is even more miniature. According to the construction suggested in Fig. 2.6.9 below, there were originally five sub-tables on the clay tablet, single multiplication tables with the head numbers

[7], 6 40, 6 on the obverse, and 7 30, 7 12 on the reverse.

All the sub-tables are of type a. The text on the tablet is so badly preserved that it cannot be decided if variant number signs are used for 19.

Note that there is something wrong with the ordering of the sub-tables on MS 3870. The head numbers do not form a decreasing sequence, as would have been expected.

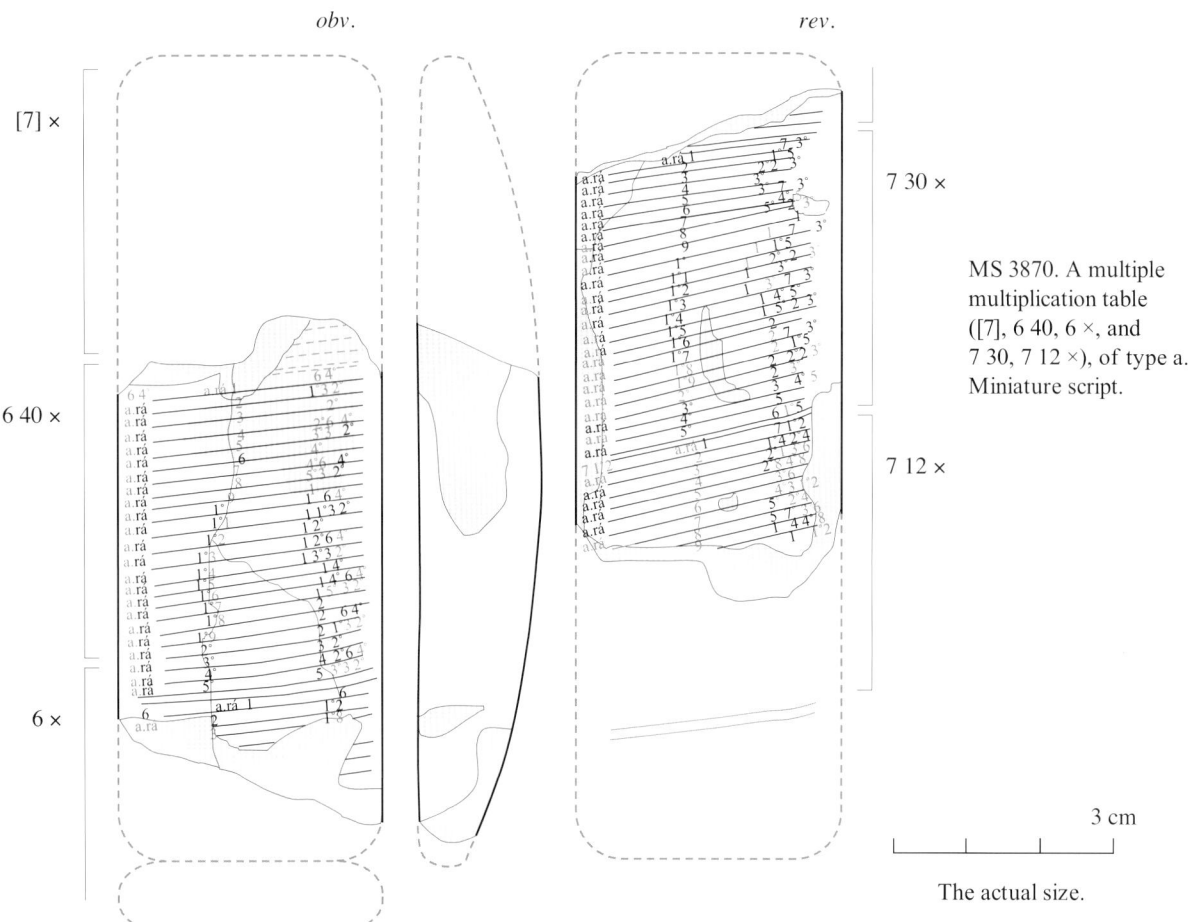

Fig. 2.6.9. MS 3870. A third example of and Old Babylonian multiple multiplication table.

2.6 d.   The Old Babylonian Combined Multiplication Table

MS 3845 (Fig. 2.6.10 below) is a substantial fragment, the lower half of a large clay tablet. According to the reconstruction presented in Fig. 2.6.11, it was inscribed, originally, with eleven sub-tables and a subscript. The first sub-table is a table of reciprocals, apparently of standard type, type a, with entries like, for instance, igi.1 21.gál.bi 44 26 40. The ten other sub-tables are single multiplication tables of type a, with entries like, for instance, 44 26 40 a.rá 1   44 26 40, a.rá 2 1 28 53 20, *etc*. The head numbers represented in MS 3845 are

50, 45, 44 26 40, 40, 36, 30, 25, 24, 22 30, and 20.

The outline of the text in Fig. 2.6.11 shows that the correctness of the reconstruction is beyond doubt, since at least some part of each one of the eleven sub-tables is preserved.

A comparison with the combined multiplication table on the cylinder **A 7897** (*MCT*, 25), and with the survey in *MKT 1*, 35, shows that in MS 3845 the head number 48 is missing. On the other hand, this head number is missing also in 18 of the 25 combined multiplication tables, which can be examined with respect to this issue, listed in *MKT 1-2* and *MCT*. Moreover, there are no *single* multiplication tables with the head number 48 in *MKT 1-2* and *MCT*.

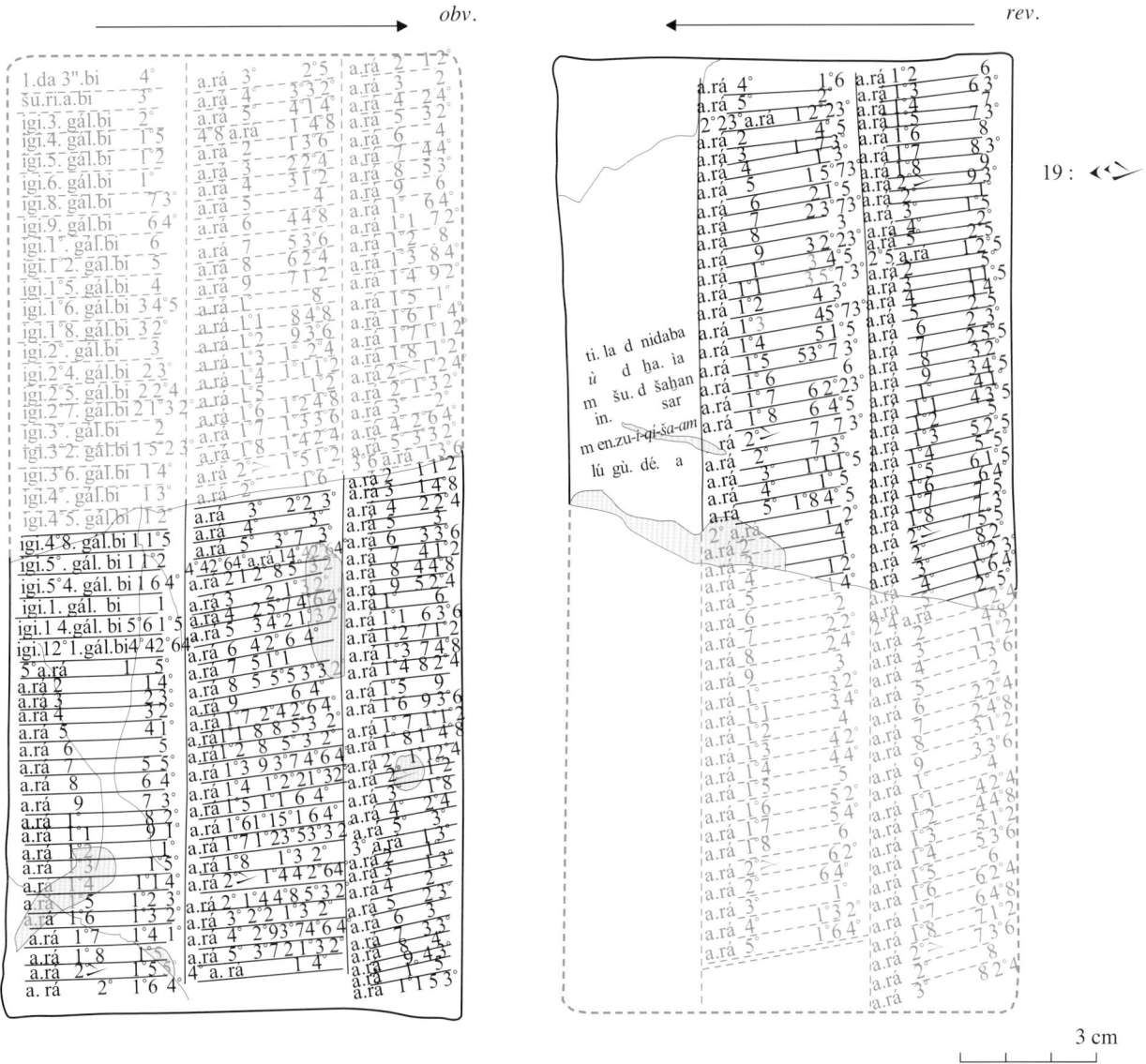

MS 3845. A large fragment of an Old Babylonian combined multiplication table of type a (11 sub-tables). The table ends with the multiplication table for 20, followed by a colophon. The arrows indicate that the columns on the obverse are to be read from left to right, but the columns on the reverse from right to left. This is a standard feature of cuneiform texts.

Fig. 2.6.10. MS 3845. The eleven first sub-tables of an Old Babylonian combined multiplication table.

In five of the sub-tables on MS 3845, the sign for 19 is preserved. In all these instances, the same form of a variant number sign for 19 is used. Standard number signs are used for the digits 4, 7, and 8.

The 11 sub-tables on MS 3845 are recorded in three columns on the obverse and two on the reverse. The sixth and last of the columns is empty except for a subscript in 6 lines:

1	ti.la ^dnisaba /	By the life of Nisaba
2	ù ^dḫa.ia /	and Haia,
3	^mšu^d.šaḫan /	Shushahan
4	in.sar /	wrote (it).
5	^men.zu-*i-qí-ša-am* /	Sueniqisham,
6	^{lú}gù.dé.a /	he of Gudea.

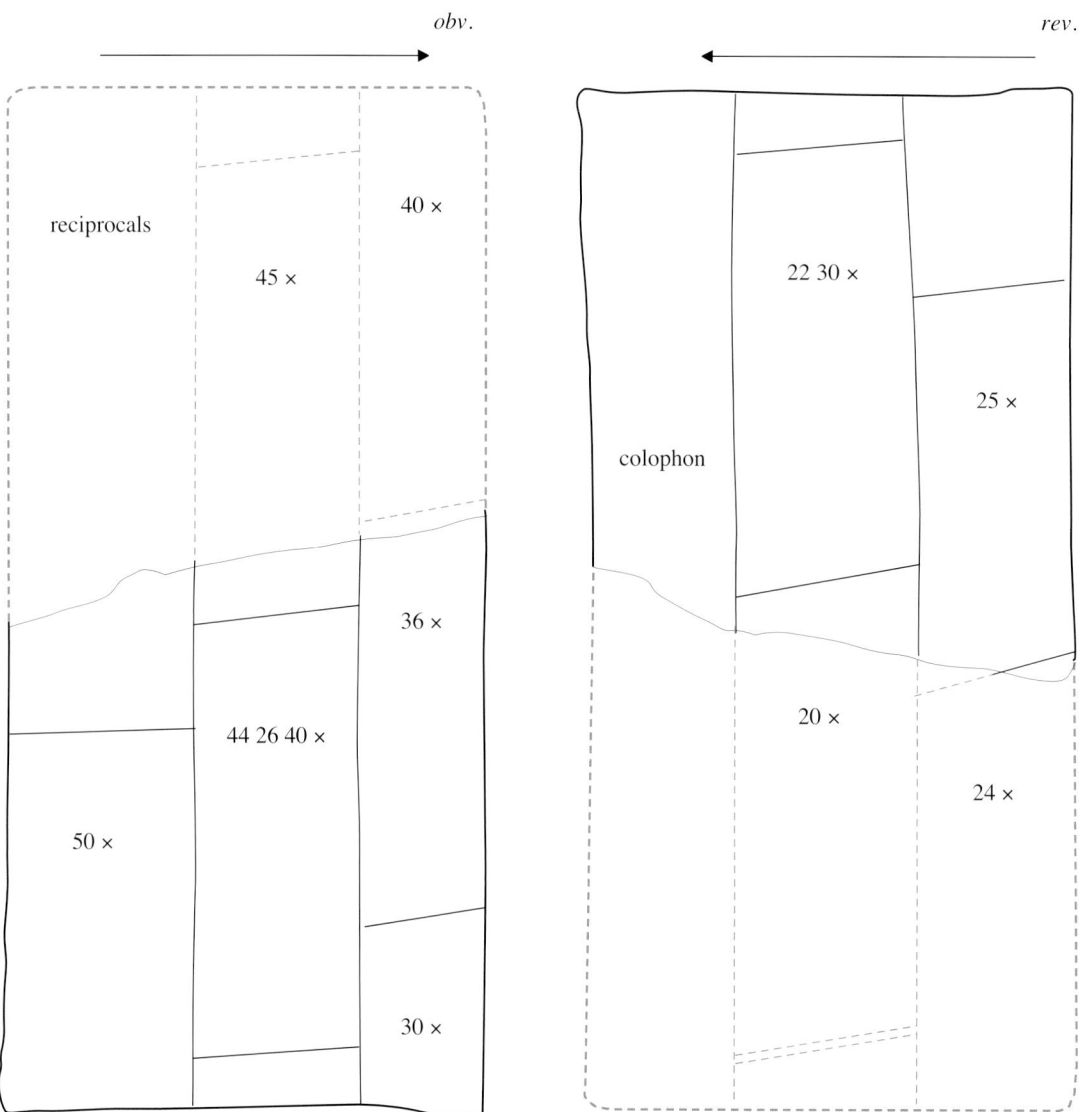

MS 3845, outline. A combined multiplication table with individual sub-tables for reciprocals and for the first 10 of maximally 43 single multiplication tables.

Fig. 2.6.11. MS 3845. The eleven first sub-tables of an Old Babylonian combined multiplication table.

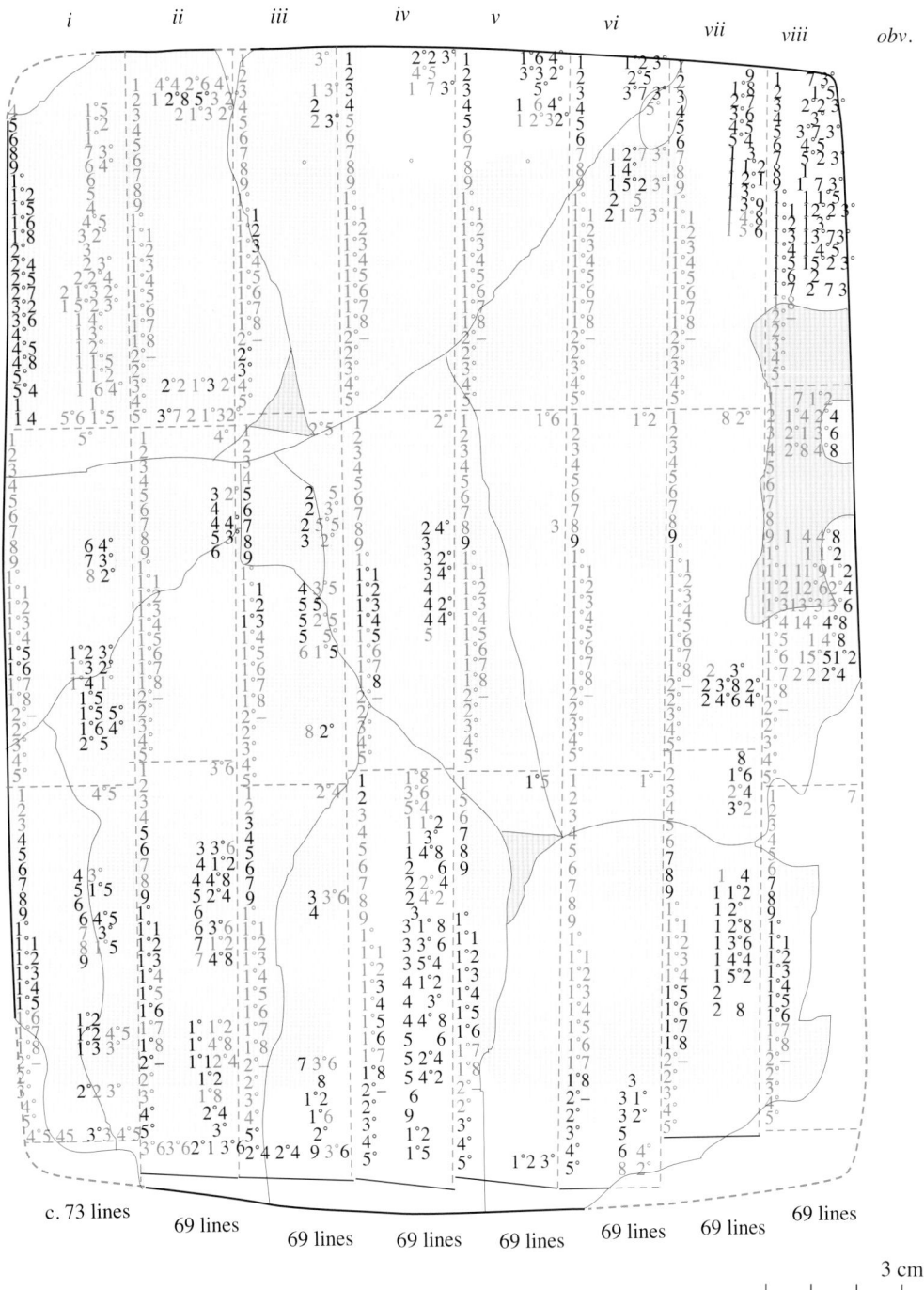

Fig. 2.6.12. MS 3974, *obv.* A nearly complete OB combined multiplication table with 40 sub-tables.

MS 3974 (Figs. 2.6.12-13 above) is a nearly *complete* combined multiplication table with 40 sub-tables, in contrast to MS 3845, which, with its 11 sub-tables, is only *the first quarter* of a complete combined multiplication table. (Cf. the survey in *MKT 1*, 35, which clearly shows that several of the known Old Babylonian combined multiplication tables can be understood as either the first or the second *half* of a complete combined multiplication table. At least two of them, Ist. O 4849 and HS 204, are a *quarter* of a complete table and end with the multiplication table for 24, just like MS 3845.)

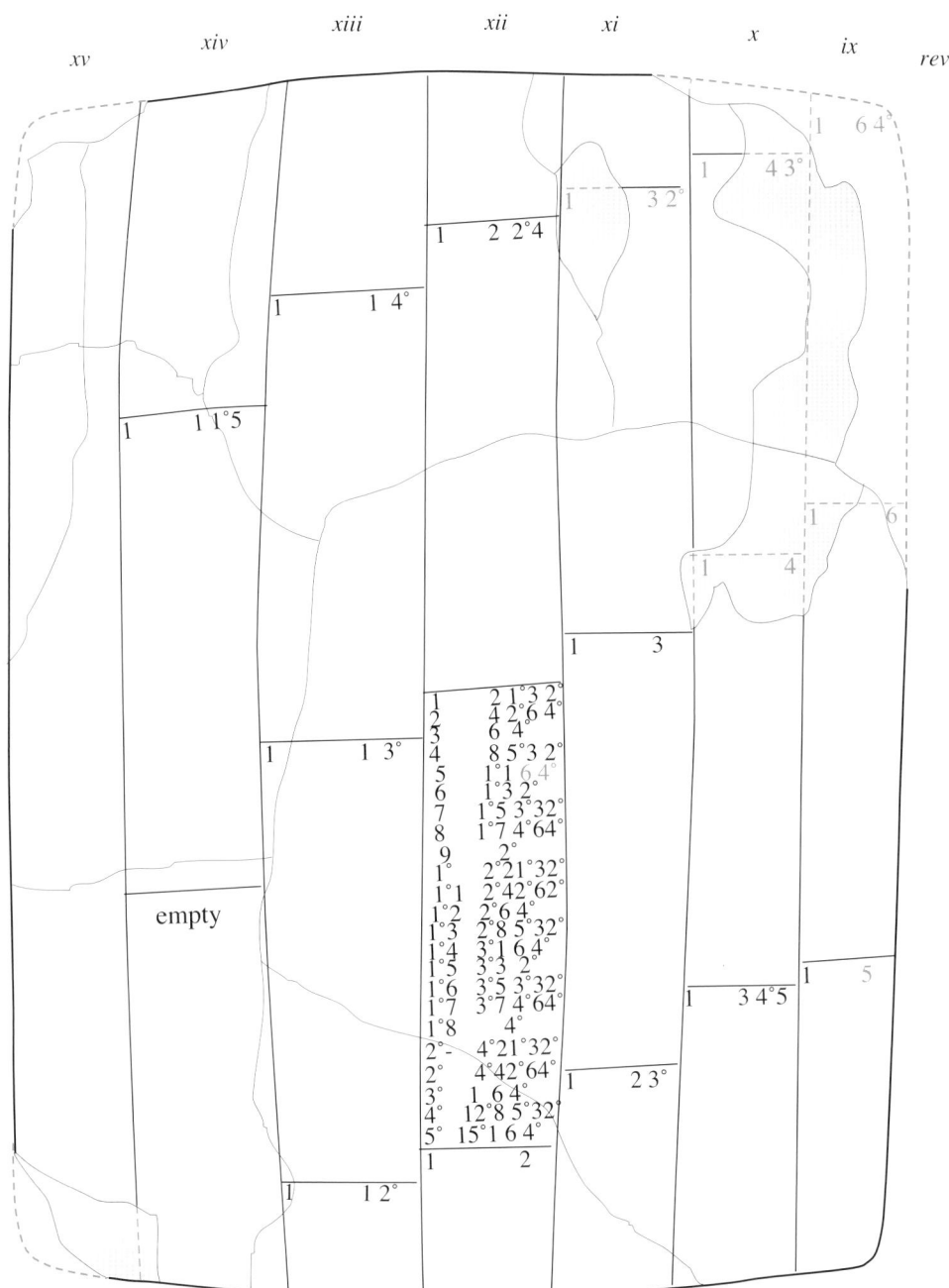

Fig. 2.6.13. MS 3974, *rev*. The standard head number 2 15 is replaced here by the new head number 2 13 20.

MS 3974 is well preserved in the sense that only small parts of the clay tablet are missing. The text on the reverse is also well preserved and readable. The text on the obverse is mostly rubbed off and almost unreadable. Fortunately, however, as a result of the particularly strong curvature of the tablet near the edges, much of the text close to the edges can still be read, so that at least a tentative reconstruction of the general layout of the text is possible (40 sub-tables in 14 columns, c. 924 lines):

Standard number notations are used in MS 3974 for the digits 4, 7, 8. For 19 a simple variant notation is used throughout, transcribed in the conform transliteration as '20 –'.

The 40 sub-tables in the 15 columns of MS 3974 are the table of reciprocals and the single multiplication tables with the following head numbers:

rec.	44 26 40	30	22 30	16 40	12 30	9	7 30	[6 40]	4 30	3 20	2 24	1 40	1 15
50	40	25	20	16	12	8 20	7 12	6	4	3	2 13 20	1 30	
45	36	24	18	15	10	8	[7]	5	3 45	2 30	2	1 20	

Just as in MS 3845, the head number 48 is missing. Other "missing" head numbers are 2 15 and 1 12. On the other hand, an unexpected "extra" head number is 2 13 20. Other unexpected features are that the initial table of reciprocals ends with the pair 1 04, 56 15, omitting the otherwise standard pair 1 21, 44 26 40, and that the three sub-tables in the lower left corner of the obverse all seem to end with lines giving the squares of the head numbers (45, 45, 3 45; 36, 36, 21 36; 24, 24, 9 36).

The head number 2 15 is missing also in 2 of 6 combined multiplication tables in *MKT 1-2* and *MCT* that can be examined with respect to this issue. Moreover, there are no single multiplication tables with the head number 2 15 in *MKT 1-2* and *MCT*. Neither is the head number 1 12 represented in any of the combined or single multiplication tables in *MKT 1-2* and *MCT*.

2.6 e. Tables of Reciprocals and Multiplication Tables in the Schøyen Collection

3 tables of reciprocals, 151 single multiplication tables, 35 different head numbers.

rec.	MS 3869/5	3874	3890		3
50 ×	MS 2826	3834/1	3849 (c)	3894/1-3	6
45 ×	MS 2710	3044/1-2	3044/3 (b*)		4
44 26 40 ×	MS 2711/1 (a')	2711/2	3858/1-2	3979	5
40 ×	MS 2810	2733	3909/3 (b*)	3909/1-2, 4-5	7
36 ×	MS 2709	3859/1-7			8
30 ×	MS 2875 (a")	2895 (a')	3877/1-3		5
25 ×	MS 2808	3857/1-6			7
24 ×	MS 2316 (a')	2825	3873/1-2		4
22 30 ×	MS 2339	3915/1-3			4
20 ×	MS 2807	3900/1-3			4
18 ×	MS 2708	3950/1, 3-4			4
16 40 ×	MS 2363	2660/1-2	2701	3856	5
16 ×	MS 2184/2 (a')	3875/1-4			5
15 ×	MS 2738	3867/1-4			5
12 30 ×	MS 2184/4, 9	2876	3851/1-3		6
12 ×	MS 2184/3 (a')	2286/2 (a')			2
10 ×	MS 2805	3923			2
9 ×	MS 2184/6	2352	3850/1-4		6
8 20 ×	MS 2184/5	2703	3872/1-5		7
8 ×	MS 2042	3043/1-2			3
7 30 ×	MS 2184/8	2702	3930		3
7 12 ×	MS 2286/3 (a')	2812	3040/1-4		6
7 ×	MS 2804/1-2	2829 (a")	3852/1-4		7
6 40 ×	MS 2184/1, 7	3934			3
6 ×	MS 3902/1-3				3
5 ×	MS 2707	3933/1-2			3
4 30 ×	MS 3861/1-3				3
4 ×	MS 3036/1-2	3878/1-2			4
3 45 ×	MS 2286/1	2809	2811	3039/1-2	5
3 20 ×	MS 2854	3853/1-2	3865		4
3 ×	MS 3957/1				1
2 30 ×	MS 3978				1
2 ×	MS 3967/1, 2, 3				3
1 30 ×	MS 3855/1-2				2
1 12 ×	MS 3866				1

Of the 151 single multiplication tables, all are of type a, except 7 of type a', 2 of type a", and 2 of type b*.
Underlined texts are those shown in Figs. 2.6.1 and 2.6.3-2.6.5.
MS 3874 (rec.), MS 3890 (rec.), MS 2184/3 (12 ×), and MS 3044/3 (45 ×) are shown in color in Appendix 10.

It is easy to check that the first eleven items in this list coincide with the eleven sub-tables on MS 3845. Furthermore, the 33 first sub-tables on MS 3974 correspond to the first 33 items on the list. (Only some of the smallest head numbers of sub-tables of MS 3845, namely 2 24, 2 13 20, 1 40, 1 20, and 1 15, do not correspond to head numbers of single multiplication tables in the Schøyen Collection.) This can hardly be a coincidence. The most obvious explanation is that there once existed a canonical Old Babylonian table text, for which a fitting name would be "the Old Babylonian combined multiplication table", of which all the tables in the list above are single excerpts, while MS 3845 and MS 3974 are longer, more complete excerpts. This explanation is further supported by the evidence provided by the double and multiple multiplication tables discussed in Secs. 2.6 b-c above. Here is first a detailed list of all double multiplication tables in the Schøyen Collection:

Nine Double Multiplication Tables

rec.	&	50 ×	MS 2877 (b*)	
25	&	24 ×	MS 3942	
22 30	&	20 ×	MS 2878	
18	&	16 40 ×	MS 3950/2,	2719
16	&	15 ×	MS 2806	
12 30	&	12 ×	MS 3038	
7 30	&	7 12 ×	MS 3964	
4	&	3 45 ×	MS 3977	
[3 20]	&	3 ×	MS 3957/2	

8 double tables of type a, and 2 of type b*.
Underlined texts are those shown in Figs. 2.6.6 and 2.6.7.

What this list suggests is that (at least in the enumerated cases) double multiplication tables are excerpts of *pairs of consecutive sub-tables* in the Old Babylonian combined multiplication table. A similar conclusion can be drawn from the following brief list of sub-tables in the multiple multiplication tables in the Schøyen Collection (other than the large texts MS 3845 and 3974):

Three Multiple Multiplication Tables

[7], 6 40, 6 × (obverse) 7 30, 7 12 × (reverse)	MS 3870 (a)
8, 7 30, 7 12, 7 ×	MS 3939 (a & b)
7, 6 40, 6 ×	MS 3891 (b*)

1 table of type a, 1 of type b*, and 1 of mixed type.
Underlined texts are those shown in Figs. 2.6.8 and 2.6.9.

According to this list, multiple multiplication tables can be regarded as excerpts of *several consecutive sub-tables* from the Old Babylonian combined multiplication table. However, in the case of MS 3870, the sub-tables on the reverse should rightly have *preceded* the sub-tables on the obverse. (Whenever an Old Babylonian cuneiform text is inscribed on a clay tablet that is flat on one side and rounded on the other, the flat side is normally the obverse and the rounded side the reverse.)

It is, of course, already a well known fact that there existed a more or less canonical Old Babylonian combined multiplication table, and that single, double, or multiple multiplication tables are excerpts from such a combined table. This was shown by Neugebauer in a series of tabular summaries in *MKT* and *MCT*. See, in particular, the surveys of head numbers in single or multiple multiplication tables in *MKT* 1, 34-35. Nevertheless, with the many new examples of multiplication tables in the Schøyen Collection, any statistical conclusion that one might want to draw about Old Babylonian multiplication tables can now be much better documented.

Attested head numbers in Old Babylonian single or multiple multiplication tables

sub-table	fraction	other explanations	MKT/MCT		MS	
rec. t.			42 rec.		3 rec. \	
50 ×	rec. 1 12	5/6	5 smt	a, a'	6 smt /	a, c
48 ×	rec. 1 15		——	–	——	–
45 ×	rec. 1 20	walking number (distance/man-day)	3 smt	a	4 smt	a, b*
44 26 40 ×	rec. 1 21		5 smt	a, b	5 smt	a, a'
40 ×	2/3		1 smt	a	7 smt	a, b*
36 ×	rec. 1 40		6 smt	a, b	8 smt	a
30 ×	1/2		3 smt	b	5 smt	a, a'
25 ×	rec. 2 24		13 smt	a, a', b	7 smt \	a
24 ×	rec. 2 30		14 smt	a, b	4 smt /	a, a'
22 30 ×	*	→ used in certain trailing part algorithms	1 smt	a	4 smt \	a
20 ×	rec. 3	1/3	1 smt	a	4 smt /	a
18 ×	rec. 3 20		6 smt	a	4 smt \	a
16 40 ×	*	→ decimal value 1,000	9 smt	a, ?	5 smt /	a
16 ×	rec. 3 45		7 smt	a, b	5 smt \	a, a'
15 ×	rec. 4		3 smt	a, b	5 smt /	a
12 30 ×	*	→ weight, rectangular mud brick (wet)	3 smt	a	6 smt \	a
12 ×	rec. 5		2 smt	a, b	2 smt /	a'
10 ×	rec. 6	weight, rectangular mud brick (sun-dried)	4 smt	b	2 smt	a
9 ×	rec. 6 40		8 smt	a, b	6 smt	a
8 20 ×	*	→ weight, rectangular mud brick (baked)	4 smt	a, b	7 smt	a
8 ×	rec. 7 30		3 smt	a, b	3 smt	a
7 30 ×	rec. 8		2 smt	a	3 smt \	a
7 12 ×	*	→ molding number, rectangular mud bricks	6 smt	a	6 smt /	a, a'
7 ×	*	→ the smallest non-regular integer	7 smt	a, b	7 smt	a, a'
6 40 ×	rec. 9		2 smt	a	3 smt	a
6 ×	rec. 10	man's load, rectangular mud bricks (sun-dried)	5 smt	a	3 smt	a
5 ×	rec. 12		3 smt	a, b	3 smt	a
4 30 ×	*	→ carrying number, rectangular mud bricks	5 smt	a, b	3 smt	a
4 ×	rec. 15		1 smt	a	4 smt \	a
3 45 ×	rec. 16		2 smt	a, b	5 smt /	a
3 20 ×	rec. 18		7 smt	a	4 smt	a
3 ×	rec. 20		1 smt	a'	1 smt	a
2 30 ×	rec. 24		8 smt	a	1 smt	a
2 24 ×	rec. 25		1 smt	?	——	–
2 15 ×	*	→ man's load, square mud bricks (sun-dried)	——	–	——	–
2 13 20 ×	rec. 27	man's load, mud; mud basket	——	–	——	–
2 ×	rec. 30		2 smt	?	3 smt	a
1 40 ×	rec. 36	decimal value 100	4 smt	a, b	——	–
1 30 ×	rec. 40		4 smt	b	2 smt	a
1 20 ×	rec. 45		——	–	——	–
1 15 ×	rec. 48		1 smt	?	——	–
1 12 ×	rec. 50		——	–	1 smt	a
42 head numbers			42 rec., 162 smt		3 rec., 148 smt	

Of the 42 *attested* head numbers, 34 are from the standard table of reciprocals, while 8 are explained in other ways.
Double multiplication tables (only MS) are indicated above by slash-backslash (\ and /).
The only single multiplication tables of type a' in *MKT/MCT* are A 1555 (50 ×), NBC 6349 (25 ×, from Larsa), and *MMAP 27*, 61 (3 ×, from Susa, early OB).
Attested in *multiple* or combined multiplication tables (*MKT/MCT* and MS) are all entries above except 1 12!
A reconstruction of the *complete* Old Babylonian combined multiplication table is presented in Appendix 2.

In the first column of the table above are listed *all head numbers* appearing in known Old Babylonian single or multiple multiplication tables. In the last two columns of the table, a comparison is made between the *MKT/MCT* corpus and the MS corpus of Old Babylonian *single* multiplication tables. The two are of about the same size, but there are a few instances when a head number is documented in one corpus and not in the other.

There are examples of single multiplication tables with the head numbers 2 24, 1 40, and 1 20 in the *MKT/MCT* corpus, but not in MS. On the other hand, there is a multiplication table with the head number 1 12, *the smallest head number known so far*, only in the Schøyen Collection!

Common for both the *MKT/MCT* and the MS texts is that there are no examples of single multiplication tables with the head numbers 48, 2 15, and 1 20. Those head numbers can be found only in some of the multiple or combined multiplication tables.

In both corpuses, a majority of the single multiplication tables are of type a. In the *MKT/MCT* corpus, there are 15 tables of type b, and 3 of type a', while in the MS corpus, there are 6 tables of type a', and no tables of type b but 2 of type b*. These small differences may or may not be significant. They possibly indicate that the multiplication tables in the two corpuses come from different sites. (Most of the single multiplication tables in *MKT 1* are from Nippur and Kish. See the map in Fig. 9.2.)

The second column in the table above was inserted there in *an attempt to answer the question why the listed head numbers, and no others, are attested in Old Babylonian single or multiple multiplication tables*. It is, of course, well known that some kind of connection exists between the Old Babylonian standard table of reciprocals and the set of head numbers used in Old Babylonian multiplication tables, but the details of this connection have never before been worked out.

Now, look at the following simplified display of the Old Babylonian standard table of reciprocals, and compare the numbers appearing in it with the attested head numbers.

3"	40		10	6		**27**	2 13 20		**54**	**1 06 40**
2'	30		12	5		30	2		**1**	**1**
3	20		15	4		**32**	**1 52 30**		**1 04**	**56 15**
4	15		16	3 45		36	1 40		1 12	50
5	12		18	3 20		40	1 30		1 15	48
6	10		20	3		45	1 20		1 20	45
8	7 30		24	2 30		48	1 15		**1 21**	44 26 40
9	6 40		25	2 24		50	1 12			

In the first column of this table of reciprocals are listed *all the 1-place regular sexagesimal numbers*

$$3, 4, 5, 6, 8, 9, 10, 12, 15, 16, 18, 20, 24, 25, 27, 30, 32, 36, 40, 45, 48, 50, 54, 1,$$

and *two 2-place regular sexagesimal numbers*:

$$1\ 04, 1\ 12, 1\ 15, 1\ 20, 1\ 21.$$

All of these are among the attested head numbers, with the exception of

$$27, 32, 54, 1, 1\ 04, \text{and } 1\ 21.$$

These non-attested head numbers are the numbers in bold style in the first column of the table above.

In the second column of the table of reciprocals are listed the reciprocals of the numbers in the first column. These reciprocals are:

$$40, 30, 20, 15, 12, 10, 7\ 30, etc.$$

All of these, too, are attested head numbers, with the exception of

$$1\ 52\ 30, 1\ 06\ 40, 1, \text{and } 56\ 15.$$

These non-attested head numbers correspond to the numbers in bold style in the second column of the table above. Thus, the bold numbers in the table of reciprocals above can be collected into *four pairs of reciprocals*:

(32, 1 52 30), (54, 1 06 40), (1, 1), (1 04, 56 15)   **four missing pairs**

and *two single numbers:*

27, 1 21   **two missing single numbers**

The result of this rapid analysis can be expressed more briefly in the following way:

Some but not all of the numbers appearing in the first or second column of the standard table of reciprocals correspond to attested head numbers in single or multiple multiplication tables.

It is also easy to check that, conversely,

Some but not all of the numbers attested as head numbers in single or multiple multiplication tables appear in the first or second column of the standard table of reciprocals.

Indeed, the following eight attested head numbers do not appear in the standard table of reciprocals:

22 30, 16 40, 12 30, 8 20, 7 12, 7, 4 30, 2 15, 2 13 20   **nine extra single numbers**

The situation is obviously more complicated than one had reason to expect.

There are a couple of additional questions that need to be answered:

Why is 50 the first head number, and why do the head numbers form a decreasing sequence?

2.6 f. An Explanation of the Head Numbers in the Combined Multiplication Table

1. Why 50 is the first head number. In the standard table of reciprocals, the number 50 is not placed in a prominent position, so there must be some other reason why 50 is the first head number in the combined multiplication table. The reason may be that the multiplication table for 50, or rather ;50 = 5/6, was *chronologically* the first multiplication table. Although this is just a conjecture, it does make sense. Indeed, the fraction 5/6 is one of only four "basic fractions" for which there existed a special cuneiform sign (6″). (See Fig. A4.2 in Appendix 4 below.) The other basic fractions, 1/3, 1/2, and 2/3 (with the signs 3′, 2′, and 3″) had been around for a long time, having been constantly used in Sumerian cuneiform texts since at least the ED IIIa period in the middle of the third millennium BC. Presumably, counting with 1/3, 1/2, and 2/3 was one of the basic skills of any Sumerian or Babylonian school boy. *The sign for 5/6, on the other hand, was not as commonly used.* According to a catalogue search at <http://cdli.ucla.edu> it is, apparently, first attested, in the form 6″.ša, in two small Old Akkadian clay tablets, **CT 50, 138** and **ITT 2, 4355**, both badly understood. It is reasonable to assume that the need was felt for a multiplication table for this uncommon basic fraction, and that such a multiplication table became one of the first successful applications of the new way of writing sexagesimal numbers in place value notation. Since this hypothetical first multiplication table can have been used only for a brief period before the invention of the more general combined multiplication table, it is easy to understand why no such primordial multiplication tables have been preserved.

Nevertheless, there does exist a text that may be an example of just such a primordial multiplication table. That text is **MS 3849** (Fig. 2.6.14), which differs from the usual kind of Old Babylonian single multiplication tables in several ways. Already the format of the clay tablet sets this text apart from all the multiplication tables discussed above. The single or multiple multiplication tables shown in Figs. 2.6.1 and 2.6.3-9 are written on one column tablets, and the combined multiplication tables in Figs. 2.6.10-13 on large tablets with several columns on each side. MS 3849, on the other hand, is a medium size two column tablet with a single multiplication table in the first column on the obverse. The rest of the tablet is empty.

In the terminology used, for instance, by Proust in her analysis of the corpus of OB mathematical cuneiform texts from Nippur (Proust, *TMN* (2004)), MS 3849, *obv.* appears to be a text of type IIa, while the one column tablets discussed above are of type III, and the tablets with combined multiplication tables of type I. (See Proust, *op. cit.*, 5.2.1-5.2.4.) As demonstrated by Proust, the overwhelming majority of all OB mathematical cuneiform texts from Nippur are either metrological lists and tables, or arithmetical tables, inscribed on the obverse and/or the reverse of tablets of type II (*op. cit.*, Annexe 1, Figs. 1 and 2). In most cases, tablets of type II are inscribed on the obverse with large number signs in two columns, but on the reverse with number signs of normal size in several columns. The text on the obverse is the part of the curriculum most recently studied by the student who wrote the text, while the text on the reverse is a repetition of other recently learned parts of the curriculum. As observed by Proust, a good example is the Nippur text **CBM 11340+** (Hilprecht, *BE 20/1* (1906), text 20, with photos of obverse and reverse on plates IV-V). An outline of CBM 11340+ is offered by Neugebauer in *MKT 2*, pl. 62. On the obverse of that text there are two columns. In the left column, someone has inscribed a multiplication table of type a with the head number 45, using unusually big cuneiform signs. In the right column, there is a clumsy and unfinished attempt to copy the content of the left column. On the reverse, there are four columns of text inscribed with an alternating series of copies of the table of reciprocals and the multiplication table with head number 50.

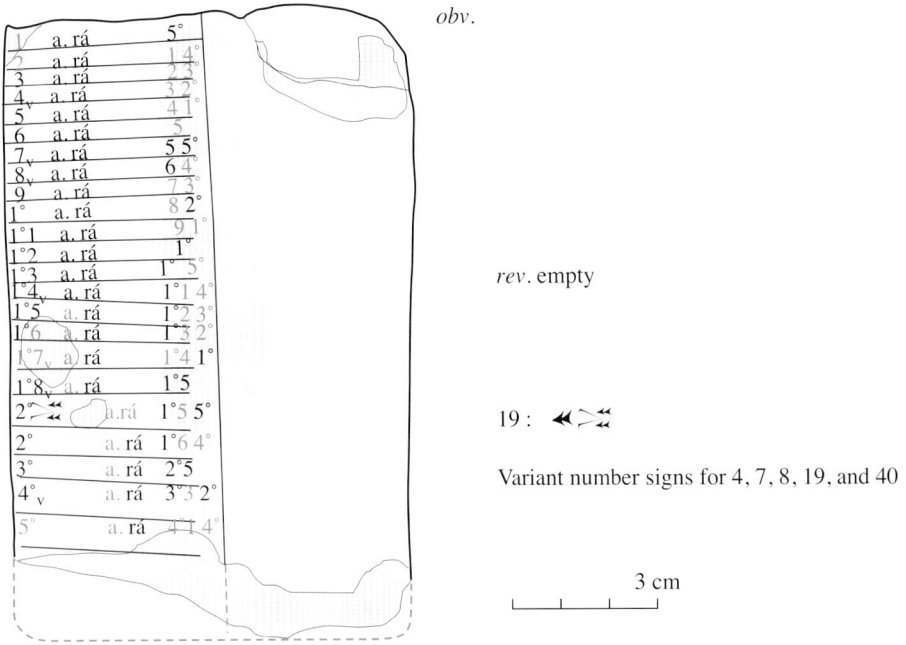

Fig. 2.6.14. MS 3849, *obv*. A single multiplication table (50 ×) of type c (new).

For some reason, on MS 3849 only the first column on the obverse is inscribed, with a multiplication table for the head number 50. There is no way of knowing what the remaining space on the clay tablet was intended for, but it is clear that this is not a tablet of the kind where the teacher's neat writing in the first column was to be copied in a student's clumsy hand in the second column. Indeed, the table in the first column was probably produced by an inexperienced student. The writing is not very elegant, and the signs are so weakly imprinted in the clay that it is very difficult to read them.

Another quite unusual feature of MS 3849 is that it makes use of variant number signs not only for 7, 8, and 19, but also for 4 and 40. The only other example of an arithmetical table text of this kind discussed above is the early Old Babylonian table of reciprocals MS 3874 in Fig. 2.5.1. Even more significant appears to be the fact that the multiplication table on MS 3849 is of an otherwise unattested kind (here called type c), with all entries of the form

$$n \text{ a.rá } p \cdot n \text{)} \qquad \text{instead of the normal equation} \qquad \text{a.rá } n \qquad p \cdot n \,.$$

Taken together, all these unusual features suggest that, possibly, MS 3849 is older than all the other clay tablets with multiplication tables discussed in this volume.

2. Head numbers borrowed from the table of reciprocals. Now, if a multiplication table for 50 was really the first multiplication table using the new sexagesimal place value numbers, then it cannot have lasted a long time before someone tried to construct other multiplication tables using the same technique. This kind of haphazard construction of new multiplication tables was by necessity only a transitory stage of the development, rapidly followed by the invention of the Old Babylonian combined multiplication table with its extensive assortment of multiplication tables associated in a systematic way with a whole series of head numbers. Presumably, the initial intention of the anonymous inventor of this combined multiplication table was to *facilitate division* through the compilation of a list of sexagesimal numbers in relative place value notation representing all numbers of the kind

$$n \cdot \text{rec.} \, m, \text{ where } m \text{ is any reasonably small regular sexagesimal number and } n \text{ any reasonably small integer.}$$

(In modern mathematical language, numbers of the form $n \cdot$ rec. m can be interpreted as rational numbers n/m. The Old Babylonian combined multiplication table can be explained as a systematically arranged list of rational numbers that can be expressed as finite sexagesimal fractions.)

So, with departure from the already established multiplication table with head number 50, the inventor of the combined multiplication table started to construct new multiplication tables with the following head numbers imported *from the end of the right column* of the standard table of reciprocals:

$$(50,)\ 48, 45, 44\ 26\ 40.$$

Automatically, the head numbers so obtained were ordered in a *decreasing* sequence.

Obviously having a lot of trouble with his construction of the multiplication table for the 3-place head number 44 26 40, the inventor of the table now somewhat rashly decided to strike from his table of reciprocals the three pairs (27, 2 13 20), (32, 1 52 30), and (54, 1 06 40), because they contained 3-place numbers. Next, he made a new start *from the beginning of the right column* of the standard table of reciprocals, constructing multiplication tables with the following head numbers:

$$40, 30, 20, 15, 12, 10, 7\ 30, 6\ 40, 6, 5, 4, 3\ 45, 3\ 20, 3, 2\ 30, 2\ 24, 2, 1\ 40, 1\ 30, 1\ 20, 1\ 15, 1\ 12.$$

The procedure stopped at 1 12, because the 3-place number 1 06 40 had already been cancelled, because 1 would be of no interest as a head number, and because 56 15 did not fit into the decreasing sequence of head numbers listed so far!

Obvious gaps in this decreasing sequence of head numbers were then filled out by adding to the list head numbers imported *from the left column* of the table of reciprocals, read *upwards*:

$$36, 30, 25, 24, 18, 16, 9, 8.$$

By turning around in this way and proceeding upwards after reaching the pair (50, 1 12), the inventor of the combined table *unintentionally* missed adding to the list of head numbers both the pair (1 04, 56 15) and the single number 1 21, since they were situated in the table *below* the pair (1, 1).

3. The extra head numbers. The procedure described above explains all attested head numbers in Old Babylonian single or multiple multiplication tables, except the following nine:

22 30, 16 40, 12 30, 8 20, 7 12, 7, 4 30, 2 15, 2 13 20   **nine extra head numbers**

These head numbers were added to the list of head numbers for various reasons.

The most obvious case is that of **16 40**, which is the sexagesimal equivalent of the decimal round number 1,000. *Since decimal numbers were used in everyday life in Babylonia, outside the school environment,* it would clearly be advantageous to include a multiplication table for 1,000 = 16 40 in the combined multiplication table. (An extra multiplication table for 100 = 1 40 was not needed, since 1 40 = rec. 36 was already included in the list of head numbers.)

The inclusion of **7** in the list of head numbers was probably motivated by completely different considerations, in particular that 7 is the smallest sexagesimally *non-regular* integer. It is also one of the "special numbers" 7, 11, 13, 14, and 19, which occupied an important niche in both pre-Babylonian and Old Babylonian mathematics, as a counterweight to the otherwise dominating counting with regular sexagesimal numbers. (See Høyrup, *JNES* 52 (1993).)

Six of the seven remaining extra head numbers are associated with Old Babylonian "brick metrology". *(For a brief, but fairly thorough account of this interesting topic, see Sec. 7.3 a below, in particular the discussion of the new brick text MS 2221.)* Old Babylonian mathematical texts dealing with bricks usually mention the following three types of bricks:

sig$_4$	'bricks'	(rectangular bricks)
sig$_4$.áb	'cow-bricks'	(half-square bricks)
sig$_4$.al.ùr.ra	'tile-bricks' (literal translation unknown)	(square bricks)

Various constants for these and other, less common, types of bricks are listed in several Old Babylonian mathematical "tables of constants". A particularly important constant for any given type of brick was its "molding number" (*naš pakum*), the number of brick-šar of bricks of that type contained in a volume-šar (1 sq. ninda · 1 cubit). A brick-šar was a convenient counting unit for bricks, equal to 12 sixties, and 1 ninda = 12 cubits, with 1 cubit = 30 fingers. Therefore, an alternative way of defining the molding number of any given type of bricks is the number of bricks of that type contained in one twelfth of one sixtieth of a volume-šar, that is in a volume equal to

$$1 \text{ volume-šar}/(12\,00) = 1 \text{ sq. ninda} \cdot 1 \text{ cubit} \cdot 1/(12\,00) = 1 \text{ sq. cubit} \cdot 6 \text{ fingers } (;01 \text{ ninda}).$$

In particular, Old Babylonian rectangular bricks of the *standard format* 1/2 cubit × 1/3 cubit × 5 fingers had the molding number 2 · 3 · 6/5 = 36/5 = 7 1/5, or simply 7 12 in sexagesimal numbers in Babylonian relative place value notation. Consequently, the "extra" head number **7 12** in the Old Babylonian combined table of constants can be explained as the molding number for rectangular bricks of the mentioned standard format. It was, of course, very useful for an Old Babylonian mathematician to have easy recourse to a multiplication table for this important constant.

The volume of a rectangular brick of the standard format can be expressed as follows:

$$1/(7;12) \cdot 1 \text{ volume-šar}/(12\,00) = ;08\,20 \cdot ;00\,05 \text{ volume-šar} = ;00\,00\,41\,40 \text{ volume-šar}.$$

This result can be compared with the text of the round hand tablet **YBC 7284** (*MCT*, 97). On the reverse of that clay tablet are mentioned the numbers 41 40, 8 20, and 12, the latter called igi.gub.ba.bi 'its constant'. On the obverse, it is stated that the weight of one (rectangular) brick is 8 1/3 mina (8 written with the variant number sign 8_v). Since 1 00 minas = 1 talent, it follows that 8 1/3 mina = ;08 20 talent. Consequently, the implication of YBC 7284 is that

the weight of a "unit brick" of the dimensions 1 sq. cubit · 6 fingers (;01 ninda) is precisely 1 talent.

This is a unitary relation of a kind that was typical for Sumerian/Old Babylonian metrology.

Another Old Babylonian text, the table of constants **RAFb = BM 36776** (see again Sec. 7.3 a below), makes it clear that the weight mentioned in YBC 7284 was the weight of a *baked* rectangular brick of standard format, and also that a sun-dried brick was assumed to be 1/5 lighter than a freshly made brick, and a baked brick 1/6 lighter than a sun-dried brick. In other words, it was assumed that

12;30 minas	is the weight of	1 *freshly made* rectangular brick
10 minas	is the weight of	1 *sun-dried* rectangular brick
8;20 minas	is the weight of	1 *baked* rectangular brick.

Apparently, the inventor of the Old Babylonian combined multiplication table thought that it would be useful to include multiplication tables for these brick constants, too. This is a plausible explanation of the extra head numbers **12 30** and **8 20**. (A possible alternative explanation of 8 20 as a head number is that 8 20 is the reciprocal of the molding number 7 12. However, there is no similar alternative explanation of 12 30, which is the reciprocal of 4 48.)

The extra head numbers **4 30** and **2 15** are explained by the mentioned text **MS 2221** discussed in Sec. 7.3 a below. In the second column on the obverse of MS 2221 are listed the three numbers 6, 4 30, and 2 15. The three numbers can be explained as follows:

6, 4;30, and 2;15 *sun-dried* bricks of the three types R1/2c, H2/3c, and S2/3c constitute a man's-load (1 talent).

Here R1/2c, H2/3c and S2/3c are convenient names for common sizes of rectangular bricks, half-square bricks, and square bricks. Clearly it was useful to have at hand multiplication tables for 6, 4 30, and 2 15, representing *man's loads of bricks of the most common formats and sizes*. (The multiplication table 6 = rec. 10 was, of course, already included in the combined multiplication table.)

The extra head number **2 13 20** has a similar explanation. It appears, as a matter of fact, right below the three numbers 6, 4 30, and 2 15 in the second column on the obverse of MS 2221, where it represents *a man's load of mud*, corresponding to the standard size of a mud basket.

As elegant as it seems to be, the explanation proposed above for the extra head numbers 4 30, 2 15, and 2 13 20 is not entirely convincing. This is shown by a look at the table presented above of attestations of head numbers in single multiplication tables. According to that table, 4 30 appears as a head number in 8 single multiplication tables in *MKT/MCT* or MS, while there are no known single multiplication tables with the head numbers 2 15 or 2 13 20. Luckily, the reason for this imbalance is not hard to find. It is that 4 30 can alternatively be explained as the *carrying number* for rectangular bricks of standard size (type R1/2c). The carrying number for any type of bricks is the product of a man's load of that type of bricks and the *walking number* for a man carrying bricks, 45 00 ninda per man-day. Thus, for rectangular bricks of standard type, the carrying number can be calculated as

$$45\ 00\ \text{ninda/man-day} \cdot 6\ \text{bricks} = 4\ 30\ (00)\ \text{brick-ninda/man-day}.$$

In the table on the obverse of MS 2221, the first row

$$45 \qquad 6 \qquad 4\ 30 \qquad \textit{libittum}\ (\text{'brick'})$$

can be explained as a calculation of the carrying number 4 30 in precisely this way

The only remaining extra head number, **22 30**, is tougher to explain, since the number 22 30 is not mentioned in any of the known Old Babylonian tables of constants. However, it is possible that it was found to be useful to have a multiplication table for 22 30 in certain systematic applications of the trailing part algorithm in combination with the doubling-and-halving-algorithm (see Sec. 1.4 above). A beautiful example is the algorithm table **CBS 1215**, exhibited and explained in Appendix 3. In that text, the starting point is the pair of reciprocals 2 05, 28 48, where 2 05 is the third power of 5. With departure from this initial pair, 20 new pairs of reciprocals are constructed through the simple device of repeatedly doubling the first number and halving the second number. For each pair of reciprocals n, m, it is shown how m can be obtained from n, and n from m, by use of the trailing part algorithm. The details of the algorithm are displayed in Fig. A3.4 in Appendix 3.

Now, in this algorithm table, each reciprocal m is computed as the product of the reciprocals of factors of the corresponding number n (and conversely). The following regular sexagesimal numbers are used in various combinations as factors in the computations of the 21 *reciprocals*:

$$24, \mathbf{22\ 30}, 18, 12, 9, 6, 3\ 45, 3, 2\ 24, 1\ 30.$$

The following regular numbers are used in the computations of the 21 *reciprocals of reciprocals*:

$$16, 10, 6\ 40, 5, 4, 2\ 30, \mathbf{2\ 13\ 20}, 2, 1\ 40, 1\ 20, 1\ 15.$$

All these regular numbers are attested head numbers in the Old Babylonian combined multiplication table, all of them except 22 30 (and possibly 2 13 20) being borrowed directly from the Old Babylonian standard table of reciprocals. The number 2 13 20 occurs only once as a factor in CBS 1215, in the eighth sub-table where it is multiplied by 2, but the number 22 30 occurs as a factor 7 times. In sub-tables 9 and 10, 22 30 is multiplied by 18 and 9, respectively, in sub-tables 13 and 14 by 3 45, in sub-tables 17 and 18 by 14 03 45, the second power of 3 45, and in sub-table 21 by 52 44 03 45, the third power of 3 45. Thus, it is clear that it must have been very useful to have recourse to a multiplication table for 22 30 when constructing an algorithm table like CBS 1215.

4. The uniqueness of the Old Babylonian combined multiplication table. The preceding discussion of the rather peculiar form of the list of head numbers in the Old Babylonian combined multiplication table, and of the convoluted way in which that list must have been constructed, clearly demonstrates the uniqueness of the combined multiplication table. It is inconceivable that a table with this complicated structure can have been composed independently by two different persons.

It is now possible to give a tentative sketch of the historical development of counting with sexagesimal numbers at the beginning of the second millennium BC: At some time late in the Ur III period the relative place value system for sexagesimal numbers had been invented, as a clever modification of the Sumerian non-positional sexagesimal system (with "sign value numbers") which had been in use practically without change during the whole preceding millennium. Among the first applications of the new number system were the table of reciprocals in its original form (as in the table texts HS 201, Ist. T. 7375, and Ist. Ni 374), and a multiplication table for the fraction 5/6 = '50' (as in MS 3849), for which now also a new sign was invented.

In an early part of the Old Babylonian period the teaching of mathematics in the scribe schools was intensified, partly because the students needed training in the use of the new positional numbers which were not as intuitively comprehensible as the Sumerian sign value numbers, and partly because it turned out to be so much easier to count with the new numbers. It is likely that at this initial stage each teacher constructed his own multiplication tables with various head numbers for specific purposes. In particular, sexagesimal multiplication tables were probably constructed for the decimal units 10, 100 = 1 40, and 1,000 = 16 40, and multiplication tables were also constructed for the reciprocals of small regular integers, in order to facilitate division. Other multiplication tables were constructed with various head numbers suitable for counting with bricks, and so on.

Very soon, some anonymous teacher of mathematics decided to construct a large table text with many sub-tables, including at the same time a new streamlined version of the table of reciprocals, multiplication tables for reciprocals of regular numbers, and several of the special multiplication tables that were in circulation at the time. The result was the Old Babylonian combined multiplication table, constructed in the way described above. (For a detailed presentation of the table, see Appendix 2.)

The new combined multiplication table became very popular, soon being used in scribe schools all over Mesopotamia. Presumably, every mathematics teacher had two or three large clay tablets containing all the sub-tables of the combined multiplication table. The students were then asked to make excerpts (one, or two, and sometimes more sub-tables) from the large texts, as writing exercises and in order to learn some of the tables by heart. Apparently, the making of duplicates was avoided. That must be why most of the head numbers, except the very last ones in the list, are represented with about the same frequency in the extant corpus of Old Babylonian single multiplication tables, and why there is little overlapping of the double multiplication tables.

It is likely that the successful compilation of the combined multiplication table was the inspiration that led to the creation also of the other canonical arithmetical table texts, the tables of squares, square sides, and cube sides. (The related case of the large combined *metrological* table texts, and the many small excerpts from them, will be discussed below, in Sec. 3.)

Note: What was said above about the uniqueness of the combined multiplication table is not the whole truth. It is worth noting that no *single* multiplication tables with the head number 48 have yet been found, and in the large multiple multiplication text MS 3845 (Figs. 2.6.10-11), the head number 48 is absent. In the survey of multiple multiplication tables in *MKT 1,* 35 this head number is present only in 4 out of 18 listed multiple or combined multiplication tables. What must have happened is that some Old Babylonian scribe making a copy of the original combined multiplication table inadvertently forgot to include the single multiplication table with this head number. After that, all copies or copies of copies of this one incomplete combined table similarly lacked the head number 48.

5. Late Babylonian multiplication tables with non-standard head numbers. Noteworthy is also that in *JCS* 22 (1968/1969) Aaboe published two "atypical" multiplication tables from Uruk. One of them (**IM 2899**) is a double multiplication table with the head numbers 2 24 (= rec. 25) and 2 13 20 (= rec. 27). We now know, of course, that this is not really an atypical multiplication table, since both 2 24 and 2 13 20 are attested elsewhere as head numbers.

The other multiplication table published by Aaboe (**Ist. U 91**) is *Late Babylonian*. It is a multiple multiplication table with the following curious list of mostly non-standard head numbers:

[...], 32, **28 48, 18 45, 11 15, 9 22 30, 6 45, 4 20, 3 30,** 2 15, 2 13 20, [...].

Two of these head numbers are, in addition, non-regular: 3 30 (= 7 · 30), and 4 20 (= 13 · 20).

BM 141493 = 1990-1-30 is another Late Babylonian clay tablet, previously unpublished,[1] with a single multiplication table for the non-regular head number 13, which is also a non-standard head number. The entries in this multiplication table are of the following type:

1 · 13	13	
2 · 13	26	
... ...	...	*Continued*

---

1. Hand copies of BM tablets in this book (in most cases produced by F. Al-Rawi) are published here with the kind permission of the trustees of the British Museum.

```
... ...          ...
29 · 13          6 17          (the hand copy shows 6 07)
29 · 30          6 30
```

Note the use in this text of *a Late Babylonian multiplication sign*, which in this text is in the form of an oblique line of three oblique wedges. (Surprisingly, three oblique wedges was also a Late Babylonian stenographic sign for '9'.)

obv.   *rev.*

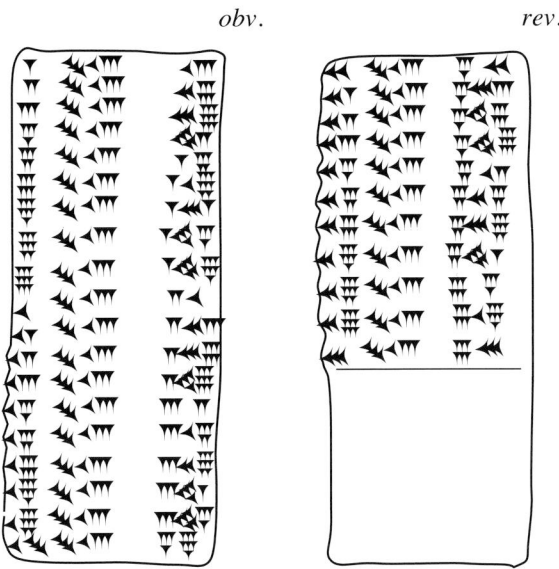

Fig. 2.6.15. BM 141493. A Late Babylonian single multiplication table with the irregular head number 13.

2.7.   Old and Late Babylonian Sexagesimal Representations of Decimal Numbers

2.7 a.   *MS 3970. An Old Babylonian Unfinished Conversion Table*

MS 3970 (Fig. 2.7.1 below) is a clay tablet with only the beginning of an arithmetical table. The writing of the text stopped abruptly in the fifth and last line when only the initial ten-sign of the sexagesimal number 17 20 had been inscribed.

The meaning of the text becomes immediately clear if the five sexagesimal numbers of the aborted table are translated into decimal numbers:

11 40	=	700
13 20	=	800
15(00)	=	900
16 40	=	1,000
1[7 20]	=	[1,100]

It is obvious that what the writer of this text wanted to write down was *the sexagesimal representations of multiples of the decimal unit 100*, starting with 700.

There is room enough on the obverse of the clay tablet for two columns of numbers. If the text had been completed, then it is likely that the right column would have listed the Akkadian *non-positional* decimal equivalents of the sexagesimal numbers in the left column, in the following way:

11 40	7 *me*	
13 20	8 *me*	
15	9 *me*	
16 40	1 *li-im*	
17 20	1 *li-im* 1 *me*	*etc.*

However, it is possible that it never was the writer's intention to include a second column in the text. It seems peculiar that in an Old Babylonian text like this the traditional (decimal) numbers would appear in the right column and the sexagesimal numbers in the left column. Compare with Old Babylonian metrological table texts where the sexagesimal numbers are always written in the right column. Another possibility is that MS 3970 is a *Late Babylonian* text, since in Late Babylonian metrological and other tables, the sexagesimal equivalents are usually written in the left column, as if the table was to be read from right to left. See Friberg, *GMS* 3 (1993). See also BM 36841 below.

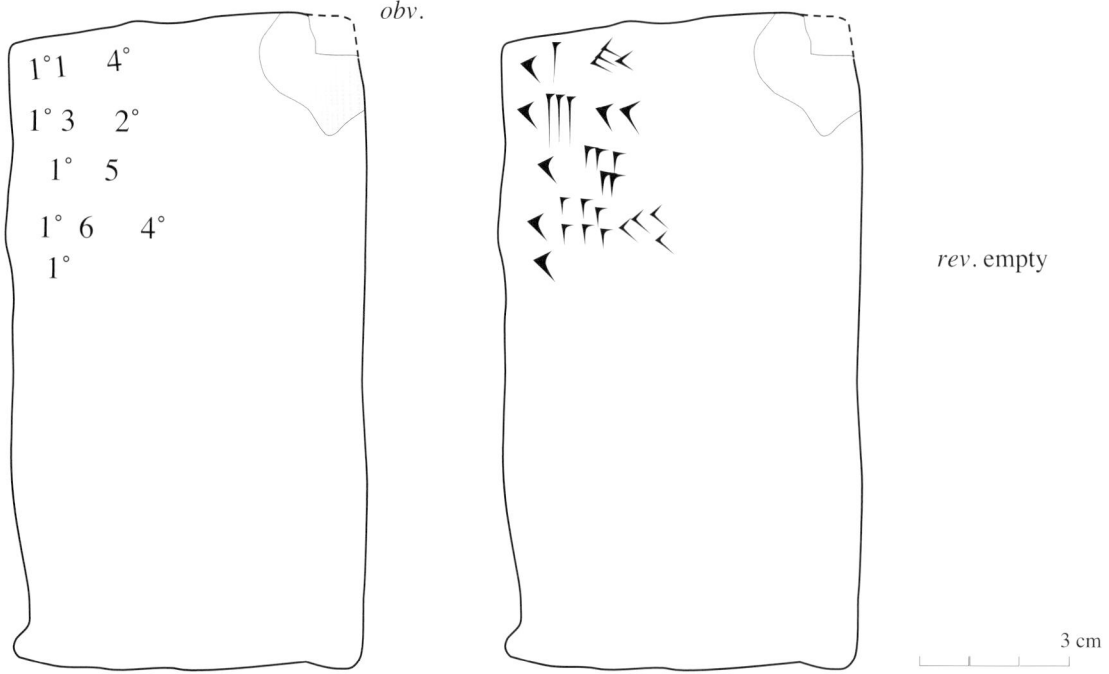

Fig. 2.7.1. MS 3970. Un unfinished text. Sexagesimal equivalents of multiples of the decimal unit 100.

2.7 b.   BM 36841. A Late Babylonian Parallel Text

In view of the fact that sexagesimal numbers were used in Mesopotamia only for calculations, not in the daily life, one might have expected to find many examples of clay tablets with sexagesimal representations of decimal numbers. For some reason, that is not the case. Indeed, MS 3970 is the only known Old(?) Babylonian example of a text of that kind, and the only known parallel text is **BM 36841** (Figure 2.7.2 below), a previously unpublished fragment of a Late Babylonian clay tablet.

There are fragments of two table texts on the obverse of BM 36841. The table to the left is a table of reciprocals, quite different from the Old Babylonian standard table of reciprocals, with entries like

4	15-*ú*	4	(is the sexagesimal equivalent of)	a 15-th	(1/15)
3 45	16-*ú*	3 45	(is the sexagesimal equivalent of)	a 16-th	(1/16)
etc.					

The table to the right on the obverse is a decimal-sexagesimal conversion table with entries like

5	[3 *me*]	5	(is the sexagesimal equivalent of)	3 hundred
6 40	[4 *me*]	6 40	(is the sexagesimal equivalent of)	4 hundred
etc.				

The conversion table continues on the reverse, and proceeds to

2 46 40	10 *lim*	2 46 40	(is the sexagesimal equivalent of)	10 thousand
[… …]	[… …]			

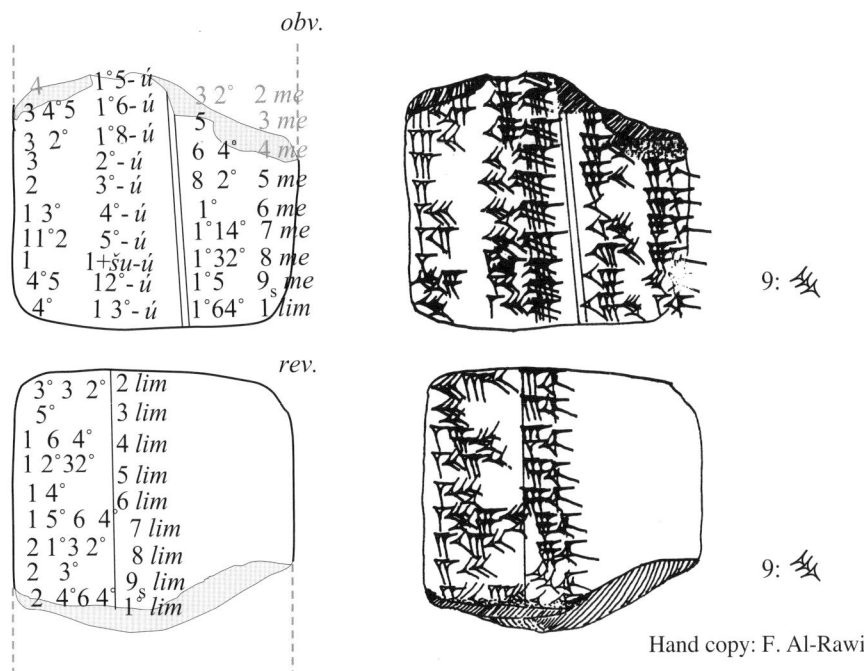

Fig. 2.7.2. BM 36841. A Late Babylonian table of reciprocals and a decimal-sexagesimal conversion table.

3
Old Babylonian Metrological Table Texts

The various systems of measures for length, area, weight, and capacity that were being used in Mesopotamia in the Old Babylonian period were essentially identical with the ones that had been used in the preceding Neo-Sumerian Ur III period. Actually, although to a varying degree, those systems of measures had roots reaching far beyond the Ur III period. The system of area measures, in particular, had been in use in Mesopotamia since at least the proto-literate period in the late fourth millennium, a distinction it shared with the system of (non-positional) sexagesimal counting numbers.

All the Sumerian/Old Babylonian systems of measures had one important property in common, namely that they were adapted to work smoothly together with sexagesimal numbers. This is apparent in the form of their respective "factor diagrams" (see below) where *all the "conversion factors" are regular sexagesimal numbers*. The connection between sexagesimal numbers and the various kinds of measures was that numerical operations involving measures, such as computing the wages due to a gang of hired workers, or the interest on a loan, or the area of a field, were normally carried out in the following way: First the given measures, expressed in the traditional way, were converted to *sexagesimal multiples of certain small "basic units"*, then the necessary computations were effected in sexagesimal arithmetic, and finally the result was converted back to traditional measures.

Once place value notation for sexagesimal numbers had been invented, processes of this kind became easier to execute, although perhaps less easy to understand intuitively. Anyway, a useful tool for converting traditional measures into sexagesimal multiples of a basic unit, and vice versa, were "metrological tables" for the various systems of measures. Clay tablets with metrological tables look very much like clay tablets with multiplication tables. They are fairly well represented in the corpus of Old Babylonian mathematical table texts in general, and in the Schøyen Collection in particular.

3.1. Old Babylonian Capacity Measures. System *C*

In the Old Babylonian period, amounts of cereals and other kinds of dry food stuff were quantified in terms of a series of *units of capacity measure*. These units were:

gín	*šiqlu*	shekel	1/60 of a sìla
sìla	*qû*	small measuring-vessel	c. 1 liter
bán	*sūtu*	large measuring-vessel	c. 10 liters
barig	*parsiktu, pānu*	bushel, basket	c. 60 liters
gur	*kurru*	barrel	c. 300 liters

It is a simplification of the actual situation to say that 1 Old Babylonian sìla was equal to about 1 liter. There was, indeed, an Old Babylonian sìla equal to a cube of side ;01 ninda = 6 fingers (c. 10 cm), which was almost exactly equal to 1 liter. On the other hand, in addition to this "cubic sìla" there were several other types of Old Babylonian sìla measures in use, for instance the "cylinder sìla", a cylinder of diameter 6 fingers and height 6 fingers, with a capacity measure of about 3/4 liter. (See Friberg, *BagM* 28 (1997) § 7 for a detailed discussion of Old Babylonian sìla types and their associated "storing numbers".)

Anyway, whatever the size of a sìla was in a given situation, the *relative sizes* of the gín, the bán, the barig, and the gur compared to the sìla never varied, at least not in the Old Babylonian period. It was always true that

1 gur = 5 00 (300) sìla, 1 barig = 1 00 (60) sìla, 1 bán = 10 sìla, and 1 gín = ;01 (1/60) sìla.

Hierarchic relations of this kind between a series of measure units can be described in a simpler and more compact way by use of what may be called a "factor diagram for system *C*":

In this factor diagram for Neo-Sumerian/Old Babylonian capacity measures, information is given about the Sumerian and Akkadian names for the series of capacity units, in suitable cases the English translations, and both absolute and relative sizes of the units. The numbers on top of the arrows, forming a "chain of conversion factors", indicate that

1 gur = 5 barig, 1 barig = 6 bán, 1 bán = 10 sìla, and 1 sìla = 60 gín.

Even if a factor diagram of the kind illustrated above contains a wealth of information, anyone wanting to actually read a metrological table of capacity measures in the original cuneiform script needs to know more: Which are the cuneiform signs for the units of capacity, and which kinds of number signs are used for what may be called "capacity numbers"? Additional information along these lines is provided by a factor diagram suitably modified in the following way:

A "sign-oriented" factor diagram of this kind gives information not only about the structure of the system of measures in the actual case (the names of the units, and the chain of conversion factors), but also about the cuneiform signs for the units and about the form of the number signs. However, even this kind of modified factor diagram is unable to provide exhaustive information about the way in which Old Babylonian capacity numbers were written. One needs to know also the rules for forming multiples of the units:

The way of writing multiples of the sìla and the shekel was straightforward enough. Capacity numbers from 1 to 9 sìla and from 1 to 19 gín were written with the usual signs for the digits 1 through 19, followed by the signs for gín or sìla. On the other hand, 20, 30, 40, and 50 gín were usually not written as signs for the tens followed by the sign for gín, since these quantities were regarded as fractions of the sìla (1/3, 1/2, 2/3 and 5/6 sìla) rather than as multiples of the gín.

The way of writing multiples of the bán, the barig, and the gur, was more idiosyncratic. There were special cuneiform number signs for the capacity numbers, from 1_{bn} = '1(bán)' to 5_{bn} = '5(bán)' and from 1_{bg} = '1(barig)' to 4_{bg} = '4(barig)'. These number signs are displayed in the upper right corner of Fig. 3.1.1 below. They could be followed by the sign še 'barley, cereal', but this was optional. A sign pi = barig after the barig-numbers was optional, too, and used only if the barig multiples stood alone and were not followed by multiples of smaller capacity units. (Note that notations of the type 1_{bn}, 1_{bg}, *etc.*, are used here only in conform transliterations, while notations of the type 1(bán), 1(barig), *etc.*, are used in standard transliterations and translations.)

For multiples of the gur, Sumerian *non-positional* sexagesimal number notations were used, with the small modification that digits from 1 to 9 were written with *horizontal* instead of vertical wedges.

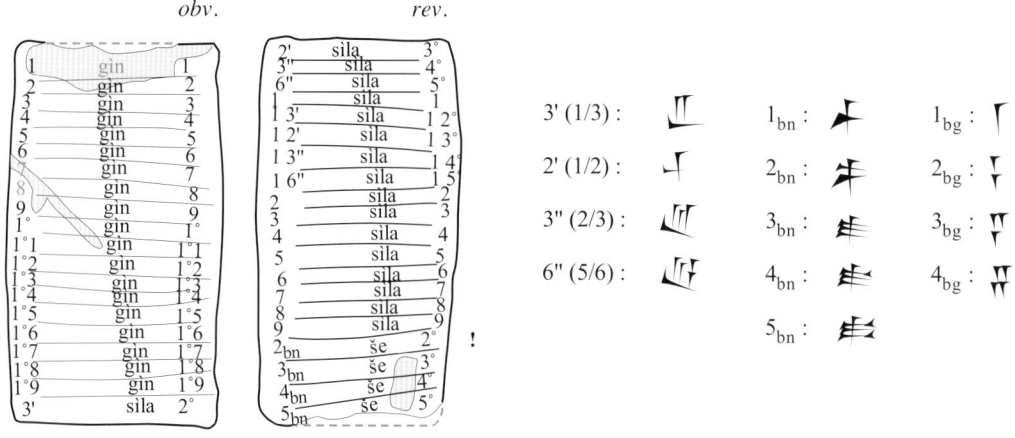

MS 2827. A metrological table for system *C*, basic unit sìla (40 lines),
from '1 shekel = 1 (· 60⁻¹ sìla)' to '5(bán) = 50 (· 1 sìla)'.

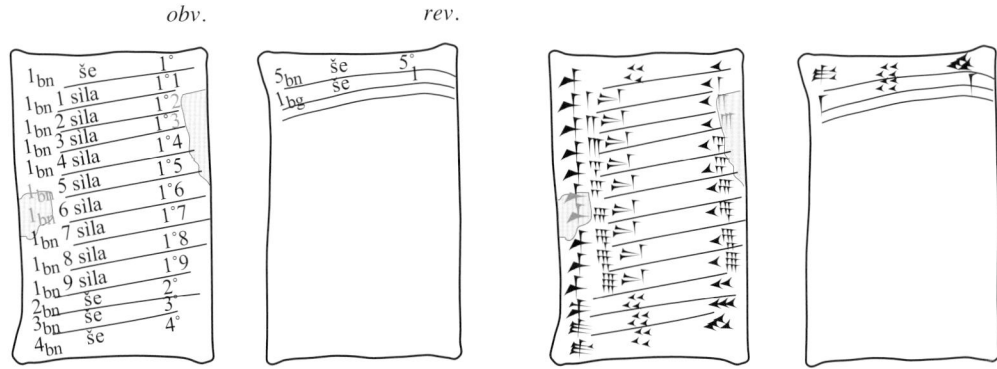

MS 3905. A metrological table for system *C*, basic unit sìla (15 lines),
from '1(bán) of barley = 10 (· 1 sìla)' to '1(barig) of barley = 1 (· 60 sìla)'.

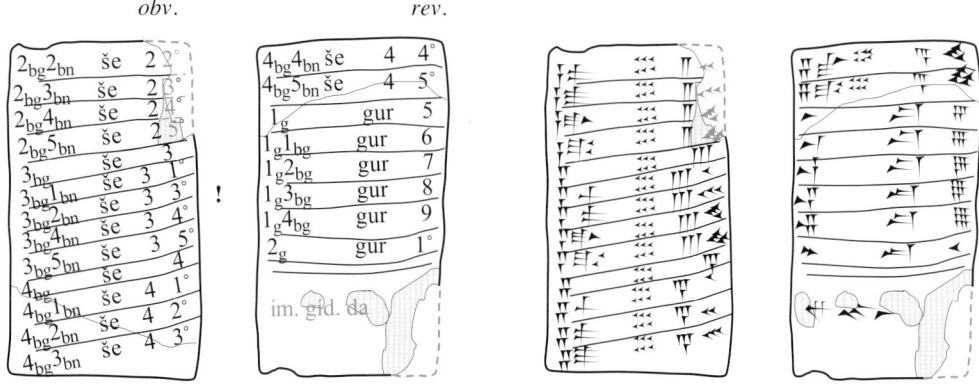

MS 2734. A metrological table for system *C*, basic unit sìla (21 lines),
from '2(barig) 2(bán) = 2 20 (· 1 sìla)' to '2(gur) = 10 (· 60 sìla)'.

Fig. 3.1.1. Three Old Babylonian metrological tables for the lower and middle ranges of system *C*.

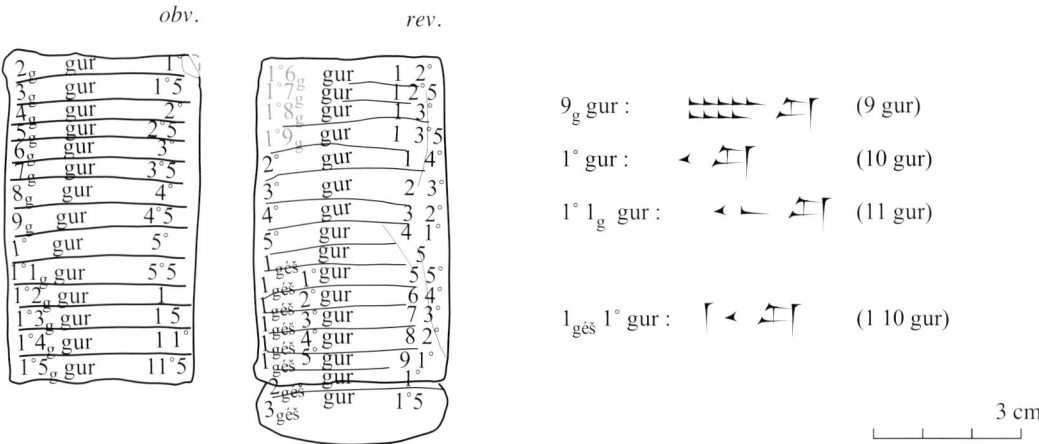

MS 2828. A metrological table for system *C*, basic unit sìla (30 lines),
from '2 gur = 10 (· 60 sìla)' to '3 (· 60) gur = 15 (· 60² sìla)'.

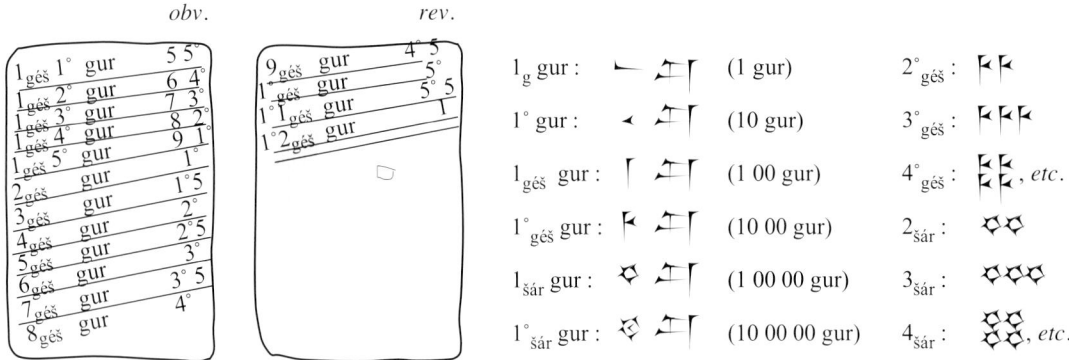

MS 3869/6. A metrological table for system *C*, basic unit sìla (16 lines),
from '1 10 gur = 5 50 (· 60 sìla)' to '12 00 gur = 1 (· 60³ sìla)'.

MS 2704 . A metrological table for system *C*, basic unit sìla (12 lines),
from '2 · 60² gur = 10 (· 60³ sìla)' to '13 · 60² gur = 1 05 (· 60³ sìla)'.

Fig. 3.1.2. Three Old Babylonian metrological tables for the middle and upper ranges of system *C*.

It should be clear from this brief account that it was *a very complicated system of notations for capacity numbers* that the students in the Old Babylonian scribe schools had to learn. The way they did master the complex system was by copying and diligently studying metrological table texts for system *C*. There are 12 such table texts in the Schøyen Collection. Seven of these twelve texts are shown in Figs. 3.1.1-3, some only in conform transliterations, others both in conform transliterations and as direct hand copies. To the extent that it was possible, six of the metrological table texts were chosen so that they would be *non-overlapping excerpts* from a "complete" metrological table for the Old Babylonian system *C*. The seventh text is just such a complete metrological table.

MS 2827 (Fig. 3.1.1, top) is an Old Babylonian metrological table for the *lower end* of system *C*. In the first column of the table are listed capacity measures from 1 shekel to 5 barig, and in the right column the values of all the listed capacity measures as multiples of 1 sìla, the "basic unit'" in all metrological tables for system *C*. *The multiples of the basic unit are expressed in terms of sexagesimal numbers in relative place value notation. The basic unit itself is not explicitly mentioned.* Here is an explanation of the meaning of the table:

1	gín	=	1	$(\cdot\,60^{-1}$ sìla$)$		2/3	sìla	=	40	$(\cdot\,60^{-1}$ sìla$)$	
2	gín	=	2	$(\cdot\,60^{-1}$ sìla$)$		5/6	sìla	=	50	$(\cdot\,60^{-1}$ sìla$)$	
3	gín	=	3	$(\cdot\,60^{-1}$ sìla$)$		1	sìla	=	1	$(\cdot\,1$ sìla$)$	
4	gín	=	4	$(\cdot\,60^{-1}$ sìla$)$		1 1/3	sìla	=	1 20	$(\cdot\,60^{-1}$ sìla$)$	
5	gín	=	5	$(\cdot\,60^{-1}$ sìla$)$		1 1/2	sìla	=	1 30	$(\cdot\,60^{-1}$ sìla$)$	
6	gín	=	6	$(\cdot\,60^{-1}$ sìla$)$		1 2/3	sìla	=	1 40	$(\cdot\,60^{-1}$ sìla$)$	
7	gín	=	7	$(\cdot\,60^{-1}$ sìla$)$		1 5/6	sìla	=	1 50	$(\cdot\,60^{-1}$ sìla$)$	
8	gín	=	8	$(\cdot\,60^{-1}$ sìla$)$		2	sìla	=	2	$(\cdot\,1$ sìla$)$	
9	gín	=	9	$(\cdot\,60^{-1}$ sìla$)$		3	sìla	=	3	$(\cdot\,1$ sìla$)$	
10	gín	=	10	$(\cdot\,60^{-1}$ sìla$)$		4	sìla	=	4	$(\cdot\,1$ sìla$)$	
11	gín	=	11	$(\cdot\,60^{-1}$ sìla$)$		5	sìla	=	5	$(\cdot\,1$ sìla$)$	
12	gín	=	12	$(\cdot\,60^{-1}$ sìla$)$		6	sìla	=	6	$(\cdot\,1$ sìla$)$	
13	gín	=	13	$(\cdot\,60^{-1}$ sìla$)$		7	sìla	=	7	$(\cdot\,1$ sìla$)$	
14	gín	=	14	$(\cdot\,60^{-1}$ sìla$)$		8	sìla	=	8	$(\cdot\,1$ sìla$)$	
15	gín	=	15	$(\cdot\,60^{-1}$ sìla$)$		9	sìla	=	9	$(\cdot\,1$ sìla$)$	
16	gín	=	16	$(\cdot\,60^{-1}$ sìla$)$		**1(bán)**	**še**	**=**	**10**	**$(\cdot\,1$ sìla$)$**	missing line
17	gín	=	17	$(\cdot\,60^{-1}$ sìla$)$		2(bán)	še	=	20	$(\cdot\,1$ sìla$)$	
18	gín	=	18	$(\cdot\,60^{-1}$ sìla$)$		3(bán)	še	=	30	$(\cdot\,1$ sìla$)$	
19	gín	=	19	$(\cdot\,60^{-1}$ sìla$)$		4(bán)	še	=	40	$(\cdot\,1$ sìla$)$	
1/3	sìla	=	20	$(\cdot\,60^{-1}$ sìla$)$		5(bán)	še	=	50	$(\cdot\,1$ sìla$)$	
1/2	sìla	=	30	$(\cdot\,60^{-1}$ sìla$)$							

Note that this metrological table text has a superficial similarity with a single multiplication table, or an abridged table of squares, both in form and content. Thus, the clay tablet on which the table is written has the same shape as clay tablets with single multiplication tables, and the multiples of 1 gín increase in steps of 1 gín from 1 to 19 gín, then in steps of 10 gín from 20 to 50 gín. However, according to *a strict convention*, expressions like '20 gín', '30 gín', *etc.*, are *not allowed* in the cuneiform script but are replaced by *basic fractions* of the sìla, that is by '1/3 sìla', '1/2 sìla', '2/3 sìla', '5/6 sìla'. This is one of many rules that the scribe students had to learn through their study of metrological table texts.

The table continues with a brief group of entries of the form '1 sìla plus a basic fraction of a sìla', but wisely refrains from a repetition, with entries of the form '2 sìla plus a basic fraction of a sìla'. Instead it proceeds in steps of 1 sìla from 2 to 9 sìla, and in steps of 1 bán (= 10 sìla) from 1 to 5 bán. (However, through a mistake of the scribe, the line '1(bán) še = 10' is missing in the text.) Here MS 2827 stops, just before the table text reaches the level '1(barig) = 1 (00)'.

An interesting observation is that the text of this metrological table can be divided into two roughly parallel parts. Indeed, the entries from 1 to 19 gín are "small fractions of the sìla" in the same way as the entries from 1 to 9 sìla are "small fractions of the barig", and the entries from 1/3 to 5/6 sìla are "basic fractions of the sìla" in the same way as the entries from 1 bán to 5 bán are "basic fractions of the barig". Also, in the same way as it is not allowed to write '20 gín', '30 gín', *etc.*, it is not allowed to write '10 sìla', '20 sìla', *etc.*

MS 3905 (Fig. 3.1.1, middle) is a brief metrological table, expanding the section from 1 to 5 bán in the preceding table by adding multiples of the sìla to 1 bán as fractions of the bán:

1(bán)	še	=	10	(· 1 sìla)
1(bán)	1 sìla	=	11	(· 1 sìla)
1(bán)	2 sìla	=	12	(· 1 sìla)
1(bán)	3 sìla	=	13	(· 1 sìla)
1(bán)	4 sìla	=	14	(· 1 sìla)
1(bán)	5 sìla	=	15	(· 1 sìla)
1(bán)	6 sìla	=	16	(· 1 sìla)
1(bán)	7 sìla	=	17	(· 1 sìla)

1(bán)	8 sìla	=	18	(· 1 sìla)
1(bán)	9 sìla	=	19	(· 1 sìla)
2(bán)	še	=	20	(· 1 sìla)
3(bán)	še	=	30	(· 1 sìla)
4(bán)	še	=	40	(· 1 sìla)
5(bán)	še	=	50	(· 1 sìla)
1(barig)	še	=	1	(· 60 sìla)

Note here again a layout reminding of the organization of single multiplication tables. The table proceeds in steps of 1 sìla from 10 to 19 sìla, and then in steps of 10 sìla from 20 to 1 00 sìla.

MS 2734 (Fig. 3.1.1, bottom) starts, for some reason, at the odd level 2 barig 2 bán, and then proceeds in steps of 1 bán to 4 barig 5 bán, and after that in steps of 1 barig from 1 gur (= 5 barig) to 2 gur (= 10 barig):

2(barig) 2(bán) še	= 2 20	(· 1 sìla)
2(barig) 3(bán) še	= 2 30	(· 1 sìla)
2(barig) 4(bán) še	= 2 40	(· 1 sìla)
2(barig) 5(bán) še	= 2 50	(· 1 sìla)
3(barig) še	= 3	(· 60 sìla)
3(barig) 1(bán) še	= 3 10	(· 1 sìla)
3(barig) 2(bán) še	= 3 20	(· 1 sìla)
3(barig) 3(bán) še	= 3 30	(· 1 sìla)
3(barig) 4(bán) še	= 3 40	(· 1 sìla)
3(barig) 5(bán) še	= 3 50	(· 1 sìla)
4(barig) še	= 4	(· 60 sìla)

4(barig) 1(bán) še	= 4 10			(· 1 sìla)
4(barig) 2(bán) še	= 4 20			(· 1 sìla)
4(barig) 3(bán) še	= 4 30			(· 1 sìla)
4(barig) 4(bán) še	= 4 40			(· 1 sìla)
4(barig) 5(bán) še	= 4 50			(· 1 sìla)
1(gur)		gur	= 5	(· 60 sìla)
1(gur) 1(barig)		gur	= 6	(· 60 sìla)
1(gur) 2(barig)		gur	= 7	(· 60 sìla)
1(gur) 3(barig)		gur	= 8	(· 60 sìla)
1(gur) 4(barig)		gur	= 9	(· 60 sìla)
2(gur)		gur	= 10	(· 60 sìla)

Note the surprising way in which 1 gur 1 barig, 1 gur 2 barig, *etc.*, are expressed as 1_g 1_{bg} gur, 1_g 2_{bg} gur, and so on. This way of writing reveals that the multiples of the barig, written 1_{bg}, 2_{bg}, 3_{bg}, 4_{bg}, were regarded as "basic fractions of the gur". Similarly, when 4 barig 1 bán, 4 barig 2 bán, *etc.*, are expressed as 4_{bg} 1_{bn} še, 4_{bg} 1_{bn} še, and so on, this reveals that the multiples of the bán, written 1_{bn}, 2_{bn}, 3_{bn}, 4_{bn}, 5_{bn}, were regarded as "basic fractions of the barig".

MS 2828 (Fig. 3.1.2, top) starts at 2 gur, proceeds in steps of 1 gur to 19 gur, and after that in steps of 10 gur from 20 gur to 1 50 gur, and in steps of 1 (00) gur from 2 (00) to 3 (00) gur. It is important to observe that what looks like, for instance, 2 gur actually is 2(géš) = 2 sixties of gur, while 2 gur is written as 2_g gur, with the kind of special number sign that is used for small multiples of the gur. Since 1 gur = 5 00 sìla, the text of MS 2878 can be explained as follows:

2(gur)	gur	=	10	(· 60 sìla)
3(gur)	gur	=	15	(· 60 sìla)
4(gur)	gur	=	20	(· 60 sìla)
5(gur)	gur	=	25	(· 60 sìla)
6(gur)	gur	=	30	(· 60 sìla)
7(gur)	gur	=	35	(· 60 sìla)
8(gur)	gur	=	40	(· 60 sìla)
9(gur)	gur	=	45	(· 60 sìla)
10	gur	=	50	(· 60 sìla)
11(gur)	gur	=	55	(· 60 sìla)
12(gur)	gur	=	1	(· 60^2 sìla)
13(gur)	gur	=	1 05	(· 60 sìla)
14(gur)	gur	=	1 10	(· 60 sìla)

17(gur)	gur	=	1 25	(· 60 sìla)
18(gur)	gur	=	1 30	(· 60 sìla)
19(gur)	gur	=	1 35	(· 60 sìla)
20	gur	=	1 40	(· 60 sìla)
30	gur	=	2 30	(· 60 sìla)
40	gur	=	3 20	(· 60 sìla)
50	gur	=	4 10	(· 60 sìla)
1(géš)	gur	=	5	(· 60^2 sìla)
1(géš) 10	gur	=	5 50	(· 60 sìla)
1(géš) 20	gur	=	6 40	(· 60 sìla)
1(géš) 30	gur	=	7 30	(· 60 sìla)
1(géš) 40	gur	=	8 20	(· 60 sìla)
1(géš) 50	gur	=	9 10	(· 60 sìla)

Continued

15(gur)	gur	=	1 15	($\cdot$ 60 sìla)
16(gur)	gur	=	1 20	($\cdot$ 60 sìla)

2 (géš) gur	=	10($\cdot$ 60 00 sìla)
3 (géš) gur	=	15($\cdot$ 60^2 sìla)

MS 3869/6 (Fig. 3.1.2, middle) is partly overlapping with MS 2828. The table starts at 1 10 gur and proceeds in steps of 10 gur to 1 50 gur, then in steps of 1 (00) gur from 2 (00) gur to 12 (00) gur:

1(géš) 10	gur	=	5 50	($\cdot$ 60 sìla)
1(géš) 20	gur	=	6 40	($\cdot$ 60 sìla)
1(géš) 30	gur	=	7 30	($\cdot$ 60 sìla)
1(géš) 40	gur	=	8 20	($\cdot$ 60 sìla)
1(géš) 50	gur	=	9 10	($\cdot$ 60 sìla)
2 (géš)	gur	=	10	($\cdot$ 60^2 sìla)
3 (géš)	gur	=	15	($\cdot$ 60^2 sìla)
4 (géš)	gur	=	20	($\cdot$ 60^2 sìla)

5 (géš)	gur	=	25	($\cdot$ 60^2 sìla)
6 (géš)	gur	=	30	($\cdot$ 60^2 sìla)
7 (géš)	gur	=	35	($\cdot$ 60^2 sìla)
8 (géš)	gur	=	40	($\cdot$ 60^2 sìla)
9 (géš)	gur	=	45	($\cdot$ 60^2 sìla)
10 (géš)	gur	=	50	($\cdot$ 60^2 sìla)
11 (géš)	gur	=	55	($\cdot$ 60^2 sìla)
12 (géš)	gur	=	1	($\cdot$ 60^3 sìla)

Note that 2, 3, 4, … sixties are written as 2, 3, 4, …, and that 10 sixties is written with the sign geš'u = *nēr*, which in some texts looks very much like 1 10. (The factor diagram for system *S*(S/OB), the Sumerian *non-positional* system of notations for sexagesimal numbers, is displayed in Appendix 4.)

MS 2704 (Fig. 3.1.2, bottom) is a brief excerpt from the upper end of the complete metrological table for system *C*. The excerpt starts at 2 šár gur and proceeds in steps of 1 šár gur to 13 šár gur. It is not clear why it stops at such an odd number. Note that, for instance, the complete metrological table on the 6-sided prism MS 2723 (Fig. 3.1.3) proceeds all the way to 2 šár.gal gur (= 2 00 00 00 gur = 10 00 00 00 00 sìla), and so does the complete table YBC 4633 (Nemet-Nejat, *JNES* 54 (1995), Fig. 8). Here is an explanation of the brief table on MS 2704:

2(šár)	gur	=	10	($\cdot$ 60^3 sìla)
3(šár)	gur	=	15	($\cdot$ 60^3 sìla)
4(šár)	gur	=	20	($\cdot$ 60^3 sìla)
5(šár)	gur	=	25	($\cdot$ 60^3 sìla)
6(šár)	gur	=	30	($\cdot$ 60^3 sìla)
7(šár)	gur	=	35	($\cdot$ 60^3 sìla)
8(šár)	gur	=	40	($\cdot$ 60^3 sìla)
9(šár)	gur	=	45	($\cdot$ 60^3 sìla)
10 (šár)	gur	=	50	($\cdot$ 60^3 sìla)
11(šár)	gur	=	55	($\cdot$ 60^3 sìla)
12 (šár)	gur	=	1	($\cdot$ 60^4 sìla)
13(šár)	gur	=	1 05	($\cdot$ 60^3 sìla)

$2 \cdot 60^2$ gur = $10 \cdot 60^3$ sìla

$12 \cdot 60^2$ gur = $1 \cdot 60^4$ sìla

MS 2723 (Fig. 3.1.3) is a well preserved 6-sided prism with a complete metrological table for the Old Babylonian capacity system inscribed on its six faces, as usual with the basic unit 1 sìla.

The table on MS 2723 starts at 1 shekel (gín) and proceeds in steps of 1 gín to 19 gín, then continues with basic fractions of a sìla, with 1 sìla plus basic fractions of a sìla, and with multiples of the sìla (basic fractions of the bán), from 2 to 9 sìla. So far the text of MS 2723 is identical with the text of 2827 (Fig. 3.1.1, top), but then it continues with 1 bán and 1 bán plus basic fractions of the bán (from 1 to 9 sìla). Note that this whole part of the bán section is missing in the text of MS 2827, probably because the one who wrote it mistakenly thought he had copied all lines as far as '1(bán) 9 sìla   19' when he had only come as far as '9 sìla 9'. The correct form of the section can be found instead in MS 3905 (Fig. 3.1.1, middle).

The next section of MS 2723, the barig section, proceeds through the basic fractions of the barig (from 1 bán to 5 bán) and 1, 2, 3, or 4 barig plus basic fractions of the barig. After that comes a brief section for 1 gur and 1 gur plus the basic fractions of the gur (from 1 barig to 4 barig). This part of the text of MS 2723 is mirrored by the text of MS 2734 (Fig. 3.1.1, bottom).

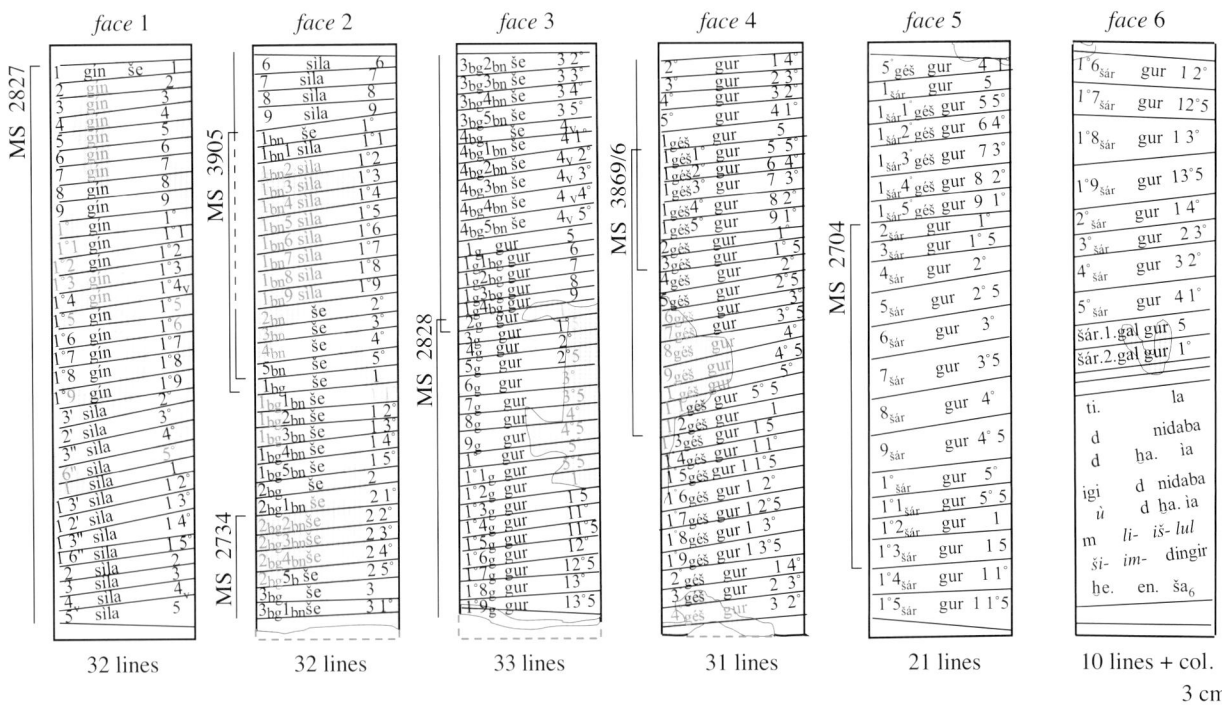

A 6-sided prism with a metrological table for system C, basic unit sìla,
from '1 shekel of barley = 1 ($\cdot 60^{-1}$ sìla)' to '$2 \cdot 60^3$ gur = 10 ($\cdot 60^4$ sìla)', altogether 159 lines.
Variant number sign for 4. 12 cm × 5.5 cm.

Fig. 3.1.3. MS 2723. A complete metrological table for Old Babylonian capacity measures.

The gur section of MS 2723 continues from 2 to 19 gur in steps of 1 gur, and from 20 to 1 50 gur in steps of 10 gur. This part of MS 2723 is mirrored by MS 2828 (Fig. 3.1.2, top). A similar sixty-gur section proceeds from 2 to 19 sixties gur in steps of sixty gur, and from 20 to 50 sixties gur in steps of 10 sixties gur. This range is partially covered by the text MS 3869/6 (Fig. 3.1.2, middle).

The next section of MS 2723 starts with 1 šár gur (1 00 00 gur), and 1 šár gur plus basic fractions of 1 šár gur (10 to 50 sixties gur), then continues from 2 to 19 šár gur in steps of 1 šár gur, and from 20 to 50 šár gur in steps of 10 šár gur. A partial excerpt of this section is MS 2704 (Fig. 3.1.2, bottom).

The brief final section of MS 2723 contains only two lines for 1 šár.gal gur (1 00 00 00 gur) and 2 šár.gal gur, written as šár.1.gal gur and šár.2.gal gur, respectively. Thus, the metrological table stops at

$$2 \text{ šár.gal gur} = 2 \,(00\ 00\ 00) \text{ gur} = 2 \,(00\ 00\ 00) \cdot 5 \,(00) \text{ sìla} = 10 \,(00\ 00\ 00\ 00) \text{ sìla.}$$

Five of the six faces of the 6-sided prism, and the upper half of the sixth face, are occupied by the metrological table for system C. The second half of the sixth face contains an interesting subscript in 8 lines (discussed below in Appendix 5, Sec. A5 g).

The analysis above of the construction of the complete metrological table for the Old Babylonian system C clearly shows that *the table, despite its apparent complexity, is exhaustive and well organized, with a repetitive structure*. Note, by the way that all the values in the right column are *at most 2-place sexagesimal numbers*, with never more than 3 digits!

MS 2723 is the *most comprehensive* metrological table for system C found so far, ranging from '1 shekel of barley = 1 ($\cdot 60^{-1}$ sìla)' to '$2 \cdot 60^3$ gur = 10 ($\cdot 60^4$ sìla)', in 159 lines. The only other known such table, sufficiently well preserved to allow a comparison, is the first section of the combined metrological table **YBC 4633** (photo in Nemet-Nejat, *JNES* 54 (1995), Fig. 8). The range of the table for system C in YBS 4633 is from '1/3 sìla = 20 ($\cdot 60^{-1}$ sìla)' to '$2 \cdot 60^3$ gur = 10 ($\cdot 60^4$ sìla)', in 140 lines.

The range of MS 2723 can also be compared with that of **VAT 2596** (Meissner, *Beiträge* (1893), 58), which ranges from '1/3 sìla of barley' to '1 · 60^3 gur', in 134 lines. However, since VAT 2596 records only the capacity numbers, not their sexagesimal values as multiples of the basic unit, VAT 2596 is only a "metrological list" for system *C*, not a metrological table. VAT 2596 is a large clay cylinder with 8 narrow columns of text.

CBM 10990+ (Hilprecht, *BE 20/1* (1906), text 29), *obv.* cols. *i-iv*, is another metrological list for system *C*, the first section of a combined metrological list. It ranges from […] to šár.gal gur (1 · 60^3 gur), immediately followed by šár.gal.šu.nu.tag gur 'the great šár the hand cannot touch' (presumably meaning 1 · 60^4 gur). CBM 10990+ is a great fragment of a large clay tablet with originally probably six columns on the obverse and seven on the reverse. (See Appendix 5, Fig. A5.5-6.)

3.2. Old Babylonian Weight Measures. System *M*

In Sumerian and Old Babylonian cuneiform texts, amounts of more or less precious *metals*, like gold, silver, and copper were measured by weight, with the corresponding "weight numbers" expressed in what may be called "system *M*". The units of this system were:

še	*uṭṭatu*	grain, barley-corn	c. 45 mg
gín	*šiqlu*	shekel	c. 8 g
ma.na	*manû*	mina	c. 500 g
gú	*biltu*	load, talent	c. 30 kg

Also amounts of certain other substances, such as bricks, wool and dates were measured in minas. The smallest weight unit, the barley-corn, could, of course, be used in a meaningful way only for gold and silver. Even then, it cannot have been practicable to weigh smaller quantities than 22 1/2 barley-corns = 1/8 shekel, or perhaps 10 barley-corns (c. 0.4 g). In the not infrequent cases when smaller quantities were recorded, they must have been *computed for the purpose of bookkeeping*, not weighed.[1]

Note: In MS 5088/1-55, a hoard of stone weights, the smallest weight is 10 barley-corns. See Sec. 4 e

It is a great simplification of the actual situation to say that the Sumerian or Old Babylonian mina was equal to about 500 g or 1/2 kg. This is not the proper place to confront the considerable complexities involved in a thorough discussion of Mesopotamian weights. The interested reader is referred instead to Powell's *RlA* article "Masse und Gewichte" (1990) Sec. V. Anyway, whatever the size of an Old Babylonian mina was in a given situation, the barley-corn, the shekel, the mina, and the talent always had the same *relative sizes*. Thus, it was always true that

1 talent = 1 00 (60) minas, 1 mina = 1 00 (60) shekels, and 1 shekel = 3 00 (180) barley-corns.

This series of relations can be expressed more concisely by use of the factor diagram for system *M*:

$$M\text{(S/OB)} \quad : \quad \underset{\underset{(c.\ 30\ kg)}{biltu}}{\underset{g\acute{u}}{talent}} \xleftarrow{\ 60\ } \underset{\underset{(c.\ 1/2\ kg)}{man\hat{u}}}{\underset{\text{ma.na}}{mina}} \xleftarrow{\ 60\ } \underset{\underset{(c.\ 8\ g)}{\check{s}iqlu}}{\underset{\text{gín}}{shekel}} \xleftarrow{\ 180\ } \underset{\underset{(c.\ 45\ mg)}{u\dot{t}\dot{t}atu}}{\underset{\text{še}}{barley\text{-}corn}}$$

Multiples of the barley-corn (from 1 to 29 barley-corns), of the shekel (from 1 to 19 shekels), and of the mina (from 1 to 59 minas), were written by use of standard number signs for sexagesimal numbers, followed by the names of the units. Larger multiples of the barley-corn were expressed as basic fractions of a shekel, and larger multiples of the shekel as basic fractions of a mina. Multiples of the talent, finally, were expressed by use of standard number signs for Sumerian non-positional sexagesimal numbers, followed by the name of the

---

1. Actually, silver was the main value indicator in the Old Babylonian period, although sometimes barley could be used as a substitute for silver, normally at a round exchange rate of '1 gur of barley for 1 shekel of silver'. In Old Babylonian mathematical cuneiform texts, the fixed daily pay for a 'hired worker' was *6 barley-corns of silver or 1 bán of barley*. The corresponding monthly pay was *1 shekel of silver or 1 gur of barley*.

unit, with the slight modification that the cuneiform number signs for from 1 and up to 9 talents were written with *horizontal* rather than vertical wedges.

There are 6 clay tablets with metrological tables for system *M* in the Schøyen Collection. They are all shown in Figs. 3.2.1-3.2.3 below and discussed individually in the text. The six tablets together fairly well cover the whole range of system *M*. However, there is no example of a complete metrological table for system *M* in the Schøyen Collection.

The four tablets displayed in Figs. 3.2.1-3.2.2 cover in various ways multiples of the barley-corn and fractions of the shekel. Then there is a small gap, since no tablet in the collection covers multiples of the shekel and fractions or small multiples of the mina. In Fig. 3.2.3, the first tablet covers multiples of the mina and multiples of the talent, up to 1 00 (60) talents. A few examples of higher multiples of the talent are considered in the second tablet in Fig. 3.2.3.

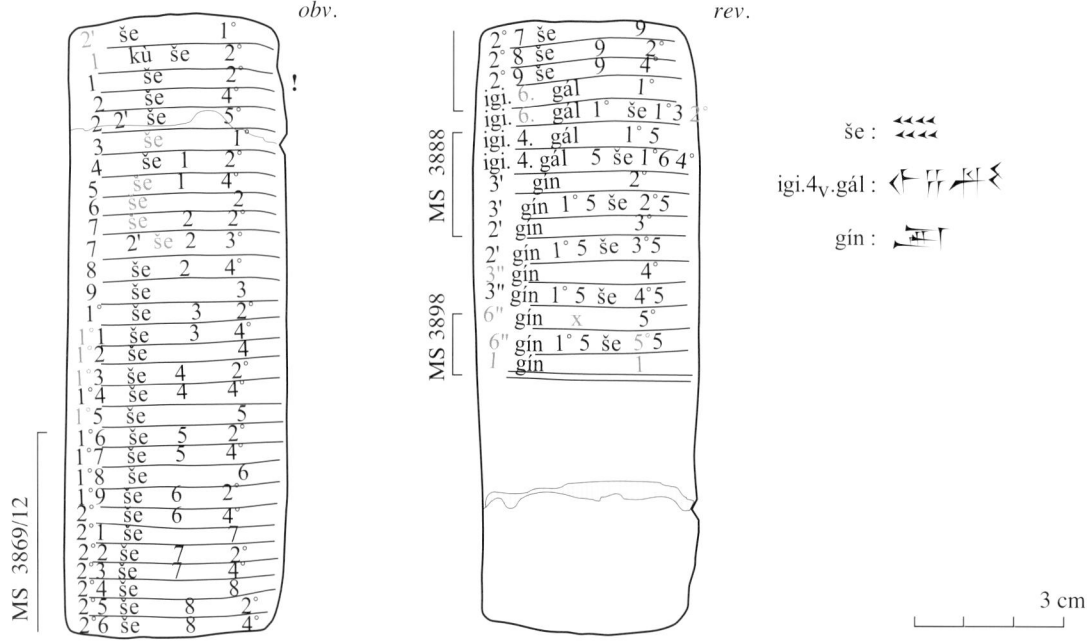

MS 2186. A metrological table for system *M*, basic unit mina (46 lines),
from '1/2 barley-corn = 10 ($\cdot$ 60^{-3} mina)' to '1 shekel = 1 ($\cdot$ 60^{-1} mina)'.

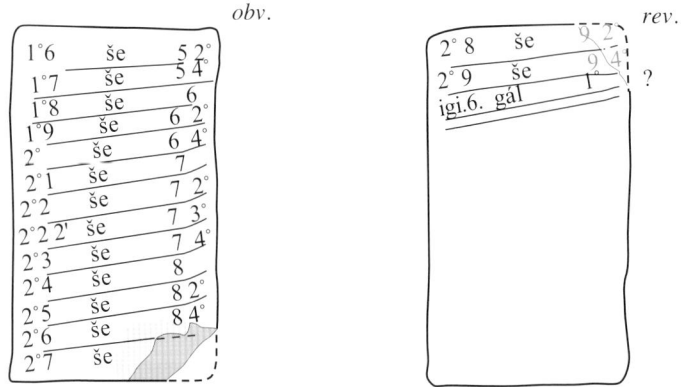

MS 3869/12. A metrological table for system *M*, basic unit mina (16 lines),
from '16 barley-corns = 5 20 ($\cdot$ 60^{-3} mina)' to '1/6 (shekel) = 10 ($\cdot$ 60^{-2} mina)'.

Fig. 3.2.1. Two Old Babylonian metrological tables for the lower range of system *M*.

MS 2186 (Fig. 3.2.1, top) is a relatively extensive metrological table for the lower range of system *M*. As in all metrological tables for system *M*, *the basic unit is the mina.*

The table begins with 1/2 barley-corn = $10 \cdot 60^{-2}$ shekel and proceeds in five steps of 1/2 barley-corn up to 3 barley-corns = $1 \cdot 60^{-1}$ shekel. The table continues from 1 to 29 barley-corns in steps of 1 barley-corn. After 29 barley-corns, the table proceeds in three steps of 10 barley-corns from '1/6', meaning '1/6 shekel' (= 30 barley-corns) to 1/3 shekel (= 60 barley-corns). There is an interpolated entry for '1/4', meaning '1/4 shekel'. The table then continues from 1/3 shekel to 1 shekel, in eight steps of 15 barley-corns. Note that in this way the table comes to include entries for all the basic fractions of the shekel (1/3, 1/2, 2/3, and 5/6 shekel)

MS 2186

1/2	še	=	10	($\cdot\ 60^{-3}$ mina)	
1 kù	še	=	20	($\cdot\ 60^{-3}$ mina)	
1 1/2	še	=	30	($\cdot\ 60^{-3}$ mina)	an error in the text is corrected here
2	še	=	40	($\cdot\ 60^{-3}$ mina)	
2 1/2	še	=	50	($\cdot\ 60^{-3}$ mina)	
3	še	=	1	($\cdot\ 60^{-2}$ mina)	
4	še	=	1 20	($\cdot\ 60^{-3}$ mina)	
5	še	=	1 40	($\cdot\ 60^{-3}$ mina)	
6	še	=	2	($\cdot\ 60^{-2}$ mina)	
7	še	=	2 20	($\cdot\ 60^{-3}$ mina)	
7 1/2	še	=	2 30	($\cdot\ 60^{-3}$ mina)	1/24 shekel
8	še	=	2 40	($\cdot\ 60^{-3}$ mina)	
9	še	=	3	($\cdot\ 60^{-2}$ mina)	
10	še	=	3 20	($\cdot\ 60^{-3}$ mina)	
11	še	=	3 40	($\cdot\ 60^{-3}$ mina)	
12	še	=	4	($\cdot\ 60^{-2}$ mina)	
13	še	=	4 20	($\cdot\ 60^{-3}$ mina)	
14	še	=	4 40	($\cdot\ 60^{-3}$ mina)	
15	še	=	5	($\cdot\ 60^{-2}$ mina)	
16	še	=	5 20	($\cdot\ 60^{-3}$ mina)	
17	še	=	5 40	($\cdot\ 60^{-3}$ mina)	
18	še	=	6	($\cdot\ 60^{-2}$ mina)	
19	še	=	6 20	($\cdot\ 60^{-3}$ mina)	
20	še	=	6 40	($\cdot\ 60^{-3}$ mina)	
21	še	=	7	($\cdot\ 60^{-2}$ mina)	
22	še	=	7 20	($\cdot\ 60^{-3}$ mina)	
23	še	=	7 40	($\cdot\ 60^{-3}$ mina)	
24	še	=	8	($\cdot\ 60^{-2}$ mina)	
25	še	=	8 20	($\cdot\ 60^{-3}$ mina)	
26	še	=	8 40	($\cdot\ 60^{-3}$ mina)	
27	še	=	9	($\cdot\ 60^{-2}$ mina)	
28	še	=	9 20	($\cdot\ 60^{-3}$ mina)	
29	še	=	9 40	($\cdot\ 60^{-3}$ mina)	
1/6 (gín)		=	10	($\cdot\ 60^{-2}$ mina)	30 b.c.
1/6 10 še		=	13 20	($\cdot\ 60^{-3}$ mina)	40 b.c.
1/4 (gín)		=	15	($\cdot\ 60^{-2}$ mina)	45 b.c.
1/4 (gín) 5 še		=	16 40	($\cdot\ 60^{-3}$ mina)	50 b.c.
1/3 gín		=	20	($\cdot\ 60^{-2}$ mina)	1 00 b.c.
1/3 gín 15 še		=	25	($\cdot\ 60^{-3}$ mina)	1 15 b.c.
1/2 gín		=	30	($\cdot\ 60^{-2}$ mina)	1 30 b.c.
1/2 gín 15 še		=	35	($\cdot\ 60^{-3}$ mina)	1 45 b.c.
2/3 gín		=	40	($\cdot\ 60^{-2}$ mina)	2 00 b.c.
2/3 gín 15 še		=	45	($\cdot\ 60^{-3}$ mina)	2 15 b.c.
5/6 gín		=	50	($\cdot\ 60^{-2}$ mina)	2 30 b.c.
5/6 gín 15 še		=	55	($\cdot\ 60^{-3}$ mina)	2 45 b.c.
1 gín		=	1	($\cdot\ 60^{-1}$ mina)	3 00 b.c.

In the barley-corn section of the table there is an interpolated line stating '7 1/2 barley-corn 2 30'. A likely reason for this interpolation is that 7 1/2 barley-corn = 1/24 shekel. (In Late Babylonian economical texts, and also in the Late Babylonian metrological tables CBS 11032 and CBS 11019, 1/24 shekel was a standard fraction of the shekel with the name *girû*, meaning 'carat' or 'bean of carob'. See Friberg, *GMS 3* (1993), 389-390.)

There is an error in line 3 of the table, which says '1 barley-corn   20' instead of the expected '1 1/2 barley-corn   30'. The sign kù in line 2 is an abbreviation for the word kù.babbar 'silver'.

MS 3869/12 (Fig. 3.2.1, bottom) is a small clay tablet with only 16 lines of text, leaving most of the reverse empty. It is inscribed with a metrological table for system *M*, proceeding from 16 barley-corns to 1/6 (shekel) = 30 barley-corns, in steps of 1 barley-corn. This brief table text is obviously identical with a small section of the larger table text MS 2186, as indicated in the conform transliteration of MS 2186 in Fig. 3.2.1.

MS 3888 (Fig. 3.2.2) is a small clay tablet with an even briefer metrological table:

1/4	(gín)	=	15	(· 60⁻² mina)	45 b.c.
1/4	(gín) 5 še	=	16 40	(· 60⁻³ mina)	50 b.c.
1/4	10 še	=	18 20	(· 60⁻³ mina)	55 b.c.
1/3	gín	=	20	(· 60⁻² mina)	1 00 b.c.
1/3	gín 5 še	=	21 40	(· 60⁻³ mina)	1 05 b.c.
1/3	gín 10 še	=	23 20	(· 60⁻³ mina)	1 10 b.c.
1/3	gín 15 še	=	25	(· 60⁻² mina)	1 15 b.c.
1/3	gín 20 še	=	26 40	(· 60⁻³ mina)	1 20 b.c.
1/3	gín 25 še	=	28 20	(· 60⁻³ mina)	1 25 b.c.
1/2	gín	=	30	(· 60⁻² mina)	1 30 b.c.

The table proceeds from 1/4 shekel to 1/2 shekel in 9 steps of 5 barley-corns. It is clearly not identical with the corresponding section of the larger metrological text MS 2186, which contains only 5 entries between 1/4 and 1/2 shekel.

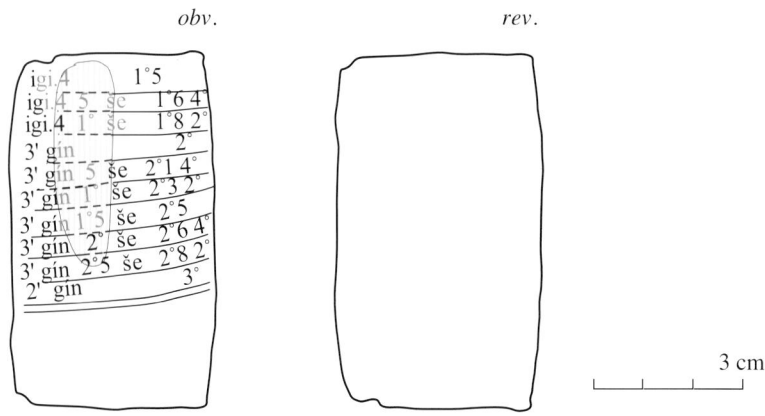

obv.   *rev.*

3 cm

MS 3888. A metrological table for system *M*, basic unit mina (10 lines),
from '1/4 (shekel) = 15 (· 60⁻² mina)' to '1/2 shekel = 30 (· 60⁻² mina)'.

Fig. 3.2.2. MS 3888. A very brief Old Babylonian metrological table for the lower range of system *M*.

MS 3898 (Fig. 3.2.3, top) is a large fragment of a clay tablet with a metrological table of 30 lines for system *M*. It proceeds simplemindedly in steps of 1 barley-corn from 5/6 shekel to 5/6 shekel 29 barley-corns, so that it stops just before it reaches 1 shekel (= 5/6 shekel 30 barley-corns). The corresponding section of the larger table on MS 2186 contains only 2 entries. Clearly, then, MS 3898 is not identical with a sub-section of MS 2186.

obv.   *rev.*

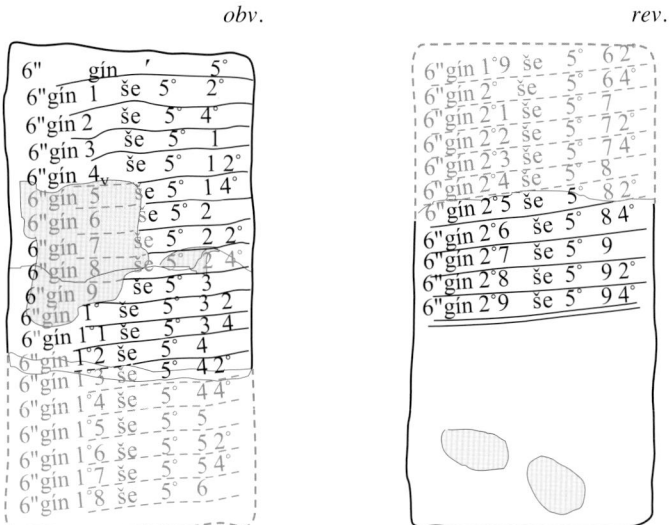

MS 3898. A metrological table for system *M*, basic unit mina (30 lines),
from '5/6 shekel = 50 (· 60^{-2} mina)' to '5/6 shekel 29 barley-corns = 59 40 (· 60^{-3} mina)'.

obv.   *rev.*

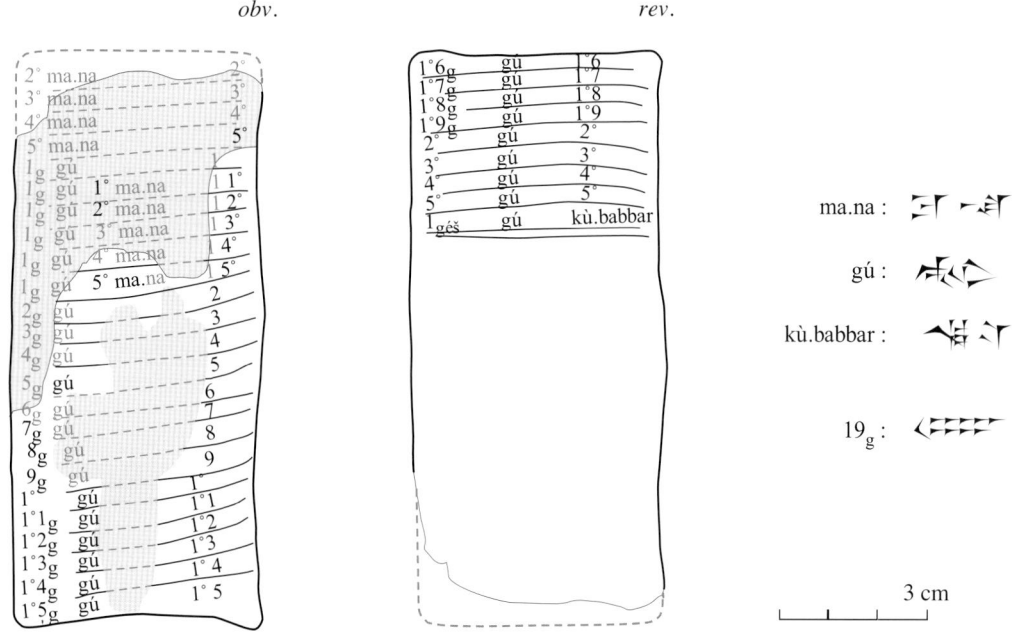

MS 3869/1. A metrological table for system *M*, basic unit mina (33 lines),
(probably) from '20 minas = 20 (minas)' to '1 (00) talents = 1 (· 60^2 minas)'.

Fig. 3.2.3. Two Old Babylonian metrological tables for the lower and upper ranges of system *M*.

MS 3869/1 (Fig. 3.2.3, bottom) is a clay tablet with a damaged obverse, on which is inscribed a metrological table for system *M*. According to the most plausible reconstruction of the damaged part of the text, the table begins with 20 minas and proceeds in 18 steps of 10 minas to 2 talents. It then continues (imitating the structure of a single multiplication table) in 18 steps of 1 talent to 20 talents, and from there in steps of 10 talents to 1 (00) talents. The last line of the table is incorrectly written, with omission of the sexagesimal value '1', as

1(géš) talents   silver

MS 3869/1

20 minas		=	20	(· 1 mina)		9 talents	=	9	(· 60 minas)
30 minas		=	30	(· 1 mina)		10 talents	=	10	(· 60 minas)
40 minas		=	40	(· 1 mina)		11 talents	=	11	(· 60 minas)
50 minas		=	50	(· 1 mina)		12 talents	=	12	(· 60 minas)
1 talent		=	1	(· 60 minas)		13 talents	=	13	(· 60 minas)
1 talent	10 minas	=	1 10	(· 1 mina)		14 talents	=	14	(· 60 minas)
1 talent	20 minas	=	1 20	(· 1 mina)		15 talents	=	15	(· 60 minas)
1 talent	30 minas	=	1 30	(· 1 mina)		16 talents	=	16	(· 60 minas)
1 talent	40 minas	=	1 40	(· 1 mina)		17 talents	=	17	(· 60 minas)
1 talent	50 minas	=	1 50	(· 1 mina)		18 talents	=	18	(· 60 minas)
2 talents		=	2	(· 60 minas)		19 talents	=	19	(· 60 minas)
3 talents		=	3	(· 60 minas)		20 talents	=	20	(· 60 minas)
4 talents		=	4	(· 60 minas)		30 talents	=	30	(· 60 minas)
5 talents		=	5	(· 60 minas)		40 talents	=	40	(· 60 minas)
6 talents		=	6	(· 60 minas)		50 talents	=	50	(· 60 minas)
7 talents		=	7	(· 60 minas)		1(géš) talents	=	<1>	(· 60^2 minas)
8 talents		=	8	(· 60 minas)					

NBC 2513 (see Fig. A.5.1 in Appendix 5) is an Old Babylonian 6-sided prism with a complete metrological table for system *M*. (There is no complete metrological table for system *M* in the Schøyen Collection.) This complete table goes from 1/2 barley-corn to 1 (00) talents, a circumstance suggesting that *MS 3869/1 is an excerpt from the very end of a complete metrological table*. The last line of the table on NBC 2513 is written as

1_g gú 1 kú.babbar   1 talent (equals) 1 silver.

Apparently, there is an error in this line, namely that the number sign preceding the gur sign is a horizontal wedge, normally used to indicate 1 talent, rather than the vertical wedge normally used to indicate sixty talents. On the other hand, the number sign 1 in front of the word kù.babbar 'silver' is the sexagesimal value of sixty talents, equated with a multiple of the mina, presumably the basic unit for system *M*. As stated above, this sexagesimal value is missing in the last line of MS 3869/1.

A comparison of the (probably) complete metrological table for system *M* exemplified by the large text NBC 2513 with the diverse sections of the complete table on the small tablets MS 2186, 3869/12, 3888, 3898, and 3869/1 in Fig. 3.2.1-3.2.3 gives the somewhat surprising result that overlapping parts of these various texts do not always coincide.(There was no corresponding difficulty in the case of overlapping metrological tables for system *C* in Sec. 3.1 above.) One possible explanation is that NBC 2513 is an *abbreviated version* of a hypothetical canonical complete metrological table for system *M*, while another equally possible explanation is that MS 2186, MS 3888, and MS 3898 are *expanded excerpts* from the complete table.

Sub-tables for system *M* in two previously published Old Babylonian combined metrological tables extend like NBC 2513 all the way to the uppermost range of the system. The two sub-tables in question are:

2N-T 530 = **UM 55-21-78** (Neugebauer and Sachs, *JCS* 36), section M   from [......]   to   1(géš) talents

YBC 4633 (Nemet-Nejat, *JNES* 54), section M   from 1/2 b.c.   to   [1(géš) talents]

Note that the last line of the sub-table for system *M* in UM 55-21-78 is deficient in the same way as the last line of the text MS 3869/1, in that it omits mention of the sexagesimal value '1'.

MLC 1854 (Clay, *BRM 4*, text 41) is an Old Babylonian complete metrological *list* (lacking the right column with sexagesimal values) for system *M*, proceeding from [1/2 b.c.] silver, 1 b.c., *etc.* to 50 talents, 1(géš) talents silver. The text may be early Old Babylonian, since variant number signs are used in it for 4, 7, 8, and 40. The last line of the list, written on the lower edge of the tablet, is

1 gú kú.babbar   1 talent (of) silver.

Maybe this last line in the complete metrological *list* for system was the model for the deficient last lines in the metrological *tables* MS 3869/1 and UM 55-21-78. If so, then this circumstance suggests that *maybe metrological lists were the predecessors of metrological tables*. This is a rather reasonable conjecture. Indeed, the extensive use of sexagesimal numbers in place value notation for the second column in the metrological tables indicates that the Old Babylonian type of metrological tables were invented in connection with the invention of place value numbers, late in the Ur III period or early in the Old Babylonian period. Metrological lists, on the other hand, do not have a second column and therefore do not need to make use of sexagesimal numbers in place value notation.

CBM 10990 (Hilprecht, *BE 20/1*, text 29) is a combined metrological list, in which the sub-table for system *M* extends from 1 b.c. all the way to 1 šár.gal gú (1 00 00 talents or c. 100,000 kg), that is beyond the range of the four metrological tables MS 3869/1, NBC 2513, UM 55-21-78, and YBC 4 633, and also beyond the range of the metrological list MLC 1854.

MS 3925 (Fig. 3.2.4 below) is an Old Babylonian "lentil", a round and fairly thick hand tablet of the kind that was used for young students' assignments. It is inscribed on the obverse with four lines of what looks like a metrological table for system *M*, and on the reverse with a second copy of those four lines. The obverse and reverse are apparently written in two different hands, as for instance the sign for kù 'metal, silver' on the reverse is of an older, more elaborate type than the corresponding sign on the obverse. This may be an example of an exercise of the not uncommon kind where a teacher's model on the obverse was copied by a student on the reverse.

As a brief Old Babylonian metrological table for system *M*, this text is atypical in several ways. It is written on a lentil, rather than on a rectangular clay tablet of the usual format, and it repeats the word 'silver' in each line of the table, after the sexagesimal values in the right column. Furthermore, the metrological table itself is unusually brief, and goes beyond the presumed range of the complete metrological table for system *M*.

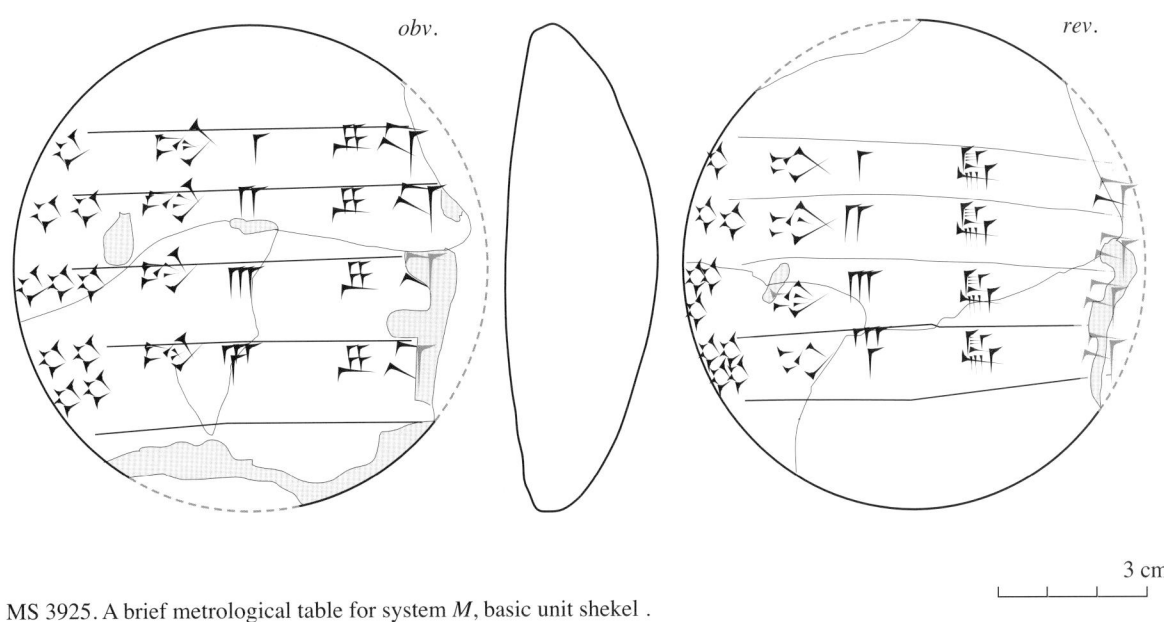

3 cm

MS 3925. A brief metrological table for system *M*, basic unit shekel .
    The 4 lines on the obverse are repeated on the reverse, in a different hand.

Fig. 3.2.4. An Old Babylonian exercise text in the form of a non-standard metrological table for system *M*.

Here follows a transliteration and translation of the four lines on the obverse of MS 3925:[2]

1(šár) gú	1	kù.babbar
2(šár) gú	2	kù.babbar
3(šár) gú	3	kù.babbar
4(šár) gú	4	kù.babbar

$1 \cdot 60^2$ talents	(equals)	$1 \ (\cdot \ 60^3$ minas of) silver
$2 \cdot 60^2$ talents	(equals)	$2 \ (\cdot \ 60^3$ minas of) silver
$3 \cdot 60^2$ talents	(equals)	$3 \ (\cdot \ 60^3$ minas of) silver
$4 \cdot 60^2$ talents	(equals)	$4 \ (\cdot \ 60^3$ minas of) silver

If this is a text with a teacher's model and a student's copy, then the question is: who wrote what? It appears that the signs on the reverse are somewhat neater than the signs on the obverse. In addition, the signs kù on the reverse have an early Old Babylonian form, while the corresponding signs on the obverse have a simplified, presumably later form. Therefore, maybe, the teacher's model was inscribed on the reverse of the clay tablet

3.3. Old Babylonian Area Measures. System *A*

In Sumerian and Old Babylonian cuneiform texts, the extent of surfaces of cultivated fields, gardens, court-yards, foils of silver, *etc.*, were measured in terms of area measure, with the corresponding "area numbers" expressed in what may be called "system *A*". The units of this system were:

gín	*šiqlu*	shekel	1/60 šar
šar	*mūšaru*	garden-plot	1 square ninda
iku	*ikū*	field	100 šar
èše	*eblu*	rope, string	6 iku
bùr	*būru*	?	3 èše
bùr'u	?		10 bùr
šár?	?		1 00 bùr
šár'u?	?		10 00 bùr
šár.gal?	?		1 00 00 bùr

The medium sized units of this system, the iku, the èše, and the bùr, had been in constant use in Mesopotamia at least since the proto-literate period around the end of the fourth millennium BC, and the remaining units since the Old Sumerian period towards the middle of the third millennium. The relations between consecutive units were always the same. They can best be described by use of the factor diagram for system *A*:

							rope	field	garden

A(S/OB) : šár.gal? $\longleftarrow$ šár'u? $\longleftarrow$ šár? $\overset{6}{\longleftarrow}$ bùr'u $\overset{10}{\longleftarrow}$ bùr $\overset{3}{\longleftarrow}$ èše $\overset{6}{\longleftarrow}$ iku $\overset{100}{\longleftarrow}$ šar $\longleftarrow$

				būru	*eblu*	*ikû*	*mūšaru*
1 00 00 b.	10 00 b.	1 00 b.	10 b.	18 i.	6 i.	1 40 sq. n.	1 sq. n.
(c. 230 km²)		(c. 4 km²)				(3,600 m²)	(36 m²)

It is not known what the Sumerian and Akkadian names were for the area units written the same way as the sexagesimal units šár, šár'u, and šár.gal. The Akkadian counterpart to the Sumerian bùr'u is also unknown.

A sign-oriented factor diagram can be used to display the special number signs used to write Old Babylonian area numbers (cf. Fig. A4.10 in Appendix 4):

A(OB) :

šár.gal	šár'u	šár	bùr'u	bùr	èše	iku	ubu

Note that there were special number signs for '1/2(iku)' and '1/4(iku)'.

---

2. The four weight numbers are of quite unrealistic sizes. Thus, for instance, 1 šár talents = 3,600 · 30 kilograms = more than 100,000 kilograms. (Remember that the main application of system *M* was to write out weight numbers for precious metals like silver and copper.)

Here follow the rules for forming multiples of the area units: There were special cuneiform number signs for the capacity numbers, from $1_i = 1(iku)$ to $5_i = 5(iku)$ and from $1_e = 1(èše)$ to $2_e = 2(èše)$. The way in which number signs for multiples of the iku were formed by horizontal wedges is demonstrated by the examples in **MS 2735**, lines 6-10 (Fig. 3.3.1, top). The sign for '2 èše' was formed as two copies of the sign for $1(èše)$.

Multiples of the bùr, from $1_b = 1(bùr)$ to $9_b = 9(bùr)$, were written as 1 to 9 copies of the sign 1_b, looking exactly like the sign $1°$ for the sexagesimal unit '10'. In a similar way, the signs $1°_b = 1(bùr'u)$, $1_š = 1(šár)$, and $1°_š = 1(šár'u)$, were repeated as many times as needed, in order to write multiples of the corresponding area units. For examples, see **MS 2768** (Fig. 3.3.1, bottom). Multiples of the šár.gal were never written with more than one gal sign.

There are only 2 clay tablets with metrological tables for system A in the Schøyen Collection, those shown in Figs. 3.3.1 below, and no example of a complete metrological table for system A.

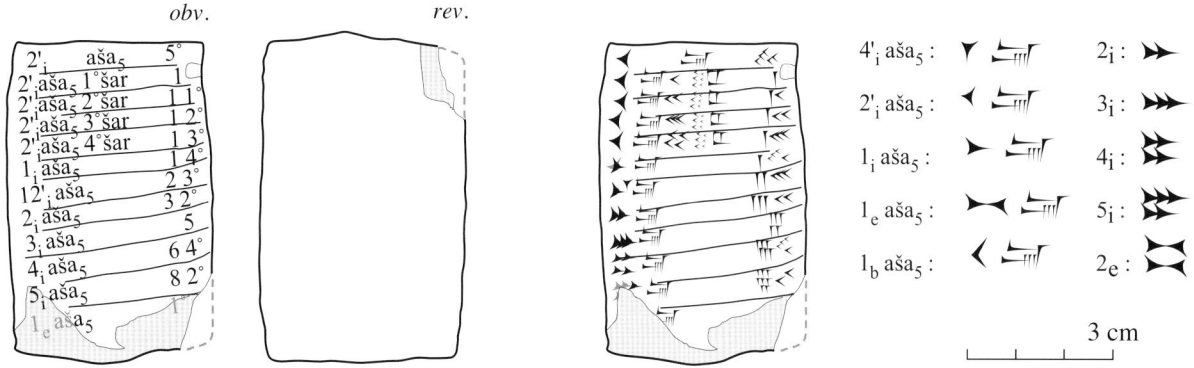

MS 2735. Metrological table for system A, basic unit šar = sq. ninda (12 lines),
from '1/2 iku = 50 (sq. ninda)' to '1 èše = 10 (· 60 sq. ninda)'.

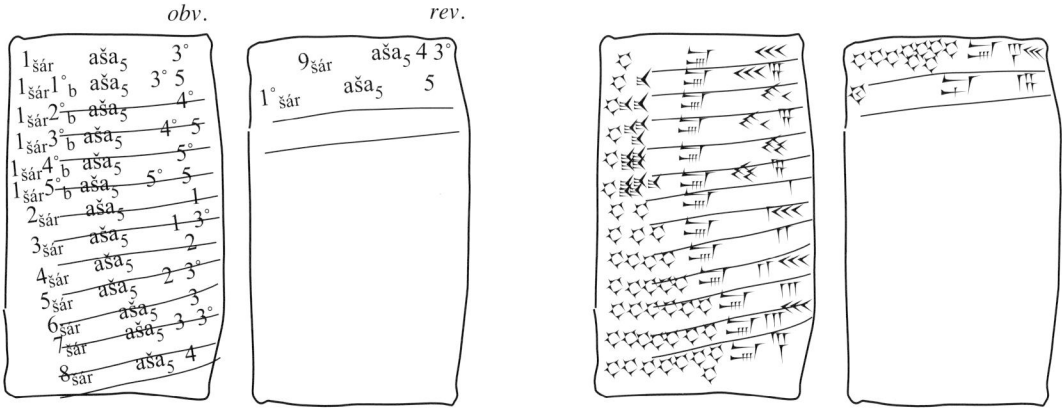

MS 2768. Metrological table for system A, basic unit šar (15 lines),
from '1 šár = 30 (· 60^2 sq. ninda)' to '10 šár = 5 (· 60^3 sq. ninda)'.

Fig. 3.3.1. One Old Babylonian metrological table for the lower, another for the upper range of system A.

The only known example of a complete metrological table for the Old Babylonian system A is **A 21948**, a clay tablet from Ishchali, near Baghdad (Greengus, *PIHANS 44* (1979); see the conform transliteration in Appendix 5, Fig. A5.2, which is followed by a standard transliteration of the table).

A badly damaged complete section for system A in the combined metrological *list* CBM 10990+ (Hilprecht, *BE 20/1* (1906), text 29, *rev.* col. iv; see Appendix 5, Fig. A5.6) ends with the two entries šár×1.gal aša$_5$, šár×1.gal šu.nu.tag aša$_5$, presumably meaning 10^2 bùr and 10^3 bùr.

3.4. Old Babylonian Length Measures. Systems *Ln* and *Lc*

In Sumerian and Old Babylonian cuneiform texts, linear dimensions (lengths, widths, and heights) are measured in terms of length measure, with the corresponding "length numbers" expressed in what may be called "system *L*". The following units of this system frequently appear in Old Babylonian mathematical cuneiform texts and Old Babylonian metrological tables:

šu.si	*ubānu*	finger	1/30 cubit	c.17 mm
kùš	*ammatu*	cubit, ell	1/12 ninda	c. 0.5 m
ninda	*nindanu*	rod?		c. 6 m
uš	*šiddu?*	length?	60 ninda	c. 360 m
danna	*bēru*	stage?	30 uš	c. 10.8 km

The ninda, the central unit of the system, had been in almost constant use since at least the proto-literate period, around the end of the fourth millennium BC.

The relations between consecutive units of system *L* can be described by use of the following factor diagram, with the ninda as the basic unit:

$$
\begin{array}{ccccccccc}
 & & \text{length?} & & \text{rod?} & & \text{cubit} & & \text{finger} \\
Ln(\text{OB}): & \text{danna} \xleftarrow{30} & \text{uš} & \xleftarrow{60} & \text{ninda} & \xleftarrow{12} & \text{kùš} & \xleftarrow{30} & \text{šu.si} \\
 & \textit{bēru} & ? & & \textit{nindanu} & & \textit{ammatu} & & \textit{ubānu} \\
 & 30\ 00\ \text{n.} & 1\ 00\ \text{n.} & & (\text{c. 6 m}) & & ;05\ \text{n.} & & ;00\ 10\ \text{n.}
\end{array}
$$

Other units of the system, appearing occasionally in Old Babylonian mathematical texts but never in OB metrological tables, are the gi 'reed' = 6 cubits or 1/2 ninda, and the éš(e) 'rope' = 60 ninda.

Multiples of the length units were written as ordinary sexagesimal numbers followed by the names of the units. Since there are no multiples of a length unit larger than 59 times the unit, no need arises for the use of any non-positional sexagesimal number signs, such as geš'u and šár.

There are five metrological tables for system *L* in the Schøyen Collection. Three of them are non-overlapping and will be presented below, in Figs. 3.4.1-2 and in the accompanying discussion.

Actually, two of the tables cover precisely the same range of length measures, but have *different basic units,* in one case the ninda, in the other case the cubit, and can therefore not be said to be overlapping. The reason why there exist metrological tables for the Old Babylonian system *L* with two different basic units is well known. In both Sumerian and Old Babylonian administrative or mathematical texts, horizontal dimensions (lengths and widths) were preferentially measured in multiples of the ninda, while vertical dimensions (heights or depths) were measured in multiples of the cubit. This practice makes sense, since in practical applications vertical dimensions of three-dimensional objects (walls, canals, *etc.*) are normally much smaller than horizontal dimensions.

To emphasize the difference between metrological tables for system *L* with one or the other basic unit, it is convenient to distinguish between "metrological tables for system *Ln*", with the ninda as the basic unit, and "metrological tables for system *Lc*" with the cubit as the basic unit.

MS 3869/11 (Fig. 3.4.1, top) is a metrological table for the lower range of system *Ln*. In its initial sub-section, it proceeds from 1 to 9 fingers in steps of 1 finger. The next sub-section is devoted to basic fractions of the cubit (1/3, 1/2, and 2/3 cubit), followed by 1 cubit and 1 cubit plus the basic fractions of the cubit. The table continues from 2 cubits to 1 ninda (= 12 cubits) in ten steps of 1 cubit, though preferring to write 1/2 ninda instead of 6 cubits. The final sub-section extends from 1 to 5 ninda in eight steps of 1/2 ninda.

MS 2705 (Fig. 3.4.1, bottom) is a brief metrological table for the upper range of system *Ln*. It proceeds from 1 uš (= 60 ninda) to 2 uš in steps of 10 ninda, then to 1/3 danna (= 10 uš) in steps of 1 uš, and ends with the basic fractions of the danna (1/3, 1/2, 2/3 danna) and finally the danna itself.

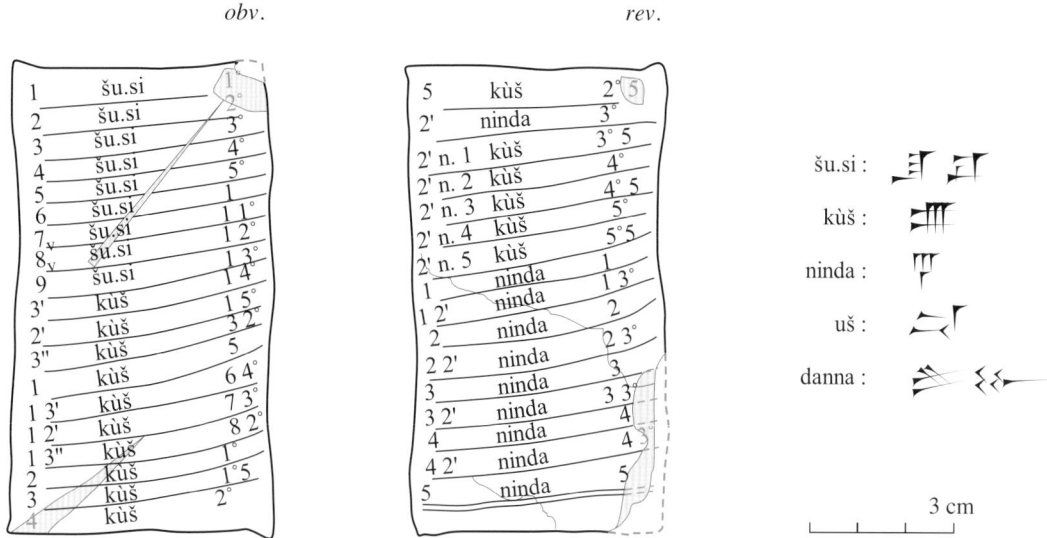

MS 3869/11. Metrological table for the OB system *Ln* (35 lines),
from '1 finger = 10 ($\cdot 60^{-2}$ ninda)' to '5 ninda = 5 ($\cdot$ 1 ninda)'.

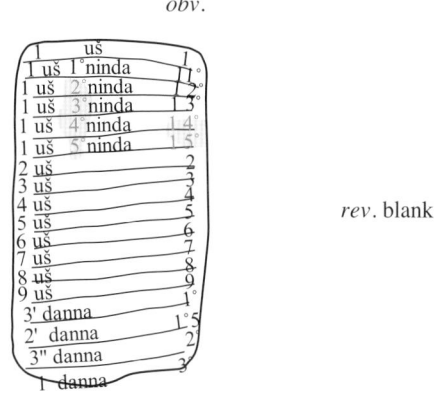

MS 2705. Metrological table for the OB system *Ln* (18 lines),
from '1 uš = 1 ($\cdot$ 60 ninda)' to '1 danna = 30 ($\cdot$ 60 ninda)'.

Fig. 3.4.1. Two Old Babylonian metrological tables for length measures, basic unit ninda (system *Ln*).

MS 2896 (Fig. 3.4.2) is a metrological table for system *Lc* (basic unit the cubit), which proceeds from 1 finger to 5 ninda in precisely the same steps as MS 3869/11 (basic unit the ninda).

The differences between the two tables show up clearly in the parallel transcriptions below:

MS 3869/11: System *Ln*

1	finger	=	10	($\cdot 60^{-2}$ ninda)
2	fingers	=	20	($\cdot 60^{-2}$ ninda)
...	...	...	...	
1/3	cubit	=	1 40	($\cdot 60^{-2}$ ninda)
1/2	cubit	=	2 30	($\cdot 60^{-2}$ ninda)
2/3	cubit	=	3 20	($\cdot 60^{-2}$ ninda)
1	cubit	=	5	($\cdot 60^{-1}$ ninda)

MS 2896: System *Lc*

1	finger	=	2	($\cdot 60^{-1}$ cubit)
2	fingers	=	4	($\cdot 60^{-1}$ cubit)
...	...	...	...	
1/3	cubit	=	20	($\cdot 60^{-1}$ cubit)
1/2	cubit	=	30	($\cdot 60^{-1}$ cubit)
2/3	cubit	=	40	($\cdot 60^{-1}$ cubit)
1	cubit	=	1	($\cdot$ 1 cubit)

Continued

```
... ... ...         ...                          ... ... ...         ...
1/2   ninda         =   30  (· 60⁻¹ ninda)       1/2   ninda         =    6  (· 1 cubit)
1/2   ninda 1 cubit =   35  (· 60⁻¹ ninda)       1/2   ninda 1 cubit =    7  (· 1 cubit)
...   ...           ...   ...                     ...   ...           ...   ...
1     ninda         =    1  (· 1 ninda)          1     ninda         =   12  (· 1 cubit)
1 1/2 ninda         =  1 30 (· 60⁻¹ ninda)       1 1/2 ninda         =   18  (· 1 cubit)
...   ...           ...   ...                     ...   ...           ...   ...
5     ninda         =    5  (· 1 ninda)          5     ninda         =    1  (· 60 cubits)
```

obv.   *rev.*

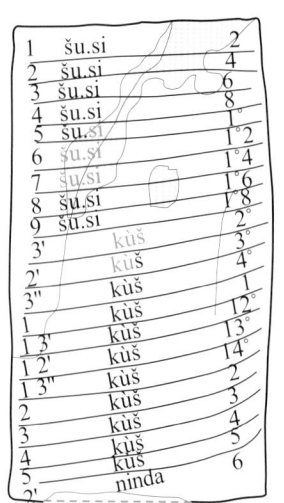 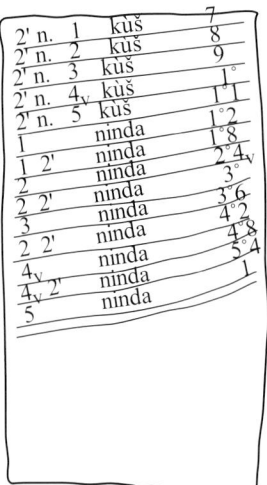

MS 2896. Metrological table for the OB system *Lc* (35 lines),
from '1 finger = 2 (· 60⁻¹ cubit)' to '5 ninda = 1 (· 60 cubits)'.

Fig. 3.4.2. An Old Babylonian metrological table for length measures, basic unit the cubit (system *Lc*).

There are no complete metrological tables for systems *Ln* and *Lc* in the Schøyen Collection. The best candidate for such a complete table is the section for systems *Ln* and *Lc* in the Nippur text **UM 55.21.78 (2N-T 530)** (Neugebauer and Sachs, *JCS* 36 (1984)), a small fragment of a combined metrological table. In that text, the table for system *Ln* starts with '1 finger   10' and ends with '[1(géš) danna]   30'. Then follows what is left of the first seven lines of the table for system *Lc*, from '[1 finger]   1' to '[7 fingers]   14'. Neugebauer and Sachs, who were unaware of the existence of metrological tables for system *Lc*, admitted in the following words (*op. cit.*, 250) that they could not understand the meaning of those seven badly preserved lines:

> "We do not know how the last seven lines are to be restored. The difficulty of making a reasonable restoration is twofold. The preserved sexagesimal numbers at the ends of the lines seem to force us to assume an otherwise completely unknown higher unit of 4,0 (that is 4 (00)) danna. On the other hand, one cannot begin a complete new system of measures after 1,0 danna since all the available systems have already been used up, in their standard arrangement: [capacity], weight, area, and length."

The sub-section for length measures in the combined table **Ash. 1923.366**, a 6-sided prism (van der Meer, *Syllabaries* (1938), 156; Robson, *SCIAMVS* 5 (2004); see Sec. A5 e in Appendix 5) is a fairly well preserved complete metrological table of 105 lines for system *Ln*, extending from 1 finger to 60 danna. Another, not quite as extensive, (nearly) complete metrological table of 95 lines for system *Ln* is Gurney, ***UET 7, 114***, which ends with an entry for 2 danna (Friberg, *RA* 94 (2000), 155; see also Fig. A5.3 in Appendix 5). A comparison of MS 3869/11 with Ash. 1923-366 shows that MS 3869/11 is an exact excerpt of the 35 first lines from the more extensive table. On the other hand, MS 3869/11 can be viewed as only an abbreviated excerpt from *UET 7*, 114, omitting the entries 1/3, or 1/2, cubit plus 1, 2, 3, 4 fingers, and 2/3 cubit plus 1, 2, ..., 9 fingers. MS

2705 is an unabbreviated excerpt from *UET 7*, 114, although it stops before the end of the complete table, omitting the final entries for 1 plus 1/3, 1/2, or 2/3 danna, and for 2 danna. *UET 7*, 114, in its turn, seems to be an imperfect version of the more extensive complete table in a section of UM 55-21-78. Not only does it omit the entries between 2 and 60 danna, it also omits the entries from 11 to 14 uš, as well as all the entries 1/2 danna plus 1, 2, 3, 4 uš and 2/3 danna plus 1, 2, …, 9 uš. Ash. 1923-366 gives a third version of what the entries between 10 uš and 2 danna should look like.

It is interesting that the table on *UET 7*, 114 ends with the subscript

[nam.uš].dagal.la.šè	for (use with) *lengths* and widths *in general.*

(Here 'in general' is an attempted translation of the prefix nam, a Sumerian mark for *abstract concepts*.) This is a way of saying that the metrological table on *UET 7*, 114 is a table for system *Ln*.

Two other extensive Old Babylonian metrological tables for length measures describe their content in similar ways. Thus, the fragment **UET 7, 115** (Friberg, *RA* 94 (2000), 156; see also Fig. A5.4 in Appendix 5) originally contained a (nearly) complete table for system *Ln*, followed by a parallel table for system *Lc*. The two tables extend as far as the entries for 20 danna, and end with the following two illuminating subscripts

nam.uš.[… … …]	*for* (use with)] lengths *and widths* in general
nam.sukud.bùr.šè	for (use with) heights and depths in general.

In a similar way, the section for length measures on the fragment **BM 92698** (*MKT 2*, pl. 61; see again Sec. A5 e in Appendix 5) contains tables for systems *Ln* and *Lc*, extending to their respective entries for 2 danna, followed by the subscripts

nam.uš.sag aša₅.šè	for (use with)] lengths and fronts of fields (areas) in general
nam.sukud.bùr.saḫar.šè	for (use with) heights and depths of mud (volumes) in general.

Note: The correct readings and interpretations of the various subscripts mentioned above were published for the first time in Friberg, *RlA 7* (1990) Sec. 5.1, contrary to what is stated in Nemet-Nejat, *JNES* 54 (1995), footnote 71.

3.5. Old Babylonian Combined Metrological Tables

Old Babylonian multiple and combined multiplication tables were discussed in Secs. 2.6 c and 2.6 d above. It was suggested that in most cases multiple multiplication tables can be explained as excerpts from a "combined multiplication table". The set of head numbers for the single multiplication tables in the combined table was shown to have such a complicated structure that it is beyond doubt that all Old Babylonian single or multiple multiplication tables must have a common origin as copies or copies of copies of the tables in one, and only one, original combined multiplication table. It is also obvious that there must have been a close relation between the invention of sexagesimal numbers in place value notation and the construction of the first combined multiplication table.

A similar observation can be made with respect to Old Babylonian metrological tables, although in this case the situation is somewhat less clear. It has long been known that in what may be called "combined metrological tables" tables for the systems *C*, *M*, *A*, and *L* always appear in this order (See, for instance, the passage cited from Neugebauer and Sachs on the preceding page, where the authors refer to the "standard arrangement" of the metrological tables. See also Friberg, *GMS* 3 (1993), where it is observed that in Late Babylonian combined metrological tables this standard order between the sub-tables can be reversed.)

A plausible conjecture is that *the first combined metrological table, with the sub-tables in the order C, M, A, L, was constructed in connection with the invention of sexagesimal numbers in place value notation and the construction of the first combined multiplication table*, and that all later Old Babylonian metrological tables were more or less direct descendants of this first table. What makes the situation somewhat complicated is that, as has been observed repeatedly above, there seems to have been no canonical version of the Old Babylonian metrological tables. When two metrological tables for the same system are overlapping, they tend to differ in details of their arrangement. The reason why this is so may be that metrological tables are simpler than multiplication tables with respect to the computational work involved. Also, a teacher in the scribe school wanting to construct his own combined metrological table text did not have to remember a complicated set of head numbers as in the case of combined multiplication tables. All he had to do was to remember a few simple rules for the construction of a combined metrological table:

the fixed order: *C, M, A, L,*
the basic units: sìla, ma.na (mina), šar, ninda, and kùš (cubit),
the smallest units: 1/3 sìla, 1/2 še (barley-corn), 1/3 šar, and 1 šu.si (finger),
the repetitive structure, with small multiples and/or basic fractions of successively larger units of the system,
the restriction that the sexagesimal equivalents in the right column should be at most 2-place sexagesimal numbers.

With these rules in mind, the actual computation of the entries offered no additional difficulties. Therefore, it is no wonder if each scribe school teacher had his own personal version of the combined metrological table.

MS 3869/13 (Figs. 3.5.1-2 below) is one of only two multiple metrological tables in the Schøyen Collection. The preserved part of MS 3869/13 is a relatively small fragment. The curvature of the fragment as viewed from the side indicates that the obverse of the fragment constitutes roughly the right half and the upper two-thirds of the obverse of the original clay tablet. Knowing that, it is possible to make an informed conjecture about the entire content of the intact text. The suggested reconstruction of the combined metrological table fits the available space nicely, except that it leaves the final one and a half column on the reverse empty. According to the reconstruction, the number of lines in the three first columns on the reverse are, respectively, 40, 38, and 37. Consequently, the remaining space in the last one and a half columns was enough for about 60 lines of text.

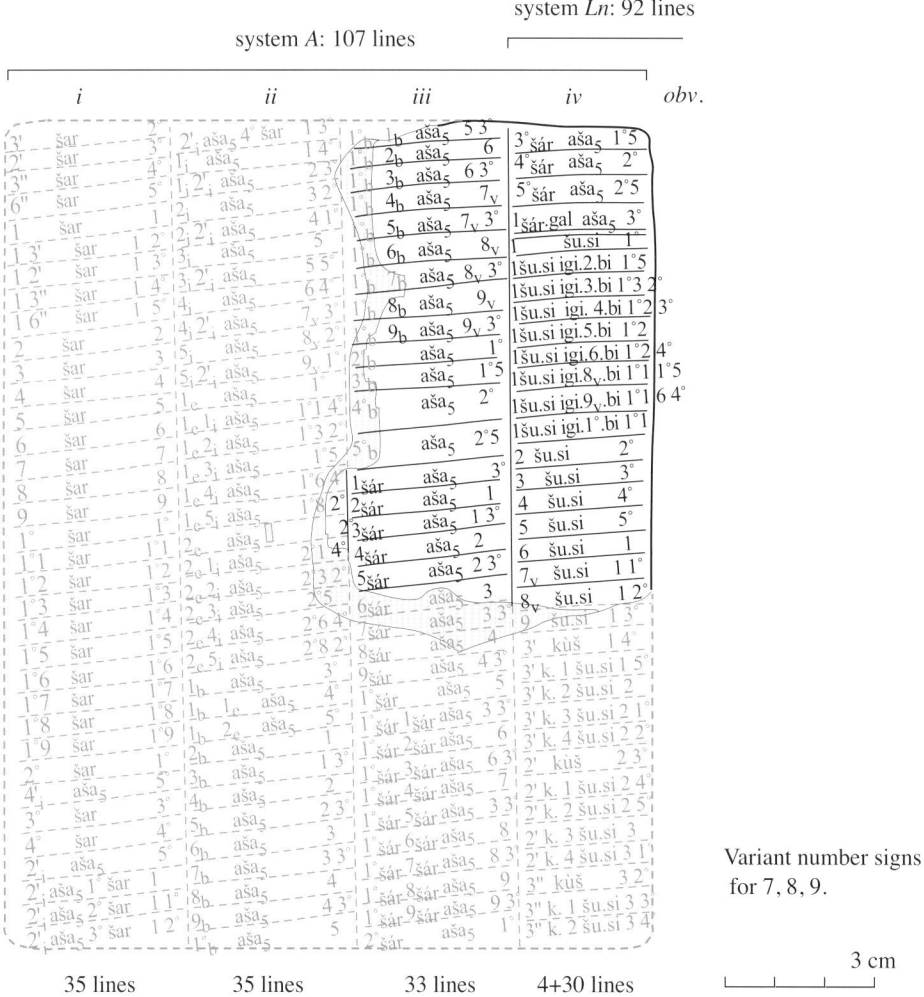

Fig. 3.5.1. MS 3869/13, *obv.* A fragment of an OB combined metrological table for systems *A, Ln*, and *Lc*.

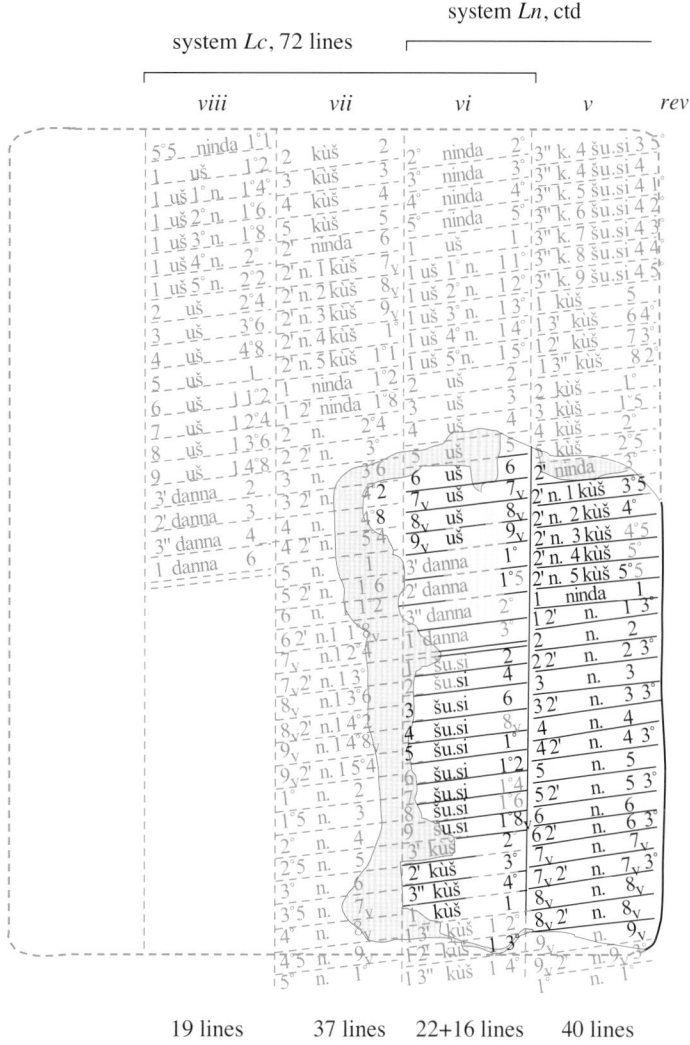

Fig. 3.5.2. MS 3869/13, *rev*. Metrological tables for systems *Ln* (continued) and *Lc*.

According to the reconstruction suggested in Fig. 3.5.1, the obverse of MS 3869/13 was originally inscribed with a nearly complete metrological table for system *A* (107 lines), and with the beginning of a metrological table for system *Ln* (92 lines). It is fortunate that the end of the table for system *A* and the beginning of the table for system *Ln* are both preserved on the obverse of the fragment. On the reverse of the fragment, the end of the table for system *Ln* is also preserved, together with the beginning of the table for system *Lc*. Only the end of the table for system *Lc* is lost.

A complete table for system *A* proceeds from 1/3 sìla = 20 ($\cdot\, 60^{-1}$ sìla) to 2 šár.gal aša$_5$ = 1 ($\cdot\, 60^4$ šar). (See Sec. A5 a in Appendix 5.) Thus, the table for system *A* on MS 3869/13 misses only the very last entry of a complete table.

A complete table for system *Ln* proceeds from 1 finger (šu.si) = 10 ($\cdot\, 60^{-2}$ ninda) to 1(géš) danna = 30 ($\cdot\, 60^2$ ninda). (See Sec. A5 b in Appendix 5.) The table for system *Ln* on MS 3869/13, on the other hand, goes from 1 finger = 10 to only 1 danna = 30. Thus it misses the last part of a complete table.

In the last two columns on the reverse of MS 3869/13, there is room enough for an additional table with about 20 + 40 = 60 lines. If there really was an additional table on the intact tablet, then it is likely that it was *a table of squares*, from 1.e 1 íb.si$_8$ to 1(00 00).e 1(00) íb.si$_8$, as on the fragment BM 92698 (see Appendix 5 f), where metrological tables for systems *Ln* and *Lc* are followed by arithmetical tables for squares, square

sides, and cube sides. Or else, it was *a table of square sides* as on the 6-sided prisms Ash. 1923.366 and AO
8865 (Appendix 5 f), where tables for systems *Ln* and *Lc* are followed by tables for square sides and cube sides.

On MS 3869/13, variant number signs are used for the digits 7, 8, and 9 (but not for the digit 4), a fact suggesting that the text is early Old Babylonian, from one of the southern sites in Mesopotamia.

MS 3869/14 (Figs. 3.5.3-4 below; color photos in Appendix9) is the second multiple metrological table in
the Schøyen Collection. It is not a fragment like MS 3869/13 but a whole tablet. On the other hand, much of
the text is more or less rubbed off and unreadable, in particular on the obverse. Therefore it is impossible to
make a complete and reliable reconstruction of the text. For that reason parts of columns *ii* and *iv* are left empty
in the conform transliteration of the obverse in Fig. 3.5.3.

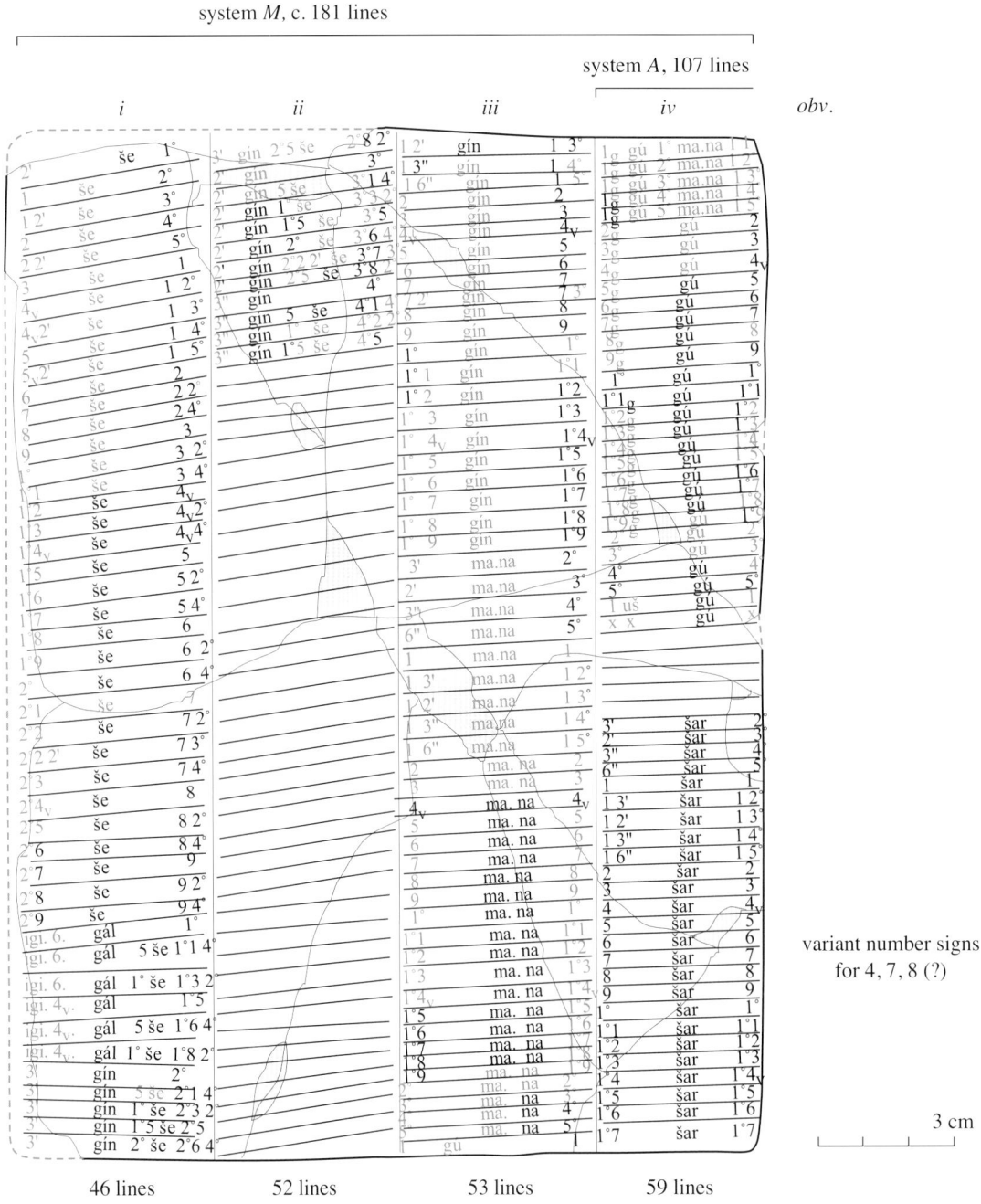

Fig. 3.5.3. MS 3869/14, *obv*. An Old Babylonian combined metrological table for systems *M* and *A*.

According to the tentative reconstruction of the text in Figs. 3.5.3-4, MS 3869/14 is inscribed with metrological tables for systems *M* and *A*. The table for system *M* appears to proceed from 1/2 barley-corn (še) = 10 (· 60^{-2} mina) to at least 60 talents (gú) = 1 (· 60^2 minas). It is possible that there are 6 additional entries at the end of this table, but it is impossible to determine precisely what they are since the text is too much rubbed off at this location.

A complete metrological table for system *M* normally seems to contain about 128 lines (see Appendix 5 b). The table for system *M* on MS 3869/14, on the other hand, apparently contains as many as 185 lines. Unfortunately, most of the extra lines are the ones in cols. *ii* and *iv* which are impossible to read!

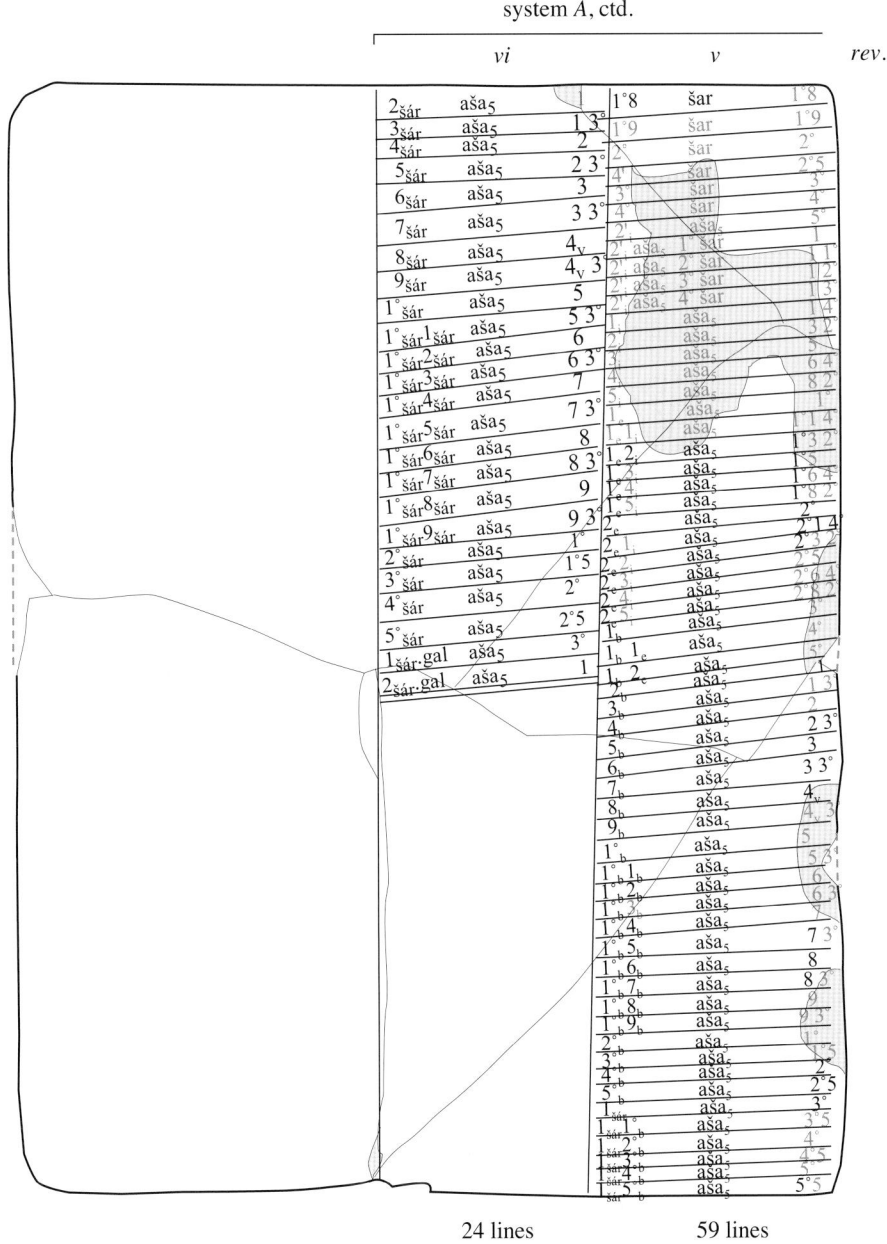

Fig. 3.5.4. MS 3869/14, *rev*. A metrological table for system *A*, continued.

The table for system *A* on MS 3869/14 (107 lines) proceeds from 1/3 šar = 20 (· 60^{-1} šar) in col. *iv* on the reverse to 2 šár.gal aša₅ = 1 (· 60^4 šar) in col. *vi* on the reverse. This is, therefore, a normal complete metrological table for system *A* (cf. Appendix 5 c).

It is somewhat surprising that the text on MS 3869/14 does not begin with a table for system *C*, and that a table for system *Ln* was not inscribed in the empty space on the reverse.

Apparently, variant number signs for 4, 7, and 8 were used in MS 3869/14, but the text is so rubbed off that it is impossible to know for sure.

4
Mesopotamian Weight Stones

4.1. *MS 4576. A Kassite Reused Talent Weight with an Inscription in Sumerian*

The way in which weight numbers are formed in Sumerian and Old Babylonian cuneiform texts is shown in Old Babylonian metrological lists and tables for system *M*. (See Sec. 3.2 above and Sec. A5 b in App. 5.) In particular, it is shown by the metrological lists and tables that the units of the Sumerian/Old Babylonian system *M* were the talent, the mina, the shekel, and the barley-corn.

The term used in cuneiform texts for the objects actually used in the weighing process is na$_4$ = *abnu* 'stone', obviously for the reason that weights normally were made out of stone. Archaeological finds have shown that such weights were made chiefly in the shape of "ducks, spindles, bombs, ellipsoids, and barrels" (Powell, *SNM* (1977), 242). The form of a number of Mesopotamian weights in the Schøyen Collection will be shown below.

MS 4576 (Fig. 4.1) is a massive object in the form of a small stone monument (a stele). Its dimensions are 38 cm × 23 cm × 15 cm. The stone is explicitly marked 1 gú '1 talent', so that it is clear that it once was a talent weight. However, it is no longer in perfect shape, since there is a small piece missing on the left shoulder, a somewhat larger piece missing in the lower right part of the rear face, and a deep hole gouged out of the lower part of the front face. (Two smaller holes placed symmetrically in the upper parts of the front and the rear may have been part of the design of the talent stone, placed there to serve as handles.) There is a long inscription on the front face, placed slightly off-center. In addition there is a large, roughly rectangular, and perfectly centered, area in the lower half of the front face, around the big hole, where the smooth surface of the stone has been ground off. This may be what remains of a completely obliterated earlier inscription on the talent weight. The big hole (right in the middle of the erased inscription!) can probably be explained as the result of the talent weight having been used for a long time as a door-socket, supporting the weight of a heavy door.

The extant inscription on the talent weight is unique; there is no similar inscription on any other known Sumerian or Babylonian weight stone. (See Powell, *op. cit.*, 205-207, 249-273, for an enumeration of known weight stones (as of 1977) with or without inscriptions. In most cases, inscriptions on weight stones are very brief, mentioning only the weight of the stone and, possibly, the name of the authority guaranteeing that the weight is correct.)

Also the form of the talent weight is quite unusual. Normally big Sumerian or Old Babylonian weight stones are in the form of ducks with the head turned back. See, again, Powell, *op. cit.*, where the heaviest of the enumerated weights are two bronze lions from Susa (4 talents) and Khorsabad (2 talents), a granite duck from Lagash/Girsu (2 talents), and three ducks in gray marble or whitish limestone from Babylon and Susa (1 talent). Weight stones similar to MS 4576 are described by Powell as "four-sided pyramids, rounded off at the top". One well known example is the *Marduk-šar-ilani* weight BM 910005 (2 minas), which has a long inscription covering the whole front face. (See the photo in Belaiew, *RA* 26 (1929).)

The direction of writing in the inscription on MS 4576 is from top to bottom, with all cuneiform signs upright, in their pictographically correct positions. (See the reference in Sec. 6 b below to the discussion in Picchioni, *OrNS* 49 (1980) of the change of direction of writing in cuneiform texts.) A small part of the inscription is damaged, the last line of the first register, and there is surface damage to the first three lines of the second register, which makes it difficult to read the text there.

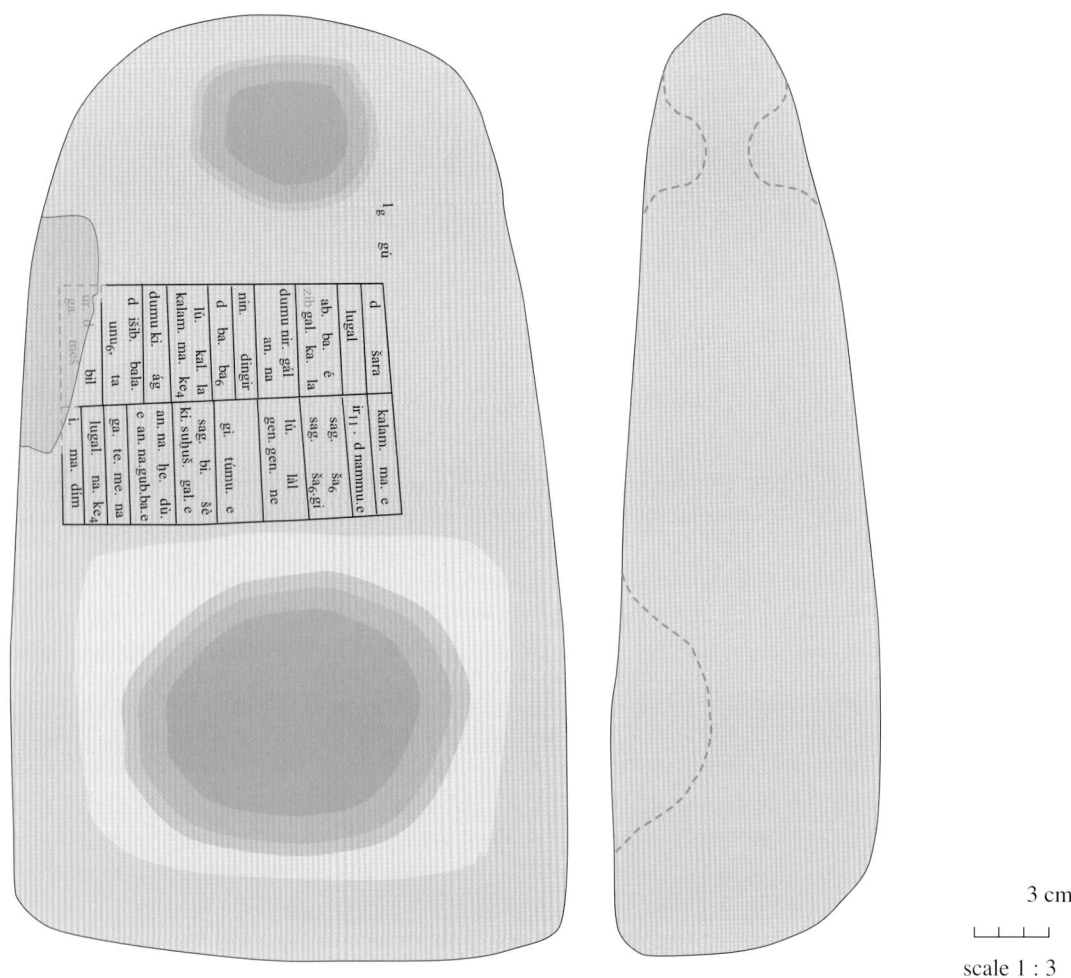

Fig. 4.1. MS 4576, a talent weight, probably reused as a door socket. Present weight 27.5 kg.

The inscription is addressed to Shara, the patron deity of the town Umma (see the map in Fig. 9.2):

		1 gú	1 talent
i	1-2	^dsára / lugal /	(For) the god Shara, (her) king,
	3	ab.ba.é.«zib.»gal.ka.la /	Abbaegalkala,
	4	dumu nir.gál.an[!].na /	daughter of Nirgalanna,
	5-6	nin.dingir / ^dba.ba₆	the priestess (of) the goddess Baba,
	7	lú.kal.la kalam.ma.ke₄	the strong man, the Sumerian (*lit.* 'of the land'),
	8-9	dumu.ki.ág / an.me.bala./unu₆.ta /	(her) beloved son from Anmebalaunu.
	10	[ur.^d]bìl./[ga.mes] /	*Ur.Gilgamesh*,
ii	11-12	kalam.ma.e / ir₁₁ ^dNammu.e /	the Sumerian, the servant of the god Nammu
	13	sag.ša₆ sag.ša₆ gi /	who established the goodness[?],
	14	lú.làl.gen.gen.ne /	the man who made sweet,
	15	nam.šul.gi túmu.e	(and) brought the rulership of Shulgi,
	16	sag.bi.šè ki.suḫuš.gal.e /	from its head to the great foundation of hell,
	17	an.na[!].ḫe.dù.e an.na.gub.ba.e /	who built and raised,
	18-19	Gá[!].te.me.na / lugal.na.ke₄ /	Gatemena, her king,
	20	ì.ma.dím	(he) fashioned (it).

The date of the inscription can be estimated from the sign shapes, the spelling, and the vulgar Sumerian. Apparently, the inscription (Fig. 4.2 below) is a Kassite imitation of an Ur III original.

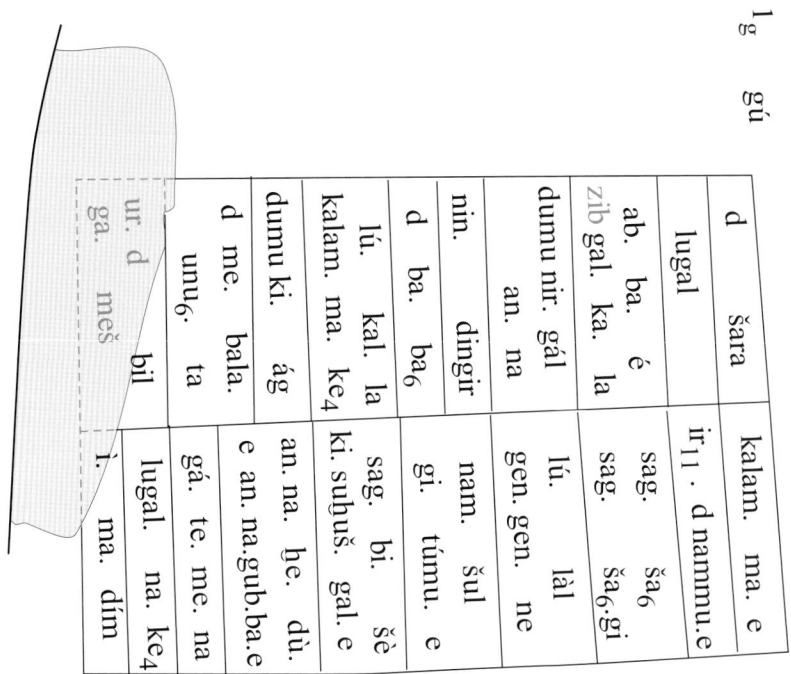

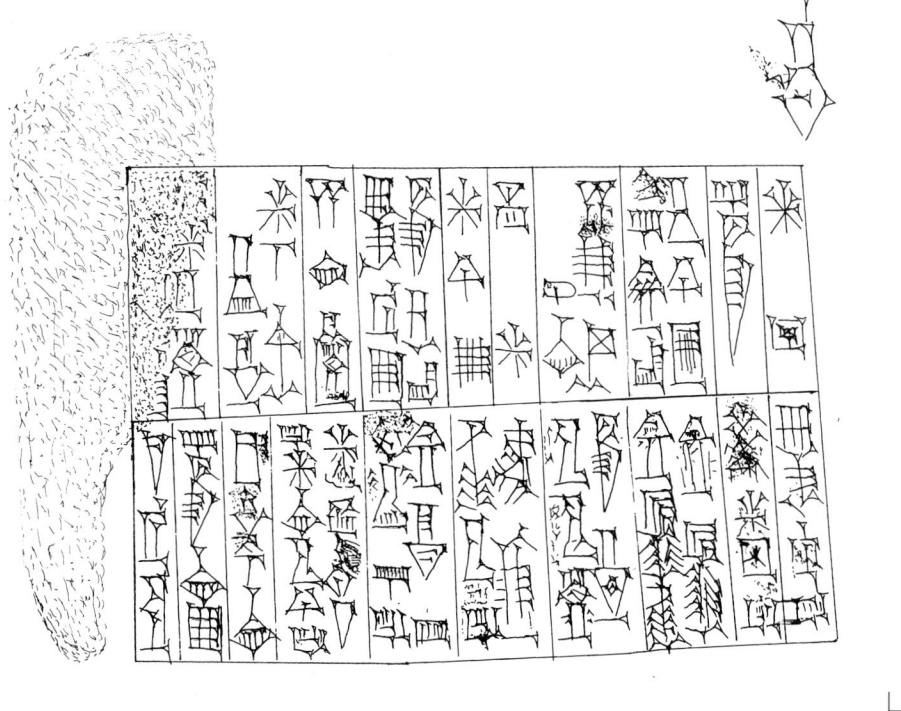

3 cm

Hand copy: F. Al-Rawi

Fig. 4.2. The inscription on MS 4576.

4.2.   MS 2481. A Barrel-Shaped 1 Mina Weight with an Inscription in 3 Lines

1 ma.na gi.	na
ša m d iškur. mu. si-*na*	
dumu m *é- reš*	d iškur

3 cm

Scale 1 : 3

Hand copy: F. Al-Rawi

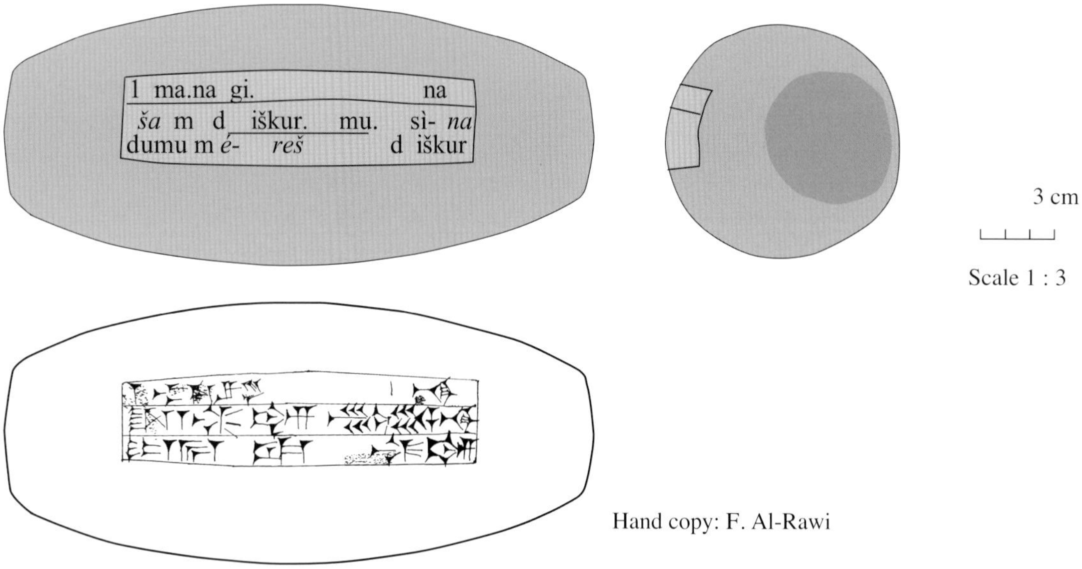

Fig. 4.3. MS 2481. A 1 mina weight in hematite, with an inscription in 3 lines. Shaped like a stretched barrel.

MS 2481 (Fig. 4.3) is 12 cm long and 5.3 cm in diameter at the center, where it is thickest. Its weight is 478.2 grams, corresponding to a shekel weighing 7.97 grams. The inscription is as follows:

1 ma.na gi.na /	1 mina true,
ša ^{md}*adad-šumu-iddina* (^{md}iškur.mu.ni si^{na}) /	that of Adad-shumu-iddina,
d u m u ^{m}*é-reš-*^{d}*adad* (^{d}iškur)	son of Eresh- Adad.

The correctness of the weight is guaranteed by a certain Adad-shumu-iddina, son of Eresh-Adad. The use of a family name of this type was not common until the first millennium BC. It is known that Adad-shumu-iddina was a vassal of the Assyrian king Tukulti-Ninurta I, c. 1230 BC. (Cf. Brinkman, *AnOr* 43 (1968), 65, 86, fn. 332, 335, 336, 375, 444, 449, 457.)

4.3.   MS 2837. An Ellipsoid-Shaped 3 Shekels Weight with a Brief Inscription

3 cm

Fig. 4.4. MS 2837. A weight in hematite, with the inscription '3 shekels'. Pierced.

MS 2837 (Fig. 4.4) is 4.1 cm long, with a diameter of 1.5 cm. Its weight is 24 grams, which corresponds to a shekel weighing 8 grams, and a mina weighing 480 grams.

4.4.   MS 2836. A Small Duck Weight in Agate with the Inscription '1/3 Shekel'

Fig. 4.5. MS 2836. A small duck weight in agate. Inscription '1/3 shekel'. Pierced lengthwise for a string.

MS 2836 (Fig. 4.5 above) is a miniature duck weight, beautifully made in agate, with a white head and a reddish-brown body. Its dimensions are 0.9 cm × 1.8 cm × 0.9 cm, and its weight is 3 grams, corresponding to a shekel weighing 9 grams.

4.5.   MS 5088. 55 Assorted Weight Stones Found Together in a Damaged Bronze Pot

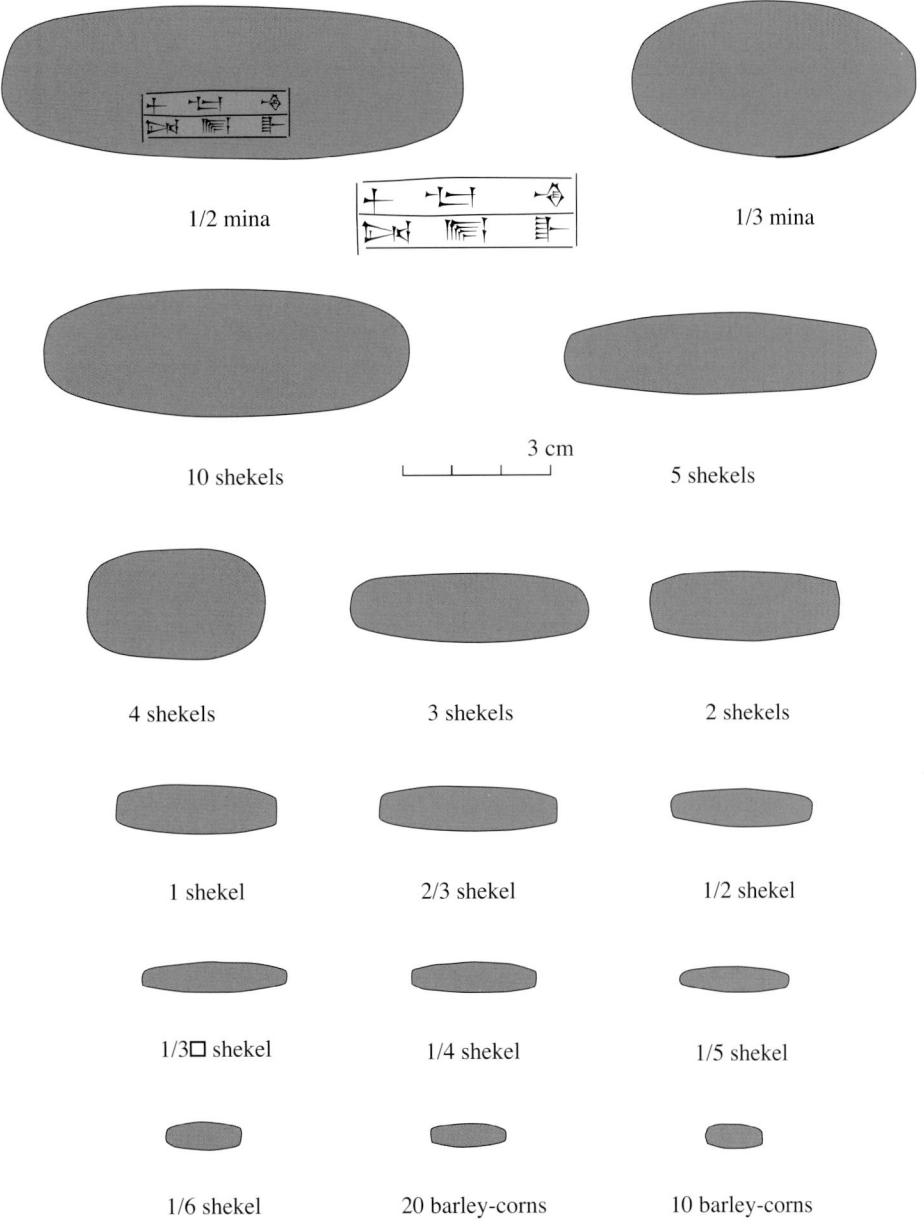

Fig. 4.6. MS 5088. Selected examples of weight stones, from 1/2 mina to 10 barley-corns.

MS 5088/1-55 (Fig. 4.6) is a set of 55 stone weights, probably Old Babylonian, allegedly found together in a damaged bronze pot, at some time before 1955. No location for the find is given, but the find spot may have been in the ruins of an ancient palace, since the biggest of the weights is a half-mina weight with the inscription

2' ma.na / na₄ é.gal   1/2 mina. Weight of the palace.

Most of the weights are of black stone (hematite), ellipsoidal, and polished to a shiny finish. (See the color photos 4 e, 1-3 in App. 10.) One 10-shekels weight is a stone of a lighter color (diorite). A 4-shekels weight is in the shape of a cylinder with rounded ends, and a 2-shekels weight is shaped like a thimble (a truncated cone). All the weights are finely worked and perfectly preserved, with the exception of a broken 2-shekels weight.

Here is a tabular summary of the mass in grams/kilograms of the 58 weights in the MS Collection. It is also shown what the corresponding weight of 1 shekel, 1 mina, or 1 talent would be in each case.

stone	1 talent	1 mina	1/2 m.	1/3 m.	10 sh.	5 sh.	4? sh.	3 sh.	2 sh.
actual weight	27.5 kg + [...]	478.2 g	245 g	139.9 g	81.6 g – 80 g	41.7 g	34.3 g	24.4 – 24.0	17.3 g – 16.3 g
weight of 1 shekel	> 7.64 g	7.97 g	8.17 g	7.0 g	8.16 g – 8.0 g	8.34 g	8.57 g	8.13 g – 8.0 g	8.65 – 8.15 g
weight of 1 mina	> 458.3 g	478.2 g	490 g	420 g	489.6 g – 480 g	500.4 g	514.5 g	488 g – 480 g	519 g – 489 g
weight of 1 talent	> 27.5 kg	28.7 kg	28.4 kg	25.2 kg	29.4 kg – 28.8 kg	30.0 kg	30.9 kg	29.3 kg – 28.8 kg	31.1 kg – 29.3 kg

stone	1 sh.	2/3 sh.	1/2 sh.	1/3 sh.	1/4 sh.	1/5 sh.	1/6 sh.	20 b.c.	10 b.c.
actual weight	8.6 g – 8.2 g	6.1 g – 5.5 g	4.4 g – 4.1 g	3.2 g – 2.6 g	2.3 g – 1.9 g	1.8 g – 1.7 g	1.5 g – 1.2 g	1.0 g – 0.8 g	0.4 g
weight of 1 shekel	8.6 g – 8.2 g	9.1 g – 8.2 g	8.8 g – 8.2 g	9.6 g – 7.8 g	9.2 g – 7.6 g	9 g – 8.5 g	9 g – 7.2 g	9.0 g – 8.0 g	7.2 g
weight of 1 mina	516 g – 492 g	549 g – 495 g	528 g – 492 g	576 g – 468 g	552 – 456 g	540 g – 510 g	540 g – 432 g	540 g – 480 g	432 g
weight of 1 talent	31.0 kg – 29.5 kg	32.9 kg – 29.7 kg	31.7 kg – 29.5 kg	34.6 kg – 28.0 kg	33.1 kg – 27.4 kg	32.4 kg – 30.6 kg	32.4 kg – 25.9 kg	32.4 kg – 28.8 kg	25.9 kg

The table shows a great variation in the (computed) values for the mass of a shekel, a mina, or a talent. Compare with the conclusion drawn by Powell in a similar situation (*op. cit.*, 207):

> "As one may see from the preceding table, the multiplicity of norms indicated by the documentary evidence is entirely confirmed by the material evidence from the inscribed weights. Even if we dismiss all the uncertain values ..., we are still left with a span of norms which precludes reduction to several 'preferred' norms."

Powell continued his survey of ancient Mesopotamian weight metrology in the very informative paper Powell, *AOAT 203* (1979). There one can learn, for instance, that "standard weights which bear the names of some guaranteeing authority ... are, in Babylonia ... almost without exception in the form of ellipsoids, elongated barrels, or ducks, which permits even unmarked specimens of these types to be classified as weights during the first sorting". Powell mentions "a few small beautifully worked ducks which have been pierced by boring", just like MS 2836, and adds that "pierced ducks have been excluded from the statistical sample ... on the grounds that they may be weight stones which have been made into amulets".

As a help for the "rating" of unmarked weights, Powell states that his thorough analysis of all available evidence, exhibited in his comprehensive Table 1A (*op. cit.*), indicates a *mean mina norm of 504 grams*, with a *standard deviation of 15 grams* (about 3%) for all specimens, and 16 grams for the shekel-fractions. In Table 1, and in an appended detailed discussion, he has collected a wealth of data for all the weights mentioned in the weight section of the encyclopedic lexical series ur$_5$.ra = ḫubullu (Hh XVI 417-452 = Landsberger *et al., MSL 10* (1970), 15-16, 49-50, 60-61):

> na$_4$.1.gú.un, na$_4$.50.ma.na, na$_4$.40.ma.na, na$_4$.30.ma.na, na$_4$.20.ma.na, na$_4$.15.ma.na, na$_4$.10.ma.na,
> na$_4$.5.ma.na, na$_4$.3.ma.na, na$_4$.2.ma.na, na$_4$.1.ma.na, na$_4$.3".ma.na, na$_4$.2'.ma.na, na$_4$.3'.ma.na,
> na$_4$.10.gín, na$_4$.5.gín, na$_4$.4.gín, na$_4$.3.gín, na$_4$.2.gín, na$_4$.1.gín, na$_4$.3".gín, na$_4$.2'.gín, na$_4$.3'.gín,
> na$_4$.igi.4.gál.la, na$_4$.igi.5.gál.la, na$_4$.igi.6.gál.la, na$_4$.22 2'.še, na$_4$.20.še, na$_4$.15.še, na$_4$.10.še,
> na$_4$.5.še, na$_4$.4.še, na$_4$.3.še, na$_4$.2.še, na$_4$.1.še, na$_4$.2'.še, na$_4$.3'.še.

Note that all these weights appear also in the Old Babylonian metrological table for system *M* (see Sec. A5 b in App. 5), with the exception of 1/3 barley-corn and 1/5 shekel.

In his Table 2 (*op. cit.*), Powell includes a detailed survey of "weight data according to provenance". Here is an abbreviated version of that table, with an extra row for the MS weights:

site	talents	minas and mina-fractions	shekels	shekel-fractions	total	smallest shekel-fraction	mean mina norm
Ur		2	16	9	27	1/2 shekel	502 g
Uruk		1	1	8	10	—	494 g
Lagash/Girsu	1	9	2		12	—	498
Adab		2	1			—	489
Nippur		4	4	10	18	1/6 shekel	525
Kish		7	18	3	28	1/2 shekel	506
Sippar			8	7	15	1/2 shekel	504
Eshnunna		5	12	3	20	15 b.c. = 1/12 sh.	509
Ishchali		2	41	16	59	1/2 shekel	506
Khafajah		1	4		5	—	512
Susa	1	6	6	1	14	1/3 shekel	508
uncertain	2	30	46	21	99	22 1/2 b.c. = 1/8 sh.	502
MS	1	3	22	33	59	10 b.c. = 1/18 sh.	502

The table shows that the weight stones in the Schøyen Collection are a sizable addition to the corpus.

4.6. YBC 4652. Weight Stones in an Old Babylonian Mathematical Theme Text

YBC 4652 is a small fragment of an early Old Babylonian mathematical theme text from some southern site, possibly Ur. (It belongs to Group 2 a, see Friberg, *RA* 94 (2000), 162.) YBC 4652 is a fairly well organized theme text written entirely in Sumerian. It originally contained 22 exercises, all on the theme na$_4$ ì.pa ki.lá nu.na.tag 'I found a stone, the weight unmarked(?)'. Of the 22 exercises, only 7, ## 7-9 and ## 19-22, are so well preserved that they allow a detailed interpretation.

The common setting for all the exercises on YBC 4652 is a thoroughly unrealistic situation where the size of the weight stone is increased and/or decreased in several steps, with the final size of the weight stone always being 1 mina. The stated problems can be reduced to "chains" of linear equations. The relatively simple solution procedures are never indicated. Here are a couple of examples:

YBC 4 652 ## 9, 19 (*MCT* (1945), text R; Melville, *HM* 29 (2002))

1	na₄ ì.pa ki.lá nu.na.tag	I found a stone, the weight unmarked(?).
	igi.7.gál ba.zi igi.11.gál bí.daḫ /	A 7th-part I tore off, an 11th-part I joined,
2	[igi.1]3.gál ba.zi ì.lá 1 ma.na	*a 13th-part* I tore off, I weighed (it): 1 mina.
	sag na₄ en.nam /	The head (initial weight) of the stone (was) what?
3	[sag] na₄ 1 ma.na 9 2' gín 2 2' še	*The head* of the stone: 1 mina 9 1/2 shekel 2 1/2 barley-corns.

1	na₄ ì.pa ki.lá nu.na.tag	I found a stone, the weight unmarked(?).
	6.bi ì.lá 2 gín [bí.daḫ-*ma*] /	6 of it I weighed, 2 shekels *I joined, then*
2	igi.3.gál igi.7.gál a.rá 24.kam tab	a 3rd-part (of) a 7th-part steps of 24 I repeated
3	bí.daḫ-*ma* / ì.lá 1 ma.na	(and) joined, then I weighed (it): 1 mina.
	sag na₄ en.nam	The head of the stone (was) what?
	sag na₄ 4 3' gín	The head of the stone: 4 1/3 shekels.

In modern notations, # 9 can be reformulated as the following *chain of three linear equations*:

$$w_1 - 1/7 \cdot w_1 = w_2, \quad w_2 + 1/11 \cdot w_2 = w_3, \quad w_3 - 1/13 \cdot w_3 = 1(00) \text{ shekels.}$$

The solution can be found in three easy steps, by repeated use of the rule of false value (see Melville, *op. cit.*, 7 for a less compact and less anachronistic version of the same solution procedure):

Set $w_3 = 13$. Then $w_3 - 1/13 \cdot w_3 = 13 - 1 = 12$, igi $12 \cdot 1(00)$ sh. = 5 sh., $w_3 = 13 \cdot 5$ sh. = 1 05 sh.
Set $w_2 = 11$. Then $w_2 + 1/11 \cdot w_2 = 11 + 1 = 12$, igi $12 \cdot 1\ 05$ sh. = 5;25 sh., $w_2 = 11 \cdot 5;25$ sh. = 59;35 sh.
Set $w_1 = 7$. Then $w_1 - 1/7 \cdot w_1 = 7 - 1 = 6$, igi $6 \cdot 59;35$ sh. = 9;55 50 sh., $w_1 = 7 \cdot 9;55$ 50 sh. = 1 09;30 50 sh.
Thus, the initial weight was 1 09;30 50 sh. = 1 mina 9 1/2 shekels 2 1/2 barley-corns.

(In the first line of this solution procedure, 13 is a false value for w_3, 12 is the corresponding false result, 5 sh. is the correction factor, and $13 \cdot 5$ sh. = 1 05 sh. is the correct value for w_3. Similarly, 11 and 7 are the false values in the second and third lines of the solution procedure, and 5;25 sh. and 9;55 50 sh. are the corresponding correction factors.)

The other example, # 19, can be reformulated as a *chain of two linear equations*:

$$6 \cdot w_1 + 2 \text{ sh.} = w_2, \quad w_2 + 24 \cdot 1/3 \cdot 1/7 \cdot w_2 = 1(00) \text{ sh.}$$

The solution can be found in two easy steps, as follows (see again Melville, *op. cit.*, 7 for a less compact and less anachronistic version of the same solution procedure):

Set $w_2 = 3 \cdot 7 = 21$. Then $w_2 + 24 \cdot 1/3$ of $1/7 \cdot w_2 = 21 + 24 = 45$, igi $45 \cdot 1(00)$ sh. = 1;20 sh., $w_2 = 21 \cdot 1;20$ sh. = 28 sh., $w_2 - 2$ sh. = 26 sh., $w_1 = $ igi $6 \cdot 26$ sh. = 4;20 sh.
Thus, the initial weight was 4 1/3 shekels.

4.7.   YBC 4669 § 1. Measuring Vessels in an OB Mathematical Theme Text

It is interesting to compare the ordered set of Old Babylonian weight stones shown in Fig. 4.6 above with the ordered set of Old Babylonian measuring vessels that appear, indirectly, in the brief theme text **YBC 4669 § 1 a-i**. Here is a transliteration and translation of the first of nine exercises in that theme text:

YBC 4 669 § 1 a (*MKT 1*, 514; *MKT 3*, pl. 3; Friberg, *BagM* 28 (1997), 307-308)

1	ᵍⁱˢba.rí.ga /	A barig-vessel.
2	3" kùš 4 šu.si dal /	2/3 cubit 4 fingers (is) the transversal (diameter).
3	sukud.bi en.nam /	The height (is) what?
4	3" kùš 2 2' šu.si sukud	2/3 cubit 2 1/2 fingers (is) the height.

The mention of a diameter makes it clear that the object of this exercise is a *cylindrical* container. Its capacity and its diameter are given as 1 barig and 2/3 cubit 4 fingers = 24 fingers, respectively, and its height is found to be 2/3 cubit 2 1/2 fingers = 22 1/2 fingers. The other eight exercises in YBC 4668 § 1 are similar, as shown in the tabular summary below, where n. = n./60 and f. = finger:

§	name	diameter d	height h	sq. $d \cdot h$	capacity C
1 a	giš.ba.ri.ga	24 f. = 4 <u>n</u>.	22 1/2 f. = 3;45 <u>n</u>.	1 00 sq.<u>n</u>. · <u>n</u>.	1 barig
1 b	giš.ba.an 3$_{bán}$	18 f. = 3 <u>n</u>.	20 f. = 3;20 <u>n</u>.	30 sq.<u>n</u>. · <u>n</u>.	3 bán
1 c	giš.ba.an 10 <sìla>	12 f. = 2 <u>n</u>.	15 f. = 2;30 <u>n</u>.	10 sq.<u>n</u>. · <u>n</u>.	1 bán
1 d	giš.ninda 1 sìla	6 f. = 1 <u>n</u>.	6 f. = 1 <u>n</u>.	1 sq.<u>n</u>. · <u>n</u>.	1 sìla
1 e	giš.ninda 2' sìla	4 1/2 f. = ;45 <u>n</u>.	5 1/3 f. = ;53 20 <u>n</u>.	;30 sq.<u>n</u>. · <u>n</u>.	1/2 sìla
1 f	giš.ninda 3' sìla	4 f. = ;40 <u>n</u>.	4 1/2 f. = ;45 <u>n</u>.	;20 sq.<u>n</u>. · <u>n</u>.	1/3 sìla
1 g	giš.ninda 10 gín	3 f. = ;30 <u>n</u>.	4 f. = ;40 <u>n</u>.	;10 sq.<u>n</u>. · <u>n</u>.	10 gín
1 h	giš.ninda 5 gín	2 f. = ;20 <u>n</u>.	4 1/2 f. = ;45 <u>n</u>.	;05 sq.<u>n</u>. · <u>n</u>.	5 gín
1 i	giš.ninda 1 gín	1 f. = ;10 <u>n</u>.	3 1/2 f. = ;35 <u>n</u>.	c. ;01 sq.<u>n</u>. · <u>n</u>.	1 gín

The 3-bán vessel was common enough to have a name in Akkadian: *ṣimdu*. The Akkadian names for barig, bán, sìla, ninda, and gín were *parsiktu*, *sūtu*, *qû*, *akalu*, and *šiqlu*.

The giš-prefixes for all the mentioned measuring vessels indicate that the vessels were fabricated (of wood). Cf. the following entries in the Nippur version of the Old Babylonian Giš List (a lexical *List of Trees and Wooden Objects*; Veldhuis, *EEN* (1997), 163), lines 515-526, where líd.ga is another Sumerian word for gur (see the discussion of the Old Sumerian division exercise *TSS* 50 in Sec. A6 g of App. 6):

> giš.líd.ga,
> giš.ba.rí.ga,
> giš.ba.an, giš.ba.an.5.sìla, giš.ba.an.2.sìla,
> giš.1.sìla, giš.3".sìla, giš.2'.sìla, giš.3'.sìla,
> giš.10.gín, giš.5.gín, giš.3.gín, giš.2.gín, giš.1.gín.

It is shown in Fig. 4.7 below what the nine measuring vessels with the data of YBC 4669 § 1 would look like. (However, it is not clear if Old Babylonian measuring vessels like the ones shown in Fig. 4.7 actually existed, or if YBC 4669 § 1 is just a cleverly devised series of theoretical examples.)

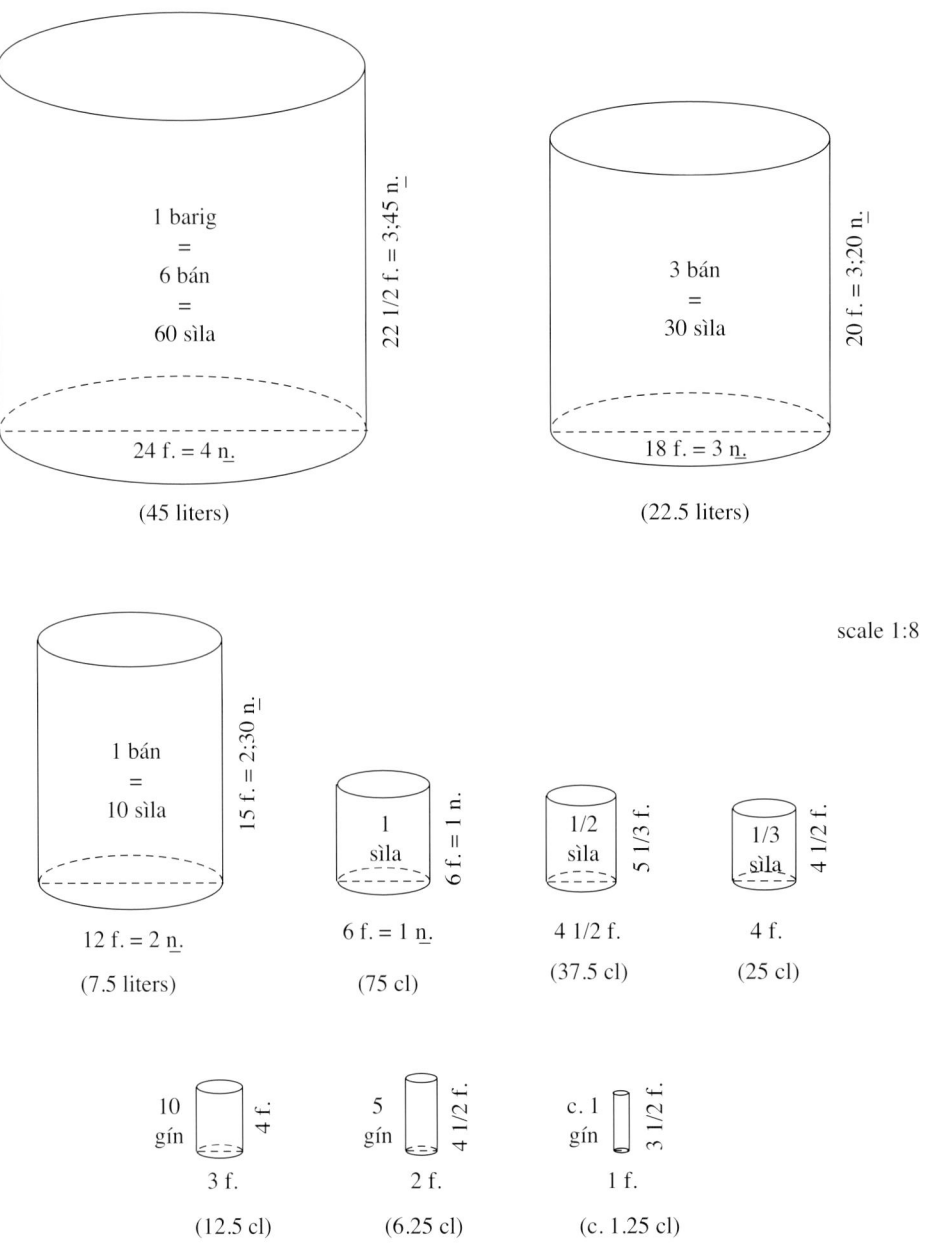

Fig. 4.7. YBC 4669 § 1 a-i. A series of Old Babylonian cylindrical measuring vessels.

5
Neo-Sumerian Field Plan Texts (Ur III)

5.1. MS 1984. A Field Plan Text from Umma with a Summary on the Reverse

MS 1984 (Fig. 5.1 below) is a square clay tablet with rounded corners, inscribed on the obverse with a detailed field plan and on the reverse with a brief summary.

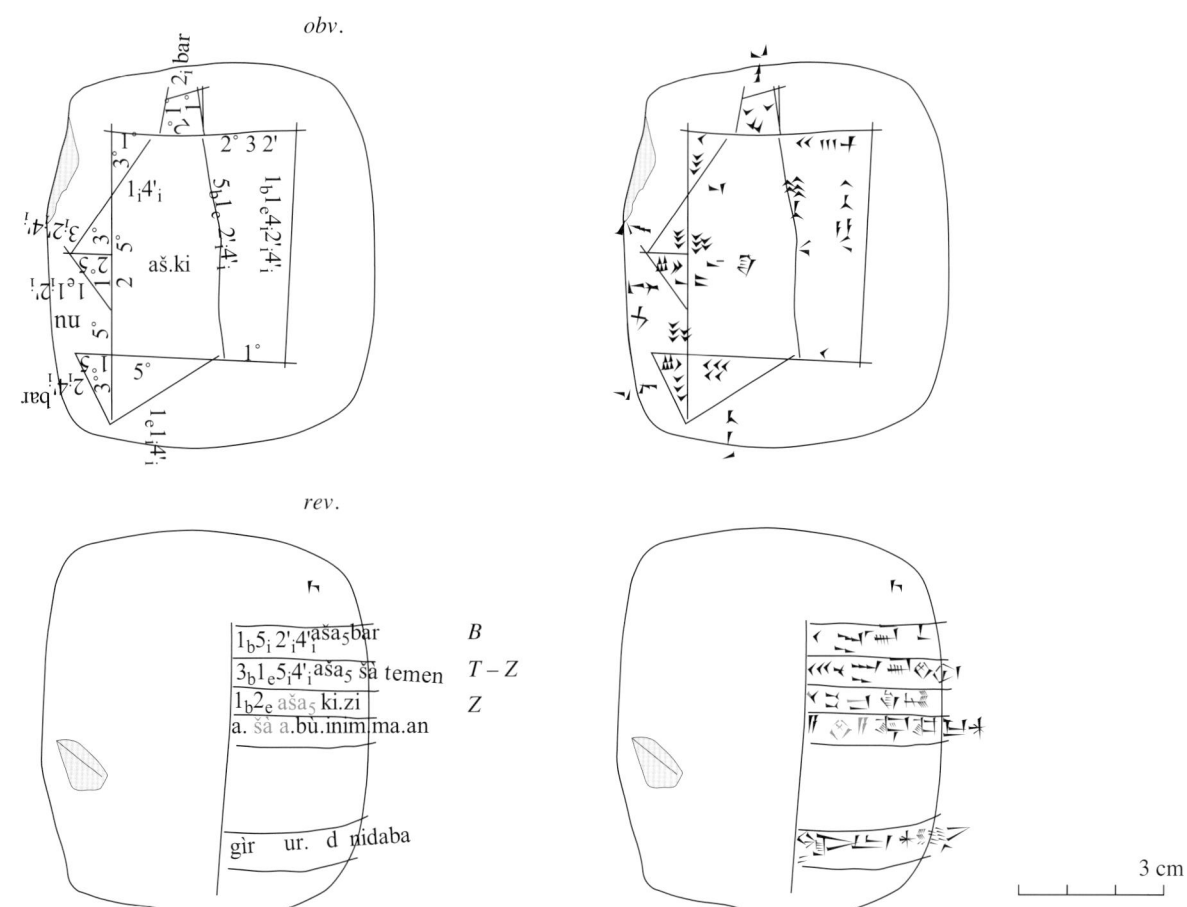

Fig. 5.1. MS 1984. A field plan text, probably from Umma.

The field outlined on the obverse of MS 1984 has an irregular, relatively complicated shape. The computation of its area is carried out in a roundabout way that would have been difficult to reconstruct without the help of the inscription on the reverse. As it turns out, the irregular field, apparently named a.šà a.bù inim.ma.an, is partly overlapping a field of a more regular, trapezoidal shape called the temen (Akk. *temennu*). The standard translation of temen is 'foundation', 'foundation document' (see below for a tentative explanation of the use of this term here). A curving line crossing the temen obliquely cuts off a large part of the temen, a roughly trapezoidal field that is outside the field a.šà a.bù inim.ma.an.

Whatever the meaning is of the term temen, it seems to be clear that the area of the irregular field was computed as the area of the temen, minus the area of the large trapezoidal field and the area of a small triangular field in the upper left corner of the temen, plus the areas of a small rectangular field and four small triangular fields outside the border of the temen.

The author of the text made himself guilty of some minor miscalculations. Immediately below follows an account of the progress of the computation of the area of the irregular field, with the errors in the text corrected. Afterwards, a separate account will be given of the errors, and explanations proposed for where they come from.

The following notations will be used (cf. Fig. 5.2 below):

A is the area of the irregular field, T the area of the rectangular temen.
B_1, B_2, B_3, B_4, and B_5 are the added areas of the rectangle and the four triangles.
Z_1 and Z_2 are the subtracted areas of the large trapezoid and the small triangle.

These notations are acronyms alluding to the Sumerian terms a.šà 'field, area', temen 'foundation document', bar 'extra, outside?', and zi 'tear out, subtract'.

Here are the successive steps of the (corrected) computation:

$B_1 = 20$ n. $\cdot$ 10 n. $= 3\ 20$ sq. n. $=$ 2 iku

$B_2 = 30$ n. $\cdot$ 25 n./2 $= 6\ 15$ sq. n. $=$ 3 1/2 1/4 iku

$B_3 = 1\ 00$ n. $\cdot$ 25 n./2 $= 12\ 30$ sq. n. $=$ 1 èše 1 1/2 iku

$B_4 = 30$ n. $\cdot$ 15 n./2 $= 3\ 45$ sq. n. $=$ 2 1/4 iku

$B_5 = 50$ n. $\cdot$ 30 n./2 $= 12\ 30$ sq. n. $=$ 1 èše 1 1/2 iku

$B_1 + B_2 + B_3 + B_4 + B_5 =$ $B = 1$ bùr   5 iku

$Z_1 = 2\ 50$ n. $\cdot$ (23 1/2 + 10) n./2 $= 2\ 50 \cdot 16;45$ sq. n. $= 47\ 27;30$ sq. n. $= 1$ bùr 1 èše 4 1/4 iku 22 1/2 šar

$Z_2 = 30$ n. $\cdot$ 10 n./2 $= 2\ 30$ sq. n. $= 1$ 1/2 iku

$Z_1 + Z_2 = Z = 1$ bùr 1 èše 5 1/2 1/4 iku 22 1/2 šar $=$ appr. 1 bùr 2 èše

$T = 2\ 50$ n. $\cdot$ (53 1/2 + 1 00) n./2 $= 2\ 50 \cdot 56;45$ sq. n. $= 2\ 40\ 47;30$ sq. n $= 5$ bùr 1 èše 1/4 iku 22 1/2 šar

$T - Z = 3$ bùr 2 èše 1/2 iku

$T - Z + B = A = 4$ bùr 2 èše 5 1/2 iku $= 5$ bùr $- 1/2$ iku $=$ appr. 5 bùr

These values can be compared with the ones appearing on the obverse and reverse of MS 1984:

$B_1 = 2$(iku) bar	*obv.*	correct	
$B_2 = 3$ 1/2 1/4(iku)	*obv.*	correct	
$B_3 = 1$(èše) 1 1/2(iku)	*obv.*	correct	
$B_4 = 2$ 1/4(iku) bar	*obv.*	correct	
$B_5 = 1$(èše) 1 1/2(iku)	*obv.*	correct	
B = 1(bùr) 5 1/2 1/4(iku) bar	*rev.*	small error	surplus 1/2 1/4(iku)
$Z_1 = 1$(bùr) 1(èše) 4 1/2 1/4(iku)	*obv.*	small error	surplus 1/4(iku) 2 1/2 šar
$Z_2 = 1$ 1/2(iku)	*obv.*	correct	
Z = 1(bùr) 2(èše) ki.zi	*rev.*	appr. correct	surplus 2 1/2 šar
$T = 5$(bùr) 1(èše) 1/2 1/4(iku)	*obv.*	small error	surplus 1/4(iku) 2 1/2 šar
T − Z = 3(bùr) 1(èše) 5 1/4(iku) šà temen	*rev.*	small error	deficit 1/2 1/4(iku)
$T - Z + B = A = 4$(bùr) 2(èše) 5(iku) not given in the text		appr. correct	deficit 1/2(iku)

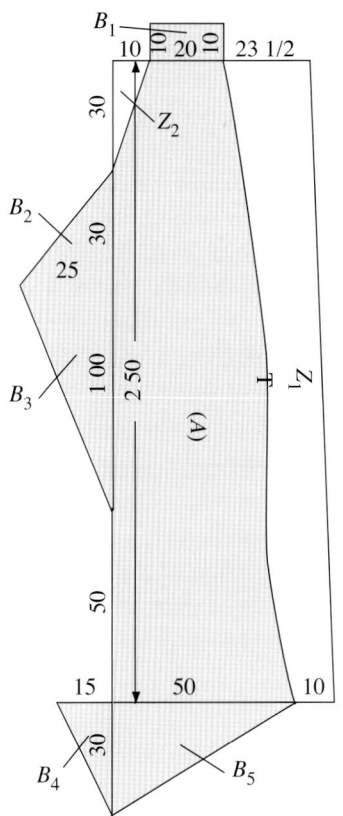

This text was first published and extensively discussed by Allotte de la Fuÿe in *RA* 12 (1915). A new hand copy appeared in Grégoire, *MVN 10* (1981), 214.

The dotted line indicates the boundary of the trapezoidal field called temen.

The field colored gray is the irregularly shaped field called a.gu$_7$.inim.ma.an.

Z_1 and Z_2 are areas to be *subtracted* from the area T of the temen.

B_1, B_2, B_3, B_4, and B_5 are areas to be *added* to the area of the temen.

The area of the field a.bù.inim.ma.an is computed as the area of the temen minus the subtracted areas plus the added areas.

The total area of the field is close to 5 bùr.

Scale: 1 : 12,000 (1 mm representing 2 ninda)

Fig. 5.2. The field plan on MS 1984, *obv.*, drawn to scale.

Here follows a transliteration of the summary and subscript on the reverse of MS 1984:

1(bùr) 5(iku) 2'(iku) a.šà bar /	1 bùr 5 1/2 iku, the field (= area) outside(?).
3(bùr) 1(èše) 5(iku) 4'(iku) a.šà šà temen /	3 bùr 1 èše 5 1/4 iku, the field inside the temen.
1(bùr) 2(èše) a.šà ki.zi /	1 bùr 2 èše the field of the ground torn off.
a.šà a.bù inim.ma.an /	The a.bù field Inim.ma.an.
gìr ur.^{d}nisaba	Inspector: Ur-Nisaba.

Clearly, the author of MS 1984 computed the area of the irregular field in the following way. First he computed *the area of the part of that field inside the temen*, explicitly named a.šà šà temen 'the field (= area) inside the temen' in line 2 of the text on the reverse. Then he *added the area of the small fields outside the temen*, named a.šà bar 'extra? field' in line 1, and subtracted the area of two parts of the temen, called a.šà ki.zi 'the field of the ground torn off' in line 3.

The error made in the computation of the sum B of the exterior fields can be explained as a simple addition error. The error made in the computation of the subtracted area Z_1 is more interesting. The scribe's intention was to compute it as

$$Z_1 = 1/2 \cdot 2\,50 \cdot 33;30 \text{ sq. n.} = 1/2 \cdot 1\,34\,55 \text{ sq. n.},$$

but apparently he made the mistake of setting $1/2 \cdot 1\,34\,55$ equal to $47\,00 + 55$ instead of $47\,00 + 27;30$. This mistake explains the surplus of $27;30$ sq. n. = $1/4$(iku) $2\,1/2$ šar.

The same surplus appears in the incorrect value for the area T of the temen. The reason is probably that T was computed as the sum of Z_1 and the area of the trapezoid with the height $2\,50$ and the parallel sides 30 and 50.

It is not clear what caused the small error in the value for $T - Z$.

A possible clue to the meaning of this text is the observation that the area of the irregularly shaped field is *a nearly round number*. Indeed, according to the corrected computation it is 5 bùr minus 1/2 iku, where 1/2 iku is 1/90 of 5 bùr, and according to the slightly incorrect computation on the clay tablet MS 1984 it is 5 bùr minus 1 iku, in both cases a very good approximation to the round number 5 bùr. Hence, the following tentative explanation:

Some high official or wealthy institution originally held the title (the temen!) to a regularly shaped piece of land measuring 5 bùr 1 èše (= 16 èše, about 345,000 sq. meters). Then something catastrophic happened so that a large part of the originally allotted land was lost, the piece to the right of the curved line across the property, and also a small piece in the upper left corner. In compensation, the title holder was allowed to add to what remained of his property several peripheral pieces of land, the rectangle and the triangles together called bar. This was done in a carefully calculated way so that after the change the total holdings came to measure almost exactly 5 bùr (= 15 èše), only slightly less than the area of the original estate.

5.2. *MS 1850. A Field Plan Text without a Summary on the Reverse*

MS 1850 (Fig. 5.3 below) is a badly broken square clay tablet with rounded corners. It is inscribed with a field plan on the obverse and only a few scattered numbers on what remains of the reverse. By sheer luck, almost no important details of the inscription were contained in the damaged parts of the obverse. Therefore, it is possible to reconstruct in its entirety the original drawing of the field plan with all its associated length and area numbers.

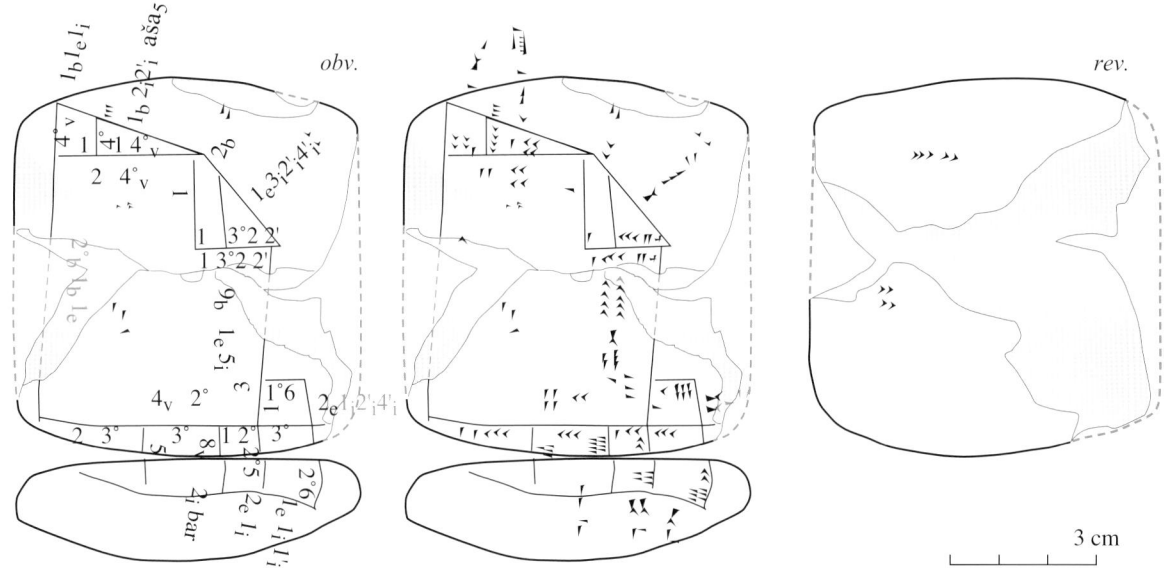

Fig. 5.3. MS 1850. A field plan with a central region and eight added fields around the border.

The anonymous field depicted on the obverse of MS 1850 consists of a large central region and eight smaller peripheral fields, triangular or trapezoidal. The central region can be divided into two nearly rectangular parts. Fig. 5.4 below shows the field plan drawn to scale.

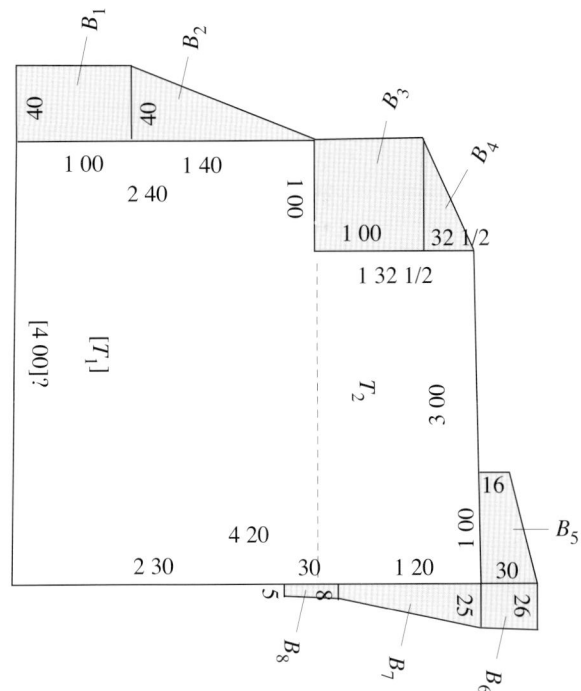

T_1 and T_2 are the areas of the two nearly rectangular parts of the central region.

$B_1, B_2, B_3, B_4, B_5, B_6, B_7, B_8$ (colored gray) are eight areas *added* to the area of the temen.

The total area of the field is close to 38 bùr.

Scale: 1 : 24,000 (1 mm representing 4 ninda).

Fig. 5.4. The field plan on MS 1850, drawn to scale.

Here are the successive steps of the computation of the areas recorded on MS 1850, *obv.*:

$T_1 = 4\ 00$ n. · $2\ 40$ n. =	$10\ 40\ 00$ sq. n. =	21 bùr 1 èše
$T_2 = 3\ 00$ n. · $(1\ 32\ 1/2 + 1\ 40)$ n./2 =	$4\ 48\ 45$ sq. n. =	9 bùr 1 èše 5 1/4 iku
$T = T_1 + T_2 =$	30 bùr 2 èše 5 1/4 iku =	31 bùr (– 1/2 1/4 iku)
$B_1 = 1\ 00$ n. · $41\ 1/2$ n. =	$41\ 15$ sq. n. =	1 bùr 1 èše
$B_2 = 1\ 40$ n. · 20 n. =	$33\ 20$ sq. n. =	1 bùr 2 iku
$B_3 = 1\ 00$ n. · $1\ 00$ n. =	$1\ 00\ 00$ sq. n. =	2 bùr
$B_4 = 1\ 00$ n. · $16\ 1/4$ n. =	$16\ 15$ sq. n. =	1 èše 3 1/2 1/4 iku
$B_5 = 1\ 00$ n. · 23 n. =	$23\ 00$ sq. n. =	2 èše 1 1/2 1/4 iku (5 šar)
$B_6 = 30$ n. · $25\ 1/2$ n. =	$12\ 45$ sq. n. =	1 èše 1 1/2 iku (15 šar)
$B_7 = 1\ 20$ n. · $16\ 1/2$ n. =	$22\ 00$ sq. n. =	2 èše 1 iku (20 šar)
$B_8 = 30$ n. · $6\ 1/2$ n. =	$3\ 15$ sq. n. =	2 iku (– 5 šar)
$B = B_1 + B_2 + \ldots + B_8 =$		7 bùr
$A = T + B = 37$ bùr 2 èše 5 1/4 iku =		38 bùr (– 1/2 1/4 iku)

Of the values listed above, the area T_1 is lost, being written in one of the destroyed parts on the obverse of the clay tablet. Apparently, the sums T, B, and $A = T + B$ were not recorded, although it is possible that the damaged parts of the obverse contained one or more of these sums.

Both the area T of the central property, and the area A of that central property plus the eight extra pieces of land are *nearly round numbers*. This striking fact supports the proposed interpretation of the field plan MS 1984. Indeed, a similar interpretation of the field plan on MS 1850 is that a wealthy person or institution originally held the title to a very extensive, regularly shaped piece of land, measuring almost exactly 31 bùr (2.4 sq. kilometers). Some higher authority then decided to allow that person or institution to expand this property by adding to it 7 bùr. This was done through expropriation of several adjoining smaller pieces of land owned by less favored individuals or institutions. In line with this interpretation, MS 1850 can be explained as a first draft for an official document recorded in some more elaborate form, perhaps on a stone tablet, to commemorate the occasion.

5.3. Four Ur III Field Plan Texts, Published in 1915, 1898, 1922, and 1962

Although the great majority of the mathematical cuneiform texts in the Schøyen Collection are new additions to the corpus, probably emanating from relatively recent excavations in Iraq, the field plan text MS 1984 has been known for a very long time. It was first published by Allotte de la Fuÿe in **RA 12 (1915)**, who presented the text as a clay tablet that had "arrived in France together with a batch (of tablets) from the tell of Djokha, situated not far from Tello", that is from the site of the ancient city Umma, near Lagaš-Girsu. As a further corroboration of the tablet's provenience from Umma, Allotte de la Fuye pointed to the circumstance that the name of the inspector mentioned in the inscription on the reverse was Ur-Nisaba, and that there apparently was in Umma a cult of the goddess Nisaba.

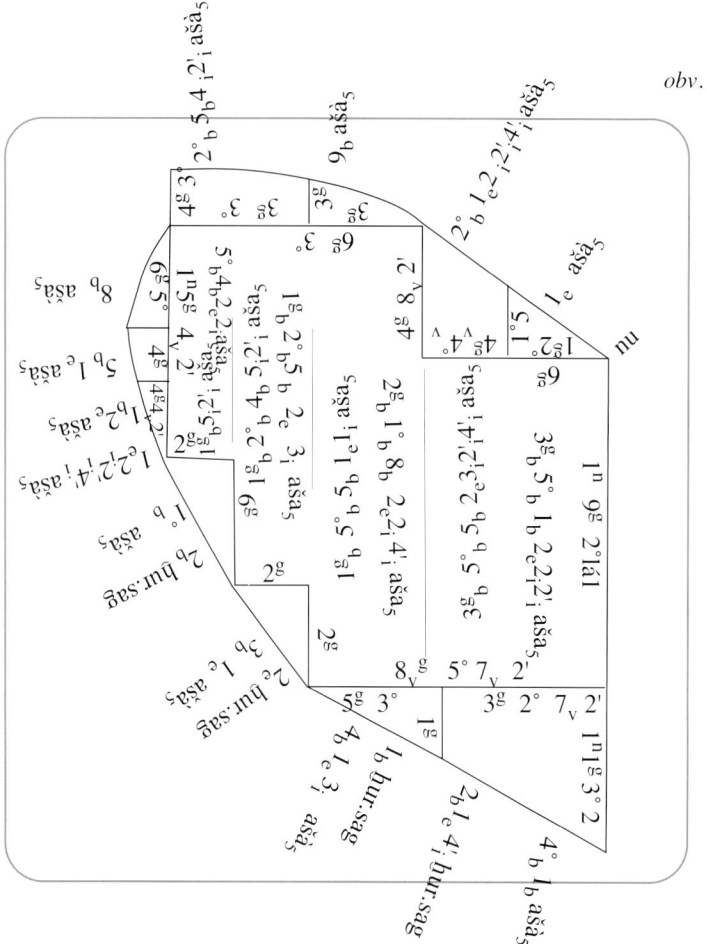

Fig. 5.5. Ist. O (MIO) 1107. A field plan with a central region and eleven added fields around the border.

A similar text is **Ist. O (MIO) 1107** (see Figs. 5.5-5.7), an Ur III text from the region around Lagaš-Girsu, published by F. Thureau-Dangin first in *RA* 4 (1897), then in *RTC* (1903) # 416. It has a field plan on the obverse, and a summary and subscript on the reverse.

The field plan on Ist. O 1107 resembles the one on MS 1850 but has a couple of extra features that make it particularly interesting. The most conspicuous extra feature is that the central region, in the summary called the temen, has been divided into four nearly rectangular sub-regions, and that for each of the four sub-regions its area has been computed twice, with different results. Thus, in each of the four nearly rectangular parts of the temen, two area numbers are recorded, one facing left, the other facing right. A convincing explanation for this strange feature was lacking for a long time but has been presented quite recently by Quillien in *RHM* 9 (2003).

Apparently, the area of the temen was computed twice. The first computation was based on the assumption that *the first three of the four sub-regions of the temen, counted from the left, are rectangular, while the fourth is trapezoidal.* The second computation was based on an identical assumption, after the field plan had been rotated to an *upside-down* position. The situation is made clear in Fig. 5.6 below, where the temen in its normal position is shown to the left, while the temen in its upside-down position is shown to the right.

The length numbers that are explicitly given in the drawing on the obverse of Ist. O 1107 are written in their correct positions along the sides of the temen in the two copies of the temen in Fig. 5.6. Computed length numbers for parts of the sides and for transversal lines are written within brackets. Thus, in the copy of the temen to the left, since the leftmost rectangular sub-region has one of its short sides given as 2 00, its other short side must be equal to 2 00, too. Similarly in the case of the second rectangular sub-region. It follows that the two short sides of the third rectangular sub-region must both be equal to 6 30 (given) minus twice 2 00 (computed). That is, they are both equal to 2 30. The fourth sub-region has then one of its parallel sides equal to 6 00 (given), while its second parallel side is 8 57 1/2 (given) minus 2 30 (computed). That is, it is equal to 6 27 1/2, as indicated.

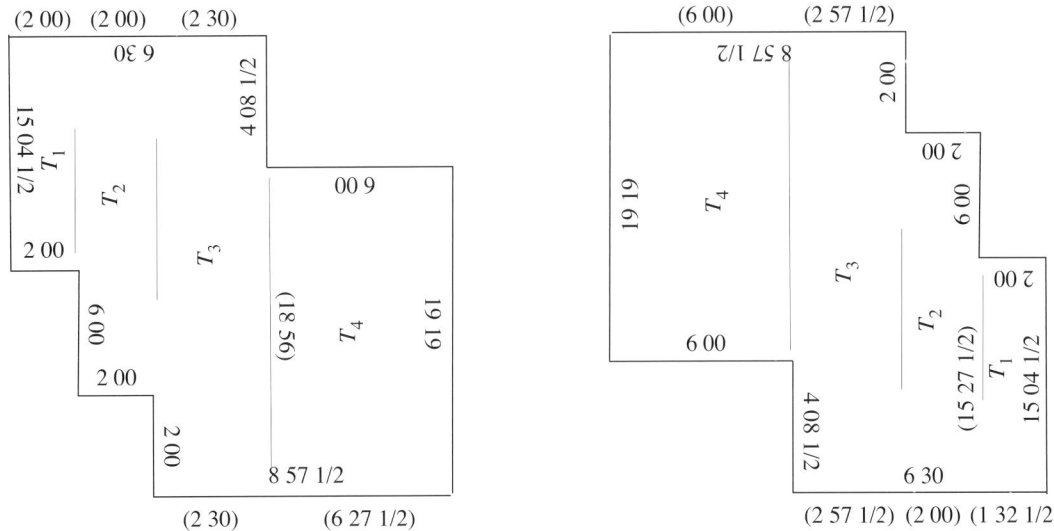

Fig. 5.6. Ist. O (MIO) 1107. The temen in its normal and upside-down positions.

The long sides of the three rectangular sub-regions to the left are clearly

$$15\ 04\ 1/2 \text{ (given)}, \ 15\ 04\ 1/2 + 6\ 00 = 21\ 04\ 1/2, \text{ and } 21\ 04\ 1/2 + 2\ 00 = 23\ 04\ 1/2.$$

The crucial observation made by Quillien is that in this first round of computations, the height of the trapezoidal sub-region to the right is assumed to have the *computed* length 23 04 1/2 – 4 08 1/2 = 18 56 rather than the *given* length 19 19! Now, with all the side lengths of the four sub-regions of the temen either given or computed, the four areas and their sum can be computed as follows:

$$T_1 = 15\ 04\ 1/2 \text{ n.} \cdot 2\ 00 \text{ n.} = 30\ 09\ 00 \text{ sq. n.} = 1\ 00 \text{ bùr } 5\ 1/2 \text{ iku } (-\ 10 \text{ šar})$$
$$T_2 = 21\ 04\ 1/2 \text{ n.} \cdot 2\ 00 \text{ n.} = 42\ 09\ 00 \text{ sq. n.} = 1\ 24 \text{ bùr } 5\ 1/2 \text{ iku } (-\ 10 \text{ šar})$$
$$T_3 = 23\ 04\ 1/2 \text{ n.} \cdot 2\ 30 \text{ n.} = 57\ 41\ 15 \text{ sq. n.} = 1\ 55 \text{ bùr } 1 \text{ èše } 1/2\ 1/4 \text{ iku}$$
$$T_4 = 18\ 56 \text{ n.} \cdot 6\ 13;45 \text{ n.} = 1\ 57\ 56\ 20 \text{ sq. n.} = \ 3\ 55 \text{ bùr } 2 \text{ èše } 3\ 1/2\ 1/4 \text{ iku } (+\ 5 \text{ šar})$$
$$T_1 + T_2 + T_3 + T_4 = T = 8\ 15 \text{ bùr } 2 \text{ èše } 3\ 1/2 \text{ iku } (-\ 15 \text{ šar}).$$

In the second round of computations, with the temen turned upside-down, the leftmost rectangle has the sides 19 19 and 6 00. The sides of the second rectangle can then be computed as

$$19\ 19 + 4\ 08\ 1/2 = 23\ 27\ 1/2 \quad \text{and} \quad 8\ 57\ 1/2 - 6\ 00 = 2\ 57\ 1/2.$$

The third rectangle has the sides

$$23\ 27\ 1/2 - 2\ 00 = 21\ 27\ 1/2 \quad \text{and} \quad 2\ 00.$$

Finally, the height and the two parallel sides of the fourth sub-region, the trapezoid, are

$$21\ 27\ 1/2 - 6\ 00 = 15\ 27\ 1/2, \quad 2\ 00, \quad \text{and} \quad 6\ 30 - (2\ 57\ 1/2 + 2\ 00) = 1\ 32\ 1/2.$$

With these given or computed side lengths for the four sub-regions of the upside-down temen, the four areas and their sum are computed as follows, beginning with the one at the left:

$$T_4 = 19\ 19\ \text{n.} \cdot 6\ 00\ \text{n.} = 1\ 55\ 54\ 00\ \text{sq. n.} = 3\ 51\ \text{bùr}\ 2\ \text{èše}\ 2\ 1/2\ \text{iku}\ (-10\ \text{šar})$$
$$T_3 = 23\ 27\ 1/2\ \text{n.} \cdot 2\ 57\ 1/2\ \text{n.} = 1\ 09\ 23\ 51\ 1/4\ \text{sq. n.} = 2\ 18\ \text{bùr}\ 2\ \text{èše}\ 2\ 1/4\ \text{iku}\ (+6\ \text{šar})$$
$$T_2 = 21\ 27\ 1/2\ \text{n.} \cdot 2\ 00\ \text{n.} = 42\ 55\ 00\ \text{sq. n.} = 1\ 25\ \text{bùr}\ 2\ \text{èše}\ 3\ \text{iku}$$
$$T_1 = 15\ 27\ 1/2\ \text{n.} \cdot 1\ 46\ 1/4\ \text{n.} = 27\ 22\ 26;52\ 30\ \text{sq. n.} = 54\ \text{bùr}\ 2\ \text{èše}\ 1\ 1/2\ \text{iku}\ (-3\ \text{šar})$$
$$T_1 + T_2 + T_3 + T_4 = T = 8\ 31\ \text{bùr}\ 3\ 1/4\ \text{iku}\ (-7\ \text{šar}).$$

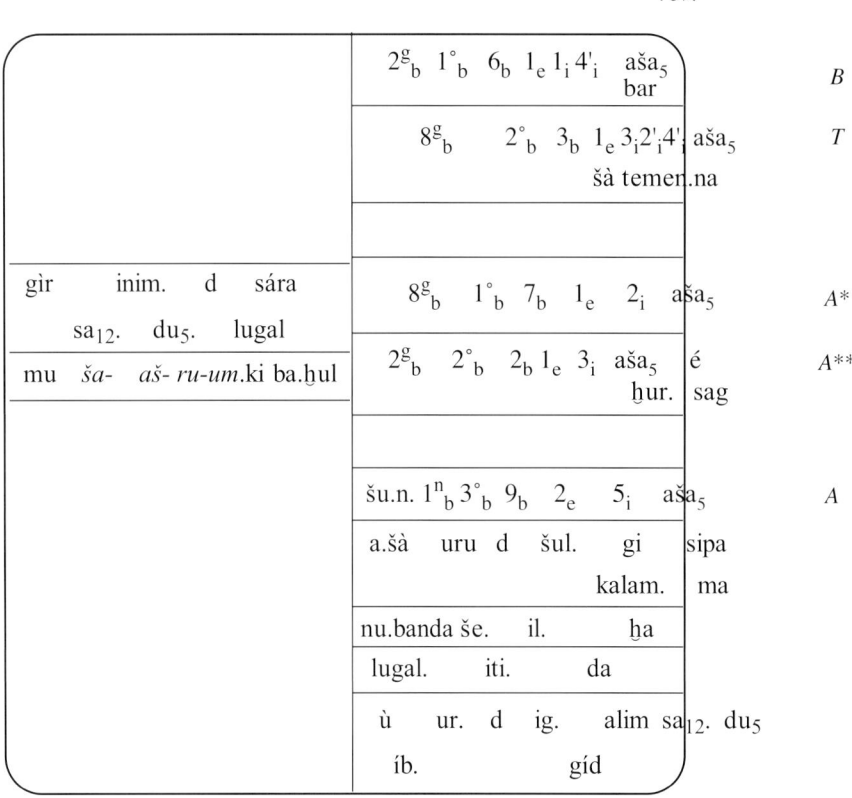

Fig. 5.7. Ist. O (MIO) 1107. The reverse with the summary and a subscript.

The reverse of Ist. O 1107 with its summary and subscript is shown in Fig. 5.7 above. The summary can be explained as follows:

$B = 2\ 16\ \text{bùr}\ 1\ \text{èše}\ 1\ 1/4\ \text{iku}$ is the sum of the areas of the eleven added fields (Sum. bar) around the temen

$T = 8\ 23\ \text{bùr}\ 1\ \text{èše}\ 3\ 1/2\ 1/4\ \text{iku}$ is the average of the two computed values for the area of the temen

$A = T + B = 10\ 39\ \text{bùr}\ 2\ \text{èše}\ 5\ \text{iku}$ is the total (Sum. šu.nígin) area of the field.

As an afterthought, the total area is split in two parts, $A^* = 8\ 17\ \text{bùr}\ 1\ \text{èše}\ 2\ \text{iku}$, called aša₅ 'field', and A^{**} = 2 22 bùr 1 èše 3 iku, called aša₅ e 'fields with houses' and ḫur.sag 'hilly terrain'. On the obverse, five of the added areas are split in the same way, but the figures don't add up. Conceivably, the explanation for the discrepancy is that Ist. O 1107 is an incomplete copy of a more complete text, where *all* the added areas were divided into their cultivated and non-cultivated constituents.

Just as in MS 1984 and MS 1850, the total area in the case of Ist. O 1107 is *very close to a round number*. Indeed,

$$A = 10\ 39\ \text{bùr}\ 2\ \text{èše}\ 5\ \text{iku} = 10\ 40\ \text{bùr} - 1\ \text{iku} = \text{appr.}\ 10\ 40\ \text{bùr}\ (= 41.5\ \text{sq. kilometers}).$$

The subscript states that the field plan depicts a town or village (Sum. uru) called Šulgi.sipa.kalam.ma '(king) Šulgi is the shepherd of the country', further mentions the names of the responsible overseer (Sum. nu.bànda), of two 'surveyors' (sa$_{12}$.du$_5$) who measured (íb.gíd) the fields, and of the 'inspector' (gìr) with the title (sa$_{12}$.du$_5$.lugal) 'royal surveyor'. The subscript ends with the year name 'the year when the city Šašrū was destroyed'.

A third Ur III field plan is **Wengler 36**, first published by Deimel in *Or* 5 ed. 2 (1930). It has a very elaborate field plan on the obverse, a summary and subscript on the reverse. According to the subscript, it is a text from Umma. A beautiful hand copy, by Maul, in Nissen/Damerow/Englund, *ABK* (1993), Figs. 58-59, is reproduced in Quillien (*op.cit.*), Fig. 8. As in the other Ur III field plans discussed above, the field plan on Wengler 36 shows a field composed of a central temen and a large number of peripheral fields (actually 48). The temen, in its turn, can be composed into six or seven rectangular or trapezoidal fields, for each of which the area is computed in two ways, in the same way as in the case of the field plan on Ist. O 1107. See the detailed analysis in Quillien (*op. cit.*). Unfortunately, although the text is only slightly damaged, it is not clear what the total area $A = T + B$ amounts to. (It seems to be an integral multiple of 1 bùr.) The text was written in 'the year when king Amar-Sîn destroyed Urbilum'.

A fourth Ur III field plan is **HSM 1659** (Dunham, *RA* 80 (1986), 34). It has a relatively simple field plan on the obverse, a summary and subscript on the reverse. According to the subscript, it is a text from Lagash. In this field plan, too, there is a temen. It is composed of a rectangle and a trapezoid. There are also 4 small trapezoidal fields and 2 small triangular fields added outside the temen. The total area $A = T + B = 3$ bùr 2 èše 1/2 iku. The text was written in "the year when king Shu-Sîn built a big ceremonial boat for the god En-Lil".

A late addition to the manuscript:

Hand copies of two further field plan texts involving a temen (**VAT 7029** and **VAT 7030**) were published by Schneider in *Or* 47-49 (1930). The hand copies are so sketchy that it is difficult to see what the precise layout actually may have been of the two field plans in these texts.

YBC 3879 (Fig. 5.8 below) is yet another field plan text from Ur III Umma. It was published by A. T. Clay as *Yale Oriental Series 1* (1915) text 24. The field plan on the obverse of the clay tablet shows a trapezoidal temen with six subtracted fields, all denoted by the term ki. The reverse contains a summary of standard type, followed by the data for *a division* into *five parallel stripes of equal area of the 'good' land*, denoted by the term sig$_5$ and meaning the temen minus the six subtracted fields.

It can be shown that the computation of the widths of the five parallel stripes required, among other things, *the calculation of the 'feed' or growth rate of the subtracted trapezoid* along the right part of the temen, as well as *the solution of a series of quadratic equations*. This is quite surprising since there are no other known examples of *pre-Old Babylonian* texts demonstrating a knowledge of the concept of the 'feed' of a trapezoid or a familiarity with quadratic equations. There are also, by the way, no previously known examples of the use of quadratic equations in a *non-mathematical* cuneiform text!

Moreover, the series of calculations on the reverse of YBC 3879 of the widths of the five parallel stripes of equal area is the earliest known example of a "geometric algorithm". (An astonishing new example of an *Old Babylonian* geometric algorithm, a "chain of trapezoids with fixed diagonals", is discussed in Friberg, *Amazing Traces* (2007), App. 1.)

A detailed discussion of YBC 3879 = *YOS 1*, 24 will appear in a separate publication with the title "A Geometric Algorithm Making Use of Quadratic Equations in a Neo-Sumerian Field Plan Text".

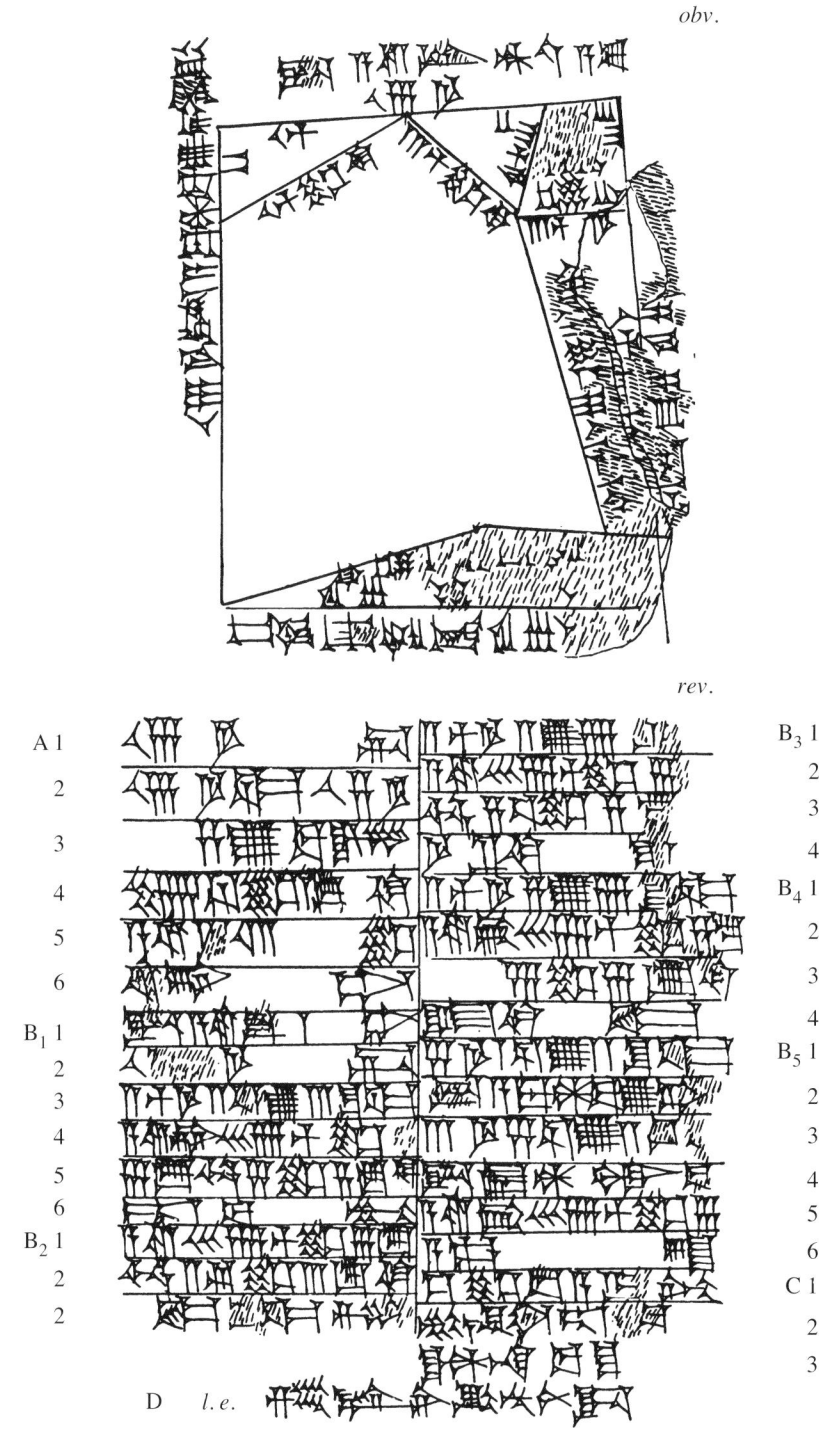

Hand copy: A. T. Clay

Fig. 5.8. *YOS 1*, 24 = YBC 3879. A field plan text from Ur III Umma with a summary and a field division.

6

An Old Sumerian Metro-Mathematical Table Text (Early Dynastic IIIa)

Shuruppak was one of the Sumerian city states, situated on the Euphrates river in south-central Mesopotamia. Excavations in 1902-03 by the Deutsche Orient-Gesellschaft and in 1931 by the University Museum of the University of Pennsylvania uncovered important remains from the Early Dynastic period, including a wealth of cuneiform documents, both administrative texts and school texts, from the Early Dynastic IIIa period (c. 2600-2500 BC). Among the school texts are some of the earliest known mathematical, or rather "metro-mathematical", texts (see below, Figs 6.1.1-6.1.2). The term "metro-mathematical" is appropriate, since all numbers appearing in these early mathematical texts are various kinds of measure numbers (length numbers, area numbers, capacity numbers, *etc.*).

6.1. Three Previously Published Metro-Mathematical School Texts from Shuruppak

As an introduction to the discussion of MS 3047 below (Fig. 6.2.1), three previously published metro-mathematical school texts from Shuruppak will be considered, all of them like MS 3047 dealing with areas of quadrilaterals. The texts will be presented here in hand copies with rotated cuneiform signs, like MS 3047 *obv.*, although they were probably written with unrotated, upright signs.

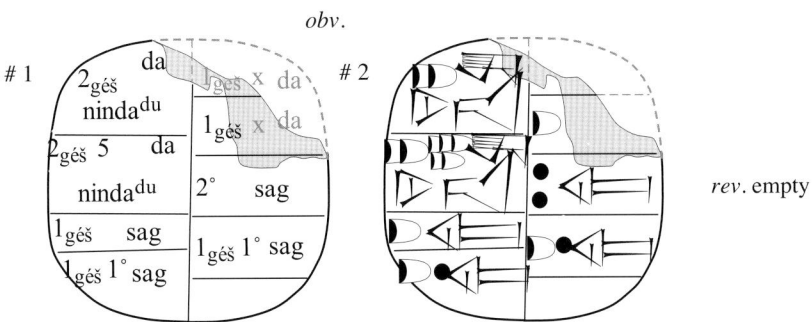

Fig. 6.1.1. *TSS* 926. The four sides of two quadrilaterals.

***TSS* 926** (Jestin *TSS* (1937)) is a small tablet inscribed on the obverse with two exercises. In the first exercise, the lengths of the sides of a quadrilateral are given in the following way:

2(géš) ninda^{du}	da	2(géš) ninda the side
2(géš) 5 ninda^{du}	da	2(géš) 5 ninda the side
1(géš)	sag	1(géš) the front
1(géš) 10	sag	1(géš) 10 the front

(An explanation of the "curviform" number signs in cuneiform texts from the third millennium can be found in Figs. A4.1 and A4.10 in App. 4.) It is surprising to find here the Sumerian da 'side' as a word for the longer, more or less parallel, sides of a quadrilateral. In later Sumerian and Babylonian texts, the word for the longer sides is normally uš 'length'. Another difference is that in this text the length measure is written as ninda^{du}, that is ninda plus a determinative du 'walk', instead of simply ninda as in later Sumerian and Babylonian texts. (Or, possibly, as ^{ninda}ninda$_x$, with a phonetic determinative.) Anyway, the lengths of the two longer sides of the quadrilateral are, in non-positional sexagesimal numbers, 2(géš) ninda and 2(géš) 5 ninda (corresponding in positional numbers to 2 00 and 2 05 ninda). The two shorter sides are 1(géš) ninda and 1(géš) 10 ninda (corresponding in positional numbers to 1 00 and 1 10 ninda). It is likely that the exercise was an *assignment*, to compute the area of the quadrilateral with the indicated sides. (See Friberg *AfO* 44/45 (1997/98) for a discussion of similar assignments in proto-cuneiform texts from the end of the fourth millennium BC. See, in particular, Figs. 8.1.5-6 below.) The area would have been computed by use of the only approximately correct "quadrilateral area rule", according to which the area in the given example is the following:

$$(2(géš) + 2(géš) 5)/2 \text{ n.} \cdot (1(géš) + 1(géš) 10)/2 \text{ n.} = 2(géš) 2 1/2 \text{ n.} \cdot 1(géš) 5 \text{ n.}$$
$$= (12 1/4 \cdot 10 \text{ n.}) \cdot 6 1/2 \cdot 10 \text{ n.}) = 1 19 1/2 1/8 \text{ sq.} (10 \text{ n.}) = 1 20 \text{ iku} (- 1/4 1/8 \text{ iku}).$$

In the second exercise on *TSS* 926, only the lengths of the shorter sides are perfectly preserved.

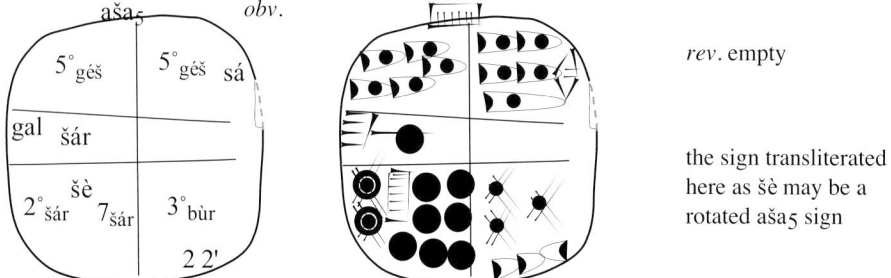

Fig. 6.1.2. *TSS* 188. The sides and the area of a very large square.

TSS 188 (Jestin *TSS* (1937)) is a small clay tablet from Shuruppak inscribed with both length and area numbers. The area number is so large that it occupies the space of *three* text boxes, instead of only one. The word aša$_5$ (or gán) 'field, area' is written on the edge of the clay tablet, so that it can be seen at a glance that this is a text with an area computation. The whole the text can be transliterated and translated as follows, with the help of the factor diagrams in App. 4, Figs. A4.1 and A4.10:

aša$_5$			field (area)		
50(géš)	50(géš)	sá	50(géš) (n.)	50(géš) (n.)	equalsided
1(šár).gal 20(šár) šè 7(šár) 30(bùr)			1(šár) 27(géš) 30(bùr)		
2 2'			2 1/2		

The meaning of the sign transliterated here as šè is not known. It seems to be part of the number sign for 20(géš) bùr, and may be a rotated aša$_5$ sign. The sign sá (or si$_8$) 'equal' probably indicates that the given length numbers are the longer and shorter sides, respectively, of a quadrilateral where the two longer sides are equal, as well as the two shorter sides. Since the same length number is given for both the longer and the shorter sides, the quadrilateral is a square. (Never mind that, from a modern point of view, a quadrilateral with all sides equal can be a parallelogram with equal sides.)

Since 1(šár).gal sq. ninda equals 2(géš) bùr, the area of the square can be computed as follows:

50(géš) n. · 50(géš) n. = 41(šár).gal 40(šár) sq. n. = 2 · 41(géš) 40 bùr = 1(šár) 23(géš) 20 bùr.

This, however, is not the answer given in the text of *TSS* 188. It is clear that the student who wrote the text made an error in his computation. How that error came about can be explained as follows: Near the lower edge of the clay tablet is inscribed a number which probably means '2 1/2'. (The sign used here for '1/2' is different from the form of '1/2' in later texts, where an upright cup-shaped sign would have been crossed by a thin wedge

See Fig. A4.1 in App. 4.) The explanation is probably that the area of the square was computed in several small steps, as follows:

sq. (50(géš) n.) = 25 · sq. (10(géš) n.) = 25 · (3(géš) 20) bùr = 2 1/2 · (33(géš) 20) bùr = 1(šár) 23(géš) 20 bùr.

The value of sq. (10(géš) n.) was either held in memory or taken from a table of areas of squares like VAT 12593 (below). Either way, the simple mistake made by the student was to use a slightly incorrect value for sq. (10(géš) n.), 3(géš) 30 bùr instead of 3(géš) 20 bùr. He then counted like this:

sq. (50(géš) n.) = 25 · sq. (10(géš) n.) = 25 · (3(géš) **30**) bùr = 2 1/2 · (35(géš)) bùr = 1(šár) 27(géš) 30 bùr.

VAT 12593 (Deimel, *SF* (1923) 82; Nissen/Damerow/Englund, *ABK* (1993), Fig. 119) is a large clay tablet from Shuruppak, with a "table of areas of squares". It progresses from larger to smaller squares, in contrast to the Old Babylonian (arithmetical) tables of squares of standard type, which always progress from smaller to larger squares.

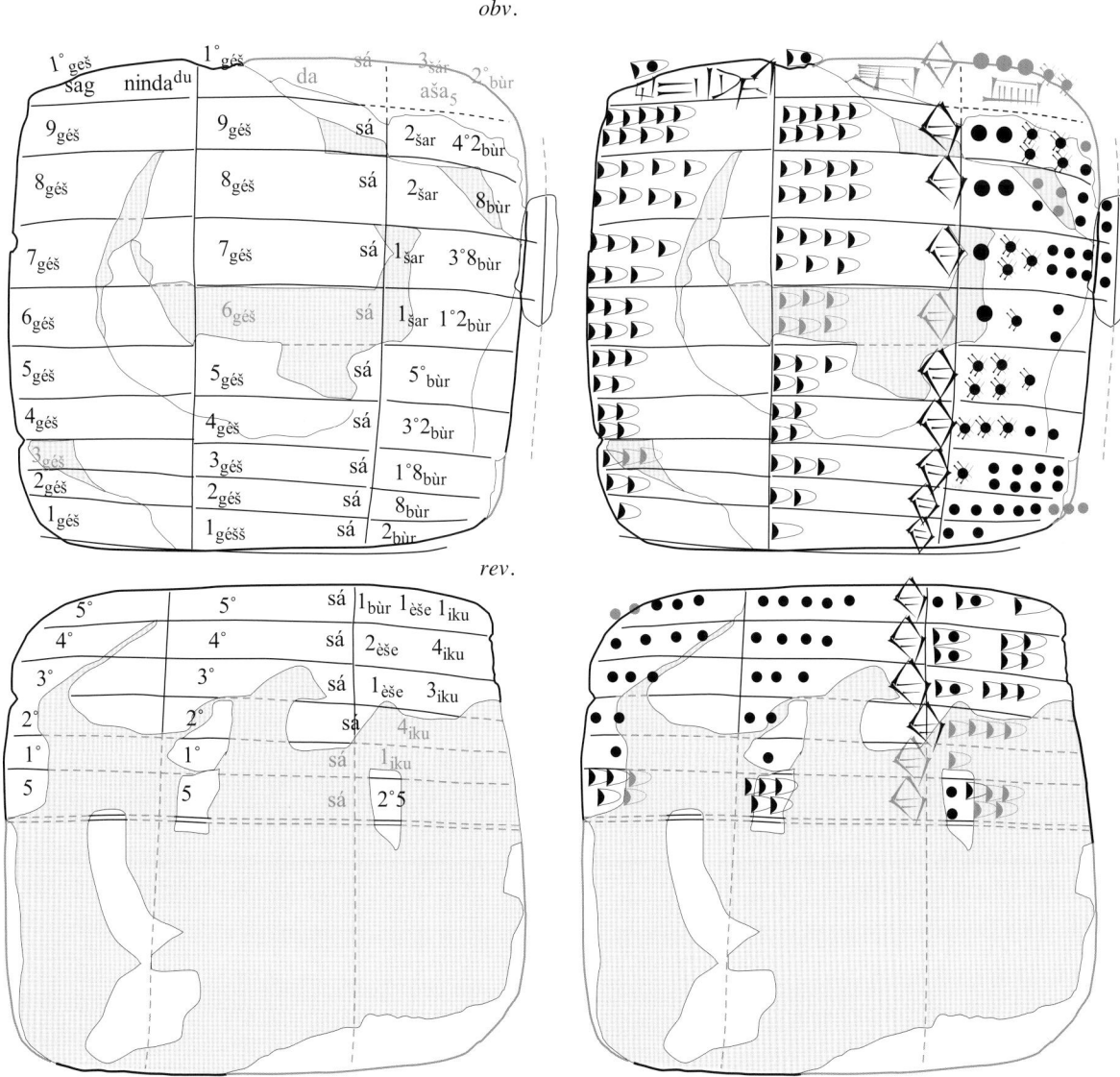

Fig. 6.1.3. VAT 12593. A descending table of areas of large squares, from sq. (10 · 60 ninda) to sq. (5 ninda).

The table begins with the area of the large square

$$\text{sq. (10(g é š) n.)} = 3(\text{g é š}) \ 20 \ \text{b ù r}.$$

(This is the value that was remembered incorrectly by the author of *TSS* 188!) This initial line of the table, which is now partly lost, was originally inscribed on the edge of the clay tablet, in the same way as the word a š a₅ is inscribed on the edge of *TSS* 188, as a quick indication of the content of the text.

The two smallest values recorded in this table of areas of squares are

sq. (10 n.) = 1(g é š) 40 sq. n. = 1(i k u)   and

sq. (5 n.) = 25 sq. n., which may have been written as 2[5 š a r], although this is not certain.

The reason why the table stops there, quite abruptly, is probably that 25 square ninda equals 1/4 i k u, and that smaller fractions of the i k u normally do not appear in cuneiform texts.

Note: CUNES 50-08-001, a previously unpublished Early Dynastic clay tablet from the collections of the department for Near Eastern Studies at Cornell University, is a metro-mathematical combined table text with a sub-table that is closely related to the table on VAT 12593. See App. 7 below.

6.2.   MS 3047. An Old Sumerian Metro-Mathematical Table Text

MS 3047 (Fig. 6.2.1) is another example of an Old Sumerian metro-mathematical text. It is likely that it, too, comes from Early Dynastic Shuruppak, in view of both its shape (a roundish square tablet) and the form of the cuneiform signs it is inscribed with. Its content is new and interesting.

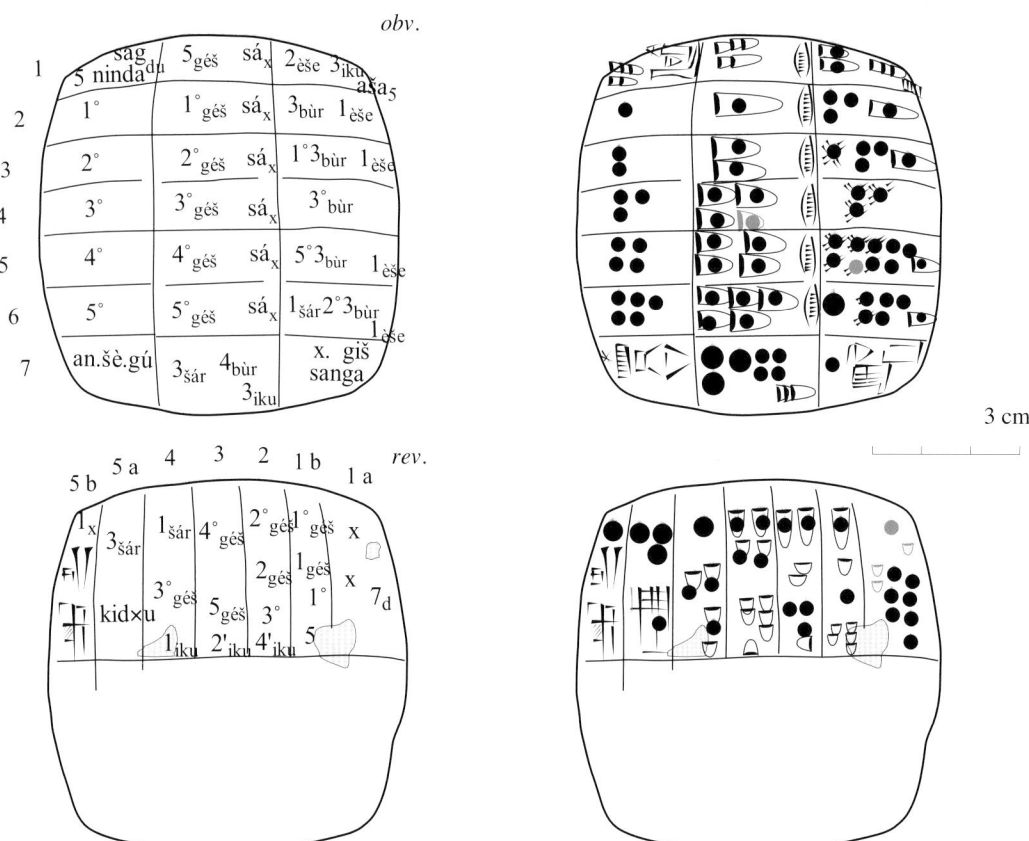

Fig. 6.2.1. MS 3047. A metro-mathematical table text, probably from Shuruppak, Early Dynastic IIIa.

Note: When this clay tablet first entered the Schøyen collection, about half its surface was covered with salt incrustations, covering the text. When the tablet had been cleaned, the appearance of a total in the last entry on the obverse came as a complete surprise. See the photos of MS 3047 in App. 10.

A puzzling detail is that the direction of writing on the reverse of MS 3047 does not agree with the direction of writing on the obverse. (Normally a clay tablet is first inscribed on the obverse, and then turned over along a horizontal or vertical axis so that it can be inscribed also on the reverse. In the case of MS 3047, the tablet has also been rotated around its center.) An effort has been made to show this anomaly in the hand copy above. Note how the longish "cup-shaped" number signs seem to be written horizontally on the obverse, but vertically on the reverse.

It is probable that the conflicting orientations are the result of a mistake. The writing on the reverse is weakly inscribed, as if the tablet after being inscribed on the obverse had been lying around for a while and become somewhat dry, before it was picked up again and inscribed on the reverse. It could then easily happen that the author of the text on the reverse did not remember to orient the reverse correctly with respect to the obverse.

In this connection it must be pointed out that it is not clear which the correct orientation of a text from Early Dynastic IIIa really is. What is known is that at some time during the course of the third millennium BC (probably close to the end of the millennium) the direction of writing in cuneiform texts was changed. Indeed, on clay tablets with proto-cuneiform script from the end of the fourth millennium, the text is written *from top to bottom in vertical text boxes*, arranged from right to left in horizontal registers, while on Old Babylonian clay tablets from the early part of the second millennium, the text has come to be written *from left to right in horizontal rows*, arranged from top to bottom in vertical columns. (See Picchioni, *OrNS* 49 (1980).) When the direction of writing changed, most of the cuneiform signs were rotated with it, from an upright to a lying down position. (However, according to an established convention in assyriological publications, hand copies of cuneiform texts are normally showing the text in the rotated position, no matter from which chronological period the text happens to be.)

6.2 a.   MS 3047, obv. *The Sum of the Areas of a Series of Similar Large Rectangles*

The survey above of earlier published metro-mathematical texts from Shuruppak concerned with the sides and areas of squares and other quadrilaterals was a necessary preparation for the following discussion of the text of MS 3047 (Fig. 6.2.1). On the obverse of MS 3047, there is the following metro-mathematical table:

sag 5 ninda^{du}	5(géš)	sá$_x$	2(èše) 3(iku)	aša$_5$
10	10(géš)	sá$_x$	3(bùr) 1(èše)	
20	20(géš)	sá$_x$	13(bùr) 1(èše)	
30	30(géš)	sá$_x$	30(bùr)	
40	40(géš)	sá$_x$	53(bùr) 1(èše)	
50	50(géš)	sá$_x$	1(šár) 23(bùr) 1(èše)	
an.šè.gú	3(šár) 4(bùr) 3(iku)		x.giš.sanga	

(The phrase an.šè.gú is a commonly occurring notation for 'sum, total' in Early Dynastic administrative texts. The phrase x.giš.sanga is probably a signature. The frequently occurring word sanga stands for 'priest/accountant'. The sign sanga itself may be the picture of a box for number tokens.)

It is clear that the first six lines of MS 2047, *obv.* is a table of sides and areas of *rectangles*. All the rectangles have the long side 60 times longer than the short side. The table is in several ways similar to the table of areas of squares on VAT 12593. In particular, the sequence 5, 10, 20, 30, 40, 50, in the first column of MS 3047, *obv.* is the same as the last six entries in the first column of VAT 12593, except for the ordering of the entries. The table on MS 3047, *obv.* is an *ascending* table of areas of rectangles, in contrast to the table on VAT 12593, which is a *descending* table of areas of squares. Note also that the unrealistic proportions of the rectangles indicate that MS 3047, *obv.* is a series of metrological exercises for use in the school rather than a practically useful reference table.

Another obvious difference between MS 3047, *obv.* and VAT 12593 is that in the second column of the former text each entry ends with the sign ki 'place, ground' while in the latter text each entry ends with the sign sá (or si$_8$) 'equal'. It is possible that ki functions here as a substitute for aša$_5$ 'field, area', and is inserted as a reminder that the product of the length numbers in the first two columns is the area number in the third column. Clearly, the two words ki and aša$_5$ are semantically related. Moreover, in proto-cuneiform texts ki sometimes has the same function as aša$_5$. (See Friberg, *AfO* 44/45 (1997/98), Figs 5.1 and 7.2.) An alternative, relatively plausible, explanation is that the ki signs are badly written versions of the normal sá signs, which is why they here are called sá$_x$.

Without recourse to sexagesimal numbers in Babylonian place value notation, the author of the table on the obverse of MS 3047 may have counted in the following way:

5 n. ·	5(géš) n. =	25(géš) sq. n. =		2(èše) 3(iku)
10 n. ·	10(géš) n. =	4 · 2(èše) 3(iku) =	3 (bùr)	1(èše)
20 n. ·	20(géš) n. =	4 · 3(bùr) 1(èše) =	13(bùr)	1(èše)
30 n. ·	30(géš) n. =	9 · 3(bùr) 1(èše) =	30(bùr)	
40 n. ·	40(géš) n. =	16 · 3(bùr) 1(èše) =	53(bùr)	1(èše)
50 n. ·	50(géš) n. =	25 · 3(bùr) 1(èše) = 1(šár)	23(bùr)	1(èše)

sum: 3(šár) 4(bùr)   3(iku)

6.2 b. *MS 3047, rev. A Geometric Progression of Areas(?)*

On the reverse of MS 3047, there is a table of a completely different type, with no known parallel. The interpretation of the table is quite difficult, not only because the table is of a new type, but also because its layout is confusing, and because some of the numbers in the first text case of the table are so weakly imprinted that they are hard to read. In addition, the meaning of the non-numerical signs in the last couple of text cases is not clear.

In order to begin to understand the table on the reverse, it is necessary to realize that there are (probably) only five or six, not seven, lines in the table, and that the entries in each line can be separated into two measure numbers. The first number in each pair of measure numbers appears to be a sexagesimal number, while the second number in each pair may (or may not) be a small area number. The following seems to be a relatively adequate transliteration of the table:

11(géš) 15	[x] 7(d) [x]	2 · 11(géš) 15	= 22(géš) 30
22(géš) 30	4'(iku?)	2 · 22(géš) 30	= 45(géš)
45(géš)	2'(iku?)	2 · 45(géš)	= 1(šár) 30(géš)
1(šár) 30(géš)	1(iku?)	2 · 1(šár) 30(géš)	= 3(šár)
3(šár) kid×u	1(šár?) x x		

In spite of this partial success, there are several remaining obscure points in this interpretation. First of all, it is only a conjecture that the second number in each entry is an area number. The second numbers in entries 2 and 3 do look like the cuneiform signs for 1/2 iku and 1/4 iku (App. 4, Fig. A4.9). Unfortunately, this conjecture cannot be supported by the form of the corresponding numbers in entries 1 a and 5 b. In addition, the number which according to this interpretation should have the meaning '1/8 iku' in entry 1 is written in the form [x] 7(d) [x] (where 7(d) stands for 7 disk-shaped number signs). If 7(d) is a sexagesimal number, it has the value 70, if 7(d) is an area number, it has the value 7(bùr), but 7(d) can also be some other kind of number, previously not attested.

It is regrettable that it seems impossible at the present time to give a complete explanation of the metro-mathematical table on the reverse of MS 3047, but the fact remains that MS 3047 is a very important text for the history of mathematics. Already the table of areas of rectangles on its obverse, with the sides of the six rectangles in the ratio 1 : 60, demonstrates that the mathematics taught in the scribe schools in Mesopotamia in the middle of the third millennium was unexpectedly sophisticated. Such a table cannot have been of any practical use; it is an example of mathematics for its own sake. The computation of the sum of the areas of the six rectangles seems to have been added as an afterthought, to make the text look more like an ordinary administrative text.

The geometric progression of sexagesimal numbers in the table on the reverse is also an unexpectedly sophisticated feature of the text. *It is the only pre-Babylonian example of a geometric progression.*

It is interesting that the combination of the number sign šár with the sign kid×u (of unknown meaning), which appears in text case 5 a on the reverse of MS 3047, also appears as an entry in Jestin's ***TSS* 190** (1937), a small lexical text on the theme ŠÁR from Early Dynastic Shuruppak (see Fig. 6.2.2 below). There is also a parallel text from Ebla (Pettinato, *MEE 3*, 72 (1981); see Sec. A6 b below, footnote 8), in which the entry šár kid×u is replaced by the simpler šár kid (that is, without the punched circular hole in the kid sign).

šár šè.gín$^?$	šár nì.búr$^?$.gu$_7$
šár giš	
[šár] kid×u	šár u$_4$.ni.gi.gu$_7$
[š]ár sá	šár u$_4$.u$_4$.u$_4$
šar dirig	šár an.[x]
šár šu nu.gi	

3 cm

Scale 1 : 1

Fig. 6.2.2. *TSS* 190. A small Early Dynastic lexical text on the theme ŠÁR.

This photo of *TSS* 190 is published here with the assistance and gracious permission of Veysel Donbaz.

7
Old Babylonian Hand Tablets with Practical Mathematics

7.1. MS 2317. Division of a Funny Number by a Non-Regular Factor

7.1 a.   Interpretation of the Three Numbers in the Text

MS 2317 (Fig. 7.1.1) is a quite small square hand tablet, inscribed on the obverse with three lines of numbers. At first sight, the text looks unpromising. The meaning of the four wedges in line 1 is not immediately clear, and the numbers 13 and 4 41 37 in lines 2 and 3 are, quite obviously, *non-regular* sexagesimal numbers.

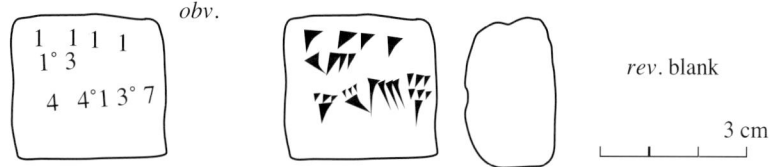

Fig. 7.1.1. MS 2317. A division exercise for a funny number.

A renewed look at the text reveals that it is unexpectedly interesting. The key to understanding what is going on here is to *work in sexagesimal arithmetic*. Indeed, a moment's reflection leads to the insight that the 3-place sexagesimal number in line 3 can be factorized as

$$4\ 41\ 37 = 4\ 37 \cdot 1\ 01.$$

For verification, note that 4 37 · 1 01 = 4 37 00 + 4 37 = 4 41 37. Hence, 4 41 37 is the product of two non-regular prime numbers, 4 37 (= 277) and 1 01 (= 61).

A connection between the number 13 in line 2 and the product 4 41 37 = 4 37 · 1 01 in line 3 is that *4 37 is an approximate reciprocal to 13*, since

$$13 \cdot 4\ 37 = 52\ 00 + 6\ 30 + 1\ 31 = 1\ 00\ 01.$$

(In absolute values, 13 · ;04 37 = 1;00 01.) Combining the observations that 4 41 37 = 4 37 · 1 01 and that 13 · 4 37 = 1 00 01, one finds that

$$13 \cdot 4\ 41\ 37 = 13 \cdot 4\ 37 \cdot 1\ 01 = 1\ 00\ 01 \cdot 1\ 01 = 1\ 00\ 01\ 00 + 1\ 00\ 01 = 1\ 01\ 01\ 01.$$

Consequently, the four wedges in line 1 of MS 2317 must be understood as the sexagesimal number 1 01 01 01, written as 1 1 1 1 in Babylonian place value notation without zeros. The whole text can then be interpreted as *a rather curious division exercise*:

What is 1 01 01 01 divided by 13?   Answer: 4 41 37.

The division exercise is curious, because 1 01 01 01, written with just four ones, is what may be called a "funny number", and because it seems rather strange that anyone would have bothered to find out that this large and non-regular sexagesimal number is not a prime number but a product of two (or more) smaller numbers.

7.1 b. A Proposed Solution Algorithm for the Division Problem

In Old Babylonian mathematics, division problems could be solved in two ways. If the set task was to divide a given number *a* by a *regular* sexagesimal number *b*, then the reciprocal igi *b* of *b* was first computed, and then *a* was multiplied by this reciprocal. Thus, the rule was that

a divided by *b* = igi *b* · *a* if *b* is a regular sexagesimal number.

(Cf. the discussion of the division exercise MS 3871 in Sec. 1.3 above. In that example, all the numbers involved are regular sexagesimal numbers.)

On the other hand, if *b* was a *non-regular* sexagesimal number, then a question like one of the following ones was asked:

mi-nam a.na *b lu-uš-ku-un ša a i-na-di-nam*   (as in YBC 4608, *MCT* D)

mi-nam a.na *b* ḫé.gar *ša a in.si*   (as in Str. 363, *MKT 1*)

In both cases, the translation would be something like

What shall I put as much as *b* that will give me *a*?

In ordinary language: What times *b* is equal to *a*? Thus, also in the case of a non-regular divisor, the division problem was transformed into an equivalent multiplication problem. No details are known about how an Old Babylonian mathematician would attack a problem of this kind. However, there are reasons to believe that a systematic approach like the "recursive division algorithm" described below may have been used, at least in more complicated cases. (See the discussion in Secs. A6 e-g in App. 6 of division problems in mathematical cuneiform texts from the third millennium BC.)

Take the concrete example when *b* = 13 and *a* = 1 01 01 01. The idea would be to start by first finding approximate solutions to the successive equations

13 · ? = 1 00 (1 géš), 13 · ? = 10 00 (10 géš), 13 · ? = 1 00 00 (1 šár),

13 · ? = 10 00 00 (10 šár), 13 · ? = 1 00 00 00 (1 šár.gal),

and then combine the results in the proper way. This could be done in the following way:

13 ·	**5**	=	1 05	=**1** 00	(+ 5)
13 ·	46	=	9 58	=**10** 00	(− 2)
13 ·	**4 37**	=	1 00 01	=**1** 00 00	(+ 1)
13 ·	46 09	=	9 59 57	=**10** 00 00	(− 3)
13 ·	**4 36 55**	=	59 59 55	=**1** 00 00 00	(− 5).

The addition of the approximate equations in lines 5, 3, and 1, gives the intermediate result that

13 · (4 36 56 + 4 37 + 5) = 1 00 00 00 (− 5) + 1 00 00 (+ 1) + 1 00 (+ 5).

Hence, the final result of the division algorithm is that

13 · 4 41 38 = 1 00 00 + 1 00 00 + 1 00 (+ 1) = 1 01 01 01, exactly!

Each step of the algorithm in the example above can be based on the result of the preceding step. When it has been shown that, for instance, 13 · 5 = 1 00 + 5, then a multiplication of both sides of this equation by 10 will show that 13 · 50 = 10 00 + 50. More precisely, 13 · 46 = 10 00 + 50 − 52 = 10 00 − 2. In the next step, multiplying both sides of the equation 13 · 46 = 10 00 − 2 by 6 will show that 13 · 4 36 = 1 00 00 − 12, or, more precisely 13 · 4 37 = 1 00 00 − 12 + 13 = 1 00 00 + 1. And so on.

From a modern, anachronistic point of view, the method is equivalent to finding successively improved approximations to 1/13 and 10/13 in terms of *sexagesimal fractions*. Indeed, it follows from the indicated computations, lines 1, 3, and 5, that successively better approximations to 1/13 are ;05, ;04 37, and ;04 36 56. Similarly, the intermediate lines 2 and 4 in the same series of computations show that successively better approximations to 10/13 are ;47 and ;46 10. It is tempting to conjecture that similar considerations prompted some anonymous Old Babylonian mathematician to compose the text of the cuneiform tablet M 10 (Free Library of Philadelphia), a curious table of approximate values, in sexagesimal fractions, for 1/7, 1/11, and 1/13, followed by approximate values for 10/14 and 10/17. See Sachs, *JCS* 6 (1952).

7.1 c.   UET 5, 121 § 2. A Parallel Text from Early Old Babylonian Ur

In the extensive corpus of known Old Babylonian mathematical texts, there are only three examples of division exercises. Two of those are MS 2317 immediately above and MS 3817 in Sec. 1.3 above.

The third example is **UET 5, 121**, a small clay tablet from Ur with a series of inheritance problems (parallel to MS 1844 in Fig. 7.4.2 below) on the obverse, and three division problems on the reverse, all pretending to be examples of practical mathematics. (See Friberg, *RA 94* (2000), 138-139.)

One of the three division problems is *UET 5, 121 § 2 b*, a *dressed up version* of the division problem in MS 2317. In *UET 5, 121*, just as in MS 2317, the answers to the division problems are given, but not the details of the solution procedures. In the dressed up version of § 2b, the prescribed data are that 1 01 01 01 (= 1 01 · 1 00 01) sheep are allotted to 13 shepherds, and the explicitly given answer is that each shepherd is allotted precisely 4 41 37 (= 1 01 · 4 37) sheep. (Note that 1 00 01 = 3,601 = 4 37 · 13.)

In § 2a of the same text, 1 01 01 01 (= 1 01 · 1 00 01) goats are divided among 13 13 (= 1 01 · 13) shepherd boys, and in § 2c, 1 01 01 sheep are divided among 7 shepherds.

UET 5, 121 §§ 2 a-c.
Sheep, shepherds, and a funny number. A dressed up division exercise.

1	1(šár'u).gal 1(šár) 1(géš) 1 ud₅.ḫá / 1(géš'u) 3 13 kab.ra	1' 01 01 01 goats, 13 13 shepherd boys.
2	kab.ra.1.e / en.nam íb.ši.ti	1 shepherd boy, what does he approach to (take)?
3	4ᵥ 37 íb.ši.ti	4 37 he approaches to.
1	1(šár'u).gal 1(šár) 1(géš) 1 udu.ḫá 13 sipa /	1' 01 01 01 sheep, 13 shepherds.
2	sipa.1.e en.nam íb.ši.ti /	1 shepherd, what does he approach to?
3	4 41 37 íb.ši.ti	4 41 37 he approaches to.
1	1(šár) 1(géš) 1 udu.ḫá 7ᵥ sipa /	1 01 01 sheep, 7 shepherds.
2	sipa.1.e en.nam íb.<ši.ti> /	1 shepherd, what does he <approach to>?
3	8 43 íb.ši.ti	8 43 he approaches to.

UET 5, 121 is interesting in several ways. It is one of a group of four small mathematical clay tablets found in what remains of a rich man's house at "1 Broad Street" in Ur. This group of clay tablets can be dated rather exactly to a relatively early part of the Old Babylonian period, among other things because the city of Ur was abandoned in 1763 BC (in the middle chronology), after Hammurabi had defeated Rim-Sîn of Larsa. The clay tablets are written almost exclusively in Sumerian, and they all use variant number signs. What is particularly interesting is that in the questions (but not in the answers) in *UET 5, 121*, large sexagesimal numbers are expressed in the Sumerian *non-positional* system, with special notations for 60, 10 · 60, and 60 · 60.

The implications of all this are vague but exciting. Did the division problems §§ 2a-c on the tablet *UET 5, 121*, form part of a Sumerian corpus of mathematics, of which virtually nothing else is known? Is the clay tablet MS 2317 with its undressed version of one of the division problems also from Ur, or is it from someplace else and from a later part of the Old Babylonian period? If it is from Ur, why is it not a round tablet, as all other known mathematical texts from Ur inscribed exclusively with numbers? If it is not from Ur, is it an example of how a problem type invented in one of the Mesopotamian cities could spread to other parts of Mesopotamia?

Note: Also *decimal* funny numbers can have interesting factorizations. Thus, for instance, the decimal funny number 1,001 (as in *Thousand and One Nights*), equal to the sexagesimal number 16 41, is the product of 7, 11, and 13. (Cf. the discussion in Sec. 7.2 c below of MS 2297, YBC 7353, YBC 11125, and VAT 7530 § 3.)

7.2.   Combined Market Rate Exercises

7.2 a.   Combined Market Rate Exercises with Regular Sexagesimal Market Rates

MS 2830 (Fig. 7.2.1, top) is a small rectangular hand tablet, similar in format to hand tablets with single multiplication tables. Two tabular arrays are inscribed on **MS 2830, rev.** The way in which these tabular arrays are constructed is fairly obvious. They are reproduced schematically below, in a somewhat more readable transliteration:

MS 2830 § 2a				MS 2830 § 2b			
1 shekel of silver				1 shekel of silver			
1	1	28 48	28 48	2	30	28 07 30	56 15
2	30	14 24	28 48	3	20	18 45	56 15
3	20	9 36	28 48	15	4	3 45	56 15
4	15	7 12	28 48	6	10	9 22 30	56 15

In the case of the first array (§ 2a), for instance, one may assume that the numbers 1, 2, 3, 4 in col. *i* are *arbitrarily prescribed data*. The numbers 1, 30, 20, 15 in col. *ii* are *the reciprocals* of the numbers in col. *i*, presumably with the values 1, ;30, ;20, and ;15, the "inverted values" of the four given numbers. *The sum of these inverted values* is

$$1 + {;}30 + {;}20 + {;}15 = 2{;}05, \text{ a } regular \text{ sexagesimal number.}$$

The number 28 48 appearing in all four lines of col. *iv* is the reciprocal of 2 05, since

$$2\ 05 \cdot 28\ 48 = 5 \cdot 5 \cdot 5 \cdot 28\ 48 = 5 \cdot 5 \cdot 2\ 24 = 5 \cdot 12 = 1 \quad \text{(in relative numbers).}$$

Hence, it is reasonable to interpret the number 28 48 recorded in all four lines of col. *iv* as *the inverted value ;28 48 of the sum 2;05 of the inverted values in col. ii.*

The numbers in the four lines of col. *iii* can now be seen to be equal to *the values in the four lines of col. ii, multiplied by the constant factor ;28 48.* Indeed,

$$\begin{aligned} {;}28\ 48 \cdot 1 \quad &= \quad {;}28\ 48, \\ {;}28\ 48 \cdot {;}30 \quad &= \quad {;}14\ 24, \\ {;}28\ 48 \cdot {;}20 \quad &= \quad {;}09\ 36, \\ {;}28\ 48 \cdot {;}15 \quad &= \quad {;}07\ 12. \end{aligned}$$

The computation described above is the solution algorithm for what may be called a "combined market rate exercise". (See Friberg, *RlA 7* Sec. 5.2 h.) Let the "market rate" *r* of a given commodity be the number of "units" of that particular commodity that *can be purchased for 1 shekel of silver.* The nature of a unit depends, of course, on the commodity considered. For *barley, etc.*, it could be the g u r (a capacity unit of 5 00 s i l a, each equal to about 1 liter), for *metals* it could be the mina (a weight unit equal to about 500 grams), for *fish* it could be a basket of sixty fishes, and so on.

Note that in a market economy *before the invention of money*, it was more convenient to operate with market rates (Sum.: g á n . b a or k i . l a m, Akk.: *maḥīrum*) than with prices!

In col. *i* of MS 2830 § 2a, four market rates are listed,

$$r = 1, 2, 3, \text{ and } 4 \text{ units, respectively, for 1 shekel of silver.}$$

The inverted value of the market rate of a given commodity may be understood as its "unit price" *p* in silver, because

if *r* units can be bought for 1 shekel of silver, then $p = 1/r$ shekels of silver is the price of 1 unit.

In col. *ii* of § 2a are inscribed the unit prices of the four given commodities:

$$p = 1, 1/2\ ({;}30), 1/3\ ({;}20), \text{ and } 1/4\ ({;}15) \text{ shekels of silver per unit.}$$

The sum of the four different unit prices can be called the "combined unit price" *P*:

$$P = 2{;}05 \text{ shekels of silver for a combination of four units, one of each kind of commodity.}$$

The inverse value of the combined unit price is the "combined market rate" *R*:

$$R = {;}28\ 48 \text{ combinations of 1 unit of each kind of commodity per shekel of silver.}$$

Another, equivalent, way of characterizing the combined market rate *R* is to say that

R is the combined market rate if the total price of *R* units of each kind of commodity is 1 shekel of silver.

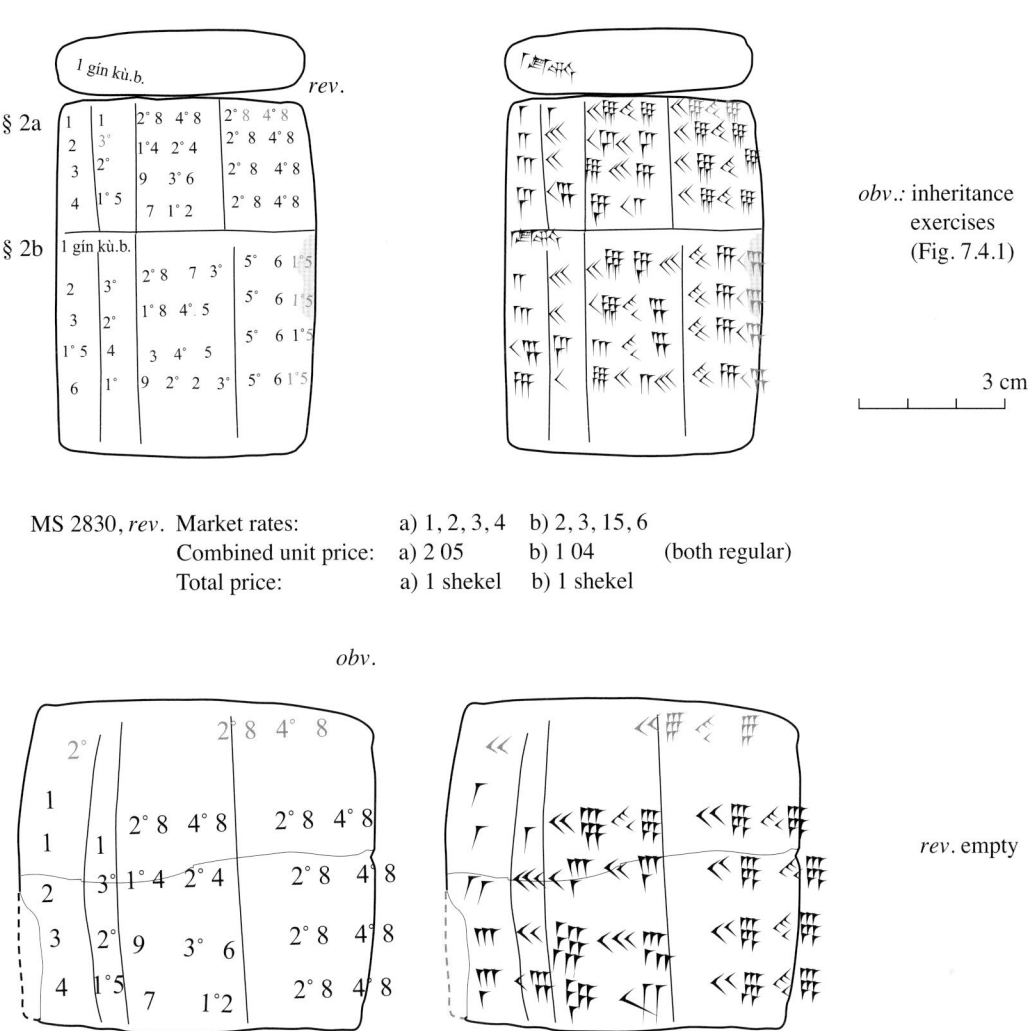

MS 2830, *rev.* Market rates: a) 1, 2, 3, 4   b) 2, 3, 15, 6
                  Combined unit price: a) 2 05   b) 1 04   (both regular)
                  Total price: a) 1 shekel   b) 1 shekel

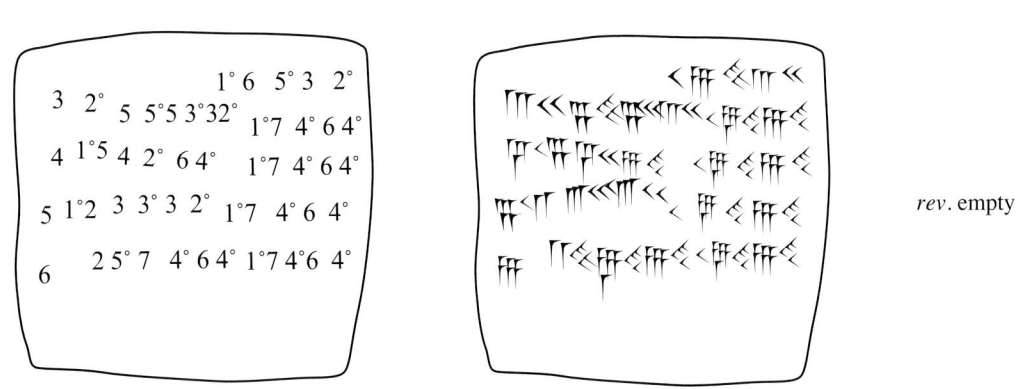

MS 2832. Market rates: 1, 2, 3, 4. Combined unit price: 2 05 (regular). Total price: 1.

MS 2299. Market rates: 3, 4, 5, 6. Combined unit price: 57 (non-regular). Total price: 16 53 20.

Fig. 7.2.1. Three Old Babylonian combined market rate exercises with regular market rates.

In col. *iii* of § 2a on MS 2830 are inscribed the prices of *R* units of each kind of commodity, when *R* = ;28 48:

;28 48 · 1 = ;28 48, ;28 48 · 1/2 = ;14 24, ;28 48 · 1/3 = ;09 36, and ;28 48 · 1/4 = ;07 12 (shekels of silver).

It is easy to check that, indeed, the total of these four individual prices is precisely 1 shekel of silver:

;28 48 + ;14 24 + ;09 36 + ;07 12 = 1 (shekel).

Col. *iv* seems to be the final result of the computation, namely that in order to get a total price equal to precisely 1 shekel of silver, one has to purchase *R* = 28 48 units of *each* kind of commodity.

Briefly, it seems to be clear that the purpose of the first tabular array (§ 2a) on MS 2830, *rev.* is *to compute a number R such that 1 shekel of silver is the total price of R units of each of four commodities with the individual market rates 1, 2, 3, and 4 units per shekel of silver.* Note that the prescribed total price, 1 shekel of silver, is inscribed above the array, on the edge of the clay tablet.

The second tabular array on MS 2830, *rev.* (§ 2b), is similar. Here the four given market rates are 2, 3, 15, and 6 units per shekel, the corresponding unit prices ;30, ;20, ;04, and ;10 shekels per unit. Consequently,

the combined unit price in § 2b is ;30 + ;20 + ;04 + ;10 shekels = 1;04 shekels.

Since 1 04 is a *regular* sexagesimal number with the reciprocal 56 15 (see Sec. 2.5), it follows that

the combined market rate *R* in § 2b is equal to ;56 15 units per shekel (the value inscribed in col. *iv*).

The individual prices of *R* units of each kind of commodity are recorded in col. *iii*:

;56 15 · ;30 = ;28 07 30, ;56 15 · ;20 = ;18 45, ;56 15 · ;04 = ;03 45, and ;56 15 · ;10 = ;09 22 30 shekels.

Again, it is easy to check that the total of these four individual prices is precisely 1 shekel of silver:

;28 07 30 + ;18 45 + ;03 45 + ;09 22 30 = 1 (shekel).

This prescribed total price, 1 shekel of silver, is recorded above the array.

MS 2832 (Fig. 7.2.1, middle) is a square clay tablet inscribed with an array of rather large number signs. In this respect, the text is reminding of the simple multiplication, squaring, and division exercises discussed in Secs. 1.1.1-1.1.3 above. As for its content, MS 2832 is an almost exact parallel to MS 2830 § 2a. The only significant difference is that the prescribed total price is written above the array as '1 shekel silver' in MS 2830 § 2a, but simply as '1' in MS 2832. There is also a half erased number, 20 above the array on MS 2832, probably having little to do with the ensuing calculation. The similarly half erased number 28 48 in the upper right corner of the obverse was probably erased because it was misplaced. The four copies of the computed combined market rate 28 48 are correctly placed in lower positions on the obverse.

MS 2299 (Fig 7.2.1, bottom) is, just like MS 2832, a "combined market rate text" on a square clay tablet inscribed with relatively large number signs. The text looks rather messy, as the cuneiform signs are badly formed and weakly imprinted. There are no ruled lines between the columns of the array, and the numbers in different columns are not adequately separated from each other. Anyway, the intended arrangement is shown in the corrected transliteration to the right below:

MS 2832				**MS 2299**			
20			28 48				
1						16 53 20	
1	1	28 48	28 48	3	20	5 55 33 20	17 46 40
2	30	14 24	28 48	4	15	4 26 40	17 46 40
3	20	9 36	28 48	5	12	3 33 20	17 46 40
4	15	7 12	28 48	6	(10)	2 57 46 40	17 46 40

Just like the numbers 1, 2, 3, and 4 in col. *i* of MS 2830 § 2a, and the numbers 2, 3, 15, 6 in col. *i* of MS 2830 § 2b, the given numbers 3, 4, 5, and 6 in col. *i* of MS 2299 are all *regular* sexagesimal numbers. The numbers in col. *ii* are their reciprocals, with the values ;20, ;15, ;12, and ;10. (By mistake, the author or copyist of the text forgot to write down the number 10 in col. *ii*, row 4.)

An added complication in MS 2299, in comparison with the examples in MS 2830 Secs. 2a and 2b, is that the sum of the reciprocals of the given market rates, that is the combined unit price P, is a *non-regular* sexagesimal number in MS 2299:

$$P = {;}20 + {;}15 + {;}12 + {;}10 = {;}57.$$

The combined market rate is the reciprocal of the combined unit price. Therefore, *when the combined unit price is a non-regular sexagesimal number, as in MS 2299, the value of the combined market rate cannot be computed exactly as a sexagesimal fraction*. For this reason, the problem stated and solved in MS 2299 is of a slightly different type compared with the corresponding problems in MS 2830 Secs. 2a and 2b. In particular, in MS 2299 the given total price is no longer 1 shekel of silver, as in the preceding problems. Instead, the given amount of silver is '16 52 30', the number recorded above cols. *iii* and *iv*.

The purpose of a tabular array such as the one on MS 2299 seems to have been *to compute a number N such that a given amount S of silver is the total price of N units of each of several commodities with given individual market rates*. The given market rates are inscribed in col. *i* of the array, the given amount S of silver is recorded above the array, and the computed number N is repeated several times in col. *iv*.

The first step of the solution algorithm, to compute the combined unit price P, gave the result in the case of MS 2299 that P has the non-regular value ;57 (shekels for 1 unit of each commodity). The second step of the solution algorithm is to find N as a solution to the linear equation

$$N \cdot P = S.$$

When P is a non-regular sexagesimal number, as in the case of MS 2299, the solution N to this linear equation is an *exact* sexagesimal number only if S is a multiple of P. It is not difficult to check that this requirement is satisfied here, since 16 53 20 = 57 · 17 46 40. Indeed, in sexagesimal arithmetic,

$$57 \cdot 17\ 46\ 40 = 16\ 09\ (00\ 00) + 43\ 42\ (00) + 38\ (00) = 16\ 53\ 20\ (00).$$

It remains to find out how an Old Babylonian student could find an answer to the question

What times 57 is 16 53 20?

The simplest way would have been to start by finding a factorization of 16 53 20, since the author of the problem was kind enough to let the given value 16 53 20 be the combined unit price 57 multiplied by a *regular* sexagesimal number. The factorization could be achieved in a series of simple steps:

$$16\ 53\ 20 = 20 \cdot 50\ 40, \quad 50\ 40 = 40 \cdot 1\ 16, \quad 1\ 16 = 4 \cdot 19, \quad \text{hence} \quad 16\ 53\ 20 = 20 \cdot 40 \cdot 4 \cdot 19.$$

Since 57 = 3 · 19, it was then easy to see that 16 52 30/57 = 20 · 40 · 4/ 3 = 17 46 40.

Remark: In the absence of any specific indications in the text of MS 2299 how the recorded numbers should be interpreted, there is more than one conceivable explanation of the problem and its solution. Instead of commodities purchased, one may think of wares produced, or work finished, *etc.*, in a given period of time. (Cf. Friberg, *RlA 7* Sec. 5.6 h.) Thus, instead of a combined market rate problem of the kind described above, the problem behind the numerical array on MS 2299 may have been, for instance, a "combined work norm problem" of the following kind:

Given four kinds of wares produced or work finished *in equal quantities* at four different work rates, namely 3, 4, 5, and 6 units per man-day, and given a total of 16;53 20 man-days (understood as 16 and 2/3 and 1/3 of 2/3 man-days). Then the combined cost in labor is ;57 man-days for 1 unit of each kind, and 17;46 40 units of each kind can be produced or finished in the given 16;53 20 man-days. The cost in labor for 17;46 40 units of the first kind is 17;46 40 · ;20 man-days = 5;55 33 20 man-days, *etc.*

7.2 b.   *YBC 7234, 7235, 7354, 7355, 7358, and 11127, Six Parallel Texts in MCT*

A number of cuneiform texts with tabular arrays were published by Neugebauer and Sachs in *MCT* (1945), 17. Although the general structure of the arrays was correctly analyzed, no explanation of the meaning of the arrays was offered. Photos of YBC 7358 and 11127, were published by Nemet-Nejat in *JNES* 54 (1995), and of YBC 7234, 7235, 7354, 7355, 7358, and 11127, again by Nemet-Nejat, in *UOS* (2002), in both cases with no explanation offered. Actually, the characterization of such hand tablets with tabular arrays as "help tables for combined market rate exercises", and the explanation of "combined market rate exercises" as parallels to the more generally understood "combined work norm exercises", appeared for the first time in Friberg, *RlA 7* (1990) Sec. 5.2 h and §§ 5.6 h-i.

Two of the tabular arrays published in *MCT* are the following:

YBC 7358

		1 42 45	
1	1	45	45
2	30	22 30	45
3	20	15	45
4	15	11 15	45
[5]	12	9	45

YBC 7235

			1 03 20	
1 40	1	1	40	1 06 40
5	3	20	13 20	1 06 40
6 40	4	15	10	1 06 40
				40

The first of these arrays, **YBC 7358**, can immediately be interpreted as the data for a combined market rate problem similar to the one in MS 2299. According to this interpretation, five given market rates are 1, 2, 3, 4, and 5 (units per shekel). The corresponding unit prices are 1, ;30, ;20, ;15, and ;12 (shekels per unit). The combined unit price (as usual, not explicitly indicated in the text) is 2;17 (shekels for 1 unit of each kind). Clearly, 2 17 is a *non-regular* sexagesimal number. As in MS 2299, the given total price in this case (recorded above the array in YBC 7358), is equal to the combined unit price 2 17 multiplied by a *regular* sexagesimal number. It is easy to see that 1 42 45 = 15 · 6 51 = 15 · 3 · 2 17 = 45 · 2 17. Consequently the number of units purchased of each commodity is 45, the number recorded five times in col. *iv*.

In **YBC 7235**, the numerical array has *five* columns, one more than the usual number. The most likely interpretation in this case is that three given market rates are 1 40 (= 100), 5 00 (= 300), and 6 40 (= 400) units per shekel. (col. *i*), and that the given total price is 1 03;20 shekels (recorded above the array). Expressed differently, the given market rates are 1, 3, and 4 *hundreds per shekel* (col. *ii*). The corresponding unit prices are 1, ;20, and ;15 *shekels per hundred* (col. *iii*). Consequently, if 1 hundred of each kind were purchased, the combined price would become 1;35 shekels. The given total price, presumably to be understood as 1 03;20 shekels, is 40 times larger. Therefore, the number 40 is recorded under the other numbers in col. *v*. The final result is that *40 hundreds* of each kind must be purchased. The number 40 hundreds = 1 06 40 (in decimal numbers 4,000) is recorded three times in col. *v*, and the corresponding individual prices 40, 13;20, and 10 (shekels) are recorded in col. *iv*. It is easy to check that, as required, 40 + 13;20 + 10 = 1 03;20 (shekels).

7.2 c.   Combined Market Rate Exercises with One Non-Regular Factor in the Data

In all the market rate tables considered in Sec. 7.2 a above, the given market rates in col. *i* were *regular* sexagesimal numbers. Otherwise it would not have been possible to compute exactly the unit prices in col. *ii* as the inverted values of the market rates.

In **MS 2268/19, *obv.*** (Fig. 7.2.2, top), on the other hand, a somewhat changed approach allows the presence of a *non-regular* factor, the same in one or more of the given market rates:

MS 2268/19, *obv.*

		1 00 05 20	
3 30	1	21 20	1 14 40
5 50	36	12 48	1 14 40
7	30	10 40	1 14 40
7 30	28	9 57 20	1 14 40
14	15	5 20	1 14 40

The given numbers in col. *i* of this text are 3 30, 5 50, 7, 7 30, and 14, presumably with the intended values 3;30 = 7 · ;30, 5;50 = 7 · ;50, 7, 7;30, and 14 = 7 · 2. Thus, all of these, except 7;30, contain the *non-regular* sexagesimal number 7 as a factor. Clearly, the sexagesimal reciprocals of the numbers with 7 as a factor do not exist. Therefore, assuming that the numbers in col. *i* are market rates, the numbers in col. *ii* cannot be the corresponding unit prices. Instead, they are the prices of 3;30 = 3 1/2 units of each kind of commodity. The number 3;30 can be thought of as the "false" (in the sense of "posited") size of the equal purchases made of each commodity, in an application of the method of false value. (See Friberg, *RlA 7* (1990) Sec. 5.7 d.) It is easy to check that

The price of 3;30 units at a market rate of	3;30	units per shekel	is	1	shekel,	
the price of 3;30 units at a market rate of	5;50	units per shekel	is	;36	shekel	(3;30/5;50 = 3/5 = ;36),
the price of 3;30 units at a market rate of	7	units per shekel	is	;30	shekel	(3;30/7 = 1/2 = ;30),
the price of 3;30 units at a market rate of	7;30	units per shekel	is	;28	shekel	(3;30/7;30 = 7/15 = ;28),
the price of 3;30 units at a market rate of	14	units per shekel	is	;15	shekel	(3;30/14 = 1/4 = ;15).

Hence, the combined price for 3 1/2 units of each commodity is

$$1 + ;36 + ;30 + ;28 + ;15 = 2;49 \text{ (shekels).}$$

Here 2 49 is a *non-regular* sexagesimal number. As could be expected, 1 00 05 20, the given total price, is a multiple of this combined price. Assuming that the intended value of 1 00 05 20 is, for instance, 1;00 05 20 shekels, one finds that the value of the total price is precisely ;21 20 times the false price 2;49 shekels, as shown by the factorization

$$1\ 00\ 05\ 20 = 20 \cdot 3\ 00\ 16 = 20 \cdot 4 \cdot 45\ 04 = 20 \cdot 4 \cdot 4 \cdot 11\ 16 = 20 \cdot 4 \cdot 4 \cdot 4 \cdot 2\ 49 = 21\ 20 \cdot 2\ 49.$$

Therefore, the "true prices" in col. *iii* are ;21 20 times the "false prices" in col. *ii*. Indeed,

;21 20 · 1		=	;21 20,
;21 20 · ;36	= ;12 + ;00 36 + ;00 12	=	;12 48,
;21 20 · ;30	= ;21 20/2	=	;10 40,
;21 20 · ;28	= ;09 20 + ;00 28 + ;00 09 20	=	;09 57 20,
;21 20 · ;15	= ;21 20/4	=	;05 20.

The sum of these true prices is, of course, equal to the given total price:

$$;21\ 20 + ;12\ 48 + ;10\ 40 + ;09\ 57\ 20 + ;05\ 20 = 1;00\ 05\ 20.$$

Finally, the true number of units that can be purchased of each commodity for this total price must be ;21 20 times the initially chosen false size of the equal purchases. In other words, it is

$$;21\ 20 \cdot 3;30 = 1;10 + ;03\ 30 + ;01\ 10 = 1;14\ 40 \text{ (units).}$$

This is, then, the value of the number 1 14 40 recorded in all 5 rows of col. *iv* of MS 2268/19.

The strange form of the given total price, 1;00 05 20 shekels = 1 shekel + 1/4 and 1/60 barley-corn, can be explained as follows. It was shown above that the combined price for 3 1/2 units of each kind is 2;49 shekels. Therefore, if the reciprocal of 2;49 had existed, the combined market price could have been computed as

$$3;30/2;49 = 3;30 \cdot \text{rec. } 2;49.$$

Since 2;49 is non-regular, the reciprocal does not exist, but ;21 20 is a good *approximative reciprocal*. Indeed,

$$;21\ 20 \cdot 2;49 = 1;00\ 05\ 20.$$

Hence, the number 1 14 40 recorded in the fourth column of MS 2268/19 can be interpreted as the *approximate combined market rate*

$$3;30 / 2;49 = \text{appr. } ;21\ 20 \cdot 3;30 = 1;14\ 40 \text{ (units of each kind per shekel).}$$

(Approximate solutions of this kind are extremely rare in Old Babylonian mathematics!)

7.2 d.   *N 3914, a Parallel Text with 10 Given Numbers*

Another text of a similar type is **N 3914**, an Old Babylonian clay tablet from Nippur, of the same size and shape as the YBC clay tablets mentioned above with numerical data for combined market rate problems. (All those clay tablets are unprovenanced). N 3914 was published by Robson in *SCIAMVS* 1 (2000), 28, as a hand copy with a transliteration, but without any valid interpretation.

obv.

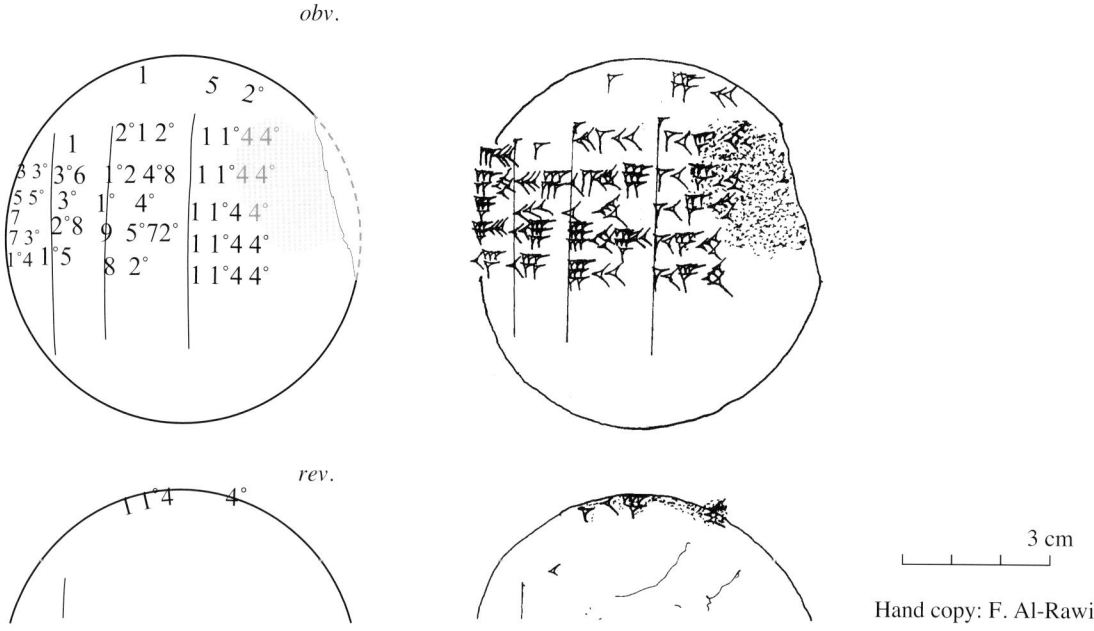

rev.

MS 2268/19, *obv.* Market rates: 3 30, 5 50, 7, 7 30, 14 (non-regular). Combined price: 2 49 for 3 30.
Total price: 1 00 05 20. Number of purchases: 1;14 40 units of each kind.
rev. The number recorded on the upper edge looks like 1 14 00 40 but may be a badly written 1 14 40.

obv.

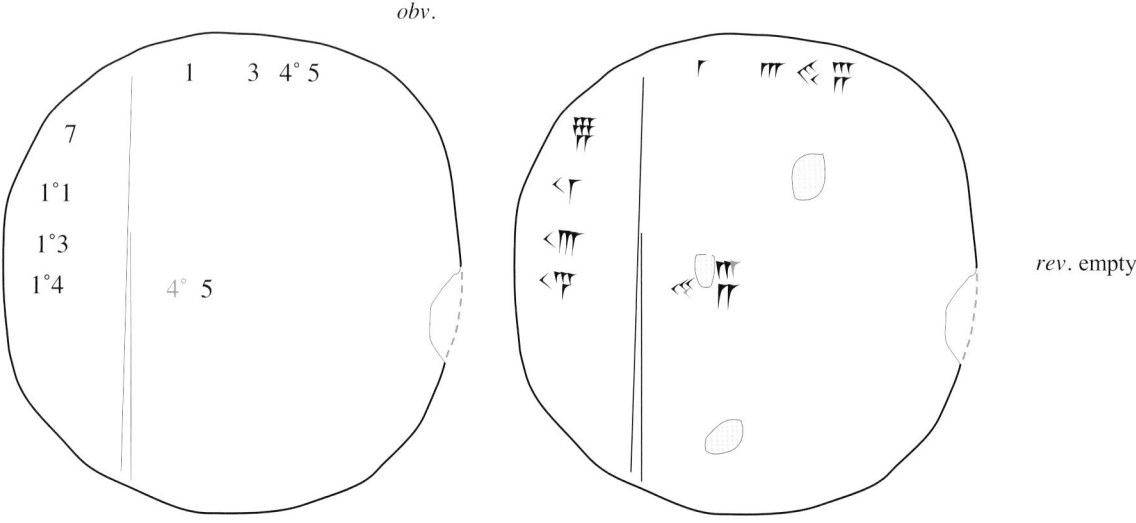

rev. empty

MS 2297. Market rates: 7, 11, 13, 14 (non-regular).
[Combined price: 6 22 30 for 16 41 units. 10 · 6 22 30 = 1 03 45.]

Fig. 7.2.2. Two Old Babylonian combined market rate tables with non-regular market rates.

In N 3914, the given market rates range from 1 to 10, and the given total price is 3 25 01 40, possibly to be understood as 3 25;01 40 shekels = 3 1/3 minas 5 shekels 5 barley-corns. One of the given market rates, 7, is *non-regular*. For that reason, the initially chosen false size of the equal purchases is 7 (units). The numbers recorded in col. *ii* are the ten false prices, the individual unit prices multiplied by 7. The combined price for 7 units of each kind is

$$7 + 3;30 + 2;20 + 1;45 + 1;24 + 1;10 + 1 + ;52\ 30 + ;46\ 40 + 42 = 20;30\ 10.$$

This combined price is one tenth of the given total price 3 25;01 40. Consequently, the correct size of the equal purchases is ten times as large as the false size. Therefore, it is $10 \cdot 7 = 1\ 10$ (units), the number repeatedly recorded in col. *iv*. The total of all the ten purchases, $11\ 40 = 10 \cdot 1\ 10$ (units), is recorded in the right margin.

Here follows a slightly amended transliteration of the text, with damaged parts reconstructed:

N 3914

3 25 01 40

1	7	1 10	1 10	11 40
2	3 30	35	1 10	
3	2 20	23 20	1 10	
4	1 45	17 30	1 10	
5	1 24	14	1 10	
6	1 10	11 40	*etc.*	
7	1	10		
8	52 30	8 45		
9	46 40	7 46 40		
10	42	7		

7.2 e. Combined Market Rate Exercises with Several Non-Regular Factors in the Data

MS 2297 (Fig. 7.2.2, bottom) is a round clay tablet inscribed with what looks like only the first column of a tabular array, plus traces of additional numbers. The sexagesimal number 1 03 45 can be seen near the upper edge, at the place where the given total price is usually recorded in combined market rate exercises. Conceivably, a student first watched a teacher's solution to a combined market rate problem, then entered on this tablet the given market rates and the given total price, intending to try later to find on his own the solution to the problem. Which he never got around to do. Here follows a transliteration of the unfinished text:

MS 2297

1 03 45

7	
11	
13	
14	45?

A further comment to this text will have to wait until some possibly parallel texts have been considered. (See the note inserted below, after the discussion of the text VAT 7530 § 3.)

Two possibly parallel texts are **YBC 7353** and **YBC 11125**, published in transliteration, but without any attempted explanation, by Neugebauer and Sachs in *MCT*, 17:

YBC 7353

3 11 15

7	2 23	1 11 30	8 20 30
11	1 31	45 30	8 20 30
13	1 17	38 30	8 20 30
14	1 11 30	35 45	8 20 30

YBC 11125

4 15

7	2 23	1 35 20	11 07 20
11	1 31	1 00 40!	11 07 20
13	1 17	51 20	11 07 20
14	1 11 30	47 40	11 07 20

In both texts, four given market rates are 7, 11, 13, and 14 (units per shekel). Since the given market rates are *non-regular* sexagesimal numbers (with three different non-regular factors, 7, 11, and 13), the corresponding unit prices cannot be expressed exactly as finite sexagesimal numbers. The difficulty is sidestepped by assuming *false equal purchases of a sufficiently large quantity,* namely

$$7 \cdot 11 \cdot 13 = 16\ 41 \text{ (in decimal notation 1,001).}$$

The corresponding false prices are then

$$16\ 41/7 = 11 \cdot 13 = 2\ 23,\ 16\ 41/11 = 7 \cdot 13 = 1\ 31,\ 16\ 41/13 = 7 \cdot 11 = 1\ 17, \text{ and } 16\ 41/14 = 1\ 11;30.$$

The resulting combined price (not indicated in the texts) is

$$2\ 23 + 1\ 31 + 1\ 17 + 1\ 11;30 = 6\ 22;30.$$

In YBC 7353, the first of the two texts, the given total price can be interpreted as, for instance, 3 11;15 (shekels), half the combined price. Then, the correct size of the equal purchases is ;30 · 16 41 = 8 20;30 (col. *iv*), and the true prices in col. *iii* are precisely half the false prices in col. *ii*.

In YBC 11125, the second of the two texts, the given total price can be interpreted as 4 15, two-thirds of the combined price. Consequently, the true size of the equal purchases is ;40 · 16 41 = 11 07;20 (col. *iv*), and the true prices in col. *iii* are precisely two-thirds of the false prices in col. *ii*.

7.2 f.   VAT 7530. A Theme Text with Combined Market Rate Problems

In Secs. 7.2 a-c above, five MS texts, 4 parallel YBC texts, and a single text from Nippur were discussed as examples of texts with tabular arrays for combined market rate problems. All those texts, with the exception of MS 2830 *rev.*, are in the form of small round or squarish clay tablets on which are recorded numerical data without any explaining text. They are, in other words hand tablets of the kind discussed in Robson, *MMTC* (1999), App. 5, and Friberg, "Mathematics at Ur", *RA* 94 (2000). The text on the reverse of MS 2830 may have been copied from two such hand tablets.

In addition to these brief and practically wordless texts, there is also one Old Babylonian problem text dealing with combined market rate problems. This is the "theme text" **VAT 7530** published by Neugebauer in *MKT 1*, 287-289, with photo and hand copy in *MKT 2,* and with a renewed, partly successful attempt to analyze its meaning in *MCT*, 18. Six combined market rate problems are formulated in VAT 7530 §§ 1-6. No answers or detailed solution algorithms are provided.

The problems formulated in §§ 1-2 of VAT 7530 are unlike previously known examples of combined market rate problems, and so damaged that it is difficult to find valid interpretations of them. The problems in §§ 5-6 are similar to the one on N 3914 (one non-regular given market rate), while the problems in §§ 3-4 are similar to YBC 7353 (several non-regular market rates). Here is, for instance, the text of VAT 7530 § 5 in transliteration and translation:

VAT 7530 § 5 (*obv.* 17-21).

1	1 ma.na.ta.àm 2 ma.na [3 ma.n]a 4 ma.n[a]	1 mina each, 2 minas, *3 min*as, 4 minas,
2	[5 ma.na] / 6 ma.na 7 ma.na 8 ma.na	*5 minas*, 6 minas, 7 minas, 8 minas,
	9 ma.na 10 m[a.n]a /	9 minas, 10 m*in*as.
3	10 gín igi.4ᵥ.gál *ù*	10 shekels a 4th-part (of a shekel) and
	igi.4ᵥ-*a-at* še kù.babbar /	a 4th part of a barley-corn.
4	kù.babbar *li-li li-ri-da-ma* /	The silver may go up (and) go down, but
5	[gán.]ba *li-im-ta-ḫi-ra*	the market rates may be equal.

In this problem, the given market rates are the same as in N 3914. Hence, just as in N 3914, if the initially chosen false size of the equal purchases is 7 (units), then the combined price is 20;30 10 (shekels). Here, however, the given total price is 10 1/4 shekel 1/4 barley-corn = 10;15 05 shekels, not 3 25; 01 40 (shekels) as in N 3914. This means that the given total price is precisely 1/2 of the combined price. Accordingly, the true size of the equal purchases is 1/2 · 7 = 3;30 (units), and so on. However, no solution algorithm is provided, and no answer to the stated problem is given.

Slightly more complicated is the problem formulated in VAT 7530 § 6:

VAT 7530 § 6 (*rev.* 1-7).

1	1 ma.na.ta.àm 1 ma.na *ù* 10 gín.ta.à[m] /	1 mina each, 1 mina and 10 shekels each,
2	2 ma.na.ta.àm 2 3' ma.na /	2 minas each, 2 1/3 minas,
3	3 ma.na 3 2' ma.na 4$_v$ ma.na 4$_v$ 3" ma.na /	3 minas, 3 1/2 minas, 4 minas, 4 2/3 minas,
4		
5	5 ma.na 5 6" ma.na 2 3' 25 še /	5 minas, 5 5/6 minas, 2 1/3 (shekels) 25 barley-corns
6	*ù* igi.4$_v$-*a-at* še /	and a 4th part of a barley-corn.
7	[kù.babbar] *li-li li-ri-da-ma* /	*The silver* may go up (and) go down, but
	[gán.]ba *li-im-ta-ḫa-ar*	*the mar*ket rates may be equal.

The total price given in lines 4-5 is 2 1/3 shekels 25 1/4 barley-corns = 2;28 25 shekels. Therefore, *the problem* stated *in words* in the text of VAT 7530 can be reformulated, *together with its solution* (not given in the text), *in a tabular array* of the following form:

2 28 25

1	7	35	35
1 10	6	30	35
2	3 30	17 30	35
2 20	3	15	35
3	2 20	11 40	35
3 30	2	10	35
4	1 45	8 45	35
4 40	1 30	7 30	35
5	1 24	7	35
5 50	1 12	6	35

Note that 7 is the only non-regular a factor in 1 10, 2 20, *etc.* (actually, 1;10 = 7 · ;10, 2;20 = 7 · ;20, *etc.*)

The problem stated in VAT 7530 § 3 is the following:

VAT 7530 § 3 (*obv.* 7-10).

1	7 ma.na.ta.àm *ù* 11 ma.na.[ta.àm] /	7 minas each and 11 minas *each*,
2	13 ma.na.ta.àm *ù* 14 ma.na.[ta.àm] /	13 minas each and 14 minas *each*.
3	1 gín 11 še igi.4$_v$-*a-at* še kù.babbar /	1 shekel 11 barley-corns, a 4th part of a barley-corn.
4	kù.babbar *li-li ù li-ri-da*	The silver may go up and go down, but
	ma-ḫi-[rum] li-im-ta-ḫar	the market *rates* may be equal.

The given total price is 1 shekel 11 1/4 barley-corns = 1;03 45 shekel. Hence, the solution to the stated problem, if given in the form of a tabular array, would have taken the following form:

1 03 45

7	2 23	23 50	2 46 50	Cf. the discussion of the arrays in YBC 7353 and 11125. With false equal
11	1 31	15 10	2 46 50	purchases of 7 · 11 · 13 = 16 41 minas, the combined price is 6 22;30 shekels,
13	1 17	12 50	2 46 50	6 00 times the given total price 1;03 45 shekel. Therefore, the true equal pur-
14	1 11 30	11 55	2 46 50	chases are 16 41 / 6 00 = 2;46 50 minas = 2 2/3 minas 6 5/6 shekels.

Note: The text of VAT 7530 § 3 contains only a question, but no answer and no solution procedure. Stated in the form of a tabular array, the question alone would take the form of only the first column and the superscript 1 03 45 in the array above. This is precisely the form of the array on MS 2297 above! Thus, it appears that MS 2297 was an assignment. The student was expected to complete the array with the missing columns and/or write a text along the lines of VAT 7530 § 3.

The problem stated in VAT 7530 § 4 is particularly interesting:

VAT 7530 § 4 (*obv.* 11-16).

1	7 ma.na.ta.àm 11 ma.na.ta.àm 13 ma.na.ta.àm /	7 minas each, 11 minas each, 13 minas each,
2	14 ma.na.ta.àm 19 ma.na.ta.àm *ma-ḫi-rum* /	14 minas each, 19 minas each is the market rate.
3	[1 ma].na 8 6" gín 11 2' še	*1 mina* 8 5/6 shekels 11 1/2 barley-corns,
	igi.4ᵥ-*a-at* še [kù.babbar] /	a 4th part barley-corn of *silver*.
4	*ma-ḫi-ir* 7 ma.na *ù* 19 ma.na [··· ···] /	The market rate of 7 minas and 19 minas ··· ··· ,
5	*ma-ḫi-ir* 11 ma.na *ù* 14 ma.na [··· ···] /	the market rate of 11 minas and 14 minas ··· ··· ,
6	*ma-ḫi-ir* 13 ma.na *li-[te]-er*	the market rate of 13 minas may go beyond.

Expressed as a tabular array, the solution to the problem stated in lines 1-3 would look like this:

			1 08 54 15
7	45 17	22 38 30	2 38 29 30
11	28 49	14 24 30	2 38 29 30
13	24 23	12 11 30	2 38 29 30
14	22 38 30	11 19 15	2 38 29 30
19	16 41	8 20 30	2 38 29 30

The meaning of lines 4-6 in VAT 7530 § 4 is far from obvious, in particular in view of the damage to the ends of lines 4 and 5 and to the middle part of the crucial inflected verb form *li-[te]-er* in line 6. Nevertheless, an interpretation will be attempted here: In more explicit form, the displayed solution to the stated problem may be rephrased as follows:

If five different commodities have the five different market rates 7, 11, 13, 14, and 19 units per shekel, and if the total amount of silver available is 1 08;54 15 shekels, then 2 38;29 30 units can bought of each kind of commodity, that is altogether 5 · 2 38;29 30 units = 13 12;27 30 units.

Therefore, a mixed bag of the five different commodities, with an equal number of each kind, may be bought at an average market rate of 13 12;27 30 units for 1 08;54 15 shekels = (approximately) 11;30 units per shekel.

This "average market rate" is, actually, the *harmonic mean* of the five given market rates. In modern mathematical terms:

If $r_1, ..., r_5$ are the five given market rates then the combined market rate is $5/(1/r_1 + ... + 1/r_5)$.

As a kind of average of the given market rates, the harmonic mean is smaller than the more familiar *arithmetical mean*, which in the case considered here is (7 + 11 + 13 + 14 + 19)/5 = 64/5 = 12;48. A sharp-eyed Babylonian mathematician may have observed that this arithmetical mean of the five market rates is close to 13, the value of the market rate in the middle, as well as to the arithmetical mean (11 + 14)/2 = 12;30 of the two nearest market rates on each side of 13, and even to the arithmetical mean (7 + 19)/2 = 13 of the two extreme market rates on each side of 13. He may then have suspected that something similar may be shown in the case of the average market rate (that is, the harmonic mean). Actually, the average market rate of 7 and 19 is 2 · 7 · 19/(19 + 7) = 7 · 19/13 = 10 3/13 = appr. 10;14, which is not far away from, but *less than* 11;30 (line 4). The average market rate of 11 and 14 is 2 · 11 · 14/(14 + 11) = 12 8/25 = appr. 12;19, which again is not far away from, yet *greater than* 11;30 (line 5). The market rate in the middle, 13, is also *greater than* 11;30 (line 6).

Of all the mentioned Old Babylonian texts concerned with combined market rate problems, MS 2830 *rev.* is almost certainly the *earliest* one, being (probably) from early Old Babylonian Ur. VAT 7530 may also be an early Old Babylonian text, from one of the southern cities in Mesopotamia. This is indicated by the use of the variant number sign 4ᵥ in certain expressions, such as 4ᵥ ma.na, igi.4ᵥ, and igi.4ᵥ-*a-at*.

It is interesting to note that MS 2830 *rev.* is also the *simplest* example of all the known texts of this kind, with only regular sexagesimal numbers used for the given market rates, and with the given total price equal to precisely 1 shekel, written explicitly as 1 gín kù.babbar, not just as '1'.

7.3. Old Babylonian Brick Types and Brick Constants

For obvious reasons, counting with bricks was a popular topic in Old Babylonian practical mathematics, exemplified by many entries in tables of constants and several different kinds of problem texts. Detailed surveys of the subject have been published by Friberg in *ChV* (2001), and by Robson in *MMTC* (1999) §§ 4-6. An effort to combine and refine the results of these two parallel but independent investigations was made in a review of Robson's *MMTC* by Friberg in *AfO* 46/47 (1999/2000). The mentioned surveys can now be completed through the addition to the corpus of a text from the Schøyen Collection with important implications for the subject.

7.3 a.   *MS 2221, obv. Walking Numbers, Loading Numbers, and Carrying Numbers for Three Kinds of Bricks, and for Mud*

MS 2221 (Fig. 7.3.1) is a small square hand tablet, with a table of constants on the left edge and the obverse, and a numerical tabular array on the reverse. The text is an exercise in the use of constants for bricks and mud.

obv.

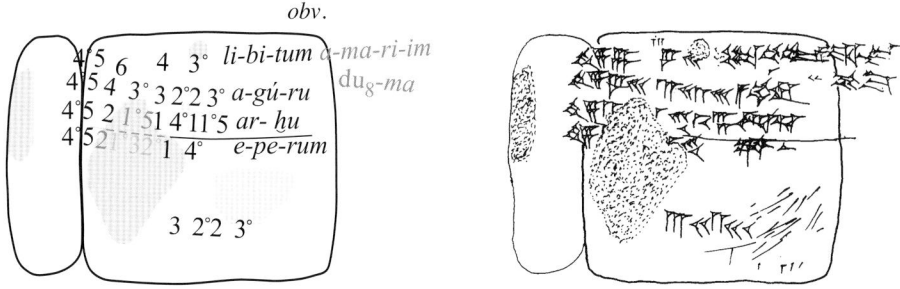

MS 2221 *obv.*: Computations of carrying numbers for three kinds of bricks and for mud.

rev.

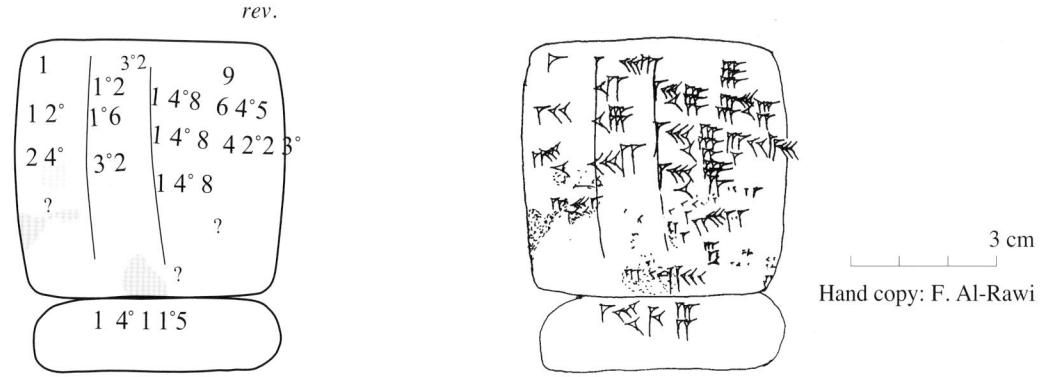

3 cm

Hand copy: F. Al-Rawi

MS 2221 *rev.*: Computation of a combined work norm for the carrying of three kinds of bricks

Fig. 7.3.1. An Old Babylonian hand tablet with brick constants and a brick problem.

Here follows a transliteration of the inscription on the left edge and on the obverse of MS 2221. (Note that the inscription on the right edge following after the first line of text on the obverse is badly readable. It seems to be a half erased leftover from an earlier inscription on the clay tablet. It will not be translated below, even if it seems to mention the molding of bricks and making a 'brick pile'.)

MS 2221, *obv.*

45	6	4 [30]	*li-bi-tum a?-ma?-ri?-im?* / du₈?-*ma*
45	4 30	3 22 30	*a-gú-ru*
45	2 [15]	[1] 41 15	*ar-ḫu*
45	2 [13 20]	[1] 40	*e-pe-rum*
		3 22 30	

1 41 15

It is impossible to understand the meaning of this inscription without a preceding discussion of Old Babylonian "brick metrology", like the one inserted below for the readers' convenience.

Listed in the last column of text on the obverse are the following Akkadian words:

libittu	(Sum. sig₄)	a word for *rectangular* bricks, with the width equal to 2/3 of the length,
agurru	(Sum. sig₄.al.ur₅.ra)	a word for *square* bricks, with the width equal to the length,
arḫu	(Sum. sig₄.áb)	a word for *half square* bricks, with the width equal to 1/2 of the length,
eperu	(Sum. saḫar)	a word for mud or earth, used for the fabrication of bricks.

For lack of more precise translations, the words '(regular) bricks', 'tile-bricks', and 'cow-bricks' will be used in the following as free translations of *libittu, agurru,* and *arḫu* (*arḫu* is Akkadian both for half square bricks and for 'cow'). In the cases attested in Old Babylonian mathematical texts and tables of constants, the length of the side of a square brick may be either 1/3, 2/3, or 1 cubit (= 10, 20, and 30 fingers), or 3 or 4 sixtieths of a ninda (= 18 and 24 fingers). The length of a rectangular brick may be either 1/2 cubit or 3 sixtieths of a ninda, and the length of a half brick may be either 2/3 cubit or 4 sixtieths of a ninda. The thickness of all these diverse types of bricks is usually 5 fingers (= 1/6 of a cubit), but it may also be 6 fingers (= 1/5 of a cubit = 1 sixtieth of a ninda).

Ever since the first discussion of brick types in Old Babylonian mathematical texts, in the comment by Neugebauer and Sachs to the metro-mathematical theme text *MCT,* 91 = YBC 4607, it has been customary to refer to brick types by number (type 1, type 2, *etc.*). A more informative kind of notation is to let the names of the various brick types refer to the *format,* with S for square bricks, H for half bricks, R for rectangular bricks, as well as to the *size,* with the tags 1/3c, 2/3c, and 1c for bricks of length 1/3, 2/3, or 1 cubit, and the tags 3n and 4n for bricks of length 3 or 4 sixtieths of a ninda. Thus, the attested brick types may be referred to as

S1/3c, S2/3c, S1c, S3n, and S4n	in the case of *square* bricks of regular thickness (5 fingers),
H2/3c and H4n	in the case of *half bricks* of regular thickness,
R1/2c and R4n	in the case of *rectangular* bricks of regular thickness.

For bricks of extra thickness (6 fingers), these notations may be augmented by a tag v for "variant", so that one writes S1/3cv, *etc.*

The dimensions of the bricks figuring in *MCT* O = YBC 4607, for example, are mentioned explicitly, which makes it easy to see that those bricks are of the four types R1/2c, R3n, S2/3c, S1c. The dimensions of the bricks figuring in Haddad 104 §§ 9-10 (Al-Rawi and Roaf, *Sumer* 43 (1984)) are mentioned explicitly, too, so that it is clear that the bricks in this latter case are of type S2/3cv.

In most other cases, the dimensions of bricks figuring in Old Babylonian mathematical texts or tables of constants are not mentioned explicitly. The brick type is specified only indirectly, often by reference to its "molding number" (Akk. *nalbanum*), which may be defined as follows:

> Bricks of a given type have the molding number L if the volume of L brick šar of such bricks is 1 volume šar.
> Here 1 brick šar = 12 00 bricks, and 1 volume šar = 1 square ninda · 1 cubit (= 1 n. · 1 n. · 1 c.).

Take, for instance, the most common type of bricks in Old Babylonian mathematical texts, rectangular bricks of type R1/2c. For such "standard bricks", one can compute the following parameters:

> Base area A = 1/2 c. · 1/3 c. = ;02 30 n. · ;01 40 n. = ;00 04 10 sq. n.,
> Volume V = A · 1/6 c. = ;00 04 10 n. · n. · ;10 c. = ;00 00 41 40 volume šar,
> Volume of 1 brick šar = 12 00 · ;00 00 41 40 volume šar = ;08 20 volume šar,
> Molding number L = 7;12 (brick šar per volume šar), since 7;12 · ;08 20 = 1.

For "variant unit bricks", bricks of type S1cv, the corresponding computations are simpler:

Base area A = 1 c. · 1 c. = ;05 n. · ;05 n. = ;00 25 sq. n.,
Volume $V = A$ · 1/5 c. = ;00 25 n. · n. · ;12 c. = ;00 05 volume šar,
Volume of 1 brick šar = 12 00 · ;00 05 volume šar = 1 volume šar,
Molding number L = 1 (brick šar per volume šar).

In other words,

The volume of 1 brick šar of variant unit bricks is 1 volume šar.

"Unitary relations" of this type are relatively frequent in Sumerian and Babylonian metrology. It is likely that this particular unitary relation is the explanation for the seemingly strange introduction of the brick šar = 12 00 bricks as a counting unit for bricks.

Another, even more striking, unitary relation involving variant unit bricks is the following:

The weight of 1 baked variant unit brick is 1 talent.

Now, the volume of L brick šar of bricks of molding number L is 1 volume šar, which is the volume of 1 brick šar of variant unit bricks. This means that a variant unit brick is L times more massive than a brick of molding number L. Consequently,

The weight of L baked bricks of molding number L is 1 talent (= 1 00 minas).
Consequently, the weight of 1 baked standard brick (type R1/2c) is 8 1/3 minas (= 1 talent / 7;12).

The weight of a (baked) standard brick is mentioned explicitly in the text *MCT* Oa = **YBC 7284**, a round hand tablet with the following inscriptions (plus an unrelated number) on obverse and reverse:

obv.	41 40		;00 00 41 40 (volume šar)	the volume of 1 brick
	8 20		;08 20 (volume šar)	the volume of 1 brick šar
	igi.gub.ba.bi 12		its constant 12(00)	bricks in 1 brick šar
rev.	1 sig$_4$		1 brick,	
	ki.lá.bi en.nam		its weight is what?	
	ki.lá.bi 8 3' ma.na		Its weight is 8 1/3 minas.	

The weights of *baked* bricks of types R1/2c, H2/3c, S2/3c, and R3n are listed in § 4' of the Old Babylonian table of constants RAFb = **BM 36776** (Fig. 7.3.2; Robson, *MMTC* (1999), 206; Friberg, *ChV* (2001), 67):

[8] 20	ki.lá	sig$_4$	ṣa-rip-ti	8;20	the weight of a	brick,	baked	RAFb 13
[11 06 40]	ki.lá	[sig$_4$].áb	ṣa-rip-ti	11;06 40	the weight of a	cow-*brick*,	baked	RAFb 14
[22 13 20]	[ki.lá]	[sig$_4$].al.ùr.ra	ki.2	22 ;3 20	the *weight* of a	tile-*brick*,	"	RAFb 15
[12]	[ki.lá]	[sig$_4$.3"]-*ti*	ki.2	12	the *weight* of a	2/3-*brick*,	"	RAFb 16

The weights of *sun-dried* bricks (also known as *mud bricks*) of the same four types are mentioned in § 3' of the same table of constants:

10	ki.lá	sig$_4$	ḫa$_5$.rá	10	the weight of a	brick,	dried	RAFb 9
13 20	ki.lá	sig$_4$.[áb]	ḫa$_5$.rá	13;20	the weight of a	cow-brick,	dried	RAFb 10
26 40	ki.lá	sig$_4$.al.ùr.ra	ḫa$_5$.rá	26;40	the weight of a	tile-brick,	dried	RAFb 11
14 24	ki.lá	sig$_4$.3"-*ti*	ḫa$_5$.rá	14;24	the weight of a	2/3-brick,	dried	RAFb 12

A comparison of the values listed in § 3' and § 4' shows that *the weight of a baked brick was assumed to be 5/6 of the weight of a dried brick*. Similarly, in § 2' of the same table of constants (RAFb 5-8), the weights of [freshly made, *wet*] bricks of the same four types are mentioned. Apparently, *the weight of a baked brick was assumed to be 2/3 of the weight of a wet brick*.

Much is lost of § 1' of BM 36776, but the paragraph can be partly reconstructed as follows:

[7 12]	[sig$_4$]	[...]	7;12	brick,	[...]	RAFb 1
[5 24]	[sig$_4$.áb]	[...]	5;24	cow-brick,	[...]	RAFb 2
2 42	[sig$_4$.al.ùr.ra]	[...]	2;42	tile-brick,	[...]	RAFb 3
5	sig$_4$.3"-*ti*	[...]	5	2/3-brick,	[...]	RAFb 4

The four sexagesimal numbers listed in § 1' stand for the *molding numbers* of bricks of the four types R1/2c, H2/3c, S2/3c, and R3n, with the values 7;12, 5;24, 2;42, and 5. They are the reciprocals of the weights of baked bricks of the four types, listed in § 4'.

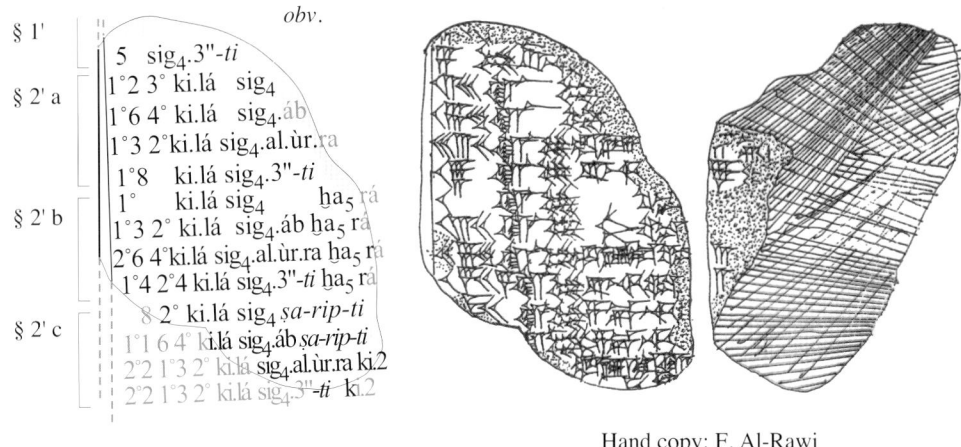

<table>
<tr><td>§ 1'</td><td>5   sig₄.3"-ti</td></tr>
<tr><td rowspan="4">§ 2' a</td><td>1°2 3° ki.lá   sig₄</td></tr>
<tr><td>1°6 4° ki.lá   sig₄.áb</td></tr>
<tr><td>1°3 2°ki.lá sig₄.al.ùr.ra</td></tr>
<tr><td>1°8   ki.lá sig₄.3"-ti</td></tr>
<tr><td rowspan="4">§ 2' b</td><td>1°   ki.lá sig₄   ha₅ rá</td></tr>
<tr><td>1°3 2° ki.lá   sig₄.áb ha₅ rá</td></tr>
<tr><td>2°6 4°ki.lá sig₄.al.ùr.ra ha₅ rá</td></tr>
<tr><td>1°4 2°4 ki.lá sig₄.3"-ti ha₅ rá</td></tr>
<tr><td rowspan="4">§ 2' c</td><td>8 2° ki.lá sig₄ ṣa-rip-ti</td></tr>
<tr><td>1°1 6 4° ki.lá sig₄.áb ṣa-rip-ti</td></tr>
<tr><td>2°2 1°3 2° ki.lá sig₄.al.ùr.ra ki.2</td></tr>
<tr><td>2°2 1°3 2° ki.lá sig₄.3"-ti ki.2</td></tr>
</table>

obv.

Hand copy: F. Al-Rawi

Fig. 7.3.2. RAFb = BM 36776. A fragment of a table of constants for four kinds of bricks.

After this excursion into Old Babylonian brick metrology, it will be easier to understand what is going on in the inscription on MS 2221. In col. *ii* of MS 2221 *obv.*, the numbers 6, 4 30, and 2 15 are mentioned as parameters for three kinds of bricks, rectangular (*libittu*), square (*agurru*), and half square (*arḫu*). The same numbers appear also together in a small cluster of entries in the table of constants Kb = **CBS 10996**, which is a post-Old Babylonian, possibly Late Babylonian, copy of an older text (see Friberg, *ChV* (2001), 65):

[6]	ᵍⁱˢmá.lá	sig₄		6	heavy boat	brick	Kb B1
[4 30]	ᵍⁱˢmá.lá	sig₄.áb		4;30	heavy boat	cow-brick	Kb B2
[2] 15	ᵍⁱˢmá.lá	sig₄.al.ùr.ra		2;15	heavy boat	tile-brick	Kb B3
4 10?	ᵍⁱˢmá.lá	sig₄.3"-ti		4;10?	heavy boat	2/3-brick	Kb B4

Cf. the following entry in the Old Babylonian table of constants NSd = **YBC 5022**:

6	*ša*	ᵍⁱˢmá.lá		6	of heavy boat	NSd 29

The four types of bricks mentioned in Kb B1-4, presumably R1/2c, H2/3c, S2/3c, and R3n, are the same as those mentioned repeatedly in §§ 1'-4' of BM 36776 above. The listed numbers 6, 4 30, 2 15, 4 10 are 5/6 of the numbers 7 12, 5 24, 2 42, 5 listed in BM 36776 § 1' (the molding numbers). At the same time, they are reciprocals of the numbers 10, 13 20, 26 40, 14 24 listed in BM 36776 § 3' (the weights of dried bricks). Hence, it is conceivable that a paragraph listing these numbers was once inscribed just before § 1' on RAFb = BM 36776, on the lost upper part of that clay tablet.

As for the meaning of this cluster of entries in a table of constants, the term ᵍⁱˢmá.lá (Akk. *malallū*) gives little information. The word normally stands for some kind of freight boat. (The translation here, 'heavy boat', is an ad hoc translation, meant to point out that lá is the Sumerian equivalent to Akk. *šaqālu* 'to weigh'.) The word ᵍⁱˢmá (Akk. *eleppu*) by itself normally means 'boat', but it appears also, out of context, in a few tables of constants (G = **IM 52916**, NSe = **YBC 7243**, and E = **IM 49949**), apparently with the same meaning there as *nalbanu* 'molding number':

7 12	*i-gi-gu-bu-ša*	*e-le-pu-um*	7;12	its constant	boat	G D 1
5		ᵍⁱˢmá	5		boat	NSe 19
4 48		ᵍⁱˢmá *e-le-pí-im*	4;48	boat	boat	E 12

The numbers 7 12, 5, 4 48 in these entries presumably stand for the molding numbers associated with bricks of the types R1/2c, R3n, and S1/2c.

The fact that ᵍⁱˢmá.lá is a variant of ᵍⁱˢmá 'boat' with the meaning 'molding number' seems to suggest that ᵍⁱˢmá.lá 'heavy boat' refers to molding numbers, 6, 4 30, *etc.*, of particularly heavy kinds of bricks, more precisely "variant" bricks that are 6/5 times more massive than bricks with the molding numbers 7 12, 5 24, *etc.*

This is the interpretation proposed in Friberg, *ChV* (2001), 85.

The numbers 6, 4;30, 2;15, 4;10 can also (or instead?) be interpreted as "loading numbers". Cf. the discussion in Robson, *MMTC* (1999) Sec. 5.4. Indeed, just as, incidentally,

> 1 talent is the weight of 7;12, 5;24, 2;42, and 5 *baked* bricks of the four types R1/2c, H2/3c, S2/3c, and R3n,

so, perhaps more importantly,

> 1 talent is the weight of 6, 4;30, 2;15, and 4;10 *sun-dried* bricks of the four types R1/2c, H2/3c, S2/3c, and R3n.

In this connection, it may be relevant that 'talent' is the modern translation (with a Greek origin) of the Sumerian word gú or gú.un (Akk. *biltum*) with the general meaning 'load'. Therefore, it is likely that 1 talent (about 30 kg) was deemed, by compassionless Mesopotamian administrators, to be a "man's-load", the weight a man could carry over long periods of time.

If this interpretation of 1 talent as a man's-load is correct, then the meaning of the alleged loading numbers may have been that

> 6, 4;30, 2;15, and 4;10 *sun-dried* bricks of the four types R1/2c, H2/3c, S2/3c, and R3n constitute a man's-load.

This interpretation is strongly supported by the explicit evidence of the table of constants FM = ***UET 5, 881*** (see the hand copy of the tablet in Friberg, *ChV* (2001), Fig. 3.1):

6	sig₄.si.sá	gú.un.lú.1.e	6	standard-bricks	load-of -1 -man	FM 1
1	sig₄.1.kùš.íb.si₈	gú.un.lú.1.e	1	brick-1-cubit-square	load-of -1 -man	FM 2
2 15	sig₄.3".kùš.ta	gú.un.lú.1.e	2;15	bricks-2/3-cubit-each	load-of -1 -man	FM 3

Line 2 can be interpreted as referring to the following *unitary relation*:

> The weight of a *sun-dried* unit brick (type S1c) is 1 talent, that is, 1 man's-load.

The talent as a large unit of weight measure was possibly originally deliberately *defined*, at some time about the middle of the third millennium BC, so that this unitary relation would hold true!

The loading number 6 is mentioned in two entries in the table of constants NSd = YBC 5022:

6	*ma-aš-šu-ú-um* [*ša*] sig₄		6 carrying tray, of bricks	NSd 41

The number '45' appears 4 times in col. *i* of MS 2221, *obv.*, on the left edge. This is the "walking number" *muttalliktum* or *tallaktum* (Sum. a.rá), which is known (cf. Robson, *MMTC* (1999), 79; Friberg, *ChV* (2001), 101) to have the value

> 45 00 ninda/man-day (around 16 km/man-day) of walking with 1 man's-load.

The significance of the numbers 4 30, 3 22 30, 1 41 15 (MS 2221 *obv.*, col. *iii*, rows 1-3) is also well known (cf. Robson, *op. cit.*, 83; Friberg, *op. cit.*, 72). These are the "carrying numbers" (Akk. *nazbalum*) for bricks of types R1/2c, H2/3c, S2/3c, with the values

> 4 30 00, 3 22 30, and 1 41 15 brick-ninda per man-day.

Here the term 'brick-ninda' stands for '1 brick carried for a distance of 1 ninda').

Until now, it has not been quite clear how these carrying numbers were understood by Old Babylonian mathematicians. The tabular array on MS 2221 *obv.* (including the numbers inscribed on the left edge) makes the situation absolutely clear, at least after an obvious error has been corrected — the words *a-gú-ru* and *ar-ḫu* in lines 2-3 must change places.

The meaning of lines 1-3 is elucidated in the following expanded version of the array:

walking number	loading number		carrying number (work norm)		brick type	
45 00 ninda/man-day	6	bricks	4 30 00	brick-ninda/man-day	bricks	(type R1/2c)
45 00 ninda/man-day	4;30	bricks	3 22 30	brick-ninda/man-day	cow-bricks	(type H2/3c)
45 00 ninda/man-day	2;15	bricks	1 41 15	brick-ninda/man-day	tile-bricks	(type S2/3c)

With the 3 first lines of the array in this expanded form, it becomes clear that what takes place here is the *computation of the carrying numbers* (work norms for the carrying of bricks) for three common types of bricks.

Indeed, it is easy to check that

$$45\ 00 \cdot 6 = 4\ 30\ 00, \quad 45\ 00 \cdot 4;30 = 3\ 22\ 30, \quad \text{and} \quad 45\ 00 \cdot 2;15 = 1\ 41\ 15.$$

What, then, is the meaning of line 4 of the tabular array on MS 2221 *obv.*, in which are mentioned the numbers 45, 2 13 20, and 1 40, followed by the word *e-pe-rum* 'mud'? The first number, 45, of course, is again the walking number. The second number, 2 13 20, appears in three different tables of constants as the constant for a *tupšikkum* (Sum. giš.dusu) 'mud basket'. (See Robson, *op. cit.*, 78; Friberg, *op. cit.*, 101). It can be shown that the value of this number is ;02 13 20 gín (= 1/27 volume shekel), and that it is simply the standardized volume of a basket used to carry mud in. Therefore, line 4 of the array on MS 2221 *obv.* can now be reformulated as:

walking number	loading number	carrying number (work norm)	kind
45 00 ninda/man-day	;02 13 20 gín (volume shekel)	1 40 gín-ninda/man-day	mud

Here the term 'gín-ninda' stands for '1 volume-shekel carried for a distance of 1 ninda'. Thus, this line is devoted to the computation of the carrying number for mud. It is, indeed, easy to check that 45 00 · ;02 13 20 = 1 40, for instance by counting in the following way:

$$45\ 00 \cdot {;}02\ 13\ 20 = 45 \cdot 2;13\ 20 = 15 \cdot 6;40 = 1\ 30 + 10 = 1\ 40.$$

The interpretation of 1 40 as the carrying number for mud is confirmed by entries in three different tables of constants mentioning 1 40 as the constant for *na-az-ba-al* saḫar 'carrying of mud' (Robson, *op. cit.*, 78; Friberg, *op. cit.*, 101.)

7.3 b.   On Carrying Numbers in Old Babylonian Metro-Mathematical Problem Texts

In the large Old Babylonian metro-mathematical recombination text **Haddad 104** (Al-Rawi and Roaf, *Sumer* 43(1984)), the last problem (# 10) is a "combined work norm problem" concerned with the carrying of mud used for the fabrication of bricks. The part of the text that is solely concerned with the carrying of mud is reproduced below:

Hadddad 104 # 10

1	… … … *iš-tu* 5 *ṣú-up-pa-am za-bi-lam-ma* /	… … … From 5 (ninda), a *ṣuppān*, bring here (mud).
	… … …	… … …
	[*pa*]-*ni* 5 *me-ṣe-tim pu-ṭur-ma* 12 *i-lí*	The opposite of 5, the distance, resolve, 12 comes up.
4	12 *a-na* 45 *mu-ta-lik-tim* / *i-ši-ma* 9 *i-lí*	12 to 45, the walking (number), lift, then 9 comes up.
5	9 *a-na* 2 13 20 *tu-up-ši-ki i-ši-ma* / 20 saḫar	9 to 2 13 20, the basket, lift, then 20, the mud
6	*iš-tu* 5 *ṣú-up-pa-am ta-az-bi-lam i-lí*	(that) from 5, a *ṣuppān*, you carried here, comes up.
	… … …	… … …

In this text, mud is carried for a distance of 5 ninda from the place where it is dug up to the place where it is used for the making of bricks. The walking number 45 00 ninda is found to be 9 00 times this distance. This means that the man carrying the mud can make 9 00 trips with his mud basket in 1 day, the standardized volume of the mud basket being ;02 13 20 volume shekel. Consequently, the work norm of carrying (that is, the carrying number for mud) is evaluated in the given situation as

$$9\ 00 \cdot 5\ \text{ninda} \cdot {;}02\ 13\ 20\ \text{volume shekel} = 20\ \text{volume shekels} \cdot 5\ \text{ninda}.$$

Note that, indeed, 20 gín (volume shekels) · 5 ninda is equal to the established carrying number for mud, which is 1 40 gín-ninda. The constant 1 40 itself is not mentioned in the text.

In a similar way, carrying numbers for various types of bricks are never mentioned explicitly in Old Babylonian metro-mathematical problem texts. Four such texts concerned with the carrying of bricks are known at present; they are all discussed in Friberg, *ChV* (2001). The texts in question are **YBC 4673** §§ 2-3 (*ibid.*, 94), **YBC 4669** § B10 (*ibid.*, 113), **YBC 10722** (*ibid.*, 141), and **AO 8862** § 2.2-3 (*ibid.*, 141, 143). All these texts deal exclusively with standard bricks (type R1/2c) and mention the work norm for carrying such bricks only indirectly, as *9 sixties of bricks carried a distance of 30* ninda *(= 3 ropes) in a man-day*. Indeed,

9 00 bricks · 30 ninda = 4 30 00 brick-ninda = the carrying number for bricks of type R1/2c.

Here is a survey of how a carrying distance of 30 ninda is presented in the mentioned texts:

a.na 30 ninda uš / lú.1.e / 9 šu-ši sig₄ íl-*ma*	(For) as much as 30 ninda of length, 1 man 9 sixties of bricks he carried, then	YBC 4673 § 2
lú.šidim.e / *i-na* 30 ninda uš / 9 šu-ši sig₄ íl.íl-*ma*	A brick worker, from 30 ninda of length, 9 sixties of bricks he kept carrying	YBC 4673 § 3
lú.1.e / 30! ninda uš / ḫé.gul.gul. ḫé.íl.íl	1 man, 30 ninda of length, may he keep demolishing and carrying	YBC 4669 § B10
ag-ra-am a-gu-ur-ma / [*i-na* 30 ninda] uš / 9 šu-ši sig₄ *iz-bi-lam*	A paid worker I hired, from 30 ninda of length, 9 sixties of bricks he carried here	YBC 10722 (MCT P)
a.na *ša-la-<ša>-aš-li-i* / *iš-te-en a-wi-lu-um* / 9 šu-ši sig₄ / *iz-bi-la-am-ma*	As much as three ropes, one man, 9 sixties of bricks he carried here, then	AO 8862 § 2.2
šum-ma a.na *ša-la-ša-aš-li* [*ši*]-*du-um-mi*	If as much as three ropes the length, say,	AO 8862 § 2.3

7.3 c. MS 2221, rev. A Combined Carrying Number Problem for Three Types of Bricks

The numbers on the reverse of MS 2221 are not properly aligned, and a separating ruled line is missing between columns 3 and 4 of the tabular array. (It is likely that the student who wrote the text copied it from someone else's clay tablet, without having a good idea of what was going on.) There are also several half erased numbers from earlier computations on the tablet. It is clear, anyway, that the intended form of the array must have been something like this:

MS 2221 *rev.*

1	12	1 48	9
1 20	16	1 48	6 45
2 40	32	1 48	4 22 30

should be 3! 22 30

This looks very much like the tabular arrays for combined market rate problems discussed in Sec. 7.2. The similarity would have been even greater if what is now col. *iv* had been placed before the other columns. Anyway, it is not difficult to find a satisfying interpretation of this text.

A good starting point for an interpretation attempt is the assumption that the text on the reverse of MS 2221, just like the text on the obverse, is concerned with carrying numbers for three common types of bricks, rectangular (R1/2c), half square (H2/3c), and square (S2/3c). If that is so, then the number 9 in row 1 of col. *iv* may be interpreted as 9 sixties of bricks (carried over a distance of 30 ninda.) Now, recall that the computation on the obverse showed that the carrying numbers for the three mentioned types of bricks are

4 30 00, 3 22 30, and 1 41 15 brick-ninda.

In agreement with the apparent convention that the typical carrying distance is 30 ninda, these numbers can be expressed in the following, alternative form:

9, 6;45, and, 3;22 30 sixties of bricks, carried over a distance of 30 ninda.

Indeed, just as 4 30 00 = 9 00 · 30, so 3 22 30 = 6 45 · 30, and 1 41 15 = 3;22 30 · 30. Thus, if the carrying distance is supposed to be the typical 30 ninda, then the numbers in col. *iv* of the tabular array above stand for the correspondingly "reduced" carrying numbers for the three types of bricks. The number written '4 22 30' can be assumed to be an error. It should be '3 22 30'.

To find the "combined carrying number" is a "combined work norm problem" of the kind discussed in Friberg, *RlA 7* (1990) Sec. 5.6 h, and Friberg, *ChV* (2001) § 6. The question to be answered is the following:

If three types of bricks have the carrying numbers 9, 6;45, and 3;22 30 sixties of bricks per 30 ninda and man-day, find a number *N* so that *N sixties of bricks of each kind* can be carried 30 ninda in one man-day.

Here is the solution to this problem suggested by the tabular array above: Begin by letting 9 be the "false" value of *N*, that is, assume that a man carries *9 sixties of each kind of brick*. To carry 9 sixties of the first type of bricks, the one with the carrying number 9 (sixties per 30 ninda and man-day), will take him 1 day. To carry 9 sixties of the second type of bricks, the one with the carrying number 6;45 (sixties per 30 ninda and man-day) will take him 1;20 days, since

$$1; 20 \cdot 6;45 = 6;45 + 2;15 = 9.$$

Similarly, to carry 9 sixties of the third type of bricks will take him 2;40 days, since

$$2;40 \cdot 3;22 \; 30 = 6;45 + 2;15 = 9.$$

Consequently, the numbers in col. *i* of the array are the multiples of 1 day that the man will need to carry 9 sixties of each type of bricks over a distance of 30 ninda. Together, the man will need

$$1 + 1;20 + 2;40 \text{ days} = 5 \text{ days to carry 9 sixties of each type of bricks (over a distance of 30 ninda).}$$

In 1 day, that is, in 1/5 of that time he will be able to carry

$$1/5 \text{ of 9 sixties} = 1;48 \text{ sixties} = 1 \; 48 \text{ of each type of bricks (over a distance of 30 ninda).}$$

The value 1 48 is inscribed in each row of col. *iii* of the array. Finally, in order to carry 1 48 of each type of bricks, the man will need

$$
\begin{aligned}
1/5 \cdot 1 &= ;12 \text{ days} &&\text{for the first type of bricks,} \\
1/5 \cdot 1;20 &= ;16 \text{ days} &&\text{for the second type of bricks,} \\
1/5 \cdot 2;40 &= ;32 \text{ days} &&\text{for the third type of bricks.}
\end{aligned}
$$

Accordingly, the numbers 12, 16, and 32 are inscribed in col. *ii* of the array.

Here is an expanded form of the tabular array, with detailed information about what is counted in each case:

carrying 9 sixties over 30 ninda	1/5 of the time	1/5 of the bricks	carrying numbers	types of bricks
1 man-day	;12 man-day	1 48 bricks	9 sixties · 30 ninda	bricks, type R1/2c
1;20 man-day	;16 man-day	1 48 bricks	6;45 sixties · 30 ninda	cow-bricks, type H2/3c
2;40 man-day	;32 man-day	1 48 bricks	3;22 30 sixties · 30 ninda	tile-bricks, type S2/3c

7.3 d.   Other Texts with Combined Work Norm Problems Involving Bricks and Mud

There are no known texts parallel to MS 2221 *obv.* with its calculation of carrying numbers for bricks and mud, or to MS 2221 *rev.* with its combined carrying number problem. On the other hand, several previously published texts with combined work norm problems for bricks and mud can now be better understood, in the light of the discussion above of both the obverse and the reverse of MS 2221. In particular, tabular arrays such as the one on MS 2221 *rev.* turn out to be very useful in presenting the solutions to such problems. Here follows a brief survey:

In **YBC 4669 § B10** (Friberg, *op. cit.*, 113), the vaguely stated problem appears to be the following: A man wrecks an old wall and carries away the bricks for a distance of 30 ninda. The answer given, without an explicit solution algorithm, is that he spends 1/5 of a day wrecking the wall and the remaining 4/5 of the day carrying away the bricks. The omitted solution algorithm could have been given in the form of a tabular array like the one below:

wrecking and carrying 25 volume shekels	1/5 of the time	1/5 of the volume	work norms	type of bricks
1 man-day wrecking	;12 man-day	5 gín	25 gín/man-day for wrecking	any type
4 man-days carrying	;48 man-day	5 gín	6;15 gín/man-day for carrying	the same type

The work norm for wrecking is not given in the text. It must have been well known. The work norm for carrying, *in terms of volume measure*, may have been computed as the product of the volume of, say, one standard brick (type R1/2c) and the carrying number for such bricks:

$$;00\ 41\ 40\ \text{gín/brick} \cdot 9\ 00\ \text{bricks} \cdot 30\ \text{ninda} = 6;15\ \text{gín} \cdot 30\ \text{ninda}.$$

The sum of the numbers in col. *i* is 5, so that a man will be able to wreck and carry 25 volume shekels in 5 days. Hence, he can wreck and carry 5 volume shekels in 1 day (cols. *ii-iii*).

YBC 4673 § 5 (Robson, *MMTC* (1999), 90-91) mentions 1 man (lú.1.e) repeatedly carrying mud (íl.íl) over a distance of 30 ninda and molding (du$_8$.du$_8$) bricks. The work norm for carrying mud (the carrying number) is, of course,

$$1\ 40\ \text{gín} \cdot \text{ninda} = 3;20\ \text{gín} \cdot 30\ \text{ninda}.$$

Since part of the answer to the stated problem is given (2 40 bricks per man-day), it is possible to count backwards from this answer to find what the work norm for molding bricks from mud must be:

2 40 bricks = ;13 20 (2/9) · 12 00 bricks = ;13 20 brick šar,
the volume of ;13 20 brick šar of regular bricks (R1/2c) is ;13 20 / 7;12 volume šar = ;01 51 06 40 volume šar,
1;51 06 40 volume gín = ;33 20 (5/9) · 3;20 volume gín.

Now, if 5/9 of the day is spent carrying mud, 4/9 of the day is spent molding bricks. Consequently, it is tacitly assumed in this text that *the work norm for molding bricks* is

$$2;15\ (9/4) \cdot 1;51\ 06\ 40\ \text{volume gín/man-day} = 4;10\ \text{volume gín/man-day}.$$

In particular,

the work norm for molding regular bricks (R1/2c) is 2;15 (9/4) · 2 40 bricks/man-day =1/2 brick šar/man-day.

It follows that this is the solution algorithm for the stated problem, in the form of a tabular array:

processing 3;20 gín	5/9 of the time	5/9 of 3;20 gín	bricks	work norms	type
1 man-day (íl.íl) ;48 man-day (du$_8$.du$_8$)	;33 20 man-day ;26 40 man-day	1;51 06 40 gín 1;51 06 40 gín	2 40	3;20 gín · 30 ninda/man-day 4;10 gín/man-day	mud R1/2c

The necessary computations are the following: The sum of the numbers in col. *i* is 1;48 (= 9/5). The reciprocal of 1;48 is ;33 20 (= 5/9), which is the first entry in col. *ii*. The second entry in col. *ii* is ;48 · ;33 20 = (4/5 · 5/9 = 4/9 =) ;26 40. The numbers in col. *iii* stand for 5/9 of 3;20 volume shekels = 1;51 06 40 volume shekels. The number of bricks in col. *iv* can be computed as the product of the volume (;01 51 06 40 volume šar) and the molding number for standard bricks:

;01 51 06 40 volume šar · 7;12 brick šar/volume šar = ;13 20 brick šar = ;13 20 · 12 00 bricks = 2 40 bricks.

In the problem text **Haddad 104 § 9** (Friberg, *op. cit.*, 110-111; Robson, *op. cit.*, 79), a combined work norm for brick making is computed, through combination of the three work norms for *alli ḫabātim* 'crushing', *alli labānim* 'molding', and *alli balālim* 'mixing'. Without going into details here, the given solution algorithm can be explained in terms of the following array:

work norms		inverted work norms	5 times the time	5 times the volume	bricks made	type
20 gín/man-day	crushing	;03 man-day/ gín	;15 man-day	5 gín	2 15	S2/3cv
20 gín/man-day	molding	;03 man-day/ gín	;15 man-day	5 gín	2 15	S2/3cv
10 gín/man-day	mixing	;06 man-day/ gín	;30 man-day	5 gín	2 15	S2/3cv

Here, the sum of the numbers in col. *ii* is ;12 man-day/gín = 1/5 man-day/gín. In col. *iii*, the times are 5 times bigger, and the sum is 1 man-day. In col. *iv*, the volume is 5 times bigger, that is 5 gín. The numbers in col. *v* are obtained as follows. First, the volume of 1 brick of type S2/3cv is

2/3 cubit · 2/3 cubit · 6 fingers = ;03 20 ninda · ;03 20 ninda · ;12 cubit = ;00 02 13 20 volume šar.

Since 2 13 20 is the reciprocal of 27, it follows that the volume of 1 brick of type S2/3cv is 1/27 of a volume shekel. Hence, the number of bricks of type S2/3cv contained in 5 volume shekels is

5 gín / 1/27 gín/brick = 5 · 27 bricks = 2 15 bricks.

Therefore, the result of the computation is that in this text

the combined wok norm for making bricks is 2 15 bricks (S2/3cv) /man-day, made out of 5 volume gin of mud.

In **Haddad 104 § 10**, a fourth work norm is added, that of carrying mud over a distance of 5 ninda. Since 1 40 gín-ninda/man-day = 20 gín/man-day · 5 ninda, the added work norm is 20 gín/man-day. Hence, the solution algorithm for the new problem may be presented in the form of the following correspondingly expanded tabular array:

work norms		inverted work norms	4 times the time	4 times the volume	bricks made	type
20 gín/man-day	crushing	;03 man-day/gín	;12 man-day	4 gín	1 48	S2/3cv
20 gín/man-day	molding	;03 man-day/gín	;12 man-day	4 gín	1 48	S2/3cv
10 gín/man-day	mixing	;06 man-day/gín	;24 man-day	4 gín	1 48	S2/3cv
20 gín/man-day	carrying	;03 man-day/gín	;12 man-day	4 gín	1 48	S2/3cv

Therefore, in this text,

the combined wok norm for carrying mud and making bricks is 1 48 bricks (S2/3cv) /man-day.

7.3 e.   An Early Text with a Brick Problem and Sexagesimal Numbers in Place Value Notation

The small hand tablet **RTC, 413** (Fig. 7.3.3 below; Thureau-Dangin 1903; Friberg, *RlA* (1990), Fig. 3) has on its obverse the numerical parameters for a brick problem and on its reverse two sexagesimal numbers in place value notation, written with oversize digits.

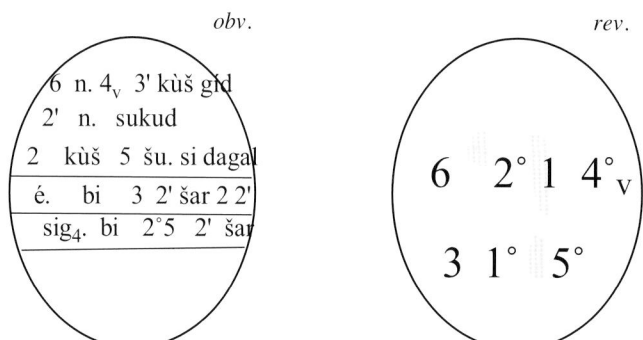

Fig. 7.3.3. *RTC* 413. A hand tablet with a brick problem and sexagesimal numbers in place value notation.

Here is a transliteration and translation of the text on the obverse:

RTC, 413

1	6 ninda 4$_v$ 3' kùš gíd /	6 ninda 4 1/3 cubits the length,
2	2' ninda sukud /	1/2 ninda the height,
3	2 kùš 5 šu.si dagal /	2 cubits 5 fingers the width.
4	é.bi 3 2' šar 2 2' /	Its house (volume) 3 1/2 šar 2 1/2 <gín>.
5	sig₄.bi 25 2' šar	Its bricks 25 1/2 šar.

In spite of its modest appearance, this is a very interesting text. It appears to be a very early Old Babylonian mathematical text. As such it may be the oldest known example of the use of sexagesimal numbers in place value notation in a mathematical cuneiform text.

Apparently, it is silently understood that the object of this exercise is a wall with a triangular cross section, built of standard mud bricks (type R1/2c). The text on the obverse of *RTC*, 413 specifies the linear dimensions of the wall, and then mentions the volume of the wall and the corresponding number of bricks.

The first step of the solution procedure is the following preliminary computation

$$1. \quad u = 6 \text{ n. } 4 \text{ 1/3 c.} = 6;21 \text{ 40 n.}, \quad s = 2;10 \text{ c.}, \quad h = 1/2 \text{ n.} = ;30 \text{ n.},$$
$$u \cdot h = 3;10 \text{ 50 sq. n.}$$

This is clearly the explanation for the numbers recorded on the reverse. The second step is the computation of the volume of the wall, which rightly should have been carried out as follows:

$$2. \quad u \cdot h \cdot 1/2 \cdot s = 1; 35 \text{ 25 sq. n.} \cdot 2;10 \text{ c.} = 3;26 \text{ 44 10 volume šar} = 3 \text{ 1/3 šar 6 2/3 gín 12 1/2 barley-corns.}$$

Instead, however, the student who wrote the hand tablet made a stupid but interesting mistake and calculated as follows:

$$2^*. \quad 1;35 \text{ 25 sq. n.} \cdot 2;10 \text{ c.} = (1;35 \text{ 25} \cdot 2 + 2;10 \text{ (sic!)} \cdot ;10) \text{ volume šar}$$
$$= (3;10 \text{ 50} + ;21 \text{ 40}) \text{ volume šar} = 3;32 \text{ 30 volume šar} = 3 \text{ 1/2 šar 2 1/2 <gín>.}$$

He then used this incorrect result in his final computation of the number of standard rectangular bricks in the wall, multiplying the volume by the molding number 7 12 for bricks of type R1/2 c:

$$3. \, 3;32 \text{ 30 volume šar} \cdot 7;12 \text{ brick šar/volume šar} = 25;30 \text{ brick šar} = 25 \text{ 1/2 brick šar.}$$

Remember that an Old Babylonian school boy making this kind of computation could make use of the multiplication table with the head number 7 12, one of the standard head numbers in the Old Babylonian combined multiplication table!

7.4. Inheritance Problems with the Shares Forming a Geometric Progression

7.4 a.   MS 2830, obv. A Theme Text with Five Inheritance Problems

MS 2830 (Fig. 7.4.1) is a small clay tablet, inscribed on both sides, but with totally unrelated texts on the two sides. Two tabular arrays on the *reverse* can be shown to be the numerical details of two combined market rate problems with regular market rates. (See Sec. 7.2 a.) The upper part of the *obverse* is badly damaged. Nevertheless, it seems to be clear that this side of the text originally was a small "theme text" with a series of five closely related "inheritance problems". The problems are written in Sumerian, and the number signs used for the numbers 4 and 7 (two wedges over two and four over three) are the kind of variant number signs that appear to be characteristic for early Old Babylonian mathematical texts from the southern part of Mesopotamia.

Actually, MS 2830 has several traits in common with a group of four important mathematical texts in the British Museum from early Old Babylonian Ur, such as the small rectangular format of the clay tablet, the almost exclusive use of Sumerian, and the use of variant number signs.(See Friberg, *R A* 94 (2000).) A plausible conjecture is that MS 2830 and the four texts from Ur have a common origin, although only MS 2830 found its way separately to the open market

The text of §§ 1d-1e is relatively well preserved, and it is obvious that at least these two problems are *variations on a common theme*. The little that remains of the text of §§ 1a-1c seems to make it clear that those three problems are variations on the same theme. It is also clear what that theme is, in spite of the damage to all lines of text on the obverse close to the right edge of the tablet, namely divisions of various amounts of silver between four brothers, with the shares always forming a geometric progression. This interpretation of the problems on the obverse of MS 2830 is supported by the existence of a parallel text, namely the first paragraph of **UET 5, 121**, one of the four early Old Babylonian mathematical texts from Ur mentioned above (reproduced in transliteration and translation in Sec. 7.4 b below). Incidentally, the second paragraph of *UET 5*, 121 is a parallel to MS 2317, the small clay tablet with a division problem for the funny number 1 01 01 01, discussed in Sec. 7.1 above.

obv.

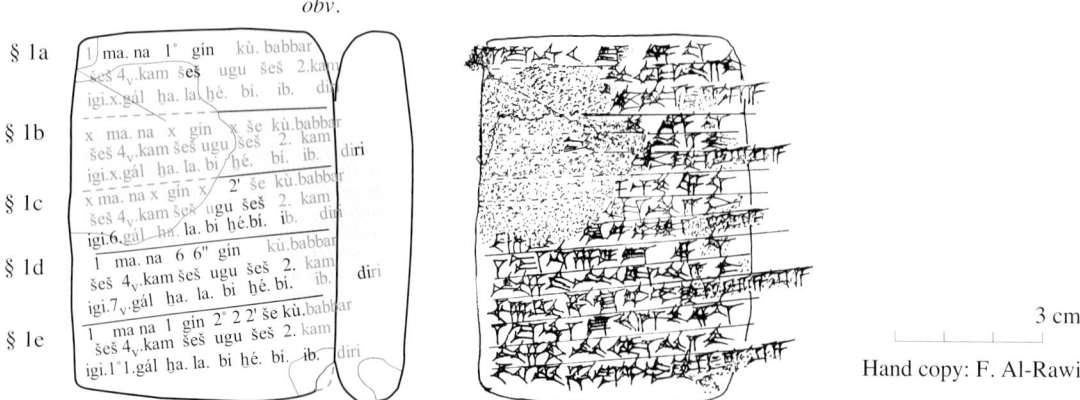

3 cm

Hand copy: F. Al-Rawi

Fig. 7.4.1. MS 2830, *obv.* A theme text with five inheritance problems.

Here follows a transliteration and translation of the text of the five problems on MS 2830, *obv.* As usual, reconstructed parts of the text are marked by straight brackets in the translation, and by italic style in the transliteration.

MS 2830 §§ 1a-1e.

1a	1	[1] ma.na 10 gín [kù.babbar] /	*1* mina 10 shekels *of silver,*
	2	[šeš 4$_v$.kam š]eš [ugu šeš 2.kam] /	*4 brothers. Brother over 2nd brother*
	3	[igi.x.gál ḫa.la.bi ḫé.bí.ib$^?$.diri]	*(by) the … of its share may it go beyond.*
1b	1	[… ma.na … gín … še kù.babbar] /	*…mina …shekels …barley-corns of silver,*
	2	[šeš 4$_v$.kam šeš ugu šeš 2.kam] /	*4 brothers. Brother over 2nd brother*
	3	[igi.x.gál ḫa.la.bi ḫé.bí.ib$^?$.di]ri	*(by) the … of its share may it go bey*ond.
1c	1	[… ma.na… g]ín 2' [še kù.babbar] /	*…mina …shekels 1/2 barley-corn of silver,*
	2	[šeš 4$_v$.kam šeš u]gu šeš 2.[kam] /	*4 brothers. Brother ov*er 2nd brother
	3	igi.6.[gál ḫa].la ḫé.bí.[ib$^?$.diri]	(by) the 6th-*part of its sh*are may it *go beyond.*
1d	1	1 ma.na 6 6" gín [kù.babba]r /	1 mina 6 5/6 shekel *silver,*
	2	šeš 4$_v$.kam šeš ugu šeš 2.[kam] /	4 brothers. Brother over 2*nd* brother
	3	igi.7$_v$.gál ḫa.la.bi ḫé.bí.[ib]$^?$.di[ri]	(by) the 7th-part of its share may it go bey*ond.*
1e	1	1 ma.na 1 gín 22 2' še kù.[babbar] /	1 mina 1 shekel 22 1/2 barley-corn *silver,*
	2	šeš 4$_v$.kam šeš ugu šeš 2.[kam] /	4 brothers. Brother over 2*nd* brother
	3	igi.11.gál ḫa.la.bi ḫé.bí.ib$^?$.[diri]	(by) the 11th-part of its share may i*t go beyond.*

If MS 2830 § 1 and *UET 5*, 121 § 1 are parallel texts, then the well preserved problem briefly stated in **§ 1e** of MS 2830 can be restated in a more intelligible form as follows:

> Silver amounting to 1 mina 1 shekel 22 1/2 barley-corns (= 1 01;07 30 shekels) is divided among 4 brothers.
> Each brother's share, less the 11th part of that share, equals the next brother's share. Find the 4 shares.

The text of MS 2830 contains neither a solution algorithm nor an answer to this problem, so that it looks like an *assignment* given to a student by his teacher in an Old Babylonian scribe school. However, in his construction of the problem, the teacher must have made himself guilty of a miscalculation. As it is stated, the problem has no simple solution. Instead of 1 mina 1 shekel 22 1/2 barley-corns = 1 01;07 30 shekels, the given amount of silver should have been

1 mina 4 1/3 shekels 22 1/2 barley-corns = 1 04;27 30 shekels.

(An explanation of the error is given below.)

If the problem had been correctly stated, the student could have argued as follows in his solution algorithm:

Let the oldest brother's "false share" be	22 11 (= 11 · 11 · 11).	The 11th part of that is 2 01.
Then the 2nd brother's false share is	22 11 – 2 01 = 20 10.	The 11th part of that is 1 50.
Then the 3rd brother's false share is	20 10 – 1 50 = 18 20.	The 11th part of that is 1 40.
Then the 4th brother's false share is	18 20 – 1 40 = 16 40.	
The "false sum" of the four shares is	22 11 + 20 10 + 18 20 + 16 40 = 1 17 21.	
The given sum is the false sum times ;00 50 shekels (1 17 21 · ;00 50 = 1 04;27 30).		
Therefore, the "true shares" must be the false shares times ;00 50 shekels, *etc.*		

Note: Instead of starting with the oldest brother's share, going from there to the shares of the younger brothers, the student could just as well have started with the youngest bother's share, going from there to the shares of the older brothers. As the false share of the youngest brother, he could have chosen, for instance, 16 40 = 10 · 10 · 10. Then the next brother's share would have been 16 40 + 16 40 · 1/10 = 16 40 + 1 40 = 18 20, and so on. More about this alternative approach later, in the discussion of MS 1844 in Sec. 7.4.b below.

The problem in § **1d** is just as well preserved as the problem in § 1e. It can be reformulated as

Silver amounting to 1 mina 6 5/6 shekels (= 1 06;50 shekels) is divided among 4 brothers.
Each brother's share, less the 7th part of that share, equals the next brother's share. Find the 4 shares.

In this case, too, the problem as it is stated has no simple solution. The given amount of silver, 1 mina 6 5/6 shekel = 1 06;50 shekel, is apparently a mistake for

1/2 mina 6 5/6 shekel = 36;50 shekels.

Indeed, with the initial data corrected in this way, the problem can be solved as follows:

Let the oldest brother's "false share" be	5 43 (= 7 · 7 · 7).	The 7th part of that is 49.
Then the 2nd brother's false share is	5 43 – 49 = 4 54.	The 7th part of that is 42.
Then the 3rd broth's false share is	4 54 – 42 = 4 12.	The 7th part of that is 36.
Then the 4th brother's false share is	4 12 – 36 = 3 36.	
The "false sum" of the four shares is	5 43 + 4 54 + 4 12 + 3 36 = 18 25.	
The given sum is the false sum times ;02 shekels (18 25 · 2 = 36 50).		
Therefore, the "true shares" must be the false shares times ;02 shekels, *etc.*		

In § **1c**, the number specifying the given amount of silver is almost completely obliterated. The problem can be partly reconstructed as follows:

Silver amounting to [... ...] 1/2 barley-corns is divided among [4] brothers.
Each brother's share, less the 6th part of that share, equals the next brother's share. Find the [4] shares.

The sign for 6 in i g i .6. g á l in this problem is damaged, but the reading is probably correct, and if it is, then the solution procedure in this case would begin as follows:

Let the oldest brother's "false share" be	3 36 (= 6 · 6 · 6).	The 6th part of that is 36.
Then the 2nd brother's false share is	3 36 – 36 = 3 (00).	The 6th part of that is 30.
Then the 3rd broth's false share is	3 (00) – 30 = 2 30.	The 6th part of that is 25.
Then the 4th brother's false share is	2 30 – 25 = 2 05.	
The "false sum" of the four shares is	3 36 + 3 (00) + 2 30 + 2 05 = 11 11.	

The given amount of silver in § 1c is probably [1 mina x shekels x] 1/2 barley-corn. It is possible that this given sum can be explained as the false sum 11 11 times ;06 30 shekel. Indeed,

11 11 · ;06 30 shekel = (1 07;06 + 5;35 30) shekels = 1 12;41 30 shekels = 1 mina 12 2/3 shekels 3 1/2 barley-corns.

Since the problems in §§ 1c and 1d are of the types 'by the 6th part beyond' and 'by the 7th part beyond', respectively, it seems to be a reasonable conjecture that the problems in §§ 1 a and 1b are of the corresponding types 'by the 4th part beyond' and 'by the 5th part beyond'. This conjecture can be tested. If, in § **1a**, a younger brother's share is equal to the next older brother's share, less 1/4 of that share, then the solution procedure for the problem in § 1a would start as follows:

Let the oldest brother's "false share" be	1 04 (= 4 · 4 · 4).	The 4th part of that is 16.
Then the 2nd brother's false share is	1 04 – 16 = 48.	The 4th part of that is 12.
Then the 3rd broth's false share is	48 – 12 = 36.	The 4th part of that is 9.
Then the 4th brother's false share is	36 – 9 = 27.	
The "false sum" of the four shares is	1 04 + 48 + 36 + 27 = 2 55.	

Now, the given amount of silver in § 1a is [x] minas 10 shekels = [xx] 10 shekels. This given sum can be explained as the false sum 2 55 times ;24 shekels. Indeed,

2 55 · ;24 shekels = 1 10 shekels = 1 mina 10 shekels.

Thus, the only problem on the obverse of MS 2830 for which no reconstruction can be suggested is the one in § **1b**, where there are no remaining traces of the data.

Remark: The numerical error in the given amount of silver in § 1d is easy to explain, since the sign for '1/2' in '1/2 mina' is written with a cuneiform sign consisting of a vertical wedge traversed by a horizontal wedge. The one who wrote or copied the text just failed to write the horizontal wedge, with the result that what should have been '1/2 mina' came to look as '1 mina' (a notational error).

The error in § 1e has a more interesting explanation. Apparently, in the construction of the data for the problem, the teacher wanted to compute the given amount of silver, in relative sexagesimal numbers, as 1 17 21 multiplied by 50. Below is shown both the correct computation, to the left, and the actual, incorrect computation, to the right:

1 17 21 · 50 =	50	1 17 21 · 50 =	50
	8 20		**5**
	5 50		5 50
	16 40		16 40
	50		50
	1 04 27 30		1 01 07 30

According to this reconstruction, the origin of the error was the incorrect multiplication 10 · 50 = 5 hundred = 5 (00), instead of 10 · 50 = 5 hundred = 8 20. In other words, what caused the error was that the one who performed the calculation was thinking in terms of decimal numbers when he was supposed to count with sexagesimal numbers. (It is well known that in the Akkadian language number words were decimal. In everyday life in Mesopotamia, decimal numbers were used for counting. Only well educated scribes knew how to count with sexagesimal numbers.)

7.4 b.   MS 1844. A Lentil with the Solution Algorithm for an Inheritance Problem

MS 1844 (Fig. 7.4.2 below) is a round hand tablet. With its diameter of 11 cm and thickness of 3.5 cm it is easily the biggest and most massive of more than ten mathematical or metrological round hand tablets in the Schøyen Collection. The reverse of MS 1844 is blank, while the obverse of the tablet is inscribed with eight lines of sexagesimal numbers in place value notation, separated by ruled lines, and followed by one line with a brief subscript. The number of sexagesimal places (double digits) in the recorded numbers decreases from the first line to the eighth, a clear indication that the text is some kind of algorithm table. The first, and longest, number is written as four sexagesimal places followed, somewhat lower down, by four additional sexagesimal places. There is a numerical error in the number recorded in line 3. It is easy to check (as will be done below) that this error is *propagated upwards*, to the numbers recorded in lines 1 and 2. This means that the numbers in the algorithm table were computed *in reverse order*, beginning with the number '2' in line 8.

7.4 c.   The Terms of a Geometric Progression

A transliteration of the text of MS 1844 is given below, to the left. Errors are indicated by bold script. Two simple *computational* or *notational* errors are 27 instead of 22 in line 3, and 15 instead of 18 in line 1. The remaining errors in lines 1 and 2 are caused by the error in line 3. To the right is a second copy of the algorithm table, with the errors corrected. The incorrect digits to the left are indicated by bold script. A key to the reconstruction of the table is the observation that the numbers in the successive lines of the table, counted *upwards* from line 8, were intended to form a geometric progression with the ratio between successive terms equal to 1 10 in relative place value notation.

The algorithm table on MS 1844.   The same text, corrected.

1	23 **15 20 36**
	12 08 53 20
2	5 02 **41 32** 35 33 20
3	4 19 **27** 02 13 20
4	3 42 18 53 20
5	3 10 33 20
6	2 43 20
7	2 20
8	2
9	igi.7.gál.bi tur.šè

23 18 09 48
10 08 53 20
5 02 35 42 35 33 20
4 19 22 02 13 20
3 42 18 53 20
3 10 33 20
2 43 20
2 20
2
igi.7.gál.bi tur.šè

The subscript can probably be translated as 'for small by the 7th part' (more about this below), and may be interpreted as meaning that *each number in the table is equal to the number in the preceding line, diminished by 1/7 of its value*. (In modern notation, each number is $1 - 1/7 = 6/7$ times the number in the preceding line.) Now, since 7 is not a regular sexagesimal number, Babylonian mathematicians would have found it difficult to *count* with a number like 1/7. On the other hand, it is known through explicit examples that they were well aware of a counting rule of the type

$$\text{if } b = a - a \cdot 1/7, \text{ then } a = b + b \cdot 1/6.$$

The correctness of this counting rule is obvious, at least in the case when a is a multiple of 7. Indeed,

$$\text{if } a = n \cdot 7 \text{ and } b = a - a \cdot 1/7, \text{ then } b = n \cdot 7 - n = n \cdot 6, \text{ and } b + b \cdot 1/6 = n \cdot 6 + n = a.$$

In view of this simple counting rule, the requirement that each number in the table shall be equal to the number in the preceding line, diminished by 1/7 of its value, can be replaced by the equivalent requirement that *each number in the table shall be equal to the number below it, increased by 1/6 of its value*. This reformulation of the requirement is a great simplification, since 6 is a regular sexagesimal number with the reciprocal 10. Therefore, increasing a given number by 1/6 of its value is equivalent to multiplying the given number by 1 10 in Babylonian *relative* place value notation, or by 1;10 in *absolute* place value notation.

This means that it was easier for the author of the text to count upwards in the algorithm table, multiplying with the factor '1 10' than to count downwards, subtracting seventh parts. That is also precisely what he did. Here follows a reconstruction of the way in which all the numbers in the algorithm table were computed, beginning with the number '2' in line 8. (An attempted explanation of the meaning of these successively computed numbers will be given below.)

2 · 1 10	=	2	+	20	=	2 20	(line 7)
2 20 · 1 10	=	2 20	+	23 20	=	2 43 20	(line 6)
2 43 20 · 1 10	=	2 43 20	+	27 13 20	=	3 10 33 20	(line 5)
3 10 33 20 · 1 10	=	3 10 33 20	+	31 45 33 20	=	3 42 18 53 20	(line 4)
3 42 18 53 20 · 1 10	=	3 42 18 53 20	+	37 03 08 53 20	=	4 19 22 02 13 20	(line 3)
4 19 22 02 13 20 · 1 10	=	4 19 22 02 13 20	+	43 13 40 22 13 20	=	5 02 35 42 35 33 20	(line 2)

Clearly, the seven numbers beginning with the number in line 8 and ending with the number in line 2 form a *geometric progression* with the *first term* 2 and the *factor* 1 10.

In the text, the mistake in line 3 (**27** instead of **22**) is propagated to line 2 above it in the following way (incorrect digits in bold type):

4 19 **27** 02 13 20 · 1 10   =   4 19 **27** 02 13 20 + 43 **14 30** 22 13 20   =   5 02 **41 32** 35 33 20.

The error in line 1, where the sum of the numbers in lines 2-8 is recorded, can be explained as the sum of the errors in lines 3 and 2 (more about this below).

The original error in line 3 is easy to explain. It is likely that the computation of 3 42 18 53 20 times 1 10 was set up on a counting board (or on another clay tablet) in the following way:

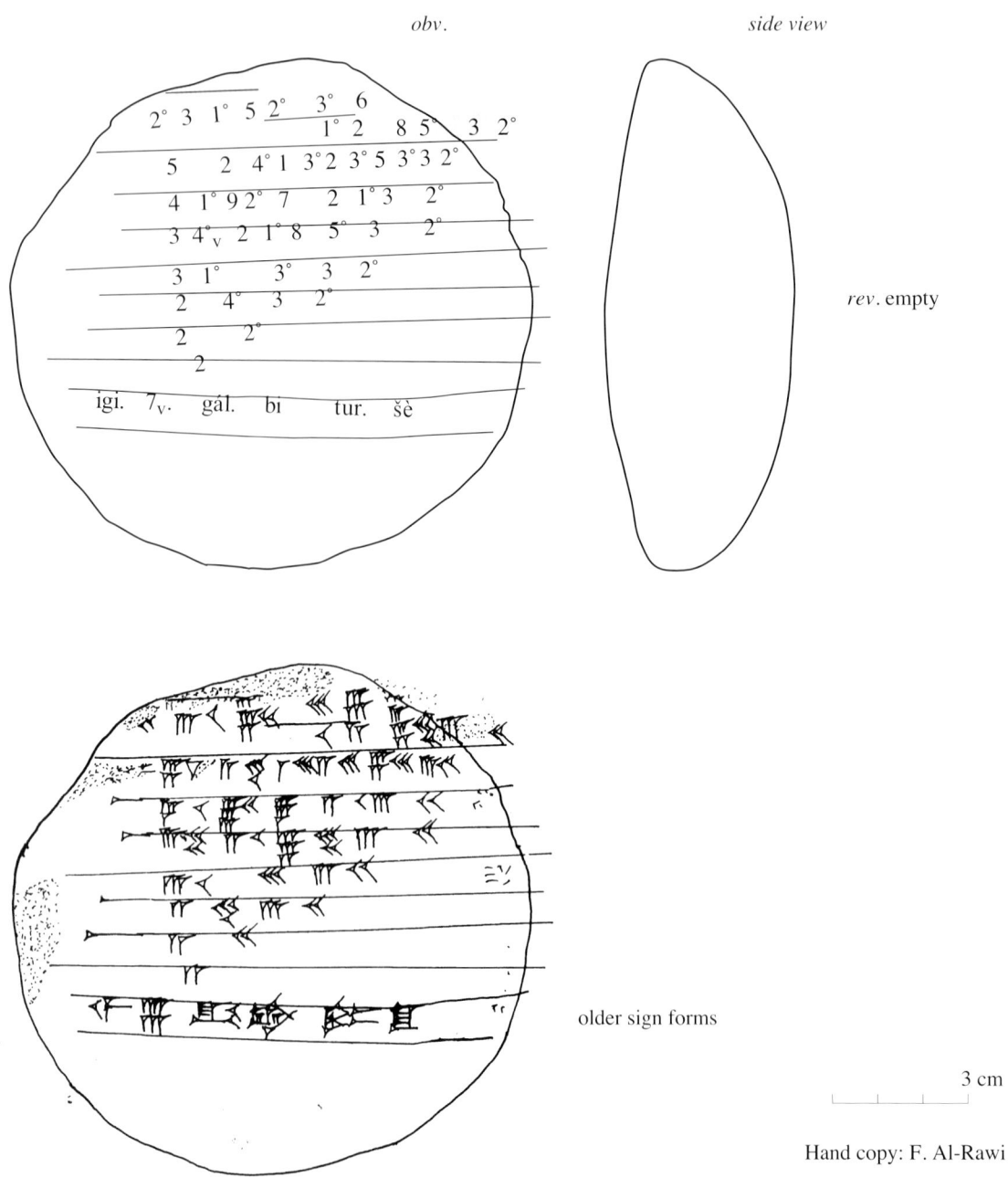

Fig. 7.4.2. MS 1844. A round hand tablet with the numerical solution algorithm for an inheritance problem.

$$\begin{array}{r}
\underline{1}\quad\;\underline{1}\;\underline{1}\quad\quad\quad\;\;\\
3\;42\;18\;53\;20\\
+\;\;37\;03\;08\;53\;20\\
\hline
4\;19\;22\;02\;13\;20
\end{array}$$

If this was done in a careless way, with poor vertical aligning of corresponding sexagesimal places, the correct addition 1+18 + 3 = 22 may have been replaced by the incorrect addition 1+ 18 + 8 = 27.

7.4 d.   The Sum of the Geometric Progression

After the author of the text had computed the seven terms of the geometric progression, it would have been easy for him to compute the sum of the seven terms as follows (addition upwards):

$$
\begin{array}{l}
\underline{23\ 18\ \mathbf{20\ 48}\ 08\ 53\ 20} \\
5\ 02\ \mathbf{41\ 32}\ 35\ 33\ 20 \\
4\ 19\ 27\ 02\ 13\ 20 \\
3\ 42\ 18\ 53\ 20 \\
3\ 10\ 33\ 20 \\
2\ 43\ 20 \\
2\ 20 \\
2
\end{array}
$$

However, this cannot be the way he did it, since in the text the expected sum 23 18 20 48 08 53 20 is split in two parts, and is written as 23 1**5** 20 36 + 12 08 53 20. (Here, 1**5** instead of 18 is a simple notational error. The cuneiform signs for 5 and 8 can easily be mistaken for each other).

The question now is *why* the sum in line 1 should be split up in this way. The most likely explanation is that the author of MS 1844 had in mind the application of the algorithm to the solution of a specific problem (see below), and that it was not meaningful to use a more than four-place sexagesimal number in the final answer to that problem. Therefore, *on his counting board* (or whatever medium he used for his calculations) he may have prepared a *round-off* by writing each more than three-place term of the geometric progression as *the sum of a four-place number and a remainder*. Moreover, this was done in such a way that, in each case, *the last place of the four-place number would be divisible by the factor 6*. The reason for this peculiar kind of round-off is that the scribe made preparations for the division by 6 which was one of the recurrent steps in his computation of the successive terms of the progression. Thus, in line 5 the computed number 3 10 33 20 was rounded off to 3 10 33 18, with the remainder 2. Similarly, in line 4, the computed number 3 42 18 53 20 was rounded off to the four-place number 3 42 18 48, remainder 5 20. In line 3, with its small computational error, the corresponding round-off was 4 19 27 02 13 20 = 4 19 27 + 2 13 20. And so on. Below is shown how the sum of the seven terms in the geometric progression may have been computed by separate addition of the four-place rounded numbers and of the remainders, respectively.

23 1**5 20 36**	+**12** 08 53 20	or, without the errors,	23 18 09 48	+ 10 08 53 20
5 02 **41 30**	+ **2** 35 33 20		5 02 35 42	+ 35 33 20
4 19 27(00)	+ 2 13 20		4 19 22(00)	+ 2 13 20
3 42 18 48	+ 5 20		3 42 18 48	+ 5 20
3 10 33 18	+ 2		3 10 33 18	+ 2
2 43 20			2 43 20	
2 20			2 20	
2			2	

It is not clear why, after this computation of the sum, the four-place round-off and the corresponding remainder was rejoined in each of the seven terms of the progression.[1]

7.4 e.   The Intended Application of the Algorithm

Inheritance problems, alternatively *division of property problems*, are usually formulated in the following way in Old Babylonian mathematical cuneiform texts: A given number of brothers, or partners, are required to share a given amount of silver, or a given piece of land, in unequal shares. The younger brothers, or the junior

---

1. A different interpretation of the data in the text is that there were no round-offs, and that the sexagesimal number in the first line was broken up in two parts only because there was not space enough on the obverse of the tablet to write the whole number in a single sequence of digits. As a matter of fact, the last two places of 12 08 53 20 are written around the edge of the tablet. However, this alternative interpretation is not very likely, because it does not explain why the fourth place of the number, which ought to be 48, was broken up in two parts and written as 36 with 12 immediately beneath.

partners, are supposed to get progressively smaller shares. More precisely, the shares are supposed to form a decreasing arithmetical or geometric progression, or some similar gradually decreasing set of weight or area numbers. An example is, of course, the theme text MS 2830, *obv.* (Fig. 7.4.1) with its series of five problems in which four brothers share given amounts of silver, with their shares forming geometric progressions.

In the case of the present text, MS 1844, the (corrected) data seem to constitute the numerical solution to an inheritance problem which (as we shall see) can be stated in the following way:

> Seven brothers, 23 minas 18 1/6 shekels of silver.
>
> Each brother's share, minus a 7th of that share, equals the next brother's share. Find the 7 shares.

In line with this interpretation, it is suggested here that the obscure sentence in the last line of the text refers to the circumstance that each share is 'a 7th less' than the preceding share.[2] The syntax of the sentence is peculiar, but maybe a clue to the correct interpretation is offered by the subscripts of three early Old Babylonian metrological tables for length measure, *UET 7*, **114-115** and **BM 92698**, two from Ur and one from Larsa (mentioned already in Sec. 3.4 above, and shown in App.5, Figs. A5.3-4). In BM 92698, for instance, two sub-tables have the following subscripts in Sumerian:

> nam.uš.sag aša₅.šè	(to be used) for lengths and fronts of fields (areas) in general
>
> nam.sukud.bùr.saḫar.šè	(to be used) for heights and depths of mud (volumes) in general.

These subscripts indicate the *purpose* of the two sub-tables: one is to be used for *horizontal* length measures, the other for *vertical* length measures. In a similar way, it may be assumed that the subscript following the algorithm table on MS 1844 indicates that *the purpose* of the algorithm text was that it should be used for the construction or solution of inheritance problems where the share of one brother is less by a seventh than the share of the preceding brother. Thus, it is proposed here that the name of problems of this type was igi.7.gál.bi tur 'small by the 7th part', and that the subscript should be translated as

> igi.7.gál.bi tur.šè	(to be used) for 'small by the 7th part'

The conjectured explanation of the intended application of the algorithm of MS 1844 will not be complete until an effort has been made to make clear what the share of the youngest brother actually was meant to be. In the text it is given just as '2', in *relative* place value notation and without any indication of the *chosen unit of measure*. Now, if the algorithm was to be used for the solution of an inheritance problem, the shares would be amounts of silver, expressed as multiples or fractions of 1 mina. In that case, the most plausible interpretation of the relative number '2' is that it stands for' 2 minas'. The computed shares of the six other brother would then be the following, with the errors corrected, and with the sexagesimal numbers conveniently rounded off:

5;02 35 40 minas	=	5 minas 2 1/2 shekels 17 barley-corns
4;19 22 00 minas	=	4 minas 19 1/3 shekels 6 barley-corns
3;42 18 50 minas	=	3 minas 42 1/6 shekels 26 1/2 barley-corns
3;10 33 20 minas	=	3 minas 10 1/2 shekels 10 barley-corns
2;43 20 minas	=	2 minas 43 1/3 shekels
2;20 minas	=	2 minas 20 shekels
2 minas		

A rounding off to multiples of ;00 00 10 mina makes sense, in view of the fact that the smallest weight measure occurring in Old Babylonian metrological tables for system *M* (Sec. 3.2) is 1/2 barley-corn = ;00 00 10 mina.

The sum of the shares computed in this way would be

> 23;18 09 50 minas=23 minas 18 shekels 29 1/2 barley-corns.

This result can be compared with the exact sum of the 7 shares, which is

> 23;18 09 58 08 53 20 minas.

Thus, the sum of the rounded shares and the exact sum of the shares are both very good approximations to the

---

2. Two more or less parallel situations are described in **BM 13901 # 10** (Neugebauer, *MKT 3*, 2), with the Akkadian phrase *mi-it-ḫar-tum a-na mi-it-ḫar-tim si-bi-a-tim im-ṭi* '(one) square side is less than (the other) square side by a seventh', and in **YBC 4714 § 6** (*MKT 1*, 487), with the Sumerian equivalent of that phrase: íb.si₈ íb.si₈.ra igi.7.gál ba.lá.

(fairly) round weight number

23;18 10 minas=23 minas 18 1/6 shekels.

For this reason, it is plausible that 23 minas 18 1/6 shekels was the given sum of the 7 shares. Apparently, the student who got as assignment to compute the solution to this inheritance problem of the type 'small by the 7th part' and the given sum of the 7 shares, seems to have started his computation by *cleverly guessing the correct size of the share of the youngest brother*. The way he reasoned may have been as follows: If the sum of the shares is about 23 minas, then the average share is about 3 minas. The smallest share must be less than the average share, so it is reasonable to assume that it is 2 minas.

Note: There is no known parallel in the corpus of Old Babylonian mathematics to this way of rounding off many-place sexagesimal numbers.

7.4 f.   UET 5, 121. A Parallel Text from Early Old Babylonian Ur

The algorithm text MS 1844 was interpreted above as the numerical solution procedure for an inheritance problem of the type 'small by the 7th part' for *seven* brothers. Similarly, the five problems stated in the damaged theme text MS 2830, *obv.*, were interpreted as inheritance problems for *four* brothers, all of the type 'by the *n*th part beyond', *n* equal to [4], [5], 6, 7, 11. These proposed interpretations are supported by the fact that there exists a known early Old Babylonian mathematical cuneiform text, written in Sumerian, in which the given problem is *explicitly* stated as an inheritance problem of the type "by the 5th part beyond" for *five* brothers. The text in question is Figulla and Martin, **UET 5, 121**, one of the mathematical texts from Ur, reproduced below in transliteration (cf. Friberg, *RA 94* (2000) § 4a):

UET 5, 121 § 1. An inheritance problem of the type "by the 5th part beyond" for five brothers.

1	26 ma.na 15 3" gín 15 [še] [kù.bab]bar /	26 minas 15 2/3 shekels 15 barley-corns of silver.
2	dumu.nita.bi 5	Its heirs are 5.
	šeš.gal šeš dumu.nita /	The big brother <over the next> brother-heir
3	igi.5.gál ḫa.la.kam ḫé.ib.diri /	<by> a 5th of the share may he be beyond.
4	šeš.gal.e	The big brother:
	7 3" ma.na 8 3" gín 15 še /	7 2/3 minas 8 2/3 shekels 15 barley-corns.
5	šeš.2.kam 6 ma.na 5 gín /	The 2nd brother: 6 minas 5 shekels.
6	šeš.3.kam 5 ma.na /	The 3rd brother: 5 minas.
7	šeš.4.kam 4 ma.na /	The 4th brother: 4 minas.
8	šeš.5.kam 3 ma.na 12 gín	The 5th brother: 3 minas 12 shekels.

It is interesting to see that the structure of this text is in a sense complementary to the structure of the algorithm text MS 1844. While the algorithm text consists exclusively of the *numerical solution procedure* for an inheritance problem of a certain kind, *UET 5* 121 consists exclusively of the *statement of the problem* and the *answer* to an inheritance problem of the same kind. The stated problem in the case of *UET 5* 121 can be formulated as follows:

Five heirs, 26 minas 15 2/3 shekels 15 barley-corns of silver.
The share of the older brother, less a 5th of that share, equals the share of the younger brother.
Find the 5 shares.

The solution procedure for this problem would proceed as follows, by use of the method of false value:

Let the oldest brother's false share be	(5 · 5 · 5 · 5 · 5 =) 10 25.	The 5th part of that is 2 05.
Then the 2nd brother's false share is	10 25 - 2 05 = 8 20.	The 5th part of that is 1 40.
Then the 3rd broth's false share is	8 20 - 1 40 = 6 40.	The 5th part of that is 1 20.
Then the 4th brother's false share is	6 40 - 1 20 = 5 20.	The 5th part of that is 1 04.
Then the 5th brother's false share is	5 20 - 1 04 = 4 16.	
The "false sum" of the five shares is	10 25 + 8 20 + 6 40 + 5 20 + 4 16 = 35 01.	

The given sum is ;26 minas 15 2/3 shekels 15 barley-corns = 26 15;45 shekels.
Thus, the given sum is the false sum times ;45 shekel (35 01 · ;45 = 26 15;45).
Therefore, the "true shares" must be the false shares times ;45 shekel, *etc.*

8
Old Babylonian Hand Tablets with Geometric Exercises

There are nine hand tablets with drawings of geometric figures in the Schøyen Collection. Five of the illustrated hand tablets are round, the other four are square or rectangular. The only text on the nine tablets consists of relative sexagesimal numbers or area numbers. Each tablet will be considered separately below, but in order to facilitate comparisons of format, size, and content, hand copies and conform transliterations of the tablets are grouped together in Figs. 8.1.1 - 2 and 8.2.2 below.

8.1. Triangles and Trapezoids

8.1 a.   MS 3042. The Area of a Triangle

MS 3042 (Fig. 8.1.1, top) is a square clay tablet with a crude drawing of a triangle on the obverse, together with some numbers. The reverse is empty. The numbers 3 and 5 40 along the sides of the triangle indicate the lengths of the short side (Sum. sag 'front') and the long side, or the height (Sum. uš 'length', 'flank'). (Old Babylonian teachers did not bother to distinguish between the height and the long side of a triangle when teaching young students elementary geometry.) These sexagesimal numbers in relative place value notation have to be understood as meaning 3 (00) ninda and 5 40 ninda, respectively. Indeed, it is an easily observed, and almost universal rule that *the sides of geometric figures in Old Babylonian mathematical texts are of sizes appropriate for cultivated fields, measured in tens or sixties of the ninda. The areas of the figures are then, correspondingly, measured in multiples of the iku (1 40 sq. ninda), the èše (10 00 sq. ninda), and the bùr (30 00 sq. ninda).*

With the mentioned values for the lengths of the sides of the triangle, the area can be computed as

$$A = 5\ 40 \cdot 3(00)/2 \text{ sq. ninda} = 8\ 30(00) \text{ sq. ninda} = 17 \text{ bùr}.$$

Hence, it is clear that the sexagesimal number 8 30 recorded in the interior of the triangle is the area of the triangle in relative place value notation. This placement of the value of the area is in agreement with what seems to be another general rule: *The area of a geometric figure is recorded in the interior of the figure, while the side lengths are recorded on the outside, along the sides they measure.*

8.1 b.   MS 2107. The Area of a Trapezoid. An Almost Round Area Number

MS 2107 (Fig. 8.1.2, top) is a round hand tablet (a lentil) with a drawing of a trapezoid on the obverse, accompanied by some numbers. The reverse is empty. The recorded numbers are

30 for the upper front, 15 for the lower front, 3 30 for the long side or height, and 1 18 45 for the area.

(In Old Babylonian cuneiform texts, the word an.ta 'upper' means to the left, and ki.ta 'lower' means to the right. This is because the direction of writing had changed from top-to-bottom to left-to-right, and when this happened the orientation of illustrating figures changed in the same way.)

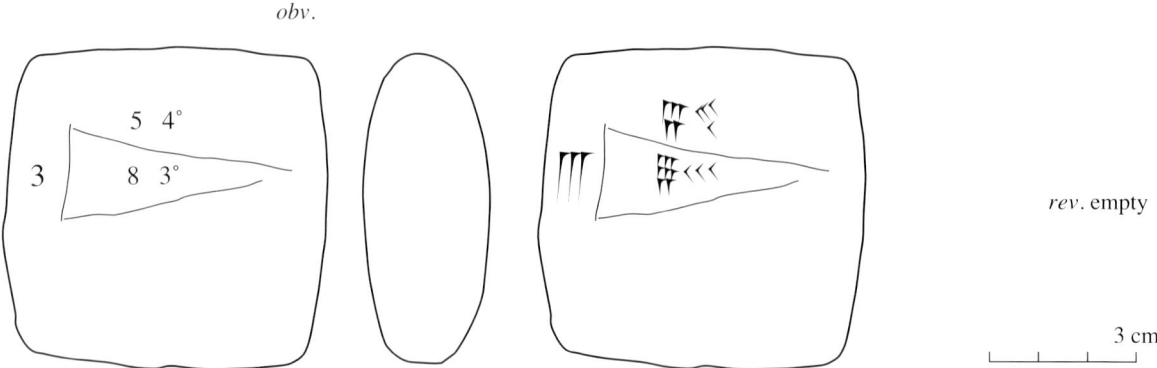

MS 3042. Computation of the area of a triangle.

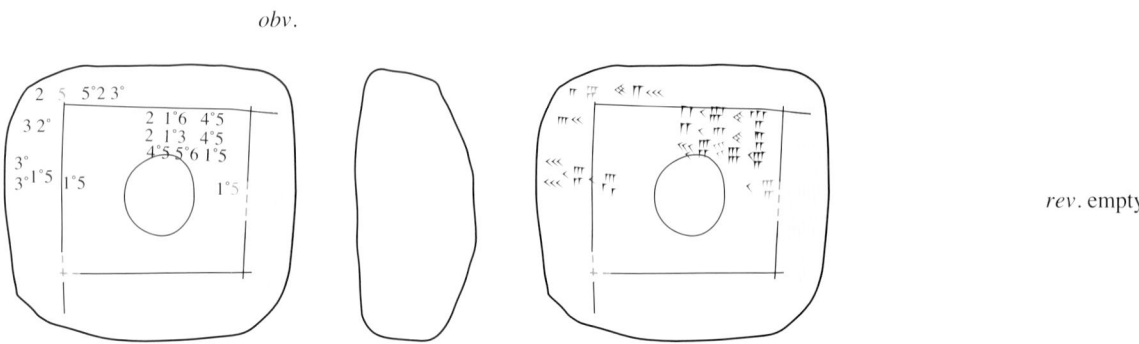

MS 2985. A problem for a circle inscribed in a square a certain distance away from the sides of the square.

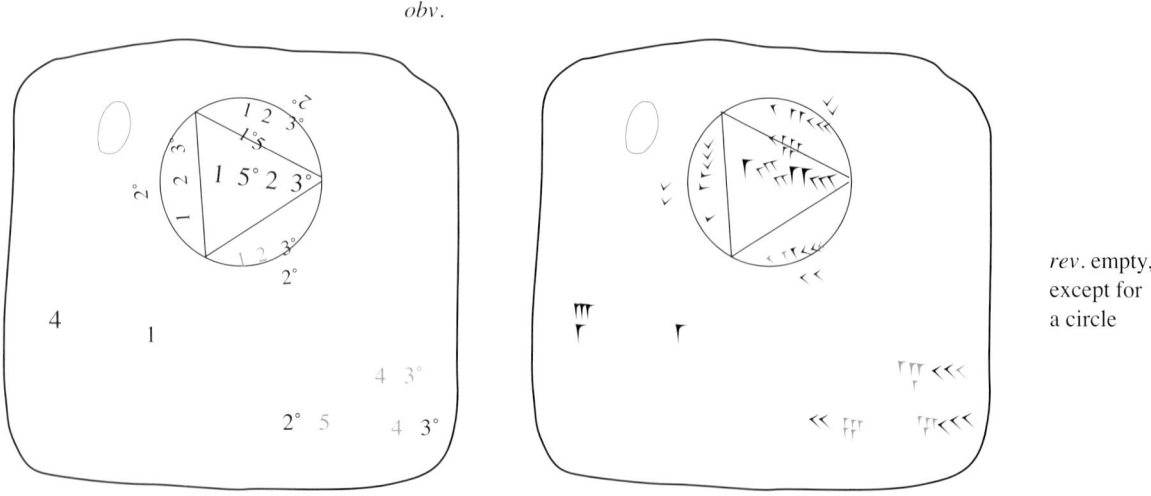

MS 3051. A problem for an equilateral triangle inscribed in a circle.

Fig. 8.1.1. Three square hand tablets with drawings of geometric figures and associated numerical data.

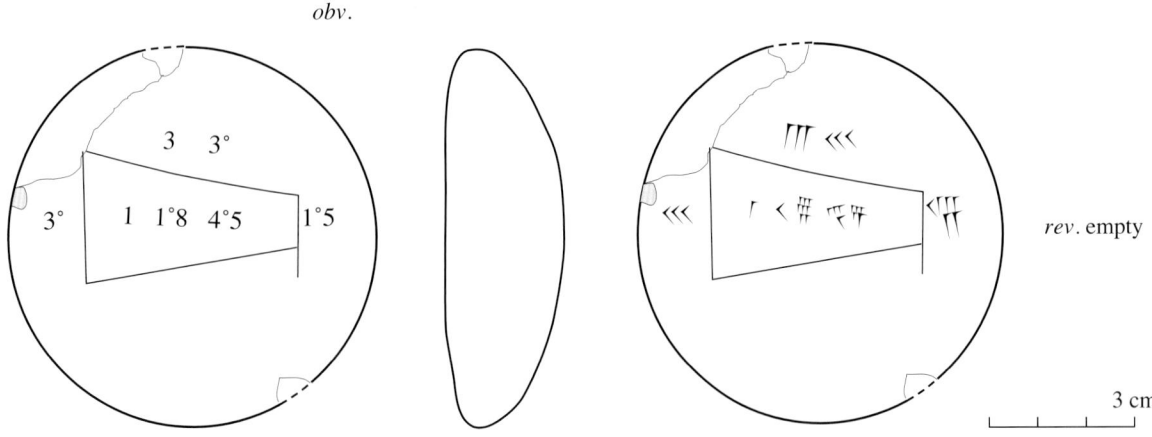

MS 2107. A trapezoid with an almost round area number.

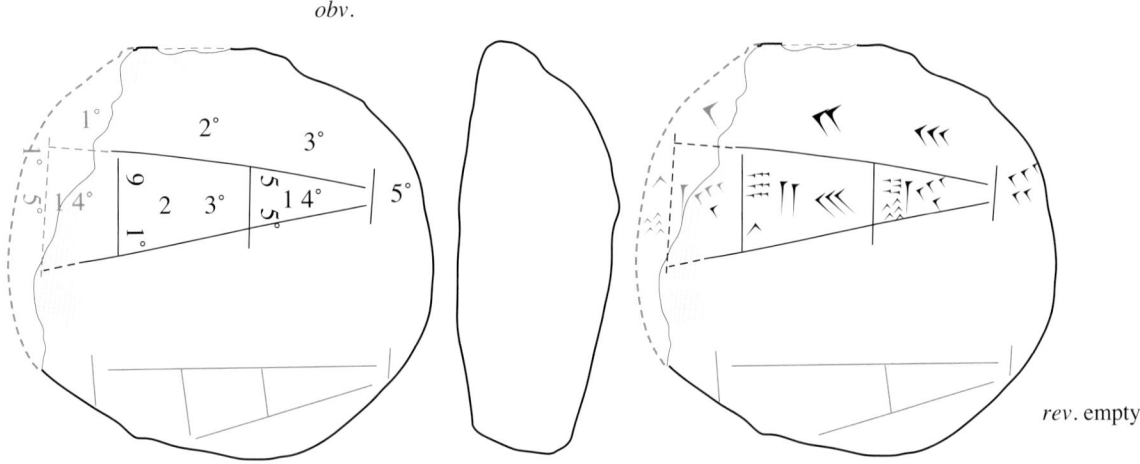

MS 3908. A trapezoid divided into three stripe,s and a complete set of associated numerical parameters.

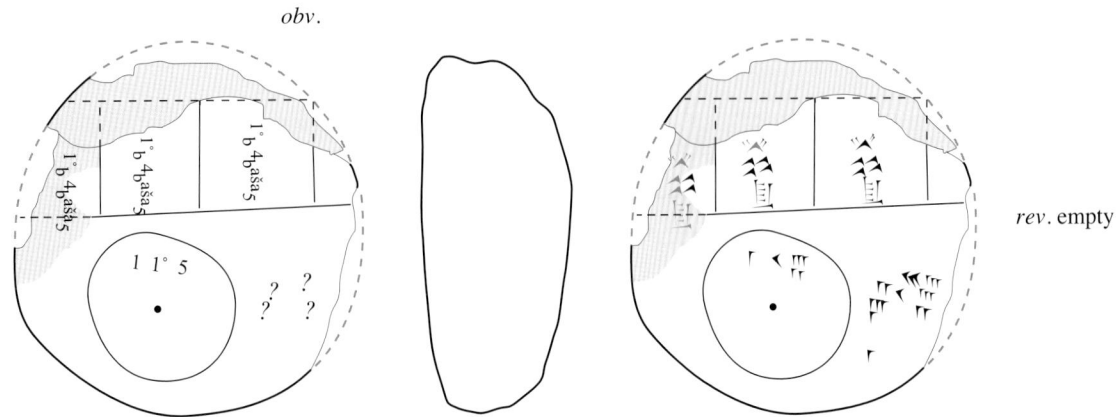

MS 3041. a) A rectangle divided into three equal parts, with area numbers. b) The area of a circle.

Fig. 8.1.2. Three round hand tablets with drawings of geometric figures and associated numerical data.

The area of the trapezoid can be computed as

$$A = 3\ 30 \cdot (30 + 15)/2 \text{ sq. ninda} = 3\ 30 \cdot 22{;}30 \text{ sq. ninda} = 1\ 18\ 45 \text{ sq. ninda} = 2 \text{ bùr } 1 \text{ èše } 5\ 1/4 \text{ iku}.$$

This computed value for the area of the trapezoid agrees with the number recorded inside the trapezoid. Thus, this geometric exercise, like the one considered above (MS 3042) does not look very exciting. However, a closer look at the numbers appearing in the exercise will tell a different story. On one hand, *the computed area number can be factorized in the following remarkable way*:

$$A = 3\ 30 \cdot 22{;}30 = (1 + 1/6) \cdot 3\ 00 \cdot (1 + 1/8) \cdot 20 = (1 + 1/6) \cdot (1 + 1/8) \cdot 1\ 00\ 00.$$

At the same time, *the computed area number is close to a round area number*:

$$A = 2 \text{ bùr } 1 \text{ èše } 5\ 1/4 \text{ iku} = 2 \text{ bùr } 2 \text{ èše } (- 3/4 \text{ iku}) = \text{appr. } 2 \text{ bùr } 2 \text{ èše} = 1\ 20\ 00 \text{ sq. ninda}$$

The deficit $3/4$ iku $= 1\ 15$ sq. ninda is as little as $1/64$ of $1\ 20\ 00$ sq. ninda. This means that the computed area number is an almost round number of the kind considered in Friberg *AfO* 44/45 (1997/98) §§ 2-3. An almost round area number is an area number which is *close to a round area number and simultaneously equal to another round area number multiplied by one or two factors of the kind* $(1 + 1/n)$, *where n is a small regular sexagesimal integer*. Such almost round area numbers typically appear in proto-literate field texts as the result of an application of what may be called the "(proto-literate) field expansion procedure", the oldest known mathematical algorithm. (The "proto-literate period" in Mesopotamia is a convenient name for the centuries just before and after 3000 BC, a period to which can be dated the oldest known clay tablets, those inscribed with a predecessor of the cuneiform script.)

The proto-literate field expansion procedure seems to have been developed as a convenient geometric method to solve a problem of the following type by use of *successive approximations*:

> To find the sides of a rectangle when the area and the ratio of the sides are given.

The way the method worked is best described by use of an explicit example.[1] Take, for instance, the following example with data borrowed from MS 2107:

> Given a rectangle of area $A = 2$ bùr 2 èše $= 1\ 20\ 00$ sq. ninda, and with the front s equal to $1/9$ of the length u.
> Find the length and the front.

It is easy to solve this problem by use of Old Babylonian metric algebra. In a typical application of the method of false value (Friberg, *RlA 7* Sec. 5.7 d), let the length initially have the false value $1\ 00$. Then the front has the corresponding false value $1\ 00/9 = 6{;}40$, and so the area of the rectangle has the false value $6\ 40$. The true value $1\ 20\ 00$ sq. ninda is equal to $6\ 40$ multiplied by 12 sq. ninda. Hence, 12 sq. ninda is the square of the "correction factor" (the ratio between the correct and the false value of the length). The correction factor itself must then be equal to $2 \cdot$ sqs. 3 ninda $=$ appr. $3{;}30$ ninda, since the Old Babylonian standard approximation to sqs. 3 was $1{;}45$ (7/4). Therefore, the correct values of the length u and the front s are found to be

$$u = \text{appr. } 1\ 00 \cdot 3{;}30 \text{ ninda} = 3\ 30 \text{ ninda, and } s = \text{appr. } 3\ 30 \text{ ninda}/9 = 23{;}20 \text{ ninda}.$$

The field expansion procedure solves the problem in a different way. The initial false value of the length is chosen as, say, the *round length number* $u_1 = 3\ 00$ ninda. The false value of the front is $1/9$ of that, that is $s_1 = 20$ ninda. The corresponding false value of the area is the *round area number*

$$A_1 = 3\ 00 \text{ ninda} \cdot 20 \text{ ninda} = 1\ 00\ 00 \text{ sq. ninda} = 2 \text{ bùr}. \qquad \text{(Fig. 8.1.3 a)}$$

Then the *deficit*, that is the difference between the wanted "true" area and the initial false area, is

$$A - A_1 = (1\ 20\ 00 - 1\ 00\ 00) \text{ sq. ninda} = 20\ 00 \text{ sq. ninda} = 1/3 \text{ of } A_1.$$

The crucial idea in this situation is to *eliminate one half of this initial deficit* by expanding the length by $1/6$ of its value, thereby expanding also the area of the rectangle by $1/6$ of its value. Then the new values of the length and the area will be:

---

1. Cf. the discussion in Sec. 1.1 c above of the fourth multiplication exercise in **MS 3955**.

$u_2 = (1 + 1/6) \cdot 3\,00$ ninda $= 3\,30$ ninda,
$A_2 = (1 + 1/6) \cdot 1\,00\,00$ sq. ninda $= 1\,10\,00$ sq. ninda $= 2$ bùr 1 èše.   (Fig. 8.1.3 b)

The new deficit is

$$A - A_2 = (1\,20\,00 - 1\,10\,00) \text{ sq. ninda} = 10\,00 \text{ sq. ninda} = 1/7 \text{ of } A_2.$$

This new deficit can be eliminated by *expanding the front by 1/7 of its value*. Since 7 is a non-regular sexage-simal number, a more appealing alternative is to expand the front by only 1/8 of its value. After this second expansion, the new values of the front and the area are

$s_2 = (1 + 1/8) \cdot 20$ ninda $= 22;30$ ninda,
$A_3 = (1 + 1/8) \cdot 1\,10\,00$ sq. ninda $= 1\,18\,45$ sq. ninda $= 2$ bùr 1 èše 5 1/4 iku.   (Fig. 8.1.3 c)

Thus, the solution to the stated problem obtained by use of the field expansion procedure is

$u_2 = (1 + 1/6) \cdot 3\,00$ ninda $= 3\,30$ ninda
$s_2 = (1 + 1/8) \cdot 20$ ninda $= 22;30$ ninda,
$A_3 = (1 + 1/8) \cdot (1 + 1/6) \cdot 1\,00\,00$ sq. ninda $= 1\,18\,45$ sq. ninda $= 2$ bùr 2 èše $- 1/2$ 1/4 iku.

This is a fairly good approximation to the correct solution. It can be compared with the approximate solution obtained by use of metric algebra and an approximation to sqs. 3, which is hardly better:

$u = 1\,00 \cdot 2 \cdot$ sqs. 3 ninda $=$ appr. $3\,30$ ninda
$s = 1/9 \cdot u = 23;20$ ninda,
$A = u \cdot s = 2$ bùr 2 èše $= 1\,21\,40$ sq. ninda $= 2$ bùr 2 èše $+ 1$ iku.

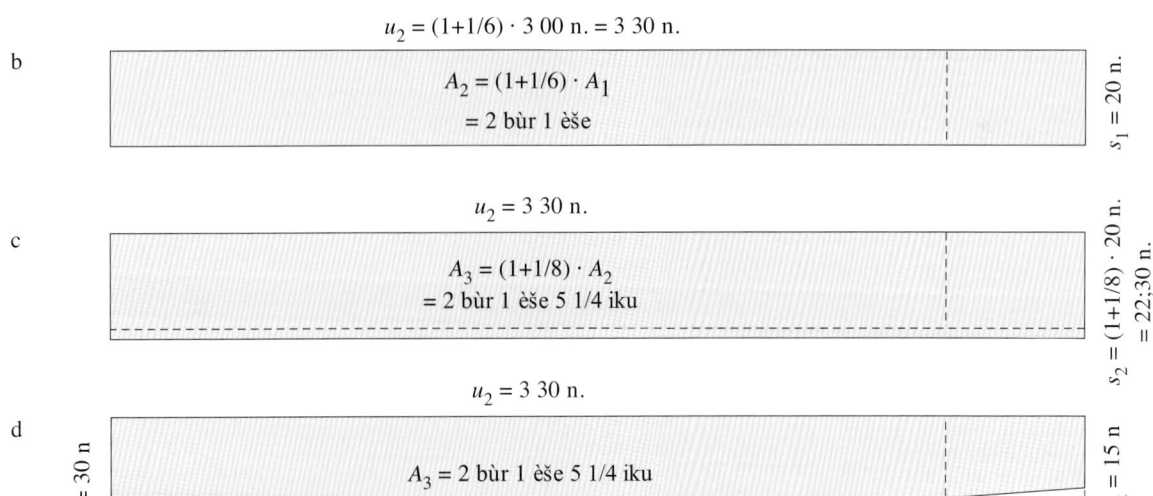

Fig. 8.1.3. MS 2107. An Old Babylonian application of the proto-literate field expansion procedure.

The geometric problem with the solution illustrated by the drawing and the numbers on MS 2107 is a more complex variant of the problem treated above. It may have been stated as follows:

Given a trapezoid with the area $A = 2$ bùr 2 èše $= 1\,20\,00$ sq. ninda,
with the half-sum of the upper and lower fronts equal to 1/9 of the length u,
and with the upper front twice as long as the lower front.
Find the length and the front.

The solution of this more complex problem would require an additional step in the field expansion procedure, as shown in Fig. 8.1.3 d.

Remark: MS 2107 has been claimed to be from the same archive as MS 1844 (Fig. 7.4.2).

YBC 7290 (Fig. 8.1.4 below; *MCT*, 44) is a parallel to MS 2107. It is a rectangular hand tablet with a drawing of a trapezoid on the obverse, accompanied by numbers indicating the area 5 03 20, the length 2 20 and the upper and lower fronts 2 20 and 2 (00).

obv.

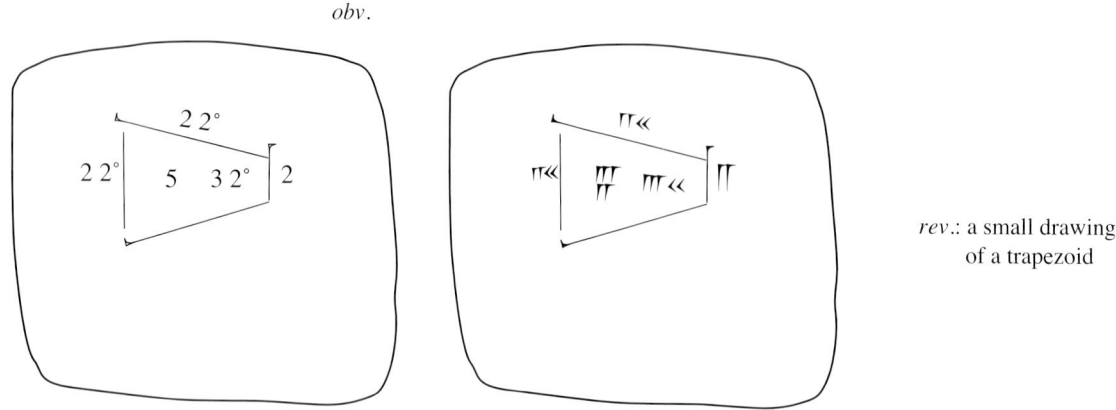

rev.: a small drawing of a trapezoid

Fig. 8.1.4. YBC 7290. A trapezoid with an almost round area number.

Just as in MS 2107, the area number in YBC 7290 is an almost round number, since

A = 5 03 20 sq. ninda = appr. 5 00 00 sq. ninda = 10 bùr, a round area number, and at the same time
A = (1 + 1/6) · 2 00 ninda · (1 + 1/12) · 2 00 ninda = (1 + 1/6) · (1 + 1/12) · 8 bùr.

8.1 c.   Examples of Proto-Literate Field-Sides and Field-Area Texts

Here follows, for comparison, a brief discussion of the application of the field expansion procedure in a couple of proto-literate texts.

W 20044, 35 (Friberg, *AfO* 44/45 (1997/98), 10; Fig. 8.1.5 below) is one of several examples of proto-literate "field texts" documenting the use of the field expansion procedure.

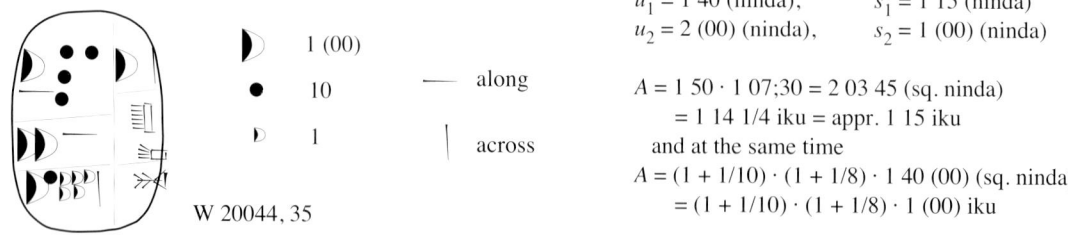

u_1 = 1 40 (ninda), s_1 = 1 15 (ninda)
u_2 = 2 (00) (ninda), s_2 = 1 (00) (ninda)

A = 1 50 · 1 07;30 = 2 03 45 (sq. ninda)
   = 1 14 1/4 iku = appr. 1 15 iku
and at the same time
A = (1 + 1/10) · (1 + 1/8) · 1 40 (00) (sq. ninda)
   = (1 + 1/10) · (1 + 1/8) · 1 (00) iku

Fig. 8.1.5. A proto-literate field-sides text (c. 3200 BC). A quadrilateral with an almost round area

In this text are recorded four length numbers, apparently the lengths of the four sides of a quadrilateral, with vertical and horizontal lines close to the numbers, indicating that 1(géš) 40 and 2(géš) stand for the sides "along", while 1(géš) 15 and 1(géš) stand for the sides "across". The silently understood length unit is the ninda, as in Old Babylonian geometric texts. W 20044, 35 can be explained as an assignment. A student was probably expected to compute the area of a quadrilateral with the given sides, using the (inexact) "quadrilateral area rule", according to which the area of a quadrilateral is (approximately) equal to the half-sum of the sides along multiplied by the half-sum of the sides across. In the case of W 20044, 35, the result would be (in modern notations, using sexagesimal numbers in place value notation):

$$A = \text{appr. } (1\ 40 + 2\ 00)/2\ \text{n.} \cdot (1\ 15 + 1\ 00)/2\ \text{n.} = 1\ 50\ \text{n.} \cdot 1\ 07;30\ \text{n.} = 2\ 03\ 45\ \text{sq. n.} = 1\ 14\ 1/4\ \text{iku} = \text{appr. } 1\ 15\ \text{iku.}$$

At the same time,

$$A = (1 + 1/10) \cdot 1\ 40\ \text{n.} \cdot (1 + 1/8) \cdot 1\ 00\ \text{n.} = (1 + 1/10) \cdot (1 + 1/8) \cdot 1\ 00\ \text{iku.}$$

Thus, clearly, the area is an almost round number, probably obtained as the result of an application of the field expansion procedure.

W 20214, 1 (Friberg, *AfO* 44/45 (1997/98), 9; Fig. 8.1.6 below) is one of a couple of proto-literate "area texts" also documenting the use of the field expansion procedure. In this text is recorded the area number

$$A = 2\ \text{bur'u}\ 4\ \text{bùr}\ 2\ \text{èše} = 24\ \text{bùr}\ 2\ \text{èše} = 1\ 14\ \text{èše} = 12\ 20\ 00\ \text{sq. ninda.}$$

The recorded area number is close to the round area number 25 bùr = 1 15 èše = 12 30 00 sq. ninda. Therefore, it is reasonable to assume that this is another example of an almost round number.

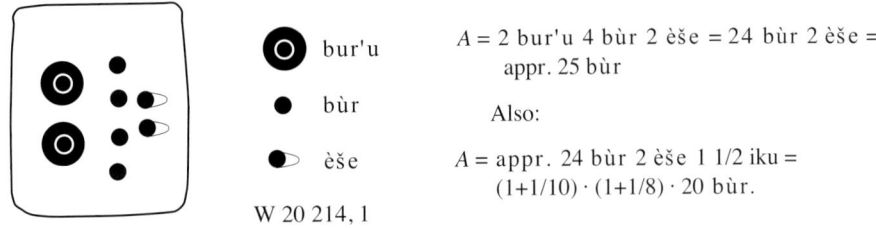

W 20 214, 1

$A = 2$ bur'u 4 bùr 2 èše = 24 bùr 2 èše = appr. 25 bùr

Also:

$A = $ appr. 24 bùr 2 èše 1 1/2 iku = $(1+1/10) \cdot (1+1/8) \cdot 20$ bùr.

Fig. 8.1.6. A proto-literate field-area text (c. 3200 BC) with an almost round area number.

There is no obvious way to factorize the recorded number exactly in the way expected of an almost round number. Nevertheless, it is not difficult to see that the recorded number may be interpreted as an almost round number, rounded off to the nearest multiple of 1 èše. Indeed,

$$(1+1/10) \cdot (1+1/8) \cdot 10\ 00\ 00\ \text{sq. ninda} = (1 + 1/10) \cdot 5\ 00\ \text{n.} \cdot (1 + 1/8) \cdot 2\ 00\ \text{n.} =$$
$$5\ 30\ \text{n.} \cdot 2\ 15\ \text{n.} = 12\ 22\ 30\ \text{sq. ninda} = 24\ \text{bùr}\ 2\ \text{èše}\ 1\ 1/2\ \text{iku} = \text{appr. } 24\ \text{bùr}\ 2\ \text{èše.}$$

8.1 d.   MS 3908. A Trapezoid Divided into Three Stripes

MS 3908 (Fig. 8.1.2, middle) is a round hand tablet with a drawing of a trapezoid, accompanied by some numbers, on the obverse. There is also a weak outline of a second trapezoid on the obverse. The reverse is empty. The trapezoid with numbers is "three-striped", that is, it is divided by two transversals, parallel with the upper and lower fronts of the trapezoid, into three stripes or sub-trapezoids. The lengths of the two transversals are given as 9 10 and 5 50, respectively. The areas of the middle and lower stripes are also given, as 2 30 and 1 40, respectively.

The upper stripe is damaged, and the numbers inscribed inside it and along its sides are not preserved. Nevertheless, it is not difficult to find a reasonable reconstruction of the lost numbers. The key observation is that the three "partial lengths" are [...], 20, 30, and that if the first number is reconstructed as [10], the three numbers will form an arithmetical progression with the sum 1 (00). Thus, the whole length of the trapezoid was probably chosen to be the round number 1 00 (ninda).

The inclination of the sides of a triangle is determined by what may be called the "growth rate" of the triangle, namely *the ratio of the front to the length*. Now, it is a generally observed convention in Old Babylonian geometry that *in every triangle the length is longer than the front*. Consequently *the growth rate of a triangle is always smaller than 1*.

The inclination of the sides of a trapezoid is similarly determined by *the growth rate of the trapezoid, defined as the ratio of the difference of the upper and lower fronts to the length. This growth rate, too, is always smaller than 1.* Therefore, if the length of the trapezoid in MS 3908 is supposed to be 1 00 (ninda), the preserved values 9 10, 5 50, and 50 for the transversals and the lower front (a decreasing sequence) must be interpreted as 9;10, 5;50, and ;50 (ninda), respectively. Consequently, the growth rates of the undamaged stripes or sub-trapezoids are

$$(9;10 - 5;50)/20 = 3;20/20 = ;10 \text{ for the middle stripe, } (5;50 - ;50)/30 = 5/30 = ;10 \text{ for the lower stripe.}$$

A naive application of a simple argument using similar triangles makes it clear that in every "striped trapezoid" like the one in the drawing on MS 3908, that is, in every trapezoid divided by a number of parallel transversals into several stripes or sub-trapezoids, *all the stripes have the same growth rate.* Many explicit examples show that this "common growth rate rule" was well known in Old Babylonian mathematics. That being the case, also the damaged upper stripe must have had the growth rate ;10. Therefore, the lost value of the upper front can be reconstructed as [10;50]. Indeed,

$$(10;50 - 9;10)/10 = 1;40/10 = ;10.$$

Since now all the lengths of the fronts, the transversals, and the partial lengths are known, the three "sub-areas" of the three-striped trapezoid can be computed. They are:

$$
\begin{array}{llll}
10 \cdot (10;50 + 9;10)/2 & = 10 \cdot 20/2 & = 1\ 40 \text{ (sq. ninda)} = 1\ \text{iku} & \text{for the upper stripe,} \\
20 \cdot (9;10 + 5;50)/2 & = 20 \cdot 15/2 & = 2\ 30 \text{ (sq. ninda)} = 1\ 1/2\ \text{iku} & \text{for the middle stripe,} \\
30 \cdot (5;50 + ;50)/2 & = 30 \cdot 6;40/2 & = 1\ 40 \text{ (sq. ninda)} = 1\ \text{iku} & \text{for the lower stripe.}
\end{array}
$$

The numbers 2 30 and 1 40 for the areas of the middle and lower stripes agree with the preserved numbers on MS 3908, while [1 40] for the upper stripe is a plausible reconstruction of a lost number.

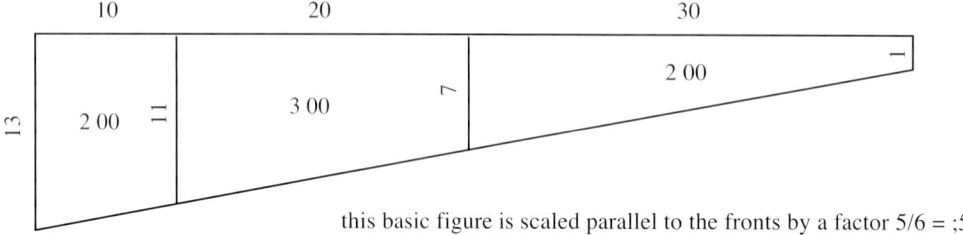

this basic figure is scaled parallel to the fronts by a factor 5/6 = ;50

Fig. 8.1.7. The basic construction for the three-striped trapezoid in MS 3908.

Two important questions concerning MS 3908 remain to be answered. *How were the data for the three-striped trapezoid constructed, and what was the purpose of the text?* A first step towards an explanation of the construction of the data is the observation that all the lengths 10;50 and ;50 of the upper and lower fronts and 9;10 and 5;50 of the upper and lower transversals are multiples of ;50. It is very likely that a simpler set of data was first constructed in some way, and then the fronts and transversals of the three-striped trapezoid were scaled by the factor ;50 (= 5/6), obviously in order to make the areas of the upper and lower stripes equal to the round area number 1 40. sq. ninda = 1 iku. That original set of data is displayed in Fig. 8.1.7 above.

It is important to understand that all the numbers displayed in Fig. 8.1.7 cannot have been prescribed freely. As a matter of fact, they must have been chosen with great care in such a way that they would not come in conflict with the *three area equations* for the areas of the three stripes:

$$
\begin{array}{llll}
A_1 & = 10 \cdot (13 + 11)/2 & = 10 \cdot 12 & = 2\ 00 & \text{the upper sub-area,} \\
A_2 & = 20 \cdot (11 + 7)/2 & = 20 \cdot 9 & = 3\ 00 & \text{the middle sub-area,} \\
A_3 & = 30 \cdot (7 + 1)/2 & = 30 \cdot 4 & = 2\ 00 & \text{the lower sub-area.}
\end{array}
$$

Neither must they conflict with the *common growth rate rule,* that is, with the *similarity equations*

$$r_1 = r_2 = r_3.$$

Actually,

$$r_1 = (13 - 11)/10 = 2/10 = ;12, \quad r_2 = (11 - 7)/20 = 4/20 = ;12, \quad r_3 = (7 - 1)/30 = 6/30 = ;12.$$

What all this means is that the 10 parameters for the three stripes, consisting of 3 partial lengths, 2 fronts, 2 transversals, and 3 sub-areas, must satisfy 3 area equations and 2 similarity equations. Therefore, *only 5 of the 10 parameters for a three-striped trapezoid can be given arbitrary values!*

The most likely situation is that the five freely prescribed parameters in Fig. 8.1.8 below were the three partial lengths 10, 20, 30 and the two the sub-areas equal to 2 00. *The remaining five parameters were then computed as the solutions to the following system of 5 linear equations for 5 unknowns*:

$$s_a + d_a = 2\ 00 \cdot 2/10 \quad = 24 \qquad \text{from the equation for the upper sub-area,}$$
$$A_m \quad = 20 \cdot (d_a + d_k)/2 \qquad \text{the equation for the middle sub-area,}$$
$$d_k + s_k = 2\ 00 \cdot 2/30 \quad = 8 \qquad \text{from the equation for the lower sub-area,}$$
$$d_k \quad = d_a - 2 \cdot (s_a - d_a) \qquad \text{from the similarity equation for the upper and middle stripes,}$$
$$s_k \quad = d_k - 3 \cdot (s_a - d_a) \qquad \text{from the similarity equation for the upper and lower stripes.}$$

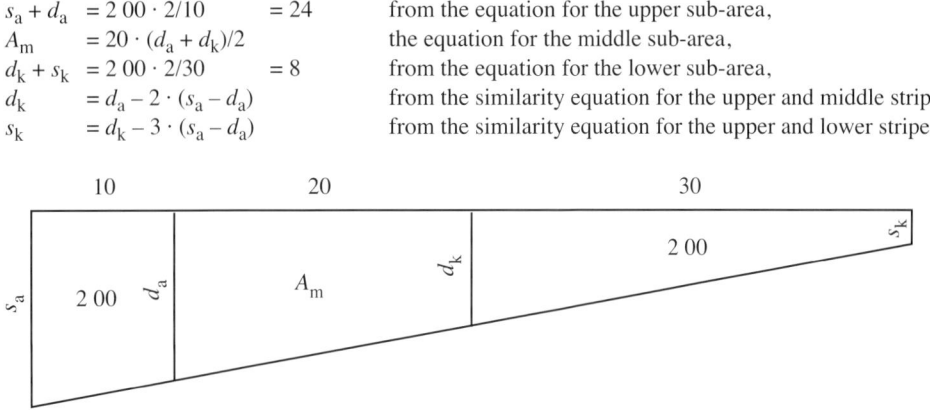

Fig. 8.1.8. The five unknowns for a three-striped trapezoid with five given parameters:
the upper and lower fronts, the upper and lower transversals, and the middle sub-area.

It is, of course, impossible to know how the one who constructed the data for MS 3908 actually solved this system of linear equations. He may have proceeded *by trial and error*. Then, again, he may have proceeded *in a systematic way,* for instance in the following series of simple steps:

1)	d_a	$= 24 - s_a$	using the first area equation,
2)	d_k	$= 3\,d_a - 2\,s_a = 1\ 12 - 5\,s_a$	using the first similarity equation and step 1,
3)	s_k	$= 6\,d_a - 5\,s_a = 2\ 24 - 11\,s_a$	using the second similarity equation, step 2, and step 1,
4)	$d_k + s_k$	$= 3\ 36 - 16\,s_a$	using step 1 and step 2,
5)	$s_a = (3\ 36 - 8)/16 = 13$		using the third area equation and step 4,
6)	d_a	$= 11, d_k = 7, s_k = 1$	using steps 1, 2, and 3,
7)	$A_m = 3\ 00$		using the second area equation and step 6.

In whichever way it was done, it was quite an accomplishment to construct the data for the striped trapezoid figuring on MS 3908. As for the purpose of the text, it may be a student's answer to an assignment. There are many different kinds of problems that a teacher could design with departure from a striped trapezoid like the one on MS 3908, simply by erasing some of the parameters and asking the student to find them again.

8.1 e. Ash. 1922.168, IM 43996, Two More Texts with Striped Trapezoids or Triangles

Although there is no known directly parallel text to MS 3908, there are several indirect parallels with striped trapezoids or triangles. Indeed, problems involving striped trapezoids or striped triangles was a popular topic in Old Babylonian mathematics. (See Friberg, *RlA 7* Sec. 5.4 i.) The most interesting example is **Str. 364** (*MKT 1*, 248 ff) with its many variations of problems for triangles with 2, 3, or 5 stripes (see Fig. 10.2.12 below). In the simplest cases, the stated problems lead to quadratic equations.

Ash. 1922.168 (Robson, *MMTC* (1999), 273-4) is round hand tablet with a drawing of a three-striped trapezoid and associated numbers. The construction of the striped trapezoid in this text is a rather close parallel to the construction of the striped trapezoid on MS 3908, but somewhat less sophisticated. The conform transliteration below, in Fig. 8.1.9, is based on Robson's hand copy.

The upper, middle, and lower lengths are 1 00, 2 00, and 3 00 (ninda) in this text, which means that they are of the same relative sizes as the corresponding values 10, 20, and 30 (ninda) in MS 3908. On the other hand, while apparently the *upper and lower areas* were arbitrarily given in MS 3908, it was probably the *upper and lower fronts*, 15 and 6;40 (ninda), that were arbitrarily given in Ash. 1922.168. With both the lengths and the fronts given, the upper and lower transversals can be computed directly by use of the common growth rate rule. First the growth rate is computed, it turns out to be $r = (15 - 6;40)/6\ 00 = ;01\ 23\ 20$. Next, the two transversals are computed as follows:

$$d_a = s_a - r \cdot u_a = 15 - ;01\ 23\ 20 \cdot 1\ 00 = 15 - 1;23\ 20 = 13;\ 36\ 40,$$
$$d_k = s_k + r \cdot u_k = 6;40 + ;01\ 23\ 20 \cdot 3\ 00 = 6;40 + 4;10 = 10;50.$$

After that, the computation of the three sub-areas is straightforward.

obv.

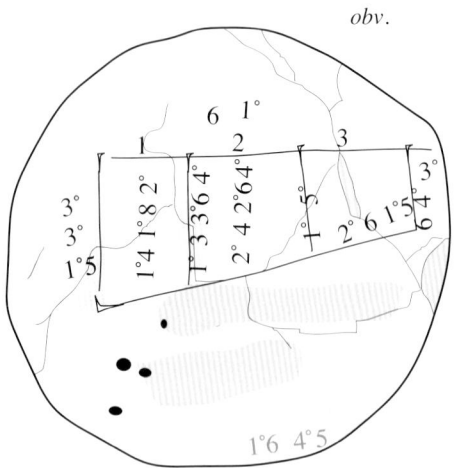

Given parameters (probably):

The upper, middle, and lower partial lengths:

   1 00, 2 00, and 3 00 (ninda).

The upper and lower fronts:

   15 and 6;40 (ninda).

(Above the trapezoid, the 10 following after 6 is the reciprocal of 6, the whole length of the trapezoid. The three numbers 30 possibly refer to the divisions by 2 in the three area equations.)

Fig. 8.1.9. Ash. 1922.168. A less sophisticated parallel text to MS 3908.

Another parallel text to MS 3908, concerned with a *triangle* rather than a trapezoid, is the square hand tablet **IM 43996** (Fig. 8.1.10; Bruins, *Sumer* 9 (1953); Friberg, *UL* (2005), Fig. 2.1.9).

obv.

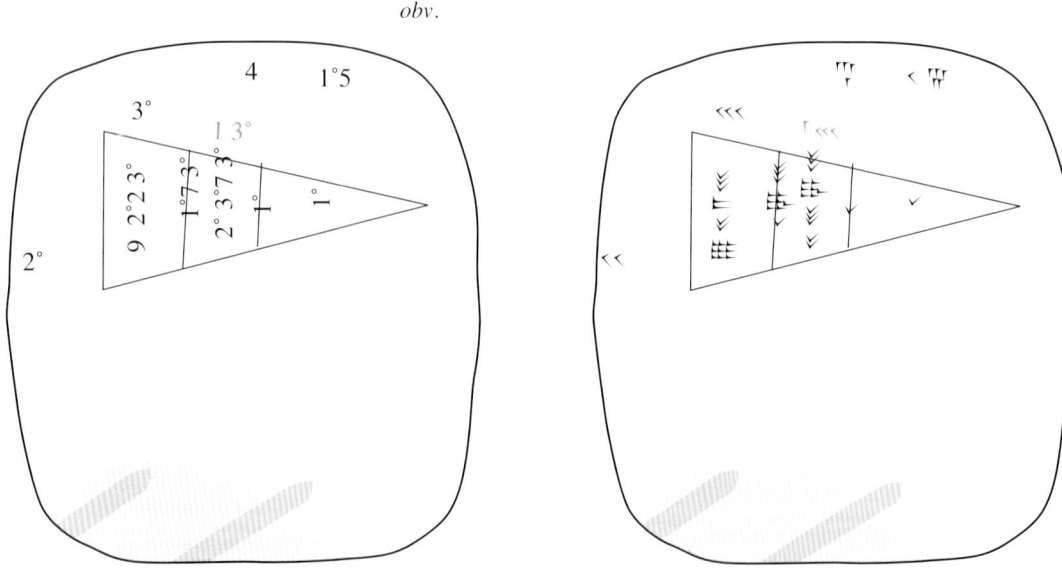

Fig. 8.1.10. IM 43996. A three-striped triangle with two erased parameters.

It is an assignment where the teacher apparently first gave the student the values of all the nine parameters for a three-striped triangle, then erased two of the values, asking the student to find them again. The text is interesting because of the way in which the values of the nine parameters must have been constructed. *In a three-striped triangle, the values for 4 of the 9 parameters can be chosen freely.* The remaining 5 values are determined by 3 area equations and 2 similarity equations.

It is likely that in the case of IM 43996 the teacher initially wanted to construct a striped triangle satisfying the following conditions:

The upper area and the lower area are both 1/2 of the middle area, the front is 20 (n.), and the upper length 30 (n.).

Then, since *the areas of similar triangles are proportional to the squares of their fronts* (a simple consequence of the common growth rate rule, with which Old Babylonian mathematicians were well acquainted), the front s and the two transversals d_a and d_k would have to satisfy the equations

$$\text{sq. } d_a - \text{sq. } d_k = 2 \text{ sq. } d_k, \quad \text{and} \quad \text{sq. } s - \text{sq. } d_a = \text{sq. } d_k.$$

Hence,

$$\text{sq. } d_a = 3 \text{ sq. } d_k, \quad \text{and} \quad \text{sq. } s = 4 \text{ sq. } d_k.$$

With the given value for the front, $s = 20$ (n.), it follows that

$$d_k = 1/2 \cdot s = 10 \text{ (ninda)} \quad \text{and} \quad d_a = \text{sqs. } 3 \cdot d_k = 10 \cdot \text{sqs. } 3 = \text{appr. } 10 \cdot 1;45 = 17;30.$$

With the given value for the upper front, $u_a = 30$ (n.), the growth rate is computed as follows

$$r = (20 - 17;30)/30 = 2;30/30 = ;05.$$

The two similarity equations then yield the values of the middle and lower lengths u_m and u_k:

$$u_m = (17;30 - 10)/r = 7;30 \cdot 12 = 1\ 30, \quad \text{and} \quad u_k = 10/r = 10 \cdot 12 = 2\ 00.$$

The last step of this construction is the straightforward computation of the areas of the three stripes. *Due to the inexactness of the approximate value used for the square side of 3*, the computed value for the upper area then is 9 22;30 instead of 10 00 and that of the middle area 20 37;30 instead of 20 00.

The problem posed to the student, with the computed values for the parameters, was *much easier* than the original construction of those values. Apparently, the student was asked to find the front s and the middle and lower lengths u_m and u_k when the areas of the three stripes were given as 9 22 30, 20 37 30, and 10, and when the two transversals were given as 17 30 and 10. All the student had to do was then to make use of the three area equations, one after the other.[2]

An interesting final example of a parallel to MS 3908 is the "geometric theme text" **MAH 16 055** (see the photo in Bruins, *Physis* 4 (1962)), a text with a series of drawings of five similar three-striped triangles. In the first of these triangles (Fig. 8.1.11), the upper and lower stripes both have the area 5 (00 sq. ninda). In the four following triangles, these areas are multiplied by 2, 3, 4, and 5, respectively, but in all the five triangles, the lengths are divided in three parts equal to 12, 12, and 36 (ninda).

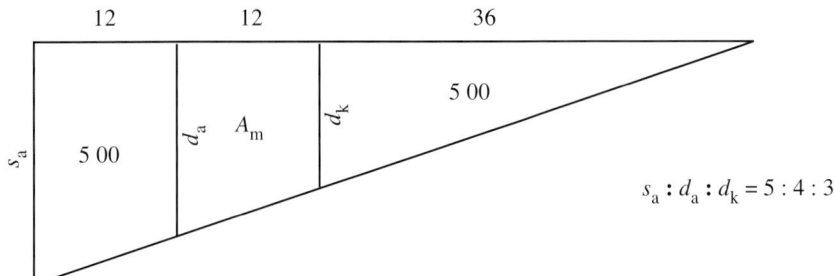

Fig. 8.1.11. The first of five similar three-striped triangles in the geometric table text MAH 16055.

The construction of the numerical values of the parameters for these three-striped triangles can be explained like this: Since the areas of similar triangles are proportional to the squares of their fronts, it follows that

If the upper and lower stripes in a three-striped triangle have the same area, then   sq. $s -$ sq. $d_a =$ sq. d_k.

---

2. The number '4' inscribed near the upper edge of the obverse, together with its reciprocal '15', apparently confirms that the first step of the solution procedure was the observation that sqs. $s = 4$ sqs. d_k, from which it follows that sq. $d_k = ;15 \cdot$ sq. $s = ;15 \cdot 6\ 40 = 1\ 40$, so that $d_k = 10$.

In other words, the front and the two transversals satisfy the indeterminate quadratic "Old Babylonian diagonal rule" (incorrectly known as the "Pythagorean equation"). The simplest and most well known solution *in integers* to that equation is, of course, the triple (5, 4, 3). Suppose, then, that the front and the two transversals have the relative values 5, 4, and 3. Then it follows from the two similarity equations that the upper, middle, and lower lengths have the relative values 5 – 4 = 1, 4 – 3 = 1, and 3. In particular, if the triangle is normalized so that its length is equal to 1 (00 ninda), then it follows that the three parts of the length have the values 12, 12, and 36 (ninda), as in MAH 16055.

Now, if the lower stripe has the area 5 (00 ninda), as in the first of the five drawings on MAH 16055, then the lower transversal must have the length 2 · 36/5 00 = 14;24 (ninda), *etc.*

8.1 f.   MS 1938/2. An Arithmetical Progression of Stripes in a Trapezoid

MS 1938/2 (Fig. 8.1.12) is a fragment of a rectangular hand tablet. The curvature of the clay tablet, as seen from the side, makes it possible to reconstruct the original shape of the tablet. The tablet is inscribed on the obverse with the picture of a striped trapezoid, accompanied by sexagesimal numbers and area numbers. There is a drawing of a circle inside a hexagon (in Old Babylonian mathematics called a 'six-front') on the reverse.

obv.

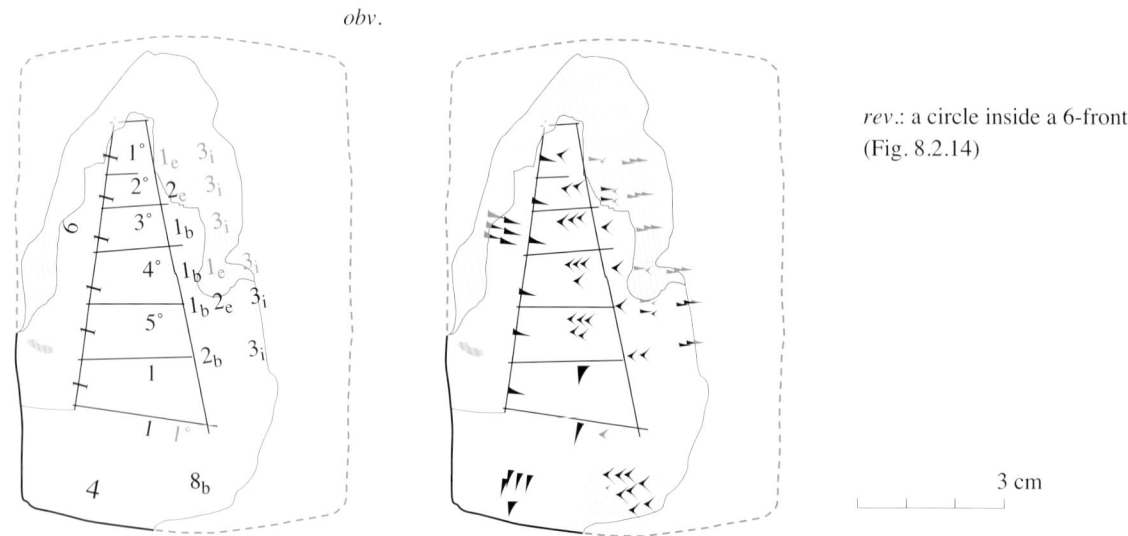

rev.: a circle inside a 6-front (Fig. 8.2.14)

3 cm

Fig. 8.1.12. MS 1938/2, *obv*. Six fields with their areas in an arithmetical progression.

The trapezoid on the obverse is divided by five evenly spaced transversals into six parallel stripes. The lengths of the two fronts and the five transversals form a decreasing arithmetical progression, proceeding in steps of 10 (ninda) from 1 10 to 10 (ninda). The the six partial lengths are all marked as '1', obviously meaning '1 00 ninda'. It is equally obvious that the number '6', written close to the left length of the trapezoid, stands for the value '6 00 ninda' of the whole length.

Alongside the trapezoid are written a series of area numbers, which can be assumed to stand for the areas of the six stripes, although the alignment of these area numbers is quite poor. The areas of the six stripes can be computed as follows:

$$
\begin{aligned}
A_1 &= 1\ 00 \cdot (1\ 10 + 1\ 00)/2 \text{ sq. ninda} &&= 1\ 05\ 00 \text{ sq. ninda} = 2 \text{ bùr} &&3 \text{ iku}, \\
A_2 &= 1\ 00 \cdot\ \ \ (1\ 00 + 50)/2 \text{ sq. ninda} &&=\ \ \ 55\ 00 \text{ sq. ninda} = 1 \text{ bùr } 2 \text{ èše } 3 \text{ iku}, \\
A_3 &= 1\ 00 \cdot\ \ \ \ \ \ \ \ (50 + 40)/2 \text{ sq. ninda} &&=\ \ \ 45\ 00 \text{ sq. ninda} = 1 \text{ bùr } 1 \text{ èše } 3 \text{ iku}, \\
A_4 &= 1\ 00 \cdot\ \ \ \ \ \ \ \ (40 + 30)/2 \text{ sq. ninda} &&=\ \ \ 35\ 00 \text{ sq. ninda} = 1 \text{ bùr } 3 \text{ iku}, \\
A_5 &= 1\ 00 \cdot\ \ \ \ \ \ \ \ (30 + 20)/2 \text{ sq. ninda} &&=\ \ \ 25\ 00 \text{ sq. ninda} = 2 \text{ èše } 3 \text{ iku}, \\
A_6 &= 1\ 00 \cdot\ \ \ \ \ \ \ \ (20 + 10)/2 \text{ sq. ninda} &&=\ \ \ 15\ 00 \text{ sq. ninda} = 1 \text{ èše } 3 \text{ iku}.
\end{aligned}
$$

These computed area numbers agree well with the preserved traces of area numbers on the clay tablet. (Cf. App.

4, Fig. A4.10.) It is easy to check that the sum of the six computed areas is equal to the area of the whole trapezoid, which can be computed as follows:

$$A = 6 \ 00 \cdot (1 \ 10 + 10)/2 \text{ sq. ninda} = 4 \ 00 \ 00 \text{ sq. ninda} = 8 \text{ bùr}.$$

The number 4, indicating this total area, is recorded in the lower left corner of the tablet. The eight oblique wedges inscribed to the right of the number 4, written in three layers of 3, 3, and 2 wedges, may stand for '8 bùr', although the normal way of writing '8 bùr' is in two layers of 4 plus 4 wedges.

Although MS 1938/2 with its drawing of a striped trapezoid and its series of area number looks like a field plan (cf. Ch. 5 above), it is more likely that the text is the answer to an assignment, possibly an inheritance problem for 'six brothers'.

Note: Normally trapezoids in drawings on Old Babylonian clay tablets are positioned with the long sides (the 'lengths') extending from left to right, and with the 'upper front' to the left and the 'lower front' to the right. However, judging from the way in which most of the numbers are written, the trapezoid on the obverse of MS 1983/2 may very well have been oriented in an unusual way, with what is normally the upper front facing downwards. MS 1983/2 is also unlike normal Old Babylonian hand tablets with geometric exercises in that the computed areas are expressed in terms of area numbers rather than sexagesimal numbers in relative place value notation. A possible explanation is that MS 1983/2 is a *Sumerian* rather than Old Babylonian geometric exercise!

For comparison is shown below a drawing of a five-striped trapezoid illustrating a mathematical problem text (of which almost nothing else is preserved) on the obverse of the fragment **IM 31248**. This a text from the site Ishchali, published by Bruins in *Sumer* 9 (1953). There is also a photo of the clay tablet in *Janus* 71 (1984). (The hand copy in Fig. 8.1.13 below is based on that photo.) The trapezoid is drawn totally out of scale, apparently in order to allow enough space to record all pertinent values. The areas of the five stripes are given as 1 57, 31, 1 09, 15, and 21, and the lengths of the four transversals as 33, 29, 17, 13. The upper front is 45, the lower front 1, and the length is divided into five parts with the lengths 3 (00), 1 (00), 3 (00), 1 (00), 3 (00). All recorded numbers are sexagesimal numbers in relative place value notation.

obv.

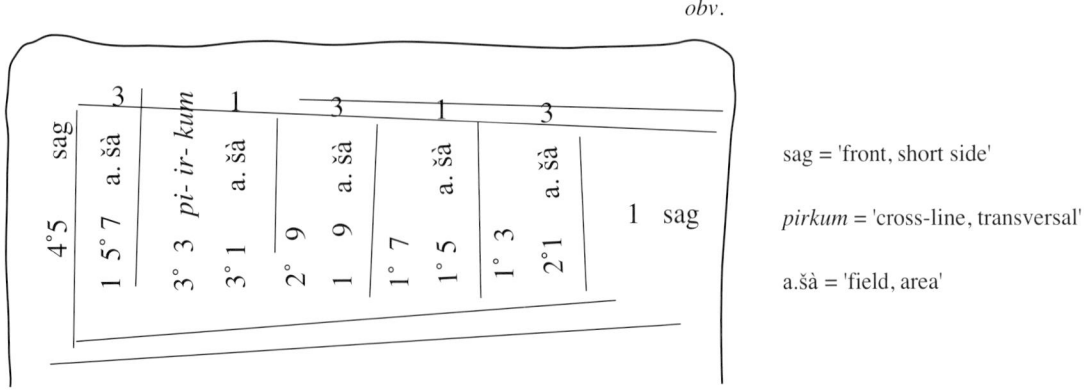

Fig. 8.1.13. IM 31248. A drawing of a five-striped trapezoid introducing an Old Babylonian problem text.

Since the text of the associated exercise is lost, there is no way of knowing what problem was illustrated by the drawing. Note, however, that the growth rate of the trapezoid is $(45 - 1)/11 \ 00 = 44/11 \ 00 = 1/15 = \ ;04$. Thus, the growth rate is a regular sexagesimal number, in spite of the fact that the length is a non-regular number.

8.1 g.   MS 3853/2. A Doodle on the Reverse of a Single Multiplication Table

MS 3853/2 (Fig. 8.1.14) is a large fragment of a clay tablet with a single multiplication table on the obverse, head number 3 20 (the reciprocal of 36). On the reverse there is a geometric doodle, a symmetric triangle divided by three vertical and two horizontal straight lines. The triangle is also divided by several further straight lines, running obliquely downwards and to the right. The full details of the drawing are not visible any more due to damage to the surface of the clay tablet. It is not clear how the missing parts of the drawing should be reconstructed.

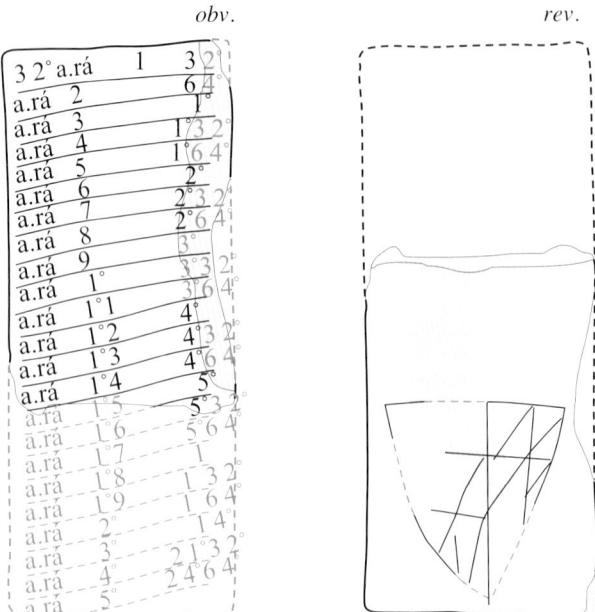

Fig. 8.1.14. MS 3853/2. A single multiplication table of type a with a geometric doodle on the reverse.

8.2. Figures Within Figures

"Figures within figures" was a popular topic in Old Babylonian geometry (see Friberg, *RlA 7* (1990), Sec. 5.4 l.). In the Schøyen Collection, there are five hand tablets with drawings of figures within figures. They will be discussed separately below.

8.2 a.   MS 2192. A Triangular Band between two Concentric Equilateral Triangles

MS 2192 (Fig. 8.2.2, top) is a round hand tablet (a lentil), with a relatively flat obverse but a strongly curved reverse. On the obverse of the lentil, there is a drawing of two "concentric" (and parallel) equilateral triangles. Each side of the inner triangle is extended in one direction. As a result, the space between the two concentric triangles, what may be called an "equilateral triangular band", is divided into a "chain" of three equal trapezoids. The lengths of all line segments in the figure are indicated by sexagesimal numbers in relative place value notation, without any indication of the unit of length, or of the absolute size of the numbers. The lentil is probably early Old Babylonian from southern Mesopotamia, since it makes use of variant number signs for 40.

The lengths indicated in the drawing can be interpreted as 10 ninda three times (the sides of the inner equilateral triangle), 43;20 ninda three times (the longer of the parallel sides in each of the three trapezoids), and 16;40 ninda six times (the non-parallel sides of the three trapezoids). Although that is not explicitly indicated in the drawing, the lengths of the shorter of the parallel sides in the three trapezoids are 10 + 16;40 = 26 40 (ninda), and the lengths of the sides of the outer equilateral triangle are all equal to 16;40 + 43;20 = 1 00 (ninda). The drawing is not true to scale (the inner triangle should be much smaller than the outer triangle).

The inscribed numbers can be explained as follows: Suppose that, initially, the lengths of the sides of the two concentric triangles were given as $m = 1\ 00$ and $n = 10$ (ninda), respectively. Let a be the length of the non-parallel sides of each of the three trapezoids. Then it is not difficult to see that the lengths of the two par-

allel sides of the trapezoids are 10 + *a* and 10 + 2 *a*, respectively, and that the
length of the side of the outer equilateral triangle is 10 + 3 *a*. Since 10 + 3 *a* = 1 00 (ninda), it follows that *a*
= 50/3 = 16;40 (ninda). Hence, the lengths of the parallel sides of the trapezoids are 10 + 16;40 = 26; 40 (ninda) and 10 + 2 · 16;40 = 43;20 (ninda), respectively.

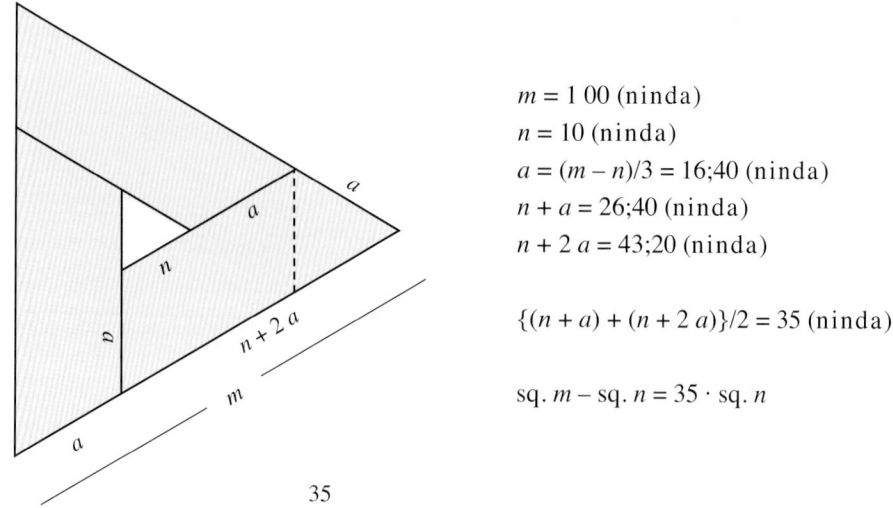

$m = 1\ 00$ (ninda)
$n = 10$ (ninda)
$a = (m - n)/3 = 16;40$ (ninda)
$n + a = 26;40$ (ninda)
$n + 2\ a = 43;20$ (ninda)

$\{(n + a) + (n + 2\ a)\}/2 = 35$ (ninda)

sq. m – sq. $n = 35 \cdot$ sq. n

Fig. 8.2.1. Explanation of the numbers associated with the drawing on MS 2192.

8.2 b.   An Assignment: To Compute the Area between the Two Equilateral Triangles

An important sub-topic of the wide topic "figures within figures" was "concentric" figures, until now represented only by *concentric circles* and *concentric (and parallel) squares*. **MS 2192** is the only known Babylonian mathematical text dealing with *concentric (and parallel) equilateral triangles*. Actually, the only previously known instance of equilateral triangles being mentioned in an Old Babylonian mathematical text is an obscure entry in a table of constants (see below).[3] In several of the known examples of figures within figures, the consideration of concentric circles or squares was associated with the problem of computing the a.šà dal.ba.na 'the field between', by which is meant the area of the circular or square "band" bounded by given concentric figures. It is likely that MS 2192 with its drawing of two concentric equilateral triangles was *an assignment, the student's task being to compute the area of the triangular band between the two triangles*. Compare the many mathematical lentils from Old Babylonian Ur with various arrays of numbers (discussed in Robson, *MMTC* (1999), App. 5, and Friberg, *RA 94* (2000)), which may all be interpreted as assignments, with the student's task being to fill in the missing details. Other examples of lentils with mathematical assignments are MS 1844 with the data for an inheritance problem (Fig. 7.4.2), MS 2268/19 with the data for a market rate problem (Fig. 7.2.2, top), MS 2107 with the area of a trapezoid (Fig. 8.1.2, top), and MS 3908 with a striped trapezoid (Fig. 8.1.2, middle).

Trivially, the area of the band between two concentric circles, squares, or triangles can be computed directly as the difference between the areas of the outer and the inner figure. If one wants to use this simple rule in order to compute the area between two concentric equilateral triangles, one has to know first the area of an equilateral triangle with sides of given length. In modern terms, if the side of an equilateral triangle is *s*, then the height *h* of the triangle is sqs. 3 /2 · *s*, and the area is

$$A = h \cdot s/2 = \text{sqs. } 3 /2 \cdot s \cdot s/2 = \text{sqs. } 3 /4 \cdot \text{sq. } s.$$

---

3. Now, however, there is also the extremely interesting problem text **MS 3876** (see Fig. 11.3.1 below), which may be interpreted as a text dealing with equilateral triangles and regular icosahedrons. See also **MS 3051** (Fig. 8.1.1, bottom.)

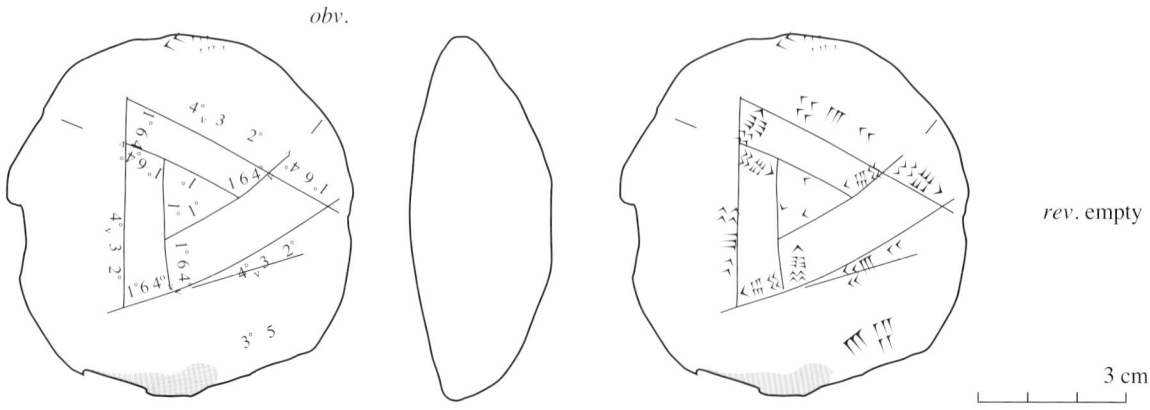

MS 2192. Beginning of a computation of the area between two concentric equilateral triangles, expressed as the sum of the areas of three trapezoids.

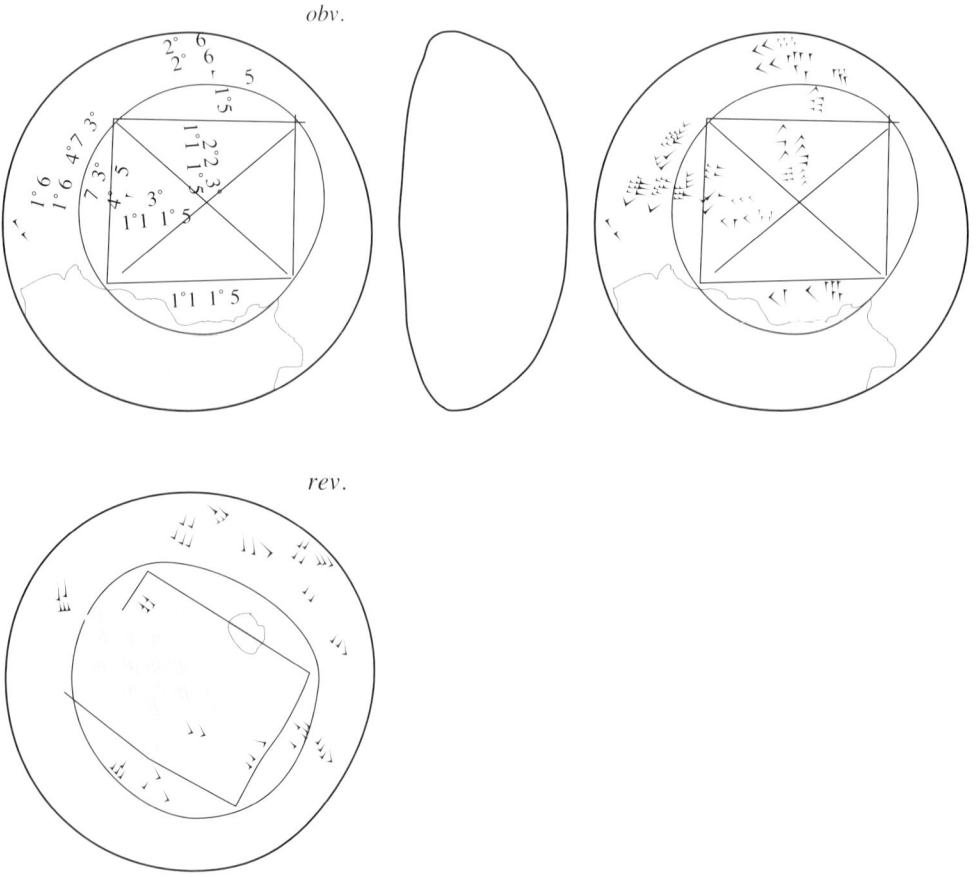

MS 3050 *obv.* A square and its diagonals inscribed in a circle, and several groups of associated numbers.
rev. A square (or rectangle) inscribed in a circle, scattered numbers and erasures.

Fig. 8.2.2. Two round hand tablets with drawings of figures within figures.

A simple approximation to sqs. 3 is 1;45 (= 7/4). The corresponding approximation to sqs. 3 /4 is

$$\text{sqs. } 3\ /4 = \text{appr. } ;26\ 15 \quad (= 7/16).$$

This approximation is mentioned in the Old Babylonian table of constants G = IM 52916, obliquely referred to there as a constant for an equilateral triangle:

sag.kak-*kum ša sa-am-na-[tu na]-ás-ḫa*	A peg-head (triangle), that with an eighth torn out,	G *rev.* 7'
26 15 *i-[gi-gu-bu-šu]*	26 15 its constant.	

(See Robson, *MMTC* (1999), 40-41.) The obscure text refers to the fact that in an equilateral triangle, with the side *s* the height *h* and the area *A* can be computed as

$$h = \text{sqs. } 3\ /2 \cdot s = 1\ 45\ /\ 2 \cdot s = 52;30 \cdot s = s - 1/8 \cdot s, \quad A = h \cdot s/2 = 26;15 \cdot \text{sq. } s.$$

Remark: In two successive exercises in the *Late* Babylonian large recombination text IM 23291, accompanied by drawings of equilateral triangles, the area of an equilateral triangle with the side 1 (00) is computed using two different approximations to the constant sqs. 3 /4. In the first exercise, the approximation used is ;26 15, in the second exercise it is ;26. (See Friberg, *BagM* 28 (1997), 285.) The reason why the Late Babylonian texts mentions two different approximate values for the same constant is probably that ;26 15 (in the first exercise) was the traditional value for sqs. 3 /4, borrowed from some Old Babylonian mathematical text, while ;26 (in the second exercise) was a new, more accurate value, the one normally used by Late Babylonian mathematicians. The Late Babylonian approximation to sqs. 3 /4 can be obtained in, for instance, the following way:

$$1\ 00 \cdot \text{sqs. } 3\ /4 = 15 \cdot \text{sqs. } 3 = \text{sqs. } (3 \cdot 3\ 45) = \text{sqs. } 11\ 15 = \text{appr. } 26, \quad \text{since} \quad \text{sq. } 26 = 11\ 16.$$

Thus, in the case of the *Old* Babylonian text MS 2192, the area of the triangular band between the two concentric equilateral triangles with the sides 1 00 and 10 could have been computed as

$$A(\text{band}) = A(\text{outer triangle}) - A(\text{inner triangle}) = \text{appr. } ;26\ 15 \cdot (\text{sq. } 1\ 00 - \text{sq. } 10) = ;26\ 15 \cdot 58\ 20 = 25\ 31;15.$$

However, this does not seem to be the method that the student who wrote MS 2192 was supposed to use. Apparently, his given task was instead to compute the area of the triangular band as the sum of the areas of the three trapezoids in the drawing. For the computation of the area of one of the trapezoids, he needed to know the height of the trapezoid. Now, it is easy to see that the height of the trapezoid is equal to the height of an equilateral triangle of side 16;40 (Fig. 8.2.1). Therefore,

$$A(\text{band}) = 3 \cdot A(\text{trapezoid}) = 3 \cdot \text{length of midline} \cdot \text{height}/2 = 3 \cdot 35 \cdot (16;40 - 2;05) = 25\ 31;15.$$

The result is, of course, the same as the difference between the areas of the outer and inner triangles.

The number '35' is inscribed on MS 2192 below the triangular band, near the edge. It is likely that this was the student's computation of the length of the midline of one of the trapezoids, the first step in his computation of the area of a trapezoid. Thus, it seems that MS 2192 is an *unfinished solution* to the problem of finding the area of the triangular band represented by the drawing.

Note: *The idea of computing the area of a triangular band as the area of a chain of trapezoids is a variation on the idea of computing the area of a square band as the area of a chain of four rectangles.* This is a simple idea, and it is likely that it was known by Old Babylonian mathematicians, although no cuneiform mathematical text has yet been found where this idea enters in an explicit way. There is, however, an interesting clay tablet from the site Tell Dhiba'i (Db$_2$-146 = **IM 67118**; Baqir *Sumer* 18 (1962)) in which the following problem is considered:

Given the diagonal (1 15) and the area (45 00) of a rectangle, find the sides of the rectangle.

Now there is also a parallel exercise, **MS 3971 § 2**, the whole text of which is reproduced in Sec. 10.1 b below. The explicit solution procedure is explained in Fig. 10.1.3.

It is well known (see, for instance, Høyrup, *LWS* (2002)) that Old Babylonian mathematicians relied on geometric models for their solutions of quadratic equations or systems of equations. A likely geometric model in the case of IM 67118 is the one illustrated to the left in Fig. 8.2.3 below, where the diagonals of the chains of rectangles forming a square ring are seen to form a square, the square on the diagonal. The basic idea of the solution procedure in IM 67118 is that if the diagonal d and the area A of a rectangle are given, and if q is the half-difference of the sides of the rectangle, then the area of the square with the side q can be computed as the area of the square with the side d minus 2 times the area A of the rectangle. (Actually, in Fig. 8.2.3, left, sq. q is seen to be equal to sq. d minus 4 times $A/2$.) In IM 67118, after the half-difference q has been computed, the square of the half-sum p is computed as the square of the half-difference q plus 4 times the area A of the rectangle.

In this connection, it is interesting to note that sq. q plus 2 times the area A is at the same time equal to sq. d (Fig. 8.2.3, left) and to sq. u + sq. s (Fig. 8.2.3, right). This is a simple and straightforward demonstration of the validity of the Old Babylonian diagonal rule, based exclusively on ideas that were well known in Old Babylonian mathematics.[4]

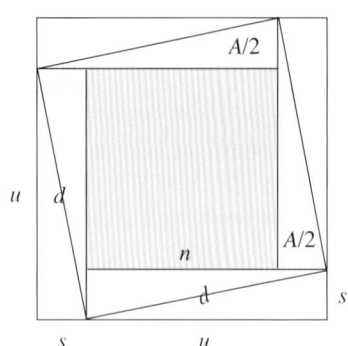 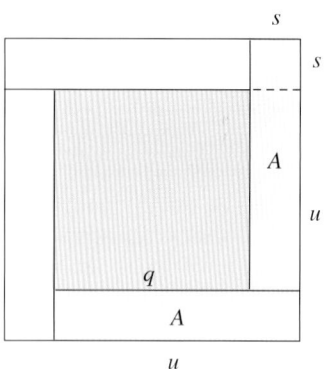

Fig. 8.2.3. Left: IM 67118. Finding the sides of a rectangle when the area and the diagonal are known.
Right and left together: A possible Old Babylonian proof of the diagonal rule.

Remark: *P.Cairo* J. E. 99127-30, 89137-43 is an Egyptian demotic mathematical papyrus from the 3rd c. BC, thus contemporary with Late Babylonian mathematical texts. It was published by Parker in *DMP* (1972). Exercises ## 34-35 in *P.Cairo* are parallels to the problem in IM 67118. One of these exercises, *P.Cairo* # 34 (Friberg, *UL* (2005), Sec. 3.1 i), asks for the sides of a rectangle with the diagonal d = 13 cubits, and the area A = 60 square cubits. The answer is given in a series of simple steps:

1) sq. d + 2 · A = 169 + 10 = 289, sqs. 289 = 17
2) sq. d − 2 · A = 49, sqs. 49 = 7
3) (17 − 7)/2 = 5 = the width of the rectangle
4) 17 − 5 = 12 = the length of the rectangle
5) verification: sq. 12 + sq. 5 = 144 + 25 = 169, sqs. 169 = 13 = the diagonal of the rectangle

An interesting interpretation of an obscure entry in the Old Babylonian table of constants BR (Bruins and Rutten, *TMS* 3 (1961), entry 30) suggests that Old Babylonian mathematicians had a name for a "chain of right triangles" like the one formed by the light grey triangles in Fig. 8.2.3, left. Such a chain of rectangles superficially resembles the cuneiform number sign šàr (= 1 00 00). (There are, for instance, several examples of this number sign in the metrological text MS 3925 shown in Fig. 3.2.4. See also the factor diagram for system *S* in Fig. A4.1.) The entry in the table of constants is:

57 36 igi.gub šà šár   57 36, constant of the šár   BR 30

The probable explanation, found by Vaiman in *VDI* (1963) is that the area of a chain of right triangles like the light grey chain in Fig. 8.2.3, left, is precisely 57 36 if the side of the chain (the diagonal of one of the right triangles) is normalized as 1 00, and if the sides of the right triangles are proportional to the triple 5, 4, 3. Then the sides of the triangles are 1 00, 48, 36, and it follows that the area of the chain of right triangles is equal to

---

4. See App. 8 below, and in particular Sec. A.8 f, for an up-to-date discussion of the role played by the diagonal rule in Old Babylonian mathematics.

the area of the oblique square with the side 1 00 minus the area of the straight square with the side 48 –36 = 12. More precisely, the area of this chain of right triangles is

$$A = \text{sq. } 1\ 00 - \text{sq. } (48 - 36) = \text{sq. } 1\ 00 - \text{sq. } 12 = 1\ 00\ 00 - 2\ 24 = 57\ 36.$$

8.2 c.   *MS 3051. An Equilateral Triangle Inscribed in a Circle*

MS 3051 (Fig. 8.1.1, bottom) is a somewhat large square hand tablet with a drawing and some numbers on the obverse. The reverse is empty, except for a weakly drawn circle. The drawing on the obverse represents an equilateral triangle inscribed in a circle. The equilateral triangle is oriented with one of its sides pointing to the left, in agreement with the observed general rule that in Old Babylonian geometric exercises *the 'front' of a triangle is always pointing to the left*. Compare with the orientation of the triangle in MS 3042 (Fig. 8.1.1, top). Compare also with the orientation of the trapezoids in MS 2107 (Fig. 8.1.2, top) and MS 3908 (Fig. 8.1.2, middle), which are oriented in agreement with the related Old Babylonian rule *that the upper front of a trapezoid is always pointing to the left*.

The drawing in MS 3051 is amazingly exact. The sides of the triangle are nearly equal. The circle passes through two of the three vertices of the triangle and passes close by the third vertex. It is clear that a compass must have been used in the construction of the figure, although there are no remaining traces of the point of the compass. It is also clear that the accurate construction of a figure of this kind would be difficult without a good understanding of basic geometric principles.

The equilateral triangle divides the circumference of the circle in three equal parts. In the drawing on MS 3051, they are all marked with the number '20', obviously meaning '20 ninda'. That means that the whole circumference of the circle is 1 00 (ninda). Now, as known from several Old Babylonian tables of constants and mathematical exercises, in Old Babylonian mathematical texts it is always assumed that *a circle of circumference a = 1 00 has the diameter d = a/3 = 20 and the area A = ;05 · sq. a = 5 00.*[5] Now, if the diameter is 20, the radius is 10. The height of the equilateral triangle is then equal to the radius 10 plus another piece, called *a* in Fig. 8.2.4 below.

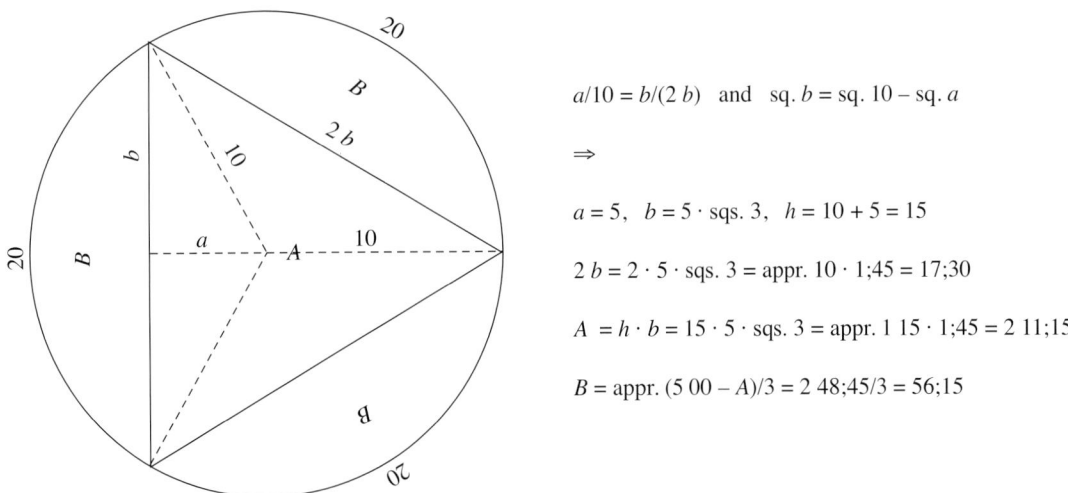

On the right side of the figure:

$a/10 = b/(2\ b)$ and sq. b = sq. 10 – sq. a

$\Rightarrow$

$a = 5$, $b = 5 \cdot$ sqs. 3, $h = 10 + 5 = 15$

$2\ b = 2 \cdot 5 \cdot$ sqs. 3 = appr. $10 \cdot 1;45 = 17;30$

$A = h \cdot b = 15 \cdot 5 \cdot$ sqs. 3 = appr. $1\ 15 \cdot 1;45 = 2\ 11;15$

B = appr. $(5\ 00 - A)/3 = 2\ 48;45/3 = 56;15$

Fig. 8.2.4. Correct parameters for an equilateral triangle inscribed in a (normalized) circle.

An Old Babylonian mathematician could compute the values of *a* and *b* (= half the side of the equilateral triangle) by use of similarity, as follows. First,

$$a/10 = b/(2\ b)\ \Rightarrow\ a = 10/2 = 4.$$

---

5. **MS 3041** (Fig. 8.1.2, bottom) is a round hand tablet with two unrelated drawings. Above, there is a drawing of a rectangle divided into three equal parts, all with the area 14 bùr, written with traditional area numbers. Below, there is a drawing of a circle, and inside the circle the sexagesimal number 1 15, which can be explained as ;05 · sq. 30, that is the area of a circle of circumference 30 (ninda).

Knowing a, he could then compute b by use of the diagonal rule:

$$\text{sq. } b = \text{sq. } 10 - \text{sq. } 5 = 1\ 40 - 25 = 1\ 15 = 3 \cdot \text{sq. } 5.$$

Hence,

$$b = \text{sq. } 1\ 15 = 5 \cdot \text{sqs. } 3 = \text{appr. } 5 \cdot 1;45 = 8;45.$$

When the Old Babylonian mathematician had computed the values of a and b, he could easily compute also the side of the equilateral triangle, $2\ b = 10 \cdot \text{sqs. } 3 = \text{appr. } 10 \cdot 1;45 = 17;30$, as well as the area of the equilateral triangle, $A = 15 \cdot 17;30/2 = 2\ 11;15$. He could then also compute the area B of any of the three circular segments outside the equilateral triangle, since $A + 3 \cdot B =$ the area of the whole circle $=$ appr. 5 00. Therefore, $B = (5\ 00 - 2\ 11;15)/3 = 2\ 48;45/3 = 56;15$.

Explicit arguments like the ones above can rarely be found in Old Babylonian mathematical texts. On the hand tablet MS 3051, there are recorded only the lengths of the three circular arcs, correctly given as '20', the area of the equilateral triangle, incorrectly given as '1 52 30', and the areas of the three circle segments, incorrectly given as '1 02 30'. In addition the height of the equilateral triangle is indicated by the number '15', misleadingly inscribed along one side of the triangle.

It is clear that the student who solved an assignment given to him by producing the hand tablet MS 3051 was guilty of some serious error. The numbers he actually recorded on his hand tablet can be analyzed as follows:

$$\text{``}A\text{''} = 1\ 52;30 = 15 \cdot 15\ /2, \quad \text{and} \quad \text{``}B\text{''} = 1\ 02;30 = (5\ 00 - 1\ 52;30)/3.$$

The mistake he made, absentmindedly thinking about more exciting things than his mathematical assignment, was to set the side of an equilateral triangle equal to the height of the triangle!

Remark: There are no parallels to MS 3051 in the known corpus of Old (or Late) Babylonian mathematics. There are, however, two parallel exercises in the Egyptian demotic mathematical text *P.Cairo*, mentioned above. One of these is *P.Cairo # 36*, where an equilateral triangle of side $s = 12$ 'divine cubits' is inscribed in a circle. The area of the circle is determined in a number of steps:

1) According to the diagonal rule, the height of the equilateral triangle is $h = \text{sqs. (sq. } 12 - \text{sq. } 6)$
   $= \text{sqs. } 108 = \text{appr. } 10\ 1/3\ 1/20\ 1/120$ d. c.
2) The area of the equilateral triangle is $A = 6 \cdot \text{sqs. } 108$ d. c. $= \text{appr. } 62\ 1/3\ 1/60$ sq. d. c.
3) The height of a circle segment is $k = 1/3$ of sqs. $108 = \text{appr. } 3\ 1/3\ 1/10\ 1/60\ 1/120\ 1/180$ d. c.
4) The area of a circle segment is $B = \text{appr. } k \cdot (s + k)/2 = 26\ 5/6\ 1/10$ sq. d. c.
5) The area of the circle is $A + 3\ B = \text{appr. } 143\ 1/10\ 1/20$ sq. d. c.

To check the result, the area of the circle is then computed in a different way:

7) The diameter of the circle is $d = h + k = \text{appr. } 13\ 5/6\ 1/45$ d. c.
8) The circumference of the circle is $a = \text{appr. } 3 \cdot d = 41\ 1/2\ 1/15$ d. c.
9) The area of the circle is $A = \text{appr. } a/3 \cdot a/4$ sq. d. c. $= 143\ 5/6\ 1/10\ 1/30$ sq. d. c.

This Late Egyptian parallel to the Old Babylonian mathematical exercise MS 3051 is particularly interesting because it also in other ways demonstrates *a close connection between an Egyptian demotic mathematical papyrus and Babylonian mathematics*. Here are some pertinent observations:

A. There are three drawings illustrating ***P.Cairo # 36***, showing the equilateral triangle inscribed in a circle, the triangle with its height, and one of the three circle segments. See Fig. 8.2.5.

In these three drawings the triangles and the circle segment are standing upright on their bases. In the drawing in the middle, the height of the triangle is shown at right angles to the base. This is precisely the way in which equilateral triangles and their heights are drawn in the Late Babylonian large mathematical clay tablet **W 23291 § 4 a-c** (Friberg, *BagM* 28 (1997)). In Old Babylonian mathematical texts, on the other hand, triangles are always drawn with the 'front' pointing to the left. In Egyptian hieratic mathematical texts, like *P.Rhind* and *P.Moscow*, triangles are oriented with the short side pointing either to the left or to the right. Thus, in this respect, the demotic *P.Cairo* is closer related to Late Babylonian mathematics than to Old Babylonian or Egyptian hieratic mathematics.

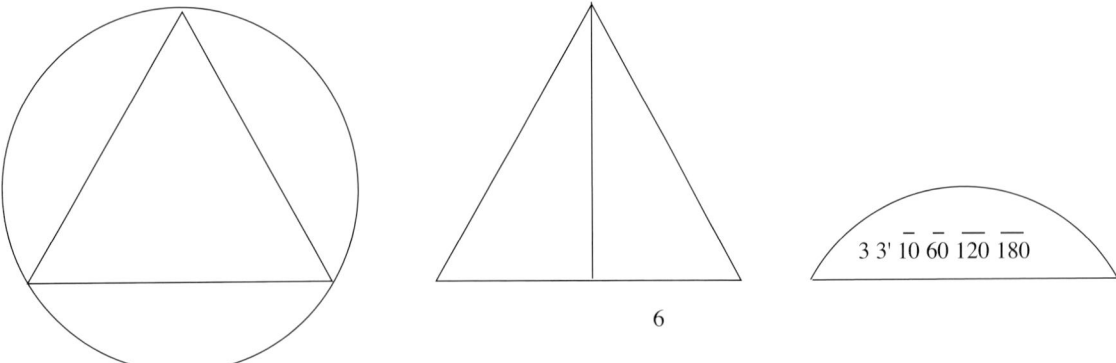

Fig. 8.2.5. Three drawings illustrating the demotic *P.Cairo* # 36.

B. In *P.Cairo* # 36, the circumference a of a circle is equal to $3 \cdot d$, where d is the diameter, and the area A is equal to $a/3 \cdot a/4$. These relations are essentially the same as in Babylonian mathematics, where $a = 3 \cdot d$ and $A = ;05 \cdot$ sq. a. (Note that $;05 = 1/12 = 1/3 \cdot 1/4$.) In the hieratic *P.Rhind*, on the other hand, the area of a circle is given by the totally different equation $A =$ sq. $(d - d/9)$.

C. The calculations in *P.Cairo* # 36 have the appearance of being carried out in terms of "unit fractions" just like all calculations in Egyptian hieratic mathematical texts. Actually, however, the calculations were almost certainly carried out using Babylonian *sexagesimal arithmetic*. This is clear from the way in which the representations of numbers as sums of unit fractions use only *unit fractions that are regular sexagesimal numbers*. Here are sexagesimal interpretations of all the numbers appearing in *P.Cairo* # 36:

10 3' $\overline{20}$ $\overline{120}$	=	10 + ;20 + ;03 + ;00 30	=	10;23 30
62 3' $\overline{60}$	=	62 + ;20 + ;01	=	1 02;21
3 3' $\overline{10}$ $\overline{60}$ $\overline{120}$	=	3 + ;20 + ;06 + ;01 + ;00 30	=	3;27;30
26 6" $\overline{10}$	=	26 + ;50 + ;06	=	26;56
143 $\overline{10}$ $\overline{20}$	=	143 + ;06 + ;03	=	2 23;09
13 6" 45	=	13 + ;50 + ;01 20	=	13;51 20
41 2' $\overline{15}$	=	41 + ;30 + ;04	=	41;34
143 6" $\overline{10}$ $\overline{30}$	=	143 + ;50 + ;06 + ;02	=	2 23;58

An Old Babylonian text that is at least a partial parallel to MS 3051 is **TMS 1**, Bruins and Rutten, (1960), a fragment of a square hand tablet from Old Babylonian Susa. On the obverse of *TMS 1* (Fig. 8.2.6, left), there is a drawing of what appears to be a "symmetric triangle" (in modern terms, an isosceles triangle) inscribed in a circle. The whole figure may or may not have been inscribed in a square, but that is of no importance. The inscription 50 uš '50, the length' indicates that the two equal sides of the triangle measured 50 ninda. The double inscription '30' and 2' sag '1/2 of the front' indicates that the front of the triangle measured 2 · 30 = 1 00 ninda.

The height of the symmetric triangle is drawn as a straight line orthogonal to the front, and is accompanied by the inscription

[40 u]š sag.kak *ga-am-ru*   40, the whole length of the peg-head (= triangle).

Note the confusing use of the same word to denote a long side of a triangle and the height of the same triangle. This is a common phenomenon in Old Babylonian geometric texts (and maybe the explanation for the fatal mistake in MS 3051, where the side of the equilateral triangle was thought to be equal to 15, where 15 was the computed length of the height of the triangle). The side of the symmetric triangle which is called '50, the length' is actually the diagonal of a triangle with the sides 50, 40, 30.

The height $h = 40$ is called 'the whole length' because the center of the circle divides it in two parts. The length of one of these parts is indicated in the drawing as 8 45. The length of the radius is also given, as 31 15 uš, where in this case the use of the word uš 'length' is explained by the fact that the radius is the long side of the small right triangle with the sides 31 15, 30, 8 45 = 1 15 · (25, 24, 7).

The radius *r* and the short side *q* of the small right triangle were probably computed as the solution to a quadratic-linear system of equations of the following kind:

$$\text{sq. } r - \text{sq. } q = \text{sq. } 30 = 15\ 00$$
$$r + q = h = 40.$$

Such systems of equations were solved routinely by Old Babylonian mathematicians, probably with more or less silent reference to geometric arguments, instances of what may be called "metric algebra". (See, for instance, the many examples of metric algebra problems in the large text MS 5112 in Sec. 11.2 below.)

Just like the drawing on MS 30 51 of an equilateral triangle inscribed in a circle, so the drawing of a symmetric triangle inscribed in a circle on *TMS* 1 is amazingly accurate. It is also made to scale so that the proportions are correct. The whole construction shows, once again, that Old Babylonian mathematicians were familiar with the basic properties of triangles and circles.

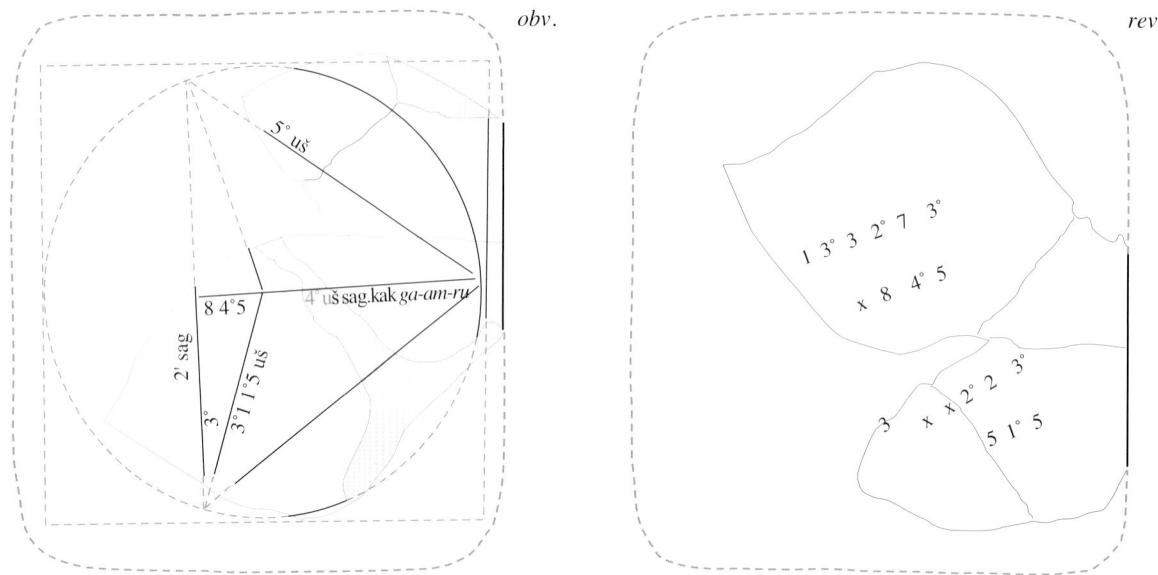

Fig. 8.2.6. *TMS* 1. A symmetric triangle inscribed in a circle. (Conform transliteration based on a photo.)

8.2 d.   MS 3050. A Square with Diagonals, Inscribed in a Circle

MS 3050 (Fig. 8.2.2, bottom) is a round hand tablet with drawings and numbers on both the obverse and the reverse. The drawing on the obverse depicts a square with diagonals, inscribed in a circle. (The drawing on the reverse is similar but partly erased.) It is not easy to make sense of the numbers associated with the drawing on the obverse. Several of the numbers are carelessly written, and their positioning does not seem to relate closely to what they stand for. Anyway, the numbers inscribed on the obverse appear to be the following:

45, 26? (twice), 22 30, 16 40?, 16?, 15, 11 15 (three times), 7 30 (twice?), 5.

Several of these numbers can be explained by assuming that *the diameter* of the circle in the drawing on the obverse is 1 00 (ninda), as in Fig. 8.2.7 below.

If the diameter of the circle is 1 00, then the diagonal of the square is also 1 00, and it follows that the area 2 · *C* of half the square is equal to sq. 30 = 15 00. This can be shown through a simple (and well known) argument where the two quarters of the square with diagonal 1 00 are seen to form together a square with the side 30. Consequently, the area *C* of the quarter-square is equal to 7 30.

Again, if the diameter of the circle is 1 00, then the circumference *a* of the circle is equal to appr. 3 00, and the area 4 · (*B* + *C*) of the circle is appr. 5 00 · sq. 3 = 45 00. The area of the semicircle is then 2 · (*B* + *C*) = 22 30, and the area of the quarter circle 1 · (*B* + *C*) is 11 15.

Thus, some of the values recorded more or less at random on the obverse of MS 3050 seem to be

45 (00)	the area of the circle with the diameter 1 00
22 30	the area of the semi-circle
11 15	the area of the quarter-circle
15 (00)	the area of the half-square
7 30	the area of the quarter-square

The area B of any one of the four circle segments is not computed, and the side of the square is not computed, either. Furthermore, there is not much that can be said about the drawing and the numbers on the reverse. Thus, it seems that the dawdling student who wrote this text did not finish his work. Note also that there are no obvious explanations for the numbers 26?, 16 40?, 16?, and 5, unless, of course, 5 is the constant for the area of a circle.

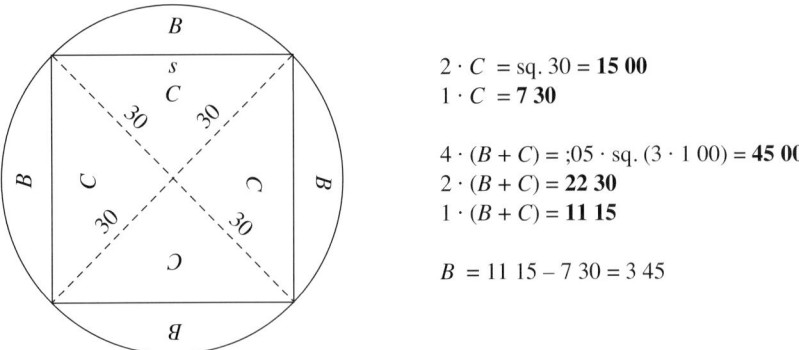

$$2 \cdot C = \text{sq. } 30 = \mathbf{15\ 00}$$
$$1 \cdot C = \mathbf{7\ 30}$$

$$4 \cdot (B + C) = {;}05 \cdot \text{sq. } (3 \cdot 1\ 00) = \mathbf{45\ 00}$$
$$2 \cdot (B + C) = \mathbf{22\ 30}$$
$$1 \cdot (B + C) = \mathbf{11\ 15}$$

$$B = 11\ 15 - 7\ 30 = 3\ 45$$

Fig. 8.2.7. Computation of some parameters for a square inscribed in a circle of diameter 1 00.

A distant parallel to MS 3050 is the well known round hand tablet **YBC 7289**, published by Neugebauer and Sachs in *MCT* (1945), 42. It has on the obverse a drawing of a square with its diagonals. There are three recorded numbers, which can be interpreted as 30 for the side of the square, 1; 24 51 10 for an excellent approximation to sqs. 2 (see Friberg, *BagM* 28 (1997), § 8 c), and 42;25 35 for the diagonal of the square. (Note that the hand copy of YBC 7289 in *MCT* is not correctly oriented. According to the observed conventions of Old Babylonian mathematics, a square should always be drawn with one of its sides facing to the left, as in Fig. 8.2.8 below, which is based on a photo of the clay tablet.)

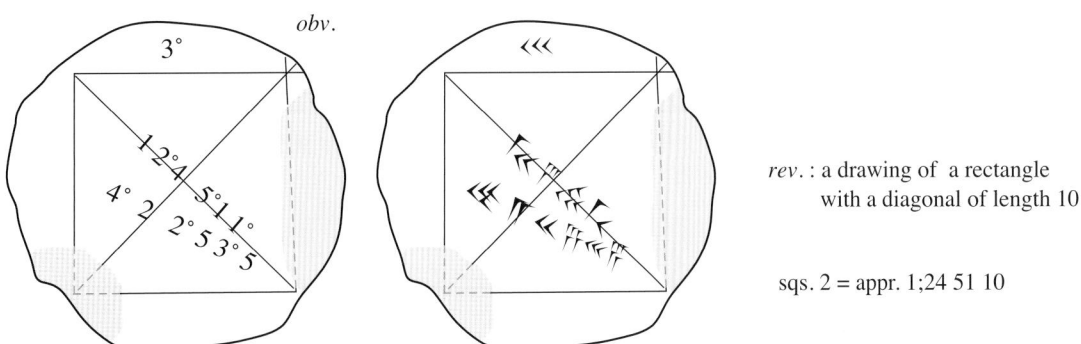

rev.: a drawing of a rectangle with a diagonal of length 10

sqs. 2 = appr. 1;24 51 10

Fig. 8.2.8. YBC 7289. A square with its diagonals. An excellent approximation to sqs. 2.

Remark: Closer parallels to MS 3050 are two exercises in the Egyptian demotic mathematical text *P.Cairo*, already mentioned twice above. One of these is *P.Cairo* # 37 (Friberg, *UL* (2005), Sec. 3.1 k), where a square is inscribed in a circle of diameter $d = 30$ cubits and area $A = 675$ (= 3/4 · sq. 30) sq. cubits. Here is a brief presentation of the computations in that exercise, with all Egyptian sums of unit fractions appearing in the text converted into equivalent sexagesimal numbers:

1) The area of the inscribed square with diagonal 30 is $B = 30 \cdot 30/2 = 450$.
2) The side of the square is $s = $ sqs. $450 = $ appr. 21 1/5 1/60 (21;13).
3) Verification: sq. 21 1/5 1/60 = (appr.) 450. (In sexagesimal numbers: sq. 21;13 = 7 30;08 49.)
4) The height of one of the four circle segments is $p = (30 - 21$ 1/5 1/60$)/2 = 4$ 1/3 1/20 1/120 (4;23 30).
5) The area of one of the four circle segments is $C = $ appr. $(s + p)/2 \cdot p = $ (sic!) 56 1/4 (56;15).
6) Verification: The area of the circle is $B + 4 \cdot C = 450 + 4 \cdot 56$ 1/4 = 675.

It is interesting that the student who made this computation cheated in step 5. He obviously *counted backwards from the expected answer* to get the exact value 56;15 for the area of a circle segment. This kind of cheating appears occasionally also in Old Babylonian mathematical texts

Remark: The attempted computations in MS 3051 and MS 3050 of the areas of circle segments cut off from a circle by an inscribed equilateral triangle or an inscribed square were not very successful. Nevertheless, it is known that at Old Babylonian mathematicians *could* compute correctly the areas of such circle segments. This is obvious in view of the following six entries from the Old Babylonian table of constants **BR = *TMS* 3**, Bruins and Rutten (1961):

13 20	igi.gub	*šà*	a.šà še	13 20	constant	of	a barley-corn field	BR 16
56 40	dal	*šà*	a.šà še	56 40	transversal	of	a barley-corn field	BR 17
23 20	*pi-ir-ku*	*šà*	a.šà še	23 20	cross-line	of	a barley-corn field	BR 18
16 52 30	igi.gub	*šà*	igi.gu₄	16 52 30	constant	of	an ox-eye	BR 19
52 30	dal	*šà*	igi.gu₄	52 30	transversal	of	an ox-eye	BR 20
30	*pi-ir-ku*	*šà*	igi.gu₄	30	cross-line	of	an ox-eye	BR 21

What is going on here is explained in detail in the commentary to problem # 1 in the Late Babylonian mathematical recombination text W 32291-x (Friberg, Hunger, and Al-Rawi, *BagM* 21 (1990).) The "barley-corn field" is an oval figure composed of two circle segments glued together, in the case when the circle segments are those cut off from a circle by an inscribed square. Similarly, the "ox-eye" is an oval figure composed of two circle segments joined together, in the case when the circle segments are those cut off from a circle by an inscribed equilateral triangle. In both cases, the figures are evidently thought of as normalized in the sense that they are bounded by circular "arcs" of length $a = 1$ (00). The "constant" of such a figure is its area, the "transversal" d is its long diameter (the common base of the joined circular segments), and the "cross-line" p is its short diameter (orthogonal to the transversal).

It can be left to the reader to check that the 6 constants BR 16-21 are correctly computed. Briefly, the first three can be explained as follows, with sqs. 2 = appr. 1;25 (17/12):

$$A = 2 \cdot 1/4 \cdot (A_{\text{circle}} - A_{\text{square}}) = \text{appr. } 2 \cdot 1/4 \cdot (;05 \cdot \text{sq. } (4 \cdot a) - 1/2 \cdot \text{sq. } (1;20\ a)) = ;13\ 20\ (2/9) \cdot \text{sq. } a$$
$$d = \text{sqs. } 2 \cdot ;40 \cdot a = \text{appr. } 1;25 \cdot ;40 \cdot a = ;56\ 40\ (17/18) \cdot a$$
$$p = 2 \cdot 1/3 \cdot (2 - \text{sqs. } 2) \cdot a = \text{appr. } ;23\ 20\ (7/18) \cdot a.$$

Similarly, the last three can be explained as follows, with sqs. 3 = appr. 1;45 (7/4):

$$A = 2 \cdot 1/3 \cdot (A_{\text{circle}} - A_{\text{triangle}}) = \text{appr. } 2 \cdot 1/3 \cdot (;05 \cdot \text{sq. } (3 \cdot a) - 3/16 \cdot 1;45 \cdot \text{sq. } a) = ;16\ 52\ 30\ (27/64) \cdot \text{sq. } a$$
$$d = \text{sqs. } 3 \cdot 1/2 \cdot a = \text{appr. } ;52\ 30\ (7/8) \cdot \text{sq. } a$$
$$p = \text{appr. } 2 \cdot (1/2 - 1/4) \cdot a = ;30\ (1/2) \cdot a.$$

8.2 e.   *MS 2985 A Circle in the Middle of a Square*

MS 2985 (Fig. 8.1.1, middle) is a square hand tablet with a drawing and numbers on the obverse. The reverse is empty. The drawing shows a square and circle inside it, a distance away from the sides of the square. The numbers are written in an inexperienced hand and are difficult to read. They seem also to be placed more or less at random on the obverse, without any obvious connection to the drawing. For this reason, it would have been difficult to find a meaningful interpretation of MS 2985 if it had not been for the existence of a couple of partially parallel texts.

The first of these partial parallels is **YBC 7359**, a square hand tablet with drawings on the obverse and the reverse of a couple of "concentric squares" with associated numbers. The drawing on the obverse is clearly a teacher's neat model, while the drawing on the reverse is a student's clumsy copy. The conform transliteration in Fig. 8.2.9 below is based on photos of the clay tablet published by Nemet-Nejat, *UOS* (2002), 275-276.

Nemet-Nejat correctly observes (*op. cit.*, 262) that the inscribed numbers are related to each other in various ways but can offer no further explanation of the text, probably because she interprets the figures as rectangles. However, a clue to the meaning of the text is the observation that the number 9 inscribed within the inner figure is the square of 3, while the number 1 40 inscribed above the outer figure is the square of 10. (The numbers 10 and 3 themselves are inscribed to the left of the drawings on the obverse and the reverse, respectively.) There-fore, it seems reasonable to start an analysis of the text with the assumption that the two figures are a couple of concentric squares with the sides 3 and 10, respectively. That this is a correct assumption becomes clear when it turns out that then the area of the region between the squares is

$$\text{sq. } 10 - \text{sq. } 3 = 1\ 40 - 9 = 1\ 31.$$

The number 1 31 is inscribed between the squares, to the left, in the drawings on both the obverse and the re-verse. Furthermore, the distance between the two squares can be computed as

$$(10 - 3)/2 = 7/2 = 3;30.$$

The number 3 30 is recorded in both drawings, between the outer and the inner square, above and below the inner square.

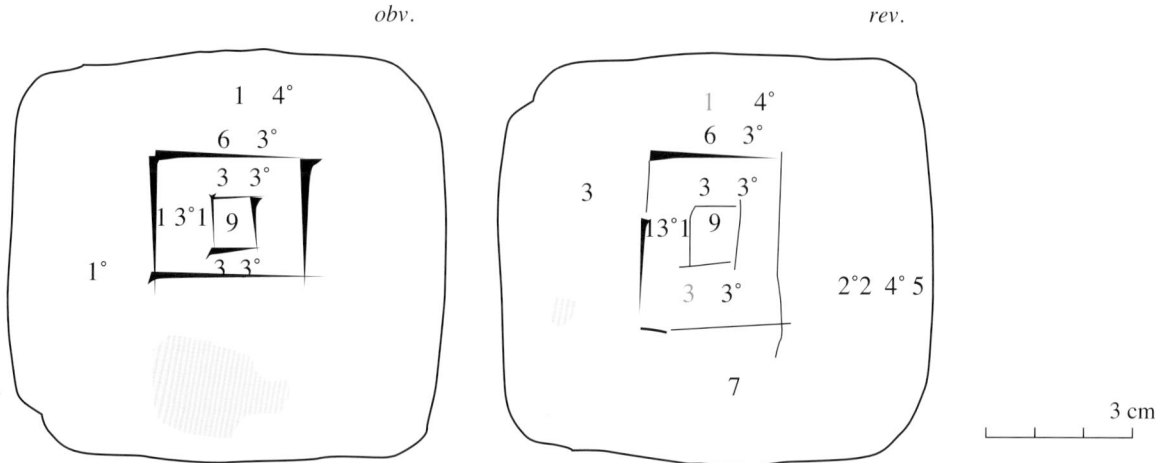

obv.   *rev.*

3 cm

Fig. 8.2.9. YBC 7359. A teacher's and a student's drawings of concentric squares.

It is likely that the drawings were meant to illustrate a metric algebra problem, which may have been of the following form:

The area *A* between two (concentric) squares is 1 31. The distance *d* between the squares is 3;30.
Find the sides and the areas of the squares.

It was known (cf. Fig. 8.2.3 above) that the region between two concentric squares with the sides *p* and *q*, re-spectively, can be divided into four rectangles, all four with the long side $(p + q)/2$ and the short side $(p - q)/2$ $= s$. Therefore, the area *A* of the region between the squares, a "square band" with the middle length *u*, can be expressed as

$$A = 4\,u \cdot s, \quad \text{where} \quad u = (p + q)/2.$$

If both *A* and *s* are known, this equation can be used to determine the value of $(p + q)/2$. Thus, with $A = 1\ 31$ and $s = 3{,}30$ as in YBC 7350, *p* and *q* can be found as the solutions to the equations

$$(p + q)/2 = A/4 \cdot 1/s = 22{;}45/3{;}30 = 6{;}30, \quad (p - q)/2 = s = 3{;}30.$$

Note that the number $A/4 = 22\ 45$ is inscribed to the right of the drawing on the reverse of YBC 7359, and that the number 6 30 appears above the drawings on both sides of the clay tablet. Hence,

$$p = 6{;}30 + 3{;}\ 30 = 10, \quad \text{and} \quad q = 6{;}30 - 3{;}30 = 3.$$

Another partial parallel to MS 2985 is the fragment **TMS 21** (Bruins and Rutten (1961)), a difficult problem text that was only recently explained by Muroi in *SCIAMVS* 1 (2000). According to Muroi, the problem is concerned with an *apsammikku*, a 'sound-hole', or what may be called a "concave square" (see Friberg, *RlA 7* (1990) Sec. 5.4 g). A concave square is the figure bounded by four circular arcs which remains when four touching quarter-circles have been removed from the four corners of a square.

In **TMS 21 a** (Fig. 8.2.10), a concave square is inscribed in the middle of a square, at a distance of 5 ninda from all the four sides of the square. The a.šà dal.ba.na, the 'field between', bounded on one side by the square and on the other side by the concave square, is given as 35 (00 sq. ninda).

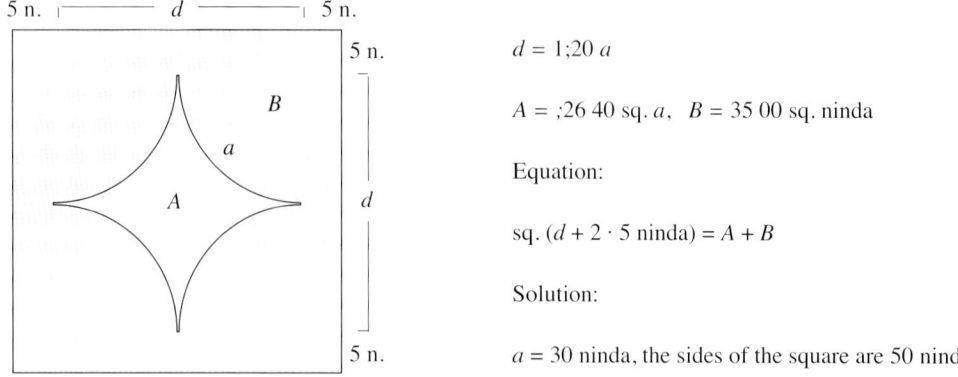

$d = 1;20\, a$

$A = ;26\ 40$ sq. a, $B = 35\ 00$ sq. ninda

Equation:

sq. $(d + 2 \cdot 5$ ninda$) = A + B$

Solution:

$a = 30$ ninda, the sides of the square are 50 ninda.

Fig. 8.2.10. *TMS* 21 a. A concave square inscribed in a square, 5 ninda away from the sides of the square.

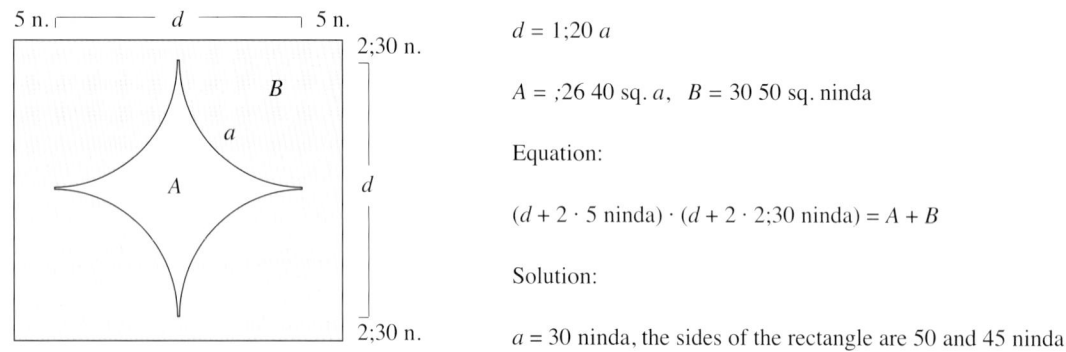

$d = 1;20\, a$

$A = ;26\ 40$ sq. a, $B = 30\ 50$ sq. ninda

Equation:

$(d + 2 \cdot 5$ ninda$) \cdot (d + 2 \cdot 2;30$ ninda$) = A + B$

Solution:

$a = 30$ ninda, the sides of the rectangle are 50 and 45 ninda.

Fig. 8.2.11. *TMS* 21 b. A concave square inscribed in a rectangle, 5 n. from the front, 2;30 n. from the length.

The following basic parameters for an *apsammikku* are given in the table of constants BR:[6]

26 40	igi.gub	šà a-pu-sà-am-mi-ki	26 40	constant	of a sound-hole	BR 22
1 20	bar.dà	šà a-pu-sà-mi-ki	1 20	cross-over (diagonal)	of a sound-hole	BR 23
33 20	*pi-ir-ku*	šà a-pu-sà-mi-ki	33 20	cross-line	of a sound-hole	BR 24

The dimensions of the square and the concave square are determined in the following way:

In a normalized concave square, the arc $a = 1\ 00$, the diameter $d = 1\ 20$, and the area $A = 26\ 40$.
In an arbitrary concave square with arc a, the diameter $d = 1;20 \cdot a$, and the area $A = ;26\ 40 \cdot$ sq. a.
Equation: sq. $(1;20 \cdot a + 2 \cdot 5$ ninda$) - ;26\ 40 \cdot$ sq. $a = 35\ 00$ sq. ninda.
The (positive) solution to this quadratic equation is $a = 30$ ninda.
Hence, $d = 1;20 \cdot a = 40$ ninda, and the side of the square is $d + 2 \cdot 5$ ninda $= 50$ ninda.

TMS 21 b (Fig. 8.2.11) is even less well preserved than *TMS* 21 a, but what remains of the text seems to suggest that in this second exercise the object considered was a concave square inscribed in a *rectangle*, 2 1/2

6. The reading bar.dà, as a variant spelling for Sum. bar.da 'cross-bar', was suggested by Muroi in *HSJ* 2 (1992). Note that the cuneiform sign bar is an upright wedge crossed over orthogonally by a horizontal wedge.

ninda away from the front (the short side) of the rectangle, and 5 ninda away from the length (the long side) of the rectangle.

It is likely that the drawing on MS 2985 is meant to illustrate a similar problem for a circle inscribed in a square a given distance away from the sides of the square. If that is so, then the situation can be described by the diagram in Fig. 8.2.12 below, which mimics the diagram in Fig. 8.2.10.

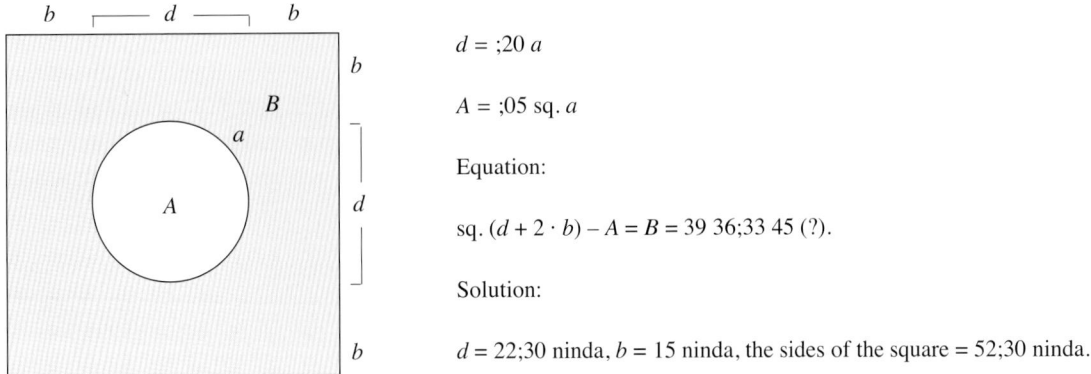

$d = ;20\ a$

$A = ;05$ sq. a

Equation:

sq. $(d + 2 \cdot b) - A = B = 39\ 36;33\ 45\ (?)$.

Solution:

$d = 22;30$ ninda, $b = 15$ ninda, the sides of the square $= 52;30$ ninda.

Fig. 8.2.12. MS 2985. A circle inscribed in a square, a distance b away from the sides of the square.

Now, according to Old Babylonian mathematical conventions, the main parameter of a circle is its 'arc' a, the circumference of the circle. The diameter d and the area A of the circle are given by the following equations:

$$d = ;20 \cdot a\ (= a/3), \quad A = ;05 \cdot \text{sq.}\ a = ;45 \cdot \text{sq.}\ d.$$

Therefore, if both the area B between the circle and the square and the distance b are given, then the value a of the arc of the circle can be found as the solution to the following quadratic equation:

$$\text{sq.}\ (;20 \cdot a + 2 \cdot b) - ;05 \cdot \text{sq.}\ a = B.$$

An equation like this would be given a geometric interpretation by Old Babylonian mathematicians and would then routinely be solved through a geometric procedure equivalent to our "completion of the square". Translated to modern symbolic notations, the essential steps of that geometric procedure, in the present case, would be the following: First, since sq. $;20 = ;06\ 40$ and $;06\ 40 - ;05 = ;01\ 40$, the quadratic equation above can be reduced to:

$$;01\ 40 \cdot \text{sq.}\ a + 1;20 \cdot b \cdot a = B - \text{sq.}\ (2 \cdot b).$$

Next, since $;01\ 40 = \text{sq.}\ ;10$, this equation can be further reduced to

$$\text{sq.}\ (;10 \cdot a + 4 \cdot b) = B + 12 \cdot \text{sq.}\ b.$$

And so on.

If this is really what is going on in the case of MS 2985 (see again Fig. 8.1.1, middle), then it should be possible to identify some of the numbers recorded on the obverse of MS 2985 as square numbers. There is one obvious candidate, since the numbers near the left side of the square are 30, 30, and 15, clearly a notation meaning that sq. 30 = 15 (00). Another likely candidate is the number 45 56 15 inscribed just above the circle, since $45\ 56;15 = \text{sq.}\ 52;30$.

Now, suppose, for instance, that the following identification can be made:

$$45\ 56;15\ \text{sq. ninda} = \text{sq.}\ (;20 \cdot a + 2 \cdot b) = \text{the area of the square containing the circle.}$$

It then follows that

$$d + 2 \cdot b = ;20 \cdot a + 2 \cdot b = \text{sqs.}\ 45\ 56\ 15\ \text{sq. ninda} = 52;30\ \text{ninda} = \text{the side of the square.}$$

As a matter of fact, the value 52 30 appears to be inscribed over the upper left corner of the square.

Furthermore, the value of b can probably be identified with the number 15 which is inscribed both to the left and to the right of the circle, inside the square, in the drawing on the obverse of MS 2985. Now, it follows from the suggested identifications that $d + 2 \cdot b = d + 30 = 52;30$ ninda, and that

$$d = 22;30 \text{ ninda}.$$

Note that this means that $2 \cdot d = 3 \cdot b$. The indicated values for d and b imply that

$$a = 3 \cdot d = 1\ 07;30 \text{ ninda}, \quad \text{and} \quad B = \text{sq.}\ 52;30 - ;05 \cdot \text{sq.}\ 1\ 07;30 = 45\ 56;15 - 6\ 19;41\ 15 = 39\ 36;33\ 45.$$

A final possible identification is that the numbers 30, 30, 15 to the left of the square in the drawing stand for the computation sq. $(2 \cdot b) = $ sq. $30 = 15$ (ninda).

Note: Although the suggested interpretation of MS 2985 appears to be quite plausible, it is disturbing that it does not take any account of four of the numbers recorded on the clay tablet (25 and 3 20 to the left of the square and 2 16 45 and 2 13 45 right above 45 56 15). The reason may be that the proposed interpretation is not correct, or it may be that the student who wrote the text made some mistakes in his calculations (although it is not clear what those mistakes can have been).

Maybe it is an unwarranted assumption that MS 2985 is related to *TMS* 21 b, the problem for a concave square in a square, which leads to a quadratic equation. An alternative, simpler explanation of MS 2985 is that the drawing was associated with the problem of finding the area between the circle and the square, when the side of the square and the distance from the circle to the sides of the square are given. This is a problem that is known from the third example on the well known large geometric theme text BM 15285. (See, most recently, the excellent hand copy of the whole text in Robson, *MMTC* (1999), App. 2.)

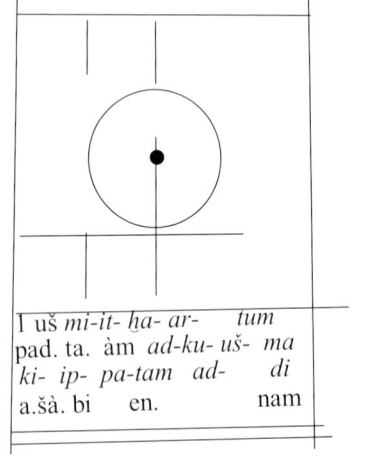

BM 15285, col. *i*, pr. 3:

1 uš (= 1 00 ninda) the equalside (= square).

A bit(?) each way I thrust (inwards),

an arc (= circle) I drew.

Its area is what?

Fig. 8.2.13. One of (probably) 41 illustrated problems on the large geometric theme text BM 15 285.

There are ten columns on BM 15285, five on the obverse and five on the reverse, with four or five small exercises in each column. The exercises are all of a common format, always with a drawing of a square divided into smaller pieces by straight lines and circular arcs, and under each subdivided square there is always a caption with a brief text specifying the construction of the small pieces, naming them, and asking for their areas. In col. *i*, the third problem (Fig. 8.2.13 above) is illustrated by a drawing of a circle inscribed in a square, a distance away from the sides of the square. The exact meaning of the text is not clear, but it seems to say that the side of the square is 1 00 ninda and that the student is asked to thrust inwards from the side of the square equally much on each side and then draw the circle. His task then is, as in all exercises on BM 15285, to compute the areas of all parts of the subdivided square.

8.2 f.   *MS 1938/2. A Circle in the Middle of a 6-Front*

MS 1938/2 is a fragment of a rectangular clay tablet, with unrelated drawings on its obverse and reverse. On the obverse, there is a drawing of a six-striped trapezoid with associated numbers (see Fig. 8.1.12). On the reverse (Fig. 8.2.14 below) there is a badly damaged drawing, which may be interpreted as what remains of a drawing originally depicting a a regular hexagon, with a circle inside it a certain distance away from the sides of the hexagon.

rev.

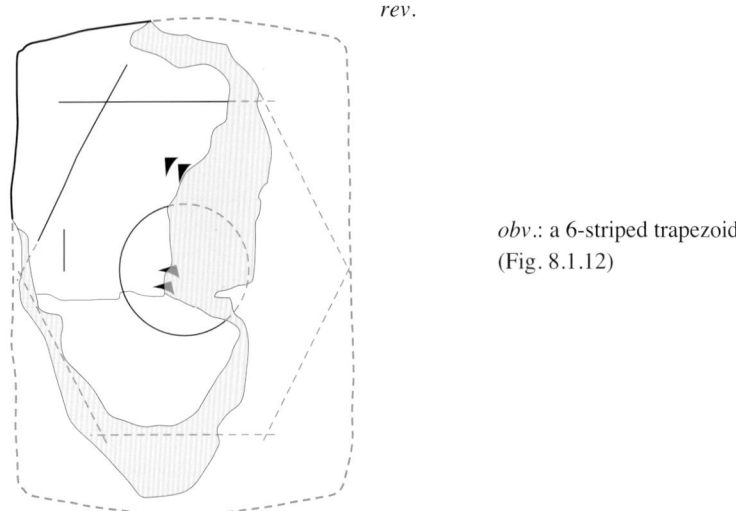

obv.: a 6-striped trapezoid
(Fig. 8.1.12)

Fig. 8.2.14. MS 1938/2, *rev.*: A circle inscribed in a regular 6-front, some distance away from the sides.

This is an assignment of the same type as the one with a circle inside a square on MS 2985 (Figs. 8.1.1 and 12). Now, if the drawing on MS 1938/2 is connected with a problem where the area between a regular hexagon and a circle is one of the given parameters, then the one who designed the problem must have known the area of a regular hexagon. The following three entries in the Old Babylonian table of constants **BR = *TMS* 3** demonstrate that, indeed, Old Babylonian mathematicians were familiar with regular polygons and had methods they used to compute the areas of such polygons:

1 40	igi.gub šà sag.5	1 40	the constant of a 5-front	BR 26
2 37 30	igi.gub šà sag.6	2 37 30	the constant of a 6-front	BR 27
3 41	igi.gub šà sag.7	3 41	the constant of a 7-front	BR 28

The area of a *normalized regular hexagon*, with the length of each side equal to 1 00, can be computed as *the sum of the areas of six normalized equilateral triangles*. Explicitly,

$$A_6 = 6 \cdot (\text{sqs. } 3/4) \cdot \text{sq. } 1\,00\,00 = \text{appr. } 6 \cdot ;26\,15 \cdot 1\,00\,00 = 6 \cdot 26\,15 = 2\,37\,30.$$

Thus, the entry in line 27 of BR = *TMS* 3 is the Old Babylonian approximation to the area of a (normalized) regular hexagon. Apparently, '6-front' was the Babylonian name for a regular hexagon.

It is just as easy to compute the area of a *normalized regular pentagon*. If the side of the pentagon is 1 00, then the circumference of the circumscribed circle is approximately equal to 5 00. Consequently, the diameter of the circle is approximately equal to 1/3 · 5 00 = 1 40. The area of the regular pentagon can therefore be computed as the sum of the areas of five symmetric triangles with one side equal to 100 and the other two equal to 50 (the radius of the circumscribed circle). The height in each triangle is easily computed and is equal to 40. Hence, the area of a normalized regular pentagon is:

$$A_5 = \text{appr. } 5 \cdot 30 \cdot 40 = 5 \cdot 20\,00 = 1\,40\,00.$$

Thus, the entry in line 26 of *TMS* 3, is the Old Babylonian approximation to the area of a (normalized) regular pentagon, called a '5-front'.

Similarly, in the case of a *normalized regular heptagon*, the circumference of the circumscribed circle is approximately equal to 7 00. The diameter is then approximately equal to 1/3 · 7 00 = 2 20, so that the radius will be 1 10. The height can then be computed as

$$h_7 = \text{sqs. (sq. } 1\,10 - \text{sq. } 30) = \text{sqs. } 1\,06\,40 = \text{appr. } 1\,00 + 6\,40/2\,00 = 1\,03;20.$$

Hence, the area of a normalized heptagon is:

$$A_7 = \text{appr. } 7 \cdot 30 \cdot 1\ 03{;}20 = 7 \cdot 31{;}40 = 3\ 41{;}40 = \text{appr. } 3\ 41.$$

Thus, the entry in line 28 of *TMS* 3, is the Old Babylonian approximation for the area of a (normalized) regular heptagon, called a '7-front'.

TMS 2 (Fig. 8.2.15 below) is a square hand tablet with drawings of a 6-front on the obverse and a 7-front on the reverse. Circumscribed circles appear to have been drawn by use of a compass as an aid for the construction, then erased when they were no longer needed. Only vague traces of the circles are now remaining.

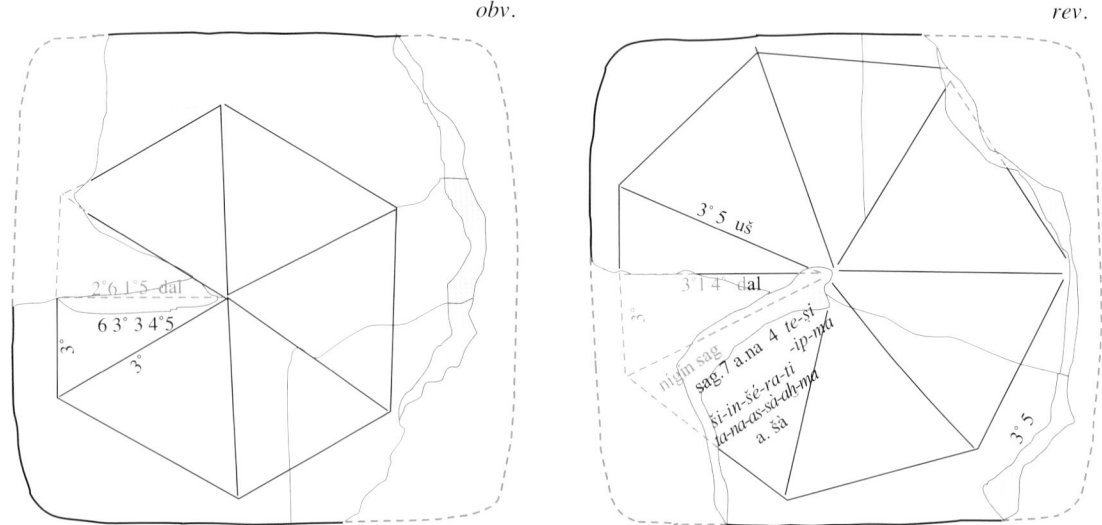

Fig. 8.2.15. *TMS* 2. A '6-front' and a '7-front', with methods for the computation of their areas. (The conform transliterations are based on Bruins' photos of the clay tablet.)

When the area of a *normalized* geometric figure was given as an entry in an Old Babylonian table of constants, it was silently understood that *the areas of similar geometric figures are proportional to the squares of their basic lengths*. In the case of an *n*-front, for instance, with $n = 3$ (an equilateral triangle), 4 (a square) 5, 6, or 7, the area is proportional to the square of the side. On the obverse of *TMS* 2, it is indicated that the length of the upper front (the left-most side) of the figure is 30, while the upper front of a *normalized* 6-front is 1 00, twice as much. Therefore, the area of the 6-front in the drawing is one fourth of the area of a normalized 6-front. Also, the area of each one of the six equilateral triangles in the 6-front is one fourth of the area of a normalized equilateral triangle:

$$A(\text{triangle}) = \text{appr. } 1/4 \cdot 26\ 15 = 6\ 33{;}45.$$

This is the value recorded inside the upper equilateral triangle in the drawing on *TMS* 2, *obv*.

The length of the upper front of the 7-front on the reverse of *TMS* 3, was also set equal to 30, although the number probably inscribed close to the upper front is no longer present. As a consequence, the length of the circumference of the circumscribed circle was equal to (approximately) $7 \cdot 30 = 3\ 30$, so that the diameter was 1 10, and the radius 35. The area of the 7-front could then be computed as the sum of the areas of 7 symmetric triangles with the front 30 and the length 35. The notation 35 uš '35, the length' is still readable close to one side of the upper triangle.

The next step of the process was to compute the height of the upper triangle as $h_7 = \text{appr. } 31{;}40 \ (= 0\ {;}30 \cdot 1\ 02{;}30)$. A notation under the central line in the upper triangle in the drawing of the 7-front, 31 40 dal '31;40, the transversal', is almost completely destroyed. Only the last half of the sign dal is preserved.

One would now expect to find the total area of the upper triangle and the area of the whole 7-front recorded in the drawing. This does not happen. Instead one finds a somewhat cryptic inscription, interpreted as follows by Robson in *MMTC* (1999), 49:

[nígin sag] sag.7 a.na *4 te-ṣi-* / *ip-ma*   *The square of the front* of the 7-front by 4 you repeat, then
ši-in-šé-ra-ti / *ta-na-as-sà-aḫ-ma* / a.šà   the twelfth you tear out, then the field (= area).

What this means is that the area of a 7-front (regular heptagon) can be computed as

$$A_7 = \text{appr. sq. } s \cdot (4 - {;}05 \cdot 4) = \text{sq. } s \cdot (4 - {;}20) = \text{sq. } s \cdot 3{;}40.$$

In other word, you get the area of the 7-front if you first multiply the square of the front by 4, then reduce the result by a twelfth of its value. This computation rule is a handy variant of the more formal computation rule

$$A_7 = \text{sq. } s \cdot 3{;}40,$$

which can be compared with the entry '3 41 the constant of a 7-front' in BR = *TMS* 3.

8.3. Labyrinths, Mazes, and Decorative Patterns

8.3 a.   *MS 4515. A Babylonian Square Labyrinth*

A labyrinth can be defined as *a convoluted path from the exterior of a square or a circle to its center*. Until now, it has been common knowledge that the idea of a labyrinth can be traced no further back in time than to the Minoan and Mycenaean civilizations on Crete and in mainland Greece, in the latter half of the second millennium BC. Now, however, two clay tablets in the Schøyen Collection suggest that, maybe, the idea of a labyrinth has a Mesopotamian origin. There is no way of dating those clay tablets, but since the overwhelmingly great majority of the mathematical clay tablets in the Schøyen Collection are unmistakably Old Babylonian, it is quite likely that the labyrinth texts, too, are Old Babylonian, hence from the first half of the second millennium BC.

In any case, it is interesting that the Babylonian labyrinths are definitely not of the same type as the "Greek labyrinth" from Crete or Greece (see below). Therefore, even if the presumably Old Babylonian labyrinths are older than the Greek labyrinth, it does not follow that the Greek labyrinth was directly inspired by its Babylonian predecessor. The two can very well have been invented independently (even if this is unlikely, since interesting ideas tend to spread rapidly).

obv.

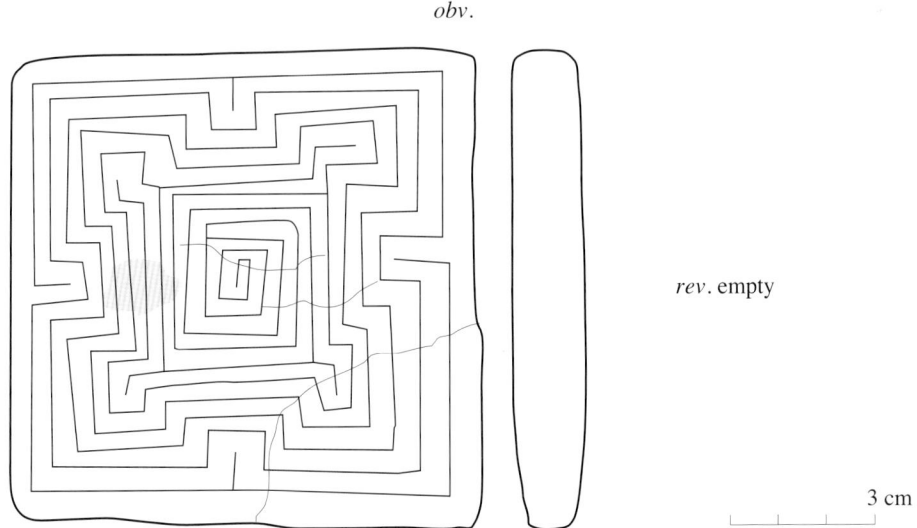

rev. empty

3 cm

Fig. 8.3.1. MS 4515. The Babylonian square labyrinth with two paths, one good and one bad.

MS 4515 (Fig. 8.3.1 above) is a square clay tablet with a drawing of a labyrinth filling out all available space on the obverse. This labyrinth, in the following called "the Babylonian square labyrinth" appears to be in the form of a fortified city with a city gate in each of the four faces of the city wall. Two of the gates are closed, the other two are open. The path that enters through one of the open gates (the right gate if the clay tablet is oriented as in Fig. 8.3.1), arrives at the center of the square after 2 clockwise loops parallel to the city walls, followed by 4 counter-clockwise loops. The path that enters through the other open gate (the left gate in Fig. 8.3.1) ends in a blind alley after 2 clockwise loops followed by 2 counter-clockwise loops. The five outermost loops of either path are formed as squares with indentations near the four gates. In Fig. 8.3.2 below, the course of the good path is shown in the diagram to the left, while the course of the bad path is shown to the right.

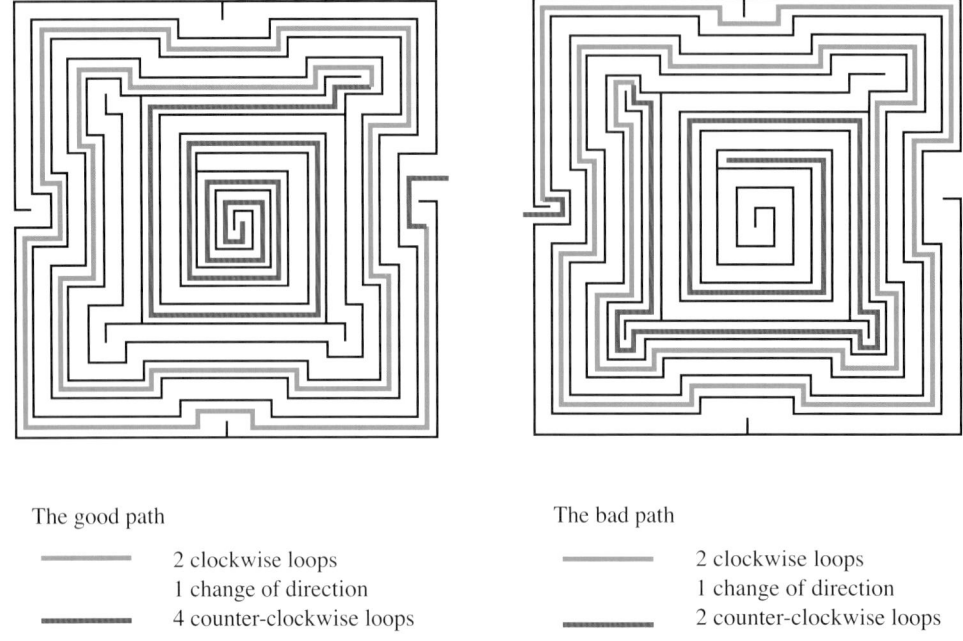

The good path

——————   2 clockwise loops
                    1 change of direction
·············   4 counter-clockwise loops

The bad path

——————   2 clockwise loops
                    1 change of direction
·············   2 counter-clockwise loops

Fig. 8.3.2. The Babylonian square labyrinth. Two paths, only one leading to the center.

It is obvious that it would be difficult to draw the Babylonian labyrinth neatly, in the way it is drawn on MS 4515, without an efficient algorithm for its construction. One such algorithm is demonstrated in a series of six diagrams in Fig. 8.3.3 below. The basic idea is to start with two separate parts of the outer wall of the square labyrinth, on either side of the open gates, and then successively continue the two parts of the wall into the interior of the square, forming segments of inner walls at a constant distance from each other, by use of *a double recursive algorithm*. Two steps at a time of the algorithm are shown in the first five of the diagrams in Fig. 8.3.3. The final result is shown in the sixth diagram.

Here follows a detailed account of how the complicated algorithm operates:

1. Start at the left gate and draw *clockwise, in black*, the upper half of the outer wall.
2. Start at the right gate and draw *clockwise, in grey*, the lower half of the outer wall. Continue to the right gate.
3. Continue the *black* line from the right gate a full loop back to the right gate.
4. Continue the *grey* line from the right gate a full loop back to the right gate.
5. Continue the *black* line from the right gate as part of a loop to the dead end in the upper right corner.
6. Continue the *grey line* from the right gate as part of a loop to the dead end in the upper left corner.
7. Start *a new black line* in the upper left corner and continue *counter-clockwise* to the upper right corner.
8. Start *a new grey line* in the upper right corner and continue *counter-clockwise* back to the upper right corner.
9. Continue the *black* line from the upper right corner a full loop back to the upper right corner.
10. Finish the *grey* line with a short segment and the dead end for the bad path.
11. Finish the *black* line with a spiral in to the center of the square.

The complexity of the construction clearly demonstrates that the drawing of a labyrinth on MS 4515 is not just a meaningless doodle. It is much more likely that this Babylonian labyrinth was the end result of a long series of experiments with various more or less advanced designs with a common objective: to produce one connected path (or two) from the periphery to the center, so convoluted that it is difficult to follow the meanderings. It is conceivable that drawings of that type could draw on experience gathered from executing more conventional and better documented types of drawings on Mesopotamian clay tablets, such as house plans, city maps, serpentine patterns, and so on.

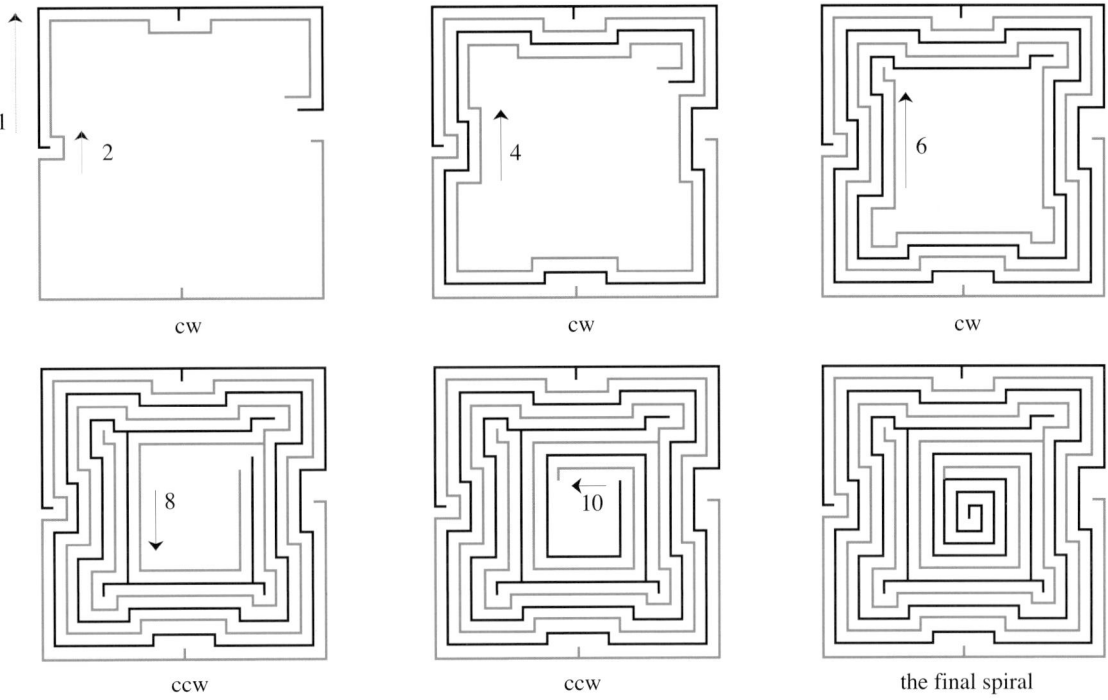

Fig. 8.3.3. The Babylonian square labyrinth. Construction in 11 steps, beginning with the outer walls.

8.3 b.   The Greek Labyrinth

The proposed algorithm for the construction of the Babylonian square labyrinth can be compared with a well known algorithm for the construction of the Greek labyrinth. That algorithm starts with a central "core" in the form of a double cross with points in the four corners formed by the outer arms of the double cross:

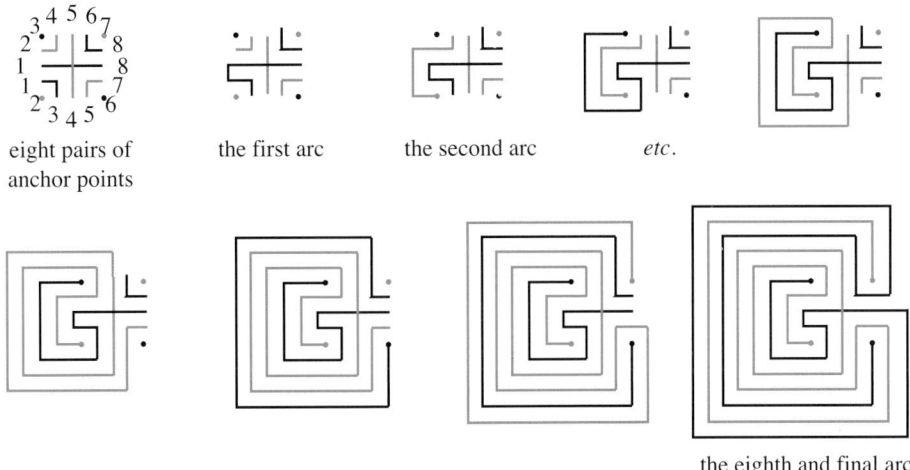

Fig. 8.3.4. The Greek labyrinth. Construction in 9 steps, beginning with the central core.

Eight successive parts of the walls of the labyrinth are then constructed, one at a time, proceeding outwards, by use of the simple idea that the eight numbered points or endpoints of cross arms in the core should be connected to the nearest available point or endpoint to the right by means of a single counter-clockwise arc.

The circumstance that the construction of the Greek labyrinth is simpler than the construction of the Babylonian labyrinth is related to the fact that there is only one path leading into the Greek labyrinth compared to the two paths leading into the Babylonian labyrinth. On the other hand, the one path in the Greek labyrinth is in a certain sense more complicated (see Figs. 8.3.5 and 8.3.12). It proceeds alternatively in four clockwise and three counter-clockwise loops. In addition, the first three loops move from the middle of the labyrinth towards the outer wall, the fourth loop stays in the middle, and the last three begin at the center and move towards the middle.

The difference between the construction of the Babylonian labyrinth on one hand and the Greek labyrinth on the other is accentuated by the drawings in Fig. 8.3.5, which show simplified versions of both labyrinths, with minimal numbers of loops. The simplified Babylonian labyrinth consists, essentially, of two parallel spirals, proceeding inwards from the two gates, but with only one of the spirals actually reaching the center, and with the endpoint of the second spiral resting on a point of the first spiral. The good path through this labyrinth is also a spiral, proceeding from one of the gates all the way to the center of the square.

The simplified Greek labyrinth consists of two spirals crossing each other at one point. There is only one path from the gate to the center of the construction, a spiral that changes direction whenever it comes near the point where the two spirals cross each other.

Fig. 8.3.5. The different basic ideas behind the constructions of the Greek and the Babylonian labyrinths.

A Greek labyrinth of a square form appears on the reverse of a clay tablet from king Nestor's palace at Pylos (**Cn 1287**, Athen's National Museum). There is an unrelated Greek inscription in Linear B on the obverse. Two Greek labyrinths of a circular form appear as decorations on two fragments of a clay bowl found at Tell Rifa'at in Syria. Both the clay tablet and the bowl are from the 13th century BC (see Kern (2000 (1982)), nos. 102-104) and consequently (probably) younger than the Babylonian labyrinth on MS 4515.

(The conform transliteration of Cn 1287 in Fig. 8.3.6 below is based on Kern's photo of the clay tablet.)

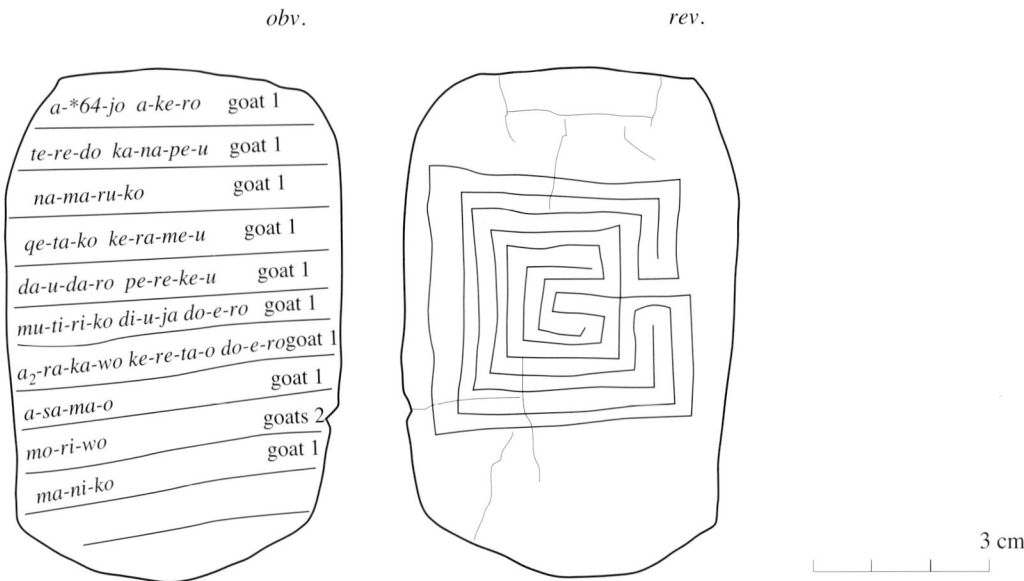

Fig. 8.3.6. Cn 1287. A Mycenaean clay tablet with the oldest known representation of the Greek labyrinth.

A drawing of a kind of double spiral on the round hand tablet **VAT 744** may be a precursor of the Babylonian labyrinth. However, according to Kern, *Labyrinthe* (2000 (1982)), the clay tablet may be Late Babylonian, and the drawing is not a labyrinth but a model of the intestines of a sacrificial animal.

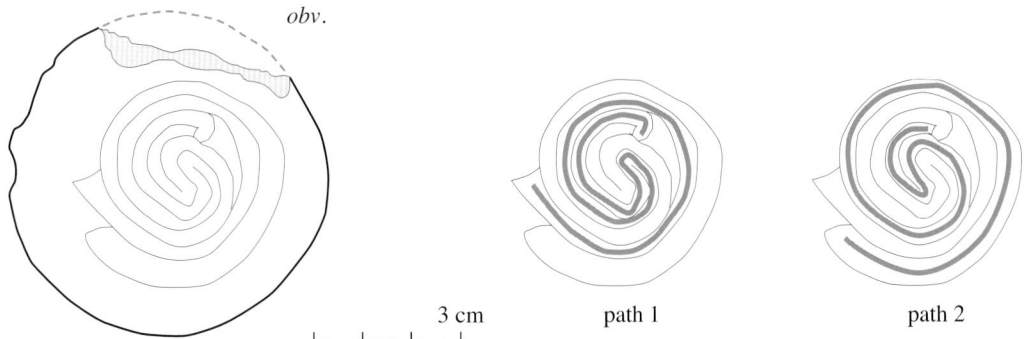

Fig. 8.3.7. VAT 744. A possible precursor of the Babylonian labyrinth.

Another possible precursor of the Babylonian labyrinth is the small clay cone **MS 3195**, inscribed with spirals on its bottom face which appear to be continued into spirals along the sides of the cone:

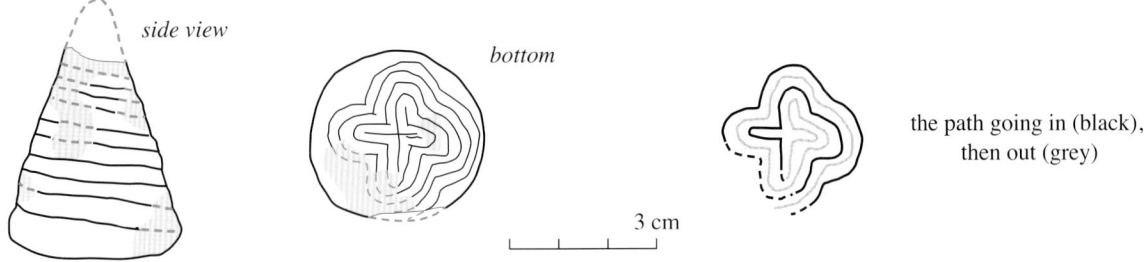

Fig. 8.3.8. MS 3195. Another possible precursor of the Babylonian labyrinth.

8.3 c.   MS 3194. A Babylonian Rectangular Labyrinth

MS 3194 (Fig. 8.3.9) is a relatively well preserved clay tablet with a drawing on the obverse of a complicated labyrinth, in the following called "the Babylonian rectangular labyrinth". Some parts of the design close
to one edge of the clay tablet are missing, and the surface of the obverse is damaged in some places, all of which
makes it difficult to reconstruct with certainty the original form of the labyrinth. Fortunately, the apparent symmetry of the design makes the task somewhat easier.

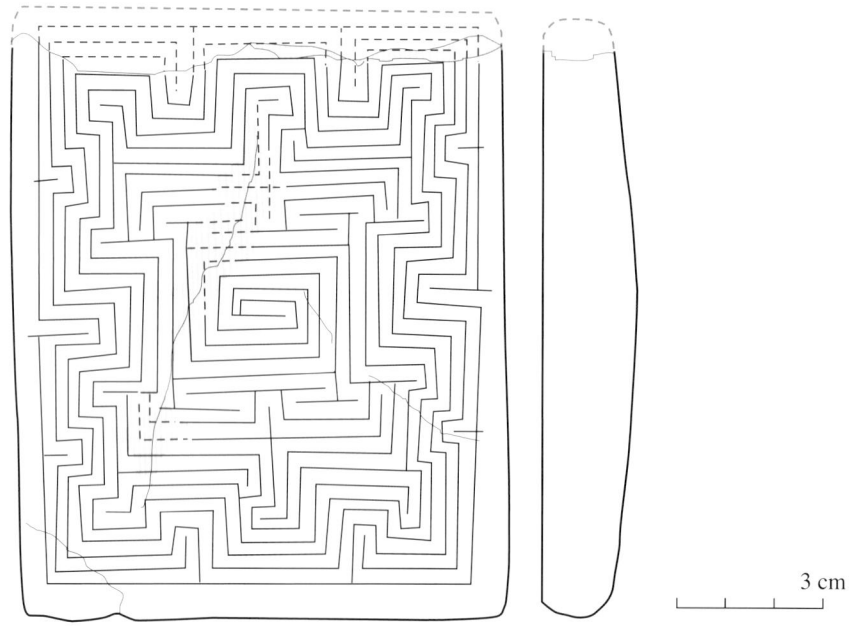

Fig. 8.3.9. MS 3194. The Babylonian rectangular labyrinth.

This Babylonian labyrinth appears to be a more elaborate variant of the Babylonian square labyrinth of MS
4515 (Fig. 8.3.1). Thus, the basic design is the same: to the left and to the right two open gates, with two paths
spiralling inwards from the gates, with only one reaching the center of the labyrinth. On the other hand, there
are also some interesting differences between the two designs. While the square labyrinth has the form of a
square city with *four* gates, two open and two closed, with one gate on each side of the square, the rectangular
labyrinth has the form of a *rectangular* city with *ten* gates, 2 open and eight closed, with three gates on each
one of the long sides of the rectangle, and two gates on each one of the short sides. As a result of the greater
complexity of the design, the "good" path leading from the outside to the center is even more convoluted in the
case of the rectangular labyrinth than it was in the case of the square labyrinth. See Fig. 8.3.12 below.

The construction of the Babylonian rectangular labyrinth is also much more complicated than the construction of the Babylonian square labyrinth. Thus, while 11 steps are needed for the construction of the square labyrinth (Fig. 8.3.3 above), 38 steps are required for the construction of the Babylonian rectangular labyrinth
(Figs. 8.3.10-11).

Yet, the basic idea of the construction is the same in both cases. One has to start from the outer wall of the
city, with the positions of the gates indicated (steps 1-2). The outer wall is divided into two separate halves by
the open gates. In the initial design, in the upper left corner of Fig. 8.3.10, the two parts of the outer wall are
drawn with grey and black lines, respectively. The construction proceeds from this initial design in a number
of simple steps. In each step, a black wall section is built parallel to and a constant distance away from the currently innermost grey wall section.

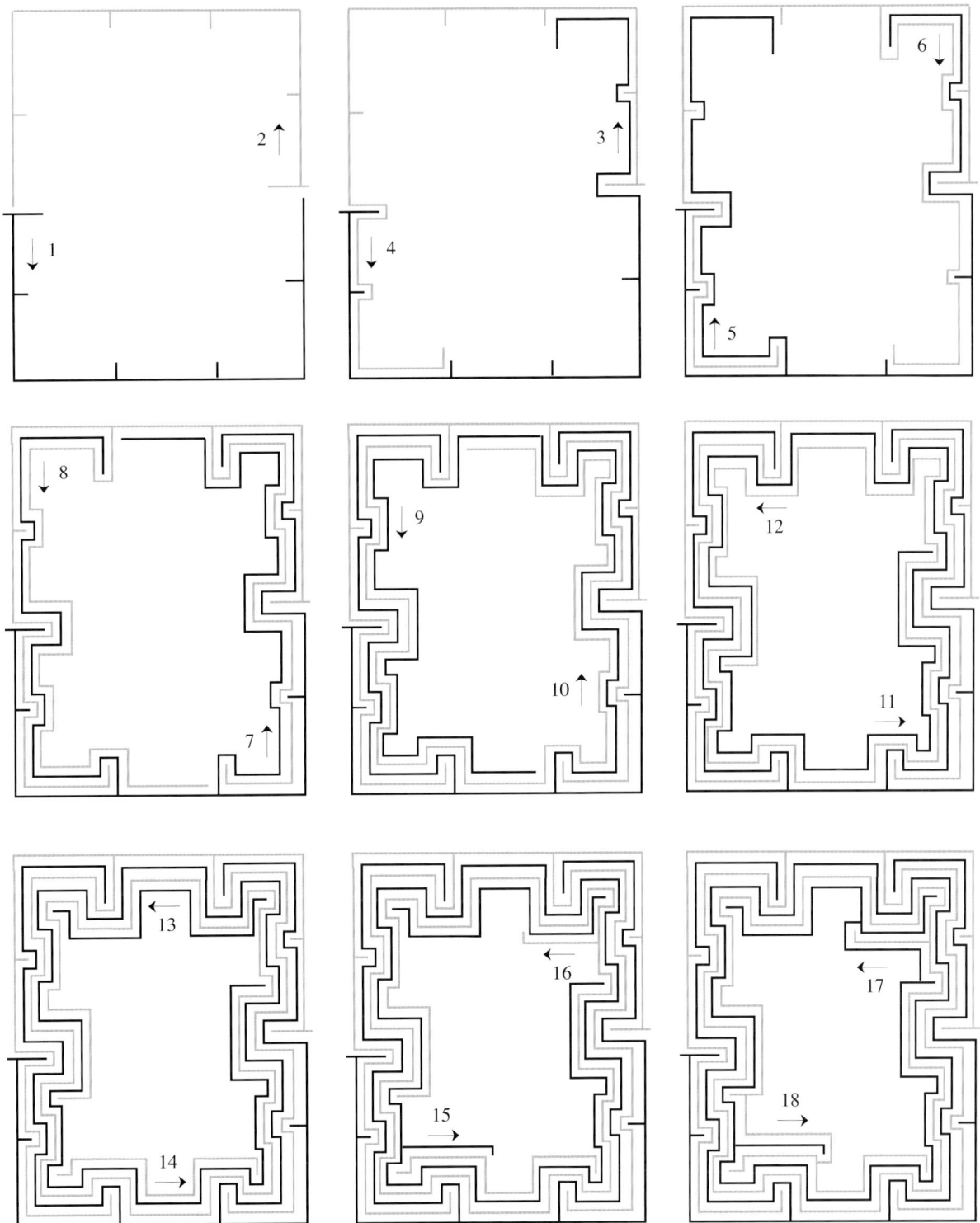

Fig. 8.3.10. The first 18 steps of the construction of the Babylonian rectangular labyrinth.

the last 6 steps

the final view

Fig. 8.3.11. The 16 final steps of the construction of the Babylonian rectangular labyrinth.

(It is easy to check that if the constant distance between the gray and black wall sections is, say, 1 ninda, then the lengths of the long and short outer walls of the rectangular labyrinth are 40 and 33 ninda, respectively, if no account is taken of the thickness of the walls, while in the case of the square labyrinth, the lengths of the four outer walls are all equal to 22 ninda.)

The sections of black and gray walls have to be drawn alternately. It is important to keep in mind that because of the requirement that there must be a constant distance of 1 ninda between the black and gray wall sections, those sections are not allowed to touch each other until the last step of the construction. Consecutive black wall sections, on the other hand (like consecutive gray wall sections), are required to touch each other. See steps 5-8, for instance.

The construction is essentially symmetrical, in the sense that the black and the gray wall sections are constructed in pairs, with each gray wall section being more or less the mirror image, with respect to the center of the labyrinth, of the corresponding black wall section.

There is one exception this rule. In steps 33-38 of the construction, the black wall is allowed to spiral inwards, and the gray wall ends by touching the black spiral. In this way, it is made sure that one of the two paths leading inwards from the open gates will end at the center of the labyrinth, while the other path will stop short of the center, at the point where the gray wall touches the black wall. (The situation is the same in the case of the square labyrinth, as shown in Fig. 8.3.3.)

Note that there are no fork points, where a path is allowed to continue through the labyrinth in more than one way. Thus, there is only one choice to be made, namely between the two open gates, and only one possible dead end, namely if one makes the wrong choice of gate to enter through.

Fig. 8.3.12. The only thread in the Greek labyrinth and the good threads in the two Babylonian labyrinths.

The forms of the good threads in the Greek and the two Babylonian labyrinths are compared in Fig. 8.3.12 above. Clearly, the Greek labyrinth is at the same time simpler and more sophisticated than the two Babylonian labyrinths. A comparison of the algorithms in Figs. 8.3.3, 8.3.4, and 8.3.10-11 for the construction of the three types of labyrinth shows that the Greek labyrinth is also easier to construct. This may be why the Greek labyrinth is the only one of the three that never lost its popularity.

8.3 d.   MS 4516. A Geometric Theme text with 8 Assorted Mazes

MS 4516 (Fig. 8.3.13 below) is a rectangular hand tablet inscribed on the obverse with four rows of drawings in two columns. There is no accompanying text. One of the eight drawings is lost, and one is so damaged that it is not possible to reconstruct all its details. The other six drawings can be reconstructed, even the ones that are quite extensively damaged.

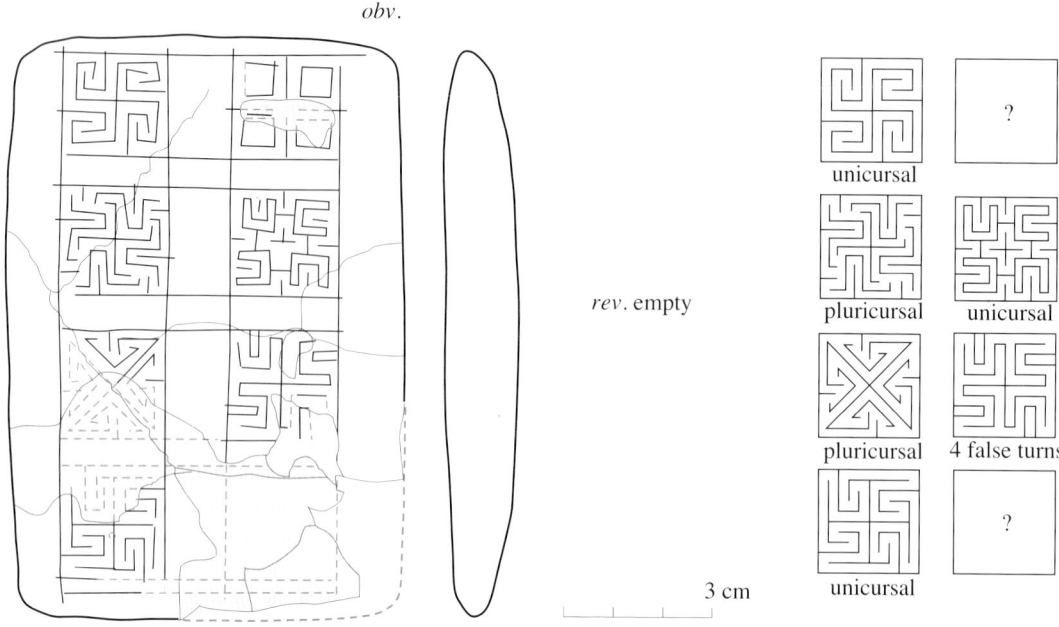

Fig. 8.3.13. MS 4516. A geometric theme text with 8 assorted mazes.

What makes the reconstruction possible in the cases when the drawings are damaged is the following observation. Each drawing consists of a square with a complex pattern of segments of straight lines inside it. All those interior patterns exhibit the same kind of *fourfold rotational symmetry*: If the pattern in, say, the upper left corner of one of the squares is rotated around the center of the square for one quarter of a full revolution, then it comes to coincide with the pattern in the lower left corner of the square. After another quarter revolution, it coincides with the pattern in the lower right corner, and after a third quarter revolution, it coincides with the pattern in the upper right corner.

The squares with their interior patterns are no labyrinths, because there is no entrance into the squares. Somewhat arbitrarily, they will here be called "mazes", which seems to be a fitting name for a pattern forming a passage way for a convoluted path with no well defined start or finish.

The first of the eight mazes can be called "unicursal" because there is only one possible path through it. The fourth and seventh mazes are also unicursal. The third maze is "pluricursal" in the sense that there is more than one path leading through it. The sixth maze differs from all the others in that there are four false turns in it leading to dead ends for the path.

There are several possible explanations for a text like MS 4516. It may be just a catalog of esthetically pleasing patterns, maybe intended to be used for decorative purposes (layouts for mosaics, tilings of floors, gardens, *etc.*). It may also have been a school text, used to teach students the basic principles of maze construction. (Cf. the well known geometric theme text BM 15 285 with its 41 illustrated exercises of the kind shown in Fig. 8.2.13 above.) Or, it may have been a preliminary exercise, a preamble to the construction of a true labyrinth.

8.3 e. MS 3940. A Pattern Superimposed on a Dense Grid of Guide Lines

MS 3940 (Fig. 8.3.14,left) is a fragment, the upper or lower half of a clay tablet, on which is drawn a dense grid of horizontal and vertical guidelines. These guidelines are used for the accurate construction of a pattern, presumably for a tiled floor or a woven fabric. The basic design in the pattern is 4 series of concentric oblique serrated squares flanked on all sides by series of oblique serrated rectangles. This basic design is repeated until all available space is filled.

This clay tablet is interesting in the present context for a couple of reasons. First, it is an excellent demonstration of the rather obvious close relation between decorative art and geometry. Recall, in particular, that concentric squares was a popular topic in Old Babylonian geometry. (Cf. Figs. 8.2.3 and 8.2.10 above.) Secondly, anyone who tries to draw for the first time a complex design like the labyrinths on MS 4515 and MS 3194 or one of the mazes on MS 4516 will soon realize that this is a difficult task unless it is done with the help of an appropriate grid of guide lines.

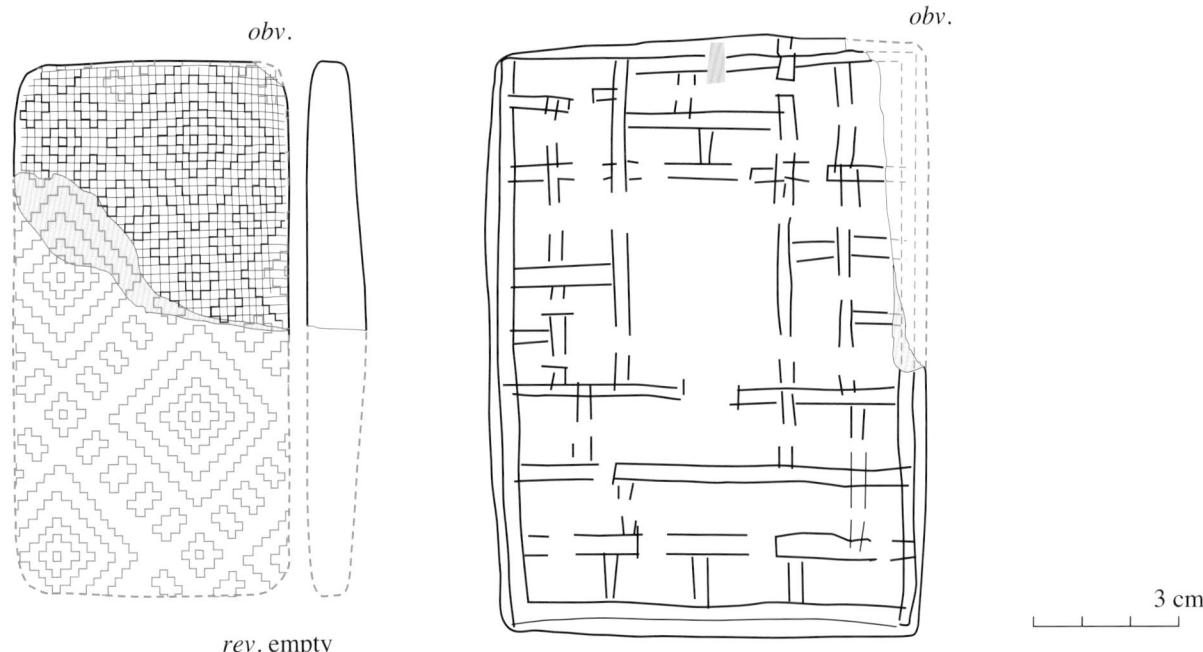

Fig. 8.3.14. MS 3940. A pattern superimposed on a dense grid of guide lines.
MS 3031. The ground plan of the palace of Nur Addad at Larsa.

8.3 f. MS 3031. The Ground Plan of a Palace

It is interesting to compare the layout of the Old Babylonian square and rectangle labyrinths with an actual labyrinthine ground plan of an Old Babylonian palace. **MS 3031** (Fig. 8.3.14, right) is a detailed drawing of the various halls, rooms, and antechambers around the central courtyard of a palace. Although there is no inscription on the clay tablet one can be sure of what the drawing depicts. The design of the palace and the proportions correspond nicely with the remains of the palace of Nur Adad at Larsa unearthed by French excavations since 1903. Besides, the clay tablet is made of a purplish clay that is said to be characteristic for clay tablets from Larsa.

So far this is the only example of a clay tablet with an identifiable house plan of a known building.

9
The Beginning and the End of the *Sumerian King List*

9.1.   The Sumerian King List

The "Sumerian King List" is the name given to a literary composition, written in Sumerian, listing a long succession of Sumerian cities alleged to have been invested for longer or briefer periods with nam.lugal 'the kingship', the names of the kings of the corresponding dynasties, and the individual lengths of their reigns. The content of the king list is known, in diverse variants, from a number of clay tablets, or fragments of clay tablets, most of them Old Babylonian (see below). Since 1999, a compiled edition of the king list is published online as a part of the *Electronic Text Corpus of Sumerian Literature* (www-etcsl.orient.ox.ac.uk), in the form of both a transliteration (c.2.1.1) and a translation (t.21.1). The main source for the compiled version is the text of **Ash. 1923.444** (also called WB 444, or the "Weld-Blundell prism"), a relatively well preserved four-sided prism with two columns on each face. An interesting, although now somewhat outdated discussion of the *Sumerian King List* can be found in Jacobsen, *AS 11* (1939).

The *end* of the *Sumerian King List*, essentially corresponding to the version known from Ash. 1923.444, is the "Ur-Isin part" of the list, giving the names and the reigns of the kings of the Neo-Sumerian Ur III dynasty and of one of the Old Babylonian dynasties succeeding Ur III:

The *Sumerian King List*, lines 341-377

úrim^{ki}.ma ur.[^dnamma] lugal.àm / mu 18 ì.ak

úrim^{ki}.ma ur.[^{d}namma] lugal.àm / mu 18 ì.ak
^{d}šul.gi dumu ur.^{d}namma.ke$_4$ / mu 46 ì.ak
^{d}amar.^{d}suen.na dumu ^{d}šul.gi.ke$_4$ / mu 9 ì.ak /
šu.^{d}suen dumu ^{d}amar.^{d}suen.na.ke$_4$ / mu 9 ì.ak /
i.bí.^{d}suen dumu šu.^{d}suen.ke$_4$ / mu 24 ì.ak /
4 lugal / mu.bi 1 48 íb.ak /
úrim^{ki}.ma ^{giš}tukul ba.an.sìg /
nam.lugal.bi ì.si.in^{ki}.šè ba.de$_6$ /
ì.si.in^{ki}.na iš.bi.^{d}èr.ra lugal.àm / mu 33 ì.ak /
šu.ì.lí.šu dumu iš.bi.^{d}èr.ra.ke$_4$ / mu 20 ì.ak /
i.din.^{d}da.gan dumu šu.ì.lí.šu.ke$_4$ / mu 21 ì.ak /
iš.me.^{d}da.g[an dumu i.din.^{d}da.gan.ke$_4$] / mu 20 ì.ak /
l[i.pí.it.eš$_4$.tár dumu iš.me.^{d}da.gan.ke$_4$] / mu [11 ì.ak] /
^{d}ur.[^{d}nin.urta mu 28 ì].ak /
^{d}bur.^{d}sue[n dumu ^{d}ur.^{d}nin.urt]a.ke$_4$ / mu 21 ì.ak /
^{d}li.pí.[it.]^{d}en.líl / dumu bur.^{d}suen.ke$_4$ mu 5 ì.ak /
^{d}èr.ra.i.mi.ti mu 8 ì.ak /
^{d}en.líl.ba.ni mu 24 ì.ak /
^{d}za.am.bi.ia mu 3 ì.ak /
^{d}i.te.er.pi$_4$.ša mu 4 ì.ak /

In Urim (Ur), Ur-Namma was king, ruled for 18 years.
  Shulgi, son of Ur-Namma, ruled for 48$^!$ years.
  Amar-Suena, son of Shulgi, ruled for 9 years.
  Shu-Suen, son of Amar-Suena, ruled for 9 years.
  Ibbi-Suen, son of Shu-Suen, ruled for 24 years.
5$^!$ kings ruled for 1 48 years.
Then Urim was smitten with weapons.
The kingship was taken to Isin.
In Isin, Ishbi-Erra was king, ruled for 33 years.
  Shu-ilishu, son of Ishbi-Erra, ruled for 10$^!$ years.
  Iddin-Dagan, son of Shu-ilishu, ruled for 21 years.
  Ishme-Dagan, son of Iddin-Dagan, ruled for 20 years.
  Lipit-Eshtar, son of Ishme-Dagan, ruled for 11 years.
  Ur-Ninurta ruled for 28 years.
  Bur-Suen, son of Ur-Ninurta, ruled for 21 years.
  Lipit-Enlil, son of Bur-Suen, ruled for 5 years.
  Erra-imitti ruled for 8 years.
  Enlil-bani ruled for 24 years.
  Zambiya ruled for 3 years.
  Iter-pisha ruled for 4 years.

^{d}ur.dul.kug.ga mu 4 ì.ak /	Ur-dulkuga ruled for 4 years.
^{d}suen.ma.gir mu 11 ì.ak /	Suen-magir ruled for 11 years.
14 lugal / mu.bi 3 23 íb.ak	**14 kings ruled for 3 23 years**.

In three of the known Nippur fragments of the *Sumerian King List,* an interesting summary of the whole king list (except the antediluvian part) follows directly after the Ur-Isin part.

According to this summary, 4 dynasties of kings in Kish, 5 in Uruk, 3 in Ur, and 1 each in Awan, in Hamazi, in Agade, (pus three others), and 1 under the foreign dominance of the Gutians, alternated as the central power in Mesopotamia "after the flood". For various reasons, this cannot be historically correct. The explanation given by Jacobsen, in *AS 11* (1939), 152-154, is that the original author of the *Sumerian King List* used as his sources a conglomerate of *local date lists* from a number of Mesopotamian cities. These local lists were cut up by him in smaller pieces, the alleged dynasties, which were then placed after each other in a linear sequence, with dynasties that were in reality more or less contemporaneous placed closely together. Jacobsen made the observation that fantasy numbers are used in the king list for the "legendary reigns" of the three first dynasties in Kish, and of the first dynasty in Uruk, while more realistic numbers are used for what he considered to be the "historical reigns" of the remaining dynasties.

The *Sumerian King List*, lines 378-397, 414-430 (summary)

šu.nígin 40.lá.[1 lugal] /	Together 40 - *1 kings*
mu.bi 4$_{šár.gal}$ […] 9 mu [3 iti 3 1/2 ud] íb.ak /	ruled for 4 šár.gal […] 9 years, *3 months, 3 1/2 days,*
a.rá 4.[kam] / šà kiš$^{[ki]}$ /	4 tim*es* in Kish.
šu.nígin 22 lugal /	Together 22 kings
mu.bi 43 30 […] 6 iti 15 ud íb.ak /	ruled for 43 30 […] years, 6 months, 15 days,
a.rá 5.kam / šà unug^{ki}.ga /	5 times in Unug (Uruk).
šu.nígin 12 lugal /	Together 12 kings
mu.bi 6 36 mu íb.ak /	ruled for 6 36 years,
a.rá 3.kam / [šà] urim$_2$^{ki}.ma /	3 times *in* Urim (Ur).
šu.nígin 3 lugal /	Together 3 kings
mu.bi 5 56 mu íb.ak /	ruled for 5 56 years,
a.rá 1.kam / šà a.wa.an^{ki} /	once in Awan.
[šu].nígin 1 lugal /	*To*gether 1 king
mu.bi 7 (00) mu [ì].ak /	ru*led* for 7 (00) years,
a.rá 1.[kam] / šà ha$^?$.[ma].zi^{ki}.a /	on*ce* in Ha*m*azi.
(16 lines missing; 3 cities, 13 kings)	16 lines missing
[šu.nígin 12] lugal /	*Together 12* kings
[mu.bi 3] 17 [mu] íb.ak /	ruled for *3* 17 *years,*
[a]./rá\ 1.kam / [šà] a.ga.dè^{ki} /	once in Agade.
šu.nígin 21 lugal /	Together 21 kings
mu.bi 2 05 mu 40 ud íb.ak /	ruled for 2 05 years, 40 days,
a.rá 1.kam / šà ugnim / gu.ti.um^{ki} /	once in the army of Gutium.
[šu.nigin] 11 lugal /	Together 11 kings
[mu].bi 2 39 mu íb.ak /	ruled for 2 39 years,
[šà i].si.in^{ki}.na /	\<once\> in Isin.
11 iri^{ki} / [iri]^{ki} nam.lugal.la íb.ak.kà /	**11 cities,** cities where the kingship was exercised.
[šu].nígin 2 14 lugal /	**Together 2 14 kings,**
[šu].nígin / mu.bi 8$_{šár.gal}$ […] 1 16	**their years 8 š á r . g a l […] 1 16.**

A study of the terminology used in various parts of the *Sumerian King List* led Jacobsen to draw the conclusion (*op. cit.*, 141) that the composition of the king list in its original form can be dated to the time of Utu-ḥegal, who reigned in Uruk just before the beginning of Ur III, and that both the "antediluvian" part at the beginning of the king list and the Ur-Isin part at the end of it were later additions.

Most recently, a privately owned Ur III version of the *Sumerian King List* was published by Steinkeller in *FS Wilcke* (2003). It can be firmly dated to between Šulgi 20 and Šulgi 48. The big and numerous discrepancies between the Ur III version (*USKL*) and the Old Babylonian standard edition (*SKL*) led Steinkeller to tentatively suggest a three-stage development of *USKL*:

> "(1) an Akkade (= the original?) version, which originated sometime under Sargon's dynasty; (2) an Uruk version from the time of Utu-ḫegal, which added the last rulers of Akkade, the 4th dynasty of Uruk, and the Gutian rulers, …; (3) a local Adab version from the time of Ur-Namma or Šulgi, …".

9.2.   *MS 1686. A New Version of the Ur-Isin King List*

MS 1686 (Fig. 9.1 below) is a small, neatly written and perfectly preserved clay tablet with the Ur-Isin king list *in mini-format*. Compared with lines 341-377 of the compiled *Sumerian King List*, MS 1686 lacks the whole narrative framework. Only the names of the kings and the years of their reigns are recorded. The names of the cities, Ur and Isin, are not mentioned. Neither is it mentioned that the kingship was transferred from one city to the other. The phrase ì.ak 'he ruled' is omitted everywhere, and the names of the fathers of the kings are not mentioned. Even the summaries are missing, so and so many kings for so and so many years.

In other ways, though, there is partial agreement between MS 1686 and the Ur-Isin part of the *ETCSL* version of the king list. The names of the kings are the same, and the lengths of the reigns are nearly the same. The two suspected errors in the *ETCSL* version, 46 instead of 48 years for Shulgi, and 20 instead of 10 years for Shu-ilishu, are corrected in MS 1686. Also, MS 1686 has, like the closely parallel versions A and B in Sollberger, *JCS* 8 (1954) 19 years for Ishme-dagan, 22 for Bur-suen, 3 for Iter-pisha, and 3 for Suen-magir, where the *ETCSL* version has 20, 21, 4, and 4 years.

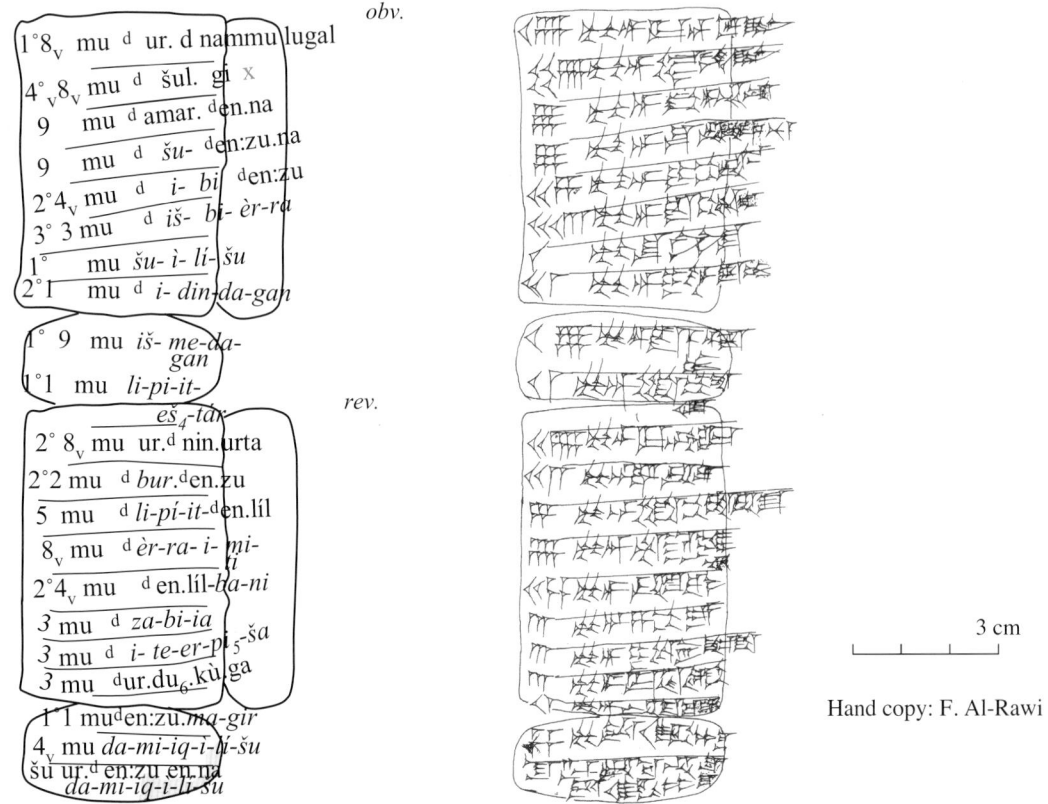

Fig. 9.1. MS 1686. An Ur-Isin king list, written in the fourth year of the reign of Damiq-ilishu of Isin.

Variant number signs are used in MS 1686 for the digits 4, 8, and 40, which seems to suggest that the tablet was written in the earlier part of the Old Babylonian period. This conclusion, though, is in conflict with the observation that MS 1686 was written by a scribe called Ur-Suen, in the fourth year of the reign of Damiq-ilishu. At least, that is what is indicated by the subscript and by the fact that Damiq-Ilishu's reign is said to have lasted only 4 year. Compare with the *ETCSL* version which ends, like Ash. 1923.444, with 11 years for Suen-magir, but does not mention Damiq-Ilishu. Compare also with **CBS 19797** (the *Sumerian King List*, variant P₅), which ends in the following way:

CBS 19797 (*PBS 5*, 5; *BE 20/1*, 47)

ᵈsuen.ma.[gi]r mu 11 ì.[ak] / da.mi.iq.ì.lí.šu dumu ᵈsuen.ma.gir mu 23 in[.ak] / [... lugal m]u.bi 3 45 iti 6 in.ak	Suen-magir ruled for 11 years. Damiq-ilishu, son of Suen-magir, *ruled* for 23 years. **... kings ruled for 3 45 years, 6 months**.

This text shows that Damiq-Ilishu's reign lasted for 19 more years after the date when MS 1686 was written. As the first published fragment of the *Sumerian King List*, CBS 19797 originally appeared as text 47 in Hilprecht's *BE 20/1* (1906), that is in the same volume as excellent reproductions of Old Babylonian mathematical and metrological table texts, a mathematical algorithm text, and a fragment of a mathematical problem text, all from Nippur. Some of those texts were mentioned above. Hilprecht could read only the reverse of the fragment, which happened to contain the entire Ur-Isin part of the king list. The inscription on the obverse, which was difficult to read because the clay tablet needed to be cleaned, was published by Poebel in *PBS 5* (1914). Ironically, the upper half of the first column on the obverse of CBS 19797 is lost, precisely the part of the text that probably contained a version of the antediluvian part of the king list.

Here is a transliteration and translation of the text on MS 1686:

MS 1686

1	18ᵥ mu ᵈur.ᵈnamma lugal	18 years Ur-Namma (was) king,
2	4ᵥ8ᵥ mu ᵈšul.gi {x}	48 years Šulgi,
3	9 mu ᵈamar.ᵈen<:zu>.na	9 years Amar-Suen,
4	9 mu ᵈšu.ᵈen:zu.na	9 years Šu-Suen,
5	24ᵥ mu ᵈi-bi-ᵈen:\ zu	24 years Ibbi-Suen,
6	33 mu ᵈiš-bi-èr-ra	33 years Išbi-Erra,
7	10 mu ⁽ᵈ⁾šu-i-lí-šu	10 years Šu-ilišu,
8	21 mu ᵈi-din-da-gan	21 years Iddin-Dagan,
9	19 mu ᵈiš-me-⁽ᵈ⁾da- / gan	19 years Išme-Dagan,
10	11 mu ᵈli-pí-it- / eš₄-tár	11 years Lipit-Eštar,
11	28ᵥ mu ᵈur.ᵈnin.urta	28 years Ur-Ninurta,
12	22 mu ᵈbur-ᵈen:zu	22 years Bur-Suen,
13	5 mu ᵈli-pí-it-ᵈen.líl	5 years Lipit-Enlil,
14	8ᵥ mu ᵈèr-ra-i-mi-ti	8 years Erra-imitti,
15	24ᵥ mu ᵈen.líl-ba-ni	24 years Enlil-bani,
16	3 mu ᵈza<-am>-bi-ia	3 years Zambiya,
17	3 mu ᵈi-te-er-pi₅-ša	3 years Iter-piša,
18	3 mu ᵈur.du₆.kù.ga	3 years Ur-dukuga,
19	11 mu ᵈen:zu-ma-gir	11 years Suen-magir,
20	4ᵥ mu da-mi-iq-i-lí-šu	4 years Damiq-ilišu.
21	šu ur.ᵈen:zu en.na / da-mi-iq-i-lí-šu	The hand of Ur-Suen, until Damiq-ilišu.

King lists like MS 1686 can be used to compile lists showing the *relative* dates of the reigns of the kings in various Mesopotamian cities. In order to arrive at the *absolute* dates one needs additional information. Recently, it has been shown by Manning, *et al.*, *Science* (2001), that a refined tree ring chronology definitely supports the so called "middle chronology", which puts the fall of Babylon at 1595 BC. (On the other hand, Gurzadyan claims in *Sky and Telescope* 100 (2000) that astronomical evidence related to eclipses mentioned in Tablets 20 and 21 of the *Enuma Anu Enlil* fixes the date of the fall of Babylon to 1499 BC, 96 years later.)

Ur III	Eshnunna	Larsa	Isin	Babylon I
Ur-Namma **18 years** 2112-2095 Šulgi **48 years** 2094-2047 Amar-Suen **9 years** 2046-2038 Šu-Suen **9 years** 2037-2029 Ibbi-Suen **24 years** 2028-2004	Ituria Ilushu-Ilia 19 years 2028-2010 Nur-Ahum 10 years 2009-1990 Kirikiri 15 years 1989-1975 Bilalama 15 years 1974-1960 Ishar-Ramashshu 15 years 1959-1945 Azuzum 5 years 1944-1940 Ur-Ninmar 5 years 1039-1935 Ur-ningizzida 5 years 1934-1930 Ibiq-Adad I 20 years 1929-1910 Sharria 10 years 1909-1900 Belakum 20 years 1899-1880 Warassa 5 years 1879-1875 Ibal-Pi-el I 15 years 1874-1860 Ibiq-Adad II 30 years 1859-1830 Naram-Suen 25 years 1829-1805 Dadusha 25 years 1804-1780 Ibal-Pi-el II 17 years 1779-1763	Naplanum 21 years 2025-2005 Emisum 28 years 2004-1977 Samium 35 years 1976-1942 Zabaya 8 years 1941-1933 Gungunum 27 years 1932-1906 Abisare 11 years 1905-1895 Sumu-el 19 years 1894-1866 Nur-Adad 16 years 1865-1850 Suen-iddinam 7 years 1849-1843 Suen-eribam 2 years 1842-1841 Suen-iqišam 5 years 1840-1836 Silli-Adad 1 year 1835 Warad-Suen 12 years 1834-1823 Rim-Suen 60 years 1822-1763	Išbi-Erra **23 years** 2017-1985 Šu-ilišu **10 years** 1984-1975 Iddin-Dagan **21 years** 1974-1954 Išme-Dagan **19 years** 1953-1935 Lipit-Eštar **11 years** 1934-1924 Ur-Ninurta **28 years** 1923-1896 Bur-Suen **22? years** 1895-1874 Lipit-Enlil **5 years** 1873-1869 Erra-imitti **8? years** 1868-1861 Enlil-bani **24 years** 1860-1837 Zambiya **3 years** 1836-1834 Iter-piša **3 years** 1833-1831 Ur-dulkuga **3 years** 1830-1828 Suen-magir **11 years** 1827-1817 Damiq-ilishu **23 years** 1816-1794	Sumu-abum 15 years 1894-1881 Sumu-la-El 36 years 1880-1845 Sabium 14 years 1844-1831 Apil-Suen 28 years 1830-1813 Suen-muballit 20 years 1812-1793 Hammurabi 43 years 1792-1750 Samsu-iluna 38 years 1749-1712 Abi-ešuh 28 years 1711-1684 Ammiditana 37 years 1683-1647 Ammisaduqa 2 years 1646-1626 Samsuditana 31 years 1625-1595

The reigns of the kings of Ur (NS), and Eshnunna, Larsa, Isin, Babylon (OB), in the middle chronology.

In the table above are listed the dates for the kings of Ur III (Neo-Sumerian) and for the kings of Eshnunna, Larsa, Isin, and the first dynasty of Babylon (all "Old Babylonian") *according to the middle chronology*.[1]

The location of Ur, Eshnunna, Larsa, Isin, Babylon, and some other important sites in ancient Mesopotamia, are shown in the map below (Fig. 9.2).

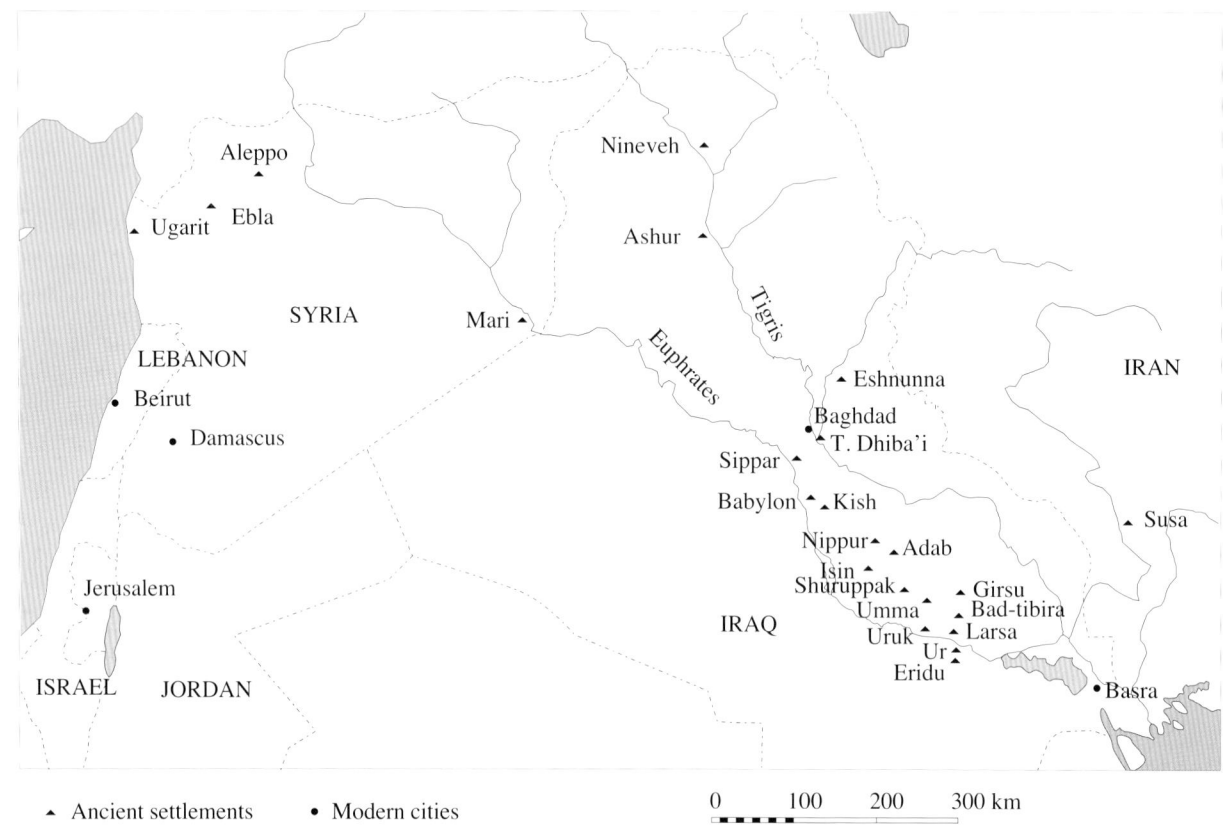

▲ Ancient settlements   • Modern cities

0   100   200   300 km

Fig. 9.2. A map of Mesopotamia and surrounding regions.

9.3.   *MS 2855. A New Version of the Antediluvian Part of the Sumerian King List*

The *beginning* of the *Sumerian King List* is reproduced below, in the Ash. 23.444 version:

The *Sumerian King List*, lines 1-39

1	[nam].lugal an.ta èd.dè.a.ba /	After the kingship descended from heaven,
2	[eri]dug^{ki} nam.lugal.la /	In **Eridu** was the kingship.
3-4	eridug^{ki}.á.lu.lim lugal / mu 8_{šár} i.ak /	In Eridu, Alulim (was) king, ruled for 8(šár) years.
5	á.làl.gar mu 10_{šár} i.ak /	Alalgar ruled for 10(šár) years.
6-7	2 lugal / mu.<bi> 18_{šár} íb.ak /	**2 kings ruled for 18(šár) years**.
8	eridug^{ki} ba.šub.<bé.en> /	Eridu <I> let fall,
9-10	nam.lugal.bi bàd.tibira^{ki}.šè / ba.de₆ /	the kingship was taken away to Badtibira.

1. In Friberg, *RA* 94 (2000), 174, the dates must be corrected. Thus, the dates of the known "early OB" mathematical texts are from around 1821 BC to 1763 BC, the year when Hammurabi defeated Rim-Suen of Larsa. The ones called "middle OB" are from 1763 BC to 1739 BC = Samsuiluna 11, the year when Ur, Larsa, and Uruk were abandoned. Similarly, the mathematical texts from Nippur cannot be from a date later than 1721 BC, the year when Nippur and Isin were abandoned.

11-12	bàd.tibira^{ki} en.me.en.lú.an.na / mu 12$_{šár}$ ì.ak /	In **Badtibira**, Enmenluana ruled for 12(šár) years.
13-14	en.me.en.gal.an.na / mu 8$_{šár}$ ì.ak /	Enmengalana ruled for 8(šár) years.
15	^{d}dumu.zi sipa mu 10$_{šár}$ ì.ak	Dumuzi, the shepherd, ruled for 10(šár) years.
16-17	3 lugal / mu.bi 30$_{šár}$ íb.ak /	**3 kings ruled for 30(šár) years.**
18	bàd.tibira^{ki} ba.šub.bé.en /	Then Badtibira I let fall,
19-20	nam.lugal.bi la.ra.ak^{ki}.<šè> ba.de$_{6}$ /	the kingship was taken away <to> Larak.
21	la.ra.ak^{ki} en.sipad.zid.an.na / mu 8$_{šár}$ ì.ak /	In **Larak**, Ensipadzidana ruled for 8(šár) years.
22-23	1 lugal / mu.bi 8$_{šár}$ íb.ak /	**1 king ruled for 8(šár) years.**
24	la.ra.ak^{ki} ba.šub.bé.en /	Then Larak I let fall,
25	nam.lugal.bi zimbir^{ki}.šè ba.de$_{6}$ /	the kingship was takenaway to Sippar.
26-27	zimbir^{ki} en.me.en.dúr.an.na / lugal.àm mu 18$_{šár}$ 50$_{géš}$ ì.ak /	In **Sippar**, Enmendurana was king, ruled for 18(šár) 50(géš) years.
28-29	1 lugal / mu.bi 18$_{šár}$ 50$_{géš}$ íb.ak /	**1 king ruled for 18(šár) 50(géš) years.**
30	zimbir^{ki} ba.šub.bé.en /	Then Zimbir I let fall,
31	nam.lugal.bi šuruppak^{ki}.<šè> ba.de$_{6}$ /	the kingship was takenaway <to> Shuruppak.
32-33	šuruppak^{ki} ubur.du.du / lugal.àm mu 5$_{šár}$ 10$_{géš}$ ì.ak /	In **Shuruppak**, Ubartutu was king, ruled for 5(šár) 10(géš) years.
34-35	1 lugal / mu.bi 5$_{šár}$ 10$_{géš}$ íb.ak /	**1 king ruled for 5(šár) 10(géš) years.**
36-38	5 iri^{ki}.me.eš / 8 lugal / mu.bi 1$_{šár.gal}$ 7$_{šár}$ íb.ak /	**In 5 cities 8 kings ruled for 1(šár).gal 7(šár) years.**
39	a.ma.ru ba.ùr.<<ra.ta>>	Then the flood swept over.

This is the so called "antediluvian" part of the *Sumerian King List*, with the names of five cities that allegedly held the kingship "before the flood". The numbers for the reigns of the eight antediluvian kings are so large that it is clear that it is non-historical and unrealistic. They are therefore traditionally referred to as "legendary numbers". In previously published discussions of the *Sumerian King List*, including the compiled *ETCSL* version, the legendary numbers are converted to decimal numbers, which in no way makes it easier to understand what they stand for. In the transliteration and translation above, the numbers are reproduced, without conversion to decimal numbers, as multiples of the g é š (60) and the š á r (sq. 60).

MS 2855, lines 1-22 (see Fig. 9.3 below) is, essentially, a close parallel to Ash. 1923.444, lines 1-39 (col. *i*). The same cities, and the same kings, are listed in both texts, in the same order, although the MS 2855 version is much less elaborate than the version in Ash. 1923.444.

MS 2855

1	eridu^{ki}<-ga> nam.lugal	In **Eridu** (was) the kingship.
2	a.lu.lim [lugal] / mu 8$_{šár}$.ì.ak /	Alul*im* (was) *king*, ruled for 8(šár) years.
3	e.làl.gar mu 10$_{šár}$ 2$_{šár}$.ì.[ak] /	Alalgar *rule*d for 12(šár) years.
4	eridu^{ki} ba.šub	Eridu fell,
5	nam.lugal.šè / bàd.tibira$^{ki\,ra}$<.šè> ba.de$_{6}$ /	as for the kingship, it was taken away <to> **Badtibira**.
6-7	am.mi.lú.an.na lugal / mu 10$_{šár}$.ì.ak /	Enmeluana (was) king, ruled for 10(šár) years.
8	en.me.gal.an.na mu 8$_{šár}$.ì.ak	Enmegalana ruled for 8(šár) years.
9	^{d}dumu.zi mu 8$_{šár}$.ì.ak /	Dumuzi ruled for 8(šár) years.
10	bàd.tibira$^{ki\,ra}$ [ba.šub] /	Badtibira *fell*,
11	nam.lugal.šè la.[ra.ak^{ki}.šè ba.de$_{6}$] /	as for the kingship, *it was taken away to* **Larak**.
12-13	en.sipa.zi.[an.na lugal] / mu 3$_{šár}$.50$_{géš}$.ì.[ak] /	Ensipazi*ana* (was) king, *ruled* for 3(šár) 50(géš) years.
14-15	la.ra.ak$^{<ki>}$ ba.šub / nam.lugal.šè zimbir^{ki}.[šè ba.de$_{6}$] /	Larak fell, as for the kingship, it was taken away to **Sippar**.
16	<en.>me.dur.an.na mu 2$_{šár}$ ì.ak /	<En>medurana ruled for 2(šár) years.
17-18	zimbir^{ki} ba.šub / nam.lugal.šè šuruppak^{ki}.šè ba.[de$_{6}$]	Sippar fell, as for the kingship, *it was taken away* to **Shuruppak**,

19	ubur.du.du <lugal> mu 10$_{šár}$.i.ak /	Ubartutu ruled for 10(šár) years.
20	[šu.níğ]in 8 lugal /	**In all 8 kings,**
21-22	[m]u.bi.me.eš 1$_{šár×1}$.gal 1$_{šár}$ 50$_{géš}$ / {x x} í.ak	**their years 1(šár).gal 1(šár) 50(géš) (that) they ruled.**
23	ši.mi.it$^?$ nam$^!$.lugal$^!$ mi.r[i.x x] /	?
24	ab za.gi im mu za ni eš lu [x] /	?
25	ki ig.bi ni im ši sá x /	?
26	da a na x [xxx] /	?
27	ù [x x x x x x]	?
28	i za [x x x x x x]	?
l. e.	ši ta za am ù diri lú diš / gù dè$^!$ <a>$^?$	?

(The subscript near the lower edge of the reverse and on the left edge is shallowly impressed, partly damaged, and totally unintelligible. The language may be neither Sumerian nor Akkadian.)

MS 2855 is a relatively small and fairly well preserved clay tablet with rounded corners and elaborate sign forms. It is probably from the early part of the Isin period. (Of all the clay tablets discussed above in this book, the only one of a similar format is the table of reciprocals HS 201, which is either early Old Babylonian or from the Ur III period. See the photo in Oelsner, in *ChV* (2001), 53-59.)

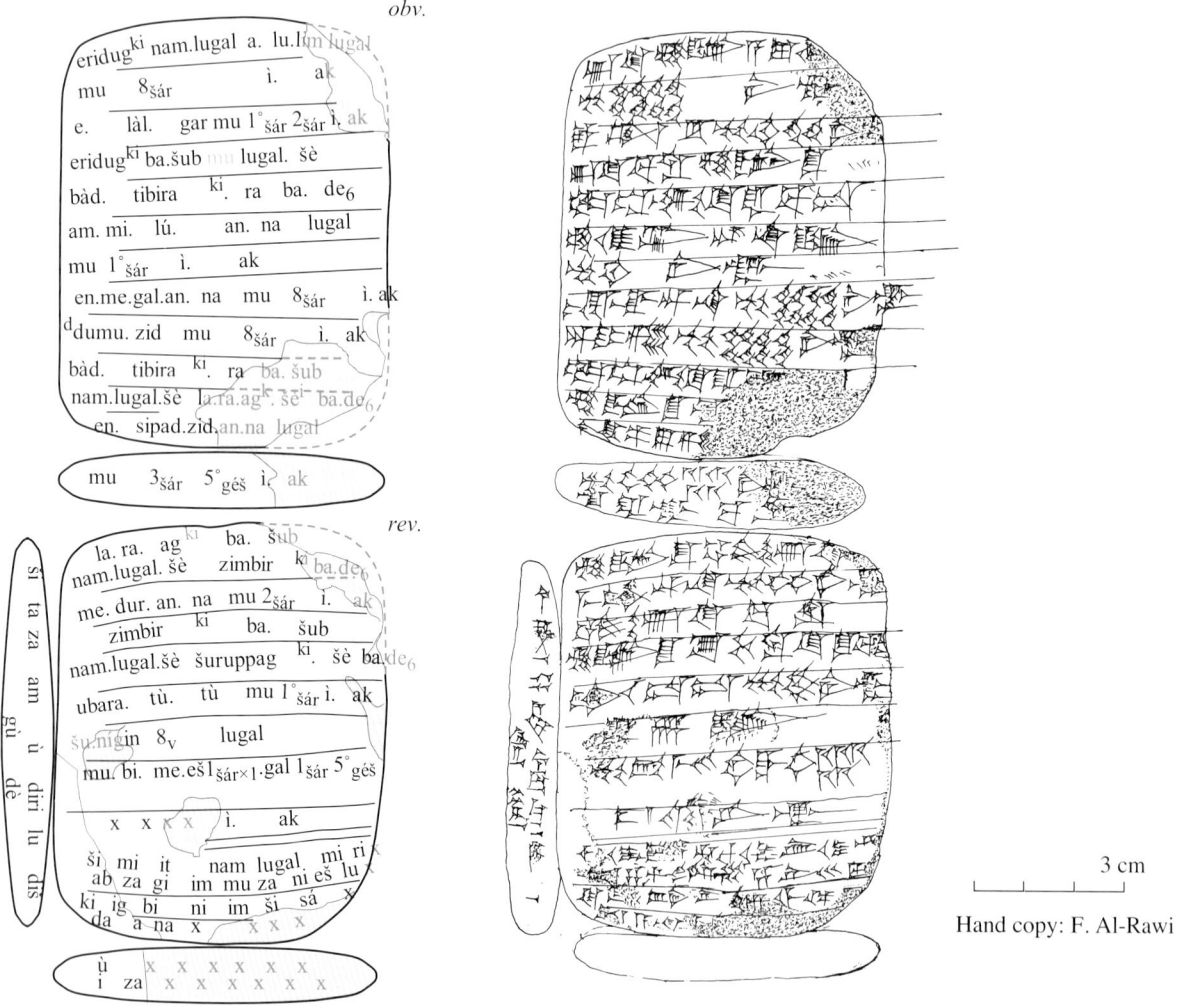

obv.

rev.

3 cm

Hand copy: F. Al-Rawi

Fig. 9.3. MS 2855. An early OB or Ur III version of the antediluvian part of the *Sumerian King List*.

If this assumption is correct, MS 2855 is older than the other known versions of the antediluvian king list, which are all Old Babylonian or later. This is somewhat curious, in view of the fact that much of the narrative framework in Ash. 1923.444 is missing in MS 2855 (see below).[2] The lengths of the reigns of the kings are also not the same in the two texts. Thus, even if it is likely that there was a common source to the two texts, one is not a copy of the other.

The reference to the kingship descending from heaven in Ash. 23.444, line 1,

| [nam].lugal an.ta èd.dè.a.ba | After the kingship descended from heaven |

and the reference to the flood in Ash. 23.444, line 39,

| a.ma.ru ba.ùr | Then the flood swept over |

have no counterparts in MS 2855. Jacobsen observed in *AS 11* (1939), 58-61, that these lines suggest that the antediluvian section of the king list was in some way derived from the Sumerian story about the beginning of the world and the flood (www-etcsl.orient.ox.ac.uk/section1/c174.htm & tr174.htm).

That story describes in segment A the creation of men and animals, then proceeds in segment B, 6-7, to tell about the descent of kingship from heaven:

6	[ud x] x nam.lugal.la an.ta èd.dè.a.ba /	After the ... of kingship had descended from heaven,
	men maḫ ^{giš}gu.za nam.lugal.la	the exalted crown and throne of kingship
7	an.ta èd.a.ba / ...	had descended from heaven ...

The story goes on, in segment B, 9-15, to describe the founding of five cities, mentioned in the same order as in MS 2855 and Ash. 1923.444: Eridu, Badtibira, Larak, Sippar, Shuruppak:

9	... [iri.bi.e].ne [sig$_4$.bi ki kug.ga im].ma.an.da.šub /	...the bricks of the cities were laid in holy places,
10	mu.bi ba.an.sa$_4$ kab dug$_4$.ga [ba.ḫal].ḫal.la /	their names were announced and the ... were distributed.
11	nesag iri.bi.e.ne eridug^{ki}	The first of the cities, **Eridu**,
	máš.sag ^{d}nu.dím.mud mi.ni.in.šúm /	was given to Nudimmud the leader.
12	2.kam.ma.šè nu.gig.ra bad.tibira^{ki} mi.ni.in.šúm /	The 2nd, **Badtibira**, was given to the Mistress.
13	3.kam.ma la.ra.ag ^{d}pa.bíl.{ḫur}.sag mi.ni.in.šúm /	The 3rd, **Larak**, was given to Pabilsag.
14	4.kam.ma zimbir^{ki} šul ^{d}utu mi.ni.in.šúm /	The 4th, **Zimbir**, was given to hero Utu.
15	5.kam.ma šuruppag ^{d}sùd^{ki}.ra mi.ni.in.šúm / ...	The 5th, **Shuruppak**, was given to Sud. ...

After a lacuna, segment C begins the story of the flood, which, interrupted by new lacunas, continues through segments D and E. The end of the flood is related in segment D in the following words:

3	ud 7.àm gi$_6$ 7.àm /	7 days, 7 nights,
4	a.ma.ru kalam.ma ba.ùr.a.ta /	the flood swept over the land,
5	^{giš}má gur$_4$.gur$_4$ a gal.la im.ḫul tuk$_4$.tuk$_4$.a.ta /	the big boat was rocked by waves and windstorms,
6	^{d}utu i.im.ma.ra.éd	Utu the sun-god came out,
	an ki.a ud gá.gá /	brightening heaven and earth with his rays.
7	zi.ud.sù.rá ^{giš}má gur$_4$.gur$_4$ ab.búr mu.un.da.bùru /	Ziusudra could break an opening in the huge boat, ...

It is interesting that Ziudsura, the Sumerian Noah, is not mentioned in Ash. 1923.444 (or MS 2855), but is mentioned in the other known versions of the antediluvian king list.

Missing in MS 2855 are, in addition to the narrative framework, the summaries for each dynasty, 'so and so many kings ruled for so and so many years', as well as the final summary 'In 5 cities, 8 kings ruled for so and so many years'. Finally, the names of the cities are never repeated in MS 2855, as they are in Ash. 23.444, lines 2-3, 10-11, 20-21, 25-26, and 31-32.

---

2. Much of the narrative framework, including the summaries, is missing also in Steinkeller's Ur III version (*USKL*) of the *Sumerian King List*, although it asserts in lines 1-2 that "The kingship descended from heaven, Kiš was king". (The antediluvian king list is not present in *USKL*, which begins with a long, uninterrupted list of kings in Kiš.)

On the other hand, MS 2855 has nearly the same formula for change of dynasty as Ash. 23.444, while other variants of the antediluvian king list have other formulas. In MS 2855, the formula is

$$A^{ki} \text{ ba.šub nam.lugal.šè } B^{ki}.\text{šè ba.de}_6$$ city A fell, this? kingship was taken away to city B

and in Ash. 23.444 it is the slightly different

$$A^{ki} \text{ ba.šub.bé.en nam.lugal.bi } B^{ki}.\text{šè ba.de}_6$$ city A I let fall, its kingship was taken away to city B

In *JCS* 17 (1963), 41-42, Finkelstein discusses the verb ba.šub.bé.en and ventures the conjecture that the use of the first person can be "explained as the form which existed in the literary source of the antediluvian tradition – in which the speaker presumably is a god (Enlil)".

The table below tries to update and make more precise the tabular surveys in Finkelstein, *op. cit*, 45-46. It demonstrates the lack of agreement between known variants of the antediluvian king list.

MS 2855	Ash. 1923.444	UCBC 9-1819	Ash. WB 62	K 11261+	Berossos
Eridu	**Eridu**	**Eridu**	**Ha.a^{ki}** (=Eridu)	**[Eridu]**	**Babylon**
Alulim	Alulim	[Alulim]	Alulim	[Alulim]$^?$	Aloros
$8_{šár}$	$8_{šár}$	$10_{šár}$	$18_{šár}$ $40_{géš}$	[...]	10 sárous
Elalgar	Alalgar	[A]lalgar	Alalgar	[Alalgar]$^?$	Alaparos
$12_{šár}$	$10_{šár}$	$3_{šár}$	$20_{šár}$	[...]	3 sárous
			Larsa		
			[...]-kidunnu		
			$20_{šár}$		
			[...]-alimma		
			$6_{šár}$		
Bad-tibira	**Bad-tibira**	**Bad-tibira**	**Bad-tibira**	**[Bad-tibira]**	**Pautibiblon**
Ammiluana	Enmenluana	Ammeluana	Dumuzi sipa	[Enmeluana]$^?$	Amelōn
$10_{šár}$	$12_{šár}$	$10_{šár}$	$8_{šár}$	[...]	13 sárous
Enmegalana	Enmengalana	Ensipadzidana	Enmenluana	[Enm]egalana	Amenōn
$8_{šár}$	$8_{šár}$	$12_{šár}$	$6_{šár}$	[...]	12 sárous
Dumuzi	Dumuzi sipa	Dumuzi sipa		[Dum]uzi sipa	Megalaros
$8_{šár}$	$10_{šár}$	$10_{šár}$		[...]	18 sárous
Larak	**Larak**		**Larak**	**Sippar**	Daōnos, shepherd
Ensipaziana	Ensipadzidana		Ensipadzidana	Enmeduranki	10 sárous
$3_{šár}$ $50_{géš}$	$8_{šár}$		$10_{šár}$	$15_{šár}$ [+?]	Euedōrachos
Sippar	**Sippar**	**Sippar**	**Sippar**	**Larak**	18 sárous
Medurana	Enmendurana	Enmenduranki	Enmedurana	Ensipaziana	**Laragchos**
$2_{šár}$	$8_{šár}$ $50_{géš}$	$1_{šár}$ $40_{géš}$	$20_{šár}$	$10_{šár}$ $20_{géš}$ [+?]	Amempsinos
Shuruppak	**Shuruppak**	**Shuruppak**	**Shuruppak**	**Shuruppak**	10 sárous
Ubartutu	Ubartutu	[Ubartutu]	Ubartutu	U[bartutu]	Otiartes
$10_{šár}$	$5_{šár}$ $10_{géš}$	[?]	$8_{šár}$	[...]	8 sárous
		[Ziusudra]$^?$	Ziusudra	Ziusudra	Xisuthros
		$5_{šár}$ [+?]	$10_{šár}$	[...]	18 sárous
5 cities, 8 kings	**5 cities, 8 kings**	**4 cities, 8$^?$ kings**	**6 cities, 10 kings**	**5 cities, 9 kings**	**3 cities, 10 kings**
$1_{šár}$·gal $1_{šár}$ $50_{géš}$	$1_{šár}$·gal $7_{šár}$	$59_{šár}$ $40_{géš}$ [+?]	$2_{šár}$·gal $6_{šár}$ $40_{géš}$	[...]	120 sárous

In addition to MS 2855 and the first column (of eight) on the four-sided prism Ash. 1923.444, the table mentions the following variants of the antediluvian section of the *Sumerian King List*:

Ash. (WB) 1923.444 (Langdon, *OECT 2* (1923), pl. 1) is a fairly well preserved four-sided prism, inscribed with two columns on each face of the prism. An elaborate version of the antediluvian king list occupies most of the first column.

UCBC 9-1819 (Finkelstein, *JCS* 17 (1963)) a small clay tablet from Khafājī, ancient Tutub, from the time between Suen-muballit and Samsuiluna of Babylon. It is a school text, with the antediluvian king list inscribed on the reverse, while there is the beginning of a typical Old Babylonian letter on the obverse.the antediluvian king list is just as concisely worded as the one on MS 2855, without much of the narrative framework of Ash. 1923.444. The term used for the termination of a dynasty is ba.gul 'it was destroyed', not ba.šub or ba.šub.bè.en as in MS 2855 and Ash. 1923.444.

Ash. WB 62 (Langdon, *OECT 2* (1923), pl. 6) is a small clay tablet, inscribed only on the obverse, with a brief version the antediluvian king list. The insertion in the list of two kings from Larsa shows that the text was written at Larsa. The end of a dynasty is indicated in this text only by a concise summary mentioning the number of kings in that dynasty.

K 11261+ (Lambert, in *Symbolae Böhl* (1973)) is a join of three small Late Assyrian fragments from Ashurbanipal's library at Niniveh. The narrative framework seems to be just as extensive as in Ash. 1923.444, with summaries after each dynasty, a final summary, and a (broken) reference at the end of the text to Enlil and the noise [of the humans]. The term used in this text for the change of dynasties is A^{ki} bala.bi ba.kúr nam.lugal.bi B^{ki}.šè ba.nigin 'the dynasty of city A changed, the kingship moved to city B'.

Berossos was a priest of Bel (Marduk) at Babylon, who wrote a history of Babylon from the mythical beginning, including the reigns of the first 10 kings, and the story of the Flood, up to the time of Alexander. The work (c. 290 BC) was dedicated to king Antiochos I and written in Greek. It is known only through excerpted fragments in the works of several Hellenistic historians. See, for instance, Schnabel, *Berossos* (1923), 261-263.

The list of kings and the years of their reigns in the antediluvian part of the *Sumerian king list* has often been likened to the list of biblical patriarchs in *Genesis* V. Here is, for comparison, the beginning and end of that list:

Genesis V.

1. This is the book of the generations of Adam. ...
   . . .
3. And Adam lived a hundred and thirty years, and begat a son in his own likeness, after his image; and called his name Seth;
4. And the days of Adam after he had begotten Seth were eight hundred years; and he begat sons and daughters;
5. And all the days that Adam lived were nine hundred and thirty years; and he died.
   . . .
28. And Lamech lived a hundred eighty and two years, and begat a son;
29. And he called his name Noah, saying ...
30. And Lamech lived after he begat Noah five hundred ninety and five years, and begat sons and daughters;
31. And all the days of Lamech were seven hundred and seventy and seven years; and he died.
32. And Noah was five hundred years old; and Noah begat Shem, Ham, and Japheth.

Including Adam, there are nine patriarchs in this list, before Noah.

The following numbers are mentioned in the list of patriarchs:

	years until the first son is born	years after the first son is born	all the years
Adam	130	800	930
Seth	105	807	912
Enos	90	815	905
Cainan	70	840	910
Mahalaleel	65	830	895
Jared	162	800	962
Enoch	65	300	365
Methuselah	187	782	969
Lamech	182	595	777
Noah	500	——	——

The biblical story of the flood follows in *Genesis* VI-VIII. See, for instance *Genesis* VII, 8:

And Noah was six hundred years old when the flood of waters was upon the earth.[3]

---

3. See Friberg, in *ABD* (1992), for, among other things, a brief account of the use of interesting numbers in the Bible.

9.4.   The Numbers in the Antediluvian King List

As mentioned already, unrealistically huge numbers are used in the king list for the "legendary reigns" of the antediluvial kings, of the kings of three first dynasties in Kish, and of kings of the first dynasty in Uruk, while more realistic numbers seem to be used for the "historical reigns" of the remaining dynasties. (Similarly, as was shown above, unrealistically large numbers are used in the list of patriarchs in *Genesis* V.) Efforts have been made to explain the origin of such fantasy numbers.

Thus, for instance, in *ASJ* 10 (1988) Steiner tried to find the "real core" of what he called the "legendary numbers" in the *Sumerian King List*. Steiner's idea was that behind the large numbers for the kings of the earliest dynasties after the flood was the ancients' concept of the average length of a *generation*, equal to 40 years. He figured that the original author of the *Sumerian King List* had equated each year of the historical reigns of the earliest kings with a generation. Accordingly, he multiplied the lengths of the reigns of the earliest kings with a factor 40. Similarly, Steiner formulated the hypothesis that the reigns of the kings before the flood had been equated with a generation of generations. The effect would be that those reigns had been multiplied with a factor of 40 times 40.

In *JNES* 47 (1988), Young made the interesting observation that many of the reigns of kings in the *Sumerian King List* can be explained as sexagesimal numbers of the kind that one often meets in Old Babylonian mathematical texts: *squares*, *sums of squares*, and *numbers with factors like 7, 11, and 13*.

Both Young and Steiner worked with the decimal equivalents of the sexagesimal numbers actually recorded in the *Sumerian King List*, which made their task more difficult than necessary. A look at a tabular survey like the one exhibited above, shows that there can be no historical background whatsoever to the reigns of the antediluvian kings, since the numbers given in the various variants of the antediluvian king list are so dissimilar.

An alternative explanation, like the one proposed by Steiner, but taking into consideration the fact that the numbers in the king list are sexagesimal, could be that the historical lengths of the reigns of the antediluvian kings were simply multiplied by 1 šár. Then the legendary numbers in MS 2855, for instance, could be interpreted as, respectively, 8, 12, 10, 8, 8, and 4 years, 3 years 10 months, 2 years, and 10 years. However, this alternative explanation, too, is contradicted by the table above showing the dissimilarity of the numbers in the known versions of the antediluvian king list.

It is difficult to see any mathematical pattern behind the numbers in any one of antediluvian king lists, just as behind the numbers in the list of patriarchs in *Genesis* V. The only observable regularity seems to be that all the numbers in the various versions of the antediluvian king list are between 1 2/3 šár and 20 šár, with fractions of the šár appearing in four of the 6 versions.

So, perhaps, the correct explanation may be that the origin of the antediluvian king list in its various versions was a popular and long lived *oral tradition*, with the lengths of the reigns imperfectly remembered, but always in the range of the šár because of a natural fascination with big numbers.

In the most elaborate versions of the *Sumerian King List* there are both summaries of the reigns of the kings of each dynasty, and final summaries. Thus, some variants of the *Sumerian King List* end with the final summary '11 cities, cities where the kingship was exercised. Together 2 14 kings, their years 8 šár.gal […] 1 16' (lines 378-430 of the *ETCSL* version). Also the Late Assyrian fragment has summaries for each dynasty and a final summary. A likely explanation is that Mesopotamian scribes were trained to write not only literary texts, including king lists, but also administrative texts and numerical accounts. So they would naturally have an ingrained habit to include summaries and final summaries in all texts they wrote with long lists of numbers.[4]

9.5.   Mesopotamian Year Names

The *Sumerian King List*, with its enumeration of kings and the durations of their reigns, may originally, at

---

4. Many proto-cuneiform accounts on proto-Sumerian clay tablets from around 3000 BC have summaries (totals) on the reverse of the numbers recorded on the obverse. There may even be final summaries (grand totals) on the reverse, as in the example *MSVO 4, 66*, an important "bread-and-beer text" from Jemdet Nasr (Friberg, *JCS* 51 (1999), Fig. 3.1).

least partly, have been compiled with departure from lists of "year names" for the various kings. Normally, a year was named after a significant event in that year or the immediately preceding year. The following few examples from the reign of the first Ur III king, Ur-Nammu, are borrowed from Sigrist and Damerow, *Mesopotamian Year Names* (2001):

mu ur.^{d}nammu lugal	Year when Ur-Nammu became king.
mu ur.^{d}nammu lugal.e sig.ta igi.nim.šè gìr si bí.sá.a	Year when Ur-Nammu the king put in order the ways from below to above.
mu ur.^{d}nammu nì.si.sá kalam.ma mu.ni.gar	Year when Ur-Nammu made justice in the land.
mu en.^{d}inanna unug^{ki}.a dumu ur.^{d}nammu lugal.a maš.e ba.pàd.da	Year when as the en-priest of Inanna in Uruk the son of king Ur-Nammu was chosen by the omens.

10
Three Old Babylonian Mathematical Problem Texts from Uruk

10.1. MS 3971. A Double-Column Mathematical Recombination Text

MS 3971 (Figs. 10.1.1-2 below) is a large fragment of a double-column clay tablet of moderate size. The text on the obverse is only partially well preserved. The text on the reverse is almost completely destroyed, with only the ends of some lines visible on the right edge of the clay tablet.

The nine mathematical exercises inscribed on the obverse can be divided into five different paragraphs, of which the last four deal with 'diagonals' (meaning rectangles). The word used for 'diagonal' is *ṣiliptum*, literally 'that which crosses over', from *ṣalāpum* (Sum. bar) 'to cross over'.

The small vocabulary of Sumerian technical terms used in MS 3971 includes the following words for mathematical operations and structuring of the text:

gar.gar	heap (add)	$du_7.du_7$	butt (square)	gaz	break (halve)
zi	tear off (subtract)	nim	lift (multiply)	in.sì	it gives (it is the result)
daḫ	join to (add to)	*n*.e *m* íb.si$_8$	*n* makes *m* equalsided (sqs. *n* = *m*)		
du$_8$	resolve (compute reciprocal)	*aššum ... amārika*	in order for you to see (as in YBC 4608, *MCT*, 149)		

Consequently, MS 3971 belongs to Goetze's "Group 3" of Old Babylonian mathematical cuneiform texts (see Friberg, *RA* 94 (2000), § 7b). As a member of Group 3, MS 3971 is from Uruk, and middle Old Babylonian, dating to before the year Samsuiluna 11 (1795 BC) (Friberg, *op. cit.*, 174).

10.1 a. MS 3971 § 1. A Broken Reed Problem of a New Type

The upper edge and the uppermost part of MS 3971 are lost. As a consequence of this circumstance, the first seven lines, or so, of § 1, including the question, are missing.

MS 3971 § 1

	… … …	… … …
8	[x x x x x x] x x sa[g gi] /	*x x x x x x* x x the original reed
9	[x x x x x x] x x sag [ki.t]a /	*x x x x x x* x x the *lower* front
10	[x x x x x a].šà il-li-ik /	*x x x x x x* the field he went
11	[x x x x x] 30-*šu at-ta-di-ma* /	*x x x x* I have given
12	[x x x x] *at-ta-an-din-nu-ma*ⁱ /	*x x x x that* I have given them, then
13	[x x] x gi.n[a a]*t-ta-an-din-ma* /	*x x x* the tr*ue x I* have given, then
14	[31] ninda 5 2' kùš sag ki.ta /	*31* ninda 5 1/2 cubits (is) the lower front.
15	6$_{bùr}$ 1$_{èše}$ 4$_{iku}$ 31 šar 15 gín a.šà /	6 bùr 1 èše 4 iku 31 šar 15 gín (is) the field.
16	sag gi e[n.n]am /	The initial reed *is what?*
17	*ki-ia-a im-ta-x*[x] /	How much x x?
18	[x x x] x [x] ninda$^?$ 1 sag gi.n[a] /	*x x x x x x* x 1 the true front.

245

19-20	*ša* [x x 1 *a*]-*na* 2 30 ki.1 daḫ / 2 31 i[n.s]i /	that *x x x 1* to the 1st 2 30 join, 2 31 it gives.
21	2 31 *ù* 2 30 / du₇.du₇-*m*[*a*] 6 17 30 in.si /	2 31 and 2 30 (make) butt (each other), 6 17 30 it gives.
22-23	2' 6 17 30 nim *ḫe-ep-pe-e-ma* / 3 08ᵥ 45 in.si /	1/2 of 6 17 30, the lifted(?), break, then 3 08 45 it gives.
24-25	*mi-na a-na* 3 08ᵥ 45 ḫé.gar! / *ša*	*What* to 3 08 45 shall I set, that
26	31 27 30 sag <ki.ta> in.si / 10 [šu.si]	31 27 40, the <lower> front, gives. 10, *a finger*.
	2 30 *a-na* 10 nim 25 /	2 30 to 10 lift, 25.
27	25 sag gi in.si /	25, the initial reed, it gives.
28	25 *a-na* 4 nim 1 40 sag an.na /	25 to 4 lift, 1 40, the upper front.
29	5-*x a-na* 25 daḫ 30 in.si /	The 5th? to 25 join, 30 it gives.
30	30 *a-na* 6 nim 3 uš gi.na	30 to 6 lift, 3, the true length.

The topic of the imperfectly preserved exercise in MS 3971 § 1 is revealed by a few telltale words:

sag gi	the head of the reed	(the initial length of the measuring stick)	(lines 16, 27)
sag ki.ta	the lower front	(of the trapezoid)	(line 14)
sag an.na	the upper front	(of the trapezoid)	(line 28)
uš gi.na	the true length	(of the trapezoid)	(line 30)
sag gi.na	the true front	(of the trapezoid)	(line 18)

These words make it likely that MS 3971 § 1 is a "broken reed problem". (See Friberg, *RlA 7* (1990), Secs. 5.4 b, 5.4 e, and 5.7 h.) The lost question, in lines 1-16 of the text, can be reconstructed through a careful analysis of the solution procedure. It appears to have been, essentially, of the following form:

> Given a trapezoid and a measuring stick (a 'reed') of unknown length.
> The reed is applied 6 00 (360) times to measure the length of the trapezoid.
> The reed is then shortened by 1/6 of its length.
> The shortened reed is applied 4 00(240) times to measure the upper front of the trapezoid.
> The shortened reed is finally applied 2 30 (150) times to measure the length of the lower front.
> For each application of the reed, a piece (always of the same unknown size) of the length of the reed is lost.
> The lower front is 31 ninda 5 1/2 cubits. (line 14)
> The area of the trapezoid is 6 bùr 1 èše 4 iku 31 šar 15 gín. (line 15)
> What was the initial length of the reed, (and what were the length, the upper front, and the lower front)? (line 16)

The solution procedure is, essentially, the following:

1. ??? (lines 17-18)
2. 2 30 + 2 29 + ... 2 + 1 = 2 31 · 2 30 · 1/2 = 3 08 45 (lines 19-23)
3. 3 08 45 · ? = 31;27 30 ninda, the lower front, ? = ;00 10 ninda = 1 finger (lines 24-26)
4. 2 30 · ;00 10 ninda (1 finger) = ;25 ninda (= 5 cubits)= the shortened reed (lines 26-27)
5. ;25 ninda · 4 00 = 1 40 ninda = the upper front (line 28)
6. ;25 ninda · (1 + 1/5) = ;30 ninda = the initial reed (line 29)
7. ;30 ninda · 6 00 = 3 00 ninda= the length (line 30)

The solution procedure starts by making use of the rule for the sum of an arithmetical progression (Friberg, *op. cit.*, Sec. 5.7 h; see also MS 5112 § 5 in Sec. 11.2 h below):

$$n + (n - 1) + ... + 2 + 1 = 1/2 · (n + 1) · n.$$

Therefore, the size of the pieces that fell off can be computed as the solution to a division problem:

> 1/2 · 2 30 · 2 31 = 3 08 45,   3 08 45 · ? = 31 ninda 5 1/2 cubits = 3 08 45 fingers.
> Answer: ? = ;00 10 ninda = 1 finger.

Next, the length of the shortened reed is computed as 2 30 times the length of one of the pieces:

> The shortened reed = 2 30 · ;00 10 ninda = ;25 ninda (= 5 cubits).
> The upper front = ;25 ninda · 4 00 = 1 40 ninda.

The length of the original reed is computed as follows:

> Since the shortened reed is 1/6 less than the original reed, the original reed is 1/5 more than the shortened reed.
> Hence, the original reed is ;25 ninda + 1/5 · ;25 ninda = ;30 ninda (= 6 cubits = 1 reed).

Finally, when the original length of the reed is known, the length of the trapezoid can be computed:

$$\text{The length} = {;}30 \text{ ninda} \cdot 6\,00 = 3\,00 \text{ ninda}.$$

10.1 b.   Other Examples of Broken Reed Problems with Arithmetical Progressions

AO 6770 # 5 (Group 1: Larsa; Thureau-Dangin, *TMB* (1938), 73)

1	gi šu.ba.an.ti	A reed I received.
	1 šu.si *im-ta-qú-ta-an-ni* /	1 finger fell off for me.
2	*a-ša-a[r ig]-ga-am-ra-an-ni* 4$_v$ kùš /	Where it was exhausted: 4 cubits.
3	*re-iš q[á]-ni-ia mi-nu-um* /	The head of my reed (was) what?
4	*re-iš qá-ni-ia* 2' kùš /	The head of my reed (was) 1/2 cubit.
5	<igi> [ig]i.gub šu.si *a-pa-ṭa-ar*	The <opposite of> the constant for a finger I resolve,
	a-na 4$_v$ kùš *a-na-aš-ši* /	to 4 cubits I lift (it),
6	[*a*]-*na ši-na e-ṣí-ip*	to two I repeat.
7	1 *wa-ṣí-ta-am* / [2']-*šu e-ḫe-pe*	1, the extension, its 1/2 I break.
	íb.si$_8$ *a-na* tab.ba bi.[da]ḫ	(Its) equalside to the doubled I join.

In this text, the stated problem is as follows:

> A reed of unknown length is applied an unknown number of times to measure a length of 4 cubits.
> For each application of the reed, a piece of length 1 finger falls off. What was the original length of the reed?

The problem in AO 6770 # 5 is complementary to the problem in MS 3971 § 1. In the former, the length of the pieces falling off is known, but not their number, while in the latter the number of pieces is known, but not their length. Exceptionally, the solution procedure in AO 6770 # 5 is *verbal and general* rather than numerical, and in the first person. It is also incomplete, simply for the reason that there was no space left on the tablet for the end of the text. Anyway, if the solution procedure had been explicit and complete, it would have run as follows (in quasi-modern terms, since it is difficult to know how an Old Babylonian writer would have expressed it):

> Let n be the number of times that the reed can be applied before it is exhausted.
> Then $1/2 \cdot (n + 1) \cdot n$ fingers = 4 cubits = ;20 ninda.
> Since 1 finger = ;00 10 ninda, 1 ninda = 6 00 fingers,
> and 4 cubits = ;20 ninda = 6 00 · ;20 ninda = 2 00 fingers.   (line 5)
> Hence, sq. $n + 1 \cdot n = 2 \cdot 2\,00 = 4\,00$, where the coefficient 1 is called *wāṣītum* the 'extension'.   (line 6)
> By completion of the square: sq. $(n + 1/2)$ = sq. 1/2 + 4 00 = 4 00;15 = sq. 15 1/2.
> The square side 15 1/2 = n +1/2. Hence, n = 15, so that the initial reed was 15 fingers = 1/2 cubit.   (line 7)

In both MS 3971 § 1 and AO 6770 # 5 the sum of the arithmetical progression is known but the length of the reed is unknown. In the next example, the length of the reed is given from the start, and the sum of the arithmetical progression has to be computed:

Str. 362 # 5 (Group 3: Uruk; Neugebauer, *MKT 1* (1935), 240; Thureau-Dangin, *TMB* (1938), 83)

1	gi.1.kùš	A 1-cubit-reed.
2	1 šu.si.ta.àm *a-di ig-ga-am-ru* / *im-ta-qú-ta-an-ni*	1 finger each time until it was exhausted fell off for me.
3	uš *ki ma-ṣi* / *al-li-ik*	How much length did I go?
4	1 ninda 2' kùš uš / *al-li-ik*	1 ninda 3 1/2 cubits the length I went.

The solution procedure is omitted in this brief text. It would, essentially, have run as follows:

$$(30 + 29 + \ldots + 2 + 1) \text{ fingers} = 1/2 \cdot 31 \cdot 30 \text{ fingers} = 7\,45 \text{ fingers}$$
$$7\,45 \text{ fingers} = 7\,45 \cdot {;}00\,10 \text{ ninda} = 1{;}17\,30 \text{ ninda} = 1 \text{ ninda } 3\ 1/2 \text{ cubits}$$

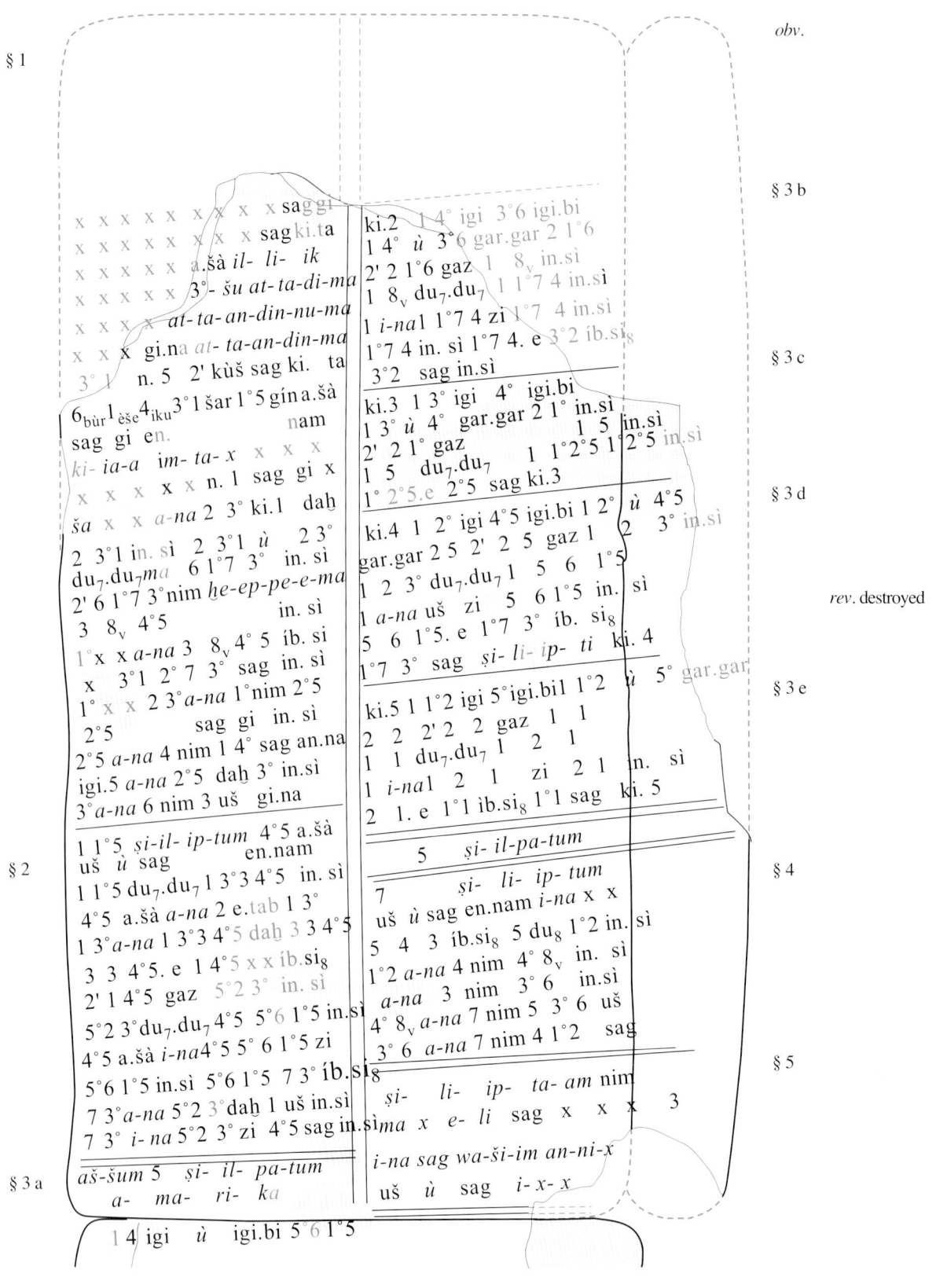

Fig. 10.1.1. MS 3971, *obv*. A large fragment of a mathematical recombination text. Five preserved themes.

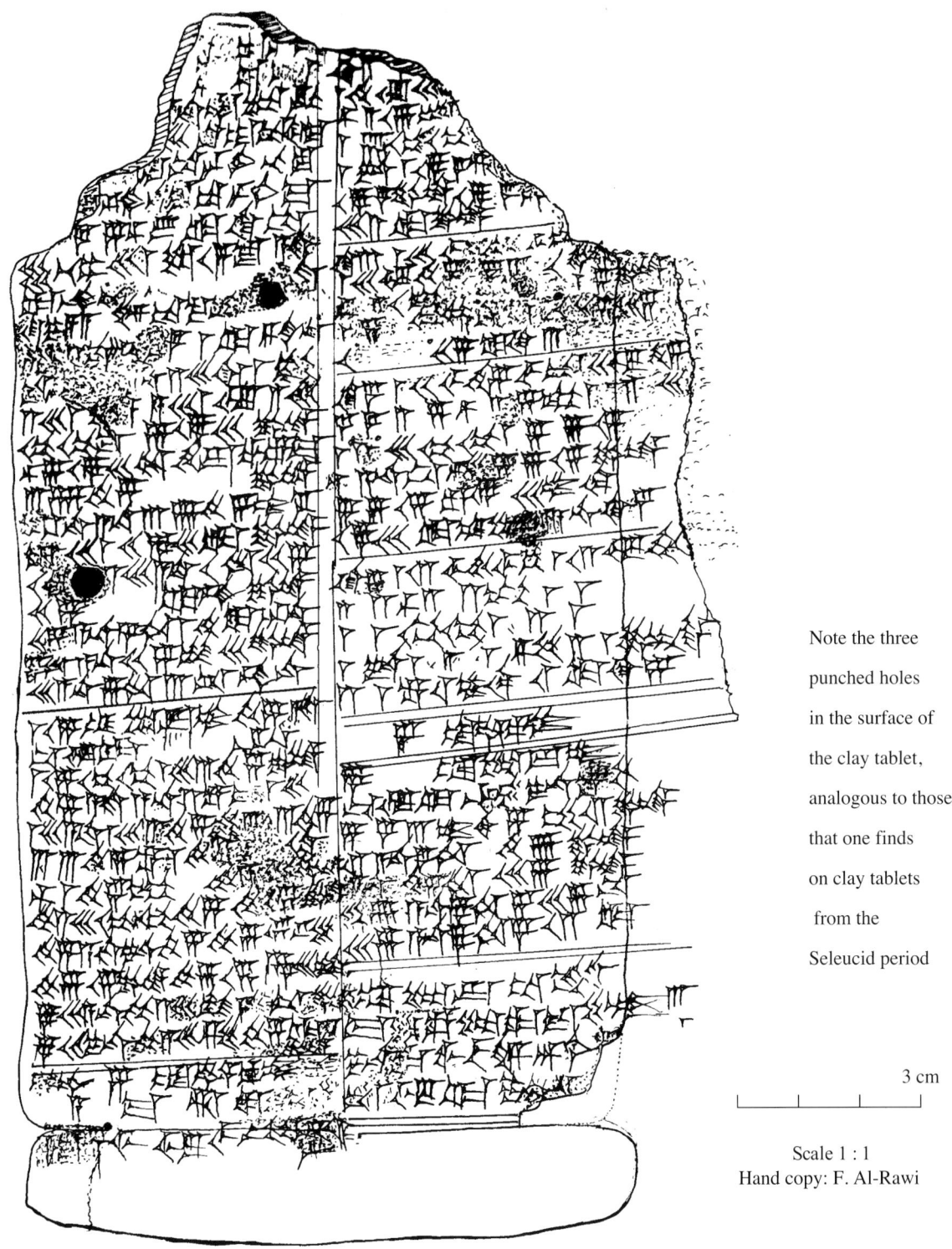

Note the three

punched holes

in the surface of

the clay tablet,

analogous to those

that one finds

on clay tablets

from the

Seleucid period

3 cm

Scale 1 : 1
Hand copy: F. Al-Rawi

Fig. 10.1.2. MS 3971, *obv*. Hand copy of the cuneiform text. (The reverse is completely destroyed.)

10.1 c.   Other Examples of Broken Reed Problems for Rectangles or Trapezoids

The simplest example of this kind is concerned with the sides of *a rectangle*:

IM 53965 (Tell Harmal; Baqir, *Sumer* 7 (1951), 39)

1	*šum-ma ki-a-am i-ša-al-[ka um-ma šu-ú-ma]* /	If somebody asks you thus:
2	*qa-na-am el-qé-e-ma*	A reed I took,
3	[*mi-in-da-su*] / *ú-ul i-de-e-ma*	its measure I do not know.
4	*šu-ši-da-am al-li-ik* / *am-ma-at aḫ-ṣú-úb-šu-ma*	Sixty-length I walked, a cubit I broke off it, then
5	*ša-la-aš pu-ta-am* / [*a*]*l-li-ik* a.šà 4 10	thirty, the front, I walked. The field (is) 4 10.
6	*ši-di ù pu-ti* / [*ki*] *ma-ṣí* *etc.*	My length and my front how much? *etc.*

The question in this exercise can be reduced to the following system of equations:

$$u = 1\ 00\ r, \quad s = 30 \cdot (1\ r - 1 \text{ cubit}), \quad u \cdot s = A = 4\ 10 \text{ sq. n., where } r \text{ is the unknown length of the reed.}$$

This system of equations can be reduced to the following rectangular-linear system of equations:

$$u \cdot s^* = 2 \cdot 4\ 10 \text{ sq. n.} = 8\ 20 \text{ sq. n.,} \quad u - s^* = 2 \cdot 30 \text{ cubits} = 5 \text{ n., where } s^* = 2 \cdot s.$$

The solution to this system of equations, computed in the usual way, is

$$u = 25 \text{ n.,} \quad s^* = 20 \text{ n., so that } s = 10 \text{ n., and } r = {;}25 \text{ n.} = 5 \text{ cubits} = \text{the original length of the reed.}$$

A more complicated example is concerned with the sides of *a trapezoid*:

VAT 7532 (Group 3: Uruk; Høyrup, *LWS* (2002), 209-213)

1	sag.ki.gud	An ox-head (trapezoid).
	gi kud	A reed I cut.
	gi *e*[*l-qé-ma i-na š*]*u-u*[*l*]*-m*[*i*]*-šu* /	The reed I to*ok in its w*hole (length),
2	1 *šu-ši* uš *al-li-i*[*k*]	1 sixty, the length, I walked.
3	[igi.6.gá]l / *iḫ-ḫa-aṣ-ba-an-ni-ma*	*The 6th-part* he cut off for me, then
	1 12 *a-na* u[š] *ú-r*[*e*]*-ed-di* /	1 12 to the length I added.
4	*a-tu-úr*	I turned back.
	igi.3.gál *ù* 3' kùš *iḫ-*[*ḫa-aṣ-ba-a*]*n-ni-ma* /	The 3rd-part and 1/3 cubit he br*oke off* for me, then
5	3 *šu-ši* sag an.na *al-li-ik* /	3 sixties, the upper front, I walked.
6	*ša iḫ-ḫa-aṣ-ba-an-ni ú-te-er-šum-*[*m*]*a* /	What he broke off for me I returned to it, then
7	36 sag <ki.ta> *al-li-ik*	36, the <lower> front, I walked.
	1$_\text{bùr}$ aša$_5$ a.šà	1 bùr (is) the field.
	sag gi en.nam *etc.*	The initial reed is what? *etc.*

The question in this exercise can be expressed as follows:

$$u = 1\ 00\ r + 1\ 12\ r^*, \quad s_\text{a} = 3\ 00\ r^{**}, \quad s_\text{k} = 36\ r^*,$$
$$u \cdot (s_\text{a} + s_\text{k})/2 = A = 1 \text{ bùr} = 30\ 00 \text{ sq. n.,}$$
$$\text{where } r^* = (1 - 1/6) \cdot r, \text{ and } r^{**} = (1 - 1/3) \cdot r^* - 1/3 \text{ cubit.}$$

In the solution procedure, this question is reduced to the following quadratic equation:

$$6\ 14\ 24 \text{ sq. } r^* - 12\ 00\ r^* = 1\ 00\ 00 \text{ sq. n.}$$

The solution to this equation is computed in the usual way. The resulting values of the unknowns are:

$$r^* = {;}25 \text{ n.} = 5 \text{ cubits,}$$
$$r = {;}30 \text{ n.} = 6 \text{ cubits, } r^{**} = {;}15 \text{ n.} = 3 \text{ cubits,}$$
$$u = 1\ 00 \text{ n.,} \quad s_\text{a} = 45 \text{ n.,} \quad s_\text{k} = 15 \text{ n.}$$

Parallel texts are VAT 7535 ## 1-3 (hand copy: *MKT 2*, pl. 47). Both VAT 7532 and VAT 7535 ##1-3 are illustrated by drawings of trapezoids, with inserted values of various parameters.

10.1 d. MS 3971 § 2. To Find a Rectangle with a Given Diagonal and a Given Area

MS 3971 § 2:

1	1 15 ṣi-i[l-i]p-tum 45 a.šà /	1 15 the cross-over (diagonal), 45 the field.
2	uš ù sag en.nam /	The length and the front are what?
3	1 15 du₇.du₇ 1 33 45 in.sì /	1 15 (make) butt (itself) 1 33 45 it gives.
4	45 a.šà a-na 2 e.ta[b] 1 30 /	45, the field, to 2 you repeat, 1 30.
5	1 30 a-na 1 33 45 d[ah] 3 03 45 /	1 30 to a 33 45 join, 3 03 45.
6	3 03 45.e 1 4[5 íb].si₈ /	3 03 45 makes 1 45 equalsided.
7	2' 1 45 gaz 5[2 30 in.sì] /	1/2 of 1 45 break, 52 30 it gives.
8	52 30 du₇.du₇ 45 5[6] 15 in.sì /	52 30 (make) butt (itself), 45 56 15 it gives.
9-10	45 a.šà i-na 45 56 15 zi / 56 15 in.sì	45, the field, from 45 56 15 tear off, 56 15 it gives.
	56 15 7 30 íb.si₈ /	56 15 <makes> 7 30 equalsided.
11	7 30 a-na 52 30 dah 1 uš in.sì /	7 30 to 52 30 join, 1, the length, it gives.
12	7 30 i-na 52 30 zi 45 sag in.sì /	7 30 from 52 30 tear off, 45, the front, it gives.

In modern language, the question in MS 3971 § 2 can be rephrased as follows:

> The diagonal d of a rectangle is 1 15, and the area A is 45 (00).
> What are the length u and the front s of the rectangle?

The successive steps of the numerical solution procedure are the following:

sq. d = sq. 1 15 = 1 33 45, $2 \cdot A = 2 \cdot 45 (00) = 1\ 30 (00)$, 1 33 45 + 1 30 (00) = 3 03 45	(lines 3-5)
sqs. 3 03 45 = 1 45 = p, 1 45 /2 = 52;30 = $p/2$, sq. 52;30 = 45 56;15	(lines 6-8)
45 56;15 – A = 45 56;15 – 45 (00) = 56;15, sqs. 56;15 = 7;30 = $q/2$	(lines 9-10)
52;30 + 7;30 = 1 (00) = u, 52;30 – 7;30 = 45 = s	(lines 11-12)

It is possible that the solution procedure in MS 3971 § 2 is best understood as consisting of two steps. In the first step, the given problem is reduced to a rectangular-linear system of equations of the following standard form:

$$u \cdot s = A = 45 (00),\quad u + s = p = 1\ 45.$$

In the second step, this rectangular-linear system is solved in the usual way.

The answer to the stated problem is, of course, that $d, u, s = 1\ 15, 1\ (00), 45$, the Babylonian standard example of a "diagonal triple". (See MS 3052 § 2 in Sec. 10.2 b below and App. 8.)

A geometric interpretation of the solution procedure is given in Fig. 10.1.3 below:

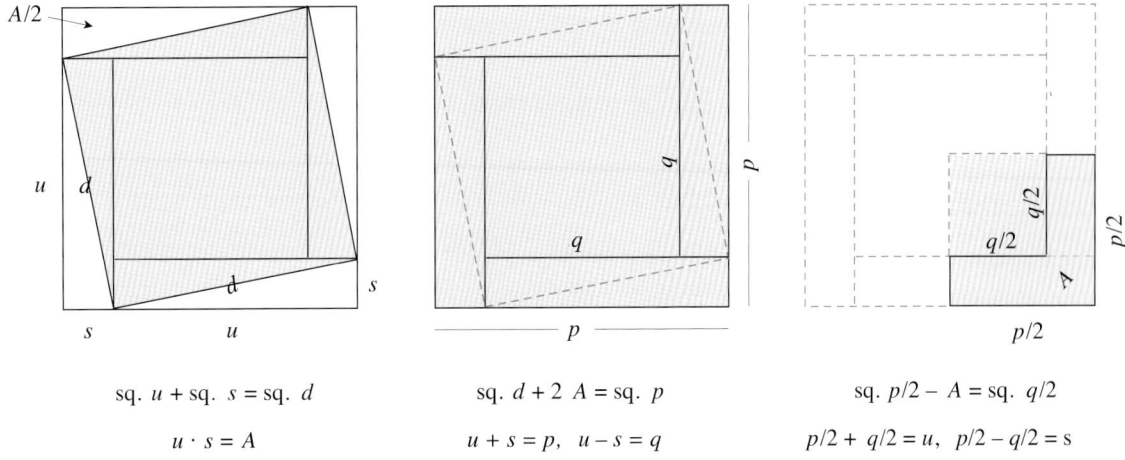

sq. u + sq. s = sq. d	sq. d + 2 A = sq. p	sq. $p/2$ – A = sq. $q/2$
$u \cdot s = A$	$u + s = p$, $u - s = q$	$p/2 + q/2 = u$, $p/2 - q/2 = s$

Fig. 10.1.3. MS 3971 § 2. Given the square on the diagonal of a rectangle and its area.

10.1 e.   A Parallel Text: IM 67118 (= Db₂-146)

IM 67118 is from Old Babylonian Eshnunna, dated to the time of Ibal-Piel II (1779-1763).

IM 67118, the first half (Høyrup, LWS (2002), 257-261)

1-2	*šum-ma ṣi-li-ip-ta-a-am i-ša-lu-ka / um-ma šu-ú-ma*	If (about) a cross-over (rectangle) he asks you thus:
	1 15 *ṣí-li-ip-tum* 45 a.šà /	1 15 the cross-over (diagonal), 45 the field.
3	*ši-di ù* sag.*ki ki ma-a-ṣí*	My length and (my) front, how much?
	at-ta i-na e-pé-ši-ka /	You, in your doing:
4	1 15 *ṣí-li-ip-ta-ka me-ḫe-er-šu i-di-i-ma* /	1 15, your cross-over, its copy lay down.
	šu-ta-ki-il-šu-nu-ti-ma	make them eat each other, then
5	1 33 45 *i-li* / 1 33 45 šu ku.u.zu? /	1 33 45 comes up. 1 33 45 the hand x x.
6	45 a.šà-*ka a-na ši-na e-bi-il-ma* 1 30 *i-li* /	45, your field, bring to two, then 1 30 comes up.
7	*i-na* 1 33 45 ḫu-ru-uṣ-*ma*	From 1 33 45 cut (it) out, then
	{1 3}3 45 *ša-pí-il-tum* /	3 45' the remainder.
8	ib.sí 3 45 *le-qe-e-ma* 15 *i-li*	The equalside of 3 45 take, then 15 comes up.
9	*mu-ta-šu* / 7 30 *i-li*	Its half-part(?) 7 30 comes up.
	a-na 7 30 *i-ši-i-ma* 56 15 *i-li* /	To 7 30 raise (it), then 56 15 comes up.
10	56 15 šu-*ka*	56 15 your hand.
11	45 a.šà-*ka e-li* šu-*ka* / 45 56 15 *i-li*	45, your field, above your hand, 45 56 15 comes up.
12	ib.si 45 56 15 *le-qe-ma* / 52 30 *i-li*	The equalside of 45 56 15 take, then 52 30 comes up.
13	52 30 *me-ḫe-er-šu i-di-i-ma* /	52 30, its copy lay down, then
14	7 30 *ša tu-uš-ta-ki-lu*	7 30 that you made eat
15-16	*a-na iš-te-en* / *ṣí-ib-ma i-na iš-te-en* / ḫu-ru-uṣ	to one add, then from one cut out:
	1 uš-*ka* 45 sag.ki	1, your length, 45, the front.

This is the same metric algebra problem as the one in MS 3971 § 2, and the solution procedure is (essentially) the same in both exercises. It is interesting to note, however, that IM 67118 is verbose and written mostly in syllabic Akkadian, while MS 3971 § 2 is concise and written mostly in terms of Sumerian logograms. Another, insignificant, difference between the two texts is that in the solution procedure in IM 67118, the area is *subtracted* from the square of the diagonal, while in MS 3971 § 2 the area is *added* to the square of the diagonal.

10.1 f.   MS 3971 § 3. Five Examples of igi-igi.bi Problems

MS 3971 § 3

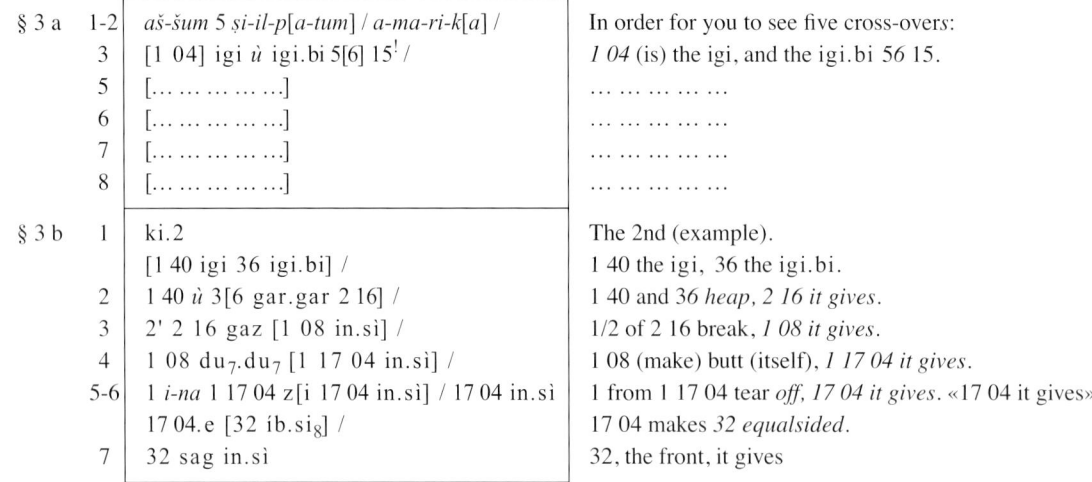

§ 3 a	1-2	*aš-šum* 5 ṣi-il-p[*a-tum*] / *a-ma-ri-k*[*a*] /	In order for you to see five cross-overs:
	3	[1 04] igi *ù* igi.bi 5[6] 15' /	1 04 (is) the igi, and the igi.bi 56 15.
	5	[... ... ...]	... ...
	6	[... ... ...]	... ...
	7	[... ... ...]	... ...
	8	[... ... ...]	... ...
§ 3 b	1	ki.2	The 2nd (example).
		[1 40 igi 36 igi.bi] /	1 40 the igi, 36 the igi.bi.
	2	1 40 *ù* 3[6 gar.gar 2 16] /	1 40 and 36 *heap, 2 16* it gives.
	3	2' 2 16 gaz [1 08 in.sì] /	1/2 of 2 16 break, *1 08* it gives.
	4	1 08 du₇.du₇ [1 17 04 in.sì] /	1 08 (make) butt (itself), *1 17 04* it gives.
	5-6	1 *i-na* 1 17 04 z[i 17 04 in.sì] / 17 04 in.sì	1 from 1 17 04 tear *off*, *17 04 it gives*. «17 04 it gives»
		17 04.e [32 íb.si₈] /	17 04 makes *32 equalsided*.
	7	32 sag in.sì	32, the front, it gives

§ 3 c	1	ki.3	The 3rd.
		1 30 igi 40 igi.bi /	1 30 the igi, 40 the igi.bi.
	2	1 30 *ù* 40 gar.gar 2 10 in.sì /	1 30 and 40 heap, 2 10 it gives.
	3	[2'] 2 10 gaz 1 05 in.sì /	*1/2 of* 2 10 break, 1 05 it gives.
	4	1 05 du₇.du₇ 1 10 25	1 05 (make) butt (itself), 1 10 25.
		<1 *i-na* 1 10 25 zi> 10 25 [in.sì] /	<1 from 1 10 25 tear off> 10 25 *it gives*.
	5	10 [25.e] 25 sag ki.3	10 *25 makes* <25 equalsided>, 25 the third front.
§ 3 d	1	ki.4	The 4th.
		1 20 igi 45 igi.bi	1 20 the igi, 45 the igi.bi.
	2	1 20 *ù* 45 / gar.gar 2 05	1 20 and 45 heap, 2 05.
		2' 2 05 gaz 1 02 30 333[in.sì] /	1/2 of 2 05 break, 1 02 30 *it gives*.
	3	1 02 30 du₇.du₇ 1 05 06 15 /	1 02 30 (make) butt (itself), 1 05 06 15.
	4	1 *a-na* uš zi 5 06 15 in.sì /	1 to' the length' tear off, 5 06 15.
	5	5 06 15.e 17 30 íb.si₈ /	5 06 15 makes 17 30 equalsided.
	6	17 30 sag *ṣi-l*[*i-i*]*p-ti* ki.4	17 30, the front of the 4th cross-over.
§ 3 e	1	ki.5	The 5th.
		1 12 igi 50 igi.bi	1 12 the igi, 50 the igi.bi.
	2	1 12 *ù* 50 [gar.gar] / 2 02	1 12 and 50 *heap*, 2 02.
		2' 2 02 gaz 1 01 /	1/2 of 2 02 break, 1 01.
	3	1 01 du₇.du₇ 1 02 01 /	1 01 (make) butt (itself), 1 02 01.
	4	1 *i-na* 1 02 01 zi 2 01 in.sì /	1 from 1 02 01 tear off, 2 01 it gives.
	5	2 01.e 11 íb.si₈ 11 sag ki.5	2 01 makes 11 equalsided. 11, the 5th front.
		5 *ṣi-il-pa-tum*	5 cross-overs.

The five exercises MS 3971 § 3 a-e differ from each other only in the initial choice of parameters, in each case a pair of reciprocals, borrowed from the Old Babylonian standard table of reciprocals (see Sec. 2.5 above). The pair is in each case referred to as igi and igi.bi 'its igi'. (Remember that every regular sexagesimal number is the reciprocal of its own reciprocal.) The five given sets of data are

§ 3 a: igi = 1 04,   igi.bi = 56 15
§ 3 b: igi = 1 40,   igi.bi = 36
§ 3 c: igi = 1 30,   igi.bi = 40
§ 3 d: igi = 1 20,   igi.bi = 45
§ 3 e: igi = 1 12   igi.bi = 50

The question may have been formulated in the initial problem, § 3 a, but if so, it is lost, as is also the solution procedure in § 3 a. However, the four other examples are relatively well preserved. Each one of them starts with the setting of the data and continues directly with the solution procedure. The missing question in each case can be reconstructed and seems to have been of the following form (in quasi-modern notations):

Let igi = u and igi.bi = s be a given pair of reciprocals,
and assume that a rectangle has the diagonal $c = (u + s)/2$
and the length $b = 1\ (00)$. Find the front a.

This is precisely the same situation as in the single exercise MS 3052 § 2. (See below, Sec. 10. 2 b.) In the case of § 3 b, for instance, the solution procedure is as follows:

$u = 1\ 40, s = 36, u + s = 2\ 16,\ \ c = (u + s)/2 = 1\ 08$ (lines 1-3)
sq. c = sq. 1 08 = 1 17 04   (line 4)
sq. c − sq. b = 1 17 04 − 1 (00 00) = 17 04, a = sqs. [sq. c − sq. b] = sqs. 17 04 = 32   (lines 5-7)

The interpretation of the solution procedure is both obvious and simple: With the diagonal c and the length b given, a is computed through an application of the Old Babylonian diagonal rule.

It is less obvious *why* the diagonal should be given as the half-sum of a regular sexagesimal number and its reciprocal in the five exercises MS 3971 § 3 a-e, as well as in the single exercise MS 3052 § 2. The interesting answer to this puzzle is given in App. 8, in connection with a renewed discussion of the famous Old Babylonian table text Plimpton 322.

10.1 g.   MS 3971 § 4. A Scaling Problem for a Rectangle with its Diagonal

MS 3971, obv.

§ 4	1-2	7 *ṣi-li-ip-tum* /	7, the cross-over (diagonal).
		uš *ù* sag en.nam *i-na* x x /	The length and the front (are) what in the x x?
	3	5 4 3 íb.si₈	5, 4, 3, the square sides.
		5 du₈ 12 in.sì /	5 release, 12 it gives.
	4	12 *a-na* 4 nim 48ᵥ in.sì /	12 to 4 lift, 48 it gives.
	5	[12 *a-na* 3] nim 36 in.sì /	*12 to 3* lift, 36 it gives.
	6	48ᵥ *a-na* 7 nim 5 36 uš /	48 to 7 lift, 5 36, the length.
	7	36 *a-na* 7 nim 4 12 sag /	36 to 7 lift, 4 12, the front.

The vaguely formulated question in this exercise can be more precisely reformulated as follows:

Find a rectangle with the diagonal 7, and with the diagonal, the length, and the front proportional to 5, 4, 3.

The solution procedure is simple enough. It begins with the following computations:

$$\text{rec. } 5 = 12, \quad 12 \cdot 4 = 48, \quad 48 \cdot 45 = 36$$

What this means is that first a "normalized" rectangle is constructed with the diagonal $d^* = 1$ and with the diagonal d^*, the length u^*, and the front s^* proportional to 5, 4, 3, so that at least one of the stated conditions is satisfied. This is achieved by setting

$$d^* = {;}12 \cdot 5 = 1, \quad u^* = {;}12 \cdot 4 = {;}48, \quad s^* = {;}45 \cdot u^* = {;}36 \qquad \text{(lines 3-5)}$$

The wanted rectangle is then obtained simply by setting

$$d = d^* \cdot 7 = 1 \cdot 7 = 7, \quad u = u^* \cdot 7 = {;}48 \cdot 7 = 5.36, \quad s = s^* \cdot 7 = {;}36 \cdot 7 = 4{;}12 \qquad \text{(lines 6-7)}$$

The two steps of the solution procedure are illustrated in Fig. 10.1.4 below:

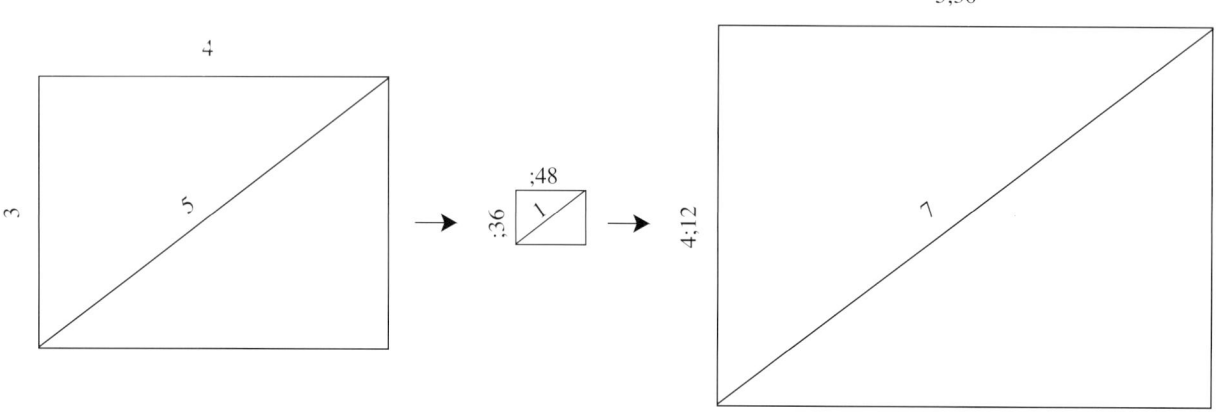

Fig. 10.1.4. Construction of a rectangle with given diagonal and given proportions.

10.2.   MS 3052. A Single-Column Mathematical Recombination Text

MS 3052 is a large single-column clay tablet with a well preserved obverse (Figs. 10.2.1-3), but an almost ruined reverse (Figs. 10.2.13-14). It is inscribed with four mathematical exercises on the obverse and four on the reverse. A subscript indicates that five of the exercises deal with partitioned mud walls, while the three remaining exercises deal with, in order, a triangle, an 'excavation', and a square. The small vocabulary of Sumerian technical terms used in the text includes the following words for mathematical operations, *etc.*:

gar.gar	heap (add)	du₇.du₇	butt (square)	gaz	break (halve)
zi	tear off (subtract)	nim	lift (multiply)	in.sì	it gives (is the result)
a-na 2 e.tab	to 2 repeat (double)	du₈	resolve (compute reciprocal)		
daḫ	join to (add to)	*n*.e *m* íb.si₈	*n* makes *m* equalsided (sqs. *n* = *m*)		
aššum … amārika	in order for you to see				

Therefore, MS 3052, just like MS 3971, appears to belong to Goetze's "Group 3" of Old Babylonian mathematical cuneiform texts. As probable member of Group 3, MS 3052 should be from Uruk, and middle Old Babylonian, dating to before Samsuiluna 11 (1795 BC). Note, however, in this connection, that no recent removals from Iraq are known to come from Uruk, a site that has remained well guarded by locals after the Kuwait war. The implication of this observation is that MS 3052 and MS 3971 were exported from Iraq and reached the antiquities market *before* the recent events in Iraq.

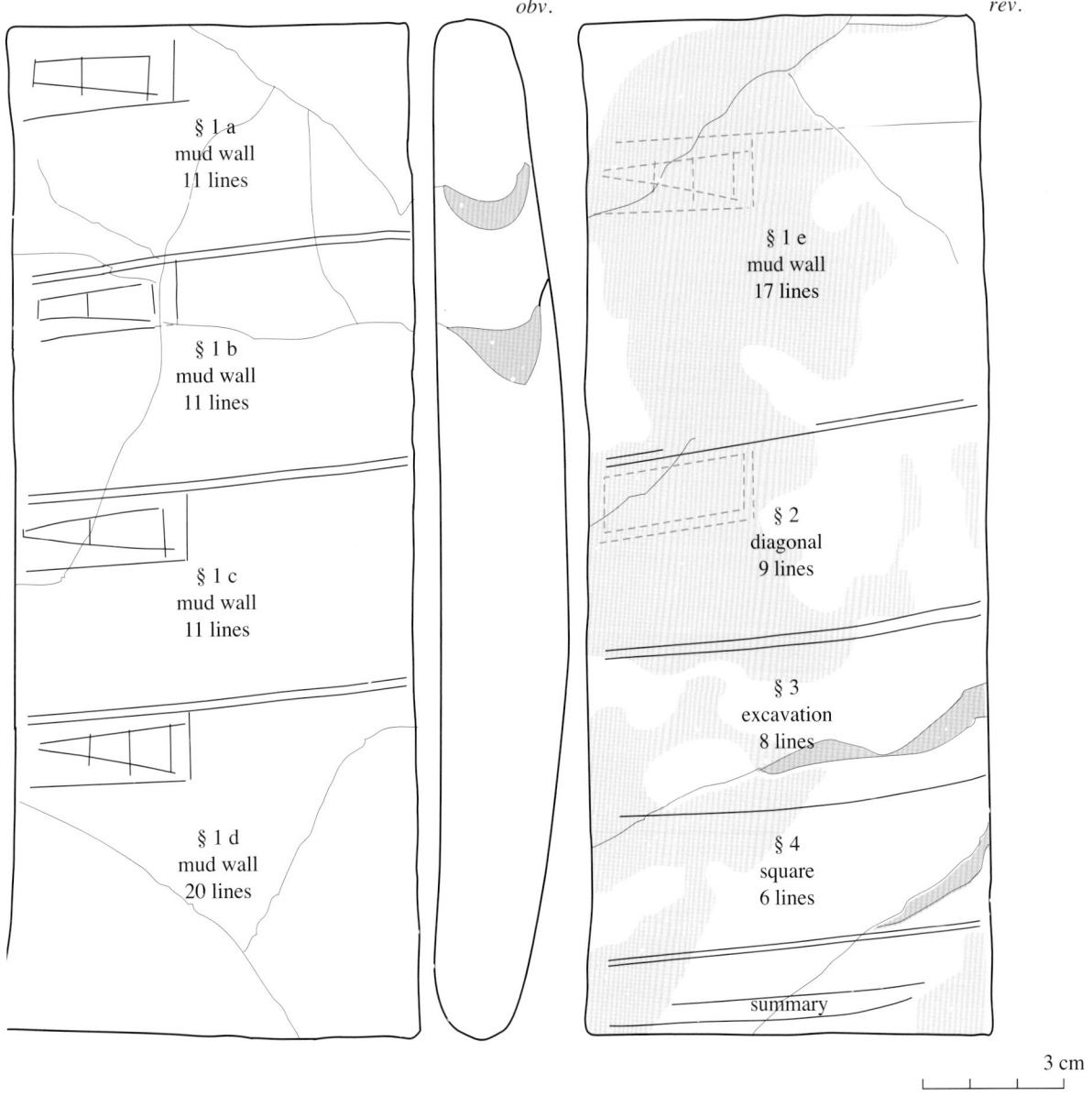

Fig. 10.2.1. MS 3052. Outline with indication of topics.

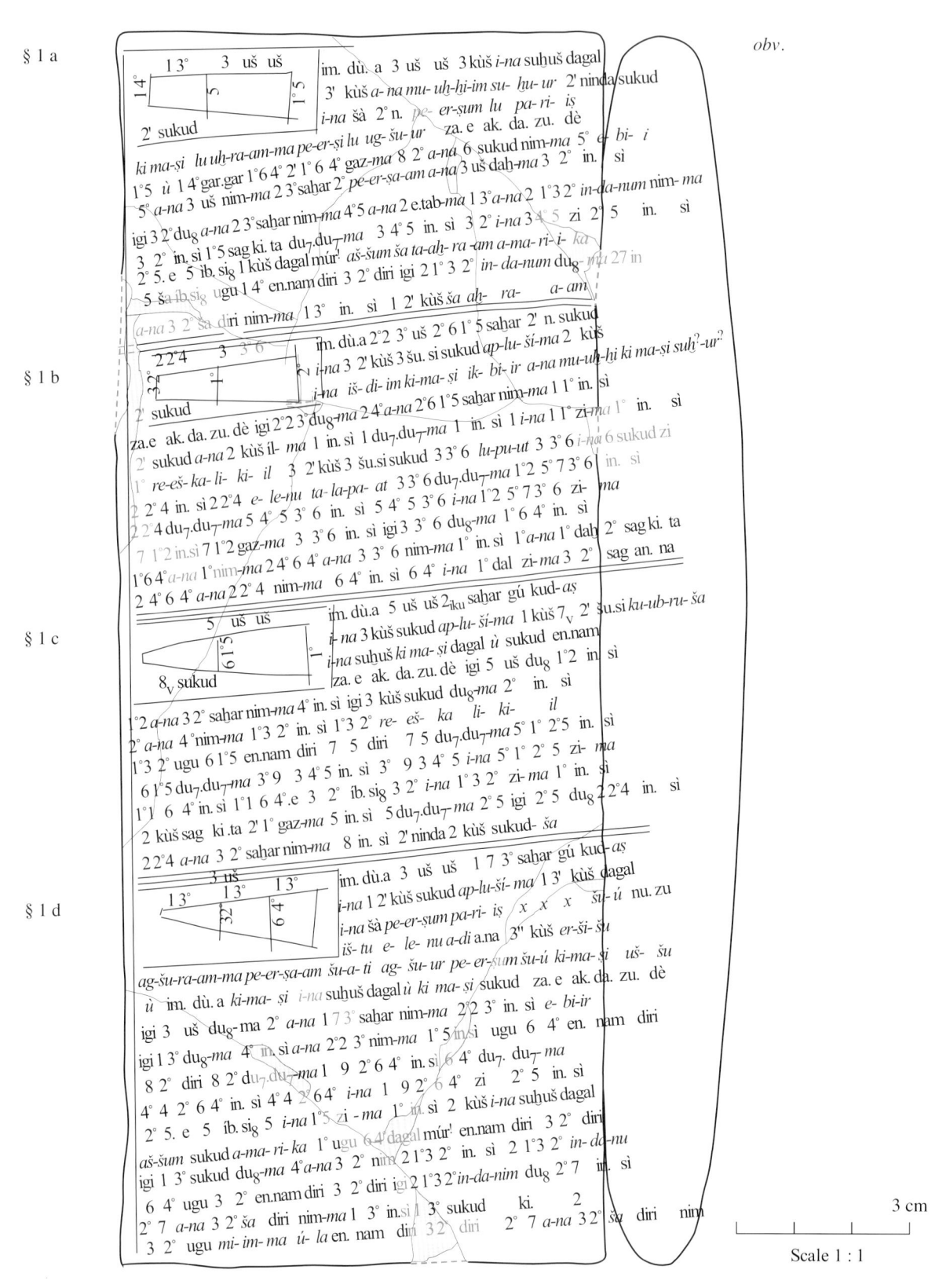

Fig. 10.2.2. MS 3052, *obv*. Conform transliteration.

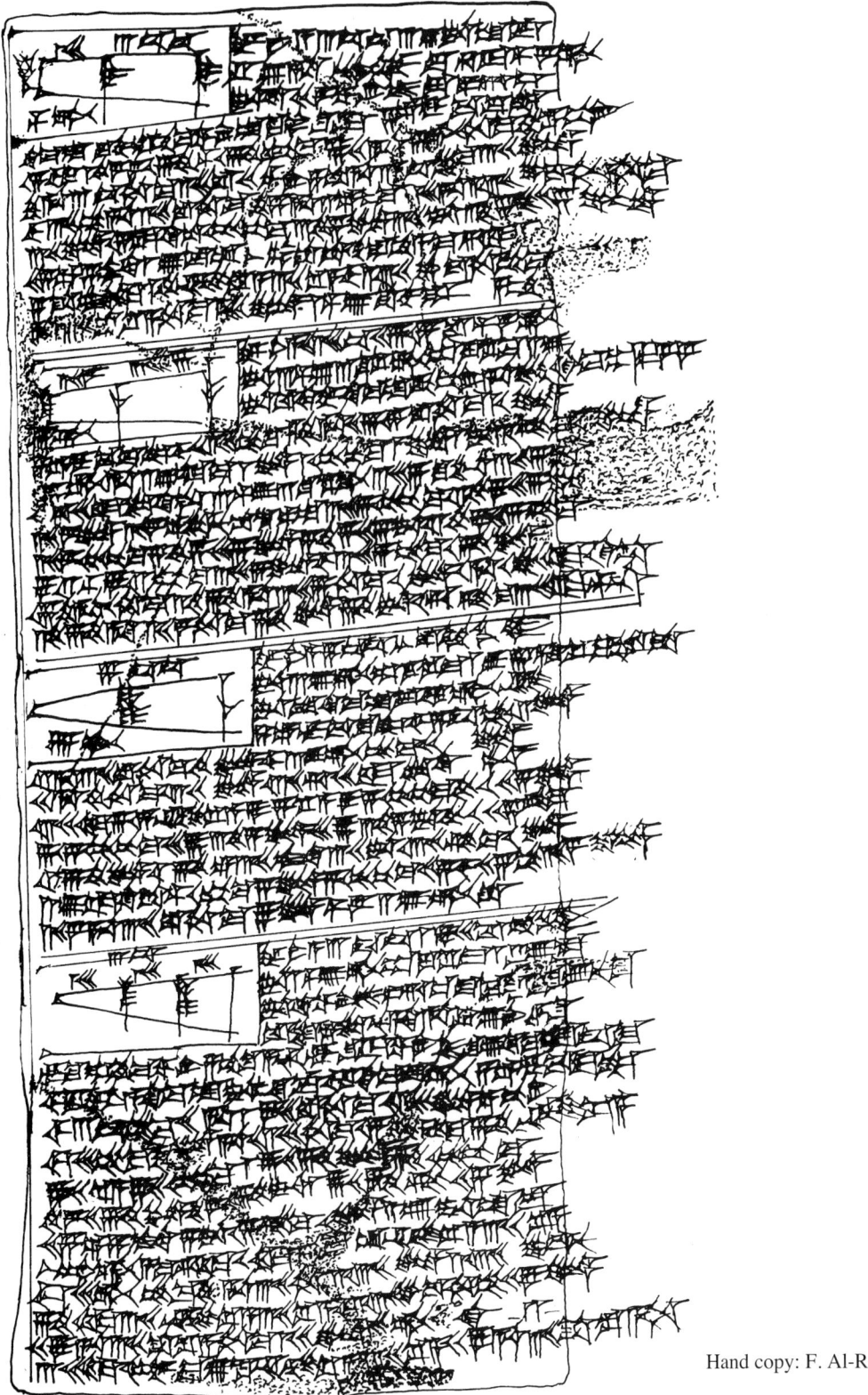

Hand copy: F. Al-Rawi

Fig. 10.2.3. MS 3052, *obv*. Hand copy of the cuneiform text.

10.2 a.   MS 3052 § 1. Mud Walls Partitioned into Two or More Separate Layers

MS 3052 § 1 a. Repairing a Breach in a Wall with Mud from the Top of the Wall

MS 3052 § 1 a

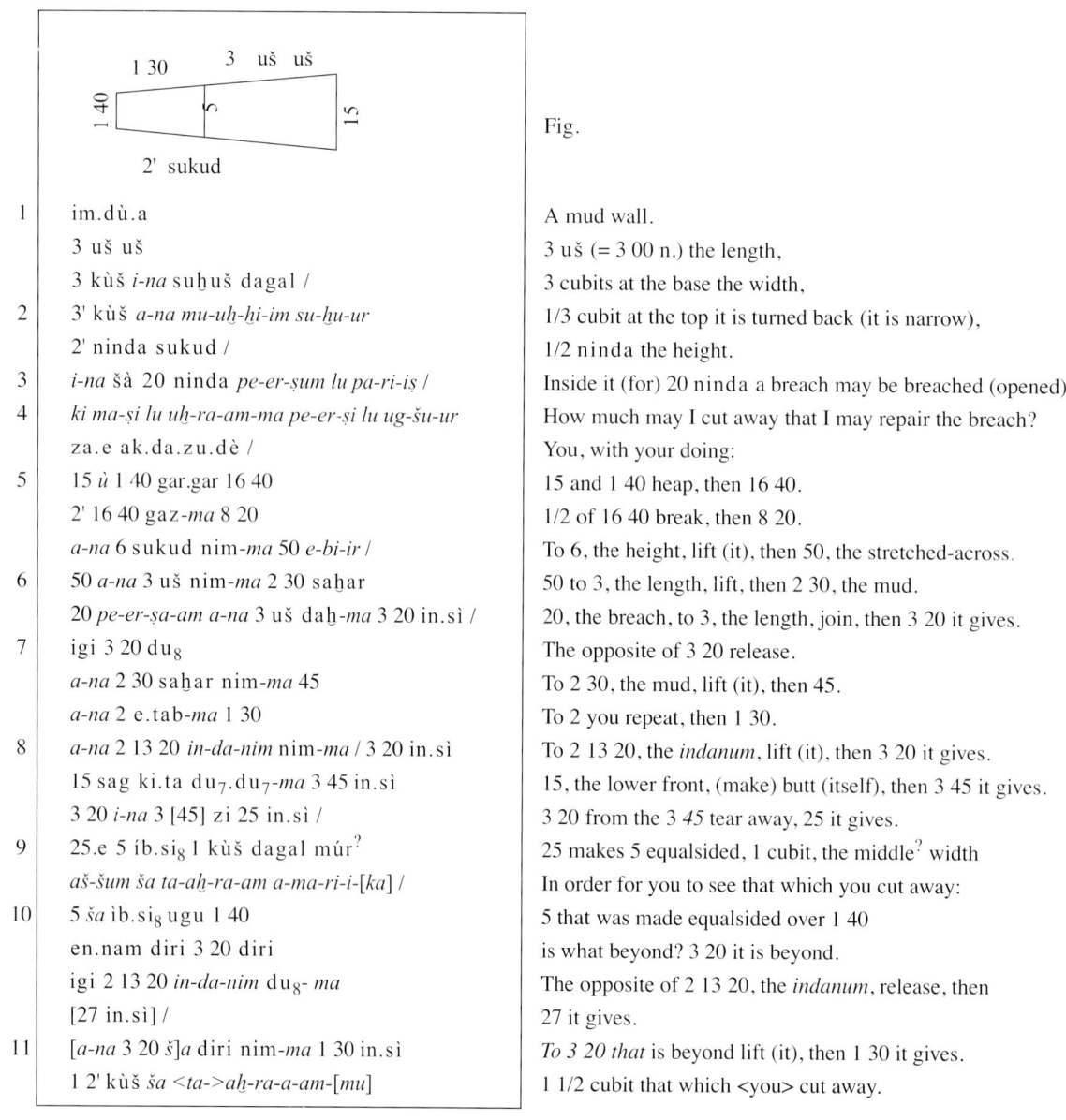

1	im.dù.a	A mud wall.
	3 uš uš	3 uš (= 3 00 n.) the length,
	3 kùš *i-na* suḫuš dagal /	3 cubits at the base the width,
2	3' kùš *a-na mu-uḫ-ḫi-im su-ḫu-ur*	1/3 cubit at the top it is turned back (it is narrow),
	2' ninda sukud /	1/2 ninda the height.
3	*i-na šà* 20 ninda *pe-er-ṣum lu pa-ri-iš* /	Inside it (for) 20 ninda a breach may be breached (opened).
4	*ki ma-ṣi lu uḫ-ra-am-ma pe-er-ṣi lu ug-šu-ur*	How much may I cut away that I may repair the breach?
	za.e ak.da.zu.dè /	You, with your doing:
5	15 *ù* 1 40 gar.gar 16 40	15 and 1 40 heap, then 16 40.
	2' 16 40 gaz-*ma* 8 20	1/2 of 16 40 break, then 8 20.
	a-na 6 sukud nim-*ma* 50 *e-bi-ir* /	To 6, the height, lift (it), then 50, the stretched-across.
6	50 *a-na* 3 uš nim-*ma* 2 30 saḫar	50 to 3, the length, lift, then 2 30, the mud.
	20 *pe-er-ṣa-am a-na* 3 uš daḫ-*ma* 3 20 in.sì /	20, the breach, to 3, the length, join, then 3 20 it gives.
7	igi 3 20 du₈	The opposite of 3 20 release.
	a-na 2 30 saḫar nim-*ma* 45	To 2 30, the mud, lift (it), then 45.
	a-na 2 e.tab-*ma* 1 30	To 2 you repeat, then 1 30.
8	*a-na* 2 13 20 *in-da-nim* nim-*ma* / 3 20 in.sì	To 2 13 20, the *indanum*, lift (it), then 3 20 it gives.
	15 sag ki.ta du₇.du₇-*ma* 3 45 in.sì	15, the lower front, (make) butt (itself), then 3 45 it gives.
	3 20 *i-na* 3 [45] zi 25 in.sì /	3 20 from the 3 45 tear away, 25 it gives.
9	25.e 5 íb.si₈ 1 kùš dagal múr?	25 makes 5 equalsided, 1 cubit, the middle? width
	aš-šum ša ta-aḫ-ra-am a-ma-ri-i-[ka] /	In order for you to see that which you cut away:
10	5 *ša* íb.si₈ ugu 1 40	5 that was made equalsided over 1 40
	en.nam diri 3 20 diri	is what beyond? 3 20 it is beyond.
	igi 2 13 20 *in-da-nim* du₈-*ma*	The opposite of 2 13 20, the *indanum*, release, then
	[27 in.sì] /	27 it gives.
11	[*a-na* 3 20 š]*a* diri nim-*ma* 1 30 in.sì	*To 3 20 that* is beyond lift (it), then 1 30 it gives.
	1 2' kùš *ša* <*ta->aḫ-ra-a-am-[mu]*	1 1/2 cubit that which <you> cut away.

The four exercises on the obverse of MS 3052 are illustrated by four drawings of what, at first sight, looks like trapezoids and triangles. However, the figures in the four drawings are broader at their right ends, while triangles and trapezoids in Old Babylonian mathematical texts normally are shown as tapering off towards the right. See, for instance, the drawing of a triangle on MS 3042 (Fig. 8.1.1), and the drawings of two trapezoids on MS 2107 and MS 3908 (Fig. 8.1.2). The explanation for the deviation from the norm is given in the first lines of the four exercises, where the objects considered are named im.dù.a 'mud wall' (Akk. *pitiqtu*). Clearly, then, what is depicted in the four drawings on the obverse of MS 3052 are pictures of the *cross sections* of four mud walls, in the usual Old Babylonian way *rotated so that the top of each mud wall is shown to the left and the base to the right*.

Another peculiarity of drawings associated with Old Babylonian mathematical exercises is that the numerical values indicated in the drawings are not always confined to the parameters given in the statement of the problem. They can include also values computed in the course of the solution procedure. Thus in MS 3052 § 1 a, the *given* values are the following:

uš	the length	3 00 ninda	(c. 1 km)
ina suḫuš dagal	the width at the base	;15 ninda = 3 cubits	(1.5 m)
ana muḫḫim suḫḫur	the 'turned back' at the top[1]	;01 40 ninda = 1/3 cubit	(16 cm)
sukud	the height	1/2 ninda = 6 cubits	(3 m)
perṣum	the breach	20 ninda	(120 m)

In the drawing associated with § 1 a, the four first of these given values are indicated, but also the following two *computed* values

dagal múr?	the middle(?) width	;05 ninda = 1 cubit	(line 9, left)
ša taḫramu	that which you cut away	1;30 cubits	(line 11, right)

The situation is clarified in Fig. 10.2.4 below, with a more accurate "modern" profile of the wall.

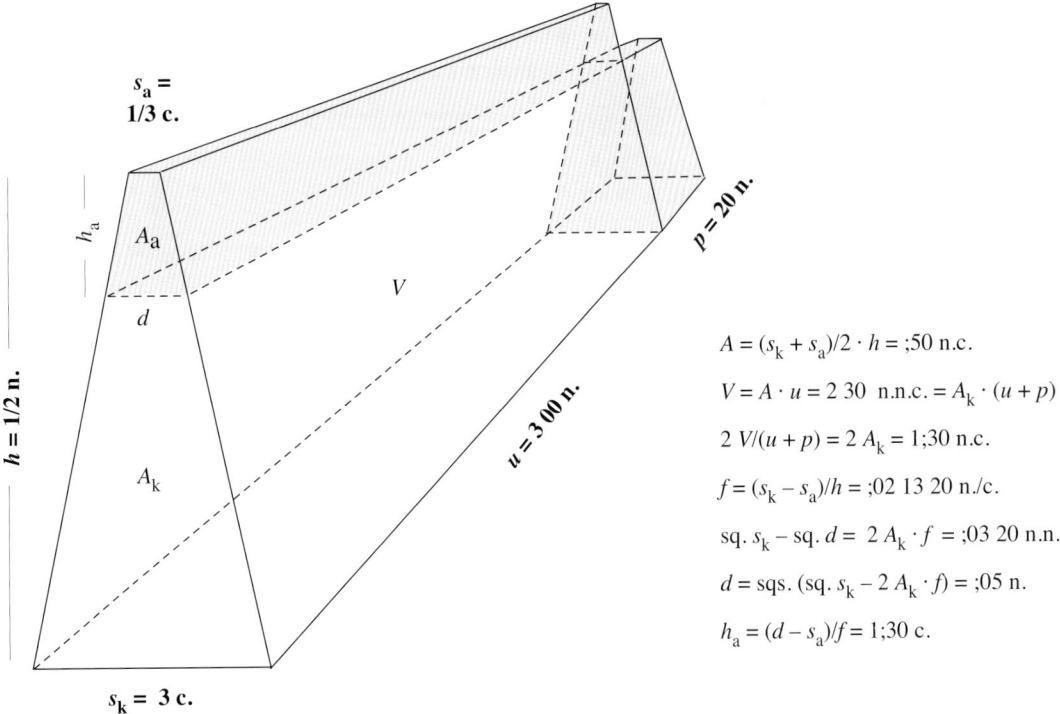

$$A = (s_k + s_a)/2 \cdot h = ;50 \text{ n.c.}$$

$$V = A \cdot u = 2\ 30 \text{ n.n.c.} = A_k \cdot (u + p)$$

$$2\ V/(u + p) = 2\ A_k = 1;30 \text{ n.c.}$$

$$f = (s_k - s_a)/h = ;02\ 13\ 20 \text{ n./c.}$$

$$\text{sq. } s_k - \text{sq. } d = 2\ A_k \cdot f = ;03\ 20 \text{ n.n.}$$

$$d = \text{sqs. } (\text{sq. } s_k - 2\ A_k \cdot f) = ;05 \text{ n.}$$

$$h_a = (d - s_a)/f = 1;30 \text{ c.}$$

Fig. 10.2.4. MS 3052 § 1 a. Repairing a breach in a wall with mud from the top of the wall.

The following standardized notations are used in Fig. 10.2.4:

u (uš)	the length	h	the height	h_a	the upper height
s_a (sag an.ta)	the upper front	s_k (sagki.ta)	the lower front	d (dagal múr)	the middle width
A_a (a.šà an.ta)	the upper area	A_k (a.šà ki.ta)	the lower area	V	the volume
p (*perṣum*)	the breach				

---

1. The tentative translation 'turned back' is based on the assumption that *su-ḫu-ur* in line 2 stands for *suḫḫur*, the stative of the G-stem of *saḫāru*. It is clear that the term here refers to the upper, narrow end of the trapezoidal cross section of the wall. A related Sumerian word is ba.an.gi₄, with the same literal translation as *suḫḫur* and apparently referring to the shorter of the parallel sides in a trapezoid. The term appears in AO 7667, an Ur III account of large numbers of bricks (Friberg, *ChV* (2001), 136-140; Robson, *MMTC* (1999), 149 ff.), and in Clay's *YOS 1* (1915), 24, an Ur III field plan where the sides of a trapezoid are given as '16 ninda uš / 16 ninda sag 14 ninda 4 kùš ba.gi₄'. (See Heimpel, *CDLJ* 2004:1.)

Note that the term sag ki.ta 'the lower front' appears in line 8 of § 1 a, although it is more appropriate for the longest parallel of a trapezoid than for the base of a wall.

In Fig. 10.2.4 is also used the notation f for what may be called the "growth rate" or "rate of increase/decrease" for the wall. Apparently this growth rate seems to have been defined as *the difference in thickness of the wall* (measured in ninda) *at two levels differing in height by 1 cubit* (this is $1/12 \cdot 2 = 1/6$ times the inverse of what we would call the *inclination* of the sides of the mud wall). The value of the growth rate in this exercise is ;02 13 20 (= 1/27) n./c. (corresponding to an inclination of 9 : 2). The term used for the growth rate, in lines 7 and 10, is *indanum* (literal meaning unknown). The term is not known from any other cuneiform mathematical text, and it is not listed in the dictionaries. A possible explanation is that it is an Akkadianization of the Sumerian word ninda(n), and that it simply stands for 'ninda (per cubit)'. One would have expected the spelling *nindanum* (listed in the dictionaries) rather than *indanum*, but compare with the following triple of entries in the Old Babylonian lexical series *Proto-Ea* (Civil, *et al.*, *MSL 13* (1979), 208-210):

208	*ni-im*	ninda
209	*gá-ar*	ninda
210	*in-da*	ninda

What this means is, presumably, that the cuneiform sign "ninda" can be read as a) níg (or nì), pronounced more or less like 'nim', b) as gar, and c) as ninda, pronounced more or less like 'inda'.

Another curious term is *ebir* 'stretched across' (stative of *ebēru* 'to cross over, stretch over, lie across'), used here, in line 5, to denote the area of the cross section of the wall.

The question in § 1 a is vaguely formulated, but the form of the solution procedure suggests that the following is an adequate, more understandable reformulation of the question:

> A mud wall with a trapezoidal cross section has the length 3 00 n., the height 1/2 n. = 3 c.,
> and the thickness at the base 3 c. = ;15 n., diminished to 1/3 c. = ;01 40 n. at the top.
> Originally, the wall was longer, but a breach of length 20 n. has been opened in the wall.
> Mud from the top of the wall is used to return the wall to its original length.
> <What is the new upper width, and> how much lower does the wall become?

The successive steps of the numerical solution procedure are the following:

$A =$		$(;15 + ;01\ 40)$ n./2 $\cdot$ 6 c. = ;50 n. c.	(line 5)
$V =$	$A \cdot u =$	;50 n. c. $\cdot$ 3 00 n. = 2 30 n. n. c.	(line 6, left)
$u + p =$		3 00 n. + 20 n. = 3 20 n.	(line 6, right)
$A_k =$	$V/(u + p) =$	2 30 n. n. c. / 3 20 n. = ;45 n. c.	(line 7, left)
$f = (s_k - s_a)/h =$	$(;15 - ;01\ 40)$ n. / 6 c. =	;13 20 n. / 6 c. = ;02 13 20 n./c.	(not mentioned)
$2\,A_k \cdot f =$		1;30 n. c. $\cdot$;02 13 20 n./c. = ;03 20 sq. n.	(lines 7 right-8 left)
sq. $s_k - 2\,A_k \cdot f =$		$(;03\ 45 - ;03\ 20)$ sq. n. = ;00 25 sq. n.	(line 8 right)
$d =$	sqs. (sq. $s_k - 2\,A_k \cdot f) =$	;05 n. = 1 c.	(line 9 left)
$h_a =$	$(d - s_a)/f =$	$(;05 - ;01\ 40)$ n. $\cdot$ 27 c./n. = 1;30 n. = 1 1/2 c.	(lines 10-11)

The solution procedure begins in a straightforward way by computing the area A of the cross section of the mud wall and the volume V of the whole mud wall. Then is computed the area of the cross section of the wall after the removal of its top and the mending of the breach. The next step of the solution procedure (cf. Fig. 10.3.4 below) makes use of the fact that, according to *the conjugate rule*,[2]

$$\text{sq. } s_k - \text{sq. } d = (s_k + d) \cdot (s_k - d) = (s_k + d) \cdot (h - h_a) \cdot f = 2\,A_k \cdot f.$$

Since the value of s_k is known, this equation can be used to compute the value of d, as in line 9. The remaining computation of the upper height h_a (called 'that which you cut away') is then easy.

MS 3052 § 1 b. Measuring the Thickness of a Wall by Drilling a Hole through It

---

2. A metric algebra demonstration of the conjugate rule can be based, for instance, on Fig. 11.2.5, right, in Sec. 11.2 d below. It is shown in that figure that sq. $s + (u + s) \cdot (u - s) = $ sq. u.

MS 3052 § 1 b

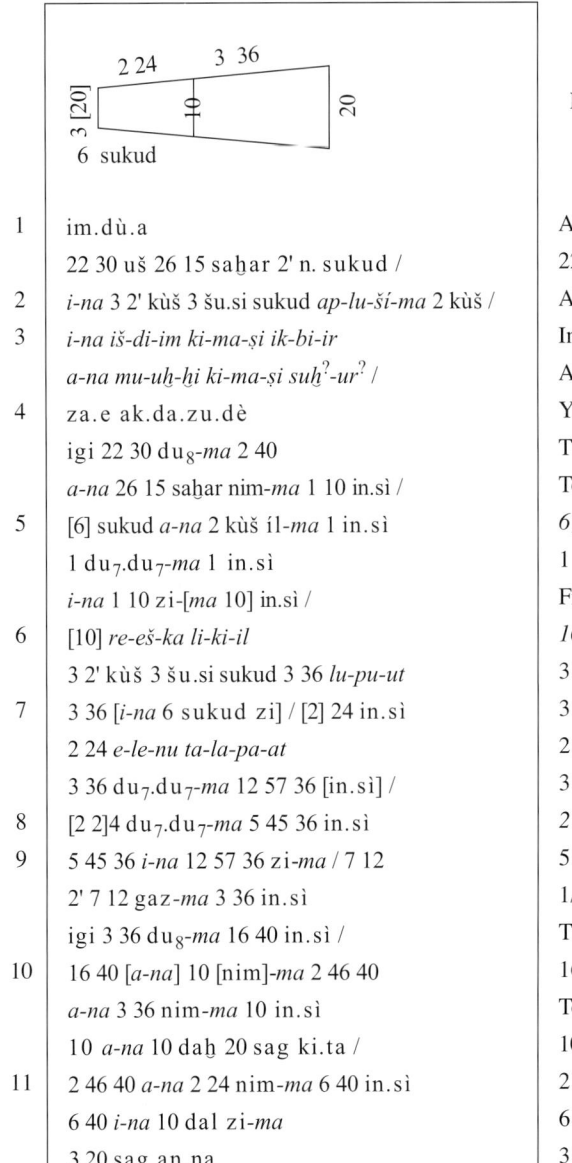

1	im.dù.a	A mud wall.
	22 30 uš 26 15 saḫar 2' n. sukud /	22 30, the length, 26 15 the mud, 1/2 ninda the height.
2	*i-na* 3 2' kùš 3 šu.si sukud *ap-lu-ší-ma* 2 kùš /	At 3 1/2 cubits 3 fingers height I drilled through, 2 cubits.
3	*i-na iš-di-im ki-ma-ṣi ik-bi-ir*	In the base, how much is it thick?
	a-na mu-uḫ-ḫi ki-ma-ṣi suḫ?-ur? /	At the top, how much is it turned back?.
4	za.e ak.da.zu.dè	You, with your doing:
	igi 22 30 du₈-*ma* 2 40	The opposite of 22 30 release, then 2 40.
	a-na 26 15 saḫar nim-*ma* 1 10 in.sì /	To 26 15, the mud, lift, then 1 10 it gives.
5	[6] sukud *a-na* 2 kùš íl-*ma* 1 in.sì	6, the height, to 2 cubits raise, then 1 it gives.
	1 du₇.du₇-*ma* 1 in.sì	1 (make) butt (itself), then 1 it gives.
	i-na 1 10 zi-[*ma* 10] in.sì /	From 1 10 tear off, *then 10* it gives.
6	[10] *re-eš-ka li-ki-il*	*10* may hold your head.
	3 2' kùš 3 šu.si sukud 3 36 *lu-pu-ut*	3 1/2 cubits 3 fingers is the height, 3 36 inscribe.
7	3 36 [*i-na* 6 sukud zi] / [2] 24 in.sì	3 36 *from 6, the height, tear off,* 2 24 it gives.
	2 24 *e-le-nu ta-la-pa-at*	2 24 above you inscribe.
	3 36 du₇.du₇-*ma* 12 57 36 [in.sì] /	3 36 (make) butt (itself), then 12 57 36 *it gives.*
8	[2 2]4 du₇.du₇-*ma* 5 45 36 in.sì	2 24 (make) butt (itself), then 5 45 36 it gives.
9	5 45 36 *i-na* 12 57 36 zi-*ma* / 7 12	5 45 36 from the 12 57 36 tear off, then 7 12 it gives.
	2' 7 12 gaz-*ma* 3 36 in.sì	1/2 of 7 12 break, then 3 36 it gives.
	igi 3 36 du₈-*ma* 16 40 in.sì /	The opposite of 3 36 resolve, then 16 40 it gives.
10	16 40 [*a-na*] 10 [nim]-*ma* 2 46 40	16 40 *to* 10 *lift,* then 2 46 40.
	a-na 3 36 nim-*ma* 10 in.sì	To 3 36 lift, then 10 it gives.
	10 *a-na* 10 daḫ 20 sag ki.ta /	10 to 10 join, 20, the lower front.
11	2 46 40 *a-na* 2 24 nim-*ma* 6 40 in.sì	2 46 40 to 2 24 lift, then 6 40 it gives.
	6 40 *i-na* 10 dal zi-*ma*	6 40 from 10, the transversal, tear off, then
	3 20 sag an.na	3 20 the upper front.

In MS 3052 § 1 b, the *given* values are the following:

uš	u	the length	22;30 ninda	(135 m)	(line 1)
saḫar	V	the volume	26;15 n. n. c.	(470 m³)	(line 1)
sukud	h	the height	1/2 ninda = 6 cubits	(3 m)	(line 1)
(sukud ki.ta)	h_k	the lower height	3 1/2 c. 3 f. = 3;36 c.	(1.8 m)	(line 2)
	d	the drilled hole	;10 ninda = 2 cubits	(1 m)	(line 2)

In the drawing illustrating § 1 b, the last three of these five given values are indicated, and also the following three *computed* values

(sukud an.na)	h_a	the upper height	2 1/3 c. 2 f. = 2;24 c.	(1.2 m)	(lines 7-8)
sag ki.ta	s_k	the lower front	;20 n. = 4 c.	(2 m)	(line 10)
sag an.na	s_a	the upper front	;03 20 n. = 2/3 c.	(33 cm)	(line 11)

The situation is clarified in Fig. 10.2.5 below.

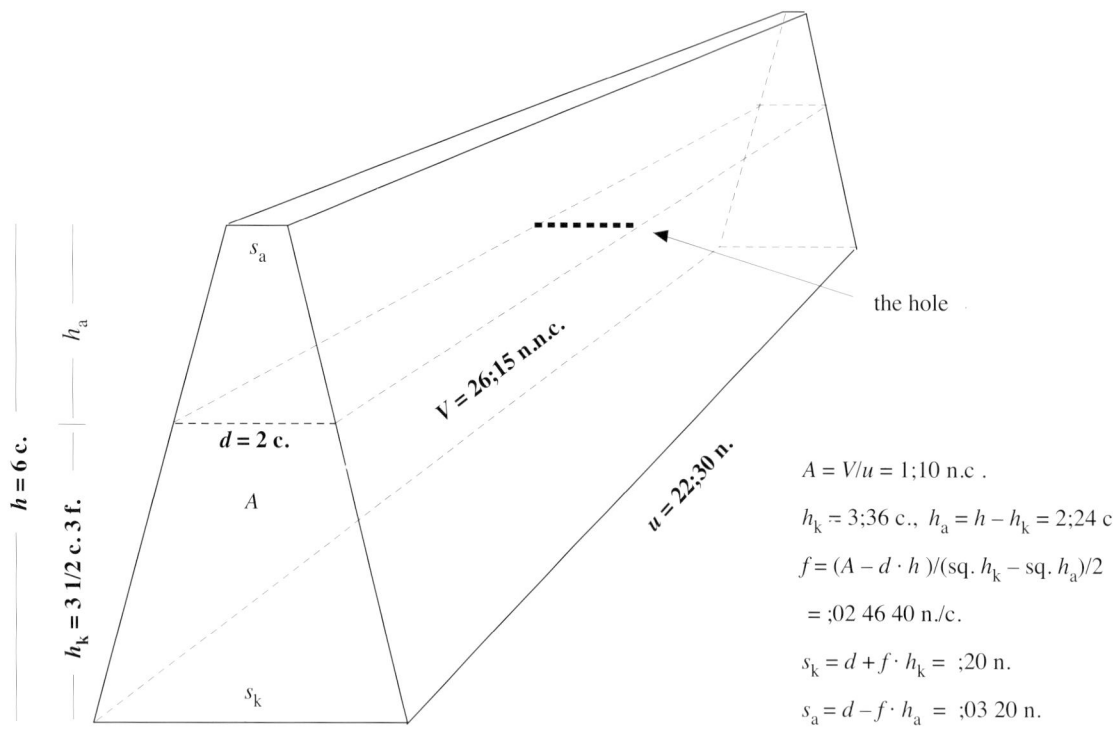

Fig. 10.2.5. MS 3052 § 1 b. A wall with a trapezoidal cross section, and with a hole drilled through it.

Note: It is interesting that in the question (lines 1-3) the transversal d is given as 2 cubits and the lower height as 3 1/2 cubits 3 fingers, while the values indicated in the drawing are 10 (meaning ;10 ninda) and 3 36 (meaning 3;36 cubits). (However, in line 6 it is said directly: 'inscribe 3 36', and in lines 10-11 the length of the drilled hole is referred to simply as '10'.) Also, in the question it is asked how thick the wall is 'in the base' and 'at the top', while in the solution procedure the answer is given in the form '20, the lower front' and '3 20, the upper front', as if the object considered were a trapezoid, rather than the cross section of a mud wall.

In line 4 of § 1 b is computed the area of the cross section of the mud wall:

$$A = V/u = 26;15 \text{ n. n. c. } / \text{ } 22;30 \text{ n. } = 1;10 \text{ n. c. } \hspace{3cm} \text{(line 4)}$$

The next step of the computation is totally unexpected and not so easy to understand. It is the computation in line 5 of an unnamed entity with the value

$$A - 1 \cdot d \cdot h = (1;10 - 1 \cdot 6 \cdot ;10) \text{ n. c. } = ;10 \text{ n. c. } \hspace{3cm} \text{(line 5)}$$

The student is asked to remember this value, until it will be needed again. Then follows, in lines 6-7, the conversion of the given value h_k = 3 1/2 cubits 3 fingers to the sexagesimal number 3;36, and the computation of the upper height as 6 – 3;36 = 2;24. The student is first asked to 'inscribe' the value 3 36, probably either in his own drawing of the cross section of the mud wall or on a hand tablet. Then he is asked to inscribe the value 2 24 'above' (namely, above 3 36).

The remembered value $A - 1 \cdot d \cdot h = ;10$ n. c. is used in the computation of

$$(A - 1 \cdot d \cdot h) / ((\text{sq. } h_k - \text{sq. } h_a)/2) = ;10 \text{ n. c. } / \{(12;57 \text{ } 36 - 5;45 \text{ } 36)/2\} \text{ sq. c. } = ;10 \text{ n. } / \text{ } 3;36 \text{ c. } = ;02 \text{ } 46 \text{ } 40 \text{ n./c.}$$

A partial explanation is provided by the final computations, where it becomes clear that ;02 46 40 (= ;25/9) n./c. is the *growth rate f* for the mud wall (corresponding to an inclination of 18 : 5). As a matter of fact, the upper and lower fronts of the trapezoid are computed as

$$s_k = d + f \cdot h_k = ;10 \text{ n. } + ;02 \text{ } 46 \text{ } 40 \text{ n./c. } \cdot 3;36 \text{ c. } = ;10 \text{ n. } + ;10 \text{ n. } = ;20 \text{ n. } \hspace{1cm} \text{(line 10)}$$
$$s_a = d - f \cdot h_a = ;10 \text{ n. } - ;02 \text{ } 46 \text{ } 40 \text{ n./c. } \cdot 2;24 \text{ c. } = ;10 \text{ n. } - ;06 \text{ } 40 \text{ n. } = ;03 \text{ } 20 \text{ n. } \hspace{1cm} \text{(line 11)}$$

It remains to explain the following curious equation for the growth rate:

$$f = (A - 1 \cdot d \cdot h) / \{(\text{sq. } h_k - \text{sq. } h_a)/2\} = {;}02\ 46\ 40 \text{ n./c.} \qquad \text{(lines 5-10)}$$

Apparently, the idea was to compare the area of a trapezoid divided by a transversal d with the area of the rectangle with base d and with the same height h as the trapezoid. It is clear from Fig. 10.2.6 below that the difference between the two areas is equal to the difference between the areas of two triangles, one with the base $s_k - d$ and the height h_k, the other with the base $d - s_a$ and the height h_a. However, *since the two triangles are similar, the same growth rate f applies to both. This common growth rate is also the growth rate for the trapezoid.* Therefore, $s_k - d = f \cdot h_k$, $d - s_a = f \cdot h_a$, and

$$A - d \cdot h = h_k \cdot (s_k - d)/2 - h_a \cdot (d - s_a)/2 = f \cdot (\text{sq. } h_k - \text{sq. } h_a)/2.$$

Consequently, as in lines 5-10 of the solution procedure in MS 3052 § 1 b,[3]

$$f = (A - d \cdot h) / (\text{sq. } h_k - \text{sq. } h_a)/2.$$

The mentioned equation for $A - d \cdot h$ may be called "the first Old Babylonian transversal rule".

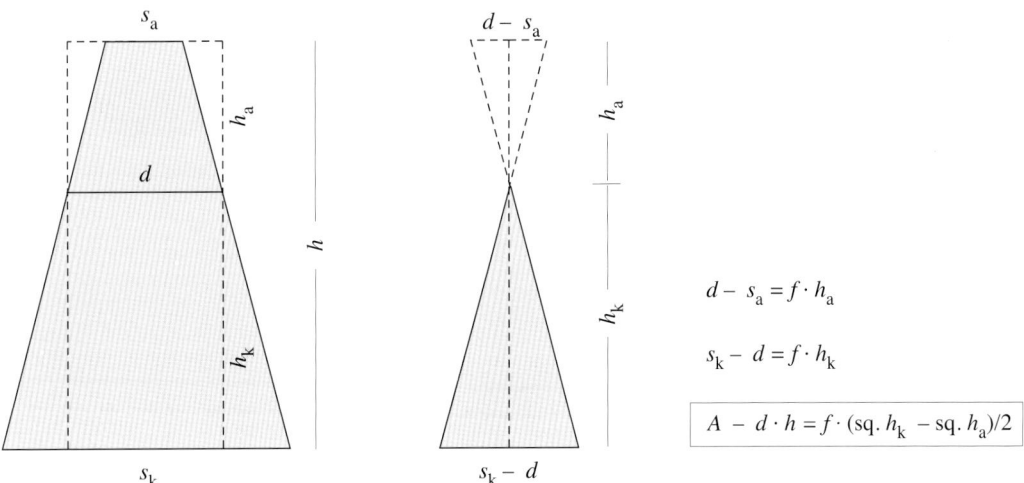

Fig. 10.2.6. MS 3052 § 1 b. The first Old Babylonian transversal rule.

MS 3052 § 1 c. Another Example of a Wall with a Hole Drilled through It

MS 3052 § 1 c

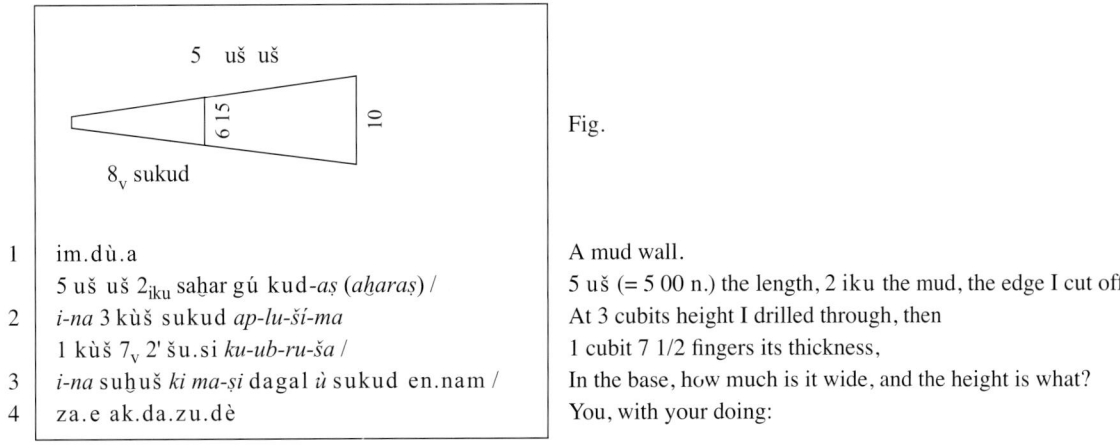

1	im.dù.a	A mud wall.
	5 uš uš 2$_{iku}$ saḫar gú kud-*aṣ* (*aḫaraṣ*) /	5 uš (= 5 00 n.) the length, 2 iku the mud, the edge I cut off
2	*i-na* 3 kùš sukud *ap-lu-ši-ma*	At 3 cubits height I drilled through, then
	1 kùš 7$_v$ 2' šu.si *ku-ub-ru-ša* /	1 cubit 7 1/2 fingers its thickness,
3	*i-na* suḫuš *ki ma-ṣi* dagal *ù* sukud en.nam /	In the base, how much is it wide, and the height is what?
4	za.e ak.da.zu.dè	You, with your doing:

---

3. Note that the squaring of 1 in line 5 is a totally unwarranted step in the solution procedure, but it does not change the result.

	igi 5 uš du$_8$ 12 in.sì /	The opposite of 5, the length, release, 12 it gives.
5	12 *a-na* 3 20 saḫar nim-*ma* 40 in.sì	12 to 3 20, the mud, lift, then 40 it gives.
	igi 3 kùš sukud du$_8$-*ma* 20 in.sì /	The opposite of 3 cubits, the height, resolve, then 20 it gives.
6	20 *a-na* 40 nim-*ma* 13 20 in.sì	20 to 40 lift, then 13 20 it gives.
	13 20 *re-eš-ka li-ki-il* / 13 20 ugu 6 15 en.nam diri	13 20 may hold your head. 13 20 over 6 15
7	7 05 diri	is what beyond? 7 05 beyond.
	7 05 du$_7$.du$_7$-*ma* 50 10 25 in.sì /	7 05 (make) butt (itself), then 50 10 25 it gives.
8	6 15 du$_7$.du$_7$-*ma* 39 03 45 in.sì	6 15 (make) butt (itself), then 39 03 45 it gives.
	39 03 45 *i-na* 50 10 25 zi-*ma* /	39 03 45 from 50 10 25 tear off, then
9	11 06 40 in.sì 11 06 40.e 3 20 íb.si$_8$	11 06 40 it gives. 11 06 40 makes 3 20 equalsided.
	3 20 *i-na* 13 20 zi-*ma* 10 in.sì /	3 20 from 13 20 tear off, then 10 it gives,
10	2 kùš sag ki.ta	2 cubits, the lower front.
	2' 10 gaz-*ma* 5 in.sì	1/2 of 10 break, then 5 it gives.
	5 du$_7$.du$_7$-*ma* 25	5 (make) butt (itself), then 25.
	igi 25 du$_8$ 2 24 in.sì /	The opposite of 25 resolve, 2 24 it gives.
11	2 24 *a-na* 3 20 saḫar nim-*ma* 8 in.si	2 24 to 3 20, the mud, lift, then 8 it gives.
	2' ninda 2 kùš sukud-*ša*	1/2 ninda 2 cubits is its height.

In this exercise (see Fig. 10.2.7 below), the *given* values are the following:

uš	u	the length	5 00 ninda	(1.8 km)
saḫar	V	the volume	2 iku = 3 20 n. n. c.	(3,600 m^3)
(sukud ki.ta)	h_k	the lower height	3 cubits	(1.5 m)
	d	the drilled hole	1 c. 7 1/2 f. = ;06 15 n.	(37.5 cm).

In the drawing illustrating § 1 c, the first and the last of these four given values are indicated, and also two *computed* values, the (width at the) base (in the solution procedure called 'the lower front'), and the height. These computed values are the following:

sukud	h	the height	1/2 n. 2 c. = 8 cubits	(4 m)
sag ki.ta	s_k	the lower front	2 c. = ;10 n.	(2 m)

The solution proceeds in a series of deceptively simple steps:

1. $A/h_k - d =$	;13 20 n. – ;06 15 n. = ;07 05 n.	(lines 4 - 7)
2. sq. $(A/h_k - d) -$ sq. $d =$	;00 50 10 25 sq. n. – ;00 39 03 45 sq. n. = ;00 11 06 40 sq. n.	(lines 7 - 9)
3. $A/h_k -$ sqs. {sq. $(A/h_k - d) -$ sq. d} =	;13 20 n. – ;03 20 n. = ;10 n. = 2 c. = s	(lines 9 - 10)
4. $V/(u \cdot s/2)$	= 3 20 n. n. c. / (5 00 · ;05) sq. n. = 8 c. = 1/2 n. 2 c. = h	(lines 10 - 11)

One peculiarity of the text is due to the fact that in the Old Babylonian relative place value notation there is no discernible difference in line 10 between u (the length) = 5 00 n. and $s/2$ (half the base) = ;05 n. Both are written simply as '5'. For this reason, the author of the text allowed himself the minor mistake of writing 'square 5', when what he meant was 'multiply 5, the length, with 5, half the base'.

The object considered in MS 3052 § 1 c is a mud wall with a *triangular* cross section. This circumstance is probably what is alluded to by the obscure phrase 'the edge I cut off' in line 1.

The first step of the solution procedure (lines 4-5) is easy to understand. It is the computation of the area of the triangular cross section of the mud wall:

$$A = V/u = 3\ 20 \text{ n. n. c. } / 5\ 00 \text{ n. } = ;40 \text{ n. c.}$$

It remains to compute the *base s* and the *height h* of the cross section of the wall, with departure from the known values for the *area A* of the triangle, the *transversal d*, and the *lower height* h_k. The solution method is based on what may be called "the second Old Babylonian transversal rule", namely that

$$d \cdot h + h_k \cdot s = 2\,A.$$

A geometric proof of the rule is easily obtained through inspection of Fig. 10.2.8 below.

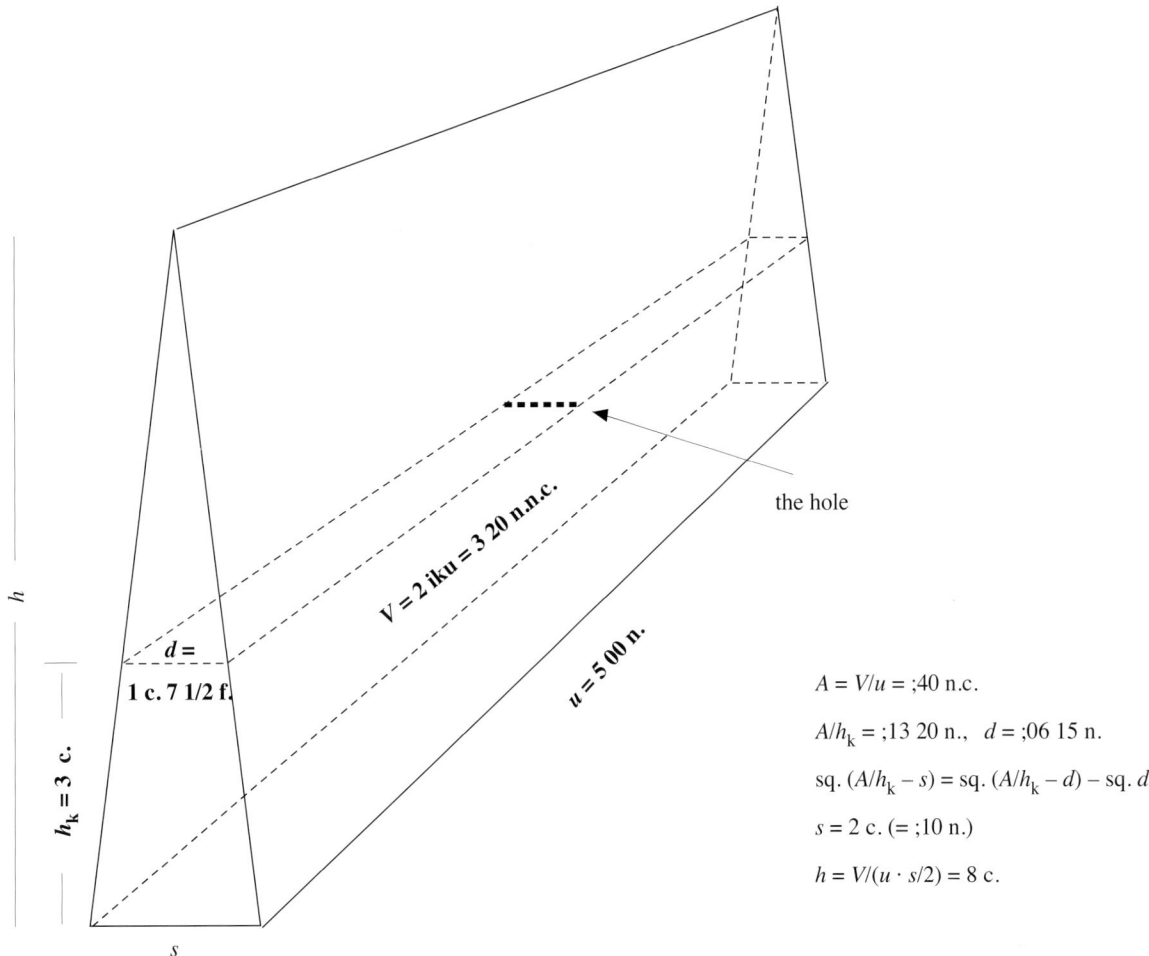

h

$d =$

1 c. 7 1/2 f.

$h_k = 3$ c.

s

$V = 2$ iku $= 3\ 20$ n.n.c.

$u = 5\ 00$ n.

the hole

$A = V/u = {;}40$ n.c.

$A/h_k = {;}13\ 20$ n., $\quad d = {;}06\ 15$ n.

sq. $(A/h_k - s) =$ sq. $(A/h_k - d) -$ sq. d

$s = 2$ c. $(= {;}10$ n.)

$h = V/(u \cdot s/2) = 8$ c.

Fig. 10.2.7. MS 3052 § 1 c. A wall with a triangular cross section, and with a hole drilled through it.

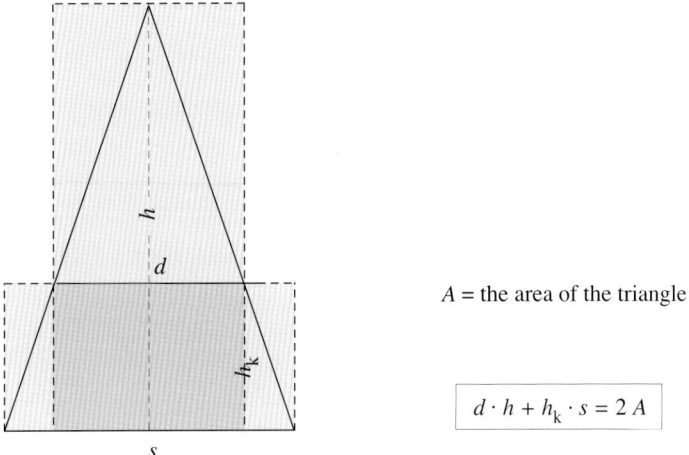

h

d

h_k

s

$A =$ the area of the triangle

$$d \cdot h + h_k \cdot s = 2\,A$$

Fig. 10.2.8. MS 3052 § 1 c. The second Old Babylonian transversal rule.

Together with the area equation the second transversal rule leads to the following *rectangular-linear system of equations* for the height h and the base s:

$$h \cdot s = 2\,A, \quad d \cdot h + h_k \cdot s = 2\,A.$$

This system of equations can easily be reduced to one of the Old Babylonian standard forms for such systems, for instance as follows: Change variables by setting $u = d \cdot h/h_k$. Then the mentioned system of equations can be replaced by the following equivalent system

$$u \cdot s = 2\,d \cdot A/h_k, \quad u \mathrel{\text{\small I}} s = 2\,A/h_k.$$

This rectangular-linear system of equations, in its turn, can be replaced by a *quadratic equation* for the single variable s. This is done most easily in the appropriate geometric model:

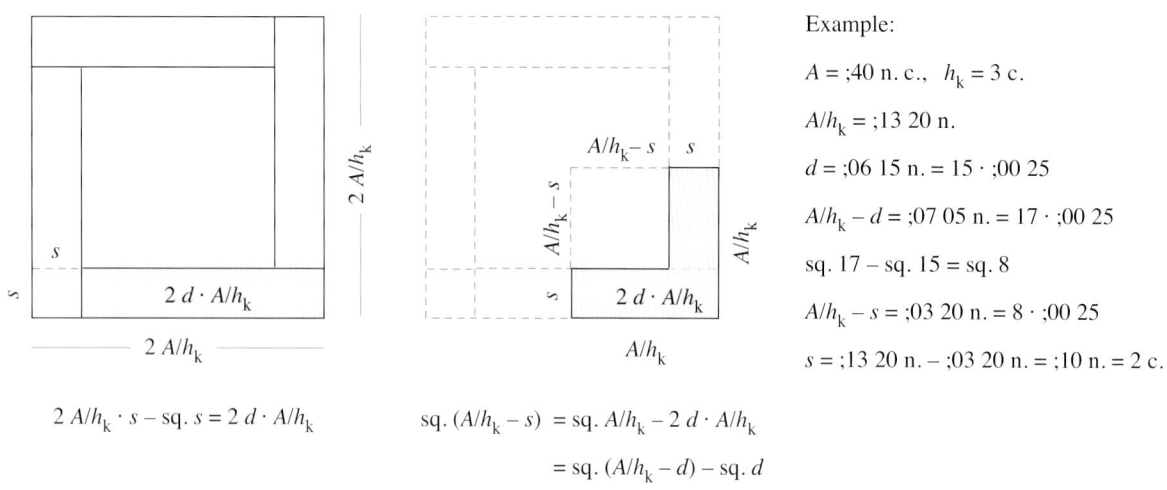

Example:

$A = {;}40$ n. c., $h_k = 3$ c.

$A/h_k = {;}13\ 20$ n.

$d = {;}06\ 15$ n. $= 15 \cdot {;}00\ 25$

$A/h_k - d = {;}07\ 05$ n. $= 17 \cdot {;}00\ 25$

sq. 17 – sq. 15 = sq. 8

$A/h_k - s = {;}03\ 20$ n. $= 8 \cdot {;}00\ 25$

$s = {;}13\ 20$ n. – ${;}03\ 20$ n. $= {;}10$ n. $= 2$ c.

$$2\,A/h_k \cdot s - \text{sq.}\,s = 2\,d \cdot A/h_k$$

$$\text{sq.}\,(A/h_k - s) = \text{sq.}\,A/h_k - 2\,d \cdot A/h_k$$
$$= \text{sq.}\,(A/h_k - d) - \text{sq.}\,d$$

Fig. 10.2.9. MS 3052 § 1 c. The quadratic equation for the base s, and its surprising solution.

The geometric model shows how the mentioned rectangular-linear system of equations for u and s can be transformed into a quadratic equation for s alone (Fig. 10.2.9, left):

$$2\,A/h_k \cdot s - \text{sq.}\,s = 2 \cdot d \cdot A/h_k.$$

The model also shows (Fig. 10.2.9, right) how this quadratic equation for s can be reduced to

$$\text{sq.}\,(A/h_k - s) = \text{sq.}\,A/h_k - 2 \cdot d \cdot A/h_k.$$

So far, the solution procedure follows the standard routine, known from a number of earlier published Old Babylonian mathematical texts. According to that routine, the expected next step in the solution procedure would be to compute the value of s as follows (in quasi-modern notations)

$$s = A/h_k - \text{sqs.}\,(\text{sq.}\,A/h_k - 2 \cdot d \cdot A/h_k).$$

Surprisingly, the solution procedure in MS 3052 § 1 c works differently. Apparently, the Old Babylonian author of the exercise had realized that he could refine the standard solution by going one step further. He saw that the expression of which the square side was computed could be expressed as the difference between two squares. Thus, the expression for s in quasi-modern notations which precisely corresponds to steps 1-4 of the solution procedure in MS 3052 § 1 c is the following:

$$s = A/h_k - \text{sqs.}\,(\text{sq.}\,(A/h_k - d) - \text{sq.}\,d).$$

This form of the solution is interesting because it shows that the triple

$$c = A/h_k - s, \quad b = d, \quad a = A/h_k - d,$$

with A, h_k, d, and s defined as in Fig. 10.2.8, is *always a solution to the diagonal equation*

$$\text{sq.}\,c = \text{sq.}\,a + \text{sq.}\,b.$$

In § 1 c, where A = ;40 n. c., h_k = 3 c., d = ;06 15 n., and s = ;10 n., the triple is

$$A/h_k - d = ;07\ 05\ \text{n.}, \quad d = ;06\ 15\ \text{n.}, \quad A/h_k - s = ;03\ 20\ \text{n.}$$

It is easy to check that

$$;07\ 05\ \text{n.} = 17 \cdot ;00\ 25\ \text{n.}, \quad ;06\ 15\ \text{n.} = 15 \cdot ;00\ 25\ \text{n.}, \quad ;03\ 20\ \text{n.} = 8 \cdot ;00\ 25\ \text{n.}$$

In other words, the triple figuring in § 1 c is a multiple of the triple 17, 15, 8, which can easily be shown to be a simple *solution in integers* to the diagonal equation.

It is no coincidence that the data for MS 3052 § 1 c were chosen so that they would lead to an *exact* solution to the problem, without approximate square sides. Indeed, it is likely that the four exercises in MS 3052 § 1 were excerpted from a large *theme text* with a series of exercises of gradually increasing complexity. One of these exercises may have been, for instance, to find the height and the area of the triangular cross section of a mud wall, when the width s at the base and the length d of a drilled hole at the height h_k over the ground are given. A problem of this kind can be reduced to finding the height h and the area A of a triangle with a given front s, divided by a transversal of given length d, a given the distance h_k from the front. The solution to this problem is

$$h = h_k \cdot s/(s - d), \quad A = h_k \cdot s/2 \cdot s/(s - d) \qquad \text{(see Fig. 10.2.8)}$$

If the value for A obtained in this way is used, together with the originally given values for d and h_k, as data for a problem of the type MS 3052 § 1 c, then it is guaranteed that there will exist a solution to that problem with exact (that is, rational) values for s and h.

It is possible to analyze the form of that solution from a modern point of view, as follows:

Set $s = d + e$. Then $A = h_k \cdot s/2 \cdot s/e$, so that $A/h_k = \text{sq. } s/(2\ e) = \text{sq. } (d + e)/(2\ e)$.
Consequently, $A/h_k - d, \ d, \ A/h_k - s = \{(\text{sq. } d + \text{sq. } e)/2, d \cdot e, (\text{sq. } d - \text{sq. } e)/2\}/e$.

The result can be reformulated as follows:

Let igi = $d/e = d/(s - d)$ and igi.bi = $e/d = (s - d)/d$.
Then $A/h_k - d, d, A/h_k - s = d \cdot [(\text{igi} + \text{igi.bi})/2, 1, (\text{igi} - \text{igi.bi})/2]$.

When d and s are *regular* sexagesimal numbers, this can be interpreted as an application the Old Babylonian rule for the construction of (rational) solutions to the diagonal equation. (See Fig. A8.4 in App. 8 below.) Note that d = ;06 15 = ;25/4 and s = ;10 *are* regular sexagesimal numbers in MS 3052 § 1 c, with igi = $d/(s - d)$ = 6 15 / 3 45 = 1;40.

MS 3052 § 1 d. A Wall with a Hole Drilled through it and a Breach Repaired

MS 3052 § 1 d

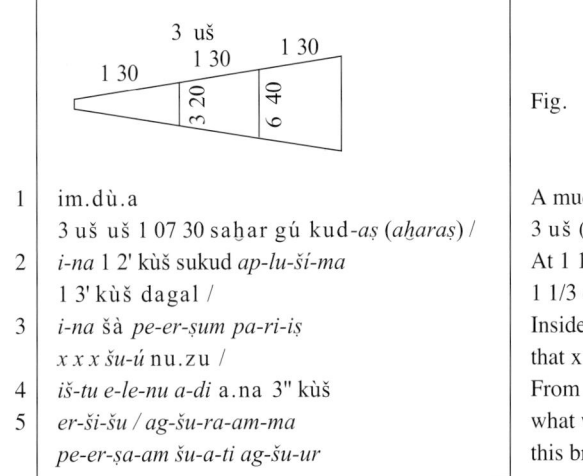

Fig.

1	im.dù.a	A mud wall.
	3 uš uš 1 07 30 saḫar gú kud-*aṣ* (*aḫaraṣ*) /	3 uš (3 00 n.) the length, 1 07 30 the mud, the edge I cut off.
2	*i-na* 1 2' kùš sukud *ap-lu-ší-ma*	At 1 1/2 cubits height I drilled through, then
	1 3' kùš dagal /	1 1/3 cubit is the width.
3	*i-na šà pe-er-ṣum pa-ri-iṣ*	Inside it a breach is breached (opened),
	x x x šu-ú nu.zu /	that x x x I do not know.
4	*iš-tu e-le-nu a-di* a.na 3" kùš	From above until as much as 2/3 cubit
5	*er-ši-šu / ag-šu-ra-am-ma*	what was asked for I repaired, then
	pe-er-ṣa-am šu-a-ti ag-šu-ur	this breach I repaired

	pe-er-ṣum šu-ú ki-ma-ṣi uš-šu /	This breach how much is its foundation,
6	*ù* im.dù.a *ki ma-ṣi i-na* suḫuš dagal	and the mud wall how much at the base is it wide,
	ù ki ma-ṣi sukud	and how much is the height?
	za.e ak.da.zu.dè /	You, with your doing:
7	igi 3 uš du₈-*ma* 20	The opposite of 3, the length, resolve, then 20.
	a-na 1 07 30 saḫar nim-*ma*	To 1 07 30, the mud, lift, then
	22 30 in.sì *e-bi-ir* /	22 30 it gives, the stretched-across.
8	igi 1 30 du₈-*ma* 4[0 in].sì	The opposite of 1 30 resolve, then 40 *it gives*.
	a-na 22 30 nim-*ma* 15 [in.s]ì	To 22 30 lift, then 15 *it gives*.
9	ugu 6 40 en.nam diri / 8 20 diri	Over 6 40 what is it beyond? 8 20 beyond.
	8 20 d[u₇.du₇]-*ma* 1 09 26 40 in.s[ì]	8 20 (make) *butt* (itself), then 1 09 26 40 it gives.
10	[6] 40 du₇.du₇-*ma* / 44 26 40 in.sì	6 40 (make) butt (itself), then 44 26 40 it gives.
	44 26 40 *i-na* 1 09 [2]6 40 zi 25 in.sì /	44 26 40 from 1 09 26 40 tear off, then 25 it gives.
11	25.e 5 íb.si₈	25 makes 5 equalsided.
	5 *i-na* 1[5 z]i-*ma* 10 [in].sì	5 from 15 *te*ar off, then 10 it gives,
	2 kùš *i-na* suḫuš dagal /	2 cubits in the base is the width.
12	*aš-šum* sukud *a-ma-ri-ka*	In order for you to see the height(s):
	10 u[gu 6 40 dagal] múr*?* en.nam diri	10 *over 6 40*, the middle*?* *width*, is what beyond?
	3 20 diri /	3 20 beyond.
13	igi 1 30 sukud du₈-*ma* 40	The opposite of 1 30, the height, resolve, then 40.
	a-na 3 20 n[im] 2 13 20 in.sì 2 13 20 *in-da-nu* /	To 3 20 lift, then 2 13 20 it gives, 2 13 20 the *ninda*.
14	6 40 ugu 3 20 en.nam diri 3 20 diri	6 40 over 3 20 what is it beyond? 3 20 beyond.
	i[gi] 2 13 20 *in-da-nim* du₈ 27 in.sì /	The opposite of 2 13 20, the *indanum*, resolve, 27 it gives.
15	27 *a-na* 3 20 *ša* diri nim-*ma* 1 30 in.[sì]	27 to 3 20 that is beyond lift, then 1 30 it *gives*.
	[1] 30 sukud ki.2 /	*1* 30, the 2nd height.
16	3 20 ugu *mi-im-ma ú-la* en.nam diri	3 20 over nothing is what beyond?
	[3 20 diri]	3 20 it is beyond.
17	27 *a-na* 3 20 *ša* diri nim / 1 30 sukud ki.3	27 to 3 20, that which is beyond, lift, 1 30, the 3rd height.
	[*aš-šum pe-er-ṣum ša pa-ri-iṣ a-ma-ri*]-*ka*	In order for you to see the breach that is breached,
	3' kùš dagal /	1/3 cubit, the width.
18	2' 3 20 gaz-*ma* [1 40]	1/2 of 3 20 *break, then 1 40.*
	[*a-na* 1 30 nim-*ma*] 2 30 in.sì /	*To 1 30 lift, then 2 30 it gives.*
19	2 30 [*a-na*] 3 [nim-*ma* 7 30 in.sì]	*2 30 to 3 lift, then 7 30 it gives.*
	[*i-na*] 1 07 [30 z]i-*ma* 1 in.sì /	*From 1 07 30 tear (it) off, then 1 it gives.*
20	1 *i-na* 1 07 30 zi-*ma* 7 30 in.sì]	*1 from 1 07 30 tear off, then 7 30 it gives.*
	[6 40 *a-na* 3] nim 20 *e-bi-ir* /	*6 40 to 3 lift, 20 the stretched-across.*
21	igi [20 du₈]-*ma* [3 in.sì]	*The opposite of 20 release*, then *3 it gives.*
	[3 *a-na* 7 30] nim-*ma* 22 30 in.sì /	*3 to 7 30 lift, then 22 30 it gives.*
22	22 30 [*pe-er-ṣum*] *ša pa-ri-iš*	22 30, *the breach* that is breached.

In MS 3052 § 1 d, the cross section of the mud wall is again triangular, as indicated by the phrase 'the edge I cut off' in line 1. The *given* values are the following:

uš	the length	3 00 ninda	(1.08 km)	(line 1)
saḫar	the volume	1 07;30 n. n. c.	(202.5 m³)	(line 1)
	the 1st height	1 1/2 c.	(75 cm)	(line 2)
	the drilled hole	1 1/3 c. = ;06 40 n.	(67 cm)	(line 2)
	the new upper width	2/3 c. = ;03 20 n.	(33 cm)	(line 4)

In the drawing associated with § 1 d, all of these five given values are recorded, except that of the volume. In addition, the last two of the following five *computed* values are indicated:

ebir	the area of the cross section	;22 30 n. c.	(1.125 m²)	(line 7)
ina suḫuš dagal	the width at the base	2 c. = ;10 n.	(1 m)	(line 11)
indanum	the growth rate	;02 13 20 n./c.	(inclination 9 : 2)	(line 13)
sukud ki.2	the 2nd height	1 1/2 c.	(75 cm)	(line 15)
sukud ki.3	the 3rd height	1 1/2 c.	(75 cm)	(line 17)

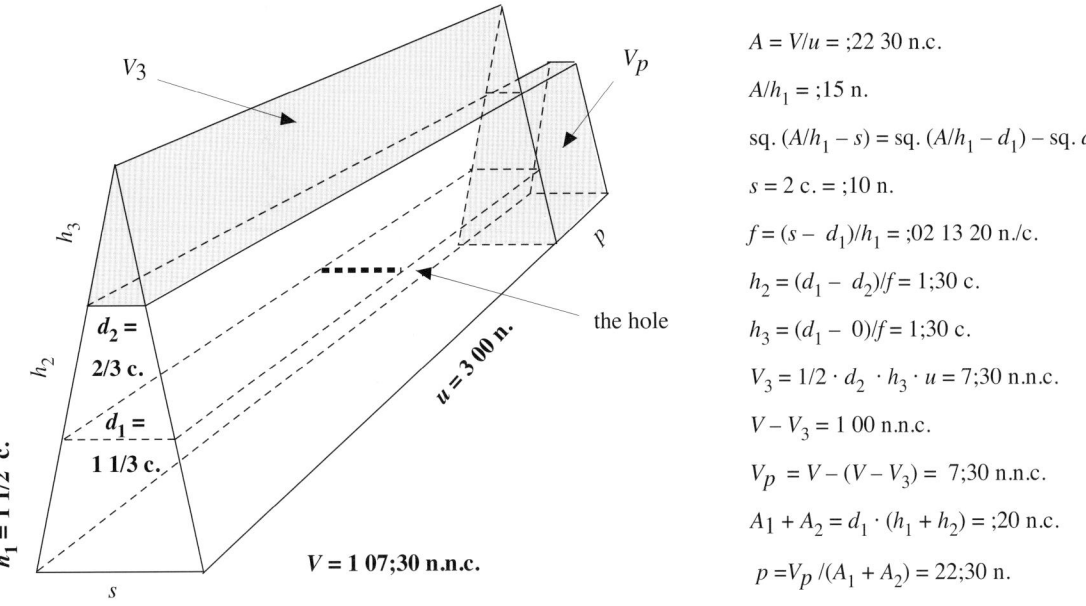

$$A = V/u = ;22\ 30 \text{ n.c.}$$

$$A/h_1 = ;15 \text{ n.}$$

$$\text{sq. } (A/h_1 - s) = \text{sq. } (A/h_1 - d_1) - \text{sq. } d_1$$

$$s = 2 \text{ c.} = ;10 \text{ n.}$$

$$f = (s - d_1)/h_1 = ;02\ 13\ 20 \text{ n./c.}$$

$$h_2 = (d_1 - d_2)/f = 1;30 \text{ c.}$$

$$h_3 = (d_1 - 0)/f = 1;30 \text{ c.}$$

$$V_3 = 1/2 \cdot d_2 \cdot h_3 \cdot u = 7;30 \text{ n.n.c.}$$

$$V - V_3 = 1\ 00 \text{ n.n.c.}$$

$$V_p = V - (V - V_3) = 7;30 \text{ n.n.c.}$$

$$A_1 + A_2 = d_1 \cdot (h_1 + h_2) = ;20 \text{ n.c.}$$

$$p = V_p /(A_1 + A_2) = 22;30 \text{ n.}$$

Fig. 10.2.10. MS 3052 § 1 d. Repairing a breach in a wall with a triangular cross section and with a drilled hole.

The first step of the solution procedure is the computation of the 'stretched-across', the area of the cross section of the mud wall:

1. $A = V/u = 1\ 07;30$ n. n. c. / 3 00 n. = ;22 30 n. c.   (line 7)

Next are computed the *width of the base s*, and the *middle and upper heights h_2 and h_3*, with departure from the known values of the *area A* of the triangular cross section of the wall, the *length of the drilled hole d_1*, the new upper width d_2, and the *lower height h_1*.

The solution procedure in § 1 d is based on the same ideas as the solution procedure in § 1 c. Thus, after the computation of the area of the cross section, the solution continues as follows:

2. $A/h_1 = ;22\ 30$ n. c. / 1;30 c. = ;15 n., $A/h_1 - d_1 = ;15$ n. – ;06 40 n. = ;08 20 n.   (lines 8 - 9)
3. sq. $(A/h_1 - d_1) = ;01\ 09\ 26\ 40$ sq. n., sq. $d_1 = ;00\ 44\ 26\ 40$ sq. n.   (lines 9 - 10)
4. sq. $(A/h_1 - d_1) -$ sq. $d_1 = ;01\ 09\ 26\ 40$ sq. n. – ;00 44 26 40 sq. n. = ;00 25 sq. n.   (line 10)
5. $A/h_1 -$ sqs. (sq. $(A/h_1 - d_1) -$ sq. d_1) = ;15 n. – ;05 n. = ;10 n. = 2 c. = s (line 11)
6. $(s - d_1)/h_1 = (;10 - ;06\ 40)$ n. / 1;30 c. = ;02 13 20 n./c. = f (lines 12 - 13)
7. $(d_1 - d_2)/f = (;06\ 40 - ;03\ 20)$ n. · 27 c./n. = 1;30 c. = h_2 (lines 14 - 15)
8. $(d_2 - 0)/f = ;03\ 20$ n. · 27 c./n. = 1;30 c. = h_3 (lines 16-17)

In steps 2 - 5 of this solution procedure, the base s is computed in precisely the same way as the base was computed in § 1 c, without any notion taken of the "upper" transversal. After that, the middle and upper heights are easily computed, in steps 6 - 8, by use of the growth rate f.

Note the previously unknown term for 'zero' used in line 16 of this text, namely *mimma ula*, the Akkadian phrase for 'nothing', composed of the words *mimma* 'anything' and *ula* 'not'.

The computation of the upper height h_3, ends in line 1 of the badly damaged reverse of MS 3052. After that, the last six lines of § 1 d are devoted to the computation of the 'foundation' (meaning the length) of the breach. It is computed as follows. First is computed the volume V_3 of the mud removed from the top of the clay wall, and the volume $V - V_3$ of what then remains of the wall:

9. $V_3 = 1/2 \cdot d_2 \cdot h_3 \cdot u = 1/2 \cdot ;03\ 20$ n. · 1;30 c. · 3 00 n. = 7;30 n. n.c.   (lines 18-19)
10. $V - V_3 = (1\ 07;30 - 7;30)$ n. n.c. = 1 00 n. n.c.   (line 19)

(See Fig. 10.2.10.) Next is computed the volume of the mud used to repair the breach:

11. $V_p = V - (V - V_3) = (1\ 07;30 - 1\ 00)$ n. n.c. = 7;30 n. n. c.   (line 20)

(It would have been simpler to say directly that the volume of the mud used to repair the breach was the same as the volume of the mud removed from the top of the wall.) Then the volume is computed of the cross section of the wall with the top removed:

12. $A_1 + A_2 = d_1 \cdot (h_1 + h_2) = {;}06\ 40\ \text{n.} \cdot 3\ \text{c.} = {;}20\ \text{n.c.}$ (line 20)

In the final step of the computation, the length of the breach is computed as follows:

13. $p = V_p/(A_1 + A_2) = 7{;}30\ \text{n. n.c.}/{;}20\ \text{n.c.} = 22{;}30\ \text{n.}$ (lines 21-22)

Earlier Published Parallel Texts: YBC 4673 § 12, Str. 364, and VAT 8512

Only one previously published Old Babylonian mathematical text is concerned with a partitioned mud wall. It is **YBC 4673 § 12**, an isolated exercise in the middle of a large mathematical recombination text, without solution procedures (Muroi, *SBM* 2 (1992), Robson, *MMTC* (1999), 89: Friberg, *ChV* (2000), 106). The transliteration and translation of that text is quite problematic.

YBC 4673 §12 (Group 2, hence from a southern site, possibly Ur)

1-2	im.dù.[a] / 5 uš uš.bi /	A clay wall. 5 uš (5 00 n.) its length,
3-4	2 kùš dagal / 2' ninda sukud.bi /	2 c. the width, 1/2 n. its height.
5	i-na 1 kùš 3' kùš gu₇ ì.gu₇-ma /	In 1 c., 1/3 c. fodder it eats.
6	guruš? gul.gul-ma /	A worker? demolishes.
7	1 2' kùš sukud ur-dám? /	1 1/2 c. it has come down?.
8	saḫar? en.nam ḫé.kur.ru?	Of mud?, what shall he accomplish??

The wall in this exercise resembles the wall in MS 3052 § 1 d. Both walls have a triangular cross section, both have the same width at the base, and both have 1 1/2 cubit torn off at the top. Only the heights are different. Note also that different expressions are used in MS 3052 § 1 and in YBC 4673 for the inclination of the wall. In YBC 4673 § 12, the curious phrase *ina* 1 kùš 3' kùš gu₇ ì.gu₇, here tentatively translated 'in 1 cubit it eats 1/3 cubit of fodder', seems to be a graphic description of the circumstance that as the wall goes upward by 1 cubit its width decreases by 1/3 cubit.

Even if there are no more known direct parallels to MS 3052 § 1, there is one text whose theme is closely, although indirectly, related to the theme of MS 3052 § 1. After all, the theme of MS 3052 § 1 is only superficially "mud walls". The real theme is "striped triangles". (See Friberg, *RlA 7* (1990) Sec. 5.4 i.) In this respect, **Str. 364** (*MKT 1*, 248-256; Fig. 10.2.12 below) is a parallel to MS 3052 § 1.

Str. 364 is a fairly well preserved medium size clay tablet, probably belonging to Group 3, hence a text from Uruk. The first and last exercises are lost, as well as important parts of the second exercise. Unfortunately, the second (and presumably also the first) exercise is where a contrived explanation of the striped triangles was given, in terms of canals, dams, and dikes. The other exercises are formulated entirely in terms of metric algebra, that is as numerical problems for geometric figures.

All the exercises in Str. 364 have the form of a question, without solution procedure and answer, but illustrated by a drawing of a triangle with one or several transversals parallel to the front. The triangles are oriented in the Old Babylonian standard way, with the front facing left.

Str. 364 is a well organized topic text with eight related themes. It is likely that it is a condensed version of much larger theme text, complete with solution procedures and answers. (Cf. the way in which YBC 4657 (*MCT*, G) is a condensed copy, with only questions and answers, of the complete theme texts YBC 4663 (MCT, H) and YBC 4662 (MCT, J). Those three texts have the same combined theme: "digging", "work norms", and "expenses in terms of man-power or silver".)

The text of **Str. 364 § 3** is perfectly preserved (see below). The situation is essentially the same as in MS 3052 § 1 c. Given are the area of the whole triangle, $A = 1\ \text{bùr}\ 2\ \text{èše} = 50\ 00\ \text{sq. n.}$, the transversal, $d = 40\ \text{n.}$, and the upper length $u_a = 33{;}20\ \text{n.}$ Therefore, the second Old Babylonian transversal rule (Fig. 10.2.8) is applicable, so that the front s and the length u (actually the height) of the triangle can be computed as the solutions to the rectangular-linear system of equations

$$u \cdot s = 2\,A = 2 \cdot 50\;00 \text{ sq. n.}$$
$$d \cdot u + u_a \cdot s = 2\,A, \quad \text{that is} \quad 40 \text{ n.} \cdot u + 33;20 \text{ n.} \cdot s = 1\;40\;00 \text{ sq. n.}$$

As shown in the discussion of MS 3052 § 1 c, this system of equations can be solved as follows:

$$A/u_a = 50\;00 \text{ sq. n.} / 33;20 \text{ n.} = 1\;30 \text{ n.}$$
$$s = 1\;30 \text{ n.} - \text{sqs. (sq. } (1\;30 \text{ n.} - 40 \text{ n.}) - \text{sq. } 40 \text{ n.}) = 1\;30 \text{ n.} - 30 \text{ n.} = 1\;00 \text{ n.}$$
$$u = 1\;40\;00 \text{ sq. n.} / 1\;00 \text{ n.} = 1\;40 \text{ n.}$$
$$u_k = 1\;40 \text{ n.} - 33;20 \text{ n.} = 1\;06;40 \text{ n.}$$

Thus, in this exercise, the triple

$$A/u_a - d = 50 \text{ n.}, \quad d = 40 \text{ n.}, \quad A/u_a - s = 30 \text{ n.}$$

is equal to the simple diagonal triple 5, 4, 3 multiplied by 10 ninda. The data for the exercise may have been constructed with departure either from this simple diagonal triple, or from the simple condition that the transversal divides the length of the triangle in two parts in the ratio 1 : 2.

Str. 364 § 3

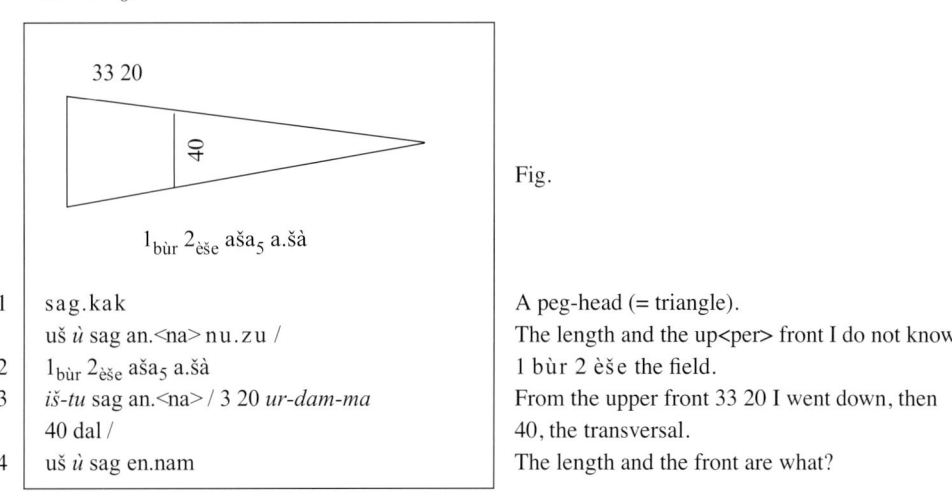

Fig.

1	sag.kak	A peg-head (= triangle).
	uš *ù* sag an.<na> nu.zu /	The length and the up<per> front I do not know.
2	1_bùr 2_èše aša₅ a.šà	1 bùr 2 èše the field.
3	*iš-tu* sag an.<na> / 3 20 *ur-dam-ma*	From the upper front 33 20 I went down, then
	40 dal /	40, the transversal.
4	uš *ù* sag en.nam	The length and the front are what?

Str. 364 § 4 a is concerned with another example of a striped triangle:

Str. 364 § 4 a

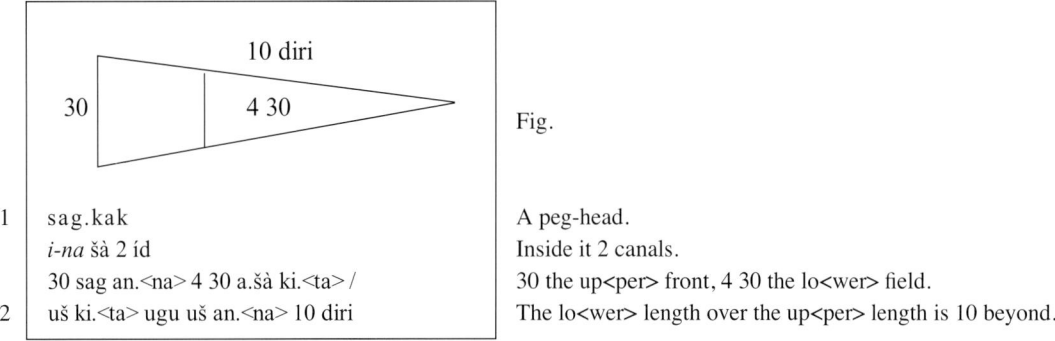

Fig.

1	sag.kak	A peg-head.
	i-na šà 2 íd	Inside it 2 canals.
	30 sag an.<na> 4 30 a.šà ki.<ta> /	30 the up<per> front, 4 30 the lo<wer> field.
2	uš ki.<ta> ugu uš an.<na> 10 diri	The lo<wer> length over the up<per> length is 10 beyond.

Here, the given parameters for the striped triangle are the lower area $A_k = 4\;30$ sq. n., the front $s = 30$ n., and the difference between the lower and upper lengths (actually heights), $u_k - u_a = 10$ n. The parameters to be computed are the transversal d, the partial lengths u_a and u_k, and the upper area A_a. If two of these unknown parameters can be found, for instance and u_k and d, then it will be easy to find also the values of the remaining two parameters. Now, the product of u_k and d is known, since it is equal to twice the lower area. Therefore, what is needed is to find a linear relation between u_k and d, so that u_k and d can be found as the solutions to a rectangular-linear system of equations.

A geometric derivation of such a linear relation is shown in Fig. 10.2.11 below. (In order to avoid unnecessary complications of the arguments, in the figure the triangle is assumed to be a right triangle.) The rule demonstrated in Fig. 10.2.11 may be called "the third Old Babylonian transversal rule". So far, it is only a conjecture that Old Babylonian mathematicians were familiar with this rule, but it seems to be a fairly reasonable conjecture.

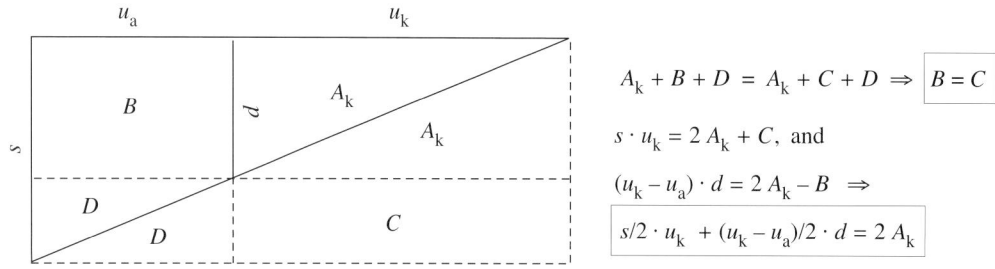

Fig. 10.2.11. Str. 364 § 4 a. The (alleged) third Old Babylonian transversal rule.

The linear relation $s/2 \cdot u_k + (u_k - u_a)/2 \cdot d = 2\,A_k$ between the unknown lower length u_k and the unknown transversal d can be obtained through a simple geometric argument as shown in Fig. 10.2.11 above. It follows that u_k and d can be computed as the solutions to the following rectangular-linear system of equations

$$u_k \cdot d = 2\,A_k, \quad s/2 \cdot u_k + (u_k - u_a)/2 \cdot d = 2 \cdot A_k.$$

This is a system of equations of the same type as the system $u \cdot s = 2\,A$, $d \cdot u + u_a \cdot s = 2\,A$ for the unknowns u and s in Str. 363 § 3. Clearly, then, the unknowns u_k and d play the same role in Str. 364 § 4 a as the unknowns u and s in § 3, and the coefficients $2\,A_k$, $s/2$ and $(u_k - u_a)/2$ play the same role in Str. 364 § 4 a as the coefficients $2\,A$, d and u_a in § 3. Therefore, in Str. 364 § 4 a

$$d = A_k /(u_k - u_a)/2 - \text{sqs. } \{\text{sq. } (A_k /(u_k - u_a)/2 - s/2) - \text{sq. } s/2\}, \quad u_k = 2\,A_k/d.$$

With the given values $A_k = 4\ 30$ sq. n., $s = 30$ n., and $(u_k - u_a) = 10$ n., it follows that

$$A_k /(u_k - u_a)/2 = 4\ 30 \text{ sq. n. } / 5 \text{ n. } = 54 \text{ n.}$$
$$d = 54 \text{ n. } - \text{sqs. (sq. } 39 \text{ n. } - \text{sq. } 15 \text{ n.}) = 54 \text{ n. } - 36 \text{ n. } = 18 \text{ n.}$$
$$u_k = 9\ 00 \text{ sq. n. } / 18 \text{ n. } = 30 \text{ n.}, \quad \text{hence} \quad u_a = 30 \text{ n. } - 10 \text{ n. } = 20 \text{ n.}$$

This solution to the problem in Str. 364 § 4 a is associated with the diagonal triple

$$A_k/(u_k - u_a)/2 - s/2 = 39 \text{ n.}, \quad A_k/(u_k - u_a)/2 - d = 36 \text{ n.}, \quad s/2 = 15 \text{ n.}$$

This, of course, is a multiple of the well known diagonal triple $13, 12, 5$.

The problem in **Str. 364 § 5 a** is concerned with a similar problem for the same triangle, but with the transversal in a different position, so that $u_a - u_k = 10$ instead of $u_k - u_a = 10$ as in Str. 364 § 4 a. There is also a new value for the lower area, $A_k = 2\ (00$ sq. n.). The problem leads to the following rectangular-linear system for the unknowns u_k and d, resembling the one in the case of Str. 364 § 4 a:

$$u_k \cdot d = 2\,A_k$$
$$s/2 \cdot u_k - (u_a - u_k)/2 \cdot d = 2 \cdot A_k \qquad \text{(see again Fig. 10.2.11 above).}$$

With the given values $A_k = 2\ (00$ sq. n.), $s = 30$ (n.), and $u_a - u_k = 10$ (n.) inserted, the solution is

$$d = \text{sqs. } \{\text{sq. } (A_k/(u_a - u_k)/2 + s/2) - \text{sq. } s/2\} - A_k/(u_a - u_k)/2 = \text{sqs. (sq. } (24 + 15) - \text{sq. } 15) - 24 = 36 - 24 = 12 \text{ n.}$$
$$u_k = 2\,A_k /d = 20 \text{ n.}$$

This solution to the problem in Str. 364 § 5 a is associated with the diagonal triple

$$A_k/(u_k - u_a)/2 + s/2 = 39 \text{ n.}, \quad A_k/(u_k - u_a)/ + d = 36 \text{ n.}, \quad s/2 = 15 \text{ n.}$$

This is, again, a multiple of the well known diagonal triple $13, 12, 5$.

The remaining exercises on the obverse of Str. 364 are routine variations on the theme. The exercises on the reverse with five-striped triangles look formidable but each one of them can be reduced to a pair of simpler exercises, one for a two-striped trapezoid, and another for a two-striped triangle.

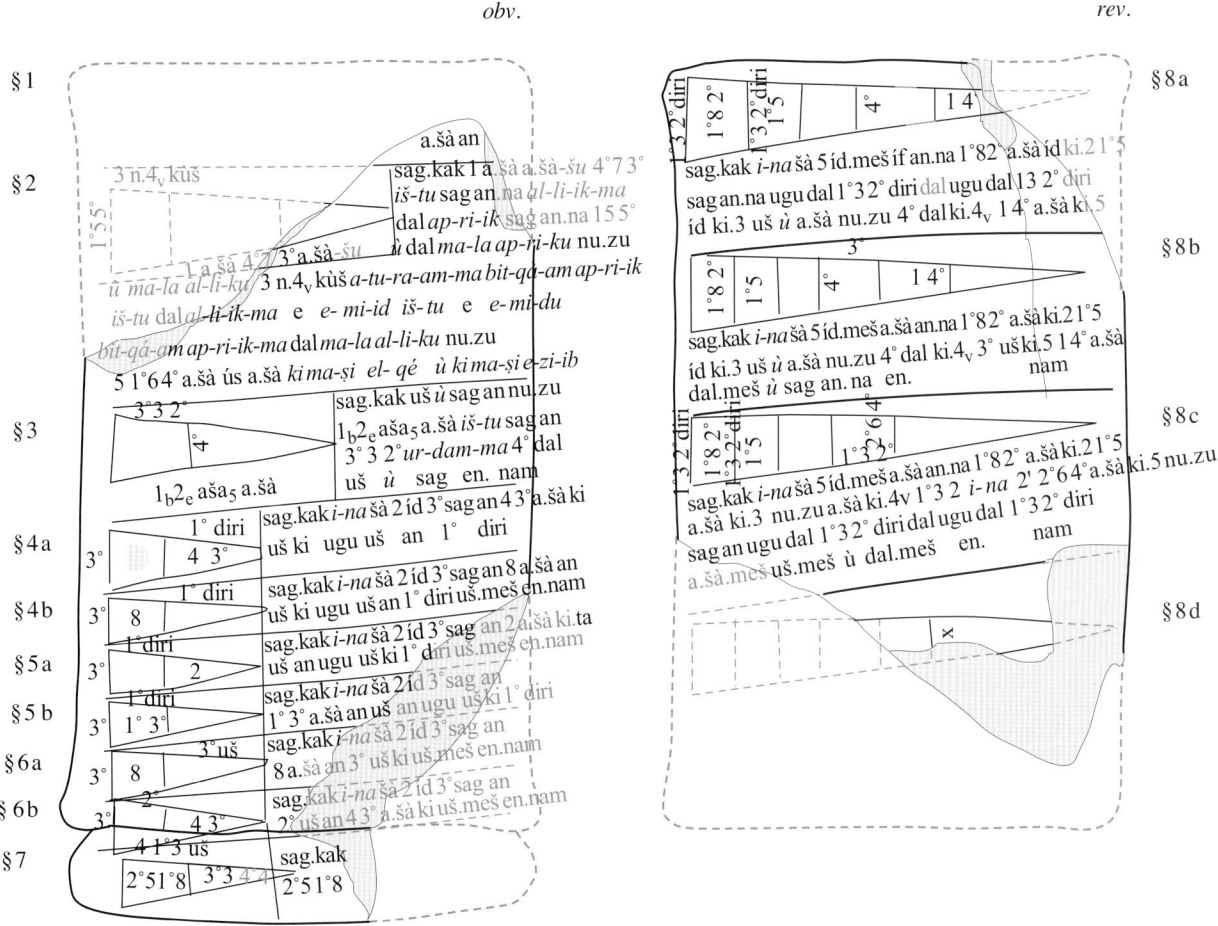

Fig. 10.2.12. Str. 364. A theme text with metric algebra problems for striped triangles.
(The conform transliteration here is based on Neugebauer's hand copy of the text in *MKT*.)

VAT 8512 (Høyrup, *LWS* (2002), 234-238; Group 4 a, hence a southern text; cf. Robson, *HM* 28 (2001), 183, fn. 21) is a problem text with a single exercise that is closely related to the theme of Str. 364 § 4 a, hence also to the theme of MS 3052 §§ 1 c-d. In VAT 8512 is considered a striped triangle with a single transversal. With the usual notations, the given parameters are

$$s = 30 \text{ n.}, \quad u_k - u_a = 20 \text{ n}, \quad A_a - A_k = 7 \,(00) \text{ sq. n.}$$

The solution is given explicitly by an elegant solution procedure. In modern notations, the first step of the solution procedure is the computation of the transversal, called *pirkum* 'cross-line', as

$$d = \text{sqs. } \{(\text{sq. }(s + p) + \text{sq. } p)/2\} - p, \quad \text{where} \quad p = (A_a - A_k)/(u_k - u_a) = 7\,00 \,/\, 20 = 21, \quad \text{so that} \quad d = 18.$$

This equation may have been found as follows: If f is the growth rate for the triangle, then

$$s - d = f \cdot u_a \quad \text{and} \quad d = f \cdot u_k, \quad \text{hence} \quad f = (2\, d - s)/(u_k - u_a).$$

In addition,

$$\text{sq. } s - \text{sq. } d = f \cdot 2\, A_a \quad \text{and} \quad \text{sq. } d = f \cdot 2\, A_k, \quad \text{so that}$$

$$\text{sq. } s - 2 \text{ sq. } d = f \cdot 2\, (A_a - A_k) = 2\,(d - s)/((u_k - u_a) \cdot 2\,(A_a - A_k).$$

It follows that

$$\text{sq. } s - 2 \text{ sq. } d = 2\,p \cdot (2\,d - s), \quad p = (A_a - A_k)/(u_k - u_a).$$

Therefore, the value of d can be found as a solution to the quadratic equation

$$\text{sq. } d + 2\,p \cdot d = (\text{sq. } s + 2\,p \cdot s)/2.$$

It is, of course, tempting to complete the squares on both sides of this equation. The result is the reformulated quadratic equation

$$\text{sq. } (d + p) - \text{sq. } p = \{\text{sq. } (s + p) - \text{sq. } p\}/2,$$

or, finally,

$$\text{sq. } (d + p) = \{\text{sq. } (s + p) + \text{sq. } p\}/2.$$

The solution given explicitly in the text of VAT 8512 follows immediately from this equation for d. Note: An ingenious *geometric* derivation of the equation was found by Gandz (1948) and Huber (1955). See the references to these authors in Høyrup (*op. cit.*) and, in particular, Fig. 59 on p. 235.

Note the following interesting similarity between MS 3052 § 1 c and VAT 8512: The explicit solution in MS 3052 § 1 c, was given in a form that made it clear that the triple

$$c = A/h_k - d, \quad b = d, \quad a = A/h_k - s \quad \text{is a solution to the } \textit{diagonal equation} \quad \text{sq. } c = \text{sq. } a + \text{sq. } b.$$

Similarly, the explicit solution in VAT 8512 is given in a form that makes it clear that the triple

$$c = d + p, \quad b = s + p, \quad a = p \quad \text{is a solution to the } \textit{equipartitioned trapezoid equation} \quad \text{sq. } c = (\text{sq. } a + \text{sq. } b)/2.$$

With the given values in VAT 8521, the triple, $s + p, d + p, p = 51, 39, 21 = 3 \cdot (17, 13, 7)$. Equipartioned trapezoids was a popular topic in Old Babylonian mathematics. (See Friberg, *RlA 7* (1990) Sec. 5.4 k.) The idea is the following: If a, b, c is an equipartioned trapezoid triple, then any trapezoid with the lower front a, and the upper front b is divided into *two sub-trapezoids of equal area* by a parallel transversal of length c. Common Old Babylonian examples of equipartitioned trapezoid triples are 7, 5, 1 or 17, 13, 7 and multiples of these triples.

MS 3052 § 1 e. A Badly Preserved Exercise Dealing with a Mud Wall with a Breach

The text of **MS 3052 § 1 e** is inscribed on the badly damaged reverse of MS 3052. Too little is preserved of the text of this exercise to allow a meaningful reconstruction. See Fig. 10.2.13 below. Anyway, since a subscript on the reverse of MS 3052 mentions '8 hand tablets (= exercises)', and, more specifically, '[x] mud walls, 1 diagonal (= rectangle), 1 excavation, and 1 square', it is clear that there were, originally, five exercises for mud walls on MS 3052. It follows that § 1 e must be the fifth exercise for a mud wall, probably one with a trapezoidal cross section, a breach, and a hole.

The word *ag-šu-ur* 'I repaired' in line 2 of § 1 e (cf. line 5 of § 1 d) shows that also § 1 e is concerned with the reparation of a breach in a wall. The fact that the numbers 3 4[5], 5, 6 15, 25, 2 36 15, 12 30, and 7 30 are mentioned in lines 6-12 of § 1 e suggests that the solution procedure in § 1 e in some way makes use of the two diagonal triples

6 15, 5, 3 45, with sq. 6 15 – sq. 5 = 39 03 45 – 25 = 14 03 45 = sq. 3 45. and
12 30, 7 30, 10 with sq. 12 30 – sq. 10 = 2 36 15 – 56 15 = 1 40 = sq. 10.

10.2 b.   MS 3052 § 2. A 'Diagonal'. The Basic igi-igi.bi Problem

The text of this exercise is badly preserved, but the fact that MS 3971, § 3 a-e are closely parallel exercises (see above, Sec. 10.1 c) makes it easy to reconstruct most of the text.

MS 3052 § 2

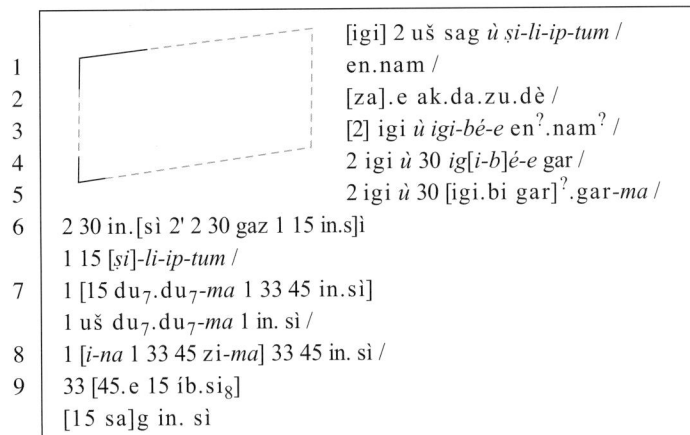

	[igi] 2 uš sag *ù ṣi-li-ip-tum* /	*The igi* (is) 2.
1	en.nam /	Length, front, and cross-over, what?
2	[za].e ak.da.zu.dè /	Fig.*You* **with your doing:**
3	[2] igi *ù igi-bé-e* en$^?$.nam$^?$ /	[2] the igi and the igi.bi what?
4	2 igi *ù* 30 ig[*i-b*]*é-e* gar /	2, the igi, and 30, the ig*i.b*i, set.
5	2 igi *ù* 30 [igi.bi gar]$^?$.gar-*ma* /	2, the igi, and 30, *the igi.bi, he*ap, then
6	2 30 in.[sì 2' 2 30 gaz 1 15 in.s]ì	2 30 it gives. *1/2 of 2 30 break, 1 15 it gi*ves.
	1 15 [*si*]-*li-ip-tum* /	1 15 (is) the cross-over.
7	1 [15 du$_7$.du$_7$-*ma* 1 33 45 in.sì]	1 *15 (let) butt (itself), then 1 33 45 it gives.*
	1 uš du$_7$.du$_7$-*ma* 1 in. sì /	1, the length, (let) butt (itself), then 1 it gives.
8	1 [*i-na* 1 33 45 zi-*ma*] 33 45 in. sì /	1 *from 1 33 45 tear off, then* 33 45 it gives.
9	33 [45.e 15 íb.si$_8$]	33 45 makes 15 equalsided.
	[15 sa]g in. sì	15, the front, it gives.

The drawing accompanying this text is badly preserved. However, the few traces of the drawing that remain seem to indicate that what was depicted was a rectangle *without* any diagonals. On the other hand, the question is almost completely preserved. It asks for the sizes of *the length, the front, and the diagonal* of the rectangle, given that a number called the igi is equal to 2.

The first step of the solution procedure is the computation of the igi.bi, the reciprocal of the igi, in the present case obviously equal to $1/2 = ;30$. The diagonal c, the length b, and the front a of the rectangle are then computed as follows:

$$c, b, a = (\text{igi} + \text{igi.bi})/2, 1, \text{sqs. (sq. } d - \text{sq. } u) = 2;30/2, 1, \text{sqs. (sq. } 2;30/2 - \text{sq. } 1) = 1;15, 1, ;45.$$

Another possibility is that the length called '1' was interpreted as 1 (00). In that case, the computed triple has to be understood as

$$c, b, a = 1\ 15, 1\ (00), 45.$$

Whichever the case may be, it is easy to see that the rectangle computed in this way is *similar* to a rectangle in which the diagonal, the length, and the front form the well known "Pythagorean" triple 5, 4, 3. (Compare with the exercise MS 3971 § 4, where it is shown that 7, 5;36, 4;12 are the diagonal, the length, and the front of another rectangle similar to the 5, 4, 3 rectangle.)

Actually, the triple 1 45, 1 (00), 45 is one of the triples listed in the famous Old Babylonian table text Plimpton 322. See App. 8 for a further discussion of this fascinating topic.

10.2 c.   MS 3052 § 3. An 'Excavation' of the igi-igi.bi Type

The text of § 3 of MS 3052 is so damaged that no reconstruction can be attempted. The statement of the problem and the answer are both lost, except the phrase igi uš ù igi.bi sag in line 1, which probably means that the floor of the excavation is a rectangle with the sides u and s, where $u \cdot s = 1$. (square ninda). Cf. **BM 85200+, ## 15-18** (Høyrup, *LWS* (2002), 158; Sec. A8 b in App. 8 below). Of the solution procedure all that remains are the computations $5 \cdot 18 = 1\ 30$ in line 3 and sq. $45 + 1 = $ sq. $1\ 15$ in lines 4-5.

10.2 d.   MS 3052 § 4. A 'Square'. Another Badly Damaged Exercise

The text of § 4 is not as extensively damaged as the text of § 3. A transliteration of the text is given below.

rev.

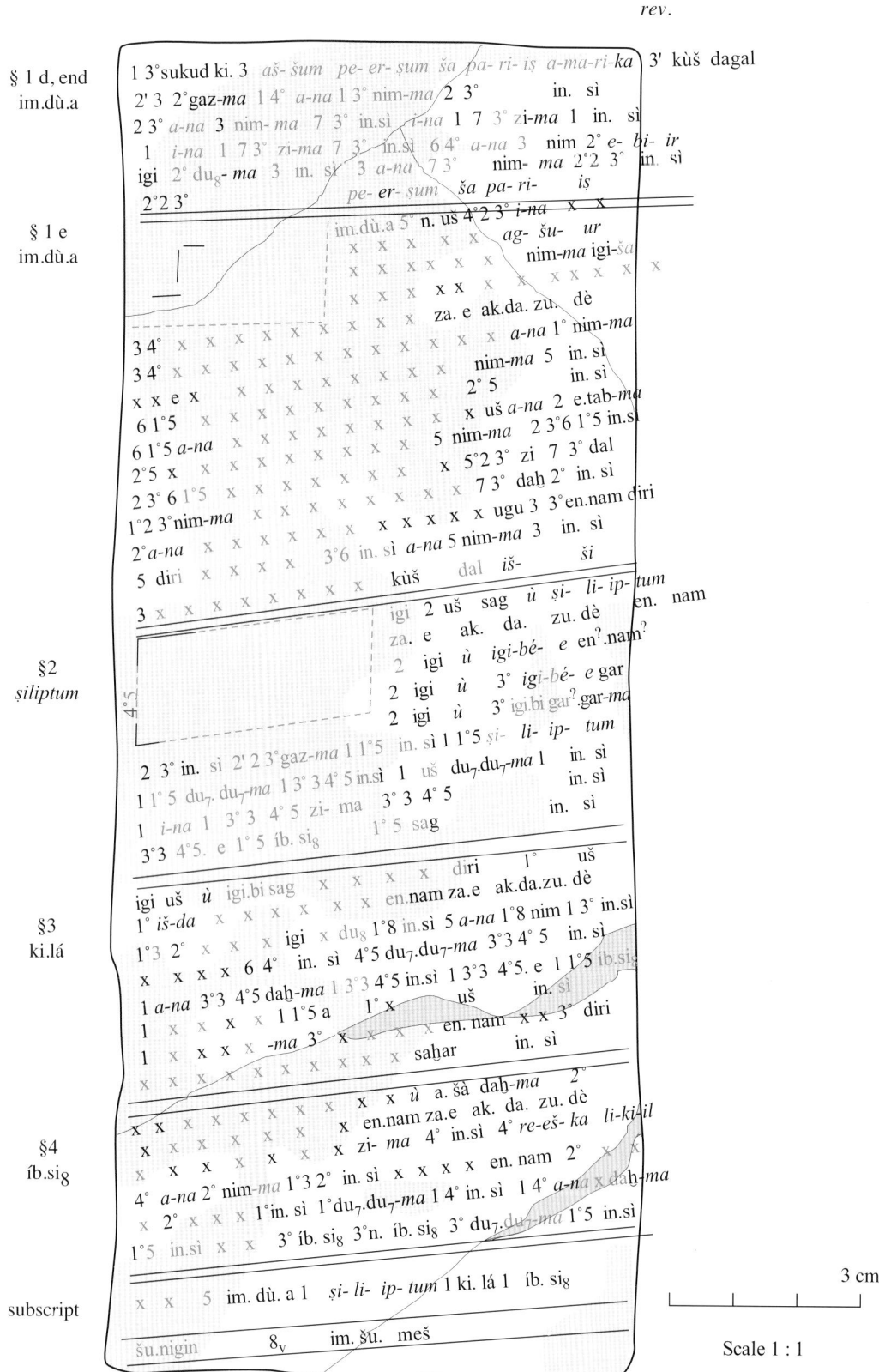

Fig. 10.2.13. MS 3052, *rev.* Conform transliteration.

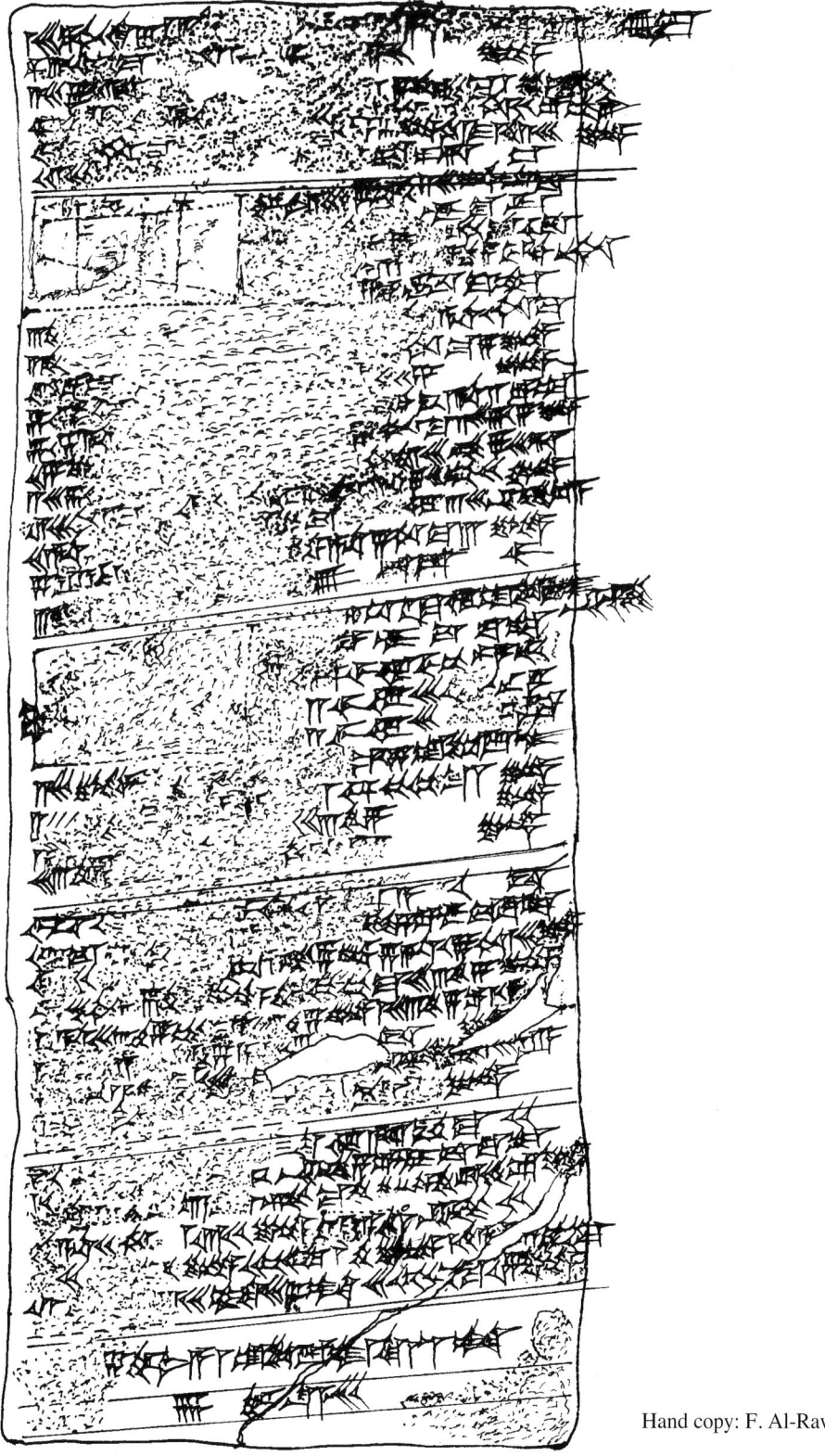

Fig. 10.2.14. MS 3052, *rev*. Hand copy of the cuneiform text.

MS 3052 § 4

1	x x [x x x x x x x x x]	x x x x x x x x x x x x
	x x *ù* a.šà daḫ-*ma* 20 /	x x and the field join, then 20.
2	x [x x x x x x x x x] en.nam	x x x x x x x x x x (are) what?
	za.e ak.da.zu.dè /	You, with your doing:
3	[x] x x [x] x [x] x zi-*ma* 40 in.sì	x x x x x x x tear off, then 40 it gives.
	40 *re-eš-ka li-ki-il* /	40 may hold your head.
	40 *a-na* 20 nim-[*ma*] 13 20 in.sì	40 to 20 lift, *then* 13 20 it gives.
4	x x x x en.nam 20 [x x] /	x x x (is) what? 20 the *x x*.
	[x] 20 [x x x] 10 in.sì	*x* 20 *x x x*, 10 it gives.
5	10 du₇.du₇-*ma* 1 40 in.sì	10 (let) butt (itself), 1 40 it gives.
	1 40 *a-n*[*a* 13 20 da]ḫ-*ma* /	1 40 to *13 20* join, then
	1[5 in.sì x x] 30 íb.si₈	15 *it gives x x*, 30 the equalside.
6	30 ninda íb.si₈	30 ninda (is) the equalside.
	30 du₇.du₇-*ma* 15 aša₅	30 (let) butt (itself), 15 the field.

Most of the question is lost in this exercise. What remains is in many places not clearly legible. Apparently, something added to the area (of a square?) makes 20. The solution procedure begins with a subtraction, with the remainder 40, which is to be remembered. This number 40 is multiplied with 20, which gives $40 \cdot 20 = 13\ 20$. Then some other operations yields the result 10, and the solution procedure seems to end with the computation sq. $10 + 13\ 20 = 15 = $ sq. 30, where $30 = 30$ ninda is said to be the side of the square. It is not at all clear what all this means.

10.2 e.   *MS 3052, Subscript. A List of the Separate Topics in the Text*

MS 3052, summary

1	[x x 5] im.dù.a 1 *ṣi-li-ip-tum* 1 ki.lá 1 íb.si₈ /	*x x* 5 mud walls, 1 cross-over, 1 excavation, 1 equalside
2	[šu.nígin] 8ᵥ im. šu.meš	*Together* 8 hand tablets (assignments).

This kind of subscript with a detailed summary of the topics in the text is an almost unique feature of MS 3052. The only known parallel is the subscript of MS 3049. (See Fig. 11.1.4 below.)

10.3.   MS 2792. Two Exercises Dealing with a Divided Ramp

10.3 a.   *MS 2792 # 1. A Layer on Top of a Ramp Divided Equally along the Length*

MS 2792 (Figs. 10.3.1-2 and 10.3.5-6 below) is a large, fairly well preserved single-column clay tablet, inscribed with one long mathematical exercise on the obverse and another on the reverse. The common topic of the two exercises is an *arammu* 'ramp'. Both exercises are accompanied by drawings showing a rectangle or trapezoid divided by three transversals into four sub-rectangles or sub-trapezoids.

The small vocabulary of Sumerian technical terms used in MS 2792 includes the following terms for mathematical operations:

gar.gar	heap (add)	du₇.du₇	butt (square)	gaz	break (halve)
zi	tear off (subtract)	nim	lift (multiply)	in.sì	it gives (the result)
a-na 2 e.tab	to 2 you repeat (double)	du₈	resolve (compute the reciprocal)		
n.e *m* íb.si₈	*n* makes *m* equalsided (sqs. *n* = *m*)			gar.ra	set (make a note of)

There is also, as in MS 3971 (Sec. 10.1) and MS 3052 (Sec. 10.2), the following Akkadian phrase, marking a new section of the solution procedure:

aššum … amārika   in order for you to see.

Consequently, MS 2792 belongs to "Group 3" of Old Babylonian mathematical cuneiform texts, just like MS 3971 and MS 3052. As a member of Group 3, MS 2792, too, can be classified as a text from the ancient southern Mesopotamian city Uruk. Note, by the way, that in MS 2792 and MS 3052, but not in MS 3971, the solution procedures begin with the phrase

za.e ak.da.zu.dè You, with your doing:

MS 2792 # 1 (the obverse):

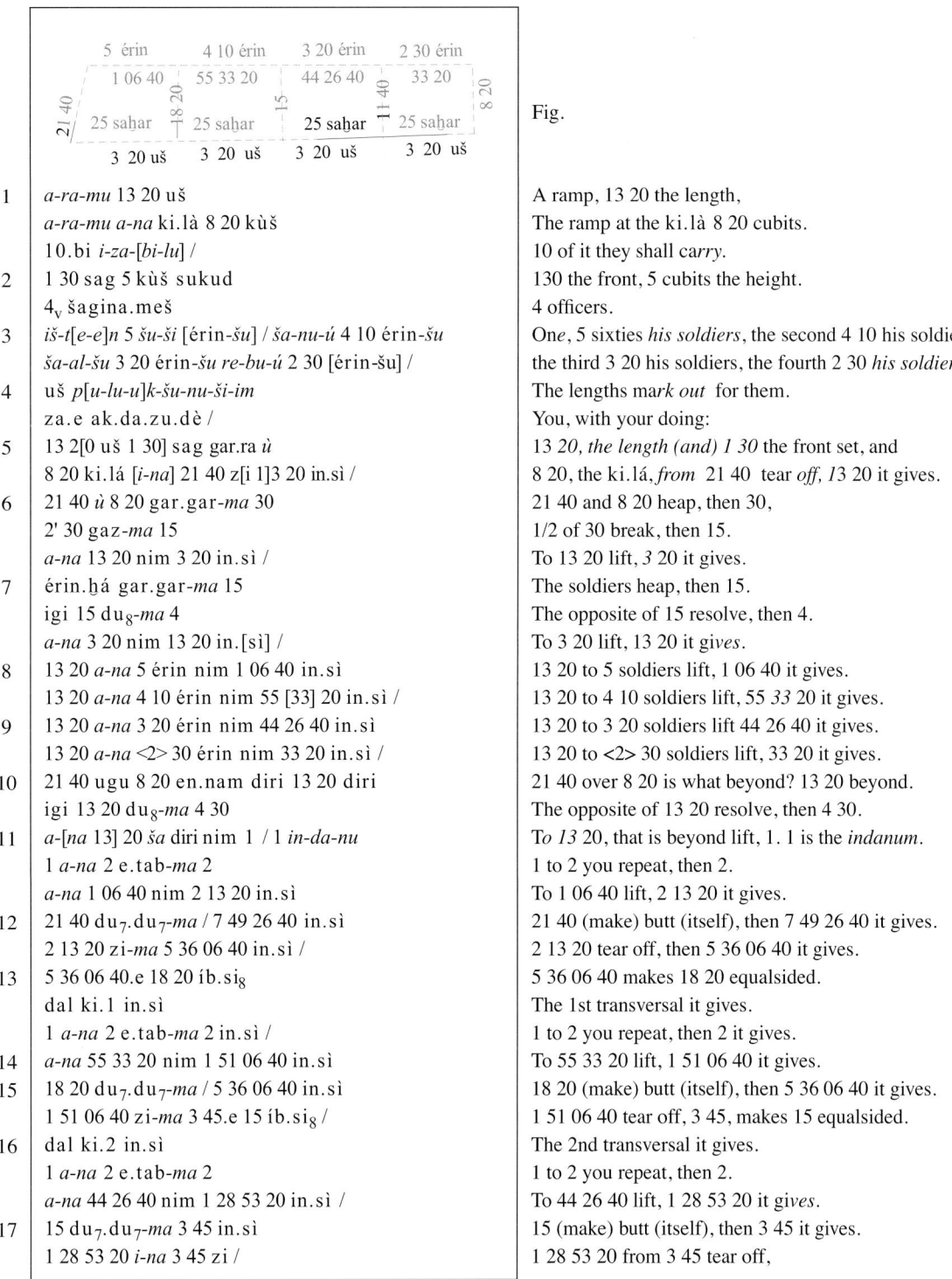

Fig.

1	*a-ra-mu* 13 20 uš	A ramp, 13 20 the length,
	a-ra-mu a-na ki.là 8 20 kùš	The ramp at the ki.là 8 20 cubits.
	10.bi *i-za-[bi-lu]* /	10 of it they shall ca*rry*.
2	1 30 sag 5 kùš sukud	130 the front, 5 cubits the height.
	4ᵥ šagina.meš	4 officers.
3	*iš-t[e-e]n* 5 šu-ši [*érin-šu*] / *ša-nu-ú* 4 10 érin-šu	On*e*, 5 sixties *his soldiers*, the second 4 10 his soldiers,
	ša-al-šu 3 20 érin-šu *re-bu-ú* 2 30 [*érin-šu*] /	the third 3 20 his soldiers, the fourth 2 30 *his soldiers*.
4	uš *p[u-lu-u]k-šu-nu-ši-im*	The lengths ma*rk out* for them.
	za.e ak.da.zu.dè /	You, with your doing:
5	13 2[0 uš 1 30] sag gar.ra *ù*	13 *20, the length (and) 1 30* the front set, and
	8 20 ki.lá [*i-na*] 21 40 z[i 1]3 20 in.sì /	8 20, the ki.lá, *from* 21 40 tear *off, 13* 20 it gives.
6	21 40 *ù* 8 20 gar.gar-*ma* 30	21 40 and 8 20 heap, then 30,
	2' 30 gaz-*ma* 15	1/2 of 30 break, then 15.
	a-na 13 20 nim 3 20 in.sì /	To 13 20 lift, *3* 20 it gives.
7	érin.ḫá gar.gar-*ma* 15	The soldiers heap, then 15.
	igi 15 du₈-*ma* 4	The opposite of 15 resolve, then 4.
	a-na 3 20 nim 13 20 in.[sì] /	To 3 20 lift, 13 20 gi*ves*.
8	13 20 *a-na* 5 érin nim 1 06 40 in.sì	13 20 to 5 soldiers lift, 1 06 40 it gives.
	13 20 *a-na* 4 10 érin nim 55 [33] 20 in.sì /	13 20 to 4 10 soldiers lift, 55 *33* 20 it gives.
9	13 20 *a-na* 3 20 érin nim 44 26 40 in.sì	13 20 to 3 20 soldiers lift 44 26 40 it gives.
	13 20 *a-na* <2> 30 érin nim 33 20 in.sì /	13 20 to <2> 30 soldiers lift, 33 20 it gives.
10	21 40 ugu 8 20 en.nam diri 13 20 diri	21 40 over 8 20 is what beyond? 13 20 beyond.
	igi 13 20 du₈-*ma* 4 30	The opposite of 13 20 resolve, then 4 30.
11	*a-[na* 13] 20 *ša* diri nim 1 / 1 *in-da-nu*	To *13* 20, that is beyond lift, 1. 1 is the *indanum*.
	1 *a-na* 2 e.tab-*ma* 2	1 to 2 you repeat, then 2.
	a-na 1 06 40 nim 2 13 20 in.sì	To 1 06 40 lift, 2 13 20 it gives.
12	21 40 du₇.du₇-*ma* / 7 49 26 40 in.sì	21 40 (make) butt (itself), then 7 49 26 40 it gives.
	2 13 20 zi-*ma* 5 36 06 40 in.sì /	2 13 20 tear off, then 5 36 06 40 it gives.
13	5 36 06 40.e 18 20 íb.si₈	5 36 06 40 makes 18 20 equalsided.
	dal ki.1 in.sì	The 1st transversal it gives.
	1 *a-na* 2 e.tab-*ma* 2 in.sì /	1 to 2 you repeat, then 2 it gives.
14	*a-na* 55 33 20 nim 1 51 06 40 in.sì	To 55 33 20 lift, 1 51 06 40 it gives.
15	18 20 du₇.du₇-*ma* / 5 36 06 40 in.sì	18 20 (make) butt (itself), then 5 36 06 40 it gives.
	1 51 06 40 zi-*ma* 3 45.e 15 íb.si₈ /	1 51 06 40 tear off, 3 45, makes 15 equalsided.
16	dal ki.2 in.sì	The 2nd transversal it gives.
	1 *a-na* 2 e.tab-*ma* 2	1 to 2 you repeat, then 2.
	a-na 44 26 40 nim 1 28 53 20 in.sì /	To 44 26 40 lift, 1 28 53 20 gi*ves*.
17	15 du₇.du₇-*ma* 3 45 in.sì	15 (make) butt (itself), then 3 45 it gives.
	1 28 53 20 *i-na* 3 45 zi /	1 28 53 20 from 3 45 tear off,

18	2 16 06 40.e 11 40 íb.si₈	2 16 06 40 makes 11 40 equalsided,
	dal ki.3 in.sì /	the 3rd transversal it gives.
19	*aš-šum* uš *a-ma-ri-ka*	In order for you to see the length:
	21 40 ugu 18 20 en.nam diri 3 20 diri /	21 40 over 18 20 is what beyond? 3 20 beyond.
20	igi 1 *in-da-nim* du₈	The opposite of 1, the *indanum*, release, 1.
	1 *a-na* 3 20 *ša* diri nim 3 20 uš ki.1 /	To 3 20 that is beyond lift, 3 20 the 1st length.
21	18 20 ugu 15 en.nam diri 3 20 diri	18 20 over 15 is what beyond? 3 20 beyond.
	igi 1 du₈-*ma* 1	The opposite of 1 release, then 1.
22	*a-na* 3 20 nim-*ma* / 3 20 uš ki.2	To 3 20 lift, then 3 20, the 2nd length.
	15 ugu 11 40 en.nam diri 3 20 diri	15 over 11 40 is what beyond? 3 20 beyond.
	igi 1 du₈-*ma* 1 /	The opposite of 1 release, then 1.
23	*a-na* 3 20 nim-*ma* 3 20 uš ki.3	To 3 20 lift, then 3 20, the 3rd length.
24	11 40 ugu 8 20 en.nam diri / 3 20 diri	11 40 over 8 20 is what beyond? 3 20 beyond.
	igi 1 du₈-*ma* 1	The opposite of 1 release, then 1.
	a-na 3 20 nim 3 20 uš ki.4ᵥ /	To 3 20 lift, 3 20 the 4th length.
25	*aš-šum* saḫar *a-ma-ri-ka*	In order for you to see the mud (volumes):
	1 30 sag *a-na* 3 20 uš nim 5 in.sì /	1 30, the front, to 3 20, the length, lift, 5 it gives.
26	5 *a-na* 5 sukud nim-*ma*	5 to 5, the height, lift, then
	25 saḫar ki.1 in.sì	25, the 1st mud it gives.
	1 30 sag *a-na* 3 20 uš ki.2 nim /	1 30, the front, to 3 20 the 2nd length lift, 5.
27	5 *a-na* 5 sukud nim-[*ma*]	To 5, the height, lift, then
	25 saḫar.ki 2 in.sì	25, the 2nd mud it gives.
	1 30 sag *a-na* 3 20 uš ki.3 nim / [5]	1 30, the front, to 3 20 the third length lift, *5*.
28	[*a-na* 5] sukud nim-[*ma*]	*To 5*, the height, lift, then
	25 saḫar.ki 3 in.sì	25, the 3rd mud it gives.
	1 30 sag *a-na* 3 20 uš ki.4 nim / [5]	1 30, the front, to 3 20 the 4th length lift, *5*.
29	[*a-n*]a 5 sukud nim-*ma*	*To 5*, the height, lift, then
	[2]5 saḫar ki.4ᵥ in.sì	25, the 4th mud it gives.

The question in this exercise is vaguely stated, and some pieces of it are lost. The question mentions a 'ramp' (*arammu*), of length 13;20 (ninda), a height? (probably at the end of the ramp), equal to 8;20 cubits, a distance? of 10 (ninda?), a 'front' equal to 1;30 (ninda), and another 'height' equal to 5 cubits. It also mentions 4 šagina (written gìr.níta) (Akk. *šakkanakkum*) 'officers, military governors', and the sizes of their respective troops of men. With departure from these data, the 'lengths' for the four troops are to be computed.

The ensuing computations give some clues to the meaning of this vaguely stated question. Thus, a likely interpretation seems to be that a ramp is already present, built by the four troops of men. Its height has to be increased by an additional 5 cubits, and the question is what the lengths (and volumes) are of the added top layers on the four sections of the ramp. The situation is clarified in Fig. 10.3.3 below, with a drawing of the ramp in perspective. The notations used in Fig. 10.3.3 are the following:

E	(érin)	soldiers	d	(dal)	transversal
A	(a.šà)	area	u	(uš)	length
s	(sag)	front	h		height

The drawing associated with the exercise MS 2792 # 1 is badly damaged. Only the line of numbers below the drawing remains intact. However, since the related drawing illustrating the exercise MS 2792 # 2, on the reverse of the clay tablet, is fairly well preserved, it is easy to reconstruct the drawing on the obverse in its original form, with all the numbers inscribed in and around it.

The values recorded in the (reconstructed) drawing on the obverse can be interpreted as:

man-days for section 1	E_1 = 5 00 'soldiers'	(300)
man-days for section 2	E_2 = 4 10 'soldiers'	(250)
man-days for section 3	E_3 = 3 20 'soldiers'	(200)
man-days for section 4	E_4 = 2 30 'soldiers'	(150)

the height of the ramp at its upper end	$h_a =$	21 2/3 cubits	(10.83 m)
the first transversal	$d_1 =$	18 1/3 cubits	(9.17 m)
the second transversal	$d_2 =$	15 cubits	(7.5 m)
the third transversal	$d_3 =$	11 2/3 cubits	(5.83 m)
the height of the ramp at its lower end	$h_k =$	8 1/3 cubits	(4.17 m)
the area of the side of section 1	$A_1 =$	1 06;40	(ninda · cubits)
the area of the side of section 2	$A_2 =$	55;33 20	(ninda · cubits)
the area of the side of section 3	$A_3 =$	44;26 40	(ninda · cubits)
the area of the side of section 4	$A_4 =$	33;20	(ninda · cubits)
the length of section 1	$u_1 =$	3;20	(ninda) (20 m)
the length of section 2	$u_2 =$	3;20	(ninda)
the length of section 3	$u_3 =$	3;20	(ninda)
the length of section 4	$u_4 =$	3;20	(ninda)
the volume of the top layer of section 1	$V_1 =$	25	(sq. ninda · cubits)
the volume of the top layer of section 2	$V_2 =$	25	(sq. ninda · cubits)
the volume of the top layer of section 3	$V_3 =$	25	(sq. ninda · cubits)
the volume of the top layer of section 4	$V_4 =$	25	(sq. ninda · cubits)

In line 1 of the text, the number '8 20' is specified by the phrase *arammu ana* ki.lá 'the ramp at the ki.lá'. It is not clear what that means. Normally, the meaning of ki.lá (Akk. *kalakkum*) is 'excavation, pit, trench'. Here, it may refer to the lower, abruptly ending, part of the ramp. The related number '21 40' is given without any specification in the text. Cf. **BM 85194 # 17** (Neugebauer, *MKT 1*, 143]), where the lower end of a ramp is qualified by the phrase *i-na* suḫuš saḫar.ḫá 'at the base of the mud', while the upper end is said to be

i-na pa-ni a-bu-li-im 'in front of the gate'.

Not recorded in the drawing on the obverse of MS 2792 are three more values given in the text:

the total length of the ramp	$u =$	13;20 (ninda)	(80 m)	(line 1)
the front of the ramp	$s =$	1;30 (ninda)	(9 m)	(line 2)
the height of the added top layer	$h =$	5 cubits	(2.5 m)	(line 2)

Now the vaguely stated question in MS 2792 # 1 can be rephrased in the following more precise way:

Given a ramp with the length 13;20 n., the front 1;30 n., the lower height 8;20 c., and the upper height 21;40 c., divided in four sections that it took 5 00, 4 10, 3 20, and 2 30 'soldiers', respectively, to build, and with an added top layer of height 5 c., find the lengths of the four sections and the volumes of the added layers on top of them.

The solution procedure begins with the computation of the area of the vertical side of the ramp, which has the form of a trapezoid:

$A = 13;20$ n. $\cdot$ (21;40 + 8;20)/2 c. = 13;20 n. $\cdot$ 15 c. = 3 20 n. c.   (line 6)

Next, the areas of the vertical sides of the four sections of the ramp are computed, by use of the obvious circumstance that these four areas are proportional to the four given numbers of 'soldiers':

$E = (5\ 00 + 4\ 10 + 3\ 20 + 2\ 30)$ 'soldiers' = 15 00 'soldiers' (meaning *man-days*)	(line 7)
$A/E = 3\ 20$ n. c. / 15 00 md = ;13 20 n. c./md.	(line 7)
$A_1 = A/E \cdot E_1 =$;13 20 n. c./md. $\cdot$ 5 00 md. = 1 06;40 n. c.	(line 8)
$A_2 = A/E \cdot E_2 =$;13 20 n. c./md. $\cdot$ 4 10 md. = 55;33 20 n. c.	(line 8)
$A_3 = A/E \cdot E_3 =$;13 20 n. c./md. $\cdot$ 3 20 md. = 44;26 40 n. c.	(line 9)
$A_4 = A/E \cdot E_4 =$;13 20 n. c./md. $\cdot$ 2 30 md. = 33;20 n. c.	(line 9)

The inclination of the ramp is then computed as

$f = (21;40 - 8;20)$ c. / 13;20 n. = 13;20 c. / 13;20 n. = 1 c./n.   (inclination 1 : 12)   (lines 10-11)

In this text, as in MS 3052 (Sec. 10.2 above), the Akkadian word *indanum* is used for 'growth rate'. The value f of the growth rate is needed in the next step of the solution procedure, the computation of the three transversals in the trapezoid formed by the vertical cross section of the ramp.

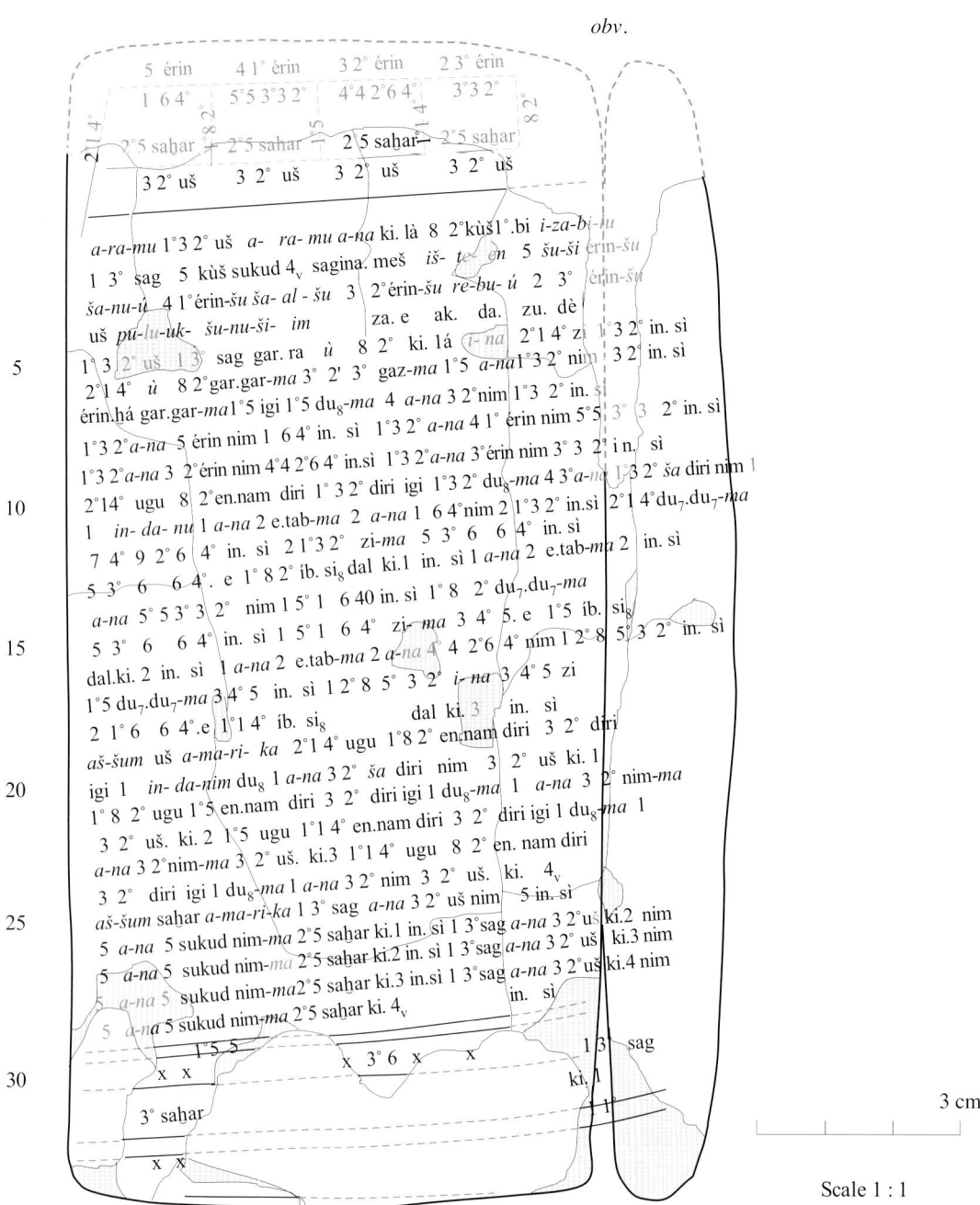

Fig. 10.3.1. MS 2792, *obv*. Conform transliteration. An equally divided ramp with an added layer on top of it.

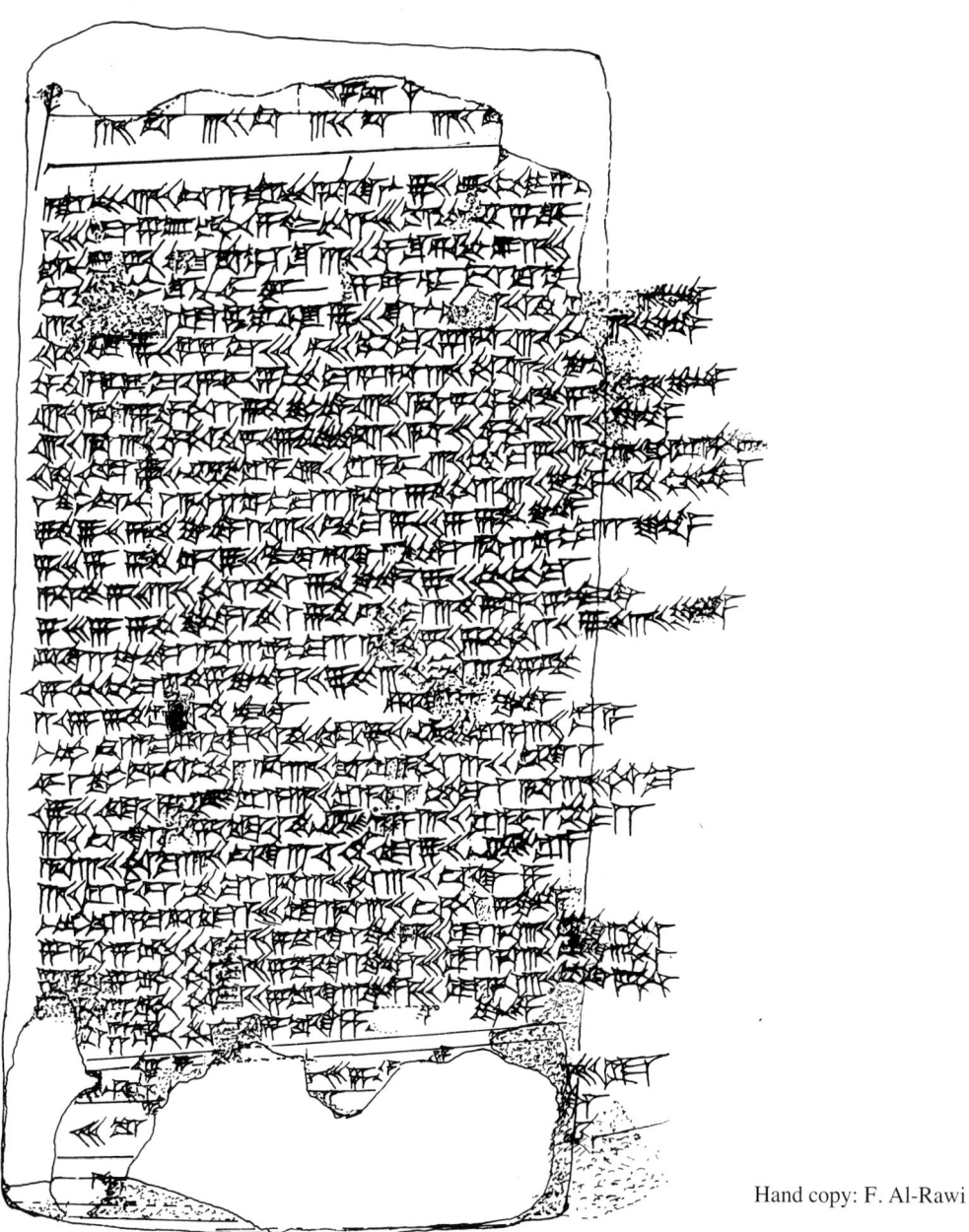

Fig. 10.3.2. MS 2792, *obv*. Hand copy of the cuneiform text.

Hand copy: F. Al-Rawi

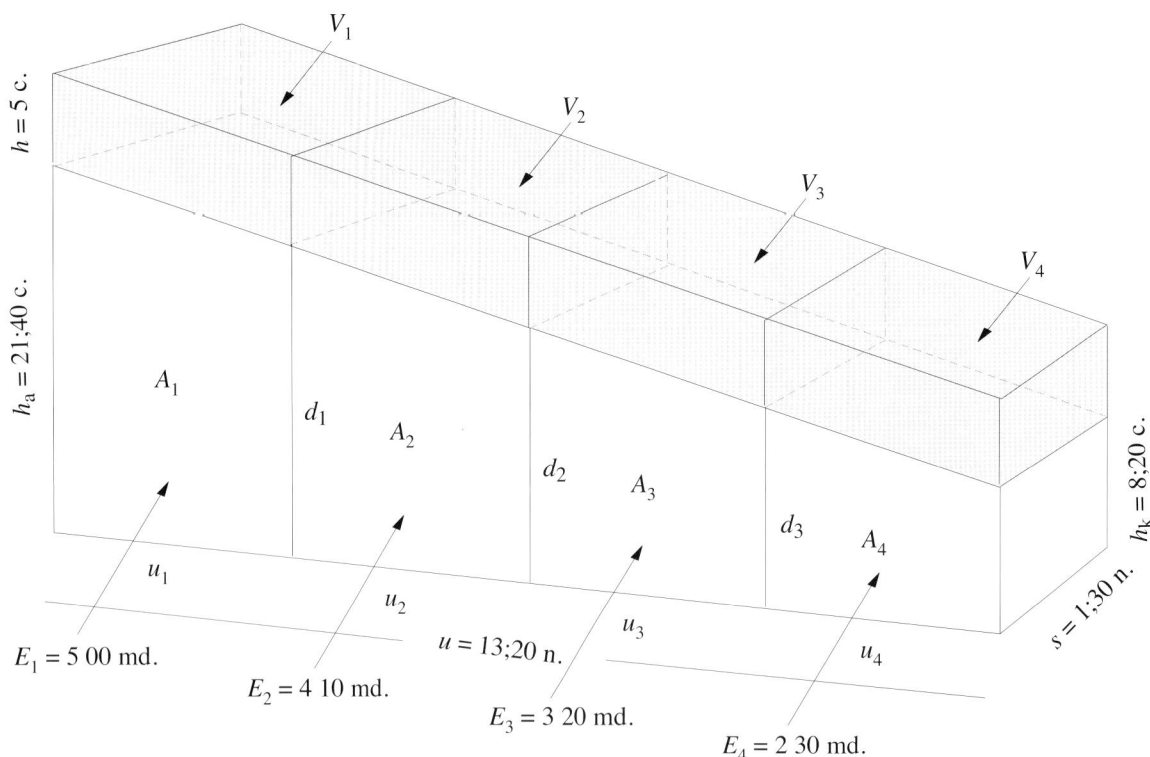

Fig. 10.3.3. MS 2792 # 1. A layer on top of a ramp, built by four troops of soldiers in sections of equal length.

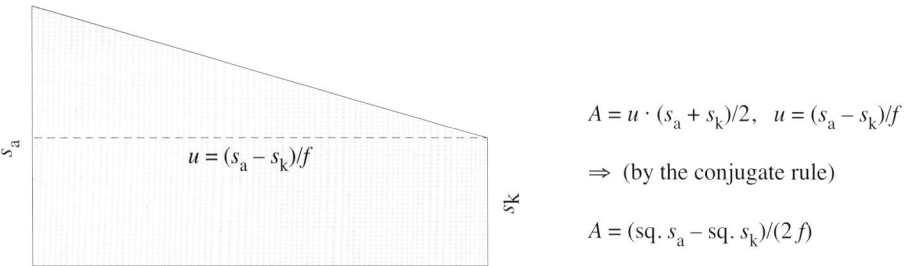

$$A = u \cdot (s_a + s_k)/2, \quad u = (s_a - s_k)/f$$

$\Rightarrow$ (by the conjugate rule)

$$A = (\mathrm{sq.}\ s_a - \mathrm{sq.}\ s_k)/(2\,f)$$

Fig. 10.3.4. MS 2792 # 1. The Old Babylonian alternative trapezoid area rule.

The Old Babylonian "alternative trapezoid area rule", derived as in Fig. 10.3.4, states that

$$A = (\mathrm{sq.}\ s_a - \mathrm{sq.}\ s_k)/(2\,f).$$

This is the explanation for the following computations of the three trapezoid transversals (see Fig. 10.3.3) in the solution procedure of MS 2792 # 1:

sq. s_a – sq. d_1 =	$(2\,f) \cdot A_1 = (2 \cdot 1\ \mathrm{c./n.}) \cdot 1\ 06{;}40\ \mathrm{n.\,c.} = 2\ 13{;}20$ sq. c.	(lines 11-13)
sq. s_a =	sq. (21;40 n.) = 7 49;26 40 sq. n.	
sq. d_1 =	sq. s_a – (sq. s_a – sq. d_1) = sq. s_a – 2 $f \cdot A_1$ = (7 49;26 40 – 2 13;20) sq. n. = 5 36;06 40 sq. n.	
d_1 =	sqs. (5 36;06 40 sq. n.) = 18;20 c.	

sq. d_1 – sq. d_2 =	$(2\,f) \cdot A_2 = (2 \cdot 1\ \mathrm{c./n.}) \cdot 55{;}33\ 20\ \mathrm{n.\,c.} = 1\ 51{;}06\ 40$ sq. c.	(lines 13-16)
sq. d_1 =	sq. (18;20 n.) = 5 36;06 40 sq. n.	
sq. d_2 =	sq. d_1 – (sq. d_1 – sq. d_2) = sq. d_1 – 2 $f \cdot A_2$ = (5 36;06 40– 1 51;06 40) sq. n. = 3 45 sq. c.	
d_2 =	sqs. (3 45 sq. n.) = 15 c.	

sq. d_2 – sq. d_3 = $(2\,f) \cdot A_3 = (2 \cdot 1\ \mathrm{c./n.}) \cdot 44{;}26\ 40\ \mathrm{n.\,c.} = 1\ 28{;}53\ 20$ sq. c.(lines 16-18) sq. d_2 = sq. (15 n.) = 3 45 sq. n.

sq. d_3 = sq. d_2 – (sq. d_2 – sq. d_3) = sq. d_2 – 2 $f \cdot A_3$ = (3 45 – 1 28;53 20) sq. n. = 2 16;06 40 sq. c.

d_3 = sqs. (2 16;06 40 sq. n.) = 11;40 c.

The solution procedure continues with the phrase

| *aššum* uš *amārika* | in order for you to see the length(s), |

meaning something like 'since you also want to know the lengths'.[4] The four partial lengths (see Fig. 10.3.3) are computed as follows:

$u_1 =$	$(s_a - d_1)/f =$	(21;40 – 18;20) c. / (1 c./n.) =	3;20 n.	(20 m)	(lines 19-24)
$u_2 =$	$(d_1 - d_2)/f =$	(18;20 – 15) c. / (1 c./n.) =	3;20 n.	(20 m)	
$u_3 =$	$(d_2 - d_3)/f =$	(15 – 11;40) c. / (1 c./n.) =	3;20 n.	(20 m)	
$u_4 =$	$(d_3 - d_4)/f =$	(11;40 – 8;20) c. / (1 c./n.) =	3;20 n.	(20 m)	

Thus, in this problem the partial lengths are equal.

The final part of the solution procedure is preceded by the phrase

| *aššum* saḫar *amārika* | in order for you to see the mud (the volumes), |

clearly meaning that it remains to compute the volumes of the added layers on top of the four sections of the ramp. The simple computations are as follows:

$V_1 =$	$s \cdot u_1 \cdot h =$	1;30 n. · 3;20 n. · 5 c. = 25 n. n. c.	(lines 25-29)
$V_2 =$	$s \cdot u_2 \cdot h =$	1;30 n. · 3;20 n. · 5 c. = 25 n. n. c.	
$V_3 =$	$s \cdot u_3 \cdot h =$	1;30 n. · 3;20 n. · 5 c. = 25 n. n. c.	
$V_4 =$	$s \cdot u_4 \cdot h =$	1;30 n. · 3;20 n. · 5 c. = 25 n. n. c.	

10.3 b.   *MS 2792 # 2. A Layer on Top of a Ramp Divided Unequally along the Length*

The problem considered in problem # 2 on the *reverse* of MS 2792 (Fig. 10.3.5)is of the same type as problem # 1 on the obverse, only with partly different data. Thus, in this new problem, a ramp is again divided into four parts, the length of the ramp is 13;20 n. and the front is 1;30 c. as before. See Fig. 10.3.7 below.

However, the height of the added layer is now 8 c. instead of 5 c., the upper height of the ramp is 23;20 c. instead of 21;40 c., and the lower height of the ramp is 10 c. instead of 8;20 c. Nevertheless, the difference between the upper and the lower height is unchanged, so that also the growth rate f remains the same.

In the exercise MS 2792 # 1, the numbers of 'soldiers' needed for the construction of the four sections of the ramp form an arithmetical progression, but in the exercise on the reverse, the corresponding numbers of 'soldiers' are chosen more at random:

man-days for section 1	$E_1 =$ 4 30 'soldiers'	(270)
man-days for section 2	$E_2 =$ 8 00 'soldiers'	(480)
man-days for section 3	$E_3 =$ 6 40 'soldiers'	(400)
man-days for section 4	$E_4 =$ 7 30 'soldiers'	(450)

In spite of the differences between problem # 1 on the obverse and problem # 2 on the reverse, there is no visible difference between the drawing on the obverse and the one on the reverse.

---

4. Similar phrases are used also in MS 3052 §§ 1 a and 1 d (Sec. 10.2 a above), and in MS 3971 § 3 a (Sec. 10.1 c above). Cf. Neugebauer and Sachs, *MCT*, p. 50, text D = **YBC 4608** (Group 3; Uruk): *aš-šum* sag an.na *ù* sag ki.ta *a-ma-ri-i-ka* (lines 21-22), and *aš-šu* a.šà *a-ma-ri-i-ka* (line 28). Cf.also in **AO 6555**, a text from the first millennium BC (Friberg, *BagM* 28 (1997), 298) § 4, 1-2: *mín-da-a-ti ki-gal-li* / é.te.me.en.an.ki / uš *ù* sag *a-na* igi.du₈.[a-*ka*] / 3(1.šu) uš 3(1.šu) sag / *ina*1 kùš *as₄-lum* 'Measures of the base of Etemenanki (the ziggurat of Babylon). In order for you to see the length and the front: 3 sixties the length, 3 sixties the front, in the *aslu* cubit.' and in § 5, 1-2: *ša-ni-iš* ka+x.meš ki.gal é.te.me.en.an.ki uš *ù* sag *a-na a-ma-ri-*[*ka*] / [10 ninda u]š 10 ninda sag *i-na* 1 kùš a.rá-*e* 'Secondly. Measures(?) of the base of Etemenanki. In order for you to see the length and the front: *10 n. the length*, 10 n. the front, in the *arû* cubit.'

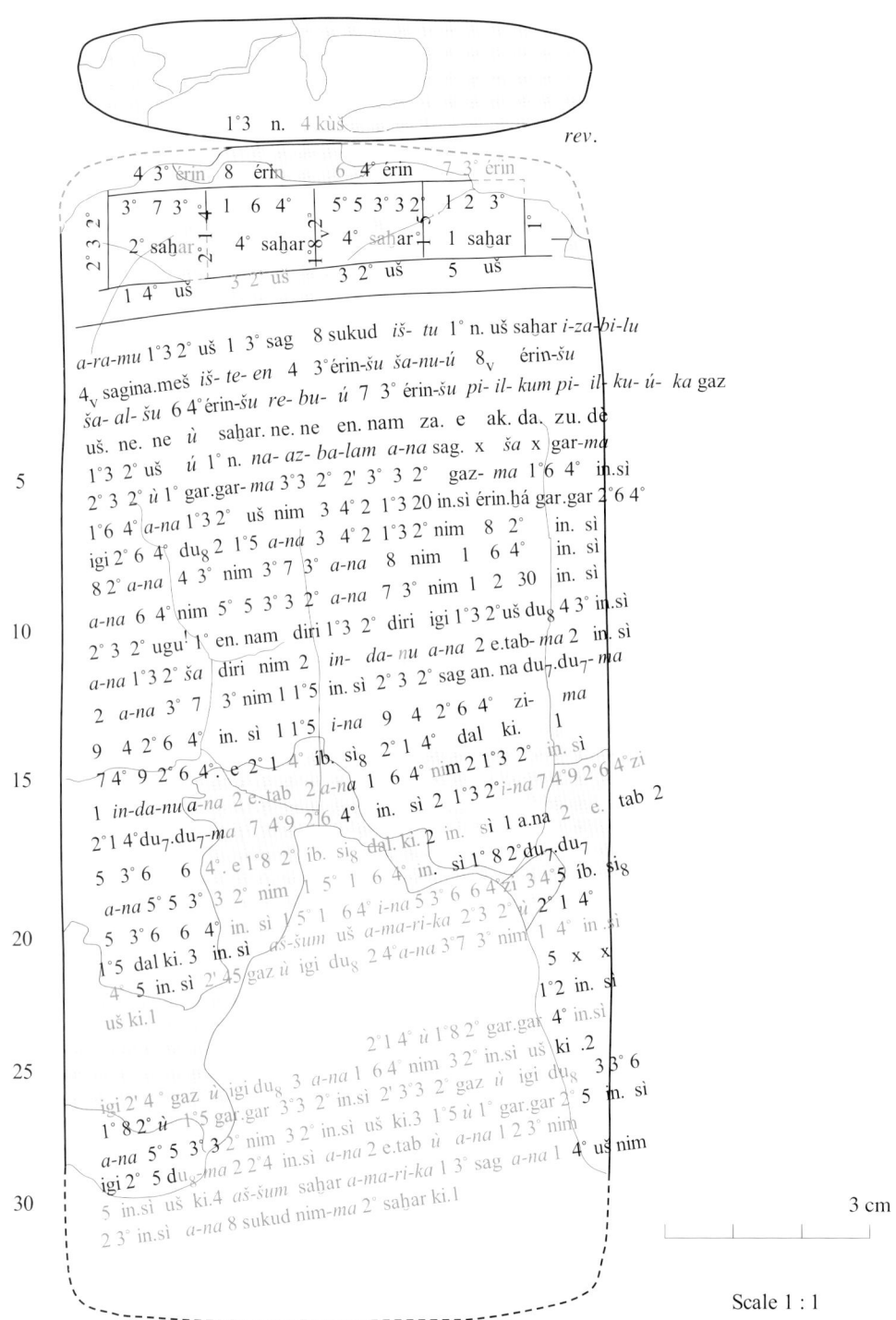

Fig. 10.3.5. MS 2792, *rev.* Conform transliteration. An unequally divided ramp with an added layer on top of it.

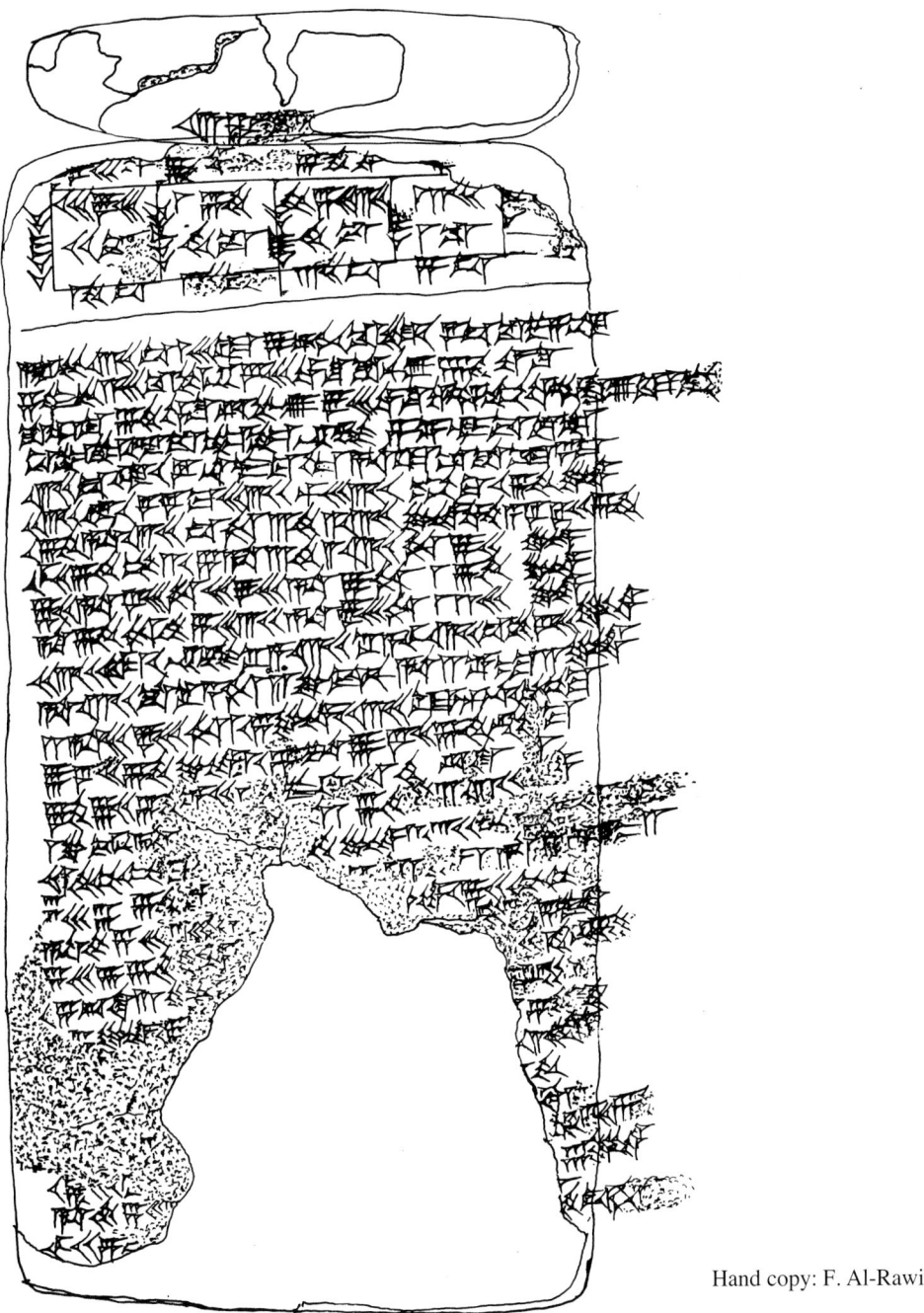

Hand copy: F. Al-Rawi

Fig. 10.3.6. MS 2792, *rev*. Hand copy of the cuneiform text.

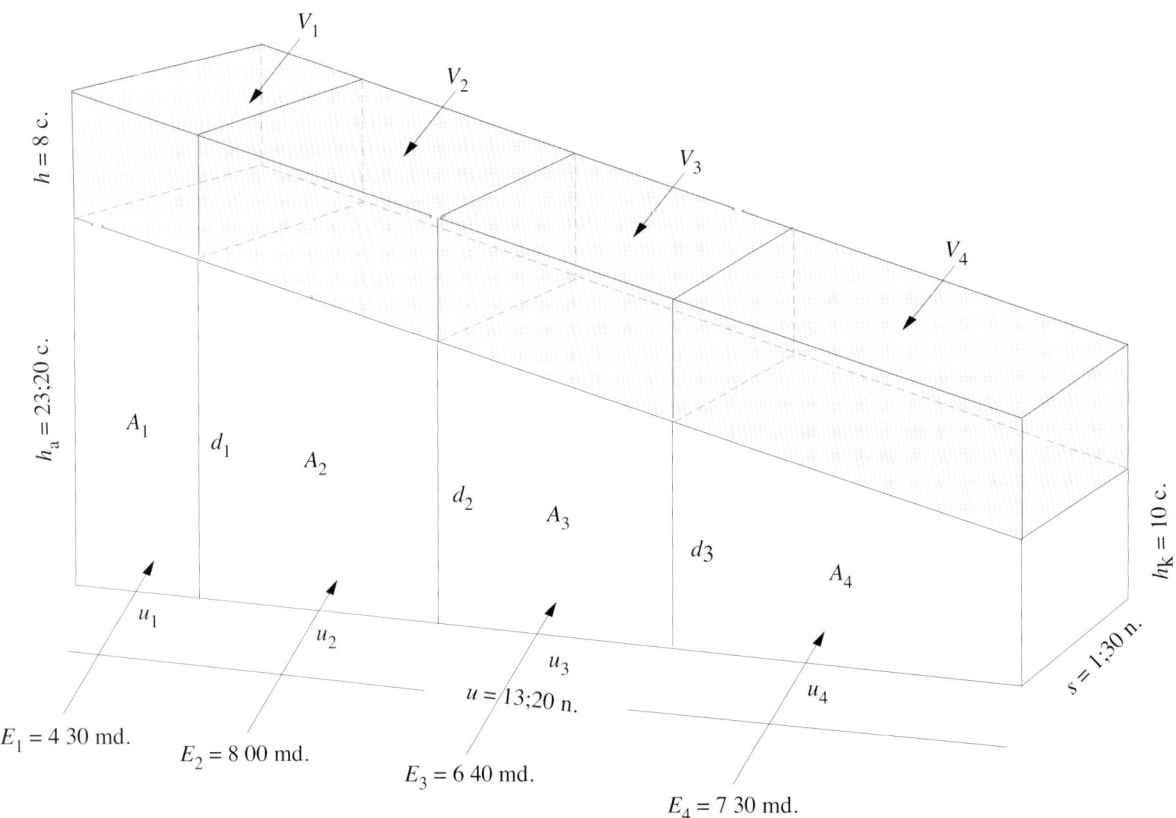

Fig. 10.3.7. MS 2792 # 2. A layer on top of a ramp, built by four troops of soldiers in sections of unequal length.

MS 2792 # 2 (the reverse)

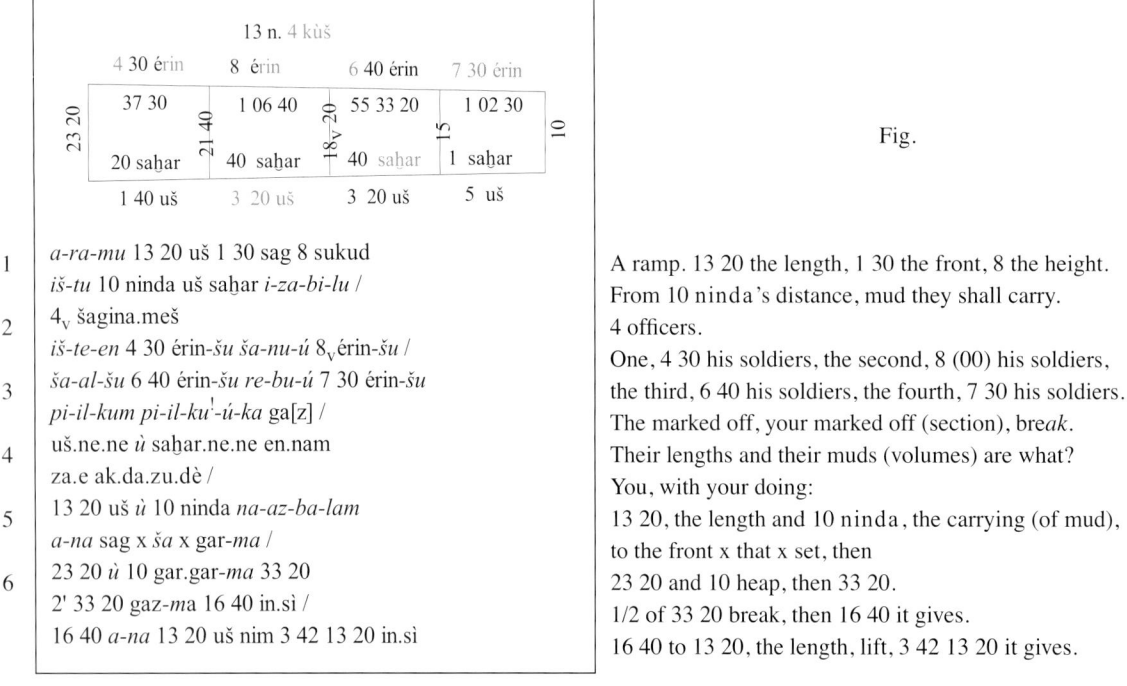

1	*a-ra-mu* 13 20 uš 1 30 sag 8 sukud *iš-tu* 10 ninda uš saḫar *i-za-bi-lu* /	A ramp. 13 20 the length, 1 30 the front, 8 the height. From 10 ninda's distance, mud they shall carry.
2	4ᵥ šagina.meš *iš-te-en* 4 30 érin-*šu ša-nu-ú* 8ᵥ érin-*šu* /	4 officers. One, 4 30 his soldiers, the second, 8 (00) his soldiers,
3	*ša-al-šu* 6 40 érin-*šu re-bu-ú* 7 30 érin-*šu* *pi-il-kum pi-il-ku'-ú-ka* ga[z] /	the third, 6 40 his soldiers, the fourth, 7 30 his soldiers. The marked off, your marked off (section), bre*ak*.
4	uš.ne.ne *ù* saḫar.ne.ne en.nam za.e ak.da.zu.dè /	Their lengths and their muds (volumes) are what? You, with your doing:
5	13 20 uš *ù* 10 ninda *na-az-ba-lam* *a-na* sag x *ša* x gar-*ma* /	13 20, the length and 10 ninda, the carrying (of mud), to the front x that x set, then
6	23 20 *ù* 10 gar.gar-*ma* 33 20 2' 33 20 gaz-*ma* 16 40 in.sì / 16 40 *a-na* 13 20 uš nim 3 42 13 20 in.sì	23 20 and 10 heap, then 33 20. 1/2 of 33 20 break, then 16 40 it gives. 16 40 to 13 20, the length, lift, 3 42 13 20 it gives.

7	érin.ḫá gar.gar 26 40 /	The soldiers heap, 26 40.
	igi 26 40 du₈ 2 15	The opposite of 26 40 resolve, 2 15.
8	a-na 3 42 13 20 nim 8 20 in.sì /	To 3 42 13 20 lift, 8 20 it gives.
	8 20 a-na 4 30 nim 37 30	8 20 to 4 30 lift, 37 30,
9	a-na 8 nim 1 06 40 in.sì /	to 8 lift, 1 06 40 it gives.
	a-na 6 40 nim 55 33 20	To 6 40 lift, 55 33 20,
10	a-na 7 30 nim 1 02 30 in.sì /	to 7 30 lift, 1 02 30 it gives.
	23 20 ugu 10 en.nam diri 13 20 diri	23 20 over 10 is what beyond? 13 20 beyond.
11	igi 13 20 uš du₈ 4 30 in.sì /	The opposite of 13 20, the length, resolve, 4 30 it gives.
	a-na 13 20 ša diri nim 2 in-da-[n]u	To 13 20, what is beyond, lift, 1! the *indanum*.
12	a-na 2 e.tab-ma 2 in.sì /	To 2 you repeat, then 2 it gives.
	2 a-na 37 30 nim 1 15 in.sì	2 to 37 30 lift, 1 15 it gives.
13	23 20 sag.an.na du₇.du₇-ma /	23 20 the upper front (make) butt (itself), then
	9 04 26 40 in.sì	9 04 26 40 it gives.
14	1 15 i-na 9 04 26 40 zi-ma /	1 15 from 9 04 26 40 tear off, then
	7 49 26 40.e 21 [40] íb.si₈	7 49 26 40, makes 21 *40* equalsided.
15	21 40 dal ki.1 /	21 40 the 1st transv*ersal*.
	1 in-da-nu a-[na 2.e.tab 2]	1, the *indanum*, *to 2 repeat, 2.*
16	[a-n]a 1 06 40 [ni]m 2 13 20 [in.s]ì /	*To 1 06 40 lift, 2 13 20 it gives.*
	21 40 du₇.du₇-m[a 7 49 26] 40 in.sì	21 40 (make) butt (itself), the*n 7 49 26 40 it gives.*
17	2 13 20 [i-na 7 49 26 40 zi] /	2 13 20 *from 7 49 26 40 tear off,*
	5 36 06 [40.e 18 20 íb.si₈]	5 36 06 *40, makes 18 20 equalsided,*
18	[dal ki.]2 in.sì	the *2nd transversal* it gives.
	1 a-na [2 e].tab 2 /	1 to *2 you repeat, 2.*
	a-na 55 3[3 20 nim 1 51 06 40 in].sì	To 55 *33 20 lift, 1 51 06 40 it gives.*
19	18 20 du₇.du₇ / 5 36 06 40 [in.sì]	18 20 (make) butt (itself), 5 36 06 40 *it gives.*
20	[1 51 06 40 i-na 5 36 06 40 zi 3 4]5	*1 51 06 40 from 5 36 06 40 tear off, 3 45.*
	íb.si₈ / 15 dal ki.3 i[n.sì]	Equalside 15, the 3rd transversal *it gives.*
21	[aš-šum uš a-ma-ri-ka]	In order for you to see the length:
	[23 20 ù 2]1 40 [gar.gar / 45 in.sì]	*23 20 and 21 40 heap, 45 it gives.*
22	[2' 45 gaz ù igi du₈ 2 40]	1/2 of 45 break and the opposite release, 2 40.
	[a-na 37 30 nim 1 40 in.sì / uš ki.1]	To 37 30 lift, 1 40 it gives, the 1st length.
23	[… … … … …] 5 x x /	… … … … … 5 x x.
	[… … … … … … … … …] 12 in.sì /	… … … … … … … …, 12 it gives.
24	[… … … … …]	… … … … …
25	[21 40 ù 18 20 gar.gar] 40 [in.sì] /	*21 40 and 18 20 heap, 40 it gives.*
	[2' 40 gaz ù igi du₈ 3]	1/2 of 40 break and the opposite release, 3.
26	[a-na 1 06 40 nim 3 20 in.sì] uš ki.2 /	To 1 06 40 lift, 3 20 it gives, the 2nd length.
	18 20 ù [15 gar.gar 33 20 in.sì]	18 20 and 15 heap, 33 20 it gives.
27	[2' 33 20 gaz ù igi du₈] 3 36 /	1/2 of 33 20 break and the opposite release, 3 36.
	a-na 55 33 [20 nim 3 20 in.sì uš ki.3]	To 55 33 20 lift, 3 20 it gives, the 3rd length.
28	[15 ù 10 gar.gar 2]5 in.sì /	15 and 10 heap, 25 it gives.
	igi 25 d[u₈-ma 2 24 in.sì]	The opposite of 25 release, then 2 24 it gives. To 2 repeat and
29	[a-na 2 e.tab ù a-na 1 02 30 nim] /	to 1 02 30 lift,
	[5 in.sì uš ki.4]	5 it gives, the 4th length.
30	[aš-šum saḫar a-ma-ri-ka]	In order for you to see the mud:
	[1 30 sag a-na 1] 40 uš nim / [2 30 in.sì]	1 30 the front to 1 40 the length lift, 2 30 it gives.
31	[a-na 8 sukud nim-ma 20 saḫar ki.1]	to 8 the height lift, then 20, the 1st mud.
	etc.	*etc.*

The solution procedure for the problem on the reverse of MS 2792 (# 2) is essentially identical with the solution procedure for the problem on the obverse. The computed values are the following:

$A =$	3 42;13 20 n. c.			(line 7)
$E =$	26 40 'soldiers'			(line 8)
$A/E =$	;08 20 n. c./md.			(line 9)
$A =$	37;30 n. c.,	$A_2 = 1\ 06;40$ n. c.,	$A_3 = 55;33\ 20$ n. c.,	$A_4 = 1\ 02;30$ n. c	(lines 10-11)
$f =$	1 c./n.			(lines 12-13)
$d_1 =$	21;40 c.,	$d_2 = 18;20$ c.,	$d_3 = 15$ c.		(lines 14-22)
$[u_1 =$	1;40 n.,	$u_2 = 3;20$ n.,	$u_3 = 3;20$ n.,	$u_4 = 5$ n.]	(lines 22-30)
$[V_1 =$	20 n. n. c.,	$V_2 = 40$ n. n. c.,	$V_3 = 40$ n. n. c.,	$V_4 = 1\ 00$ n. n. c.]	(lines 30 ff)

While the solution procedure in this exercise is relatively clear, the statement of the question in lines 1-4 is obscure and incomplete. It begins by mentioning 'the length 13 20', 'the front 1 30' and 'the height 8'. These are apparently the dimensions of the added layer on top of the ramp. The question continues by mentioning that something is carried 'here', presumably to the construction site, from a distance of 10 ninda. (Something similar is mentioned, even more cryptically, in the question at the beginning of the exercise on the obverse.) There is nothing in the solution procedure that explains what this is about. Next, the question gives the four numbers of 'soldiers', and it ends by asking for the 'lengths' and 'volumes', presumably of the added layers on top of the four sections of the ramp.

10.3 c.   The Work Norm for Building a Ramp

It might be a good idea to investigate what the ratio is in MS 2792 ## 1-2 between the number of 'soldiers' needed to build the ramps, and the volumes of the ramps. In # 1, the volume of the ramp can be computed as

$$V = 13;20 \text{ n.} \cdot 1;30 \text{ n.} \cdot (21;40 + 8;20)/2 \text{ c.} = 20 \text{ sq. n.} \cdot 15 \text{ c.} = 5\ 00 \text{ sq. n.} \cdot \text{c. (volume-šar)}.$$

Hence, the number of volume-units per 'soldier' in this case is

$$V/E = 5\ 00 \text{ volume-šar} / 15\ 00 \text{ 'soldiers'} = ;20 \text{ volume-šar/'soldier'}.$$

In view of what work-norms look like in other Old Babylonian mathematical texts (see Friberg, *RlA 7* (1990) Sec. 5.6 h), this almost certainly means that the word 'soldier' must be interpreted as a word for 'man-day', a unit of labor corresponding to 1 man working for a full day. In other words, in this text, *the work norm for building a ramp is 20 volume-shekels per man-day*, a volume-shekel being a sixtieth of a volume-šar.

In # 2, the work norm ought to be the same. However, in that case

$$V = 13;20 \text{ n.} \cdot 1;30 \text{ n.} \cdot (23;20 + 10)/2 \text{ c.} = 20 \text{ sq. n.} \cdot 16;40 \text{ c.} = 5\ 33;20 \text{ sq. n.} \cdot \text{c. (volume-šar)}.$$
$$V/E = 5\ 33;20 \text{ volume-šar}/26\ 40 \text{ 'soldiers'} = ;12\ 30 \text{ volume-šar/'soldier'}.$$

The conflict between the two implicitly given values for the work norm is probably due to a simple mistake on the part of the author of the text. In exercise # 1, the work norm is '20', the man-days (or 'soldiers') '15', and the volume '13 20' (= 20 · 15). In # 2, on the other hand, the man-days are '26 40' and the volume '5 33 20', where 5 33 20 can be factorized as 20 · 16 40. Therefore, it is likely that a mistake made by the author of the text was to let the number of 'soldiers' be 26 40 instead of 16 40!

10.3 d.   The Construction of the Data for the Two Problems

It is now obvious how the data for exercise # 2 were constructed with departure from the data for exercise # 1. The author of the text wanted the problem in # 2 to be a more complicated version of the problem in # 1. For that purpose, he added the same amount, 1;40 c., to both the upper and the lower height of the ramp, keeping the rate of decrease of the ramp the same as before. In doing so he changed the volume from 5 00 = 20 · 15 in # 1 to 5 33 20 = 20 · 16 40 in # 2. Assuming that he had not made the mentioned mistake, he would then have let the number of 'soldiers' be 16 40.

Another change made in the construction of the problem on the reverse of MS 2792 is that in this second problem the length of the ramp is divided into four *unequal* parts, whereas in the problem on the obverse the length was divided into four equal parts. In spite of the differences, however, both problems are solved in essentially the same way.

Actually, it is hard to know precisely how the partial lengths and the added volumes were computed in MS 2792 #2, since more than a third of the text is lost on the reverse of the tablet, perhaps as much as the fifteen last lines of the text. (See Figs. 10.3.5-6.) Repeated attempts to reconstruct the lost text have been only partly successful. (In particular, it is not at all clear what is going on in lines 23-25.) One of the difficulties is that the author of the text apparently chose not to use the same solution procedure in # 2 as in # 1 for the computation of the lengths of the four sections of the ramp. Thus, in # 1, lines 19-24, the four partial lengths are computed as follows:

$$u_1 = (s_a - d_1)/f, \; etc.$$

In # 2, lines 22-23 and 25-30, on the other hand, the four partial lengths are computed as

$$u_1 = 2\, A_1 / (s_a + d_1), \; etc.$$

There are clearly also other changes to the solution procedure, but what they are is not clear.

In order for the mentioned modified solution procedure in # 2 to work, it is imperative that the sums $s_a + d_1$, $d_1 + d_2$, etc., are *regular* sexagesimal numbers. It is not a trivial task to choose the data for a problem like the one in MS 2792 # 2 so that this requirement is satisfied. For that reason, it is interesting to try to explain how the author of MS 2792 actually constructed the data appearing in the text.

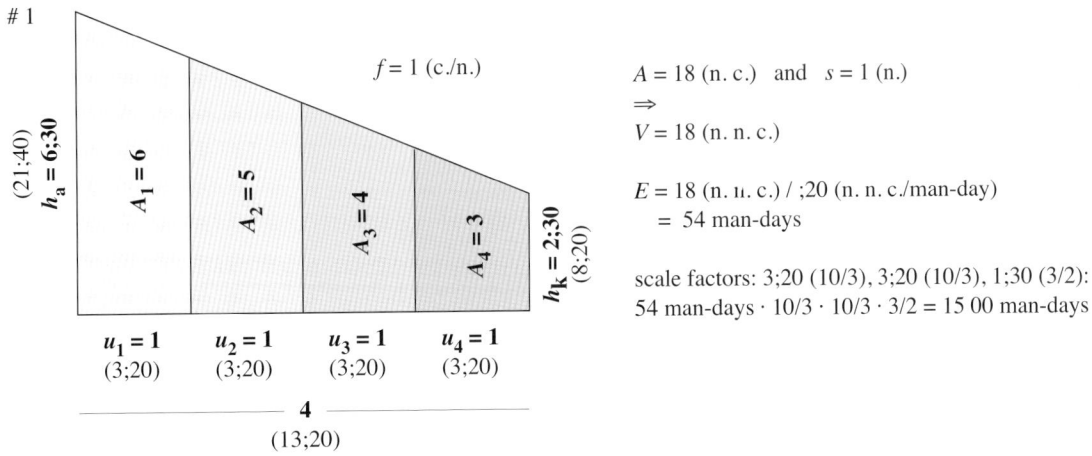

Fig. 10.3.8. A possible construction of the data for MS 2792 # 1. (The figure is not drawn to scale.)

It is likely that the construction of the data for MS 2792 #1 started with a "basic" trapezoid like the one depicted in Fig. 10.3.8 above. In this trapezoid, the length is divided into four equal pieces, all four of length 1. The growth rate for the trapezoidal cross section of the ramp is assumed to be 1, and the upper and lower fronts are set equal to 6;30 and 2;30, respectively. The lengths of the three transversals are then, of course, equal to 5,30, 4;30, and 3;30, and the areas of the four sub-trapezoids are 6, 5, 4, and 3. Therefore, the total area of the cross section is $A = 6 + 5 + 4 + 3 = 18$ (ninda · cubits). If, in addition, the width of the ramp in this basic construction is 1 (ninda), it follows that the volume is $V = 18$ (square ninda · cubits). Thus, finally, if the work norm for building the ramp is 20 volume-shekels per man-day, the needed man power for the whole ramp is

$$E = 18 \; n.\,n.\,c. \,/\; ;\!20\; n.\,n.\,c./man\text{-}day = 54 \; man\text{-}days.$$

In order to get more realistic numbers for the ramp, the author of the problem now decided to scale up the horizontal and vertical dimensions of the cross section by the factor 3;20 (10/3), and the width of the ramp by the factor 1;30 (3/2). The result was a ramp with the length $u = 3;\!20 \cdot 4 = 13;\!20$ (n.), the heights at the two ends equal to $h_a = 3;\!20 \cdot 6;\!30 = 21;\!40$ (c.) and $h_k = 3;\!20 \cdot 2;\!30 = 8;\!20$ (c.), and the width s = 1;30 · 1 = 1;30 (c.). As a result of the change of scale, the needed man power then became

$$E = 10/3 \cdot 10/3 \cdot 3/2 \cdot 54 \; man\text{-}days = 15\,00 \; man\text{-}days.$$

In Fig. 10.3.8, the scaled up numbers are within brackets. Note that $f = 1$ (c./n.) remains the same.

Compare the way in which the sides of a trapezoid are scaled up by a factor 3 in the drawing on the Old Babylonian hand tablet **YBC 11126** (Neugebauer, *MCT* (1945), 44; Fig. 10.3.9 below:

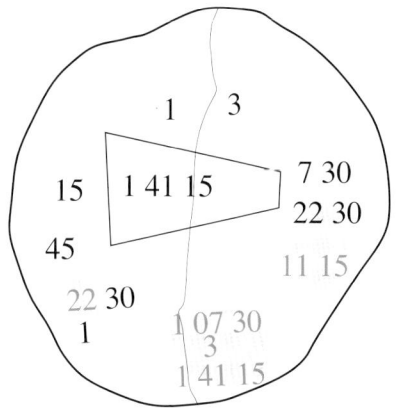

Question (reconstructed):
The given area of a trapezoid is 1 41 15.
The upper front is 1/4 of the length,
and the lower front is half the upper front.
What are the length and the fronts?
Solution by use of the rule of false value:
If the (false) length is 1(00), then the fronts
are 15 and 7;30, and the area 11 15.
The desired "true" area, 1 41 15, is 9 times larger.
Hence the correction factor is sqs. 9 = 3,
the true length is 3 (00), and the true
fronts are 3 · 15 = 45 and 3 · 7;30 = 22;30.

Fig. 10.3.9. YBC 11126. A hand tablet with scaled up numbers for the sides of a trapezoid.

The construction of the data for MS 2792 # 2 was probably similar to the construction of the data for # 1. Presumably, it started with a "basic" divided trapezoid like the one in Fig. 10.3.10 below, an 8-striped trapezoid with the length 8 and the fronts 14 and 6 and, consequently, with the growth rate $f = 1$ (c./n.). The lengths of the 7 transversals are then, of course, equal to $14 - 1 = 13, 13 - 1 = 12$, *etc.* Hence, the area of the first sub-trapezoid is $A_1 = (14 + 13)/2 \cdot 1 = 13;30.$, and the area of each sub-trapezoid is less by 1 than the area of the preceding sub-trapezoid. This means that the 8 sub-areas are

$$13;30 = 27/2, 12;30 = 25/2, 11;30 = 23/2, 10;30 = 21/2, 9;30 = 19/2, 8;30 = 17/2, 7;30 = 15/2, \text{ and } 6;30 = 13/2.$$

Here 27, 25, and 15 are regular sexagesimal numbers, but 23, 21, 19, 17, 14, and 13 are non-regular. Therefore, not all the sub-areas in the basic trapezoid are equal to regular sexagesimal numbers, and neither are all the eight sums

$$s_a + d_1 = 2 A_1/1, \quad d_1 + d_2 = 2 A_2/1, etc.$$

The author of the text cleverly circumvented this difficulty by bunching together some of the sub-rectangles, so that the basic trapezoid became divided in four parts, with the areas equal to

$$A_1 = 13;30 = 27/2, \quad A_2 + A_3 = 12;30 + 11;30 = 24, \quad A_4 + A_5 = 10;30 + 9;30 = 20, \quad \text{and}$$
$$A_6 + A_7 + A_8 = 8;30 + 7;30 + 6;30 = 22;30 = 45/2.$$

In this way he could construct a basic trapezoid divided in four parts, with all the areas of the sub-trapezoids equal to regular sexagesimal numbers. The total area of this basic trapezoid was 1 20 (ninda · cubits). With the width assumed to be 1 (ninda), the volume of the ramp in this basic form would then be 1 20 (square ninda · cubits), corresponding to 1 20 / ;20 = 4 00 man-days of labor. After application of the scale factors 1;40, 1;40, 1;30, the scaled up ramp had the same length (13;20 n.) and width (1;30 n.) as the scaled up ramp in # 1, and the needed man power had increased to

$$1;40 \cdot 1;40 \cdot 1;30 \cdot 4\ 00 \text{ man-days} = 16\ 40.$$

The corresponding numbers of man-days for the four parts of the divided ramp could then be computed by use of proportionality. In this final step of the construction, the author of the problem by mistake counted with proportional parts of 26 40 instead of 16 40.

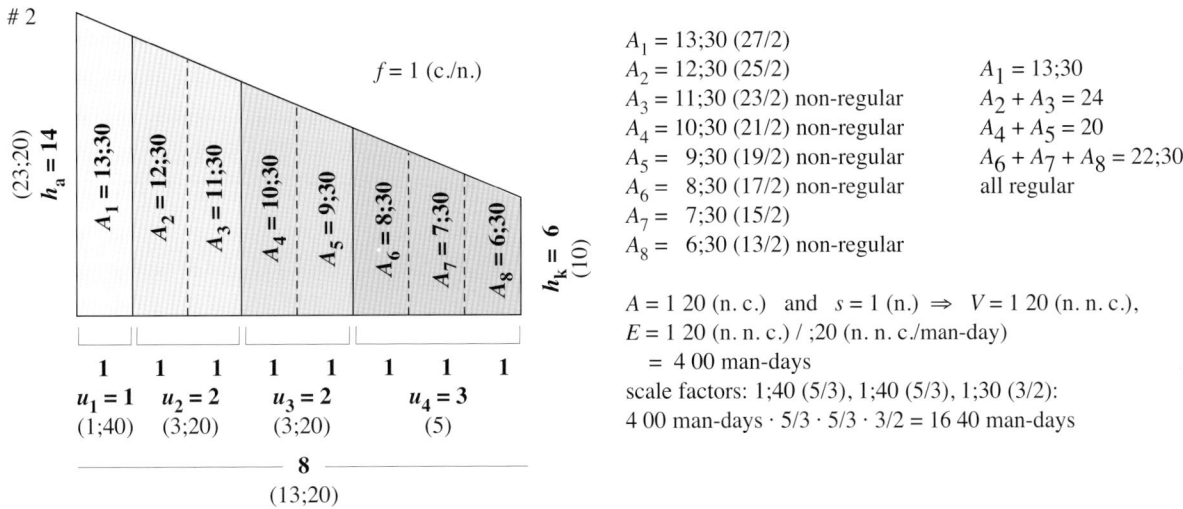

Fig. 10.3.10. A proposed construction of the data for MS 2792 # 2. (The figure is not drawn to scale.)

A Related Text: Str. 362 # 6. A Combined Work Norm for Carrying and Building

Str. 362 (*MKT 1*, 239) is a small Old Babylonian clay tablet, probably from Uruk, with six miscellaneous mathematical exercises. The last exercise is a problem for a ramp built in three sections:

Str. 362 # 6

1	*a-ra-mu*	A ramp.
	10 ninda uš 1 2' ninda sag /	10 n. the length, 1 1/2 n. the front.
2	3 šagina.meš	3 officers.
	3 ninda 4 kùš uš *iš-ba-tu* /	3 n. 4 c. they took.
3	*iš-te-en* 1 *šu-ši* érin	One, 1 sixty soldiers,
	ki.2 1 20 érin /	the 2nd, 1 20 soldiers,
4	ki.3 1 40 érin	the 3rd, 1 40 soldiers.
	iš-tu 5 [*ṣu-up-pa-am*] / saḫar.ḫá *iz-za-bi-*[*lu-nim*] /	From 5 (n.), a *ṣuppān*, mud *they* shall carry *here*.
5	*a-ra-mi šu-up-li* /	The depth of the ramp
6	*ù* saḫar.ḫá *ki ma-*[*ṣi iz-za-bi-lu-nim*]	and mud how m*uch shall they carry here?*

In this problem, a ramp is built by 3 troops of 'soldiers' under the command of three officers. Just as in MS 2792 # 1, the front is 1 1/2 n. (= 9 m), and each troop builds 3;20 n.(= 20 m) along the length of the ramp. The total length of the ramp is then 3 · 3;20 n. = 10 n. The soldiers in the first troop yield 1 00 (60) man-days of work, those in the second troop 1 20 (80) man-days, and those in the third troop 1 40 (100) man-days. The mud to build the ramp is brought from a specified distance, 5 ninda. The purpose of the exercise is to compute the height of the ramp, apparently the same everywhere, and the volume of the ramp. The solution procedure, not given in the text (nor in the commentary in Neugebauer, *MKT 1* (1935)), would probably have been as follows:

Assume that in this text, as in MS 2792 # 1 (and # 2!), the work norm for building a ramp out of mud is ;20 (= 1/3) volume-šar/man-day. However, since the mud is to be carried to the ramp from some place 5 n. away, the work norm has to be correspondingly reduced. The standard work norm for carrying mud in Old Babylonian mathematical texts is 1;40 volume-šar · ninda/man-day. (See § 7.3 a in connection with the discussion there of the table of constants on MS 2221, *obv*.) That work norm is the same as ;20 volume-šar · 5 ninda per man-day. Consequently, what is required in order to carry mud to the construction site over a distance of 5 n. and then build the ramp is 6 man-days per volume-šar, namely *3 days for the carrying* and *3 days for the building*. Now, in Str. 362 # 6, the given number of man-days is 1 00 + 1 20 + 1 40 = 4 00. Hence, the volume of the ramp must be 4 00 man-days · ;10 volume-šar/man-day = 40 volume-šar (sq. n. · c.). Since the length of the ramp is 10 n. (60 m) and the front 1 1/2 n. (9 m), the constant height of the ramp must be 40 sq. n. · c. / 15 sq. n. = 2;40 c. = (1 1/3 m).

The obvious similarities between Str. 362 # 6 and MS 2792, make it fairly reasonable to conjecture that both texts are (copies of) extracts from an original large theme text dealing with divided ramps and single or combined work norms. This would help to explain some peculiar features of MS 2792 ## 1-2, namely that the statements of the questions are so incomplete, and that they mention the carrying (of mud) from some distance, although this information is never used in the solution procedures.

Another, somewhat more distant, relative to MS 2792 is **BM 85194 # 1** (*MKT 1*, 143; from Sippar), an isolated exercise in a large recombination text. In that exercise, an *a-ra-am-mu-um* 'ramp' of a rather complicated form has the length 10 ninda and different trapezoidal cross sections of given dimensions at the lower end ('at the base of the mud') and at the upper end ('in front of the gate'). The work norm is ;10 volume-šar/man-day, and is explicitly referred to as 10 éš.kàr. The stated question is the following (cf. MS 2792 # 1, l. 4):

saḫar.ḫá en.nam *a-na* 1 lú uš *pu-lu-uk*   the mud is what? to 1 man, (his) length mark off.

In the solution procedure, it is shown that the volume of the ramp is 50 volume-šar, and that the needed amount of labor is 5 00 man-days (expressed in the form 5 érin '5 soldiers'). The final answer is computed as 10 n. / 5 00 érin = ;02 n. (= 20 cm) / érin and is expressed in the following form:

2 1 lú *i-ṣa-ba-at*   2 is (what) 1 man takes.

This is, of course, nonsense, and also mathematically incorrect. (Cf. Thureau-Dangin, *TMB* (1938), 22, fn. 1: "Ce n'est qu'une moyenne, car, dans le calcul, il n'a pas éte tenu compte de la différence de hauteur d'une extrémité à l'autre de l'ouvrage.")

The small work norm ;10 volume-šar in BM 85194 # 1, can be explained as follows: This exercise is an excerpt from a large theme text. It is possible that it was stated in some of the preceding exercises in the theme text that the mud for the construction of the ramp had to be carried there from a distance of 5 ninda, precisely as in Str. 362 # 6. In that case, 6 man-days would be needed for the construction of 1 volume-šar, and the *combined* work norm would be ;10 volume-šar/man-day.

The work norm ;20 volume-šar/man-day for some kind of construction work, possibly the building of a ramp, is behind the following curious series of entries in an Old Babylonian table of constants (G = **IM 52916**; Goetze (1951); cf. Friberg, *ChV* (2001), 6.5):

na-az-ba-al saḫar	1 40 *i-gi-gu-bu*	carrying of mud,	1 40 the constant	G *rev.* 23'
a.na 40 ninda *a-za-bi-il*	2 13 20 *al-lu-um*	for 40 ninda I shall carry,	2 13 20 the *allum*	G *rev.* 25'
a.na 20 ninda *a-za-bi-il*	4 *al-lu-um*	for 20 ninda I shall carry,	4 the *allum*	G *rev.* 26'
a.na 15 ninda [*a-za-bi-il*]	[5] *al-lu-um*	for 15 ninda *I shall carry*,	5 the *allum*	G *rev.* 27'
[a.na] 10 ninda *a-za-[bi-il*]	6 40 *al-lu-um*	*for* 10 ninda I shall carry,	6 40 the *allum*	G *rev.* 28'
[a.na 5] ninda *a-za-[bi-il*]	10 *al-lu-um*	*for* 5 ninda I shall carry,	10 the *allum*	G *rev.* 29'

The last of these entries, for instance, must be interpreted as stating that the combined work norm for carrying from a distance of 5 ninda and building, say, a ramp, is ;10 volume-šar/man-day. This is the case in Str. 362 # 6, and also, possibly, in BM 85194 # 1. In MS 2792, on the other hand, where the distance is 10 ninda, hence, according to this table, the combined work norm would be ;06 40 volume-šar/man-day.

11
Three Problem Texts Not Belonging to Any Known Group of Texts

11.1. MS 3049. A Fragment of a Mathematical Recombination Text

11.1 a. MS 3049 § 1 a. Computing the Length of a Chord in a Circle

obv.

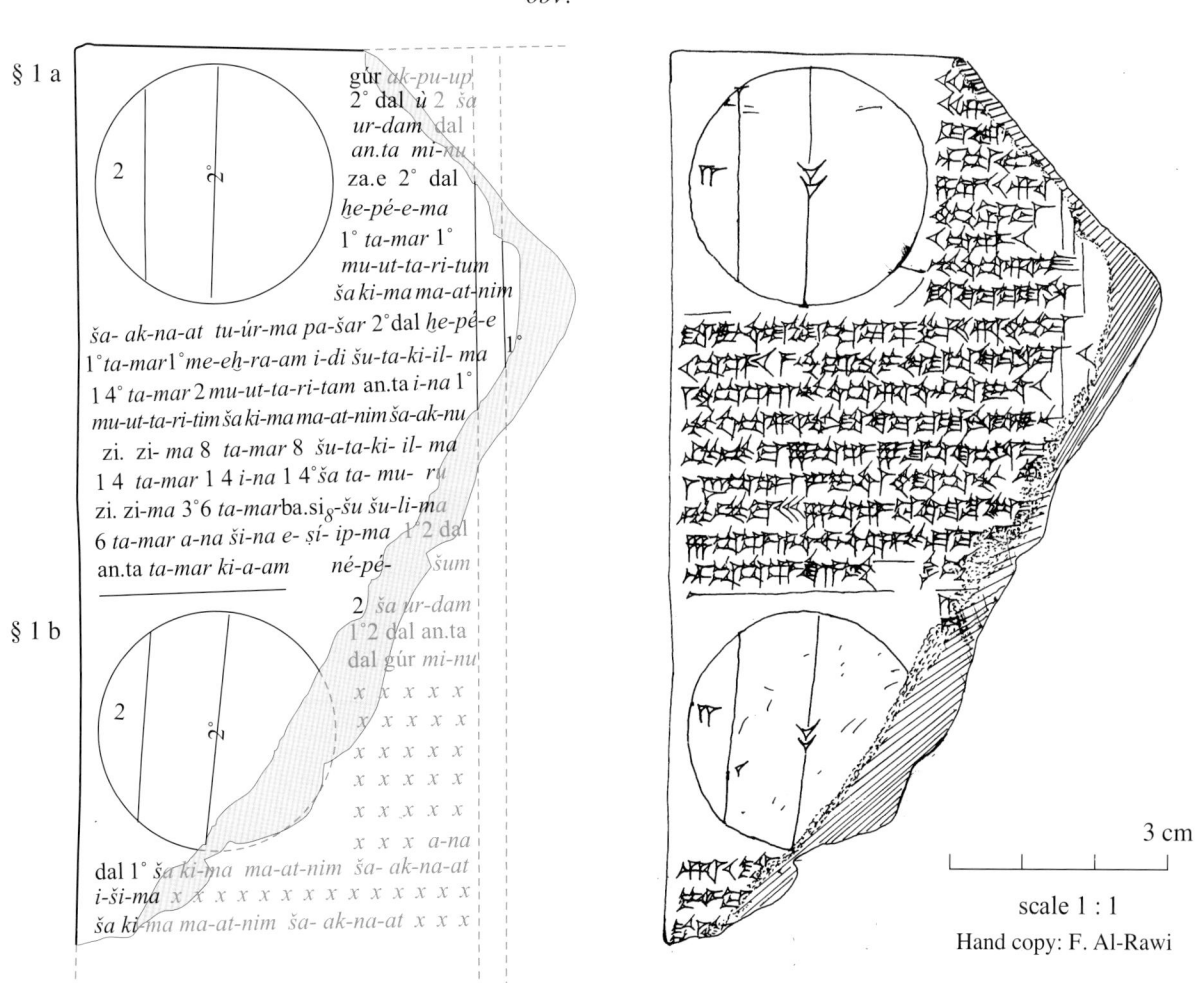

Fig. 11.1.1. MS 3049, *obv*. § 1 a: Computing the length of a chord in a circle; § 1 b: Computing the diameter(?).

MS 3049 (Figs. 11.1.1, 11.1.4, and 11.1.6) is a fragment of a large mathematical cuneiform text. By luck, a notation on the edge has not been lost (Fig. 11.1.4). It mentions the number of exercises in the text:

šu.nigin mu.bi 16   together its name(s) 16

What this means is that the text originally comprised 16 individual mathematical exercises. On the obverse, the first of these 16 exercises is preserved, as well as a small part of the second exercise. The two exercises, here called §§ 1 a-b, are accompanied by identical illustrations, drawings of a circle with a diameter, a chord, and a couple of numbers. Only a few words of the text of § 1 a are lost, those near the ends of the first four, very short lines of text, and those near the ends of the last five lines, which are of full length. Unfortunately, this means that most of the question is lost, the statement of the problem. On the other hand, almost the whole solution procedure is preserved, and it is possible to reconstruct what is lost of the question.

MS 3049 § 1 a

1	gúr¹ [ak-pu-up] /	An arc I curved,
2-3	20 dal [ù 2 ša] / ur-dam	20 the transversal, and 2 that which I went down.
4	[dal] / an.ta mi-[nu] /	The upper transversal (is) what?
5	za.e	You:
6-7	20 dal / ḫe-pé-e-ma / 10 ta-mar	20, the transversal, break, then 10 you see,
8	10 / mu-ut-ta-ri-tum /	10, the descent,
9-10	ša ki-ma ma-at-nim / ša-ak-na-at	that like a string is set.
	tu-úr-ma pa-šar	Turn back, then solve(?).
11	20 dal ḫe-pé-e / 10 ta-mar	20, the transversal, break, 10 you see.
	10 me-eḫ-ra-am i-di	10, a copy, lay down,
12	šu-ta-ki-il-ma / 1 40 ta-mar	let (them) eat each other, then 1 40 you see.
	2 mu-ut-ta-ri-tam an.ta	2, the upper descent,
13	i-na 10 / mu-ut-ta-ri-tim	from 10, the descent
	ša ki-ma ma-at-nim ša-ak-nu /	that like a string is set,
14	zi.zi-ma 8 ta-mar	tear off, then 8 you see.
15	8 šu-ta-ki-il-ma / 1 04 ta-mar	8 let eat itself, then 1 04 you see.
	1 04 i-na 1 40 ša ta-mu-r[u] /	1 04 from 1 40 that you saw
16	zi.zi-ma 36 ta-mar	tear off, then 36 you see.
17	ba.si₈-šu šu-li-m[a] / 6 ta-mar	Its likeside let come up, then 6 you see.
	a-na ši-na e-ṣí-ip-ma	To two repeat, then
18	[12 dal] / an.ta ta-mar	12, the upper transversal, you see.
	ki-a-am né-pé-[šum]	Such is the doing.

The text is predominantly written in Akkadian. There are only a few Sumerian terms present:

gúr	arc	circle
dal	transversal	diameter, chord
an.ta	upper	left
za.e	you	
zi.zi	tear off	subtract
ba.si₈	likeside (lit. 'it is equal')	square side
šà.bar	interior divider	inner diagonal (of three-dimensional object)

There is one previously unknown Akkadian mathematical term in the text of § 1 a:

muttarrittum ša kīma matnim šaknat   the descent that like a string is set   a perpendicular radius

Here muttarrittum is the usual term for a perpendicular line, drawn, according to the Old Babylonian convention, as a horizontal line, since 'up' is to the left, and 'down' is to the right. (The noun muttarrittum 'descent' and the verbal form urdam 'I went down' in line 3 of the text are both derived from the same verb warādu 'to go down, descend'.) That the perpendicular line is 'set like a string' may refer to the fact that a circle can be drawn by use of a taut piece of string rotated around a point where one end of the string is fixed.

Therefore, the 'descent that is set like a string' is a fitting name for a perpendicular radius, the *vertical distance* from the top of the circle (to the left in Fig. 11.1.2 below) to the center of the circle. The chord is clearly assumed to be orthogonal to this radius.

Although most of the question in § 1 a is lost, the accompanying drawing shows that a circle is given with the diameter $d = 20$ (ninda), hence (approximately) with the circumference 1 00 (ninda), and that a chord of the circle is given, associated in some way with the number '2'.

The solution procedure begins by computing the radius of the circle as half the diameter = 10 (ninda). By mistake(?), the radius is computed twice. Then the square of the radius is computed, as sq. 10 = 1 40 (sq. ninda). In the next step of the computation, the given number '2' is mentioned as

muttarrittum an.ta   the upper descent.

Evidently, this 'descent' stands for the *vertical distance* from the top of the circle to the chord. In Fig. 11.1.2, the acrophonic notations m, m_a, and m_k are used for *muttarrittum* 'the descent', *muttarrittum* an.ta 'the upper descent', and *muttarrittum* ki.ta 'the lower descent'.

The square of the difference $m_k = m - m_a = 8$ is subtracted from the square of the radius:

sq. m – sq. $(m - m_a) = 1\ 40 - 1\ 04 = 36$ (sq. ninda).

In the final steps of the solution procedure, the square side 6 of 36 is doubled, $2 \cdot 6 = 12$ (ninda), and the result is called

dal an.ta   the upper transversal.

It is clear from Fig. 11.1.2 that this 'upper transversal' is the chord 2 ninda below the top of the circle.

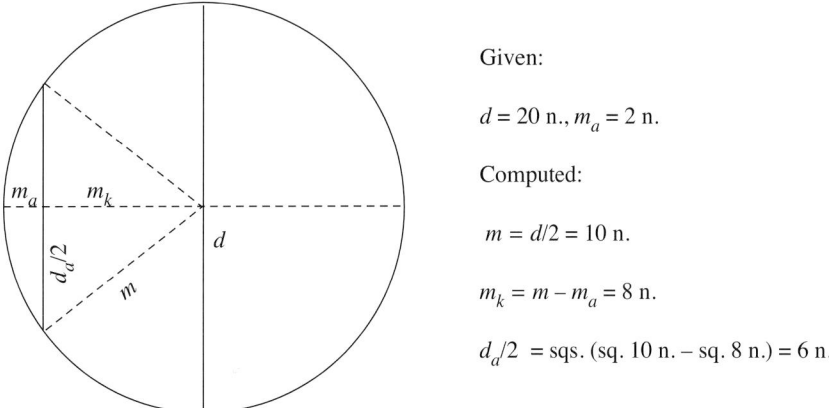

Given:

$d = 20$ n., $m_a = 2$ n.

Computed:

$m = d/2 = 10$ n.

$m_k = m - m_a = 8$ n.

$d_d/2$ = sqs. (sq. 10 n. – sq. 8 n.) = 6 n.

Fig. 11.1.2. MS 3049 § 1 a. Computing the length of a chord by use of the diagonal rule.

Too little is preserved of the text of MS 3049 § 1 b to indicate clearly what that problem was about. It must have been some kind of variation of the question presented and solved in § 1 a.

MS 3049 § 1 b

1	2 [*ša ur-dam*] / [12 dal an.ta]	*2 that I went down, 12 the upper transversal.*
2	[dal gúr *mi-nu*]	The transversal of the arc is what?
3-8	[.........................]	....
9	[x x x x x x x x x x x *a-na*] /	x x x x x x x x x x to
10	dal 10 *ša* [*ki-ma ma-at-nim ša-ak-na-at*] /	the transversal 10 that *like a string is set*
11	*i-ši-ma* [x x x x x x x x x x x]	raise, then *x x x x x x x x x x*
12	*ša k*[*i-ma ma-at-nim ša-ak-na-at* x x]	that li*ke a string is set* x x
...	....	....

A clue to the content of § 1 b exists in the form of a couple of parallel Old Babylonian exercises dealing with circles and chords:

A Couple of Parallel Texts in the Mathematical Recombination Text BM 85194

BM 85194 (Group 6 a, from Sippar) is a large Old Babylonian mathematical problem text with 35 exercises of mixed origin, a typical recombination text. It was first published as a hand copy by King in *CT 9* (1900), then republished by Neugebauer in *MKT 1* (1935) in the form of a transliteration with translation and commentary. In BM 85194, ## 21-22 (Høyrup, *LWS* (2002), 272-275) are parallels to MS 3049 § 1 a, appearing totally out of context in the recombination text.

The brief text of the two exercises is reproduced below (with a slightly improved transliteration and translation):

BM 85194 ## 21-22

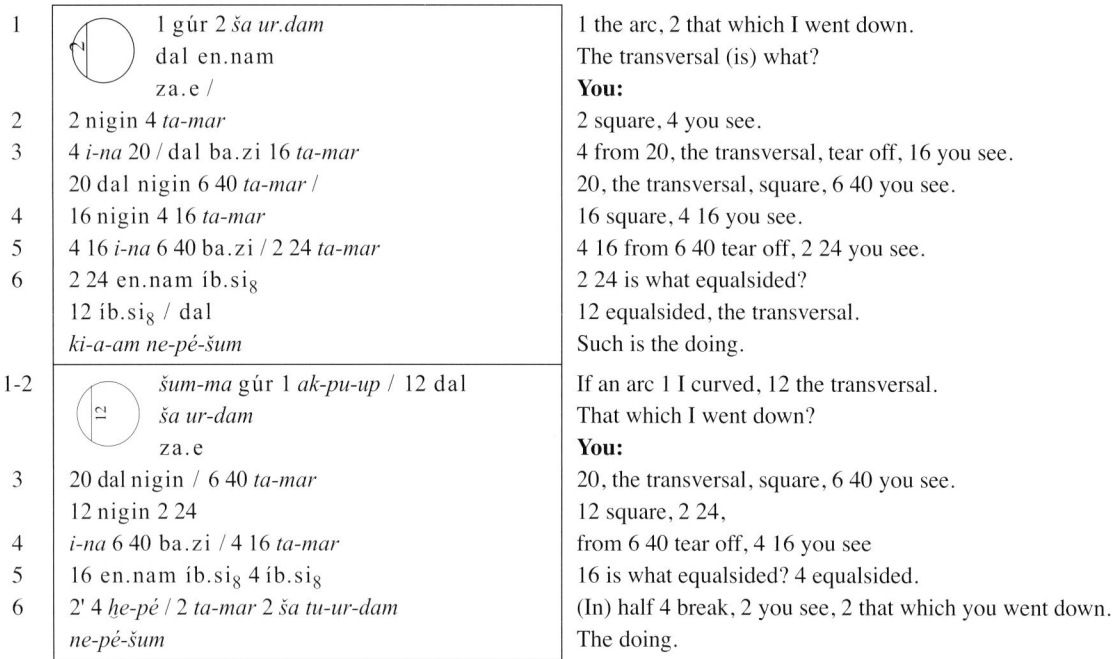

1	1 gúr 2 *ša ur.dam*	1 the arc, 2 that which I went down.
	dal en.nam	The transversal (is) what?
	za.e /	**You:**
2	2 nigin 4 *ta-mar*	2 square, 4 you see.
3	4 *i-na* 20 / dal ba.zi 16 *ta-mar*	4 from 20, the transversal, tear off, 16 you see.
	20 dal nigin 6 40 *ta-mar* /	20, the transversal, square, 6 40 you see.
4	16 nigin 4 16 *ta-mar*	16 square, 4 16 you see.
5	4 16 *i-na* 6 40 ba.zi / 2 24 *ta-mar*	4 16 from 6 40 tear off, 2 24 you see.
6	2 24 en.nam íb.si₈	2 24 is what equalsided?
	12 íb.si₈ / dal	12 equalsided, the transversal.
	ki-a-am ne-pé-šum	Such is the doing.
1-2	*šum-ma* gúr 1 *ak-pu-up* / 12 dal	If an arc 1 I curved, 12 the transversal.
	ša ur-dam	That which I went down?
	za.e	**You:**
3	20 dal nigin / 6 40 *ta-mar*	20, the transversal, square, 6 40 you see.
	12 nigin 2 24	12 square, 2 24,
4	*i-na* 6 40 ba.zi / 4 16 *ta-mar*	from 6 40 tear off, 4 16 you see
5	16 en.nam íb.si₈ 4 íb.si₈	16 is what equalsided? 4 equalsided.
6	2' 4 *ḫe-pé* / 2 *ta-mar* 2 *ša tu-ur-dam*	(In) half 4 break, 2 you see, 2 that which you went down.
	ne-pé-šum	The doing.

The question in the first of these two exercises is only superficially different from the question in MS 3049 § 1 a. The given numbers are essentially the same in the two texts, if one concedes that saying that the arc or circumference of the circle is 1 (00) is almost the same thing as saying that the transversal or diameter is 20. The solution procedures are also essentially the same, the only difference being that when MS 3049 § 1 a operates with *half* the diameter and *half* the chord, BM 85194 # 21 operates with the *whole* diameter and the *whole* chord. Compare the explanations of the two solution procedures in Figs. 11.1.2 and 11.1.3.

The question in BM 85194 # 22 is obtained from the question in # 21 through "permutation of the data", in the following sense. In # 21, (the circumference of) the circle and the upper descent are given, and the upper transversal (the chord) is computed. In # 22, the circle and the upper transversal are given, and the upper descent is computed (although the computation is nonsensical in lines 5 and 6).

It is likely that the question in the partially preserved § 1 b of MS 3049 was a similar permutation of the question in § 1 a, possibly asking for the diameter $d = 2 m$ of the circle when the upper descent m_a and the upper transversal d_a are given. With the notations used in Fig. 11.1.2, this question can be transformed into a *quadratic-linear system of equations:*

$$\text{sq. } m - \text{sq. } m_k = \text{sq. } d_a/2, \quad m - m_k = m_a.$$

This quadratic-linear system of equations has a simple solution not involving any square side, since it is equivalent to the following simple *system of linear equations*:

$$m + m_k = (\text{sq. } d_a/2)/m_a, \quad m - m_k = m_a.$$

However, it is not obvious how to reconcile the few remnants of the text in lines 10-12 of § 1 b with this interpretation attempt.

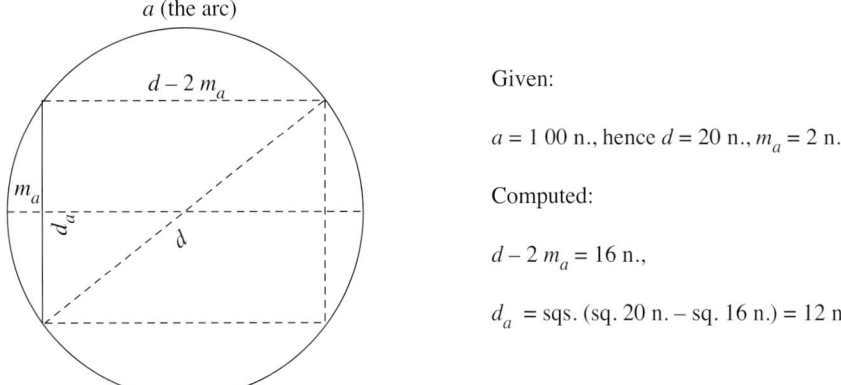

Given:

$a = 1\ 00$ n., hence $d = 20$ n., $m_a = 2$ n.

Computed:

$d - 2\ m_a = 16$ n.,

$d_a = \text{sqs. (sq. } 20$ n. $-$ sq. 16 n.$) = 12$ n.

Fig. 11.1.3. Explanation of the solution procedure in BM 85194 # 21. A parallel to MS 3049 § 1 a.

11.1 b. MS 3049, Subscript. A List of the Separate Topics in the Text

The reverse and edge of the fragment MS 3049 (Fig. 11.1.4) contain most of § 5, a small part of § 4 c, an even smaller part of what may be § 4 a, and a subscript mentioning 5 geometric(?) topics.

The subscript details the content of MS 3049, obviously a recombination text, by mentioning that it contains 6 exercises concerning circles, 5 concerning squares(?), 1 possibly concerning a triangle (it is not clear what the term recorded here is; it may be a carelessly written s a g . k a k 'peg-head', the usual term for 'triangle'), 3 exercises dealing with some 3-dimensional object related to bricks (a 'brick mold'), and 1 final exercise in which the 'inner diagonal of a gate' is computed:

MS 3049, subscript

1	mu.bi 6 gúr.meš /	Its name: 6 arcs (circles)
2	mu.bi 5 nígin$^?$-ša$^?$ /	Its name: 5 its$^?$ square$^?$
3	1 sag$^?$.kak$^?$ /	1 peghead$^?$ (triangle)
4	3 na-al-ba-tum /	3 brick molds (rectangular prisms?)
5	1 šà.bar ká	1 inner cross-over (diagonal) of a gate

In the discussion above of § 1 a and of what remains of § 1 b, it was argued that § 1 b probably was a variant of § 1 a, obtained through permutation of the data in the question (letting the unknown parameter in the case of § 1 a change places with one of the given parameters). In this sense §§ 1 a-b may have been two variations on the same theme. In the same way, it can be assumed that §§ 1 c-f were 4 further variations on the same theme, still dealing with circles.

11.1 c. MS 3049, § 4 c. A Small Fragment of a Problem for a 'Brick Mold'

Only the left two-thirds of the last five lines of § 4 c are preserved (see Fig. 11.1.4, top), which makes it nearly impossible to know what this problem is about. Fortunately, however, the text of § 4 c ends with a summary stating that § 4 c is the last of three exercises dealing with 'brick molds'.

MS 3049 § 4 c

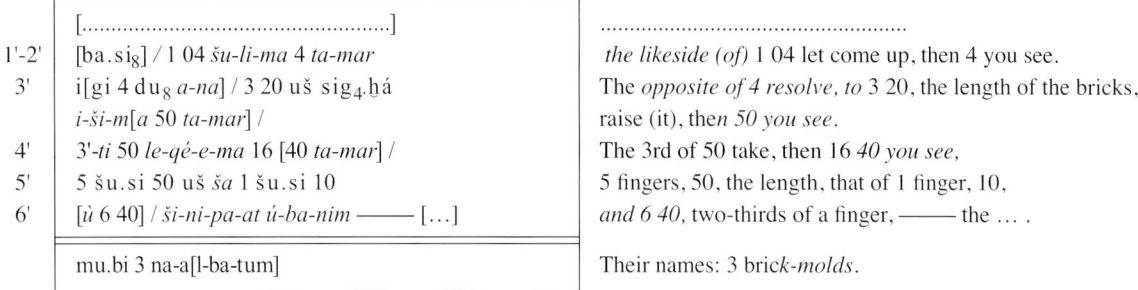

	[.....................................]
1'-2'	[ba.si₈] / 1 04 *šu-li-ma* 4 *ta-mar*
3'	i[gi 4 du₈ *a-na*] / 3 20 uš sig₄.ḫá
	i-ši-m[*a* 50 *ta-mar*] /
4'	3'-*ti* 50 *le-qé-e-ma* 16 [40 *ta-mar*] /
5'	5 šu.si 50 uš *ša* 1 šu.si 10
6'	[*ù* 6 40] / *ši-ni-pa-at ú-ba-nim* ──── [...]

mu.bi 3 na-a[l-ba-tum]	

....
the likeside (of) 1 04 let come up, then 4 you see.
The *opposite of 4 resolve, to* 3 20, the length of the bricks,
raise (it), then *50 you see.*
The 3rd of 50 take, then 16 *40 you see,*
5 fingers, 50, the length, that of 1 finger, 10,
and 6 40, two-thirds of a finger, ──── the ... .

Their names: 3 brick-*molds*.

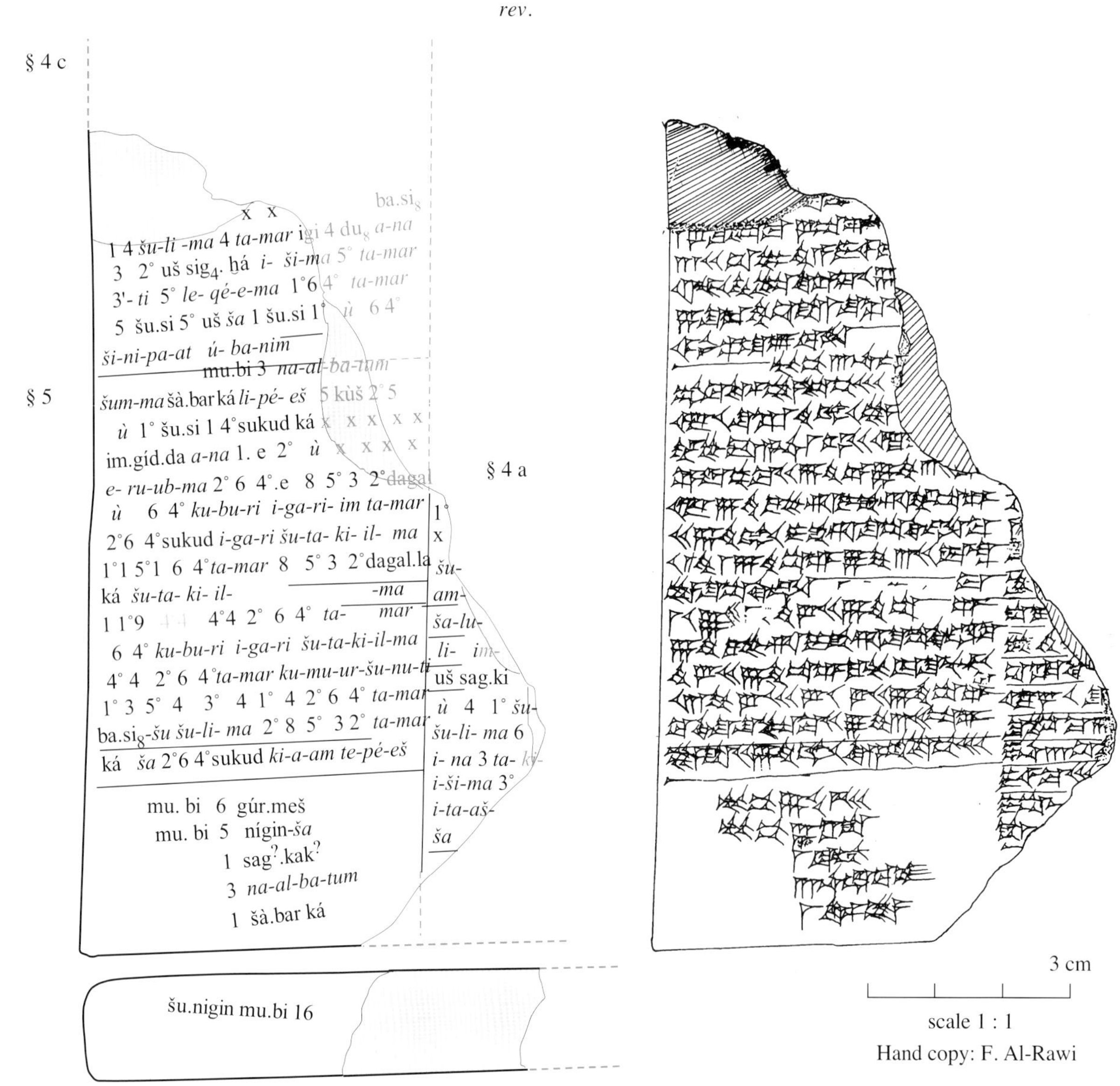

rev.

§ 4 c

x x   ba.si₈
1 4 *šu-li-ma* 4 *ta-mar* igi 4 du₈ *a-na*
3 2° uš sig₄. ḫá *i- ši-m*[a 5° *ta-mar*
3'- *ti* 5° *le- qé-e-ma* 1°6[4° *ta-mar*
5 šu.si 5° uš *ša* 1 šu.si 1° *ù* 6 4°
ši-ni-pa-at ú- ba-nim
mu.bi 3 *na-al-ba-tum*

§ 5

šum-ma šà.bar ká *li-pé- eš* 5 kùš 2° 5
ù 1° šu.si 1 4° sukud ká x x x x x
im.gíd.da *a-na* 1. e 2° *ù* x x x x
e- ru-ub-ma 2° 6 4°.e 8 5° 3 2° dagal
ù 6 4° *ku-bu-ri i-ga-ri- im ta-mar*
2° 6 4° sukud *i-ga-ri šu-ta- ki- il- ma*
1° 1 5° 1 6 4° *ta-mar* 8 5° 3 2° dagal.la
ká *šu-ta- ki-il-*   -*ma*/-*mar*
1 1° 9 4° 2° 6 4° *ta-*
6 4° *ku-bu-ri i-ga-ri šu-ta-ki-il-ma*
4° 4 2° 6 4° *ta-mar ku-mu-ur-šu-nu-ti*
1° 3 5° 4 3° 4 1° 4 2° 6 4° *ta-mar*
ba.si₈-*šu šu-li- ma* 2° 8 5° 3 2° *ta-mar*
ká *ša* 2° 6 4° sukud *ki-a-am te-pé-eš*

§ 4 a

1°
x
šu-
am-
ša-lu-
li- im
uš sag.ki
ù 4 1° *šu-*
šu-li- ma 6
i- na 3 *ta- k[*
i-ši-ma 3°
i-ta-aš-
ša

mu. bi 6 gúr.meš
mu. bi 5 nígin-*ša*
1 sag?.kak?
3 *na-al-ba-tum*
1 šà.bar ká

šu.nigin mu.bi 16

Fig. 11.1.4. MS 3049, *rev.* Parts of three exercises, a subscript, and a notation on the edge of the clay tablet.

3 cm

scale 1 : 1

Hand copy: F. Al-Rawi

The preserved text of § 4 c starts with an instruction to compute the cube side of 1 04, which is found to be 4. Then the side of a brick, measuring ;03 20 n. = 2/3 cubit, is multiplied by the reciprocal of 4. The result is

50. In the final step of the computation, 1/3 of 50 is declared to be 16 40. Then follows the explanation that 50 = 5 fingers, and that 16 40 can be split into 10 = 1 finger and 6 40 = 2/3 fingers, so that 1 2/3 fingers is the […].

In modern terms, this means that 50 has to be understood as ;00 50 ninda = 5 fingers, since 1 ninda = 12 cubits and 1 cubit = 30 fingers, so that 1 ninda = 6 00 fingers and 1 finger = ;00 10 ninda. For the same reason, 16 40 has to be understood as ;00 16 40 ninda = 1 2/3 fingers. Thus, the result of the computation is that some object (a "brick mold"), which is in some way related to square(?) bricks with the side 2/3 cubit, has the side 5 fingers (about 8 cm) and the thickness(?) 1 2/3 fingers (about 3 cm). The object in question is probably three-dimensional, in view of the extraction of a cube side in lines 1'-2', perhaps in some application of the rule of false value. (There are only two known examples of extractions of cube sides in Old Babylonian mathematical texts. One is in the fragment **CBM 12648**, an early Old Babylonian text from Nippur, in a calculation of the dimensions of a brick. (See Friberg, *Survey* (1982), 13, 49; Muroi, *Centaurus* 31 (1989).) The other example is **BM 96954+**, *rev. ii*, **6-12**, a calculation of the dimensions of a cone (Friberg, *PCHM* 6 (1996); Friberg, *UL* (2005), 262).

In other circumstances, "brick mold" seems to be the name of a two-dimensional figure. Thus, the table of constants G = **IM 52916**, a text from Shaduppum published by Goetze in *Sumer* 7 (1981), contains, among other things, the following brief list of problems for "figures within figures":

[na]-al-ba-tam i-na li-bu na-al-ba-tim e-pé-ša-am	to do a brick mold within a brick mold	G *rev.* 14'
ki-pa-[tam i]-na li-bu ki-pa-tim e-pé-ša-am	to do an arc (circle) within an arc	G *rev.* 18'
i-na li-bu na-al-ba-tim ki-pa-ta-am <e-pé-ša-am>	within a brick mold an arc	G *rev.* 20'
i-na li-bu ki-pa-tim na-al-ba-ta-am <e-pé-ša-am>	within an arc a brick mold	G *rev.* 21'
[na]-al-[ba]-at-ta-[am i-na i-in] al-pí-im e-pé-ša-am	to do a brick mold within an ox-eye	G *rev.* 30'

Here 'to do a circle within a circle' is probably some kind of problem for a circular band bounded by two concentric circles. Cf. the problem for two concentric circles on the Old Babylonian hand tablet **Böhl 1821** (Leemans and Bruins, *RAI* 2 (1951)). Cf. also the problem for two concentric squares on **YBC 7359** (Fig. 8.2.9 above). Consequently, 'to do a brick mold within a brick mold' can be assumed to be some problem for the region between two "concentric" brick molds. The other three mentioned "figures within figures" problems, may have been problems of the same kind as *TMS* 21 a-b, two problems for an *apsammikku*, a "concave square" in a square or a rectangle, some distance away from the sides of the square or rectangle (see Figs. 8.2.10-11), which were successfully interpreted by Muroi in *SCIAMVS* 1 (2000). Note that *īn alpim* (Sum. igi.gu₄) 'ox-eye' is a name for the oval figure composed of two circle segments, which was briefly discussed in Sec. 8.2 c above.

All this does not help much to explain the meaning of the term *nalbattum* 'brick mold', except for showing that it is sometimes a plane geometric figure, sometimes a solid figure. Perhaps a *rectangular plane figure* in the former case, and a *rectangular solid figure* in the latter case? In that case, the entry "to do a brick mold within a brick mold" in Goetze's table of constants may refer to a problem for two concentric rectangles, while the entries "within a brick mold a circle" and "within a circle a brick mold" may refer to problems for a circle within a rectangle or a rectangle within a circle, and so on.

11.1 d.   MS 3049, § 5. The Inner Diagonal of a Rectangular Gate in a Wall

Explanation of the Calculations in the Text of § 5

The text of § 5 is relatively well preserved (see Fig. 11.1.4). Only the last quarters of the first three or four lines are lost. Unfortunately, that means that much of the crucial beginning of the solution procedure is lost. This is doubly unfortunate, since there is no known parallel text in the Old Babylonian mathematical corpus, and since the terminology used is partly new and hard to understand.

MS 3049 § 5

1	*šum-ma* šà.bar ká *li-pé-eš*	If the inner cross-over (diagonal) of a gate he shall do,
2	[5 kùš 25] / *ù* 10 šu.si 1 40	*5 cubits, 25*, and 10 fingers, 1 40,
	sukud ká	the height of the gate.
3	[x x x x x] / im.gíd.da	*x x x x x* the table,
4	*a-na* 1.e 20 *ù* [x x x x] / *e-ru-ub-ma*	to this one 20 and *x x x x* enter, then
	26 40.e 8 53 20 [dagal] /	this 26 40, 8 53 20 *the width*,
5	*ù* 6 40 *ku-bu-ri i-ga-ri-im ta-mar* /	and 6 40, the thickness of the wall, you see.
6	26 40 sukud *i-ga-ri šu-ta-ki-il-ma* /	26 40, the height of the wall, let eat itself, then
7	11 51 06 40 *ta-mar*	11 51 06 40 you see.
8	8 53 20 dagal.la / ká *šu-ta-ki-il-ma* /	8 53 20, the width of the gate, let eat itself, then
9	1 19··· 44 26 40 *ta-mar* /	1 19··· 44 26 40 you see.
10	6 40 *ku-bu-ri i-ga-ri šu-ta-ki-il-ma* /	6 40, the thickness of the wall, let eat itself, then
11	44 26 40 *ta-mar*	44 26 40 you see.
12	*ku-mu-ur-šu-nu-ti* / 13 54 34 14 26 40 *ta-mar* /	Heap them, 13 54 34 14 26 40 you see.
13	ba.si₈-*šu šu-li-ma* 28 53 20 *ta-mar* /	Its likeside let come up, then 28 53 20 you see
14	ká *ša* 26 40 sukud	(for) the gate that (has) 26 40 (as its) height.
	ki-a-am te-pé-eš	So you do.

In spite of the mentioned difficulties, the question and the solution procedure are clear enough. In lines 1-2 are mentioned the šà.bar, literally 'inside cross-over', and the sukud 'height' of a ká 'gate', the height explicitly given as [5 cubits] and 10 fingers = [;25 ninda] + ;01 40 ninda. In the damaged lines 3-4, the student is instructed to consult a table of some kind (called im.gíd.da 'long clay tablet'). It is not at all clear what is happening here. Anyway, the result of the consultation is that in lines 4-5 the dimensions of the gate, in addition to the height 26 40, can be given as the dagal 'width' 8 53 20, and the *kuburi igāri* 'thickness of the wall' 8 53 20. In lines 6-11, the squares of these three numbers are computed. The first number is now renamed sukud *igāri* 'the height of the wall', while the second number is called dagal.la ká 'the width of the gate'. The third number is again referred to as 'the thickness of the wall'. Apparently, the object considered is a gate in a wall. The gate passes through the wall, of course, so that its depth is equal to the width of the wall. The gate is also as high as the wall, so that its height is equal to the height of the wall. The width, on the other hand, is exclusively a property of the gate. The only reasonable *absolute* values for the three parameters are:

;26 40 ninda (appr. 2.67 meters)	the height of the gate (and of the wall)
;08 53 20 ninda (appr. 0.9 meters)	the width of the gate
;06 40 ninda (appr. 0.67 meters)	the depth of the gate (and the thickness of the wall).

These values cannot be very realistic. In the course of the explanation below, it will become apparent that the three numbers were chosen solely in order to facilitate the computation.

The squares of the three numbers are (see lines 7-11):

sq. (;26 40 ninda)	= ;11 51 06 40 sq. ninda
sq. (;08 53 20 ninda)	= ;01 19 00 44 26 40 sq. ninda
sq. (;06 40 ninda)	= ;00 44 26 40 sq. ninda.

Note that in line 9 the place of the double zero is indicated by a *gap* in the number. In the gap, there are faint traces of an erased number 44, showing that the one who wrote (or copied) the text corrected himself, making a clear gap for the missing double digit where he initially had left no gap.

In line 12, the sum of the three squares is given in the form 13 54 34 14 26 40. This sum can be computed, in *relative* and *absolute* numbers, respectively, as

11 51 06 40	;11 51 06 40
1 19 00 44 26 40	;01 19 00 44 26 40
44 26 40	;00 44 26 40
13 54 34 04 26 40	;13 54 34 04 26 40

To get a sum like this right when counting only with *relative* sexagesimal numbers takes some skill!

In line 13, the square side of the sum is computed. The computation of the square side of a 6-place sexagesimal number in relative place value notation like 13 54 34 04 26 40 takes some skill, too. It can be done in a few simple steps, but *the method requires that the one doing the calculation has some preliminary notion about what the result should be in absolute numbers*. (Cf. the discussion of the additive and negative " s" in Friberg, *BagM* 28 (1997) § 8.) Here are the steps of the calculation, in *relative* numbers,

1. sq. 29 = 14 01, 13 54 = 14 01 − 7 (13 54 being the first couple of sexagesimal places in 13 54 | 34 04 | 26 40).
2. sqs. (13 54) = sqs. (sq. 29 − 7) = appr. 29 − 7/(2 · 29) = appr. 29 − 7/(2 · 30) = 28 53.
3. sq. 28 53 = 13 54 14 49, 13 54 34 04 = 13 54 14 49 + 19 15.
4. sqs. (13 54 34 04) = sqs. (sq. 28 53 + 19 15) = appr. 28 53 + 19 15 / (2 · 28 53)
$$= \text{appr. } 28\ 53 + 20/(2 \cdot 30) = 28\ 53\ 20.$$
5. sq. 28 53 20 = 13 54 34 04 26 40, the given number.

It is clear that in *absolute* numbers, the computed square side must be equal to ;28 53 20, which is comparable in size to, but slightly more than ;26 40, the largest of the three given numbers.

The Diagonal Rule for a Rectangular Prism

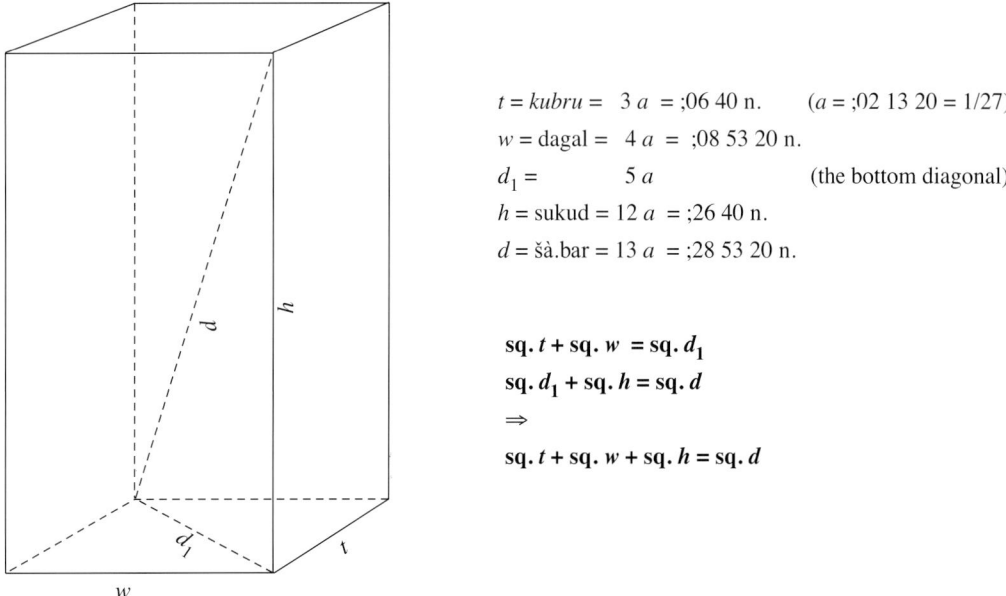

$t = kubru = \quad 3\,a = ;06\ 40$ n. $\quad (a = ;02\ 13\ 20 = 1/27)$

$w = dagal = \quad 4\,a = ;08\ 53\ 20$ n.

$d_1 = \qquad\quad 5\,a$ (the bottom diagonal)

$h = sukud = 12\,a = ;26\ 40$ n.

$d = \text{šà.bar} = 13\,a = ;28\ 53\ 20$ n.

sq. t **+ sq.** w **= sq.** d_1

sq. d_1 **+ sq.** h **= sq.** d

⇒

sq. t **+ sq.** w **+ sq.** h **= sq.** d

Fig. 11.1.5. MS 3049 § 5. The construction of a solution in integers to the three-dimensional diagonal equation.

The gate considered in MS 3049 § 5, has the form of a rectangular prism with the dimensions

kubru	'thickness'	;06 40 n.
dagal	'width'	;08 53 20 n.
sukud	'height'	;26 40 n.
šà.bar	'inner cross-over'	;28 53 20 n.

These are not arbitrarily chosen dimensions, as is obvious from the fact that the three numbers are proportional to the integers 3, 4, 12, and 13. It is easy to find the corresponding factorizations:

kubru	'thickness'	;06 40 n.	=	3 ·	;02 13 20 n.	(;02 13 20 = 1/27)
dagal	'width'	;08 53 20 n.	=	4 ·	;02 13 20 n.	
sukud	'height'	;26 40 n.	=	12 ·	;02 13 20 n.	
šà.bar	'inner cross-over'	;28 53 20 n.	=	13 ·	;02 13 20 n.	

Consider a rectangular prism like the one depicted in Fig. 11.1.5 above, and let t, w, h, d denote the thickness, the width, the height, and the inner diagonal of the prism. Then it follows from the Babylonian diagonal rule in *two* dimensions that

$$\text{sq. } t + \text{sq. } w = \text{sq. } d_1 \qquad (d_1 = \text{the diagonal of the bottom rectangle})$$

and that, at the same time,

$$\text{sq. } d_1 + \text{sq. } h = \text{sq. } d.$$

A combination of these two equations gives the Babylonian diagonal rule in *three* dimensions:

$$\text{sq. } t + \text{sq. } w + \text{sq. } h = \text{sq. } d.$$

Through MS 3049 § 5 it is demonstrated that Old Babylonian mathematicians had found a simple way to construct *exact solutions without square sides* (solutions in integers) to the *three-dimensional* diagonal equation. The method was to find two solutions t, w, d_1 and d_1, h, d to the two-dimensional diagonal equation, and then combine them to a solution to the three-dimensional diagonal equation. In § 5, the two solutions to the two-dimensional equation are 5, 4, 3 and 13, 12, 5, and the combined solution is 13, 12, 4, 3. This solution was then multiplied by a suitable constant (;02 13 20 n., where ;01 13 20 = 1/27 is a regular sexagesimal number), partly in order to get fairly realistic dimensions for the gate, partly in order to make the computations in the solution procedure more complicated.

BM 96957 + VAT 649 §§ 5-7. A Related Theme Text for the Dimensions of a Gate

The only known Old Babylonian mathematical text that is, in a restricted sense, a parallel to MS 3049 § 5, is §§ 5-7 of the recombination text **BM 96957 + VAT 6598** (Friberg, *BagM* 28 (1997) § 8 d; Robson, *JCS* 49 (1997); Robson, *MMTC* (1999), App. 4; Høyrup, *LWS* (2002), 268-272).

BM 96957 + VAT 6598 § 5 a

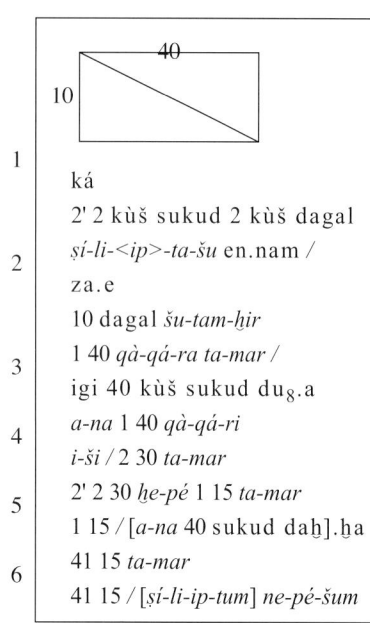

Fig.

1	ká	A gate.
	2' 2 kùš sukud 2 kùš dagal	1/2 <n.> 2 cubits the height, 2 cubits the width.
2	*ṣí-li-<ip>-ta-šu* en.nam /	Its cross-over (diagonal) is what?
	za.e	You:
	10 dagal *šu-tam-ḫir*	10, the width make equalsided,
3	1 40 *qà-qá-ra ta-mar* /	1 40, the ground, you see.
	igi 40 kùš sukud du₈.a	The opposite of 40, the height, release,
4	*a-na* 1 40 *qà-qá-ri*	to 1 40, the ground,
	i-ši / 2 30 *ta-mar*	raise, 2 30 you see.
5	2' 2 30 *ḫe-pé* 1 15 *ta-mar*	1/2 of 2 30 break, 1 15 you see.
	1 15 / [*a-na* 40 sukud daḫ].ḫa	1 15 *to 40, the height, add* on,
6	41 15 *ta-mar*	41 15 you see.
	41 15 / [*ṣí-li-ip-tum*] *ne-pé-šum*	41 15, *the cross-over.* The doing.

In BM 96957+ § 5 a, the face of a rectangular gate has the height $h = 1/2$ ninda 2 cubits = ;40 n. (about 4 meters), and the width $w = 2$ cubits = ;10 n (about 2 meters). The diagonal of the face of the gate is computed as follows:

$$d = \text{sqs. (sq. } h + \text{sq. } w) = \text{appr. } (h + \text{sq. } w /(2 h)) = (;40 + \text{sq. };10 / 1;20) \text{ n.} = ;41\ 15 \text{ n.}$$

This is an *explicit* application of the Old Babylonian additive square side rule

$$\text{sqs. (sq. } a + b) = \text{appr. } a + b / (2 a).$$

(The exact value is $d = \text{sqs. } (;26\ 40 + ;01\ 40) \text{ n.} = \text{sqs. };28\ 20 = ;10 \text{ n.} \cdot \text{sqs. } 17$. Setting $d = \text{appr. };41\ 15$, as in

BM 96957+ § 5 a, corresponds to setting sqs. 17 = appr. 4;07 30 = 4 1/8.)

Note: Above, in § 5 a, line 1, the phrase 2 kùš dagal, is transliterated as '2 cubits the width', which means that the width is equal to 2 cubits or ;10 ninda. In § 5 b, line 1, the similar phrase 40 kùš sukud, is transliterated as '40, cubits, the height' since it means that the height is equal to 40 in relative numbers but to ;40 ninda in absolute numbers, ;40 ninda being "in the range of cubits". Actually, ;40 ninda = 8 cubits. This confusing way of indicating how relative numbers should be translated into absolute numbers is a well known peculiar trait of Old Babylonian mathematical texts.

BM 96957 + VAT 6598 § 5 b

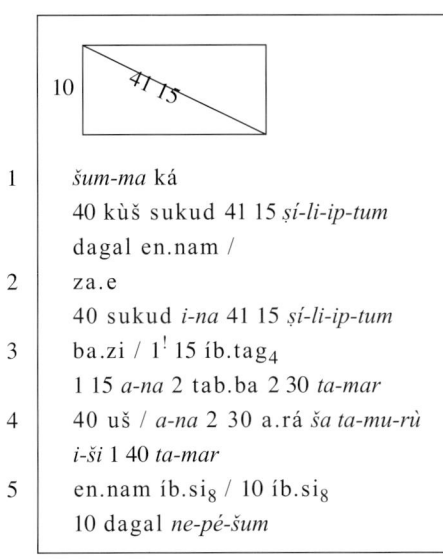

1	*šum-ma* ká	If a gate,
	40 kùš sukud 41 15 *ṣí-li-ip-tum*	40, cubits, the height, 41 15 the cross-over.
	dagal en.nam /	The width is what?
2	za.e	You:
	40 sukud *i-na* 41 15 *ṣí-li-ip-tum*	40, the height, from 41 15, the cross-over,
3	ba.zi / 1¹ 15 íb.tag₄	tear off, 1 15, the remainder.
	1 15 *a-na* 2 tab.ba 2 30 *ta-mar*	1 15 to 2 repeat, 2;30 you see.
4	40 uš / *a-na* 2 30 a.rá *ša ta-mu-rù*	40, the length, to 2;30, the steps, that you saw,
	i-ši 1 40 *ta-mar*	raise, 1 40 you see.
5	en.nam íb.si₈ / 10 íb.si₈	What is it equalsided? 10 it is equalsided.
	10 dagal *ne-pé-šum*	10, the width. The doing.

In BM 96957 + § 5 b, the height $h = ;40$ n. and the diagonal $d = ;41$ 15 n. are given. It is required to find the width w. The obvious way of doing this would be to compute $w = $ sqs. $($ sq. $d - $ sq. $h)$, but this is not the way it is done in the text. Instead, *through a reversal of the additive square side rule*, w is computed as the solution to the equation $d = $ (appr.) $h + $ sq. $w / (2 h)$. Thus,

sq. $w = (d - h) \cdot 2 h = ;01$ 15 $\cdot 2 \cdot ;40$ sq. n. $= ;01$ 40 sq. n., so that $w = ;10$ n. (exactly!).

BM 96957 + VAT 6598 § 5 c

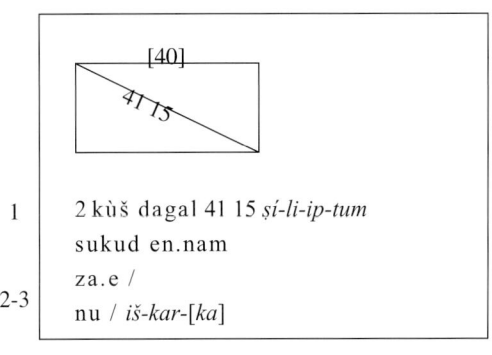

1	2 kùš dagal 41 15 *ṣí-li-ip-tum*	2 c. the width, 41 15 the cross-over.
	sukud en.nam	The height is what?
2-3	za.e /	You:
	nu / *iš-kar-[ka]*	Not. *Your* task.

In BM 96957 + § 5 c, the width w and the diagonal d of the (face of the) gate are given, and the height is unknown. The idea is, apparently, to compute the height as a solution to the equation in § 5 a, $d = $ (appr.) $h + $ sq. $w / (2 h)$, in the same way as w was computed in § 5 b as a solution to this equation when d and h were given. However, this equation for h is a quadratic equation which is considerably *more difficult* to solve than the equation for h obtained through a direct application of the diagonal rule, that is the equation $h = $ sqs. (sq. d – sq. w). The author of the text seems to have realized this, so he abandoned the project and wrote laconically nu *iš-kar-[ka]* 'not (here), your task'.

BM 96957 + VAT 598 § 6 a

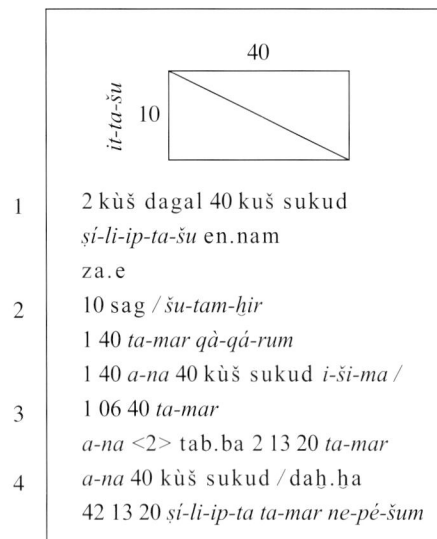

	2 kùš dagal 40 kuš sukud	2 c. the width, 40, cubits, the height.
1	*ṣí-li-ip-ta-šu* en.nam	The cross-over is what?
	za.e	You:
2	10 sag / *šu-tam-ḫir*	10, the front, make equalsided,
	1 40 *ta-mar qà-qá-rum*	1 40 you see, the ground.
	1 40 *a-na* 40 kùš sukud *i-ši-ma* /	1 40 to 40, cubits, the height, raise,
3	1 06 40 *ta-mar*	1 06 40 you see.
	a-na <2> tab.ba 2 13 20 *ta-mar*	To 2¹ repeat, 2 13 20 you see.
4	*a-na* 40 kùš sukud / daḫ.ḫa	To 40, cubits, the height, add on,
	42 13 20 *ṣí-li-ip-ta ta-mar ne-pé-šum*	42 13 20, the cross-over, you see. The doing.

Fig.

An interesting observation is that in line 2 of § 5 the square of the width is called *qaqqaru* 'ground, base', an indication that the writer had in mind not just a two-dimensional doorway but really a three-dimensional gate through a wall, just like the gate in MS 3049. This impression is strengthened by the fact that in § 6 a the square of the width is again called *qaqqaru*, and that the second time the width appears in this exercise it is called sag 'front' or 'side', probably because it is thought of as the side of the square floor of the gate.

Although the text of the solution procedure is corrupt, it is likely that what is referred to as the 'diagonal' in § 6 a is meant to be the *inner* diagonal of a gate with a square floor, that is of a gate with its thickness equal to its width. (This explanation was boldly proposed already in Friberg, *BagM* 28 (1997) § 8 d.) Note that, although the width and the height of the gate are the same in § 6 a and in § 5 a, the diagonal is not the same; in § 5 a it is 41 15, but in § 6 a it is 42 13 20. In the drawing preceding the text of § 6, the notation *it-ta-šu* next to the number 10 for the width of the gate may mean precisely that the width was the side of a square, although the correct translation of the phrase is not known. Note, however, that *itû* 'boundary, side' has the feminine plural *itâtu*, so that the meaning of *it-ta-šu* may be 'its sides', referring to the sides of the square roof of a gate!

Let $h = $;40 n. = 8 cubits be the height of the gate in § 6 a, let $s = $;10 n. = 2 cubits be the side of the square roof or bottom of the gate, and let d be the inner diagonal of the gate. Then by the Old Babylonian *three-dimensional* diagonal rule, combined with the Old Babylonian square side rule

$$d = \text{sqs. (sq. } h + 2 \cdot \text{sq. } s) = \text{appr. } h + 2 \cdot \text{sq. } s / (2 \, h) = \text{;40 n.} + \text{;03 20} / 1\text{;20} = \text{;40 n.} + \text{;02 30 n.} = \text{;42 30 n.}$$

Strangely enough, in the text of § 6 a it is stated instead that

$$d = \text{appr. } h + 2 \cdot \text{sq. } s \cdot h = \text{;40 n.} + \text{;02 13 20 n.} = \text{;42 13 20 n.}$$

Apparently, the author of the text has by mistake multiplied by h when he should have divided by $2 \, h$. A plausible explanation for this curious mistake is that, accidentally or by design, the square of the inner diagonal d is in this exercise is

$$\text{sq. } d = \text{sq. } h + 2 \cdot \text{sq. } s = \text{(;26 40} + \text{;03 20) sq. n.} = \text{;30 sq. n.} = 1/2 \text{ sq. n.}$$

Therefore, it is true that, in *relative* numbers,

$$2 \cdot \text{sq. } d = 1, \quad \text{so that} \quad d = 1/(2 \, d).$$

The author of § 6 a was probably well aware of this coincidental relation, having used it in some other, related exercise (in the original larger theme text from which § 6 a was borrowed) in order to simplify his computations. Then, inadvertently, he used the same shortcut in § 6 a, but with *h* instead of *d*, which did not work so well!

The text of the next exercise, BM 96957+ § 6 b, is almost completely destroyed. Only part of the illustrating drawing is preserved, showing a rectangle and its diagonal with the numbers 40 and [42 1]3 20, indicating that the given numbers in this exercise were h = ;40 n. and d = [;42 1]3 20 n. Presumably, the value of the side s was then computed (in the lost part of the text) by reversal of the (incorrect) equation d = appr. $h + 2 \cdot$ sq. $s \cdot h$, that is, as the solution to the following equation:

$$\text{sq. } s = (d - h) / (2 h) = (42\ 13\ 20 - 40) / 1\ 20 = 2\ 13\ 20 \cdot 45 = 1\ 40.$$

This would, of course, give the correct value $s = 10$, meaning $s = $;10 n.

There are traces of three additional exercises following § 6 b, but too little is left of the text to allow any meaningful reconstruction attempts. There is also a notation *iš-kar-ka* 'your task' just after the text of the last exercise. Presumably the first of the destroyed exercises was § 6 c, the computation of the width s of the gate through reversal of the equation d = appr. $h + 2 \cdot$ sq. $s \cdot h$. It is likely that the next three exercises, including the one that the teacher did not bother with, was a further set of similar exercises, §§ 7 a-c.

Here is a reconstructed outline of the text on MS 3049, based on the information given in the subscript:

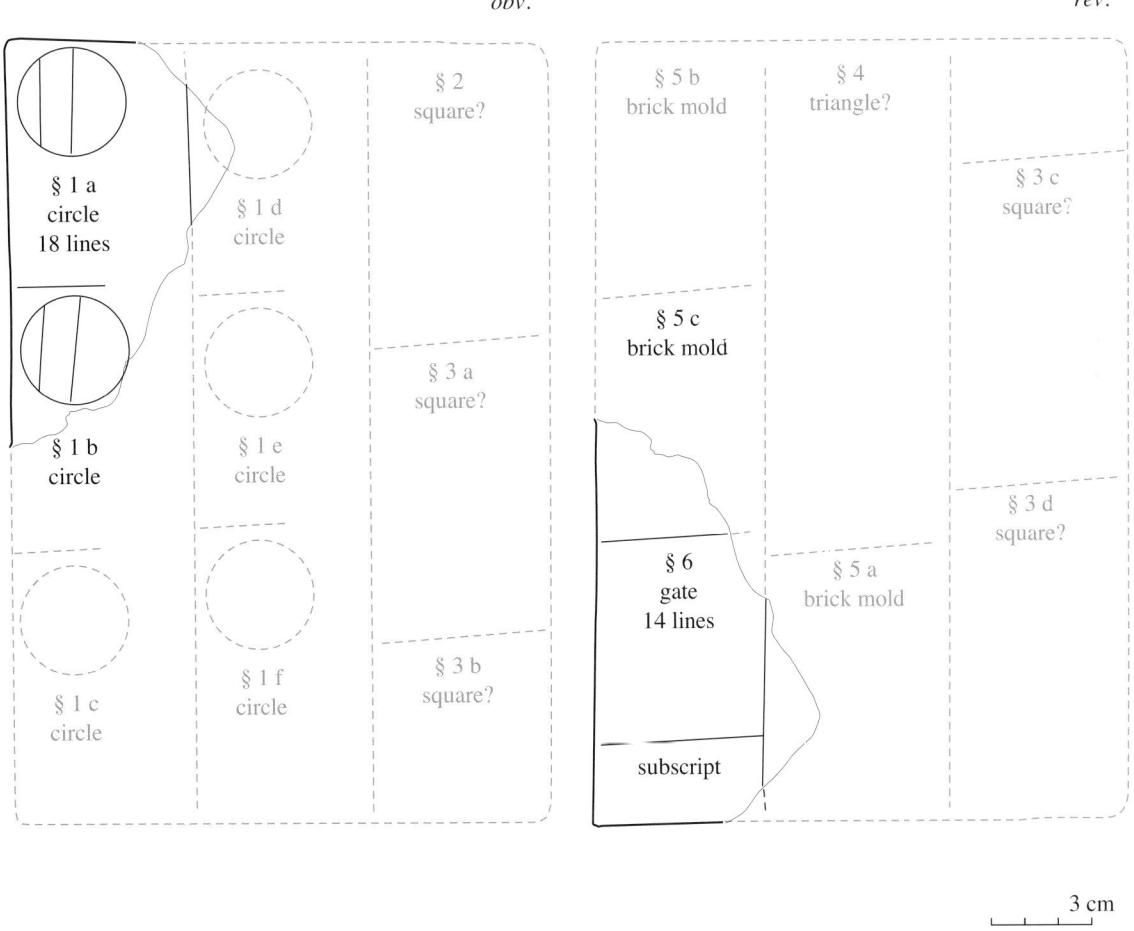

Fig. 11.1.6. MS 3049. A reconstructed outline of the text (5 themes, 16 exercises).

11.2. MS 5112. A Text with Equations for Squares and Rectangles

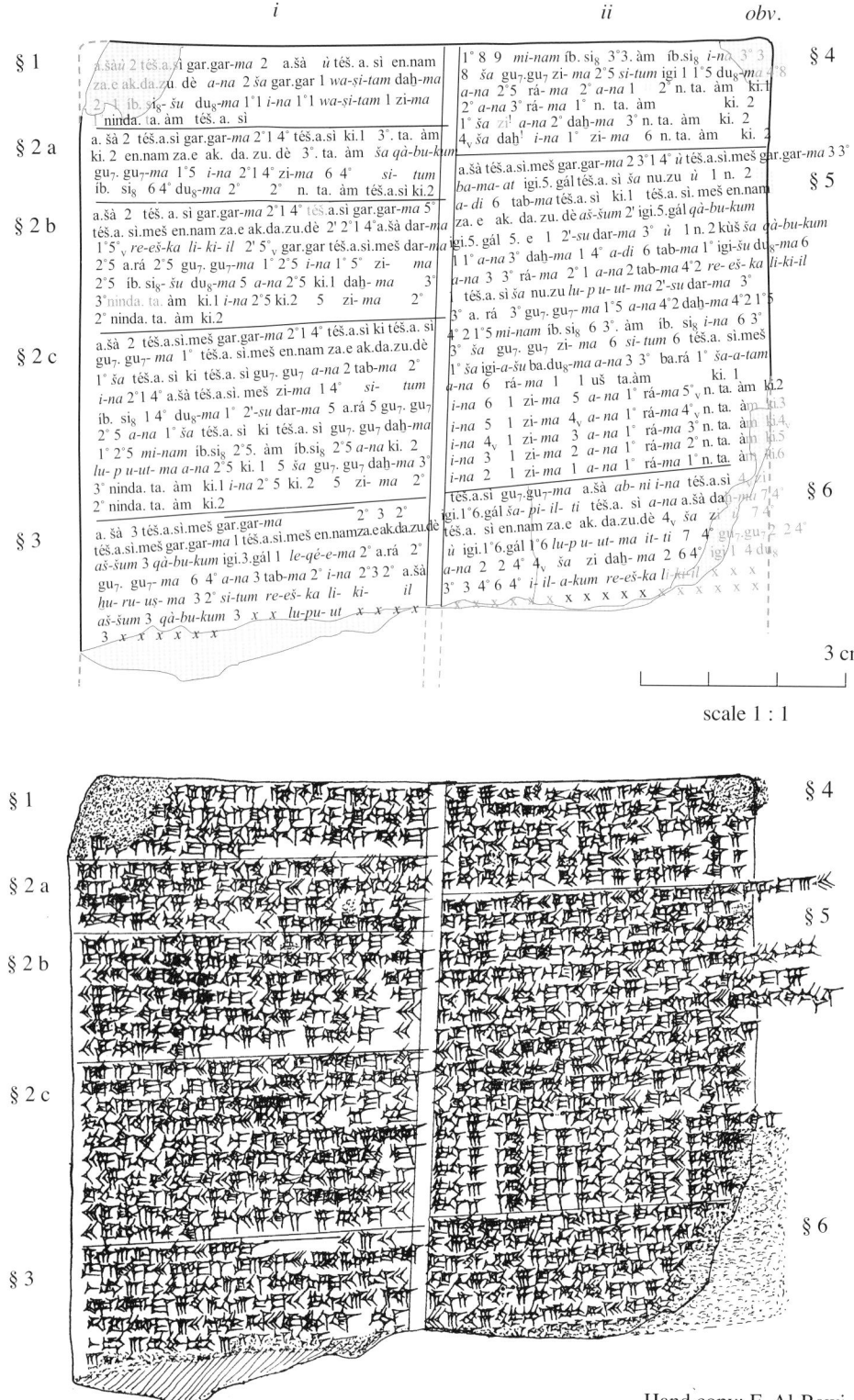

Hand copy: F. Al-Rawi

Fig. 11.2.1. MS 5112, *obv.* Eight (originally ten?) metric algebra problems for one or more squares.

MS 5112 is a substantial fragment of a large clay tablet with mathematical exercises. According to a subscript on the edge, the text originally contained 23 im.šu.meš 'hand tablets, assignments'. On the obverse of the fragment (Fig. 11.2.1 above), there are 5 well preserved exercises, and parts of 3 more. On the reverse (Fig. 11.2.9 below), there are 6 more or less completely preserved exercises and parts of 2 more. Thus, the fragment can be estimated to contain about 60% of the original text of the clay tablet.

MS 5112, obv. Metric Algebra Problems for One or More Squares

MS 5112 is a mathematical recombination text where all the exercises have a common theme: metric algebra. The exercises on the *obverse* (see the outline in Fig.11.2.17) are problems for one, two, or several *squares*. (§§ 1 and 6 are problems for 1 square, §§ 2 a-c and § 4 are problems for 2 squares, § 3 is a problem for 3 squares, and § 5 a problem for 6 squares.) The exercises on the *reverse* are problems for a *rectangle*.

11.2 a.   MS 5112 § 1. Old Babylonian Metric Algebra: Completing the Square

MS 5112 § 1 consists of a single exercise with four lines of text. A part of the text of this exercise, in the broken upper left corner of the clay tablet, is lost but can easily be reconstructed.

MS 5112 § 1 (*obv. i*: 1-4)*ff*

1	[a.šà *ù* 2 téš.a.s]ì gar.gar-*ma* 2	*The field (= area) and 2 sam*esides (I) heaped (= added together), then 2.
	a.šà *ù* téš.a.sì en.nam /	The field and the sameside (= squareside) are what?
2	[za.e ak.da.z]u.dè	*You with your do*ing:
	a-na 2 *ša* gar.gar	To 2 of the heap (= sum),
3	1 *wa-ṣi-tam* daḫ-*ma* / [2 01]	1 the extension add, then *2 01*.
	[íb.s]i₈-*šu* du₈-*ma* 11	Its *equal*side (= squareside) resolve, then 11.
	i-na 11 *wa-ṣi-tam* 1 zi-*ma* /	From 11 the extension 1 tear off, then
4	[10] ninda.[t]a.àm téš.a.sì	10 ninda each way is the sameside.

Already the text of this brief first exercise demonstrates that the mathematical terminology used in MS 5112 is *partly well known, partly novel*. Thus, the terms

gar.gar, daḫ, zi	three verbs for basic arithmetical operations (additions, subtraction)
za.e ak.da.zu.dè	the initial phrase of the solution procedure

are known to be used in Old Babylonian mathematical texts belonging to Group 3, that is in texts from Uruk (Friberg, *RA* 94 (2000), 165). The term

wāṣītum	(something) sticking out, extension

is used, repeatedly, as a technical term in the Old Babylonian metric algebra text BM 13901 (Høyrup, *LWS* (2002), 50). That text is the only one belonging to Group 1c and is from Larsa or some other southern city (Friberg, *op. cit.*, 161). On the other hand, the term

téš.a.sì (can also be read ur.a.sì)	given together (or equal)	square, square side

is new. Compare with *a b* téš.a.ta sì '*a* and *b* are given equal' in some series texts of Group Sa (Friberg, *RA* 94 (2000), 172). Compare also with téš.bi (Akk. *ištēniš*) 'together', and with téš.a, frequently used as a term for 'square' in the Late Babylonian mathematical text W 23291-x (Friberg, Hunger, Al-Rawi, *BagM* 21 (1990)). See also the discussion of the related terms ur, ur.ur, and ur.ka (or téš, téš.téš and téš.ka) in Høyrup, *LWS* (2002), 347, fn. 409.

New is also the following construction:

a íb.si₈-*šu* du₈	*a*, its equalside resolve	compute the square side of *a*

In Old Babylonian mathematical texts, the term du₈ is normally used only in the phrase igi *a* du₈ 'resolve (= compute) the opposite (= reciprocal) of *a*'. See, for instance, several examples below in MS 5112 §§ 4, [6], 9,

10, 12, 13. See also Høyrup, *LWS* (2002), 28, fn. 47, where it is emphatically pointed out that in Old Babylonian mathematical texts an íb.si₈ with the meaning 'square side' can be 'made come up' (*šūlûm*), or be 'taken' (*leqûm*), or it can be asked 'what' it is (*mīnûm*), but it is never 'detached' (du₈).

Another peculiarity of MS 5112 § 1, is that the cuneiform sign for -*ma* in this text is of an unusually elaborate form.

Thus, it is clear that MS 5112 cannot belong to any one of the already established groups of Old Babylonian mathematical texts. On the other hand, it is equally clear that MS 5112 must be from some southern Mesopotamian site. (See the map in Fig. 9.2 above.) The text may, for instance, be from Uruk, but if so, certainly not from the local "school" that produced the texts belonging to Group 3 (Str 362-3, 366-7, YBC 4608, VAT 7532, 7535, 7620, MS 3971, 3052, 2792).

What the statement of the problem (the "question") in line 1 of § 1 appears to mean is that

The area (of a square) and two sides of the square are added. The result is 2. What are the area and the sides?

In quasi-modern *abstract* notations the question can be formulated as a *quadratic equation*:

$$\text{sq.}\, p + 2 \cdot p = 2\,(00), \quad p = ? \qquad \text{(in modern notations: } p^2 + 2\,p = 2\text{).}$$

A Babylonian mathematician, on the other hand, would almost certainly interpret the question *literally*, as a question about a square field, and would think of the solution procedure, as well, in geometric terms. (See Høyrup, *LWS* (2002), Ch. 2, and Friberg, *BagM* 28 (1997) § 1.) When the text says that the area of a square and two sides of the square together are equal to 2, what this really means is that a square field is extended by a unit in two directions, as shown in Fig. 11.2.2 below, and that the area of the extended field is 2 (· 60 square ninda). The extended field consists of the initial square field with sides of some unknown length p, plus two rectangular fields, both with the sides p and $e = 1$. Hence, the area of the extended field is sq. $p + 2 \cdot p \cdot 1 =$ sq. $p + 2 \cdot p$, which can then with some justification be referred to as "the area and two sides (of a square)". In the text, the unit extension is called *wāṣītum*, a noun derived from the verb *waṣûm* 'to go out, to stick out'. (The term is discussed by Muroi in *ASJ* 20 (1998). Compare with footnote 3 below, under the text of MS 5112 § 10.)

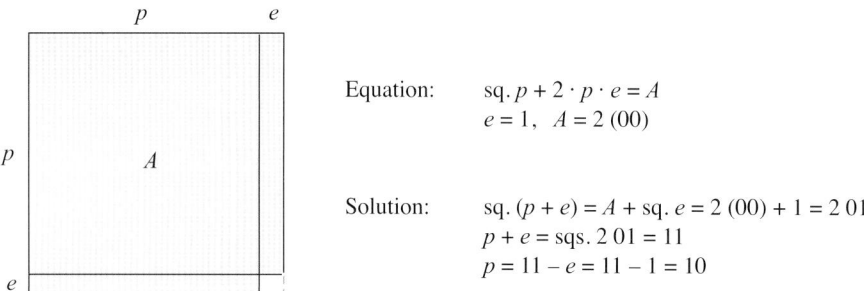

Fig. 11.2.2. MS 5112 § 1. A two ways extended square. Geometric solution procedure: Completing the square.

The solution procedure in § 1 begins by adding '1, the extension' to the area 2. This appears to be a slip on the part of the author of the text, who should have written, '1, the square of the extension'. Indeed, the natural first step of the geometric solution procedure is to add a small square with the side 1 to the extended field, thereby "completing the square" in a truly geometric sense. (See again Fig. 11.2.2 above.) Then, it is clear that the area of the completed square must be

$$(\text{sq.}\, p + 2 \cdot p \cdot e) + \text{sq.}\, e = 2\,(00) + 1 = 2\,01 \ (\text{sq. ninda}),$$

as indicated in line 2 of the text.

Now, since the side of the completed square is $p + e$, it follows that

$$\text{sq.}\,(p + e) = 2\,01, \quad \text{so that} \quad p + e = \text{sqs.}\, 2\,01 = 11, \quad \text{and} \quad p = 11 - 1 = 10 \ (\text{ninda}).$$

This interpretation agrees with lines 3-4 of the text.

There exists a clear and obvious difference between Babylonian and classical Greek (Euclidean) geometry. In Babylonian geometry, metric and metrological considerations are all-important, while in Greek geometry explicit consideration of numbers and measures is meticulously avoided. In Babylonian geometry, sides of a plane figure are always given in terms of their lengths, and the contents of plane figures in terms of their areas, while in Greek geometry lengths and areas are never mentioned. (Thus, Greek geometry in not "geometry" 'land-measuring' in the literal sense of the word!)

It is precisely in order to avoid confusion between Babylonian metric geometry and Greek abstract geometry that in the following *the branch of Babylonian mathematics concerned with numerical solutions of geometric problems* will be called "metric algebra", rather than, as in many earlier publications by other authors, "geometric algebra", "cut-and-paste geometry", or "naive geometry".

It is probably no coincidence that the first exercise in MS 5112 is the mentioned problem for a the area and two sides of a square. As will be shown below, MS 5112 is a mathematical recombination text with metric algebra as its general topic, and the area and two sides problem in § 1 can very well be said to be *the basic example of Old Babylonian metric algebra*!

A final observation: Just as there is a clear and obvious difference between Babylonian and Greek geometry, there is also a clear-cut difference between Babylonian metric algebra and the modern way of solving quadratic equations. In the present example, for instance, the modern solution to a quadratic equation of the type $p^2 + a\,p = B$ is *the algebraic formula*:

$$p = \sqrt{(B - (a/2)^2)} - a/2.$$

In contrast to this, the solution to the corresponding Babylonian metric algebra problem is in the form of *the successive steps of a metric and numerical solution procedure*, for which a suitable name would be a "metric algebra algorithm".

11.2 b. *BM 13901 # 23. A Related Text, with a Four Ways Extended Square Field*

This interesting exercise is inserted, totally out of context, near the end of **BM 13901**, a large mathematical recombination text with metric algebra as its topic. The uniqueness of the exercise, in several respects, is pointed out by Høyrup in *LWS* (2002), 222-226.

BM 13901 # 23

1	a.šà-*lam*	(About) a (square) field.
	p[a]-a-[at er-bé-et-tam ù a.š]à-*lam ak-mur-ma* 41 40 /	The *four fronts* and the *fi*eld I heaped, then 41 40.
2	4 *pa-at* er-[bé-e]*t-tam t*[a-la-p]a-at	4, the fo*ur* fronts, yo*u inscribe.*
	igi.4.gál.bi 15 /	Its 4th-part, 15.
3	15 *a-na* 41 40 [ta-n]a-ši-ma 10 25 *ta-la-pa-at* /	15 to 41 40 *you ra*ise, then 10 25. You inscribe (it).
4	1 *wa-ṣi-tam tu-ṣa-ab-ma*	1, the extension, you add to (it).
	1 10 25.e 1 05 íb.si₈ /	1 10 25 makes 1 05 equalsided.
5	1 *wa-ṣi-tam ša tu-iṣ-bu*	1, the extension that you added to it,
	ta-na-sà-aḫ-ma 5	you tear off, then 5.
6	*a-na ši-na* / *te-ṣi-ip-ma* 10 ninda *im-ta-ḫa-ar*	To two you repeat, then 10 ninda makes equal.

What this means, in normal language, is

The sum of the four sides and the area (of a square) is 41 40. What is the side (of the square)?

In quasi-modern abstract notations, this question can be expressed as a quadratic equation:

$$4\,(00) \cdot p + \text{sq. } p = 41\ 40.$$

The corresponding interpretation of the solution procedure in the text is as follows:

$$1/4 \cdot (4\ 00 \cdot p + \text{sq. } p) = 2\ 00 \cdot p/2 + \text{sq. } p/2 = 1/4 \cdot 41\ 40 = 10\ 25,$$
$$\text{sq. } (1\ 00 + p/2) = 1\ 10\ 25,$$
$$1\ 00 + p/2 = \text{sqs. } 1\ 10\ 25 = 1\ 05, \quad p = 2 \cdot (1\ 05 - 1\ 00) = 2 \cdot 5 = 10.$$

It is obvious that something is not quite right with this way of explaining the exercise. Why does "the four sides and the area" have to be interpreted as 4 00 p + sq. p, instead of just as 4 p + sq. p, and why do the terms of the original quadratic equation have to be divided by 4 before the start of the standard solution procedure?

The answer to these disturbing questions is that, just as in the case of MS 5112 § 1 above, a Babylonian mathematician would interpret the question in BM 13901 # 23 *literally*, as a question about a square field, and would think of the solution procedure, as well, in geometric terms. Thus, when the text says that four sides and the area of a square together are equal to 41 40, what this really means is that a square field is extended by a unit in four directions, as shown in Fig. 11.2.3 below, and that the area of the extended field is 41 40 (square ninda). The extended field consists of the initial square field with all sides of some unknown length p plus four rectangular fields, all four with the sides p and e = '1', as in Fig. 11.2.3, left. By mistake, or by design, the unit '1' in this exercise has the value 1 (00), not just 1 as in MS 5112 § 1.

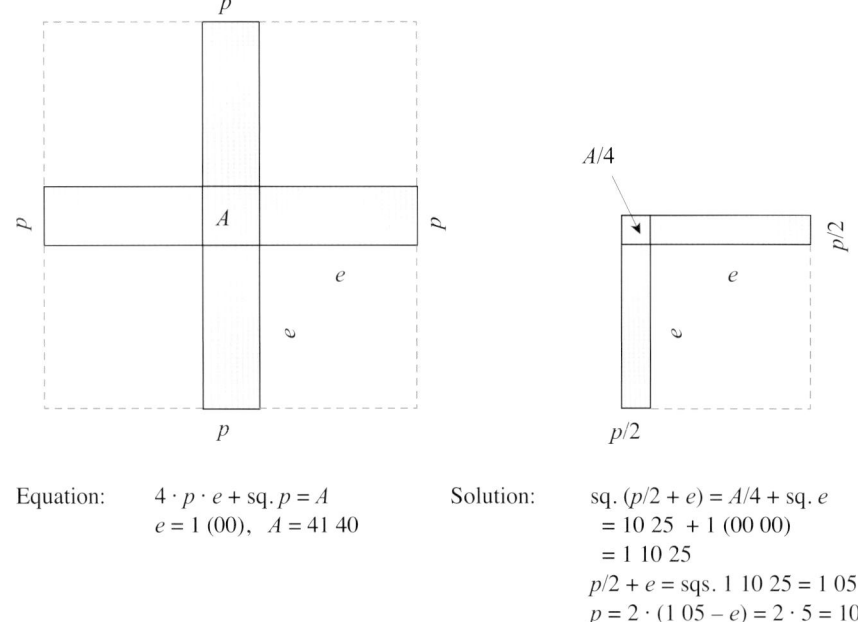

Equation: $4 \cdot p \cdot e +$ sq. $p = A$
              $e = 1\ (00),\quad A = 41\ 40$

Solution: sq. $(p/2 + e) = A/4 +$ sq. e
              $= 10\ 25\ + 1\ (00\ 00)$
              $= 1\ 10\ 25$
              $p/2 + e =$ sqs. $1\ 10\ 25 = 1\ 05$
              $p = 2 \cdot (1\ 05 - e) = 2 \cdot 5 = 10$

Fig. 11.2.3. Reduction through quartering of a four ways extended square to a two ways extended square.

The first step of the solution procedure in BM 13901 # 23 is to divide the combined area 41 40 by 4. Geometrically, this step can be interpreted as a *quartering* of the four ways extended square. The result is a two ways extended quarter-square, as in Fig. 11.2.3, middle. In this way, the question in BM 13901 # 23 has been reduced to a variant of the question in MS 5112 § 1! Therefore, the remainder of the solution procedure is practically the same as the solution procedure in that exercise.

According to Høyrup (*op. cit.*, 225), one of the unusual features of BM 13901 # 23, is that the square in that exercise has the side 10, while normally squares in Old Babylonian mathematical exercises have the side 30. Therefore, it is interesting to note that also the square in MS 5112 §1 has the side 10. Furthermore, in both exercises it is stated explicitly that the side of the square is not just '10', but actually 10 ninda.

Another unusual feature of BM 13901 # 23 is that in the question the four sides of the square are mentioned before the area of the square. It is possible that this is so because the collected area of the four extensions of the square is larger than the area of the square. (See fig. 11.2.3, left.)

11.2 c.   MS 5112 § 2 a. A trivial Problem for Two Squares

The text of § 2a is undamaged. It covers four lines of writing:

MS 5112 § 2 a (*i*: 5-8)

1	a.šà 2 téš.a.sì gar.gar-*ma* 21 40	The fields of 2 samesides (I) heaped, then 21 40.
	téš.a.sì ki.1 30.ta.àm /	The 1st sameside, 30 each way.
2	ki.2 en.nam	The 2nd is what?
	za.e ak.da.zu.dè	You with your doing:
	30.ta.àm *ša qà-bu-kum* /	30 each way that is said to you
3	gu₇.gu₇-*ma* 15	(make) eat (itself) (= square), then 15.
	i-na 21 40 zi-*ma* 6 40 *si-tum* /	From 21 40 tear (it) off, then 6 40 the remainder.
4	íb.si₈ 6 40 du₈-*ma* 20	The equalside of 6 40 resolve, then 20.
	20 ninda.ta.àm téš.a.sì ki.2	20 ninda each way is the 2nd sameside.

In this second exercise of MS 5112, the terminology continues to have unusual features. The following terms have no or few counterparts in the known corpus of Old Babylonian mathematics:

ša qabûkum	that is said to you	citing what is said in the question
a gu₇.gu₇	(make) *a* eat (itself)	squaring *a*
sittum	remainder	
ki.1, ki.2	first, second (number 1, number 2)	

The closest parallel to *ša qabûkum* is *aššum qabûkum* 'since it is said to you' in several of the mathematical Susa texts (Bruins and Rutten (1961), *TMS* 10, 16, 17, 19, 24). The phrase *aššum qabûkum* appears also below, in MS 5112 §§ 3, 5, 7 a-b, 10, and 12.

The term gu₇.gu₇ is used here, apparently, as a logogram for *tuštakal* or *šutakil* 'you make (it) eat itself', meaning 'you square (it)'.[1] See the discussion of the latter terms in Høyrup, *LWS* (2002), 6, 23, *etc.*, and compare with the more convincing arguments put forward in Muroi, *HSJ* 12 (2003). The only other known text with a similar use of gu₇.gu₇ is the Susa text *TMS* 26 (Muroi, *HSJ* 10 (2001)). Other close parallels are i.gu₇.gu₇ in YBC 4713 (Group Sa; *RA* 94, 172) and BM 85200+ (Group 6a). Similar reduplicated words for squaring or "rectangularization" are ur.ur.a (or rather téš.téš.a), possibly meaning 'bring (them) together' (Group 2 a; *RA* 94, 62) and du₇.du₇, meaning 'make (them) butt each other' (Group 3; *RA* 94, 65; examples in MS 3052 in § 10.2 above).

An indirect parallel to *si-tum* (*sittum*) here is *ši-ta-tum* (with the same meaning) in the Tell Harmal text IM 52301 (Gundlach and von Soden, *AMSH* 26 (1963)).

In normal language, the question in MS 5112 § 2 a, can be reformulated as follows:

> The sum of the areas of two squares is 21 40. The side of the first square is 30.
> What is the side of the second square?

This is a simple initial example of a "quadratic-linear" system of equations for two unknowns in the small *theme text* consisting of §§ 2 a-c.

The solution procedure is also exceedingly simple. It works, essentially, in the following way:

> Remove the square of the first side from the sum of the squares of the two sides.
> The square side of the remainder is the second side.

In quasi-modern symbolic notations, the question and the solution procedure can be reformulated as follows: Let p and q be symbolic notations for the two square sides. Then

$$\text{sq.}\, p + \text{sq.}\, q = 21\ 40\ (\text{sq. ninda}),\quad p = 30\ (\text{ninda}) \Rightarrow$$
$$\text{sq.}\, q = 21\ 40 - \text{sq.}\, 30 = 21\ 40 - 15\ 00 = 6\ 40,\quad q = \text{sqs.}\ 6\ 40 = 20\ (\text{ninda}).$$

---

1. gu₇.gu₇ is a reduplicated form of gu₇ 'eat' (Akk. *akālum*, with the Št-stem *šutākulum* 'to make -- eat each other'). The puzzling use of *eating each other* as a metaphor for squaring or rectangularization can possibly be explained as follows: If a counting board of some kind was used for the multiplication of two given numbers with each other, then presumably both numbers were initially recorded on the device. As the multiplication algorithm proceeded, first one number and then the other was deleted, and at the end of the process only the product of the given numbers remained.

11.2 d.   MS 5112 § 2 b. A Standard Quadratic-Linear System of Equations

MS 5112 § 2 b (*i*: 9-15)

1	a.šà 2 téš.a.sì gar.gar-*ma* 21 40	The fields of 2 samesides (I) heaped, then 21 40.
	[téš].a.sì gar.gar-*ma* 50 /	The *sam*esides (I) heaped, then 50.
2	téš.a.sì.meš en.nam	The samesides are what?
	za.e ak.da.zu.dè	You with your doing:
3	2' 21 40 a.šà dar-*ma* / 10¹ 50	1/2 of 21 40 the field crush, then 10 50.
	re-eš-ka li-ki-il	May it hold your head.
4	2' 50 gar.gar téš.a.sì.meš dar-*ma* / 25	1/2 of 50 the heap of the samesides crush, then 25.
	a.rá 25 gu₇.gu₇-*ma* 10 25	Steps of 25 (make) eat (itself) (= multiply), then 10 25.
5	*i-na* 10 50 zi-*ma* / 25	From 10 50 tear off, then 25.
	íb.si₈-*šu* du₈-*ma* 5	Its equalside resolve, then 5.
	a-na 25 ki.1 dah-*ma* 30 /	To the 1st 25 add, then 30.
6	30 [ninda.ta].àm ki.1	30 *ninda each* way, the 1st.
	i-na 25 ki.2 5 zi-*ma* 20 /	From the 2nd 25 tear off, then 20.
7	20 ninda.ta.àm ki.2	20 ninda each way, the 2nd.

The new terms in this exercise are:

a *rēška likīl*	may *a* hold your head	remember *a*
2' *a* dar	1/2 of *a* crush	halve *a*, compute 1/2 of *a*
a a.rá *a* gu₇.gu₇	*a* steps of *a* make eat each other	multiply *a* with *a*, square *a*

Here dar (Akk. *pa'āṣum*) 'to crush' is a new term for halving. The standard term is *ḫepûm* 'to break', replaced by the corresponding logogram gaz in texts of Group 3.

The phrase *rēška likīl* is commonly occurring in Old Babylonian mathematical texts. It is usually translated 'your head may hold (the number)', but as Muroi has pointed out in *HSJ* 12 (2003), the correct literal translation ought to be, instead, 'may it (the number) hold your head'. This change in perspective does not seem to be important, since it is clear that, in either case, the actual meaning of the phrase is 'remember this number'. However, as will be shown below (in Sec. 11.2 n), the new interpretation may be important for the correct explanation of the difficult term *takīltum*.

In this more interesting example of a quadratic -linear system of equations, the question is:

> The sum of the areas of two squares is 21 40. The sum of the sides of the two squares is 50.
> What are the sides of the two squares?

The successive steps of the numerical solution procedure are the following:

> 1/2 · 21 40 = 10 50, sq. (1/2 · 50) = sq. 25 = 10 25, 10 50 – 10 25 = 25, sqs. 25 = 5,
> 25 + 5 = 30 (ninda) = the magnitude of the first side,
> 25 – 5 = 20 (ninda) = the magnitude of the second side.

The result is correct, since sq. 30 + sq. 20 = 15 00 + 6 40 = 21 40, and 30 + 20 = 50.

In quasi-modern symbolic notations, the solution procedure in § 2 b takes the following form:

> Let p and q be the sides of the two squares in § 2 b. Then
> sq. p + sq. q = 21 40 (sq. ninda), $p + q$ = 50 (ninda) $\Rightarrow$
> (sq. p + sq. q)/2 = 10 50, sq. {($p + q$)/2} = sq. 25 = 10 25,
> (sq. p + sq. q)/2 – sq. {($p + q$)/2} = 10 50 – 10 25 = 25 = sq. 5 = sq. {($p - q$)/2},
> ($p + q$)/2 + ($p - q$)/2 = 25 + 5 = 30, p = 30 ninda,
> ($p + q$)/2 – ($p - q$)/2 = 25 – 5 = 20, q = 20 ninda.

Thus, it appears that an essential part of the solution procedure in § 2 b is a mathematical identity of the following form:

$$\text{(sq. } p + \text{sq. } q)/2 = \text{sq. } \{(p + q)/2\} + \text{sq. } \{(p - q)/2\}.$$

However, it is unlikely that the Old Babylonian mathematicians knew the identity in this algebraic form. It is much more likely that they knew a more obvious geometric version of the identity:

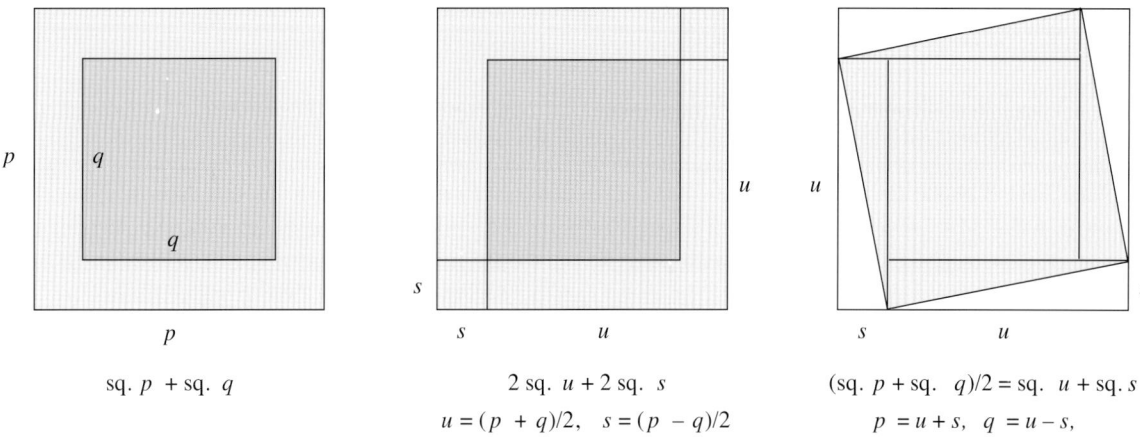

sq. p + sq. q	2 sq. u + 2 sq. s	(sq. p + sq. q)/2 = sq. u + sq. s
	$u = (p + q)/2, \quad s = (p - q)/2$	$p = u + s, \quad q = u - s,$

Fig. 11.2.4. MS 5112 § 2 b. The derivation of a useful geometric identity.

In Fig. 11.2.4 above, left, the sum of the areas of two concentric squares with the sides p and q is shown to be equal to the area of the smaller square, counted twice, plus the area of the "square band" between the two squares. In Fig. 11.2.4, middle, it is shown that *the sum of the areas of the two squares with the sides p and q is equal to twice the sum of the areas of two new squares with the sides u and s, where u is the "half-sum" of p and q, while s is the corresponding "half-difference"*. The algebraic form of this geometric identity is:

$$\text{sq. } p + \text{sq. } q = 2 \text{ sq. } \{(p + q)/2\} + 2 \text{ sq. } \{(p - q)/2\}.$$

Now, half the sum of the areas of the two squares with sides p and q is equal to the area of the smaller square, plus half the area of the square band. If the square band is divided into a chain of four equal rectangles, all with the length u and the front s, as in Fig. 11.2.4, right, then half the area of the square band is equal to the area of four right triangles, each such triangle equal to half a rectangle. Consequently, *half the sum of the areas of the two squares with the sides p and q is equal to the area of the square on the diagonal of one of the rectangles, which in its turn is equal to the sum of the areas of the squares with the sides u and s.*

The complicated algebraic explanation of the mentioned solution procedure can now be replaced by the following much more intuitive geometric explanation:

> The sum of the areas of two squares with the unknown sides p and q is known and equal to 21 40.
> The sum of the sides of the two squares is also known and equal to 50.
> Let u and s be the half-sum and the half-difference, respectively, of p and q.
> Then sq. u + sq. s = 21 40 / 2 = 10 50, and u = 50/2 = 25, are also known.
> Hence, sq. s = 10 50 – sq. 25 = 10 50 – 10 25 = 25 is also known, and s = sqs. 25 = 5.
> Therefore, $p = u + s = 25 + 5 = 30$, and $q = u - s = 25 - 5 = 20$ are also known.

Note that in this solution procedure u and s are computed in precisely the same way as the sides of the squares in MS 5112 § 2 c. Therefore § 2 b is a natural continuation of § 2 c!

A Parallel Text: BM 13901 # 8

BM 13901 # 8 (Høyrup, *LWS* (2002), 67)

1	[a.šà *mi-it-ḫa-ra-ti-ia ak-mur-ma*] 21 40 /	*The fields of my equalsides I heaped, then* 21 40,
2	[*ù mi-it-ḫa-ra-ti-ia ak-mur-ma* 50]	and *my equalsides I heaped, then* 50.
	[*ba-ma-at* 21] 40 *te-ḫe-pe* /	*The half-part of* 21 40 *you break,*
3	[10 50 *ta-la-pa-at*]	10 50 *you inscribe.*
	[*ba-ma-at* 50 *te-ḫe-pe*]	*The half-part of* 50 *you break,*
	[25 ù 25 *tu-u*]*š-ta-kal* /	25 *and* 25 *you* m*ake hold,*

4	[10 25 *lib-bi* 10 50 *ta-na-sà-ah-ma*]	10 25 off 10 50 you tear off, then
	[25.e 5] íb.si$_8$ /	*25 makes 5* equalsided.
5	5 *a-na* 25 *iš-te-en tu-ṣa-ab-*[*ma*]	5 to one 25 you add, *then*
	[30 *mi-it-ḫar-tum iš-ti-a-at*] /	30, one equalside.
6	5 *lib-bi* 25 *ša-ni-tim ta-na-sà-aḫ-m*[*a*]	5 off the second 25 you tear off, th*en*
	[20 *mi-it-ḫar-tum ša-ni-tum*]	20, the second equalside.

In **BM 13901 # 8**, the question and the solution procedure are identical with the question and the solution procedure in MS 5112 § 2 a, although with a totally different terminology, mostly in syllabic Akkadian, with only a couple of words in Sumerian. Actually, the two texts are so similar that it is conceivable that one is a translation of the other, from one jargon to another! (Most of the text of BM 13901 # 8 is destroyed, but the existence of a well preserved variant of the exercise, BM 13901 # 9, makes it possible to reconstruct the lost parts of the text of # 8. The only difference between # 8 and # 9 is that the sum of the sides in # 8 is replaced by the difference of the sides in # 9.)

11.2 e.   MS 5112 § 2 c. A Quadratic-Rectangular System of Equations

MS 5112 § 2 c (*i*: 16-25)

1	a.šà 2 téš.a.sì.meš gar.gar-*ma* 21 40	The fields (= areas) of 2 samesides (I) heaped, then 21 40.
2	téš.a.sì ki téš.a.sì / gu$_7$.gu$_7$-*ma* 10	Sameside with sameside (I made) eat (each other), then 10.
	téš.a.sì.meš en.nam	The samesides are what?
	za.e ak.da.zu.dè /	You with your doing:
3	10 *ša* téš.a.sì ki téš.a.sì gu$_7$.gu$_7$	10 that sameside with sameside (were made) eat
	a-na 2 tab-*ma* 20 /	to 2 repeat, then 20.
4	*i-na* 21 40 a.šà téš.a.sì.meš	From 2140, the fields of the samesides
	zi-*ma* 1 40 *si-tum* /	tear off, then 1 40 is the remainder.
5	íb.si$_8$ 1 40 du$_8$-*ma* 10	The equalside of 1 40 resolve, then 10.
	2'-*su* dar-*ma* 5	Its 1/2 crush, then 5.
6	a.rá 5 gu$_7$.gu$_7$ / 25	Steps of 5 (make) eat, 25.
	a-na 10 *ša* téš.a.sì ki téš.a.sì gu$_7$.gu$_7$	To 10 that sameside with sameside were made eat
7	daḫ-*ma* / 10 25	add, then 10 25.
	mi-nam íb.si$_8$ 25<.ta>.àm íb.si$_8$	What is it equalsided? 25 each way it is equalsided.
8	25 *a-na* <<ki.>> 2 / *lu-pu-ut-ma*	25 to 2 inscribe, then
	a-na 25 ki.1 5 *ša* gu$_7$.gu$_7$ daḫ-*ma* 30 /	to the 1st 25 5 that (was made) eat add, then 30.
9	30 ninda.ta.àm ki.1	30 ninda each way, the 1st.
	i-na 25 ki.2 5 zi-*ma* 20 /	From the 2nd 25 5 tear off, then 20.
10	20 ninda.ta.àm ki.2	20 ninda each way, the 2nd.

Remark: In line 6, the normal writing 2'-*šu* = *maš-šu* appears to be softened to 2'-*su* = *mas-su*.

The new terminology in § 2 c is:

a-na 2 tab	to 2 repeat	multiply by 2, double
a-na 2 *lu-pu-ut*	to two inscribe	write down in two places
b mi-nam íb.si$_8$ *a*.àm íb.si$_8$	*b* (makes) what equalsided? *a* it is equalsided	what is sqs. *b*? it is *a*
a ki *b* gu$_7$.gu$_7$	*a* with *b* (make) eat	multiply *a* by *b* (rectangularity)

The phrase *a-na* 2 tab in § 2 c can be compared with *a-na* 2 e.tab in texts of Group 3 (*RA* 94, 165). Compare also with *a-na* 2 tab.ba in Susa texts of Group 8a (*RA* 94, 171).

The phrase *a-na* 2 *lu-pu-ut* is used, in the same way as in § 2 c, in YBC 466 2 (= *MCT*, J) and 4663 (= *MCT*, H) (both of Group 2a). See also the general discussion of the meaning of the term *luput* in Robson, *MMTC* (1999), 30.

Parallels to the phrase *mi-nam* íb.si$_8$ can be found in Nippur texts of Group 6a (en.nam íb.si$_8$) and in Susa texts of Group 8a (*mi-na* íb.si$_8$). It is remarkable that different expressions for computing square sides are used in MS 5112 § 1 (*a íb.sig$_8$-šu* du$_8$) and in MS 5112 § 2 c.

It is interesting to note that the phrase téš.a.sì ki téš.a.sì gu$_7$.gu$_7$-*ma* 10 in lines 1-2 of § 2 c is referred to by the phrase 10 *ša* téš.a.sì ki téš.a.sì gu$_7$.gu$_7$ in lines 3 and 6, while the different phrase 5 a.rá 5 gu$_7$.gu$_7$ in line 5 is referred to by the phrase 5 *ša* gu$_7$.gu$_7$ in line 8. Apparently this is not done arbitrarily, since the multiplication in the former case is a rectangularization (the computation of the area of a rectangle), while the multiplication in the latter case is a squaring.

The question in § 2 c of MS 5112 can be formulated as follows:

> The sum of the areas of two squares (= sq. p + sq. q) is 21 40.
> The product of the sides of the two squares (= $p \cdot q$) is 10 (00).
> What are the sides of the two squares?

This is a "quadratic-rectangular" system of equations for two unknowns. The successive steps of the numerical solution procedure are the following:

$$2 \cdot 10 \, (00) = 20 \, (00), \quad 21 \, 40 - 20 \, (00) = 1 \, 40, \quad \text{sqs. } 1 \, 40 = 10, \quad 1/2 \cdot 10 = 5 \, (= s),$$
$$\text{sq. } 5 = 25, \quad 10 \, (00) + 25 = 10 \, 25, \quad \text{sqs. } 10 \, 25 = 25 \, (= u),$$
$$25 + 5 = 30, 30 \, \text{ninda} \, (= p), \quad 25 - 5 = 20, \quad 20 \, \text{ninda} \, (= q).$$

To try to give a full algebraic explanation of the solution procedure in § 2 c in quasi-modern abstract notations would be counterproductive. Nevertheless, it may be useful to point out that the solution procedure is based on two identities that can be expressed in algebraic form as follows:

$$(\text{sq. } p + \text{sq. } q) - 2 \, p \cdot q = 4 \, \text{sq. } [(p - q)/2], \quad \text{and} \quad \text{sq. } [(p - q)/2] + p \cdot q = \text{sq.}[(p + q)/2].$$

A geometric interpretation and explanation of these identities is offered in Fig. 11.2.5 below:

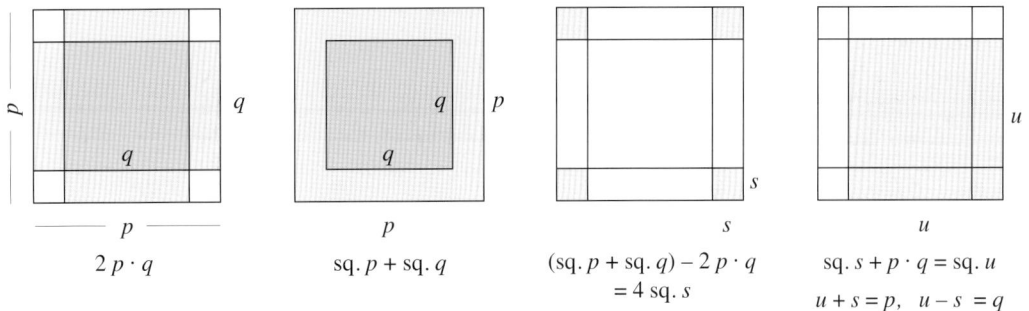

$$
\begin{array}{cccc}
p & p & s & u \\
2 \, p \cdot q & \text{sq. } p + \text{sq. } q & \begin{array}{c}(\text{sq. } p + \text{sq. } q) - 2 \, p \cdot q \\ = 4 \, \text{sq. } s\end{array} & \begin{array}{c}\text{sq. } s + p \cdot q = \text{sq. } u \\ u + s = p, \quad u - s = q\end{array}
\end{array}
$$

Fig. 11.2.5. MS 5112 § 2 c. A geometric interpretation of the solution procedure.

The first step of the solution procedure (line 3) is to compute $2 \cdot 10 \, (00) = 20 \, (00)$, twice the product of the two square sides. The geometric interpretation in Fig. 11.2.5, left, is the sum of the areas of two rectangles, both with the length p and the front q. As the figure shows, this sum is equal to twice the area of the smaller square, plus the areas of four small rectangles, all with the length q and the front $s = (p - q)/2$. Next (line 4), the result of this first operation is subtracted from the sum of the areas of the two squares, shown in Fig. 11.2.5, left middle. Numerically, the result is $21 \, 40 - 20 = 1 \, 40$, while geometrically the result is the sum of the areas of four small squares, all with the side s, as shown in Fig. 11.2.5, right middle. These four small squares can be combined into a square with the side $2 \, s = 10$, and half of this square side is $s = 5$ (line 5).

Next, it is observed that the area of the rectangle with the length p and the front q is equal to the sum of the areas of the square with side q and two rectangles with the length q and the front s (Fig. 11.2.5, right). Therefore, the sum of the area of a square with the side s and a rectangle with the length p and the front q is equal to the area of a square with the side u. Numerically, this combined area is $25 + 10 \, (00) = 10 \, 25$, and the side $u = 25$ (lines 6-7). Finally, with u and s known, p and q can be computed as the sum and the difference, respectively, of u and s (lines 8-10).

Remark: From an *algebraic* point of view, the questions and solution procedures in MS 5112 § 2 c and in Db$_2$-146 = IM 67118 (see Sec. 10.1 b above) are nearly identical, although with different numerical data. From a *geometric* point of view, on the other hand, both the questions and the solution procedures in the two exercises are different from each other. After all, MS 5112 § 2 c is a problem for two squares, while IM 67118 is a problem for a rectangle.

11.2 f.   MS 5112 § 3. Three Squares With their Sides in an Arithmetical Progression

The last few lines of § 3 are lost, but enough of the text remains so that it is clear what the question is and how the solution procedure begins.

MS 5112 § 3 (*i*: 26-32)

1	a.šà 3 téš.a.sì.meš gar.gar-*ma* 23 20 /	The fields of 3 samesides (I) heaped, then 21 40.
2	téš.a.sì.meš gar.gar-*ma* 1	The samesides (I) heaped, then 1.
	téš.a.sì.meš en.nam	The samesides are what?
	za.e ak.da.zu.dè /	You with your doing:
3	*aš-šum* 3 *qà-bu-kum*	Since 3 it is said to you,
	igi.3.gál 1 *le-qé-e-ma* 20	the 3rd-part of 1 take, then 20.
4	a.rá 20 / gu₇.gu₇-*ma* 6 40	Steps of 20 (make them) eat (each other), then 6 40.
	a-na 3 tab-*ma* 20	By 3 repeat, then 20.
5	*i-na* 23 20 a.šà / *ḫu-ru-uṣ-ma*	From 23 20 the fields cut off,
	3 20 *si-tum*	then 3 20 the remainder.
	re-eš-ka li-ki-il /	May it hold your head.
6	*aš-šum* 3 *qà-bu-kum*	Since 3 it is said to you,
	3 x x *lu-pu-ut-ma* /	3 x x inscribe, then
7	3 x x [x x x x x x x]	3 x x *x x x x x x x* ....

The new terminology in this exercise consists of the following two phrases

igi.*n*.gál *a le-qé-e*	the *n*th-part of *a* take	divide *a* by *n*
a i-na b ḫuruṣ	*a* from *b* cut off	subtract *a* from *b*

The *n*-th part of a given number is 'taken' (*leqûm*) in BM 85200+, BM 85210, and VAT 672 (Group 6), as well as in Str. 366 (Group 3). Examples: igi.7.gál 7 *le-qé* 1 *ta-mar* in BM 85200+, *rev. i*:3 (Høyrup, *LWS*, 144), and igi.3.gál 2 05 *te-le-qé-e-ma* 41 40 in Str 366, *obv.* 8.

The verb *ḫarāṣum* 'to cut off' is used to denote subtraction in mathematical texts belonging to Groups 1 (Larsa), 2 (Ur?), 6 (Sippar), and 7 (Eshnunna).

The question in MS 5112 § 3 is stated in the following way:

> The sum of the areas of three squares is 23 20. The sum of their sides is 1 (00). What are the sides?

Obviously something is missing here, because two equations are not enough to fix the values of three unknowns. The missing third requirement is that

> The square sides form an arithmetical progression. In other words,
> the difference between the first square side and the second is equal to the difference between the second and the third.

What the author of the text had in mind was almost certainly three concentric squares with a constant distance between the squares, as in Fig. 11.2.6, left.

If the sides of the three squares are called p, q, r, and if

$$d = p - q = q - r,$$

then it is clear that

$$p + q + r = (q + d) + q + (q - d) = 3\,q,$$

and that

$$\mathrm{sq.}\,p + \mathrm{sq.}\,q + \mathrm{sq.}\,r = \mathrm{sq.}\,(q + d) + \mathrm{sq.}\,q + \mathrm{sq.}\,(q - d) = 3\,\mathrm{sq.}\,q + 2\,\mathrm{sq.}\,d.$$

The second identity can be verified in various ways, for instance by use of Fig. 11.2.6, middle. Whichever way it was done, these two identities were probably used by the author of § 3.

What remains of the solution procedure can be explained as follows: Since the sum of the areas of the three squares is 23 20 and the sum of the sides 1 (00), it follows that

$$3\,q = 1\,(00), \quad \text{and} \quad 3 \text{ sq. } q + 2 \text{ sq. } d = 23\,20.$$

Hence,

$$q = 1/3 \cdot 1\,(00) = 20 \qquad \text{(line 3)}$$
$$3 \text{ sq. } q = 3 \cdot 6\,40 = 20\,(00) \qquad \text{(line 4)}$$
$$2 \text{ sq. } d = 23\,20 - 20\,(00) = 3\,20 \qquad \text{(line 5)}$$

Only a few traces remain of the continuation of the solution procedure after this point. The obvious continuation would be the successive computation of

$$\text{sq. } d = 1/2 \cdot 3\,20 = 1\,40, \quad \text{hence} \quad d = 10, \quad \text{and consequently} \quad p, q, r = 30, 20, 10.$$

For some reason, the remaining traces of the text in lines 6-7 seem to indicate that the solution procedure in the text may have ended in some other way.

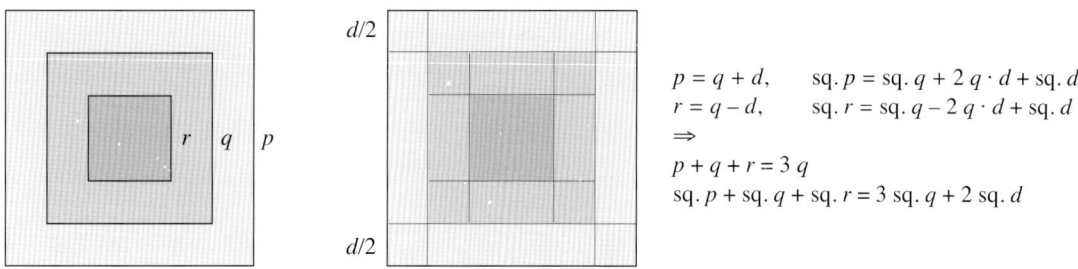

Fig. 11.2.6. MS 5112 § 3. An arithmetical progression of three concentric squares.

11.2 g.   MS 5112 § 4. A Quadratic-Linear System of Equations for Two Squares

MS 5112 § 4 (*ii* 1-6)

..	....	....
1'	18 09 *mi-nam* íb.si₈ 33.àm íb.si₈	18 09 (makes) what equalsided? 33 each way equalsided.
2'	*i-n*[*a* 33] / 8 *ša* gu₇.gu₇ zi-*ma*	From 33 the 8 that was (made) eat tear off, then
	25 *si-tum*	25 is the remainder.
	igi 1 15 du₈-*m*[*a* 48] /	The opposite (= reciprocal) of 1 15 resolve, then *48*.
3'	*a-na* 25 rá-*ma* 20	To 25 go, then 20.
	a-na 1 (erasure) 20 ninda.ta.àm ki.1 /	To 1 <go, then> 20 ninda each way, the 1st.
4'	20 *a-na* 30 rá-*ma* 10 ninda.ta.àm ki.2 /	20 to 30 go, then 10 ninda each way, the 2nd.
5'	10 *ša* [zi] *a-na* 20 daḫ-*ma*	10 that you *tore off* to 20 add, then
	30 ninda.ta.àm ki.1¹ /	30 ninda each way, the 1st.
6'	4ᵥ *ša* daḫ¹ *i-na* 10 zi-*ma*	4 that you added from 10 tear off, then
	6 ninda.ta.àm ki.2	6 ninda each way, the 2nd.

The new terminology in this exercise consists exclusively of the following phrase:

$$a\ a\text{-}na\ b\ \text{rá} \qquad a \text{ to } b \text{ go} \qquad \text{multiply } a \text{ with } b$$

In previously published Old Babylonian mathematical texts, this phrase is used only in the northern Group 6b (Friberg, *RA 94* (2000), 169). It is used in § 4 for *non-geometric* multiplication. Compare with the similar use of *našûm*, íl, nim, *etc.*, in other texts (Høyrup, *LWS* (2002), 22).

In this paragraph, the question and the beginning of the solution procedure are lost. What remains of the solution procedure are the final steps of the solution of some quadratic-linear or rectangular-linear system of equations for two unknowns, followed by a change of variables. Apparently, the equations were set up, not for the sides p and q of two squares, but for the altered sides

$$p* = p - 10 \text{ n.} \quad \text{and} \quad q* = q + 4 \text{ n.}$$

The computed values (in lines 3'-6') are

$$p* = 20 \text{ n.}, \quad q* = 10 \text{ n.}, \quad p = 30 \text{ n.}, \quad q = 6 \text{ n.}$$

It is also clear (from line 4') that the equation for $q*$ is

$$q* = 1/2 \cdot p*.$$

The computation of $p*$ (see lines 1'-2') is more complicated:

$$p* = 1 / 1\,15 \cdot p**,$$

where $p**$ (= 25) is computed as a solution to the quadratic equation

$$\text{sq. } p** + 16 \cdot p** = 17\,05.$$

Accidentally, just enough of the solution procedure is preserved so that it is possible to reconstruct the lost question. It must have been, essentially, the following *quadratic-linear system of equations for two unknown squares*:

$$\text{sq. } p + \text{sq. } q = 15\,36, \quad q + 4 \text{ ninda} = 1/2 \cdot (p - 10 \text{ ninda}).$$

Accordingly, the solution procedure can be reconstructed as follows (in quasi-modern notations):

$$\text{Let} \quad p* = p - 10 \text{ ninda}, \quad q* = q + 4 \text{ ninda} = 1/2 \cdot p*.$$

Then

$$p = p* + 10 \text{ ninda}, \quad q = 1/2 \cdot p* - 4 \text{ ninda}, \quad \text{and}$$
$$(1 + 1/4) \cdot \text{sq. } p* + (20 - 4) \text{ ninda} \cdot p* = (15\,36 - 1\,40 - 16) \text{ sq. ninda}.$$

Therefore, $p*$ can be found as the solution to the following *quadratic equation*

$$1;15 \text{ sq. } p* + 16 \text{ ninda} \cdot p* = 13\,40 \text{ sq. ninda}.$$

A geometric illustration of the metric algebra solution to an equation of this type is shown in Fig. 11.2.7 (cf. Fig. 11.2.8 below):

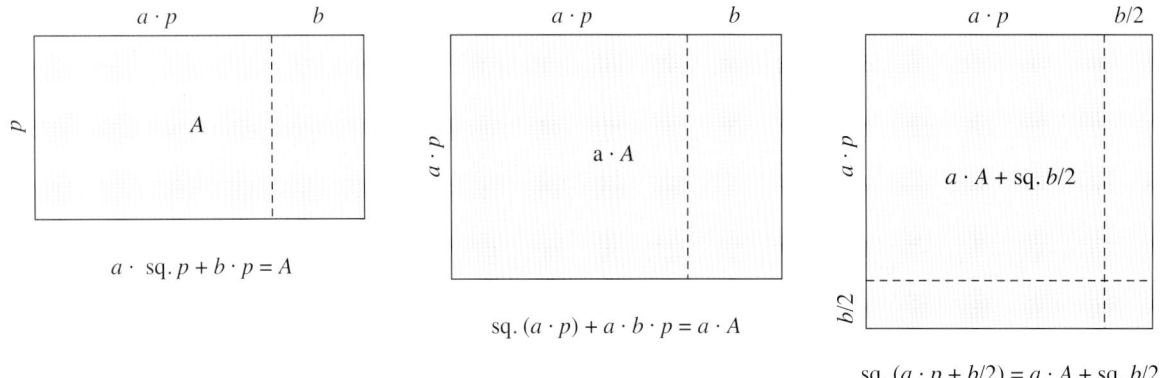

Fig. 11.2.7. The metric algebra solution to a quadratic equation of the type $a \cdot \text{sq. } p + b \cdot p = A$.

In the present case, the solution procedure continues as follows:

sq. $(1;15\,p* + 8 \text{ n.}) = 1;15 \cdot 13\,40$ sq. n. + sq. 8 n. = $(17\,05 + 1\,04)$ sq. n. = 18 09 sq. n.	
sqs. 18 09 = 33	(line 1')
$1/1;15 \cdot (33 - 8) = ;48 \cdot 25 = 20$	(lines 2'-3')
$p* = 1 \cdot 20$ n. = 20 n., $\quad q* = ;30 \cdot 20$ n. = 10 n.	(lines 3'-4')
$p = 20$ n. + 10 n. = 30 n., $\quad q = 10$ n. - 6 n. = 4 n.	(lines 5'-6')

11.2 h.   MS 5112 § 5. *Finding the Number of Terms in an Arithmetical Progression*

MS 5112 § 5 (*ii*: 7-24)

1	a.šà téš.a.sì.meš gar.gar-*ma* 2 31 40	The fields of the samesides (I) heaped, then 2 31 40,
	ù téš.a.sì.meš gar.gar-*ma* 3 30 /	and the samesides (I) heaped, then 3 30.
2	*ba-ma-at* igi.5.gál téš.a.sì	The halfpart of the 5th-part of the sameside
	ša nu.zu *ù* 1 ninda 2 <kùš> /	that is not known and 1 ninda 2 <cubits>
3	*a-di* 6 tab-*ma* téš.a.sì ki.1	until 6 (I) repeated, then the 1st sameside.
	téš.a.sì.meš en.nam /	The samesides are what?
4	za.e ak.da.zu.dè	You with your doing:
	aš-šum 2' igi.5.gál *qà-bu-kum* /	Since 1/2 of the 5th-part it is said to you,
5	igi.5.gál 5.e 1	the 5th-part of this 5 is 1.
	2'-*su* dar-*ma* 30	Its 1/2 crush, then 30,
6	*ù* 1 ninda 2 kùš *ša qà-bu-kum* / 1 10	and 1 ninda 2 cubits that it is said to you 1 10.
	a-na 30 daḫ-*ma* 1 40 *a-di* 6 tab-*ma* 10	To 30 add, then 1 40. Until 6 repeat, then 10.
	igi-*šu* du₈-*ma* 6 /	Its opposite resolve, then 6.
7	*a-na* 3 30 rá-*ma* 21	To 3 30 go, then 21.
	a-na 2 tab-*ma* 42	To two repeat, then 42.
	re-eš-ka li-ki-il /	May it hold your head.
8	1 téš.a.sì *ša* nu.zu *lu-pu-ut-ma*	1, the sameside that is not known, inscribe.
	2'-*su* dar-*ma* 30 /	Its 1/2 crush, then 30.
9	30 a.rá 30 gu₇.gu₇-*ma* 15	30 steps of 30 (make) eat (each other), then 15.
	a-na 42 daḫ-*ma* 42 15 /	To 42 add, then 42 15.
10	42 15 *mi-nam* íb.si₈ 6 30<.ta>.àm íb.si₈	42 15 (makes) what equalsided? 6 30 each way equalsided.
11	*i-na* 6ˌ 30 / 30 *ša* gu₇.gu₇ zi-*ma*	From 6 30 30 that was eaten tear off, then
	6 *si-tum* 6 téš.a.sì.meš /	6 is the remainder. 6 samesides.
12	10 *ša* igi-*a-šu* ba.du₈-*ma*	10 whose opposite was resolved, then
	a-na 3 30 ba.rá	to 3 30 went,
13	10 *ša-a-tam* / *a-na* 6 rá-*ma* 1	that same(?) 10 to 6 go, then 1.
	1 uš.ta.àm ki.1 /	1 uš (= 1 00 ninda) each way, the 1st.
14	*i-na* 6 1 zi-*ma* 5 *a-na* 10 rá-*ma*	From 6 1 tear off, 5. To 10 go, then
	50ᵥ ninda.ta.àm ki.2 /	50 ninda each way, the 2nd.
15	*i-na* 5 1 zi-*ma* 4ᵥ *a-na* 10 rá-*ma*	From 5 1 tear off, 4. To 10 go, then
	40ᵥ ninda.ta.à[m ki.3] /	40 ninda each way, the *3rd*.
16	*i-na* 4ᵥ 1 zi-*ma* 3 *a-na* 10 rá-*ma*	From 4 1 tear off, 3. To 10 go, then
	30 ninda.ta.à[m ki.4] /	30 ninda each way, the *4th*.
17	*i-na* 3 1 zi-*ma* 2 *a-na* 10 rá-*ma*	From 3 1 tear off, 2. To 10 go, then
	20 ninda.ta.à[m ki.5] /	20 ninda each way, the *5th*.
18	*i-na* 2 1 zi-*ma* 1 *a-na* 10 rá-*ma*	From 2 1 tear off, 1. To 10 go, then
	10 ninda.ta.à[m ki.6]	10 ninda each way, is the *6th*.

The new terminology in § 5 is the following:

ba-ma-at	halfpart	
a-di (*n*) tab	until *n* repeat	multiply by *n*
ša nu.zu	that is not known	unknown

According to Høyrup (*LWS*, 31) *bamtum* stands for "a natural half", used "in places where something is broken in two necessarily equal parts", in contrast to the "normal half" written 2' or *mišlum*. However, in line 2 of MS 5112 § 5, *bamtum* seems to be used as a normal half.

The phrase *a-di n* tab in § 5 is a variant of the phrase *a-na n* tab used in §§ 2 c and 3 above.

The phrase nu.zu, finally, appears elsewhere only in "southern" Old Babylonian mathematical texts, those belonging to Groups 1-3 and Sa.

The question in MS 5112 § 5 can be rephrased as follows, in normal language:

The sum of the areas of a number of squares is 2 31 40. The sum of the sides of the squares is 3 30.
The first square side is equal to six times half the fifth of the unknown square side plus 1 ninda 2 cubits.

The successive steps of the numerical solution procedure are the following:

'The unknown square side' = 5,   1/5 · 5 = 1,   1/2 · 1 = ;30,   (lines 4-5)
1 ninda 2 cubits = 1;10 (ninda),   1;10 + ;30 = 1;40,   6 · 1;40 = 10,   1/10 = ;06   (lines 5-6)
;06 · 3 30 = 21,   2 · 21 = 42   (line 7)
'The unknown square side' = 1,   1/2 · 1 = ;30,   sq. ;30 = ;15,   42 + ;15 = 42;15   (lines 8-9)
sqs. 42;15 = 6;30,   6;30 – ;30 = 6 = the number of squares   (lines 10-11)
10 (from lines 6-7) · 6 = 1 00 ninda = the 1st length   (lines 12-13)
6 – 1 = 5,   5 · 10 = 50 ninda = the 2nd length   (line 14)
5 – 1 = 4,   4 · 10 = 40 ninda = the 3rd length   (line 15)
4 – 1 = 3,   3 · 10 = 30 ninda = the 4th length   (line 16)
3 – 1 = 2,   2 · 10 = 20 ninda = the 5th length   (line 17)
2 – 1 = 1,   1 · 10 = 10 ninda = the 6th length   (line 18)

The form of the answer in lines 13-18 shows that it is silently assumed in this exercise that the *sides* of a series of squares form *a decreasing arithmetical progression* of the following form:

$$k \cdot d, \quad (k-1) \cdot d, \quad ......, \quad 2 \cdot d, \quad 1 \cdot d \quad \text{where the smallest term } d \text{ and the number } k \text{ of terms are unknown.}$$

Old Babylonian mathematicians were familiar with the summation rule for an arithmetical progression of this kind. See, for instance, the discussion in Sec. 10.1 a of the broken reed problem in MS 3971 § 1, or the explanation in Robson, *MMTC* (1999), 80-81, of **BM 85196 # 13**, an exercise in which 360 sheaves, arranged in a line with a constant distance between sheaves of 5 ninda, are to be carried to the beginning of the line.

According to the summation rule, the sum of all the terms in the mentioned progression is equal to

$$(k + (k – 1) + ... + 2 + 1) \cdot d = k \cdot (k + 1)/2 \cdot d.$$

In the first part of the solution procedure in MS 5112 § 5, in lines 4-6, the value of d, the smallest term in the progression (incorrectly called the '1st square side'), is computed as

$$d = 6 \cdot (1/2 \cdot 1/5 \cdot 5 + 1;10) = 10 \text{ (ninda)}.$$

No explanation is given for the choice of 5 as the value of 'the unknown square'. In the next part of the solution procedure, in lines 6-11, the value of k is computed as the solution to a quadratic equation:

$$k \cdot (k + 1)/2 \cdot d = 3\ 30 \text{ (ninda)}, \quad d = 10 \text{ (ninda)} \quad \Rightarrow \quad k \cdot (k + 1) = (3\ 30 / 10) \cdot 2 = 42.$$

(A parallel text is MS 3971 § 1. See Sec. 10.1 above.)

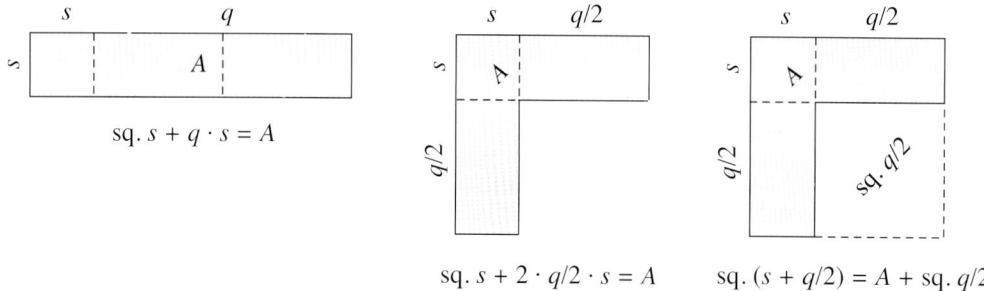

Fig. 11.2.8. MS 5112 § 5. Solving a quadratic equation geometrically by balancing and completing the square.

A quadratic equation of the type sq. $s + q \cdot s = A$ can be interpreted geometrically as the problem *to find the side of a square when the square extended a given distance q in one direction has a given area A* (see Fig. 11.2.8, left). The first step of the solution procedure, which may be called "balancing", is to replace the square extended a distance q in one direction by the same square extended the distance $q/2$ in two directions (Fig. 11.2.8, middle). Balancing the extended square does not change the area. The second step of the solution procedure is "completing" the square (Fig. 11.2.8, right). The third and fourth steps are to compute the side of the completed square and to subtract the distance $q/2$ to get the side of the original square.

Note that the first step of the solution procedure, that of balancing the one way extended square reduces the given problem to a problem of the type considered in MS 5112 § 1, and in Fig. 11.2.2!

In the quadratic equation for k in MSS 5112 § 5, $A = 42$ and $q = 1$. Therefore, the area of the balanced and completed square is $42 + $ sq. ;30 $= 42;15$ (line 9), the side of the balanced and completed square is sqs. 42;15 $= 6;30$ (line 10), and the side of the original square is $6;30 - ;30 = 6$ (line 11). This number 6 is, therefore, the number k of squares in the given progression.

When now the number of squares k and the smallest square side d, which is also the constant difference in the arithmetical progression, both are known, it is easy to find the sides of all the squares. In the text, the largest square side is computed first, its length being $6 \cdot d = 1\ 00$ (ninda). The largest square side is computed first in this exercise for the reason that, normally, arithmetical (and geometric) progressions in Old Babylonian mathematics are *decreasing*[2]. Arithmetical or geometric progressions are most often thought of as shares allotted to brothers or partners, and the largest share, the one allotted to the oldest brother or senior partner, is of course always mentioned first. Interesting examples are MS 2830, *obv.*, and MS 1844, both in Sec. 7.4 above.

In the present case, however, the arithmetical progression of square sides was probably not thought of as the shares of six brothers, since the most likely configuration is that of six *concentric* squares. On the other hand, the areas *between* the squares form a decreasing arithmetical progression that can be thought of as the shares of six brothers. (Compare the Late Babylonian mathematical recombination text **W 23291-x § 1** (Friberg, *et al.*, *BagM* 21 (1990)), where the areas bounded by five concentric circles form the decreasing arithmetical progression 1 48, 1 24, 1 00, 36, and 12 sq. ninda.)

Something is obviously wrong with this exercise. The computation of d in lines 4-6 does not make sense, in particular since it seems to be assumed that k is already known and has the value 5, not 6! In addition, no use is made of the statement that the sum of the areas of the squares is 2 31 40. It appears that the student who wrote down the teacher's statement of the problem got it wrong: the '5th part' in line 2 should be a '6th part'. Then he ignored the sum of the areas of the squares and instead invented a nonsensical computation of d, since he *knew in advance* that d should be equal to 10. Knowing d, he could then find k as a solution to a quadratic equation.

The teacher's intention was probably that the problem should be solved in the following way, using the rules for the sums of the first k integers or their squares. The given equations for k and d are then:

(sq. $k + $ sq. $(k-1) + \ldots\ldots + $ sq. $2 + $ sq. $1) \cdot$ sq. $d = k \cdot (k+1)/2 \cdot (2k+1)/3 \cdot$ sq. $d = S_2 = 2\ 31\ 40$ (cf. Fig. 2.4.5)
$(k + (k-1) + \ldots\ldots + 2 + 1) \cdot d = k \cdot (k+1)/2 \cdot d = S_1 = 3\ 30$ (cf. Fig. 2.4.4)
$d = 6 \cdot (1/6 \cdot 1/2 \cdot k + 1$ ninda 2 cubits)   (corrected)

There is no known text confirming that the mentioned rule for the sum of the squares of the first k integers was known by Old Babylonian mathematicians. On the other hand, the rule is easy to find by trial and error, and the rule appears in the Seleucid exercise AO 6484 § 2 (Neugebauer, *MKT 1* (1935), 97, 99, 103) in the following form, expressed both generally and in the special case when $k = 10$:

$$S_2 = (1/3 \cdot 1 + 2/3 \cdot k) \cdot S_1 = (;20 + ;40 \cdot 10) \cdot 55 = 7 \cdot 55 = 6\ 25.$$

It follows from the mentioned three equations for k and d in the statement of MS 5112 § 5 that

$$(2k+1) \cdot d = 3 \cdot S_2 / S_1 = 3 \cdot 2\ 31\ 40 / 3\ 30 = 2\ 10 \quad \text{and} \quad d = 1/2 \cdot k + 7.$$

This rectangular-linear system of equations for k and d can be reduced to a quadratic equation for k:

$$(2k+1) \cdot (1/2 \cdot k + 7) = 2\ 10 \quad \text{or} \quad \text{sq. } k + 14;30 \cdot k = 2\ 03.$$

The solution to this quadratic equation is $k = 6$. Consequently, $d = 1/2 \cdot k + 7 = 10$. And so on.

11.2 i.   MS 5112 § 6. A Quadratic Equation with Incorrect Data

---

2. An exception is the metric algebra exercise BM 13901 # 18 (Høyrup, *LWS* (2002), 108), where the sides 10, 20, 30 of three squares form an *increasing* arithmetical progression.

MS 5112 § 6 (*ii* 25-31)

1	téš.a.sì gu₇.gu₇-*ma* a.šà *ab-ni*	The sameside (I made) eat (itself), then a field I built.
	i-na téš.a.sì [4ᵥ zi] /	From the sameside *4 (I) tore out.*
2	igi.16.gál *ša-pi-il-ti* téš.a.sì	The 16th-part of the remainder of the sameside.
	a-na a.šà da[ḫ-*ma* 7 40] /	to the field (I) ad*ded, then 7 40.*
3	téš.a.sì en.nam	The sameside is what?
	za.e ak.da.zu.dè	You with your doing:
	4ᵥ *ša* z[*i ù* 7 40] /	4 that was torn *out, 7 40,*
4	*ù* igi.16.gál 16 *lu-pu-ut-ma*	and (for) the 16th-part 16 inscribe, then
	it-ti 7 40 [gu₇.gu₇ 2 02 40] /	with 7 40 *make it eat, 2 02 40.*
5	*a-na* 2 02 4 4ᵥ *ša zi* daḫ-*ma* 2 06 40	To 2 02 40 the 4 that was torn out add, then 2 06 40.
	[*it-ti* 16gu₇.gu₇] /	With 16 make it eat,
6	33 46 40 *i-il-a-kum*	33 46 40 comes up for you.
	re-eš-ka l[*i-ki-il*]	May *it hold* your head.
	[x x x x] /	x x x x x x x
7	....	....

The new terms in § 6 are:

a.šà *abni*	a field I built	(I constructed a rectangle)
ill(i)akkum	it comes up for you	

šapiltum   rest, remainder

Parallels to the phrase téš.a.sì gu₇.gu₇-*ma* a.šà *ab-ni* in § 6 can be found, for instance, in AO 8862 (Group 1 a; *LWS*, 164) and in the Susa text *TMS* 17, where the corresponding phrases are, respectively, uš *ù* sag *uš-ta-ki-il₅-ma* a.šà-*lam ab-ni* and uš ki sag *uš-ta-ki-il-ma* a.šà *ab-n*i.

The term *šapiltum* with the meaning 'rest, remainder' can be found also in AO 8862 and 6770 (both Group 1a), as well as in the Eshnunna text IM 54464 (Group 7a; Baqir, *Sumer* 7 (1951)).

The term *illiakum*, finally, occurs in texts of Group 5 (as in the phrase *ša i-li-kum* in YBC 6967; *LWS*, 56), but also in Eshnunna texts of Group 7a (as in 10 *i-li* 10 *ša i-li-a-ku-um* in IM 53961 (Baqir, *op. cit.*)).

The question in § 6 is damaged, and only the beginning of the solution procedure is preserved. The missing parts of the question can be reconstructed, but it is difficult to reconstruct the lost part of the solution procedure. Anyway, the question in this exercise appears to have been as follows, in quasi-modern algebraic notations:

$$\text{sq. } p + (p - 4)/16 = 7\ 40.$$

The author of this exercise must have made a simple numerical mistake (a "positional error") when he computed his given value 7 40. Indeed, suppose that the *intended* solution to the problem was $p = 20$. (It is difficult to find any other reasonable candidate.) With this value for p,

$$\text{sq. } p + (p - 4)/16 = 6\ 40 + 1 = 6\ 41, \quad \text{but the author of the exercise seems to have thought of 6;40 + 1 = 7;40.}$$

Without the mistake, the solution would have proceeded as follows (in quasi-modern notations):

$16 \cdot \text{sq. } p + p - 4 = 16 \cdot 6\ 41 = 1\ 46\ 56$	(cf. $16 \cdot 7\ 40 = 2\ 02\ 40$ in line 4)
$16 \cdot \text{sq. } p + p = 1\ 46\ 56 + 4 = 1\ 47\ 00$	(cf. $2\ 02\ 40 + 4 = 2\ 06\ 40$ in line 5)
$\text{sq. } 16 \cdot \text{sq. } p + 16 \cdot p = 16 \cdot 1\ 47\ 00 = 28\ 32\ 00$	(cf. $16 \cdot 2\ 06\ 40 = 33\ 46\ 40$ in lines 5-6)
$\text{sq. } (16\ p + ;30) = 28\ 32\ 00 + ;15 = 28\ 32\ 00;15$	
$16 \cdot p + ;30 = \text{sqs. } 28\ 32\ 00;15 = 5\ 20;30, \quad \text{so that, finally,} \quad p = 5\ 20 / 16 = 20$	

With the incorrect value 7 40 instead of 6 41 for the given number, it must have been difficult to find an acceptable solution to the stated problem. (Remember that, with rare exceptions, Old Babylonian mathematical problems were constructed so that they would have simple numbers, often small integers, as solutions.) Then, how could the author of MS 5112 § 6 finish his solution procedure, which he apparently did? Clearly, by cheating. The next step of the procedure (destroyed) should be his computation of $16 \cdot p + ;30$ as the square side of 33 46 40;15. If he had actually tried to compute this square side, he would have run into trouble. Instead, he probably just stated the result as 5 20;30, because he could count backwards from the correct solution $p = 20$!

iv *iii* *rev.*

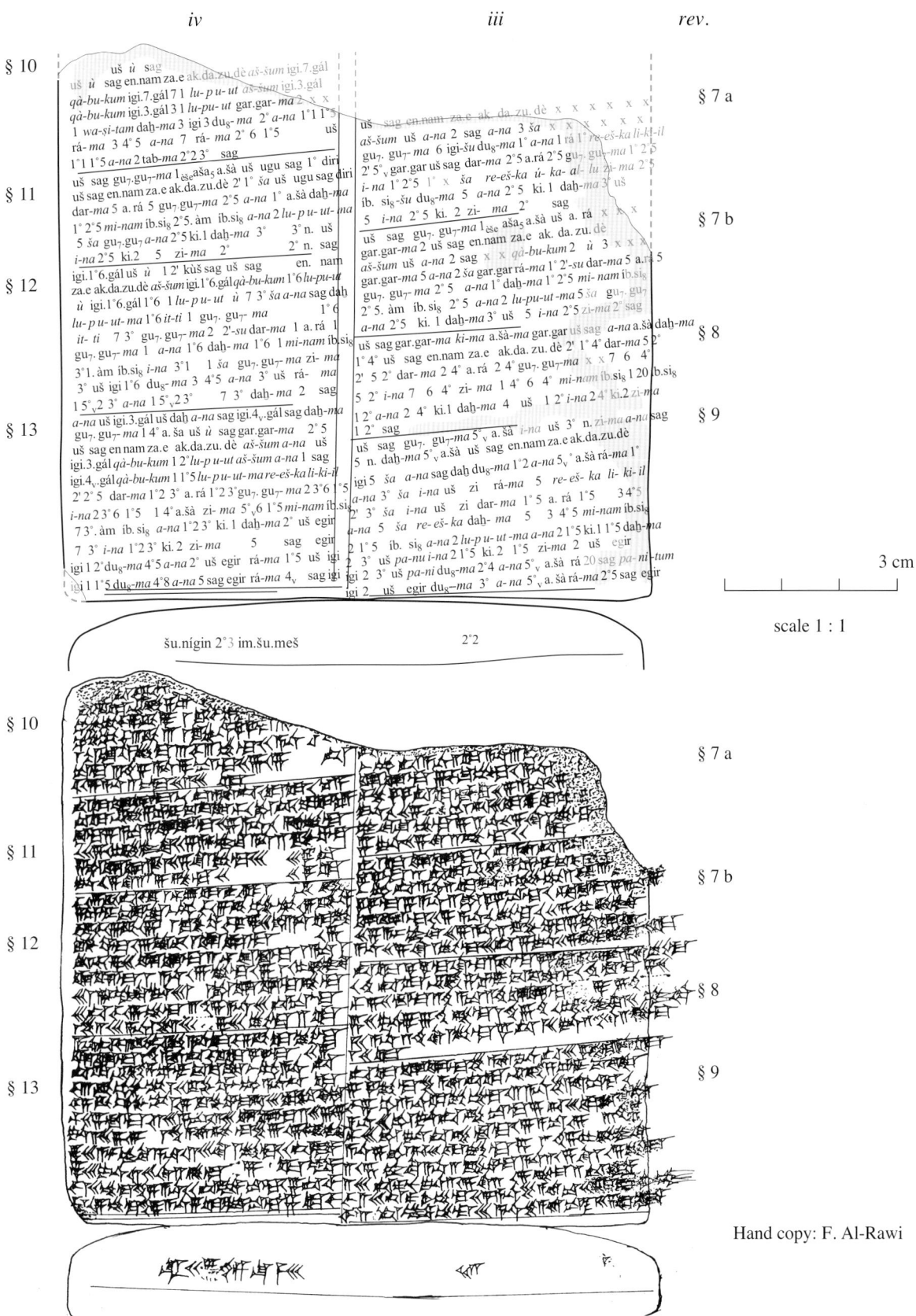

§ 10

uš ù sag
uš ù sag en.nam za.e ak.da.zu.dè aš-šum igi.7.gál
qà-bu-kum igi.7.gál 7 1 lu-pu-ut aš-šum igi.3.gál
qà-bu-kum igi.3.gál 3 1 lu-pu-ut gar.gar-ma 2˚ x x
1 wa-si-tam daḫ-ma 3 igi 3 du₈-ma 2˚ a-na 1˚1˚1˚5 uš
rá-ma 3 4˚5 a-na 7 rá-ma 2˚ 6 1˚5
1˚1˚1˚5 a-na 2 tab-ma 2˚2˚3˚ sag

§ 11

uš sag gu₇.gu₇-ma 1 ˌešˌ aša₅ a.šà uš ugu sag diri
uš sag en.nam za.e ak.da.zu.dè 2˚1˚ ša uš ugu sag diri
dar-ma 5 a.rá 5 gu₇.gu₇-ma 2˚5 a-na 1˚ a.šà daḫ-ma
1˚2˚5 mi-nam ib.si₈ 2˚5. àm ib.si₈ a-na 2 lu-pu-ut-ma
5 ša gu₇.gu₇ a-na 2˚5 ki.1 daḫ-ma 3˚ 3˚ n. uš
i-na 2˚5 ki.2 5 zi-ma 2˚ 2˚ n. sag

§ 12

igi.1˚6.gál uš ù 1 2˚ kùš uš sag en. nam
za.e ak.da.zu.dè aš-šum igi.1˚6.gál qà-bu-kum 1˚6 lu-pu-ut
ù igi.1˚6.gál 1˚6 1 lu-pu-ut ù 7 3˚ ša a-na sag daḫ
lu-pu-ut-ma 1˚6 it-ti 1 gu₇. gu₇-ma 1˚6
it-ti 7 3˚ gu₇-ma 2 2˚-su dar-ma 1 a.rá 1
gu₇.gu₇-ma 1 a-na 1˚6 daḫ-ma 1˚6 1 mi-nam ib.si₈
3˚1.àm ib.si₈ i-na 3˚1 1 ša gu₇.gu₇-ma zi-ma
3˚ uš igi 1˚6 du₈-ma 3 4˚5 a-na 3˚ uš rá-ma
1 5˚ᵥ2 3˚ a-na 1 5˚ᵥ2 3˚ 7 3˚ daḫ-ma 2

§ 13

a-na uš igi.3.gál uš daḫ a-na sag igi.4ᵥ.gál sag daḫ-ma
gu₇. gu₇-ma 1 4˚ a. ša uš ù sag gar.gar-ma 2˚5
uš sag en.nam za.e ak.da.zu.dè aš-šum a-na uš
igi.3.gál qà-bu-kum 1 2˚ lu-pu-ut aš-šum a-na 1 sag
igi.4ᵥ.gál qà-bu-kum 1 1˚5 lu-pu-ut-ma re-eš-ka li-ki-il
2˚2˚5 dar-ma 1˚2 3˚ a. rá 1˚2 3˚ gu₇-ma 2 3˚6 1˚5
i-na 2 3˚6 1˚5 1 4˚ a.šà zi-ma 5˚ᵥ6 1 1˚5 mi-nam ib.si₈
7 3˚. àm ib.si₈ a-na 1˚2 3˚ ki.1 daḫ-ma 2˚ uš egir
7 3˚ i-na 1˚2 3˚ ki.2 zi-ma 5 sag egir
igi 1 2˚du₈-ma 4˚5 a-na 2˚ uš egir rá-ma 1˚5 uš igi
igi 1 1˚5 du₈-ma 4˚8 a-na 5 sag egir rá-ma 4ᵥ sag egir

§ 7 a

uš sag en.nam za.e ak.da.zu.dè x x x x x x
aš-šum uš a-na 2 sag a-na 3 ša re-eš-ka li-ki-il
gu₇. gu₇-ma 6 igi-šu du₈-ma 1˚ a-na 1 rá 1 re-eš-ka li-ki-il
2˚5ᵥ gar.gar uš sag dar-ma 2˚5 a.rá 2˚5 gu₇. gu₇-ma 1˚2˚5
i-na 1˚2˚5 1˚ ša re-eš-ka ú-ka-al lu zi-ma 2˚
ib. si₈-šu du₈-ma 5 a-na 2˚5 ki.1 daḫ-ma 3˚ uš
5 i-na 2˚5 ki.2 zi-ma 2˚ sag

§ 7 b

uš sag gu₇. gu₇-ma 5 ˌešˌ aša₅ a.šà uš a. rá 2
gar.gar-ma 2 uš sag en.nam za.e ak. da. zu.dè
aš-šum uš a-na 2 sag x x qà-bu-kum 2˚ ù 3 x x
gar.gar-ma 5 a-na 2 ša gar.gar-ma 1˚ 2˚-su dar-ma 5 a.rá 5
gu₇. gu₇-ma 2˚5 a-na 1˚ dar-ma 1˚2˚5 mi-nam ib.si₈
2˚5. àm ib. si₈ 2˚5 a-na 2 lu-pu-ut-ma 5 i-na 2˚5 zi-ma 2˚ sag
a-na 2˚5 ki.1 daḫ-ma 3˚ uš 5 i-na 2˚5 zi-ma 2˚ sag

§ 8

uš sag gar.gar-ma ki-ma a.šà-ma gar.gar uš sag a-na a.šà daḫ-ma
1˚ 4˚ uš sag en.nam za.e ak.da. zu. dè 2˚1˚ 4˚ dar-ma 5 2˚
2˚ 5 2˚ dar-ma 2 4˚ a.rá 2 4˚ gu₇. gu₇-ma x x 7 6 4˚
5 2˚ i-na 7 6 4˚ zi-ma 1 4˚ 6 4˚ mi-nam ib.si₈ 1 2˚ ib.si₈
1 2˚ a-na 2 4˚ ki.1 daḫ-ma 4 1˚ 1 2˚ i-na 2 4˚ ki.2 zi-ma
1 2˚ sag

§ 9

uš sag gu₇. gu₇-ma 5˚ᵥ ša i-na uš 3˚ n. zi-ma a-na sag
5 n. daḫ-ma 5˚ᵥ a.šà uš sag en.nam za.e ak.da.zu.dè
igi 5 ša a-na sag daḫ du₈-ma 1˚2 a-na 5˚ᵥ a.šà rá-ma 1˚
igi 5 ša i-na uš zi rá-ma 1˚2 re-eš-ka li-ki-il
a-na 3˚ ša i-na uš zi dar-ma 1˚5 a. rá 1˚5 3 4˚5
2˚ 3˚ ša i-na uš zi dar-ma 1˚5 3 4˚5 mi-nam ib.si₈
a-na 5 ša re-eš-ka daḫ-ma 1˚ a-na 2 1˚5 ki.1 1˚5 daḫ-ma
2 1˚5 ib. si₈ a-na 2 lu-pu-ut-ma a-na 2 1˚5 ki.2 zi-ma
2˚ 3˚ uš pa-nu i-na 2 1˚5 ki.2 1˚5 zi-ma 2 uš egir
2˚ 3˚ uš pa-ni du₈-ma 2˚4 a-na 5˚ᵥ a.šà rá 20 sag pa-ni-tum
igi 2˚ uš egir du₈-ma 3˚ a-na 5˚ᵥ a. šà rá-ma 2˚5 sag egir

šu.nígin 2˚3 im.šu.meš 2˚2

3 cm

scale 1 : 1

Hand copy: F. Al-Rawi

Fig. 11.2.9. MS 5112, *rev*. Metric algebra problems for the length and front of a rectangle.

MS 5112, rev.: Metric Algebra Problems for the Length and Front of a Rectangle

Although the general topic of MS 5112 is metric algebra, there is a clear distinction between the problems on the two sides of the clay tablet. All the problems on the *obverse* are formulated in terms of the sides and areas of one or more squares, while all the problems on the *reverse* are formulated in terms of the sides and the area of a single rectangle. See the outline of the contents in Fig. 11.2.17, right.

11.2 j.   MS 5112 § 7 a-b. Two Badly Preserved Rectangular-Linear Systems

MS 5112 § 7 a (*iii: 1'-8'*)

	… … …	… … …
1'	uš [sag en.nam]	The length and the front are what?
	[za.e ak.da.zu.dè] /	You with your doing:
2'	[x x x x x] *aš-šum* uš *a-na* 2 sag	x x x x x Since the length to 2 fronts,
3'	*a-na* 3 *ša* [x x x x x x] / gu₇.gu₇-*ma* 6	to 3 that x x x x x (make) eat (each other), then 6.
	igi-*šu* du₈-*ma* 10	Its opposite resolve, then 10.
	a-na 1 [rá 10 *re-eš-ka li-ki-il*] /	To 1 *go* 10. *Let it hold your head.*
4'	2' 50 gar.gar uš sag dar-*ma* 25	1/2 of the heap of the length and the front crush, then 25.
	a.rá 25 [gu₇.gu₇-*ma* 10 25] /	Steps of 25 *go, then 10 25.*
5'	*i-na* 10 25 [10 x] *ša re-eš-ka ú-ka-a*[*l-lu*]	From 10 25 the *10, x,* that held your head
	[*zi-ma* 25] /	tear off, then 25.
6'	íb.si₈-*šu* du₈-*ma* 5	Its equalside resolve, then 5.
	a-na 25 ki.1 daḫ-*ma* [30 uš] /	To the 1st 25 add it, then *30, the length.*
7'	5 *i-na* 25 ki.2 zi-*ma* 20 sag	5 from the 2nd 25 tear off, then 20, the front.

MS 5112 § 7 b (*iii: 9'-15'*)

1	uš sag gu₇.gu₇-*ma* 1 ₑ̀šₑ aša₅ a.šà uš	The length (and) the front (I made) eat (each other), then 1 èše field.
	a.rá [x x x] /	The length steps of x x x
2	[x x x x] / gar.gar-*ma* 2	x x x heap, then 2.
	uš sag en.nam	The length (and) the front are what?
	za.e ak.da.zu.[dè] /	You with your do*ing:*
3	*aš-šum* uš *a-na* 2 sag [x x *qà*]-*bu-kum*	Since the length to 2 fronts x x it is said to you:
4	2 *ù* 3 [x x x x] / gar.gar-*ma* 5	2 and 3 x x x heap, then 5.
	a-na 2 *ša* gar.gar rá-*ma* 10	To 2 that you heaped go, then 10.
	2'-*su* dar-*ma* 5	Its 1/2 crush, then 5.
5	a.[rá 5] / gu₇.gu₇-ma 25	Steps of *5* (make) eat, then 25.
	a-na 10 daḫ-*ma* 10 25	To 10 add, then 10 25
6	*mi-nam* í[b.si₈] / 25.àm íb.si₈	(makes) what eq*ualsided?* 25 each way it is equalsided.
	25 *a-na* 2 *lu-pu-ut-ma*	25 to two inscribe.
	5 *ša* gu₇.[gu₇] /	5 that you mad*e eat*
7	*a-na* 25 ki.1 daḫ-*ma* 30 uš	to the 1st 25 add, then 30, the length.
	5 *i-na* 25 [zi-*ma* 20 sag]	5 from 25 *tear off, then 20, the front.*

In § 7 a, the whole question is lost, and some essential parts of the solution procedure. In § 7 b, a crucial part of the question is lost, and some parts of the solution procedure. As a result, it seems to be impossible to reconstruct fully the questions and the first halves of the solution procedures in both these exercises. The well preserved second halves of both solution procedures, on the other hand, are routine applications of metric algebra.

In § 7 a, the question can be concluded to have consisted of some complicated rectangular-linear system of equations, after some initial manipulations reduced to a system of the form

$$u \cdot s = 10 \ (00), \quad u + s = 50.$$

The solution to this system of equations is given in lines 5'-8' of § 7 a. The situation is illustrated in Fig. 11.2.10 below, where it is shown that any rectangular-linear system of the type

$$u \cdot s = A, \quad u + s = p$$

can be solved, in terms of metric algebra, by use of a geometric analogue to the algebraic identity

$$\text{sq. } q/2 = \text{sq. } p/2 - A, \quad \text{where} \quad q = u - s.$$

The basic idea is that if $u \cdot s = A$ and $u + s = p$ are known, then $u - s = q$ can be found by use of a geometric procedure that can be understood as "balancing, and subtracting a corner".

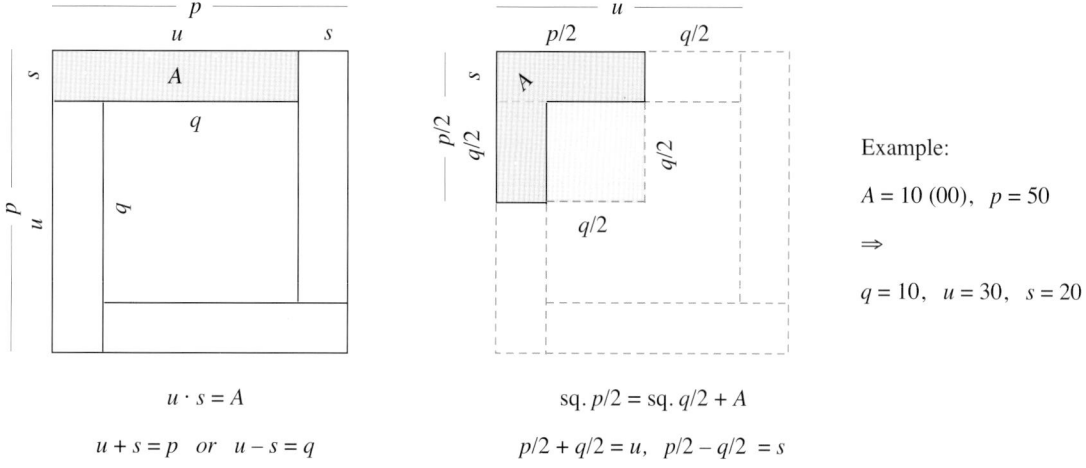

$$u \cdot s = A$$
$$u + s = p \quad or \quad u - s = q$$

$$\text{sq. } p/2 = \text{sq. } q/2 + A$$
$$p/2 + q/2 = u, \quad p/2 - q/2 = s$$

Fig. 11.2.10. The basic OB rectangular-linear systems of equations. Adding/subtracting a corner.

In § 7 b, line 1, it is mentioned that the product of the length and the front (of a rectangle) is 1 èše (= 10 00 sq. ninda). A second equation for u and s is badly preserved. What remains is the statement that the sum of two combinations of u and s is equal to 2. It does not seem possible to reconstruct the question. Anyway, as in § 7 a, it is clear that the result of the first half of the solution procedure was that the given system of equations was reduced to a simpler system of equations, in this case

$$u \cdot s = 10 \, (00), \quad u - s = 10.$$

In Fig. 11.2.10 above, it is shown that any rectangular-linear system of the type

$$u \cdot s = A, \quad u - s = q$$

can be solved, in terms of metric algebra, by use of a geometric analogue to the algebraic identity

$$\text{sq. } p/2 = \text{sq. } q/2 + A, \quad \text{where } p = u + s.$$

The basic idea is that if $u \cdot s = A$ and $u - s = q$ are known, then $u + s = p$ can be found by use of a procedure that can be understood geometrically in two ways, either as "balancing, and adding a corner", or as "balancing and completing the square". (Cf. Fig. 11.2.8).

11.2 k. MS 5112 § 8. A Rectangle where the Area is Equal to the Length Plus the Front

MS 5112 § 8 (iii: 16'-21')

1	uš sag gar.gar-*ma ki-ma* a.šà-*ma*
2	gar.gar [uš sag] *a-na* a.šà daḫ-*ma* / 10 40
	uš sag en.nam
	za.e ak.da.zu.dè
	2' 10 40 dar-*ma* 5 20 /

The length (and) the front (I) heaped, then like the field, then the heap of *field (and) length* to the front I joined, then 10 40.
The length (and) the front are what?
You with your doing:
1/2 of 10 40 crush, then 5 20.

3	2' 5 20 dar-*ma* 2 40	1/2 of 5 20 crush, then 2 40.
	a.rá 2 40 gu$_7$.gu$_7$-*ma* [x x] 7 06 40 /	Steps of 2 40 (make) eat (each other), then *x x* 7 06 *40*.
4	5 20 *i-na* 7 06 40 zi-*ma* 1 46 40	5 20 from 7 06 40 tear off, then 1 46 40.
	mi-n[*am* íb.]si$_8$ 1 20 íb.si$_8$ /	What *is it equalsided?* 1 20 equalsided.
5	1 20 *a-na* 2 40 ki.1 da$\underline{h}$-*ma* 4 uš	1 20 to the 1st 2 40 add, then 4, the length.
6	1 20 *i-na* 2 [40 ki.]2 [zi]-*ma* / 1 20 sag	1 20 from the 2*nd 2 40 tear off*, then 1 20, the front.

The question in this exercise can be rephrased as follows:

> The sum of the length and the front of a rectangle is equal to the area of the rectangle.
> The sum of the length, the front, and the area is 10 40.
> What are the length and the front?

In abstract algebraic notations, the question is this:

$$u \cdot s = u + s, \quad u \cdot s + (u + s) = 10\ 40.$$

The successive steps of the numerical solution procedure are as follows:

$$1/2 \cdot 10\ 40 = 5\ 20, \quad 1/2 \cdot 5\ 20 = 2\ 40, \quad \text{sq. } 2\ 40 = 7\ 06\ 40, \quad 7\ 06\ 40 - 5\ 20 = 1\ 46\ 40, \quad \text{sqs. } 1\ 46\ 40 = 1\ 20,$$
$$2\ 40 + 1\ 20 = 4\ (00) = u, \quad 2\ 40 - 1\ 20 = 1\ 20 = s.$$

It is clear that the solution to this problem was based on the observation that if the product and the sum of the sides of a rectangle are equal, then the sum of the product and the sum is equal both to twice the product, and to twice the sum. Therefore, the given rectangular-rectangular system of equations can be reduced to a simpler but equivalent rectangular-linear system of equations:

$$u \cdot s = 10\ 40 / 2 = 5\ 20, \quad u + s = 10\ 40 / 2 = 5\ 20.$$

This is one of the basic rectangular-linear systems of equations, solved, routinely, by metric algebra as in Fig. 11.2.10.

A Parallel Text: AO 8862 § 1 d

AO 8862 § 1 d (Høyrup, *LWS* (2002), 169, 174)

1	uš sag	Length (and) front (= a rectangle).
2	uš *ù* sag / *uš-ta-ki-il$_5$-ma*	The length and the front I made hold each other,
	a.šà *ab-ni* /	a field I built.
	a-tu-úr	I turned back.
3	uš *ù* sag *ak-mu-ur-ma* /	The length and the front I heaped, then
4	*it-ti* a.šà *mi-it-ḫa-ar* /	with the field they are equal.
5-6	uš sag *ù* a.šà *ak-mu-ur-ma* / 9	The length, the front, and the field I heaped, then 9.
	uš sag *ù* a.šà *mi-nu-um*	The length, the front, and the field are what?

This is an isolated exercise in a mathematical recombination text from Larsa (early Old Babylonian). The question is essentially identical with the question in MS 5112 § 8, except for the numerical data (9 instead of 10 40), and except for the fact that in AO 8862 § 1 d all words other than uš, sag, and a.šà are written in syllabic Akkadian. Moreover, there is no solution algorithm, and no answer. (It is easy to see that the answer would have been that $u = 3$, $s = 1;30$.)

Another Related Text: AO 6770 # 1

This is an isolated exercise in a small text with mixed content, like AO 8862 belonging to Group 1 a, hence from Larsa, and early Old Babylonian.

AO 6770 # 1 (Høyrup, *LWS* (2002), 179-181)

1	uš *ù* sag *ma-la* aša₅ *li-im-ta-ḫa*[r] /	The length and the front as much as the field I let be equal.
2	*at-ta i-na e-pe-ši-i-ka* /	You, in your doing:
3	*a-ra-am a-na ši-ni-šu ta-ša-ka-an* /	The step to its two you set.
4	*i-na li-ib-bi-im* 1 *ta-na-as-sà-aḫ* /	Out of it 1 you tear off.
5	*i-*[g]*i₄-a-am ta-pa-aṭ-ṭa-a*[r] /	The opposite you resolve.
6	*it-ti a-re-e-em ša ta-aš-ku-nu* /	With the step that you set
7	*tu-uš-ta-ka-al-ma* / sag *i-na-ad-di-ik-kum*	you let (them) hold each other. The front it gives you.

In this text, like in AO 8862, all words other than uš, sag, and aša₅ (= gán) are written in syllabic Akkadian. The text is almost unique in the Old Babylonian mathematical corpus in that it is not a mathematical exercise, but a general rule expressed verbally, without an explicit numerical example. It can be explained in quasi-modern abstract terms as saying that

$$\text{If } u \cdot s = u + s, \text{ then } s = u \cdot 1 / (u - 1).$$

This statement is trivially true if one is allowed to argue using algebraic manipulations of symbols. What if only geometric arguments are allowed? What, by the way, is the geometric meaning of the equation 'the length and the front are equal to the area'? In Fig. 11.2.11 below, the area of a rectangle with the length u and the front s is shown in grey. The sum of the length and the front, both multiplied by line segments of length 1, is shown as the sum of the areas of two smaller rectangles. The inner sides of these two rectangles divide the given rectangle in four parts whose respective areas can be called $P, Q, R,$ and S. It is then clear that

$$u \cdot s = P + Q + R + S, \text{ while } u \cdot 1 = P + Q \text{ and } s \cdot 1 = P + R.$$

Hence,

$$\text{If } u \cdot s = u \cdot 1 + s \cdot 1 \text{ then } P = S, \text{ so that } (u - 1) \cdot s = Q + S = Q + P = u \cdot 1, \text{ and } s = u \cdot 1 / (u - 1).$$

This is the rule stated in AO 6770 # 1.

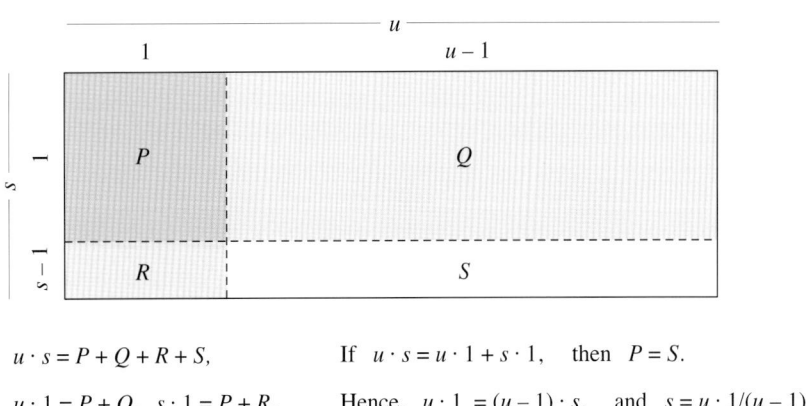

$$u \cdot s = P + Q + R + S, \qquad \text{If } u \cdot s = u \cdot 1 + s \cdot 1, \quad \text{then } P = S.$$
$$u \cdot 1 = P + Q, \; s \cdot 1 = P + R. \qquad \text{Hence, } u \cdot 1 = (u - 1) \cdot s, \quad \text{and } s = u \cdot 1/(u - 1).$$

Fig. 11.2.11. AO 6770 # 1. A geometric interpretation of 'the length and the front equal to the area'.

Other Texts Mentioning the Sum of the Length, the Front, and the Area

In the question of MS 5112 § 8, the second of the stated conditions is that *the sum of the length, the front, and the area of a rectangle* is a given number (10 40). A parallel can be found in the following question in **IM 43993**, a small clay tablet with a single exercise:

IM 43993 (Friberg and Al-Rawi, forthcoming)

1	*ši-ni-pa-at ši-di-im-mi pu-ta-am* /	Two-thirds of the length (I made) the front.
2	a.šà *ši-di ù pu-ti-mi* /	(My) field, my length and my front
3-4	*ak-mu-ur-ma* / 1-*mi*	I heaped, then 1.
	etc.	*etc.*

The question can be interpreted as a rectangular-linear system of equations for two unknowns:

$$u \cdot s + u + s = 1 \ (00\ 00), \quad s = ;40 \cdot u.$$

This system of equations, in its turn, can be reduced to a quadratic equation for a single unknown;

$$;40 \cdot \text{sq. } u + 1 \cdot u + ;40 \cdot u = 1. \qquad \text{Solution: } u = 30, \quad s = 20.$$

*The Susa text **TMS 9** a-c* is another Old Babylonian mathematical text concerned with the sum of the area, the length, and the front of a rectangle. It is an explicitly didactic text of a very unusual kind, first correctly explained by Høyrup (see *LWS* (2002), 89-95). It begins by explaining in great detail what it means that "1 length" is added to the area of the rectangle. Next it explains how "the Akkadians" would transform the situation when the sum of the area, the length, and the front of a rectangle is given into something more manageable. In the final section, this preliminary result is used in order to find the solution of a rectangular-linear system of equations.

TMS 9 (Høyrup, *LWS* (2002), 89-93)

a 1	a.šà *ù* 1 uš ul.gar 4[0] [30 uš 20 sag] /	The field and 1 length (I) heaped, 40. 30 the length, 20 the front.
2	*i-nu-ma* 1 uš *a-na* 10 [a.šà daḫ] /	When 1 length to 10 *the field is added*,
3	*ú-ul* 1 ki.du.du *a-na* 20 [sag daḫ] /	either 1, the place-mover to 20, *the front, add*,
...	etc.	*etc.*
b1	[a.šà uš *ú* sag u]l.gar 1 *i-na ak-ka-di-i* /	*The field, the length, and the front* (I) heaped, 1. In the Akkadian (way):
2	[1 *a-na* uš daḫ] 1 *a-na* sag daḫ	*1 to the length add*, 1 to the front add.
3	*aš-šum* 1 *a-na* uš daḫ / [1 *a-na* sag d]aḫ 1 *ù* 1 nigin 1 *ta-mar* /	Since 1 to the length is added, *1 to the front* is added, 1 and 1 frame (= multiply), 1 you see.
4	[1 *a-na* ul.gar uš] sag *ù* a.šà daḫ 2 *ta-mar* /	*1 to the heap of length*, front, and field add, 2 you see.
5	[*a-na* 20 sag 1 da]ḫ 1 20 *a-na* 30 uš 1 daḫ 1 30	*To 20, the front 1 ad*d, 1 20, to 30 the length add, 1 30,
...	etc.	*etc.*
c1	a.šà uš *ù* sag ul.gar 1 a.šà 3 uš 4 sag ul.gar /	The field, the length, and the front (I) heaped, 1, the field. 3 lengths and 4 fronts (I) heaped,
2	[17]-*ti-šu a-na* sag daḫ 30	its 17th to the front (I) added, 30.
...	etc.	*etc.*

A geometric explanation of *TMS* 9, a-b (cf. again Høyrup, *LWS*, 93-94) is offered in Fig. 11.2.12 below. In *TMS* 9 c, the question can be interpreted as a rectangular-linear system of equations:

$$u \cdot s + u + s = 1, \quad s + 1/17 \cdot (3\ u + 4\ s) = ;30.$$

In the solution procedure, an application of the "Akkadian" method developed in part b of *TMS* 9 makes it possible to transform the given system of equations to the following standard rectangular-linear system of equations:

$$u * \cdot s* = 2\ 06, \quad u* + s* = 32\ 30, \quad \text{where} \quad u* = 21 \cdot (s + 1), \quad s* = 3 \cdot (u + 1).$$

The solution obtained is

$$u* = 16\ 15 + 11\ 45 = 28, \quad s* = 16\ 15 - 11\ 45 = 4;30, \quad \text{hence} \quad u = ;30, \quad s = ;20.$$

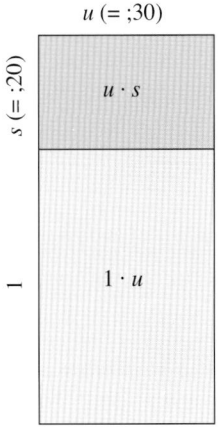

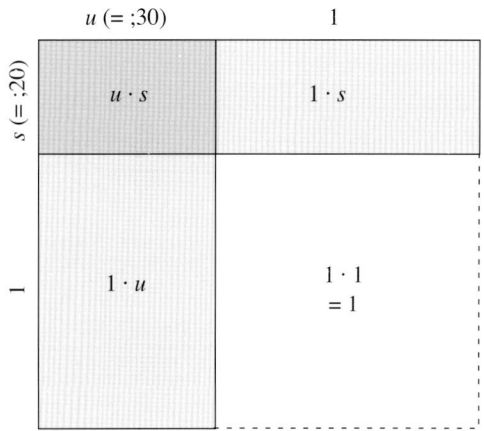

The equation

$u \cdot s + 1 \cdot u = 10 + 30 = 40$

can be replaced by the equation

$u \cdot (s + 1) = {;}30 \cdot 1{;}20 = {;}40$

The equation

$u \cdot s + u + s = 10 + 30 + 20 = 1$

can be replaced by the equation

$(u + 1) \cdot (s + 1) = 1{;}30 \cdot 1{;}20 = 2$

Fig. 11.2.12. *TMS* 9 a-b. Two examples of two kinds of a geometric change of variables.

A text with equations similar to the equation in IM 43993, but for circles instead of for a rectangle, is **BM 80209**, a small mathematical "catalog text" from Nippur (Group 6 a). The seventh and last paragraph of that text is a list of four equations of a common type but with different given numbers:

BM 80209 § 7 a-d (Friberg, *JCS* 33 (1981))

a.šà gúr dal gúr *ù si.ḫi-ir-ti* gúr ul.gar-*ma B*	The field of an arc (circle), the transversal of an arc, and the surrounding of an arc (I) heaped, then *B*.

In normal language:

The area *A*, the diameter *d*, and the circumference *a* of a circle are together equal to *B*.

Here *B* takes the four values 8 33 20, 1, 1 55, and 3 06 40. With the known Old Babylonian constants for a circle inserted, the stated condition can be interpreted as a quadratic equation for *a*:

$${;}05 \cdot \text{sq. } a + {;}20 \cdot a + a = B. \qquad \text{Solutions: } a = 10\,(00), 20, 30, 40.$$

The author of the text probably wanted the first solution to be *a* = 10 but made a positional error when he computed the corresponding value of *B*.

A third text with an equation of a related type, but for a "concave square" (*apsammikku*), is the Susa text ***TMS* 20 § 1**. The following question is stated in the broken first line of that text:

***TMS* 20 § 1** (Bruins and Rutten, *TMS* (1961))

a.šà uš *ù* bar.[dà] [*a-pu-sà-am-mi-ik-ki* 1 16 40] /	The field, the length, and the cross-over of an apsammikku, 1 16 40.

With the known Old Babylonian constants for a concave square inserted (see Fig. 8.2.10 above), the stated condition can be interpreted as a quadratic equation for the "length" *u*:

$${;}26\ 40 \cdot \text{sq. } u + u + 1{;}20 \cdot u = 1\ 16\ 40. \qquad \text{Solution: } u = {;}30.$$

Note: *There is only one kind of circle, one kind of square, and one kind of concave square, but many kinds of rectangles,* depending on the ratio between the sides. That is why all parameters of a circle can be found when the sum of the area, the diameter, and the circumference is given (as in BM 80209 § 7), and why all parameters of a concave square can be found when the sum of the area, the length, and the diagonal is given (as in *TMS 20* § 1), while all parameters of a rectangle *cannot* be found when the sum of the area, the length, and the front are given. A second condition is needed, either the ratio between the front and the length of the rectangle, as in IM 43993, or some other equation, as the additional rectangular-linear equation in MS 5112 § 8, or the additional linear equation in *TMS* 9!

11.2 l.   MS 5112 § 9. *Changing the Form of a Rectangle while Keeping the Area*

MS 5112 § 9 (*iii*: 22'-31')

1	uš sag gu₇.gu₇-*ma*	The length (and) the front (I made) eat (each other), then
	50ᵥ a.šà	50, the field.
	i-na uš 30 ninda [zi-*ma*]	From the length 30 ninda (*I*) *tore out, then*
2	*a-n[a]* sag / 5 ninda daḫ-*ma* 50ᵥ a.šà	to *the front* 5 ninda (I) added, then 50, the field.
	uš sag en.nam	The length (and) the front are what?
	za.e ak.da.zu.dè /	You with your doing:
3	igi 5 *ša a-na* sag daḫ du₈-*ma* 12	The opposite of 5 that to the front was added release, then 12.
	a-na 50 a.šà rá-*ma* 10 /	To 50 the field go, then 10.
4	*a-na* 30 *ša i-na* uš zi rá-*ma* 5	To 30 that from the length was torn out go, then 5.
	re-eš-ka li-ki-il /	May it hold your head.
5	2' 30 *ša i-na* uš zi dar-*ma* 15	1/2 of 30 that from the length was torn crush, then 15.
	a.rá 15 3 4[5] <gu₇.gu₇-*ma*> /	*Steps of* 15, 3 45.
6	*a-na* 5 *ša re-eš-ka* daḫ-*ma* 5 03! 45	To 5 that <held> your head add, then 5 03 45
	mi-nam íb.si₈ / 2 15 íb.si₈	(makes) what equalsided? 2 15 equalsided.
7	*a-na* 2 *lu-pu-ut-ma*	To two (= twice) inscribe, then
	a-na 2 15 ki.1 15 daḫ-*ma* /	to the 1st 2 15, 15 *add, then*
8	2 30 uš *pa-nu*	2 30, the earlier length.
	i-na 2 15 ki.2 15 zi-*ma*	From the 2nd 2 15, 15 tear off, then
	2 uš [eg]ir /	2, the later length.
9	igi 2 30 uš *pa-ni* du₈-*ma* 24	The opposite of 2 30, the earlier length, resolve, then 24.
	a-na 50ᵥ a.šà rá	To 50, the field, *go,*
	[20 sag *pa*]-*ni-tum* /	20, the earlier front.
10	igi 2 uš egir du₈-*ma* 30	The opposite of 2, the later length, resolve, then 30.
	a-na 50ᵥ a.šà rá-*ma* 25 sag egir	To 50, the field, go, then 25, the later front.

The only new terminology in § 9 is the pair

$$\textit{pānûm} — \text{egir} \qquad\qquad \text{earlier} — \text{later}$$

The question in this exercise can be rephrased as follows:

A rectangle has the area 50 (00 sq. n.).
The length of the rectangle is diminished by 10 n., and the front augmented by 5 n.
The new area is 50 (00 sq. n.). What are the length and the front?

These are the successive steps of the numerical solution procedure in § 9:

1/5 (added) · 50 (00) (area) · 30 (subtracted) = 5 (00 00)	(lines 3-4)
30 (subtracted) / 2 = 15, 5 (00 00) + sq. (15) = 5 03 45, sqs. 5 03 45 = 2 15	(lines 5-6)
2 15 + 15 = 2 30 = the earlier length, 2 15 – 15 = 2 (00) = the later length	(lines 7-8)
50 (00) / 2 30 = 20 = the earlier front, 50 (00) / 2 (00) = 25 = the later front	(lines 9-10)

This solution procedure appears to have been based on a geometric argument of the kind illustrated in Fig. 11.2.13 below. The original rectangle, with the sides u and s, has the area 50 (00) sq. n., which is also the area of the transformed rectangle, with the sides $u^* = u - 30$ n. and $s^* = s + 5$ n. Therefore, the small rectangle cut off from the original rectangle must has the same area as the rectangle added underneath the original rectangle (both in light grey in Fig. 11.2.13). This means that

$$u* \cdot 5 \text{ n.} = 30 \text{ n.} \cdot s, \quad \text{so that} \quad u* = 30/5 \cdot s.$$

(It would have been possible, although more laborious and less intuitive, to find this relation algebraically, without any reference to a geometric illustration.) Thus, the original system of equations

$$u \cdot s = 50 \text{ (00 sq. n.)}, \quad (u - 30 \text{ n.}) \cdot (s + 5 \text{ n.}) = 50 \text{ (00 sq. n.)}$$

can be replaced by the following simpler system of equations:

$$u \cdot u* = 30/5 \cdot 50 \text{ (00) (sq. n.)} = 5 \text{ (00 00) (sq. n.)}, \quad u - u* = 30 \text{ n.}$$

This is a "rectangular-linear" system of equations for the two unknowns u (in the text called "the earlier length") and $u*$ (called "the later length"). Such a system of equations, with the product and the difference of two unknowns given, is one of the Old Babylonian basic forms for a rectangular-linear system of equations. (See Fig. 11.2.10.) In the present case, the solution to the system of equations is

$$u = 2 \, 30, u* = 2 \text{ (00)}.$$

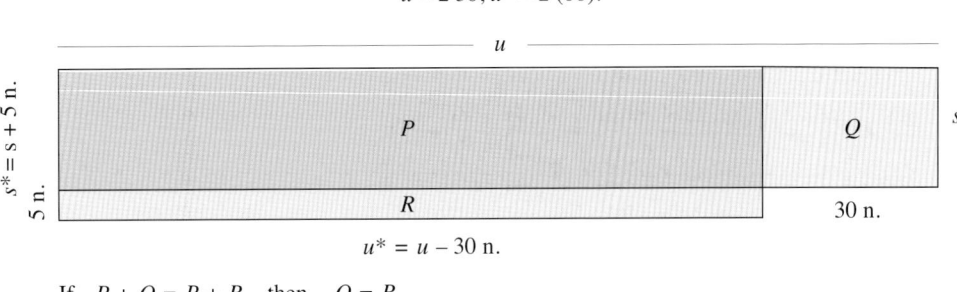

$$u* = u - 30 \text{ n.}$$

If $P + Q = P + R$,   then $Q = R$.

In other words, if $u \cdot s = (u - 30 \text{ n.}) \cdot (s + 5 \text{ n.})$,   then $(u - 30 \text{ n.}) \cdot 5 \text{ n.} = 30 \text{ n.} \cdot s$

Fig. 11.2.13. MS 5112 § 9. Changing the length and front of a rectangle without changing the area.

When the values of u and $u*$ have been determined, it is easy to find also the values of s ("the earlier front") and $s*$ ("the later front"), as follows:

$$u \cdot s = 50 \text{ (00 sq. n.)}, \quad u = 2 \, 30 \text{ n.} \quad \Rightarrow \quad s = 50 \, 00 \text{ sq. n.} / 2 \, 30 \text{ n.} = 20 \text{ n.}$$
$$u * \cdot s* = 50 \text{ (00 sq. n.)}, \quad u* = 2 \text{ (00) n.} \quad \Rightarrow \quad s* = 50 \, 00 \text{ sq. n.} / 2 \, 00 \text{ n.} = 25 \text{ n.}$$

11.2 m.   MS 5112 § 10. A System of Linear Equations for the Length and the Front

MS 5112 § 10 (*iv*: 1'-7')

1'	[x x x x] uš *ù* s[ag x x x x x x]	*x x x x x x x* the length and the front *x x x x x x x*
	[x x x x x x x x x x x x x x]	*x x x x x x x x x x x x x x x*
2'	[uš] *ù* sag en.nam	*The length* and the front are what?
	za.e [ak.da.zu.dè]	You *with your doing:*
3'	[*aš-šum* igi.7.gál] / *qà-bu-kum*	*Since the 7th-part* it is said to you,
	igi.7.gál 7 1 *lu-pu-ut*	the 7th-part of 7, 1, inscribe.
4'	[*aš-šum* igi.3.gál] / *qà-bu-kum*	*Since the 3rd-part* it is said to you,
	igi.3.gál 3 1 *lu-pu-ut*	the 3rd-part of 3, 1, inscribe.
	gar.gar-*ma* [2 x x] /	Heap, then *2 x x*
5'	1 *wa-ṣi-tam* daḫ-*ma* 3	1, the extension, add, then 3.
	igi 3 du₈-*ma* 20	The opposite of 3 resolve, then 20.
6'	*a-na* 11 1[5] / rá-*ma* 3 45	To 11 *15* go, then 3 45.
	a-na 7 rá-*ma* 26 15 uš /	To 7 go, then 26 15, the length.
7'	11 15 *a-na* 2 tab-*ma* 22 30 sag	11 15 to two repeat, then 22 30, the front.

In § 10, the question is completely destroyed, while most of the solution procedure is preserved. The solution procedure is both brief and vague, which makes it difficult to reconstruct the question. It is clear, anyway, that § 10 is *not* concerned with quadratic or rectangular equations, although the two unknowns are called uš and sag, as in all the other exercises on the reverse of MS 5112.

The successive steps of the numerical solution procedure in § 10 seem to be:

$$1/7 \cdot 7 = 1, \quad 1/3 \cdot 3 = 1, \quad 1 + 1 = 2 \qquad\qquad \text{(lines 3'-4')}$$
$$2 + 1 \ (w\bar{a}ṣ\bar{\imath}tum) = 3, \quad 1/3 \cdot 11\ 15 = 3\ 45 \qquad\qquad \text{(lines 5'-6')}$$
$$7 \cdot 3\ 45 = 26\ 15 = \text{the length } u, \quad 2 \cdot 11\ 15 = 22\ 30 = \text{the front } s \qquad \text{(lines 6'-7')}$$

A tentative interpretation of § 10 is that the given problem was a system of linear equations:

$$1/7 \cdot u + 1/3 \cdot s = 11\ 15, \quad 1/3 \cdot s = 2 \cdot 1/7 \cdot u.$$

Such a system of equations can be solved by use of the Old Babylonian rule of false value (Friberg, *RlA* 7 (1990) Sec. 5.7 d):

Let $u^* = 7$ be a preliminary ("false") value for u, so that $1/7 \cdot u^* = 1$.
Then, according to the second equation, the corresponding value for s is $s^* = 6$, so that $1/3 \cdot s^* = 2$.
Consequently, $1/7 \cdot u^* + 1/3 \cdot s^* = 1 + 2 = 3$.
According to the first equation, the correct value is greater by a factor $11\ 15 / 3 = 3\ 45$.
Hence, $u = 3\ 45 \cdot u^* = 26\ 15$, and $s = 3\ 45 \cdot s^* = 22\ 30$.

This tentative interpretation fits some, but not all, of the data in the text of § 10. In particular, it does not explain the appearance of the term *wāṣītum* in line 5' of the text. The term occurs elsewhere in texts dealing with *quadratic* equations, such as MS 5112 § 1. The term occurs also, however (in its masculine form *wāṣûm*) in VAT 8391 (see below), a text dealing with systems of *linear* equations.[3]

Other Old Babylonian Mathematical Texts with Systems of Linear Equations

Relatively few published Old Babylonian mathematical texts deal with systems of linear equations. (See Friberg, *RlA* (1990) Sec. 5.7 b.) Two of them are the pair **VAT 8389** and **VAT 8391** (Høyrup, *LWS* (2002), 77-85), in which the common theme is systems of linear equations for two fields with different rents (barley per area unit). Let the areas of the two fields be A_1 and A_2, let the rents for them be R_1 and R_2, and let the corresponding rents per area unit be r_1 and r_2, where

$$r_1 = 4\,\text{gur} / \text{bùr} = {;}40\,\text{sìla} / \text{sq. n.}, \quad r_2 = 3\,\text{gur} / \text{bùr} = {;}30\,\text{sìla} / \text{sq. n.}$$

Then four related systems of equations in the two mentioned texts can be explained as

a) $A_1 + A_2 = 30\,(00\,\text{sq. n.})$, $R_1 + R_2 = r_1 \cdot A_1 + r_2 \cdot A_2 = 18\ 20\,(\text{sìla})$,
b) $A_1 + A_2 = 30\,(00\,\text{sq. n.})$, $R_1 - R_2 = r_1 \cdot A_1 - r_2 \cdot A_2 = 8\ 20(\text{sìla})$,
c) $A_1 - A_2 = 10\,(00\,\text{sq. n.})$, $R_1 + R_2 = r_1 \cdot A_1 + r_2 \cdot A_2 = 18\ 20\,(\text{sìla})$,
d) $A_1 - A_2 = 10\,(00\,\text{sq. n.})$, $R_1 - R_2 = r_1 \cdot A_1 - r_2 \cdot A_2 = 8\ 20\,(\text{sìla})$.

Two separate strategies are employed in the texts in order to find the solutions to the systems a) and b) on one hand, and to the systems c) and d) on the other.

In all four cases, the first step of the solution procedure is to find a partial solution, satisfying only the first of the two given equations. In a) and b), this partial solution is $A_1 = A_2 = 15\,(00\,\text{sq. n.})$. However, this partial solution does not satisfy the second of the given equations. Indeed,

If $A_1 = A_2 = 15\,(00\,\text{sq. n.})$, then
$R_1 + R_2 = (r_1 + r_2) \cdot 15\,(00) = 17\ 30\,(\text{sìla})$, and
$R_1 - R_2 = (r_1 - r_2) \cdot 15\,(00) = 2\ 30\,(\text{sìla})$.
Thus, in a) and b) there are deficits of $18\ 20 - 17\ 30 = 50\,(\text{sìla})$,
and $8\ 20 - 2\ 30 = 5\ 50\,(\text{sìla})$, respectively.

---

3. See the discussion of the term *wāṣītum* in Muroi, *ASJ* 20 (1998). Unfortunately, Muroi does not take into account the possibility that the term may have a geometric meaning, as suggested repeatedly here and in Høyrup, *LWS* (2002).

To eliminate these deficits, A_1 has to be increased by a suitable amount, at the same time as A_2 is decreased by the same amount. Now, it is easy to see that

> If A_1 and A_2 are increased and decreased, respectively, by 1 (00 sq. n.), then the deficit is decreased by
> $(r_1 - r_2) \cdot 1$ (00) = 40 – 30 = 10 (sìla) in case a), and by $(r_1 + r_2) \cdot 1$ (00) = 40 + 30 = 1 10 (sìla) in case b).

As in an application of the Old Babylonian rule of false value, this initial decrease of the deficit has to be compared with the desired total elimination of the deficit. It then becomes clear that

> In both a) and b), if A_1 and A_2 are increased and decreased, respectively by 5 (00 sq. n.), then the deficit is eliminated.
> Then A_1 = 15 (00) + 5 (00) = 20 (00 sq. n.), and A_2 = 15 (00) – 5 (00) = 10 (00 sq. n.).

The situation is illustrated in Fig. 11.2.14, left:

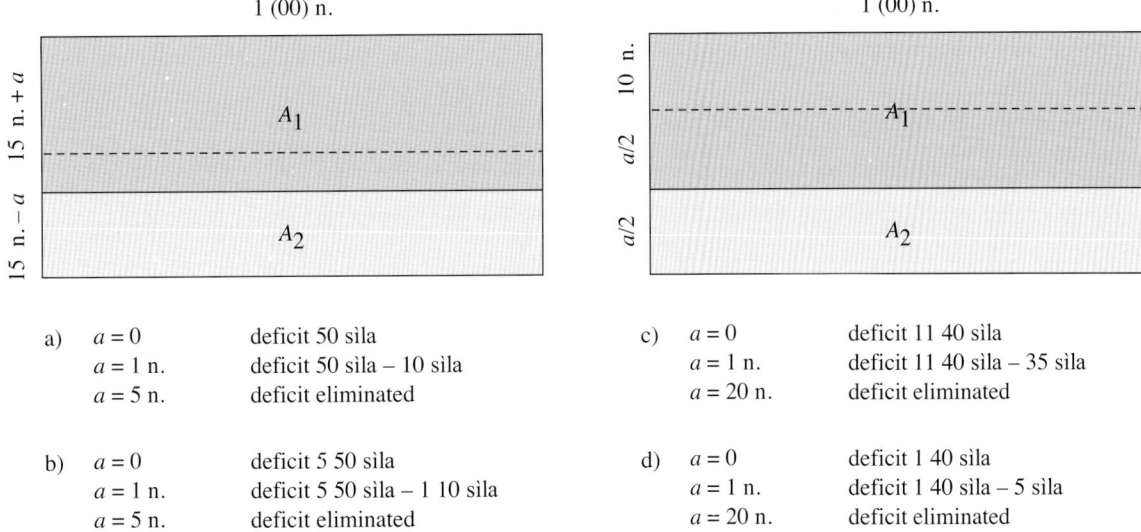

	a)	$a = 0$	deficit 50 sìla		c)	$a = 0$	deficit 11 40 sìla
		$a = 1$ n.	deficit 50 sìla – 10 sìla			$a = 1$ n.	deficit 11 40 sìla – 35 sìla
		$a = 5$ n.	deficit eliminated			$a = 20$ n.	deficit eliminated
	b)	$a = 0$	deficit 5 50 sìla		d)	$a = 0$	deficit 1 40 sìla
		$a = 1$ n.	deficit 5 50 sìla – 1 10 sìla			$a = 1$ n.	deficit 1 40 sìla – 5 sìla
		$a = 5$ n.	deficit eliminated			$a = 20$ n.	deficit eliminated

Fig. 11.2.14. VAT 8389 and VAT 8391: Systems of linear equations solved by geometric methods.

In cases c) and d), the partial solution satisfying only the first equation is A_1 = 10 (00 sq. n.), $A_2 = 0$. Once more, this partial solution does not satisfy the second of the given equations. This time,

> If A_1 = 10 (00 sq. n.), $A_2 = 0$, then $R_1 + R_2 = R_1 - R_2 = 6\ 40$ (sìla).
> Thus, in c) and d) there are deficits of 18 20 – 6 40 = 11 40 (sìla), and 8 20 – 6 40 = 1 40 (sìla), respectively.

To eliminate these deficits, A_1 and A_2 have to be increased by a common suitable amount. Now, it is easy to see that

> If A_1 and A_2 are both increased by $1/2 \cdot 1$ (00) = 30 (sq. n.), then the deficit is decreased
> by $(r_1 + r_2) \cdot 30$ = 35 (sìla) in case c), and by $(r_1 - r_2) \cdot 30$ = 5 (sìla) in case d).

If this initial decrease of the deficit is compared with the desired total elimination of the deficit, it becomes clear, as shown in Fig. 11.2.14, right, that

> In both c) and d), if A_1 and A_2 are both increased by $20 \cdot 30$ = 10 (00 sq. n.), then the deficit is eliminated.
> Then A_1 = 10 (00) + 10 (00) = 20 (00 sq. n.), and $A_2 = 0$ + 10 (00) = 10 (00 sq. n.).

Note: In c) (= VAT 8391 # 3), the initial increase 1 (00) is called *wāṣûm* (written *wa-ṣi-um*) 'extension', a fact supporting the idea that the solution procedures in VAT 8389 and VAT 8391 were based on geometric arguments, a special kind of metric algebra.

A third previously published Old Babylonian mathematical text featuring systems of linear equations is **YBC 4698**. (See the hand copy in Neugebauer, *MKT 3*, pl. 5.) This is a small mathematical recombination text with commercial problems as its very unusual topic. A new, and much improved, transliteration and translation of a "fish market problem" in # 15 of that text is presented below:

YBC 4698 # 15 a-b

1	ganba	Market rates.
	3 ku$_6$.a 5 sìla /	3 fishes (for) 5 sìla,
2	5 ku$_6$.a 1 [3" sìla] /	5 fishes (for) 1 2/3 *sìla*.
3	še gar.gar-*ma* 4$_{gur}$ 1$_{barig}$ 4$_{bán}$ /	The barley heap, then 4(gur) 1(barig) 4(bán).
4	ku$_6$.a gar.gar-ma 20$_{géš}$ /	The fishes heap, then 20(sixties).
5	še.ne ku$_6$.a.ne /	The barleys (and) the fishes (are what)?
6	ku$_6$.a íb.si$_8$-*ma*	The fishes make equal, then …

Here, the given market rates for two kinds of fish are 3 fishes for 5 sìla (of barley) and 5 fishes for 1 2/3 sìla, respectively. The corresponding unit prices (inverted market rates) are $p_1 = 1;40 (= 5/3)$ and $p_2 = ;20 (= 1/3)$ sìla/fish. The given total price, in barley, for an unknown number k_1 of fishes of the first kind and k_2 fishes of the second kind is 4 gur 1 barig 4 bán = 21 40 sìla. The total number of fishes is 20(sixties). Hence, k_1 and k_2 must satisfy the following system of linear equations:

$$k_1 + k_2 = 20 \text{ (00 fishes)}, \quad p_1 \cdot k_1 + p_2 \cdot k_2 = 21\ 40 \text{ (sìla)}, \quad \text{where} \quad p_1 = 1;40 \quad \text{and} \quad p_2 = ;20 \text{ (sìla/fish)}.$$

The solution to this problem is easy to find by use of the method displayed in Fig. 11.2.14, left:

$$k_1 = 11\ 15 \text{ (fishes)}, \quad k_2 = 8\ 45 \text{ (fishes)}.$$

This solution is not given in the text, which contains only the question. Note, by the way, that line 6 is not part of the problem in lines 1-5. Instead it mentions the possibility of an alternative question of a related kind, a combined market rate exercise (see Sec. 7.2 a above).

11.2 n.   *MS 5112 § 11. One of the Basic Rectangular-Linear Systems of Equations*

MS 5112 § 11 (*iv: 8'-13'*)

1	uš sag gu$_7$.gu$_7$-*ma*	The length (and) the front (I) made eat each other, then
	1 èše aša$_5$ a.šà	1 èše the field.
	uš ugu sag 10 diri /	The length over the front 10 beyond.
2	uš sag en.nam	The length (and) the front are what?
	za.e ak.da.zu.dè	You with your doing:
	2' 10 *ša* uš ugu sag diri /	1/2 of 10 that the length over the front is beyond
3	dar-*ma* 5 a.rá 5 gu$_7$.gu$_7$-*ma* 25	crush, then 5 steps of 5 (make) eat (each other), then 25.
4	*a-na* 10 a.šà daḫ-*ma* / 10 25	To 10 the field add, then 10 25.
	mi-nam íb.si$_8$ 25<.ta>.àm íb.si$_8$	What is it equalsided? 25 each way equalsided.
	a-na 2 *lu-pu-ut-ma* /	To two write it down.
5	5 *ša* gu$_7$.gu$_7$ *a-na* 25 ki.1 daḫ-*ma* 30	5 that was eaten to the 1st 25 add, then 30.
	30 ninda uš /	30 ninda is the length.
6	*i-na* 25 ki.2 5 zi-*ma* 20	From the second 25 the 5 tear off, then 20.
	20 ninda sag	20 ninda is the front.

The new terminology in § 11 consists of just the phrase

a ugu *b c* diri	*a* over *b* (is) *c* beyond	*a* is greater than *b* by the amount *c*

This phrase occurs commonly in theme texts and series texts belonging to Groups 2, 3, and Sa, all southern. Actually, Str. 367 (*MKT 1*, 259) is the only example from these groups of a text using the phrase without being a theme text or series text. The phrase occurs also in the large recombination texts belonging to Group 6a, a northern group, from Sippar.

The geometric meaning of the question in § 11, and of the numerical solution procedure is illustrated in Fig.

11.2.10. That such an elementary example of a rectangular-linear system of equations is placed so near the end of the text on the reverse of MS 5112 is a clear demonstration of the fact that MS 5112 is a mathematical recombination text, and not an original theme text.

Despite its simplicity, MS 5112 § 11 is interesting, for the somewhat odd reason that elementary examples of a rectangular-linear system of equations are almost non-existing in the corpus of Old Babylonian mathematical texts. The only known example, other than MS 5112 § 11, seems to be **YBC 6967**, displayed below, a text from Larsa or Uruk (see Robson, *HM* 28 (2001), 183, fn. 21).

YBC 6967, a Related Text With an igi-igi.bi Equation

YBC 6967 (Høyrup, *LWS* (2002), 55-58)

1	[igi.b]i e-*li* igi 7 *i-ter* /	*The* igi.bi *over the* igi *7 is beyond.*
2	[igi] *ù* igi.bi *mi-nu-um* /	*The* igi *and the* igi.bi *are what?*
3	a[*t-t*]a	You:
4	7 *ša* igi.bi / ugu igi *i-te-ru* /	7 that the igi.bi over the igi is beyond
5	a-na *ši-na ḫe-pé-ma* 3 30 /	to two break, then 3 30.
6-7	3 30 *it-ti* 3 30 / *šu-ta-ki-il-ma* 12 15 /	3 30 with 3 30 let them eat each other, then 12 15.
8	a-na 12 15 *ša i-li-kum* /	To 12 that came up for you
9	[1 a.šà-*l*]a-am *ṣí-ib-ma* 1 12 15 /	*1, the field,* add, then 1 12 15.
10	[íb.si₈ 1] 12 15 *mi-nu-um* 8 30 /	*The equalside* of 1 12 15 is what? 8 30.
11	[8 30 *ù*] 8 30 *me-ḫe-er-šu i-di-ma* /	8 30 *and* 8 30, its equal, lay down, then
12	3 30 *ta-ki-il-tam* /	3 30, the holder,
13	i-na *iš-te-en ú-su-uḫ* /	from one tear out,
14	a-na *iš-te-en ṣí-ib* /	to one add,
15	*iš-te-en* 12 *ša-nu-um* 5 /	one is 12, the second 5.
16	12 igi.bi 5 *i-gu-um*	12 is the igi.bi, 5 the igi.

Here the terms igi and igi.bi 'its igi' stand for a pair of reciprocal numbers, as, for instance, the pairs of reciprocals in the Old Babylonian standard table of reciprocals. Alternatively, the two numbers can be interpreted as *the length and the front, respectively, of a rectangle with the area 1 (00 00) sq. n.* This interpretation fits well together with the consistently geometric terminology of the solution procedure. Note, in particular, the phrase 1 *eqlam* (a.šà-*la-am*) *ṣí-ib-ma* in line 9, which explicitly refers to the product 1 of the igi and the igi.bi as 'the area'.

The meaning of the term *takīltum* in line 12 has been much debated. The term is commonly occurring in Old Babylonian metric algebra texts (see the many explicit examples in Muroi, *HSJ* 12 (2003)). In this connection it is important to understand that, as pointed out by Muroi (*op. cit.*), *takīltum* is the *nomen actionis* of the D-stem *kullum* 'to hold', and that the proper translation of the phrase *a reška likīl* should be 'may (the number) *a* hold your head', and *not* 'may your head hold (the number) *a*', as has usually been taken for granted. The meaning of the phrase in normal language is, of course, that 'you shall take a mental note of the number *a* for later use'. Now, with this new interpretation of *a reška likīl*, it seems to follow that the proper translation of *takīltum* should be 'that which holds' (your mind), so that *takīltum* refers to a number that you have taken a mental note of.[4] This interpretation agrees with the way in which the term is used in metric algebra texts. As a matter of fact, the phrase *takīltum* typically refers to a number used in the solution of a metric algebra problem, *a number that is first squared (and held in mind), then later used in additions or subtractions to compute the*

---

4. A more farfetched interpretation of the term *takīltum* can be found in Robson, *HM* 28 (2001), where the term is translated as 'something that has been caused to hold (something)', 'holding(-square)', which, in Robson's opinion, "suggests that the *takīltum* is conceptualized as a square configuration or frame rather than an area or a length". In the same vein, Høyrup, *LWS* (2002), 23, translates *takīltum* as 'which is made hold' (two sides of a rectangle). In Muroi, *HSJ* 12 (2003), the term is translated 'the one which contains' (two squares and two rectangles).

solutions to a quadratic equation or a rectangular-linear system of equations. In YBC 6967, for instance, the *takīltum* is the number 3 30 appearing in lines 5-7 and 12-14.

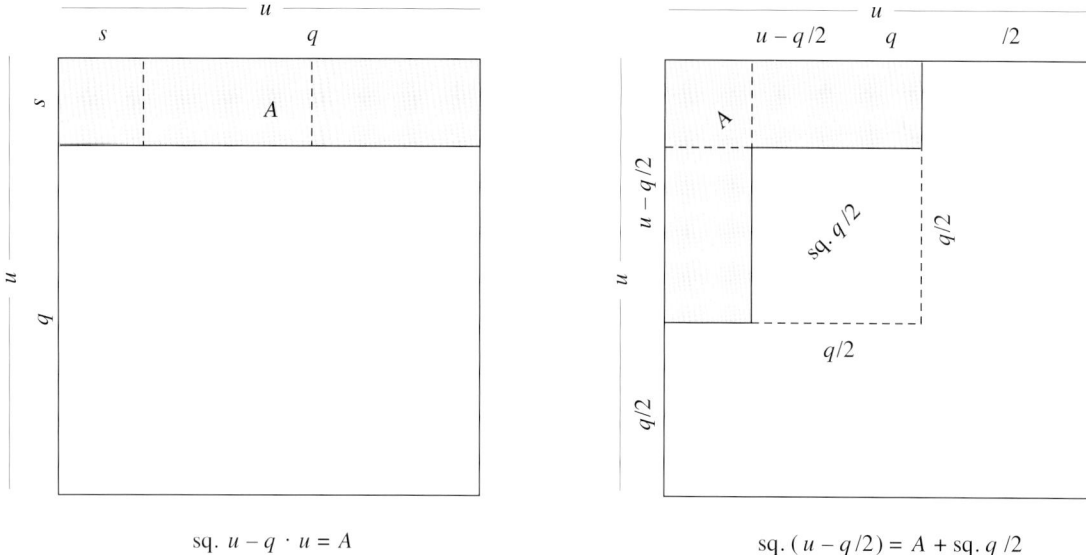

sq. $u - q \cdot u = A$ sq. $(u - q/2) = A + $ sq. $q/2$

Fig. 11.2.15. The metric algebra solution to a quadratic equation of the type sq. $u - q \cdot u = A$.

In Fig. 11.2.10, showing the metric algebra solution to two types of rectangular-linear systems of equations, $u \cdot s = A$, $u + s = p$ (or $u - s = q$), the *takīltum* is equal to $p/2$ in the first case, and to $q/2$ in the second case. In Fig. 11.2.8, showing the metric algebra solution to a quadratic equation of the type sq. $s + q \cdot s = A$, the *takīltum* is equal to $q/2$. For completeness is shown here also, in Fig. 11.2.15 above, the metric algebra solution to a quadratic equation of the type sq. $u - q \cdot u = A$.

In Fig. 11.2.15, in the case of a quadratic equation of the type sq. $u - q \cdot u = A$, precisely as in Fig. 11.2.8, in the case of a quadratic equation of the type sq. $s + q \cdot s = A$, the *takīltum* is equal to $q/2$.

11.2 o.   MS 5112 § 12. A Rectangular-Linear System of Equations

MS 5112 § 12 (*iv*: 14'-22')

1	igi.16.gál uš *ù* 1 2' kùš sag	The 16th-part of the length and 1 1/2 cubit is the front.
	uš sag en.nam /	The length (and) the front are what?
2	za.e ak.da.zu.dè	You with your doing:
	aš-šum igi.16.gál *qà-bu-kum* 16 *lu-pu-ut* /	Since the 16th-part it is said to you, 16 inscribe,
	ù igi.16.gál 16 1 *lu-pu-ut ù*	and the 16th-part of 16, 1, inscribe, and
3	7 30 *ša a-na* sag daḫ / *lu-pu-ut-ma*	7 30 that to the front was added inscribe, then
4	16 *it-ti* 1 gu₇.gu₇-*ma* 16 /	16 with 1 (make) eat, then 16.
5	*it-ti* 7 30 gu₇.gu₇-*ma* 2	With 7 30 (make) eat (each other), then 2.
	2'-*su* dar-*ma* 1	Its 1/2 crush, then 1.
6	a.rá 1 / gu₇.gu₇-*ma* 1	Steps of 1 (make) eat, then 1.
	a-na 16 daḫ-*ma* 16 01	To 16 add, then 16 01.
7	*mi-nam* íb.si₈ / 31.àm íb.si₈	What is it equalsided? 31 each way it is equalsided.
8	*i-na* 31 1 *ša* gu₇.gu₇-*ma* zi-*ma* /	From 31 the 1 that was made eat itself tear off, then
	30 uš	30, the length.
	igi 16 du₈-*ma* 3 45	The opposite of 16 release, then 3 45.
9	*a-na* 30 uš rá-*ma* / 1 50ᵥ2 30	To 30 the length go, then 1 52 30.
	a-na 1 50ᵥ2 30 7 30 daḫ-*ma* 2 sag	To 1 52 30 the 7 30 add, then 2, the front.

There is no new terminology in this exercise, except for the substitution of Akk. *itti* for Sum. ki in the phrase *a itti b* gu₇.gu₇.

The question in this exercise can be rephrased as follows:

1/16 of the length (of a rectangle) plus 1 1/2 cubit equals the front. What are the length and the front?

It is clear that something is missing here. An analysis of the ensuing solution procedure shows that what is missing is the additional condition that the area of the rectangle is equal to 1 (00 sq. n.). The reason why this condition is missing is probably that MS 5112 § 12 is a copy of a small text excerpted from a large theme text where the condition that the area should be 1 (00) was common to all the exercises but was stated only in the first exercise.

These are the successive steps of the numerical solution procedure in § 12:

record 16, 1/16 · 16 = 1, record 1, record 7 30   (lines 2-4)
16 · 1 = 16, 16 · 7 30 = 2, 1/2 · 2 = 1, sq. 1 = 1   (lines 4-6)
16 + 1 = 16 01, sqs. 16 01 = 31, 31 − 1 (that you squared) = 30 = the length   (lines 6-8)
1/16 = 3 45, 3 45 · 30 = 1 52 30, 1 52 30 + 7 30 = 2, the front   (lines 8-9)

Since 1 1/2 cubit = ;07 30 nin da, the given rectangular-linear system of equations can be interpreted as follows in quasi-modern abstract notations:

$$u \cdot s = 1 \ (00), \quad 1/16 \cdot u + ;07 \ 30 \ (\text{n.}) = s.$$

The *geometric* interpretation, on the other hand, is as follows: The situation is the one schematically depicted in Fig. 11.2.16, left: A rectangle with the ratio of the front to the length equal to 1/16 has the front extended by the amount ;07 30 n. The extended rectangle has the given area 1 (00 sq. n.).

The first step of the solution procedure can be referred to as "squaring the rectangle". The front of the extended rectangle to the left in Fig. 11.2.16 is scaled up by a factor 16. The result of the scaling is that the extended rectangle is transformed into an extended square as in Fig. 11.2.16, right (not drawn to scale). The square is extended in one direction by the amount 16 · ;07 30 = 2, and the extended square has the area 16· 1 (00) = 16 (00). In abstract notations, the result of the operation is that the given rectangular-linear system of equations has been transformed into an equivalent quadratic equation. The solution to the quadratic equation can be found in the usual way, by means of "balancing" and "completing" as in Fig. 11.2.8. The solution is u = 30 (n.). Finally, the value of s is computed by use of the linear equation $s = 1/16 \cdot u + ;07 \ 30 = 1;52 \ 30 + ;07 \ 30 = 2$ (n.)

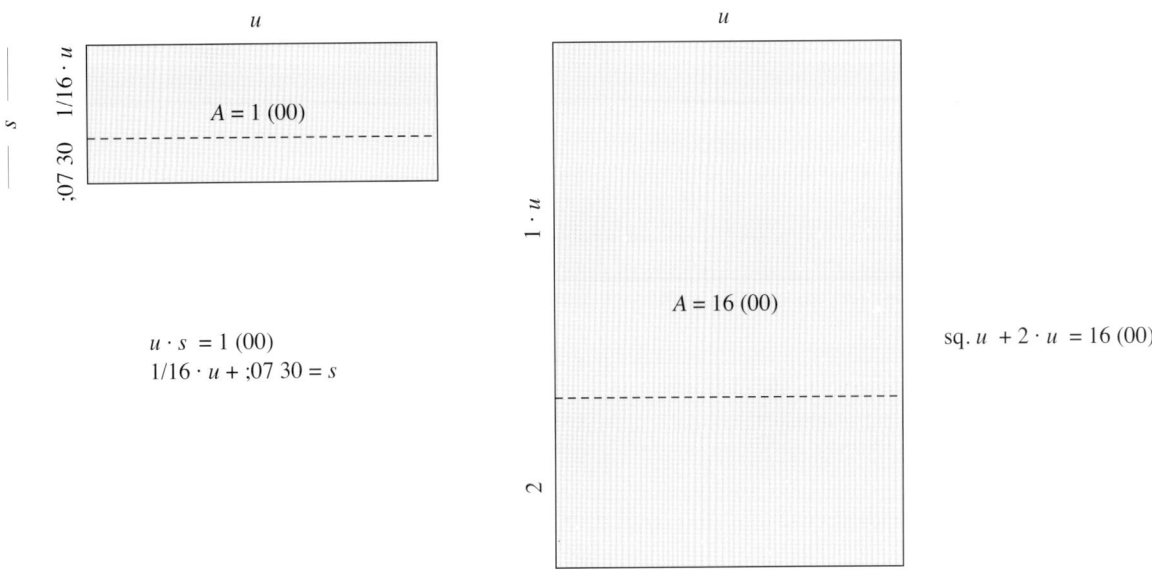

Fig. 11.2.16. A rectangular-linear system of equations reduced to a quadratic equation by scaling

11.2 p.   MS 5112 § 13. Another Rectangular-Linear System of Equations

MS 5112 § 13 (*iv*: 23'-33')

1	*a-na* uš igi.3.gál uš daḫ	To the length the 3rd-part of the length (I) added,
	a-na sag igi.4$_v$.gál sag daḫ-*ma* /	to the front the 4th-part of the front (I) added, then
2	gu$_7$.gu$_7$-*ma* 1 40 a.ša	(I made them) eat (each other), then 1 40, the field.
	uš *ù* sag gar.ga*r-ma* 25 /	The length and the front (I) heaped, then 25.
3	uš sag en.nam	The length (and) the front are what?
	za.e ak.da.zu.dè	You with your doing:
4	*aš-šum a-na* uš / igi.3.gál *qà-bu-kum*	Since to the length the 3rd-part it is said to you,
	1 20 *lu-pu-ut*	1 20 inscribe.
5	*aš-šum a-na* 1 sag / igi.4$_v$.gál *qà-bu-kum*	Since to 1 front the 4th-part it is said to *you*,
	1 15 *lu-pu-ut-ma*	1 15 inscribe.
	re-eš-ka li-ki-il /	May it hold your head.
6	2' 25 dar-*ma* 12 30	1/2 of 25 crush, then 12 30.
	a.rá 12 30 gu$_7$.gu$_7$-*ma* 2 36 15 /	Steps of 12 30 (make) eat (each other), then 2 36 15.
7	*i-na* 2 36 15 1 40 a.šà zi-*ma* 56 15	From 2 36 15, 1 40 the field tear off, then 56 15.
8	*mi-nam* íb.si$_8$ / 7 30.àm íb.si$_8$	What does it make equalsided? 7 30 each way equalsided.
	a-na 12 30 ki.1 daḫ-*ma* 20 uš egir /	To the 1st 12 30 add, then 20 the later length.
9	7 30 *i-na* 12 30 ki.2 zi-*ma* 5 sag$^!$ egir /	730 from the 2nd 12 30 tear off, then 5, the later front.
10	igi 1 20 du$_8$-*ma* 45	The opposite of 1 20 resolve, then 45.
	a-na 20 uš egir rá-*ma* 15 uš igi /	To 20, the later length, go, then 15, the earlier length.
11	[ig]i 1 15 du$_8$-*ma* 48	the opposite of 1 15 resolve, then 48.
	a-na 5 sag egir rá-*ma* 4$_v$ sag igi	to 5, the later front go, then 4, the earlier front.

The only new terminology in this exercise is the pair

$$\text{igi} - \text{egir} \qquad \text{'earlier - later'}$$

where Sum. igi. has replaced the Akk. *pānûm* used in the corresponding situation in § 9.

The question in § 13 can be rephrased as follows, in modern abstract notations:

$$(u + u/3) \cdot (s + s/4) = 1\,40, \quad (u + u/3) + (s + s/4) = 25.$$

(At least this is the form of the question which agrees with the ensuing solution procedure. It is likely, however, that the author of the text made a mistake in his solution procedure, since the second condition of the question, in line 2, right, would normally be interpreted as meaning $u + s = 25$. Accidentally, the mistake led to a considerable simplification of the solution procedure.)

The first step in the solution procedure is to replace the given equations by the equivalent

$$u^* \cdot s^* = 1\,40, \quad u^* + s^* = 25, \quad u^* = 1{;}20 \cdot u, \quad s^* = 1{;}15 \cdot s \qquad \text{(lines 4-5)}$$

This is one of the basic rectangular-linear systems of equations, solved as in Fig. 11.2.10 (lines 6-9). The solution obtained is $u^* = 20$, $s^* = 5$. Division by 1;20 and 1;15, respectively, yields the final solution, $u = {;}45 \cdot 20 = 15$, $s = {;}48 \cdot 5 = 4$ (lines 10-11).

Subscript on the edge below the last exercise

šu.nígin 23$^?$ im.šu.meš 22	Together 23$^?$ hand tablets (= exercises, assignments) 22

This subscript seems to indicate that the total number of exercises is 23, or possibly 22.

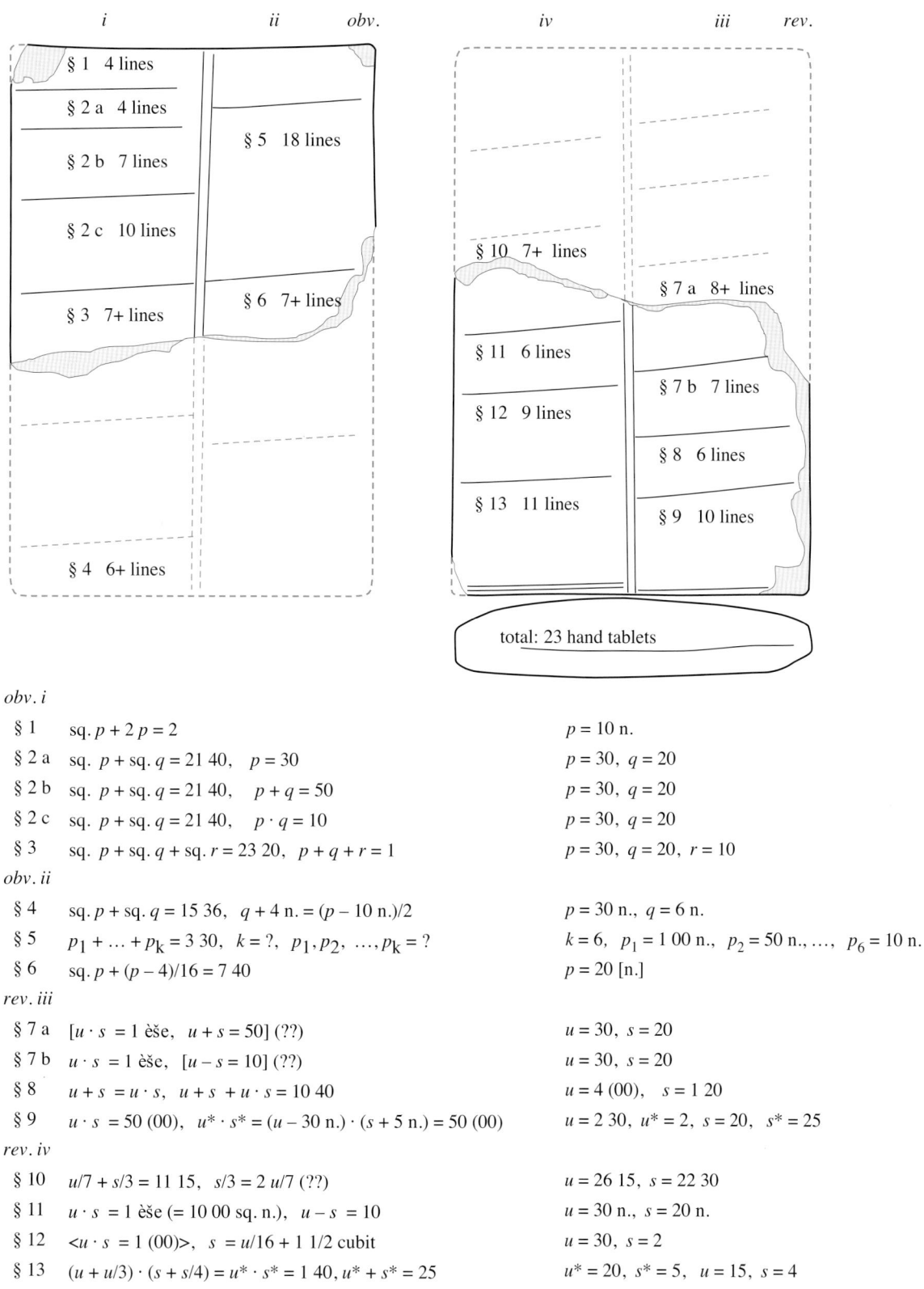

obv. i

§ 1   sq. $p + 2\,p = 2$ $p = 10$ n.
§ 2 a   sq. $p +$ sq. $q = 21\,40$, $p = 30$ $p = 30$, $q = 20$
§ 2 b   sq. $p +$ sq. $q = 21\,40$, $p + q = 50$ $p = 30$, $q = 20$
§ 2 c   sq. $p +$ sq. $q = 21\,40$, $p \cdot q = 10$ $p = 30$, $q = 20$
§ 3   sq. $p +$ sq. $q +$ sq. $r = 23\,20$, $p + q + r = 1$ $p = 30$, $q = 20$, $r = 10$

obv. ii

§ 4   sq. $p +$ sq. $q = 15\,36$, $q + 4$ n. $= (p - 10$ n.$)/2$ $p = 30$ n., $q = 6$ n.
§ 5 $p_1 + \ldots + p_k = 3\,30$, $k = ?$, $p_1, p_2, \ldots, p_k = ?$ $k = 6$, $p_1 = 1\,00$ n., $p_2 = 50$ n., $\ldots$, $p_6 = 10$ n.
§ 6   sq. $p + (p - 4)/16 = 7\,40$ $p = 20$ [n.]

rev. iii

§ 7 a   [$u \cdot s = 1$ èše, $u + s = 50$] (??) $u = 30$, $s = 20$
§ 7 b $u \cdot s = 1$ èše, [$u - s = 10$] (??) $u = 30$, $s = 20$
§ 8 $u + s = u \cdot s$, $u + s + u \cdot s = 10\,40$ $u = 4\,(00)$, $s = 1\,20$
§ 9 $u \cdot s = 50\,(00)$, $u^* \cdot s^* = (u - 30$ n.$) \cdot (s + 5$ n.$) = 50\,(00)$ $u = 2\,30$, $u^* = 2$, $s = 20$, $s^* = 25$

rev. iv

§ 10 $u/7 + s/3 = 11\,15$, $s/3 = 2\,u/7$ (??) $u = 26\,15$, $s = 22\,30$
§ 11 $u \cdot s = 1$ èše $(= 10\,00$ sq. n.$)$, $u - s = 10$ $u = 30$ n., $s = 20$ n.
§ 12 $<u \cdot s = 1\,(00)>$, $s = u/16 + 1\,1/2$ cubit $u = 30$, $s = 2$
§ 13 $(u + u/3) \cdot (s + s/4) = u^* \cdot s^* = 1\,40$, $u^* + s^* = 25$ $u^* = 20$, $s^* = 5$, $u = 15$, $s = 4$

§ 3 and § 5: arithmetic progressions

Fig. 11.2.17. MS 5112. Outline of the fragment, with a catalog of the preserved exercises.

11.3.   MS 3876. Three Problems for 20 Equilateral Triangles and a 'Horn-Figure'

11.3 a.   Computation of the Weight of a 'Horn-Figure'

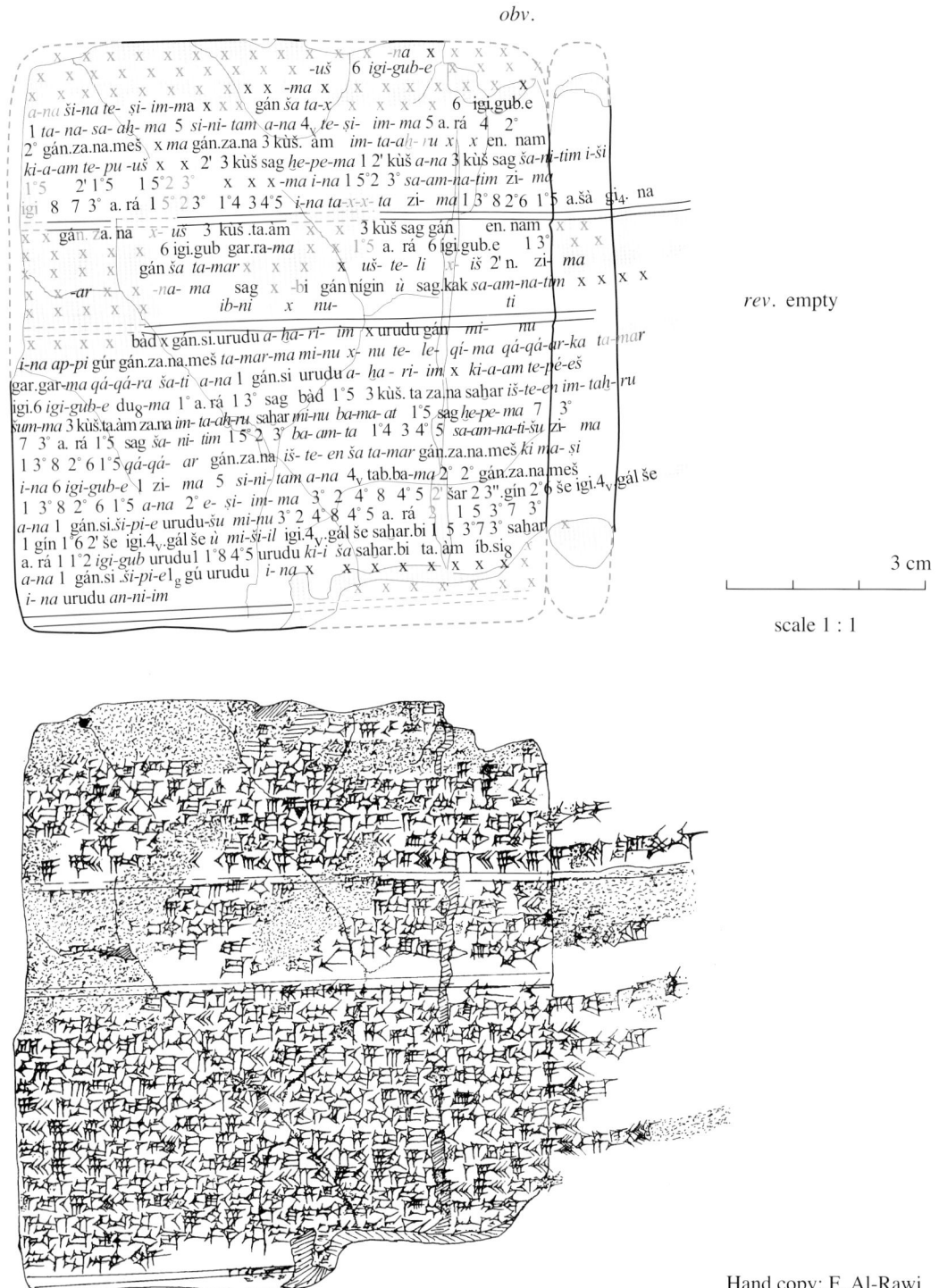

Fig. 11.3.1. MS 3876. Hand copy and conform transliteration.

MS 3876 (Fig. 11.3.1 above) is an atypical mathematical cuneiform texts in several respects, and it offers great difficulties for the interpreter. Already the format of the clay tablet is unusual. Of all clay tablets in the known corpus of mathematical cuneiform texts there is only one other clay tablet of the same format, a small and thin square tablet with only one column of text and minute script. (See Fig. 11.3.2 below.) That other text is **AO 17264** (*MKT 1*, 126; photo *MKT 2*, pl. 2), previously the only known Kassite mathematical problem text (post-Old-Babylonian, from the second half of the second millennium BC).

For what it is worth, it can be noted that also MS 4515, the clay tablet with the square labyrinth (Sec. 8.3 a), is thin and square, much like MS 3876. (Its dimensions are 95 mm × 95 mm × 15 mm.) So, if MS 3876 is Kassite like AO 17264, then maybe also MS 4515 is Kassite, and with it MS 3194, the clay tablet with the rectangular labyrinth (Sec. 8.3 c).

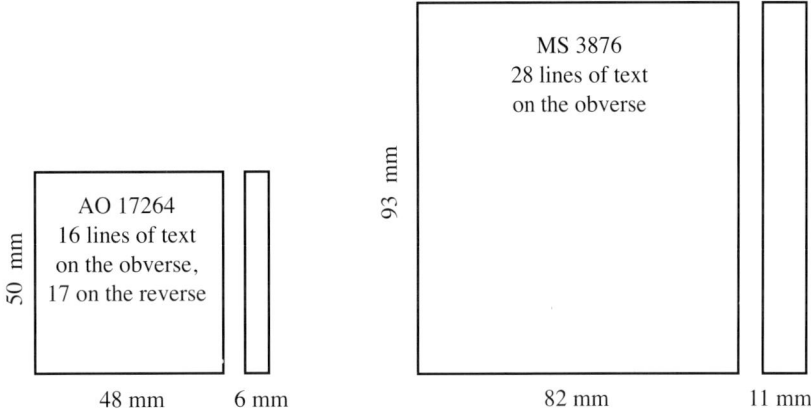

Fig. 11.3.2. The similar formats of AO 17264, MS 3876, and MS 4515.

Obstacles to the interpretation of MS 3876 are, in particular,

   a) the poor preservation of the clay tablet, with pieces missing and multiple scratches on the surface
   b) the minute, carelessly written, and shallowly punched inscription
   c) the previously undocumented topic, with previously unknown technical terms

Among the relatively few Sumerian terms for mathematical concepts in MS 3876 are:

gar.gar	heap	add together
ina a b zi	tear off *b* from *a*	subtract *b* from *a*
a a.rá *b*	*a* steps of *b*	*a* times *b*
igi *a* du₈	resolve the opposite of *a*	compute the reciprocal of *a*
gar.ra	set	
a ana b tab.ba	*a* to *b* repeat	multiply *a* by *b*
en.nam	what	

Akkadian terms for mathematical concepts in MS 3876 include the following ones:

ina a b nasāḫu	tear off *b* from *a*	subtract *b* from *a*
a ana b eṣēpu	repeat *a* to *b*	multiply *a* by *b*
a ana šina eṣēpu	repeat *a* to two	multiply *a* by 2
a ana b našû	lift (*or* carry) *a* to *b*	multiply *a* by *b*
2' (or *bamat*) *a ḫepû*	break half (of) *a*	divide *a* by 2
tamar	you see	the result is
kīam tepeš	you do (it) so	
sittum	remainder	

The phrase *kīam tepeš* 'you do it so' is used in MS 3876 to introduce solution procedures. This phrase occurs in no other known mathematical cuneiform text, with the exception of MS 3049 § 5 (Sec. 11.1 above), where it is used to conclude a solution procedure. The term *sittum* (= *šittum*) 'remainder' occurs, in an inflected form with the spelling *si-i-tam* in MS 3876 # 1, *l.* 5 and # 3, *l.* 8, and also, with the spelling *si-tum* in MS 5112 (Sec. 11.2 c above), § 2 a, *l.* 4, § 4, *l.* 2', and § 5, *l.* 11. The term is known from only one other cuneiform mathematical text, namely M 10 (Sachs, *JCS* 6 (1952)), a hand tablet with approximate reciprocals of non-regular sexagesimal numbers, in particular the pair

igi.7.bi	8 34 16 59	*si-i-tum*	(remainder, deficit)[5]	(7 · 8 34 16 59 = 59 59 58 53)
igi.7.bi	8 34 18	*diri*	(excess)	(7 · 8 34 18 = 1 00 00 06)

MS 3876 # 1

1	[x x x x x x x x x x x x x -*n*]*a* x [x x x] /	*x x x x x x x x x x x x x x x x xx x x x x x x x x x x x x x x*
2	[x x x x x x x x x x x]-*uš*	*x x x x x x x x x x x x x x*
	6 *igi-gub-e* [x x x x] /	6, the constant, *x x x x*,
3	[x x x x x x x x x] x x -*ma*	*x x x x x x x x x x x x x x x x x x*
	x [x x x x x x x x x] /	*x x x x x x x x x x*
	[*a-na*] *ši-na te-ṣi-im-ma* [x x] x	*To* 2 you repeat, then *x x* x
4	*gán ša t a-*[x x x x x x]	The field that you *x x x x x x*.
5	[*i-na*] 6 *igi-gub-e* / 1 *ta-na-sa-aḫ-ma*	*From* 6, the constant, 1 you tear off, then
	5 *si-i-tam a-na* 4ᵥ *te-ṣi-im-ma*	5, the remainder, to 4 repeat, then
6	5 *a.rá* 4 20 / 20 *gán.za.na.meš* *x-ma*	5 times 4 (is) 20, 20 gaming-piece-fields x, then
	gán.za.na 3 *kùš.àm* *im-ta-a*[*ḫ-r*]*u*	the gaming-piece-field, 3 cubits they *equa*l each other.
	x x *en.nam* /	The x x (is) what?
7	*ki-a-am te-pu-uš*	**So you do (it).**
	x [x x]	x *x* x.
	2' 3 *kùš sag* *ḫe-pe-ma* 1 2' *kùš*	1/2 (of) 3 cubits, the front, break, then 1 1/2 cubits.
	a-na 3 *kùš sag ša-ni-tim i-ši* /	To 3 cubits, the second front, raise (it).
8	[15] 2' 15 1 5[2 30] *x-x-x-ma*	*15, 1/2 15, 1 52 30* x x x, then
	i-na 1 5[2] 30 *sa-am-na-tim* zi-*ma* /	from 1 52 30 the eighth tear off, then
9	[*igi*] 8 7 30 *a. rá* [1 52 3]0 14 03 45	*the opposite* (of) 8, 7 30, steps of *1 52 30* (is) 14 03 45.
	i-na ta-[x x]-*ta* zi-*ma*	From x *x x* x tear (it) off, then
	1 38 26 15 [*a*].*šà* gi₄.*na*	1 38 26 15 the true *area*.

Unfortunately, the first few lines of this exercise are either lost or badly damaged, and with them the initial question, the presentation of the problem. There is also a constant '6' of unknown meaning and the following previously unknown technical term:

gán.za.na   gaming-piece-field, gaming-piece-figure   (equilateral triangle)

It is clear from the context (see below) that the term refers to an equilateral triangle, and one of the meanings of the word za.na (Akk. *passum*) is 'pawn, gaming-piece'. It is not inconceivable to think of an equilateral triangle as a drawing of a small gaming-piece, seen from the side.

The prefix gán (or aša₅?), meaning 'field, figure' is known, for instance, from a number of Old or Middle/Late Babylonian tables of constants, in terms like gán.gúr 'arc-field' (circle), and gán.u₄.sakar 'crescent-field' (semicircle) (NSd 20-21), gán.*pa-na-ak-ki* 'bow-field' (BR 10-12), or gán.giš.pan 'bow-field', and gán.ma.gur₈ 'boat-field' (Ka 5, 7).[6]

An equilateral triangle (inscribed in a circle) is depicted in MS 3051 (Fig. 8.1.1 above). A curious name for an equilateral triangle appears in one of the Old Babylonian tables of constants:

sag.kak-*kum ša sa-am-na-*[*tu na*]-*ás-ḫa*	A peg-head (triangle), with an eighth torn out,	
26 15 *i-*[*gi-gu-bu-šu*]	26 15 its constant.	G *rev.* 7'

The obscure phrase refers to the fact that for an equilateral triangle, normalized so that its sides are of length 1 (00), the height *h* can be computed as

$$h = \text{sqs. } 3\ /2 \cdot 1\ 00 = (\text{appr.})\ 1\ 45\ /\ 2 = 52;30 = 1\ 00 - 7;30 = 1\ 00 - 1/8 \cdot 1\ 00.$$

5. Sachs read the word in Sumerian as SI.NI.ÌB.

6. Note the use here and elsewhere in this book of convenient names for several Old Babylonian tables of constants: BR = Bruins and Rutten (1961), *TMS* 3; G = IM 52916, Goetze, *Sumer* 7 (1951); Ka = A 3553, Kilmer, *OrNS* 29 (1960), text A, NSd = YBC 5022, Neugebauer and Sachs, *MCT* (1945), text Ud; NSe = YBC 7243, Neugebauer and Sachs, *MCT* (1945), text Ue, and one Middle or Late Babylonian table of constants: Kb = CBS 10996, Kilmer, *OrNS* 29 (1960), text B.

Hence, the area *A* of an equilateral triangle with the side *s* can be conveniently computed as

$$A = \text{sqs. } 3/4 \cdot \text{sq. } s = (\text{appr.}) \; 1/2 \; \text{sq. } s - 1/8 \cdot 1/2 \; \text{sq. } s, \text{ which is the same as } ;26 \; 15 \; (7/16) \cdot \text{sq. } s.$$

Similar names for equilateral triangles appear in the Late Babylonian mathematical recombination text W 23291 §§ 4 b-c (Friberg, *BagM* 28 (1997), 285-286):

| gán.sag.kak ur.a *ša* 8-*šu na-as-ḫu* | a peghead-field, equalsided, with an 8th torn out | § 4 b |
| gán.sag.kak ur.a *ša* 10-*šu ù* 30-*šu na-as-ḫu* | a peghead-field, equalsided, with a 10th and a 30th torn out | § 4 c |

The second name for an equilateral triangle refers to a more accurate approximation to the area *A* of an equilateral triangle with the side 1 00, namely

$$A = \text{sqs. } 3/4 \cdot \text{sq. } s = (\text{appr.}) \; 1/2 \; \text{sq. } s - 1/10 \cdot 1/2 \; \text{sq. } s - 1/30 \cdot 1/2 \; \text{sq. } s, \text{ which is the same as } ;26 \cdot \text{sq. } s.$$

In MS 3876 #1, the phrase

| gán.za.na 3 kùš.àm *im-ta-a[ḫ-r]u* | a gaming-piece-figure, 3 cubits they *equ*al each other | (line 6) |

refers to an equilateral triangle with the side *s* = 3 cubits. The computation of the area of this equilateral triangle starts with the computation of *s*/2 · *s*. The steps of the computation are as follows:

$$1/2 \cdot 3 \text{ cubits} = 1 \; 1/2 \text{ cubits}, \quad 1 \; 1/2 \text{ cubits} \cdot 3 \text{ cubits} = 1/2 \cdot ;15 \; \text{n.} \cdot ;15 \; \text{n.} = ;01 \; 52 \; 30 \text{ sq. n.} \qquad \text{(line 7)}$$

Then the area is computed by use of the rule that an eighth should be subtracted from *s*/2 · *s*:

$$1/8 \cdot ;01 \; 52 \; 30 \text{ sq. n.} = ;07 \; 30 \cdot ;01 \; 52 \; 30 \text{ sq. n.} = ;00 \; 14 \; 03 \; 45 \text{ sq. n.} \qquad \text{(line 8)}$$
$$;01 \; 52 \; 30 \text{ sq. n.} - ;00 \; 14 \; 03 \; 45 \text{ sq. n.} = ;01 \; 38 \; 26 \; 15 \text{ sq. n.} \qquad \text{(line 9)}$$

The number *n* of equilateral triangles considered in this text is computed in an earlier part of the solution procedure in # 1, in the following curious way:

$$\text{with 6 being the value of a given constant, } n = (6 - 1) \cdot 4 = 5 \cdot 4 = 20 \qquad \text{(lines 4-5)}$$

What this means is, for the moment, totally obscure (but see below).

MS 3876 # 2

1	[x x] gá[n z]a.na [x x x x]-*uš*	*x x* gaming-piece-field *x x x x* x
	3 kùš.ta.àm [x x x]	3 cubits each *x x x*
	3 kùš sag? gán [x] en.n[am x x] /	3 cubits the front (of) the field, the *x x* (is) what?
2	[x x x x x x] 6 igi.gub gar.ra-*m*[*a*]	*x x x x x x*, 6, the constant, set, then
	[x x x x]	*x x x x*
	[15] a.rá 6 *igi-gub-e* 1 30 [x x] /	*15 steps of 6, the constant, (is) 1 30 x x.*
3	[x x x x x] gán *ša ta-mar*	*x x x x x* the field that you see.
	[x x x] x *uš-te-li* [*ù*] 2' ninda zi-[*ma*] /	*x x x x* x he let come up and 1/2 ninda tear off, then
4	[x x]-*ar* [x x x *n*]*a-ma* sag [x x b]i gán nígin	*x x x x x x x*, then the front *x x x x* x
	ù sag.kak *sa-am-na-tim* [x x] x x /	and the peg-head (triangle) the eighth *x x* x
5	[x x x x x x x x x] *ib-ni-šu-nu-ti*	*x x x x x x x x* x he built them.

Not much remains of the text of MS 3876 # 2, so it is far from clear what is going on here, except that again an equilateral triangle of side 3 cubits is mentioned, and that, for some reason, a new length is computed as follows:

$$\text{with 6 again being the value of a given constant, } 6 \cdot ;15 \; \text{n.} = 1;30 \; \text{n.} \; (= 18 \text{ cubits}) \qquad \text{(line 2)}$$

An eighth is again mentioned in line 4, in an obscure context, and 1/2 ninda seems to be subtracted from something in line 3.

Fortunately, the long text of MS 3876 # 3 is much better preserved:

MS 3876 # 3

1	[x x x bà]d? x gán.si urudu *a-ḫa-ri-im* x urudu x *mi-nu* /	*x x x x* the city wall x, the horn-figure, the copper behind (it) x the copper x (is) what?
2	[*i*]-*na ap-pi* gúr gán.za.na.meš *ta-mar-ma* *mi-nu* [*x*]-*nu te-le-qé-ma* *qá-qá-ar-ka t*[*a-ma*]*r* /	*At* the rim (periphery?) an arc (circle, or sphere???) of gaming-piece-fields you see, then what xx you take, then your ground (area) *you see.*
3	gar.gar-*ma qá-qá-ra ša-ti* *a-na* 1 gán.si urudu *a-ḫa-ri-im* x *ki-a-am te-pé-eš* /	Heap them, then that ground for 1 horn-figure, the copper behind (it), x, **So you do (it).**
4	igi.6 *igi-gub-e* du₈-*ma* 10 a.rá 1 30 sag bàd 15 3 kùš.ta<.àm> za.na *iš-te-en im-taḫ-ru* /	The reciprocal of 6, the constant resolve, then 10. Steps of 1 30, the front of the city wall, (is) 15. 3 cubits each (the sides of) one gaming-piece are equal.
5	*šum-ma* 3 kùš.ta.àm za.na *im-ta-aḫ-ru* saḫar! *mi-nu* *ba-ma-at* 15 sag *ḫe-pe-ma* 7 30 /	If 3 cubits each (the sides of) a gaming-piece are equal, the volume (is) what? Half (of) 15, the front, break, then 7 30.
6	7 30 a.rá 15 sag *ša-ni-tim* 1 52 30 *ba-am-ta* 14 03 45 *sa-am-na-ti-šu* zi-*ma* /	7 30 steps of 15, the second front, 1 52 30, the halved. 14 03 45, its eighth tear off, then
7	1 38 26 15 *qá-qá-ar* gán.za.na *iš-te-en* *ša ta-mar* gán.za.na.meš *ki ma-ṣi* /	1 38 26 15 (is) the ground (of) one gaming-piece-field that you see. The gaming-piece-fields how much (how many)?
8	*i-na* 6 *igi-gub-e* 1 zi-*ma* 5 *si-ì-tam* *a-na* 4ᵥ tab.ba-*ma* 20 20 gán.za.na.meš /	From 6, the constant, 1 tear off, then 5 (is) the remainder. To 4 repeat (it), then 20. 20 gaming-piece-fields.
9	1 38 26 15 *a-na* 20 *e-ṣi-im-ma* 32 48 45 [2'] šar 2 3" gín 26 še igi.4ᵥ.gál še /	1 38 26 15 to 20 repeat, then 32 48 45, *1/2* šar 2 2/3 gín 26 1/4 barley-corns.
10	*a-na* 1 gán.si *ši-pi-e* urudu-*šu mi-nu* 32 48 45 a.rá 2 1 05 37 30 /	For 1 metal-covered horn-figure, what is its copper? 32 48 45 times 2 (is)1 05 37 30,
11	1 gín 16 2' še igi.4ᵥ.gál še *ù mi-ši-il* igi.4ᵥ.gál še saḫar.bi 1 05 37 30 saḫar.bi /	2 gín 16 1/2 1/4 barley-corns and half 1/4 barley-corns (is) its mud (volume). 1 05 37 30, its volume,
12	a.rá 1 12 igi.gub urudu 1 18 45 urudu *ki-i ša* saḫar.bi 1 kùš.ta.àm íb.si₈ [x x] /	steps of 1 12, the constant of the copper, (is) 1 18 45, the copper. Instead of its mud, 1 cubit each the square side *x x*
13	*a-na* 1 gán.si *ši-pi-e* 1 g gú urudu *i-na* [x x x x x x x x x] /	for 1 metal-covered horn-figure, 1 talent, the copper in *x x x x x x x x x,*
14	*i-na* urudu *an-ni-im* [x x x x x x x]	in this copper *x x x x x x x x.*

In the imperfectly preserved first line of this exercise, the following two new terms appear:

bàd   city wall   gán.si   horn-figure   (line 1)

An amount of copper seems to be associated with these entities. Next, an unspecified number of gaming-piece-figures are said to be situated 'at the tip' of a circle (or a sphere?), and you are required to find the 'ground', presumably meaning the areas of these gaming-piece-figures and add them, in order to find the copper associated with one 'horn-figure'. All this is extremely bewildering, as long as it is not clear what is meant by a 'city wall', or a 'horn-figure', or the associated copper. At this stage of the game, one can only hope that the remainder of the exercise will manage to clarify the situation.

The solution procedure begins with the computation of the side *s* of a gaming-piece-figure:

1;30 n. is the 'front' of the 'city wall', and 6 is the given constant. Therefore,

1/6 · 1;30 n. = ;15 n. = 3 cubits = the common length *s* of the sides of the gaming-piece-figure   (line 4)

The next step is the computation of the area A_g of a gaming-piece-figure with the side 3 cubits:

If   *s* = ;15 n., then *s*/2 · *s* = ;07 30 n. · ;15 n. = ;01 52 30 sq. n.

$$A_g = (1 - 1/8) \cdot s/2 \cdot s = {;01\ 52\ 30}\ \text{sq. n.} - {;00\ 14\ 03\ 45}\ \text{sq. n.} = {;01\ 38\ 26\ 15}\ \text{sq. n.}\text{(lines 5-7)}$$

(This is a reiteration of the corresponding computation in # 1.) The third step of the solution procedure is the computation of a number n (another reiteration of a computation in # 1):

$$6 = \text{the constant of a gaming-piece-figure. Therefore, } n = (6 - 1) \cdot 4 = 5 \cdot 4 = 20 \qquad \text{(line 8)}$$

The area A_h of a horn-figure is then computed as n times the area of a gaming-piece-figure:

$$A_h = 20 \cdot {;01\ 38\ 26\ 15}\ \text{sq. n.} = {;32\ 48\ 45}\ \text{sq. n.} \qquad \text{(line 9)}$$

or, in traditional notations for area measure,

$$A_h = 1/2\ \text{šar } 2\ 2/3\ \text{shekels } 26\ 1/4\ \text{barley-corns} \qquad \text{(line 9)}$$

Now, it is silently assumed that the thickness of the horn-figure is '2', obviously meaning ;02 cubit = 1 finger. Therefore, the volume V_h of a horn-figure can be computed as

$$V_h = {;32\ 48\ 45}\ \text{sq. n.} \cdot {;02}\ \text{c.} = {;01\ 05\ 37\ 30}\ \text{sq. n.} \cdot \text{c.,} \qquad \text{(line 10)}$$

or, in traditional notations for volume measure,

$$V_h = 1\ \text{shekel } 16\ 1/2\ 1/4\ \text{and } 1/2\ \text{of } 1/4\ \text{barley-corns} \qquad \text{(line 11)}$$

The final step of the solution procedure is the computation of the weight W_h of the horn-figure:

$$W_h = {;01\ 05\ 37\ 30}\ \text{volume-shekels} \cdot 1\ 12\ \text{talents/volume-shekel} = 1;18\ 45\ \text{talents} \qquad \text{(line 12)}$$

11.3 b.   Constants for Copper and Silver in Mathematical Cuneiform Texts

The last step of the solution procedure in MS 3876 # 3 is explained in lines 12-14. Although those lines are badly preserved, it is clear that they mention a square of side 1 cubit, and a weight of 1 talent, both in connection with the constant 1 12 for copper. The interpretation that '1 12' stands for '1 12 talents per volume-shekel' is supported by the following calculation:

$$1\ \text{volume-shekel} = 1/60\ \text{sq. n. c.} = 1/60 \cdot 12 \cdot 12\ \text{sq. cubits} \cdot 30\ \text{fingers} = 1\ 12\ \text{sq. cubits} \cdot 1\ \text{finger.}$$

Therefore, the meaning of the damaged passage in lines 12-14 appears to be that

$$1\ 12\ \text{talents/volume-shekel} = 1\ \text{talent/(1 sq.cubit} \cdot 1\ \text{finger)} \qquad \text{(lines 13-14)}$$

Actually, '1 12' appears in four Babylonian tables of constants as a constant for copper:

1 12	ra-ṭù-um	ša	urudu	tube of copper	NSd 22
1 12	ra-ṭù-um	ša	urudu	tube of copper	NSe 36
1 12		šà	urudu	of copper	BR 54
1 12	igi.gub		urudu	constant, copper	Ka 18

It has not before been known what this constant stands for, but now it is clear that it means that

1 12 talents per volume-shekel is the density of copper, corresponding to
1 talent being the weight of a square sheet of copper with the side 1 cubit and the thickness 1 finger.[7]

This is a "unitary relation" of a kind that is a well known and conspicuous feature of Babylonian, Sumerian, and even proto-Sumerian, metrology.[8]

In modern terminology, this corresponds to a density for copper of about 7.2 grams per cubic centimeter, that is, 7.2 times the density of water. (The correct modern value is 8.94.)

The mentioned constants for copper in four Babylonian tables of constants can be compared with the fol-

---

7. Compare with 12 being the density of baked bricks and 1 talent being the weight of a baked square brick with the side 1 cubit and the thickness 6 fingers, or a sun-dried square brick with the side 1 cubit and the thickness 5 fingers, or a freshly made square brick with the side 1 cubit and the thickness 4 fingers. (See the discussion of MS 2221 in § 7.3 a above.)

8. Cf. Friberg, *BagM* 28 (1997), 314, fn. 70: a Late Babylonian seeding rate of 1 barig of seed barley on an area of sq. (100 cubits), or Friberg, *JCS* 51 (1999), 114, fn. 15: M2 (2 · 60) proto-literate daily food rations = 1 d of crushed barley; (*ibid.*) 123: a proto-literate seeding rate of 1 c of barley on the area 1 iku; (*ibid.*) 133: an Old Babylonian wage rate of 1 shekel of silver or 1 gur of barley per month, and a proto-Sumerian wage rate of 1 ce of silver or 1 M of barley per day.

lowing constants in the same four tables of constants:

4	*ru-uq-qu*	*ša*	kù.babbar	foil of silver	NSd 26
26 40	*ru-qú-um*	*ša*	kù.babbar	foil of silver	NSe 51
1 36		*šà*	kù.babbar	of silver	BR 58
1 36		*ša*	kù.babbar	of silver	Ka 21
1 30	*ra-ṭù-um*	*ša*	kù.babbar	tube of silver	NSd 23

(The term *ruqqum* 'metal foil' is derived from the verb *raqāqu* D 'to roll out thin'.)

The meaning of some of these constants can be deduced from the following problem text:

YBC 4669 *rev.* # 6 (Neugebauer and Sachs, *MCT* (1945), 138; Neugebauer, *MKT 3* (1937), pl. 3)

1-2	3 kùš 1 šu.si uš / 2 kùš 6 šu.si sag /	3 cubits 1 finger the length, 2 cubits 6 fingers the front.
3-4	gagar.bi en.nam / 2 3" gín 20 2' še /	Its ground (is) what? 2 2/3 shekels 20 1/2 barley-corns.
5	[4] gín kù.babbar sì-*ma* /	*4 shekels of silver is given, then*
6	3 šu.si.ta íb.si₈ /	3 fingers each square side
7	*ru-uq-qá-am im-ḫa-ṣú* /	(is) the foil they have beaten.
8-9	*i-na-an-na* / kù.babbar en.nam sì-*ma* /	*Now, (of) silver what is given, then*
10-11	2 3" gín 20 2' še gagar / ḫé.gar.ra	2 2/3 shekels 20 1/2 barley-corns of ground may be set?

This text begins (in lines 1-2) by mentioning a thin silver foil beaten out into a rectangle with the sides 3 cubits 1 finger (= ;15 10 n.) and 2 cubits 6 fingers (= ;11 n.), hence with the area

$$;15 \ 10 \ \text{n.} \cdot ;11 \ \text{n.} = ;02 \ 46 \ 50 \ \text{sq. n.} = 2 \ 2/3 \ \text{shekels} \ 20 \ 1/2 \ \text{barley-corns.}$$

Next, it mentions that 4 shekels of silver[9] can be beaten out into a square silver foil of side 3 fingers, and goes on to ask how much silver is needed for the rectangular silver foil of area 2 2/3 shekels 20 1/2 barley-corns. No answer is given in the text.

Since 4/9 = ;26 40, the information given in lines 5-6 of this text, that 4 shekels is the weight of a square silver foil of side 3 fingers, can be reformulated as a rule of the following form:

the weight *per area unit* of a silver foil is 4 shekels/(sq. (3 fingers)) or ;26 40 shekel/(sq. finger).

This is probably the meaning of the two constants 4 and 26 40 in NSd 26 and NSe 51.

Note that the thickness of the silver foil is never mentioned. Apparently, however, the silently understood *normalized thickness of a silver foil* was

1/12 finger = 1/2 barley-corn = 1 thread (= c. 1.4 mm).

(Compare with the normalized thickness of a copper sheet, which according to MS 3876 # 3 and the *rāṭum* numbers in the tables of constants seems to have been 1 finger (c. 16.7 mm).)

The small length units še 'barley-corn' = 1/6 'finger' and gu (Akk. *qû*) 'thread' = 1/2 barley-corn are mentioned in the initial section for small length measures in the large Late Babylonian combined metrological table W 23281 (Friberg, *GMS 3* (1993), 401).

Now, consider again the rule that the weight per area unit of a silver foil is ;26 40 shekel per square finger. Since 1 ninda = 12 · 30 fingers = 6 00 fingers, and 1 square ninda = 36 00 00 square fingers, that rule can be reformulated as follows:

the weight *per area unit* of a silver foil is 36 00 00 · ;26 40 shekels/sq. n. = 16 minas/area-shekel

With an assumed normalized thickness for a silver foil of 1/12 finger = 1 cubit / 6 00, the rule can again be reformulated, this time as

the weight *per volume unit* of a silver foil is 1 36 00 talents/(sq. n. · c.) = 1 36 talents/volume-shekel.

This is clearly the explanation for the constant 1 36 in BR 58 and Ka 21 (see above). In modern terminology,

---

9. Neugebauer and Sachs, who did not understand this text, erroneously read 1(?) gín instead of [4] gín in line 5. The mistake was repeated (without the question mark) in Robson, *MMTC* (1999), 126.

1 36 talents/volume-shekel corresponds to a density for silver of about 9.6 grams per cubic centimeter, that is, 9.6 times the density of water. (The correct modern value is 10.5.)[10]

It is now also easy to find also the answer to the question posed in YBC 4669 *rev.* # 6. It is

The amount of silver required to make a rectangular foil with the sides 3 c. 1 f. (c. 1.32 m) and 2 c. 6 f. (c. 1.1 m)
is 2;46 50 area-shekels · 16 minas-area-shekel = 44;29 20 minas = 44 1/3 minas 9 1/3 shekels of silver (c. 22 kg).

11.3 c. A 'Horn-Figure' Consisting of 20 Equilateral Triangles

It still remains to explain the purpose of the whole text MS 3876. In particular, what is meant by the terms bàd 'city wall' and gán.si 'horn-field' or 'horn-figure', and why are there 20 equilateral triangles in a horn-figure, with the number 20 explained as $(6 - 1) \cdot 4$?

The only explanation that comes to mind, however unlikely, is that the horn-figure (meaning a figure with many protruding "horns"?) might be an *icosahedron*, one of the five regular solids.

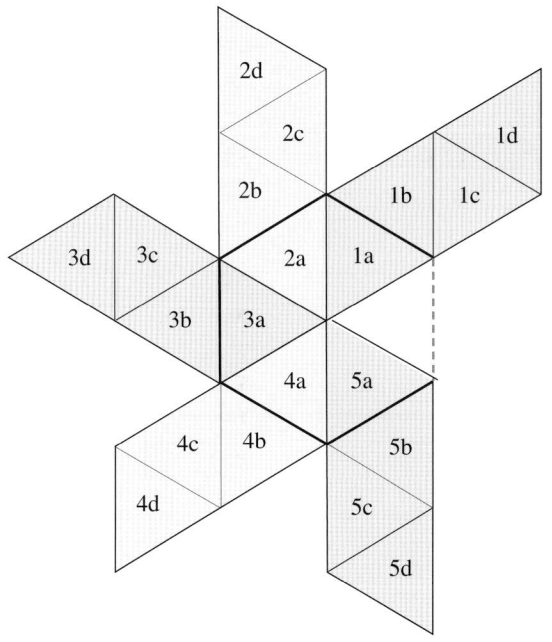

The number of equilateral triangles
in this figure is $n = (6 - 1) \cdot 4 = 20$.

Fig. 11.3.3. An unfolded horn-figure, consisting of 5 arms with 4 gaming-piece-figures in each arm.

In the light of this assumption, the various obscure terms and passages in MS 3876 # 3 can be interpreted as follows: The construction of the horn-figure starts with the 'city wall', a '6-front' (regular hexagon), divided by three transversals (diagonals) into 6 gaming-piece-figures (equilateral triangles). (See the central hexagon in Fig. 11.3.3, and compare with the figure on the obverse of *TMS* 2, Fig. 8.2.15, left.) This, then, is the meaning of the constant '6' (lines 4 and 8 of MS 3876 # 3). Assuming that the 'front' of the city wall is 1 1/2 ninda, the sides of the gaming-piece-figures are all equal to 1/6 · 1;30 n. = ;15 n. = 3 cubits (line 4).

---

10. In BR 54 and BR 57-59 are listed the constants 1 12, 124, 1 36, 1 48 for copper, tin (an.na), silver, and gold (kù.gi). If the constant 1 48 for gold in this and several other Babylonian tables of constants is interpreted in the same way as the constants 1 12 and 1 36 for copper and silver, the result seems to suggest that the density for gold was thought to be 9/8 times the density of silver. This must be an error, since the correct value for the density of gold is 19.32 grams per cubic centimeter, about 9/5 times the density of silver. The reason for the error may be that the Babylonian author of the original table of constants including these entries did not posses any actual samples of the four metals, so he could only guess how heavy they are compared to each other. Apparently, his guess was that the densities of copper, tin, silver, and gold form an arithmetical progression. (This observation is due to Robson, *MMTC* (1999), 128.) Indeed, 1 12, 1 24, 1 36, 1 48 = 12 · (6, 7, 8, 9). The constants 1 12, 1 30, 1 48 for copper, silver, and gold in the table of constants NSd can be explained in a similar way, since 1 12, 1 30, 1 48 = 18 · (4, 5, 6).

Line 7 ends by asking how many gaming-piece-figures there are (in a horn-figure). The answer, that there are (6 –1) · 4 = 20 is explained by the figure shown in Fig. 11.3.3 above. The central part of this figure is the city wall (the regular hexagon) with one sixth (one of the six gaming-piece-figures) removed. To each one of the five remaining gaming-piece-figures is then appended three more gaming-piece figures in the way shown in Fig.11.3.4. The resulting figure, built from 5 · 4 = 20 gaming-piece-figures can be folded into a horn-figure (an icosahedron). In Fig. 11.3.4, left, a front view of a horn-figure shows gaming-piece-figures belonging to arms '1', '2', and '5'.

If this interpretation of MS 3876 is correct, then what is computed in MS 3876 # 3 is the area and the weight of the outer shell of an icosahedron, with that outer shell consisting of 20 finger-thick copper sheets in the form of equilateral triangles, each with a side of 3 cubits. The fact that an icosahedron is a round object like a sphere may be what is alluded to in the cryptic phrase in line 2: *i-na ap-pi* gúr gán.za.na.meš *ta-mar* 'at the tip (*lit*. nose), an arc of gaming-piece-fields you see', or possibly instead 'at the tip of the arc, the gaming-piece-fields you see'.

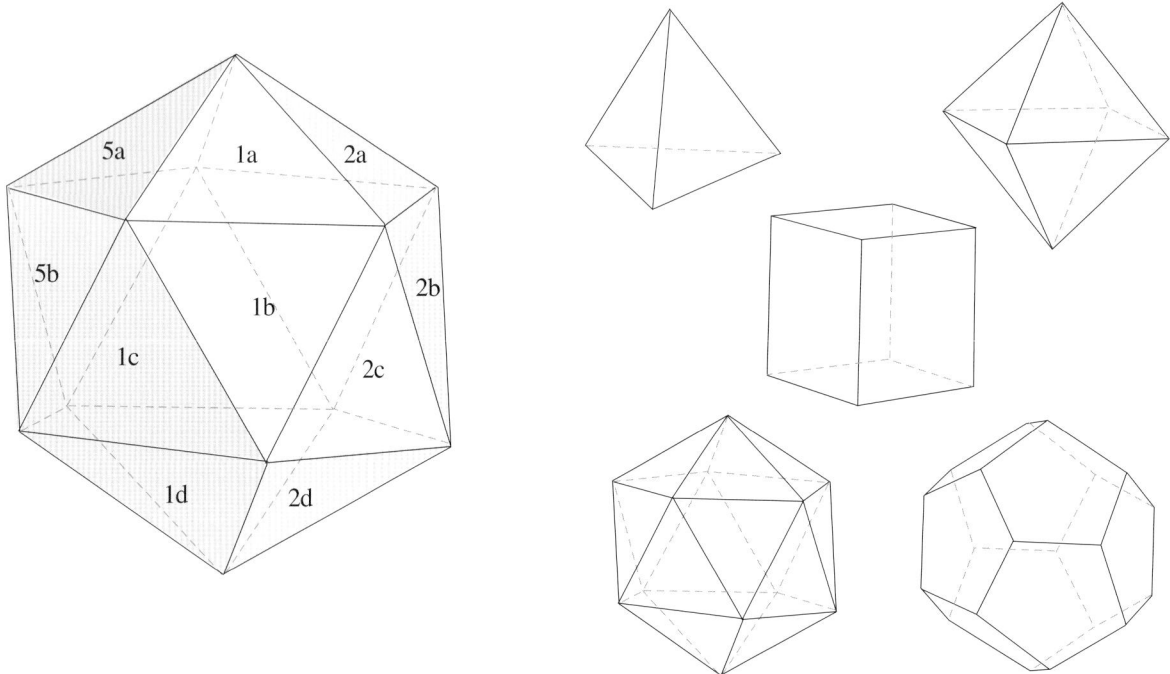

Fig. 11.3.4. Left: Ten of the twenty faces of a "horn-figure". Right: The five regular solids.

It may seem too bold to suggest that the Old Babylonians (and their Kassite successors) were familiar with the concept of an icosahedron. However, a moment's reflection leads to the conclusion that this is not so surprising, after all. Indeed, it is by now well known that Old Babylonian mathematicians were familiar with the concept of pyramids, at least pyramids with a square or rectangular base. (See the discussion of the large Old Babylonian mathematical recombination text BM 96954 + BM 102366 + SÉ 93 in Friberg, *PCHM* 6 (1996) and in Robson, *MMTC* (1999), App. 3.) They were also familiar with the concept of equilateral triangles, as shown, in particular, by the entry in the table of constants G *rev*. 7', which refers to a triangle 'with an eighth torn out', and with the constant '26 15'. It will not have been difficult for them to put these concepts together and start experimenting with pyramids with an equilateral triangle as a base. In that way they may have discovered the *tetrahedron*, a regular solid bounded by four equal equilateral triangles. This is the simplest of the five regular solids. (See Fig. 11.3.4, right.) It is even likely that Old Babylonian mathematicians could compute both the surface area and the volume of a tetrahedron. In particular, to find the volume of a tetrahedron, they would first have to find the radius of the circle circumscribed around the base of the tetrahedron. This they could do as in the related example treated in *TMS* 1, *obv*. (Fig. 8.2.6 above). Then they could compute the height of the tetrahedron by use of the three-dimensional diagonal rule (cf. Fig. 11.1.5 above). Finally, they could compute the volume of the tetrahedron by guessing (or showing; cf. the discussion in Friberg, *PCHM* 6 (1996)) that the rule for the computation of the volume of a square pyramid can be generalized to a similar rule for the volume of a tetrahedron.

Joining two square pyramids together at their bases, some Old Babylonian mathematician may also have discovered the *octahedron*, and computed both its surface area and its volume. Although no cuneiform mathematical texts are known mentioning either a tetrahedron or an octahedron, this circumstance does not mean much, in view of the fact that it was not known either until recently, and hardly even suspected, that the Old Babylonians were familiar with and could compute the volumes of both square pyramids and ridge pyramids with a rectangular base.

The *cube*, the third of the five regular solids, was also known, although its importance may have been overlooked in Old Babylonian mathematics where different length units were used for horizontal and vertical line segments. There are three known examples of a cube occurring in an Old Babylonian mathematical texts. One of the examples is a brief question in **UET 5, 289**, one of the mathematical texts from Ur, where the bottom area and volume of a cube are given (see below). The intended solution procedure was probably to divide the volume by the bottom area in order to find the depth: 40;30 n. n. c. / 2;15 n. n. = 18 c. The length and the front would then both be 1 1/2 ninda.

UET 5, 829 § 2 (Friberg, *RA* 94 (2000), 144)

1	2 šar 15 gín a.šà /	2 šar 15 shekels the field,
2	40 2' šar saḫar	40 1/2 šar the mud.
	uš sag *ù* bùr.bi íb.si₈ /	Length, front, and depth are equal.
3	uš sag *ù* bùr.bi en.nam	Length, front, and depth are what?

Another Old Babylonian example of a cube occurring in an Old Babylonian mathematical text is **IM 54478**, where the question is like this:

IM 54478 (Baqir, *Sumer* 7 (1951); Proust, *TMN* (2004), 220)

1	*šum-ma ki-a-am i-ša-al-ka um-ma šu-ú-m*a /	If someone asks you, saying thus:
2	*ma-la uš-ta-am-ḫi-ru ù-ša-pí-il-ma* /	As much as I made square-sided, I dug down, then
	mu-ša-ar ù zu-uz₄ mu-ša-ri / *e-pé-ri a-su-uḫ*	a šar and half a šar of soil I tore out.
3	*ki-ia uš uš-tam-ḫir* / *ki-ma-ṣí ù-ša-pí-il*	What did I make equalsided, and how much did I tear out?

The solution procedure given in the text is explained by Proust as follows: The given volume is compared to that of a *unit cube* with the horizontal and vertical sides equal to 1 ninda, 1 ninda, and 1 cubit, respectively. Since the volume of the unit cube is 12 volume-šar, the ratio of the given volume to the volume of the unit cube is 1;30 · igi 12 = ;07 30. The cube side of ;07 30 is ;30. Consequently, the given cube has the sides ;30 · 1 ninda = 1/2 ninda, ;30 · 1 ninda = 1/2 ninda, and ;30 · 12 cubits = 6 cubits.

A Seleucid mathematical recombination text from the late first millennium BC contains a series of exercises concerning cubes. They are all of the following type:

AO 6484 § 6 a (*MKT 1* (1935), 97)

1	1 kùš uš 1 kùš sag.ki 1 kùš sukud [...] /	1 cubit the length, 1 cubit the front, 1 cubit the height.
2	5 kùš *na-si-ik-ti* • 5 kùš *na-si-ik-t[i* rá-*ma* 25 :]	5, 1 cubit, the prone · 5, 1 cubit, the prone, go then 25.
3	[25 • 1] / [*z*]*a-qip-ti* rà-*ma* 25 :	25 · 1, the upright, go then 25.
	25 • 6 i.du[b rá-*ma* 2 30]	25 · 6, the storing (number) go then 2 30

What this means is that a cube with all sides equal to 1 cubit has the bottom area ;00 25 sq. n and the volume ;00 25 sq. n. c. = ;00 25 volume-šar. The "grain measure" of the cube is then computed as

$$;00\ 25\ \text{volume-}\check{s}\text{ar} \cdot 6\ 00\ \text{sìla/volume-}\check{s}\text{ar} = 2\ 30\ (150)\ \text{sìla (with the storing number '6').}$$

In the four other exercises in the same paragraph of AO 6484 are computed the grain measures of four cubes with their sides equal to 2, 3, 4, and 5 cubits, respectively.

How likely is it then that the Old Babylonian mathematicians discovered also the icosahedron? The picture of an icosahedron in Fig. 11.3.4 shows that the top of an icosahedron has the form of a pyramid with a regular pentagon as a base. That Old Babylonian mathematicians knew about the regular pentagon, which they called a '5-front' is shown by the following entry from a table of constants:

| 1 40 | igi.gub | *šà* | sag.5 | 140 | the constant of a 5-front | BR 26 |

(Cf. the discussion of *TMS* 2 (Fig. 8.2.15), and BR 27-28, the constants for the 5-front, the 6-front, and the 7-front in Sec. 8.2 e above. See also Fig. A6.20 in App. 6, which shows a marble plaque from Babylon with five entangled heros joined into a pentagram and enclosing a regular pentagon.) In view of the Old Babylonian mathematicians' habit to make the most of every idea they had by finding variations of it of all possible kinds, it is likely that if they were familiar with pyramids and with regular pentagons, then they would also consider pyramids with a pentagon as their base. The step from there to finding the icosahedron is not very great, in view of the relatively simple structure of the icosahedron. (Compare with the dodecahedron which cannot be found in this simple way, since the top of a dodecahedron does not have the form of a pyramid.)

The constant for the 5-front was, apparently, computed under the simplifying assumption that if the circumference of the 5-front is 5(00), then the circumference of the circumscribed circle is also approximately 5(00), so that the radius of that circle is approximately 50. Consequently, it follows from the diagonal rule that the 5-front is composed of 5 triangles, each with the base 1(00) and the height (approximately) 40, so that the area of a 5-front with the side 1(00) is approximately $5 \cdot 30 \cdot 40 = 1\ 40(00)$. This naive argument can be contrasted with Euclid's *Elements* XIII:11, a proposition stating that in a circle with rational diameter an inscribed regular pentagon has a side that is an irrational line segment of the type called 'minor'. In modern terminology, this means, more precisely, that if s the side of the regular pentagon and r the radius of the circumscribed circle, then

$$s = r/2 \cdot \text{sqs.}\ (10 - 2\ \text{sqs.}\ 5), \quad \text{or, alternatively,} \quad s = r/2 \cdot [\text{sqs.}\ (5 + 2\ \text{sqs.}\ 5) - \text{sqs.}\ (5 - 2\ \text{sqs.}\ 5)].$$

See Heath, *Euclid* (1956), vol. 3, 466. It is not very likely that the Old Babylonians found anything corresponding to this precise relation between s and r. However, if they ever did find such a relation, then their standard method for approximations of square roots would have yielded the result that $s = r \cdot 1;11\ 15$, which does not differ much from the result $s = r \cdot 1;12$ yielded by the naive argument.

To find an approximate value for the volume of an icosahedron would have been even harder (although not impossible) for an Old Babylonian mathematician, in particular in view of the documented poor quality of pictures of three-dimensional objects in Old Babylonian mathematical texts. See, for instance, the four drawings of mud-walls as trapezoids or triangles in MS 3052 § 1 a-d, or the two drawings of divided ramps as divided trapezoids in MS 2792 ## 1-2.

11.4. On the Dating of the Texts in Chapter 11

The three mathematical problem texts discussed in Ch. 10 above, MS 3971, 3052, and 2792, all belong to Group 3 from Uruk, in terms of the Goetze/Høyrup/Friberg classification of unprovenanced mathematical cuneiform texts. As members of Group 3, the mentioned texts are probably "middle" Old Babylonian, dating to between 1763 BC and 1739 BC = Samsuiluna 11, the year when the southern cities Ur, Larsa, and Uruk were abandoned. Also MS 3845, the large fragment of a combined metrological table, and MS 2723, the hexagonal prism with a metrological table for system C, both with interesting subscripts, appear to be from this period.

Two of the mathematical texts discussed in the present chapter, the metric algebra text MS 5112 and the icosahedron text MS 3896, can be dated to the Late Kassite period, from the 14th to 13th century BC, as indicated by the form of the cuneiform signs, the vocabulary, and the grammar. They are probably imitations of Old Babylonian mathematical texts, just as the talent weight MS 4576 appears to be a Kassite imitation of a Sumerian weight stone. As for the provenance of these texts, Al-Rawi suggests that they were looted from

Tell Muhammad, in Baghdad Al-Jadidah or, perhaps, from Haddad, ancient Meturan. The opinion that MS 3896 is Late Kassite agrees well with the mentioned circumstance that the format of the clay tablet is similar to the format of AO 17264, the only previously known Kassite mathematical text (Fig. 11.3.2 above).

The dating of MS 3049 is less obvious, but also in this text part of the terminology is unusual, and MS 3049 is the only mathematical text other than MS 3896 in which the phrase *kīam tepeš* 'so you do it' appears. For this reason, MS 3049 can be suspected of being a post-Old-Babylonian text. On the other hand, MS 3049 ends with a summary, and the only other known mathematical cuneiform text ending with a similar summary is MS 3052, an Old Babylonian text from Uruk.

Appendix 1
Subtractive Notations for Numbers in Mathematical Cuneiform Texts

A1.1. Hilprecht's List of Signs for '19' in Multiplication Tables from Nippur

In Fig. 2.6.2 above was displayed a selection of strange ways of writing '19' in single multiplication tables from the Schøyen Collection. It is interesting to note that a similar list of strange ways of writing '19' was compiled by Hilprecht in his pioneering work *BE 20/1* (1906), the first major publication of Old Babylonian mathematical and metrological table texts, all from Nippur. Here is a slightly amended copy of Hilprecht's list:

HS 222 a	*BE 20/1* 1	Ni. IX	2 ×	type a	
CBM 6063	4		18 ×	type a	
HS 214 a	5		18 ×	type a	
Ist. Ni. 1143	9		1 40 ×	type a	
Ist. Ni. 927	10		2 30 ×	type a	
CBM 10190	11		2 30 ×	type a	
HS 217 a	15		9 ×	type b	
CBM 10219	17		cmt	type b	
HS 210	19		cmt	type b	
CBM 11340	20	Ni. V	cmt	type a, b	
CBM 11368	21		cmt	type a, b	
CBM 19790	23		cmt	type b	

Fig. A1.1. Signs for '19' in Old Babylonian multiplication tables from Nippur.

A1.2. Ist. T 7375. An Ur III Table of Reciprocals with Subtractive Number Notations

It may have been a playful tradition that every Old Babylonian school boy should invent his own form of the sign for '19' when he had reached the level in his school work when he started writing mathematical table texts. Whichever the correct explanation of the phenomenon is, the fact remains that the origin of the various strange ways of writing '19' was that in early cuneiform table texts the use of subtractive formations of number signs was quite common. Well known examples are two of the earliest known tables of reciprocals, **Ist. T 7375** (Fig. A1.2 below; Delaporte, *RA* 8 (1911)) and **HS 201** (Oelsner, *ChV* (2001)), both from the Neo Sumerian Ur III period (or very early Old Babylonian). Here below is presented a hand copy of Ist. 7375, based on Delaporte's hand copy in the original publication of the text:

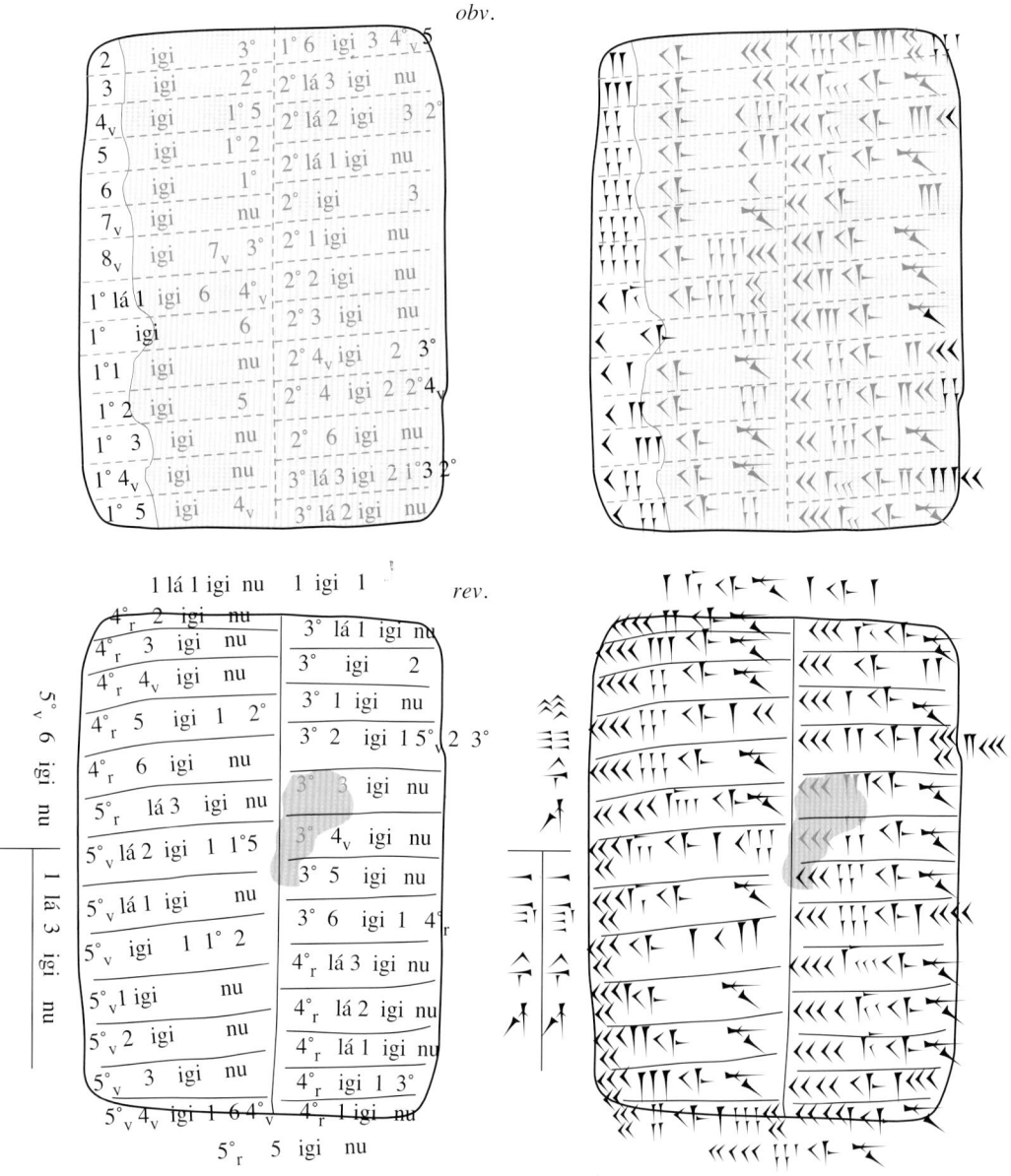

Fig. A1.2. Ist. T 7375. An Ur III table from Telloh of reciprocals for all whole numbers from 2 to 1 00.

The text on the obverse of this clay tablet is almost totally obliterated, but the text on the reverse and the edges is essentially intact. The table of reciprocals has several features which are either unique or shared only by HS 201. The most conspicuous of these features is that the table asks for the reciprocals of *all* numbers *n* between 2 and 1 (00). When *n* is non-regular, the text states that

n igi nu   *n*, the reciprocal (does) not (exist).

Variant notations are used for 4, 7, 8, 40, and 50 (4_v, 7_v, 8_v, 40_r, 50_r, 50_v). In addition, subtractive formation is used for the numbers 9, [17, 18, 19, 27, 28], 29, 37, 38, 39, 47, 48, 49, and 57, 58, 59, as shown in the list below (Fig. A1.3). (Unfortunately, the numbers 17, 18, 19, 27, 28 are not preserved in Ist. T 7375, but they are present in the parallel text HS 201.)

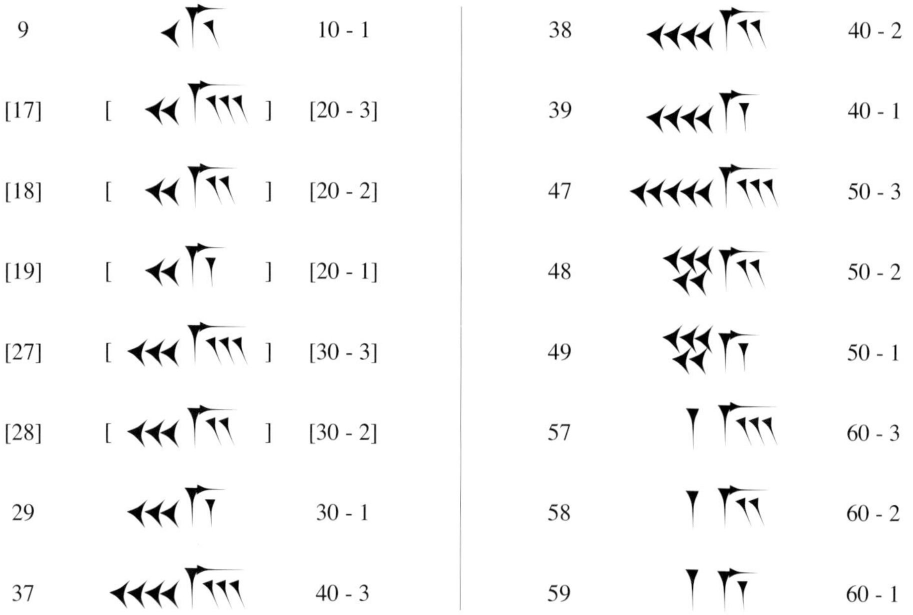

9	⟨cuneiform⟩	10 - 1		38	⟨cuneiform⟩	40 - 2
[17]	[⟨cuneiform⟩]	[20 - 3]		39	⟨cuneiform⟩	40 - 1
[18]	[⟨cuneiform⟩]	[20 - 2]		47	⟨cuneiform⟩	50 - 3
[19]	[⟨cuneiform⟩]	[20 - 1]		48	⟨cuneiform⟩	50 - 2
[27]	[⟨cuneiform⟩]	[30 - 3]		49	⟨cuneiform⟩	50 - 1
[28]	[⟨cuneiform⟩]	[30 - 2]		57	⟨cuneiform⟩	60 - 3
29	⟨cuneiform⟩	30 - 1		58	⟨cuneiform⟩	60 - 2
37	⟨cuneiform⟩	40 - 3		59	⟨cuneiform⟩	60 - 1

Fig. A1.3. Subtractive formation of number signs in the Ur III table of reciprocals Ist. T 7375.

A1.3. A 681. A Table Text from ED IIIb Adab with Subtractive Number Notations

The use of subtractive formations of number signs can be traced even further back in time.

Thus, two examples from the Old Akkadian period (c.2340-2160 BC) are the "square-side-and-area exercises" *DPA* 37 (App. 6, Fig. A6.5) and A 5446 (App. 6, Fig. A6.7). In Pul 28, the side of a square is given as 1 šár 5 géš (ninda) – 1 seed-cubit, and in A 5446 the side of another square is given as 28 šár (ninda) – 1 šu.dù.a, where 1 šu.dù.a = 1/3 cubit = 1/36 ninda.

An even older example of the use of subtractive formations of number signs is **A 681** (Luckenbill, *OIP 14*, text 70), an Old Sumerian "table of areas of small squares" from ED IIIb Adab (Fig. A1.4). (The ED III period lasted from c. 2600 to 2350 BC.) With is combination of mathematics and metrology, it can be called a metro-mathematical table text. As such it is quite unlike the Old Babylonian tables of squares of standard type, which mention only sexagesimal numbers in place value notation.

A 681 is a table of areas of small squares with sides from 1 to 11 cubits, ending with the area of a square with the side 3 reeds (1 reed = 6 cubits). The first entry in the table says that

sq. (1 cubit) = 1 sa_{10}.ma.na 15 gín.

This equation can be compared with how the area of square with the side 1 cubit would have been computed in the Old Babylonian period:

sq. (1 cubit) = sq. (;05 n.) = ;00 25 šar = ;25 shekel = 1/3 shekel 15 barley-corns.

Thus, there is the following relation between Old Sumerian and Old Babylonian shekel fractions:

$1 \, sa_{10}.ma.na = 1/3 \, gín \, (shekel) = 1/180 \, ma.na \, (mina), \quad and \quad 1 \, <sa_{10}.>gín = 1 \, še \, (barley-corn).$

A surprising explanation is given in Friberg, *JCS* 51 (1999), 132-134, namely that the Sumerian/Old Babylonian system of weight measures was invented in the Early Dynastic III period (Old Sumerian) as a combination of a system of units for weighing copper, *etc.* (talents, minas, shekels) and another system of units for weighing silver (minas, shekels). Since at that time silver was 180 times more precious than copper, the silver minas were 180 times smaller than the copper minas, and could be called "exchange minas" (sa_{10} is a Sumerian verb with the meaning 'to exchange, to buy'). Soon, the shekels, the silver minas and the silver shekels came to be interpreted as *small fractions in general*. (Ultimately, the 'silver shekel' was replaced by the še 'barley-corn' = 1/180 shekel.)

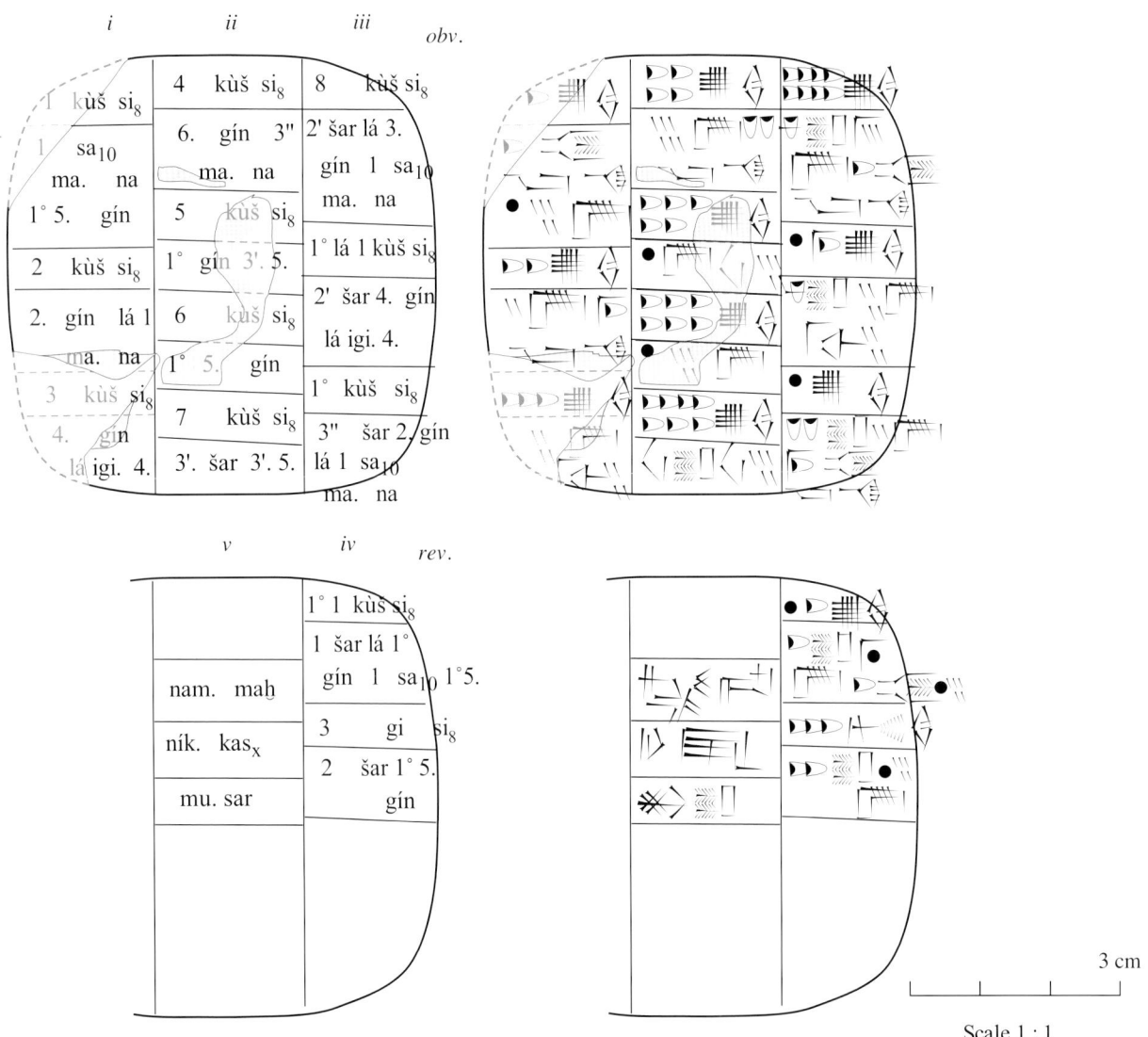

Fig. A1.4. A 681 (*OIP 14*, 70). An Old Sumerian table of areas of small squares, from Adab. (The computer-aided reproduction here of the text is based on Luckenbill's hand copy.)

The actual computation of the entries in the table A 681 is quite complicated. It must have begun with the computation of the first entry, perhaps in the following way:

$$
\begin{aligned}
1 \, sq. \, reed &= 1/4 \, šar = 15 \, gín = 45 \, sa_{10}.ma.na \\
1 \, reed \cdot 1 \, cubit &= 1 \, sq. \, reed / 6 = 2 \, 1/2 \, gín = 7 \, 1/2 \, sa_{10}.ma.na \\
1 \, sq. \, cubit &= 1 \, reed \cdot 1 \, cubit / 6 = 1 \, 1/4 \, sa_{10}.ma.na = 1 \, sa_{10} \, ma.na \, 15 \, <sa_{10}.>gín.
\end{aligned}
$$

With departure from this result, all the entries in the table could be computed as follows:

sq. 1 cubit	=	1 1/4 sa_{10}.ma.na	=	1 sa_{10} ma.na 15. $<sa_{10}.>$gín
sq. 2 cubits	=	5 sa_{10}.ma.na	=	2. gín – 1 $<sa_{10}.>$ma.na
sq. 3 cubits	=	11 1/4 sa_{10}.ma.na	=	4. gín – igi.4. $<$gín$>$
sq. 4 cubits	=	20 sa_{10}.ma.na	=	6. gín 3" $<sa_{10}.>$ma.na
sq. 5 cubits	=	31 1/4 sa_{10}.ma.na	=	10 gín 3'. $<$gín$>$ 5.
sq. 6 cubits	=	45 sa_{10}.ma.na	=	15. gín
sq. 7 cubits	=	1(géš) 1 1/4 sa_{10}.ma.na	=	3'. šar 3'. $<$gín$>$ 5.
sq. 8 cubits	=	1(géš) 20 sa_{10}.ma.na	=	2' šar – (3 gín 1 sa_{10}.ma.na)
sq. 10 – 1 cubits	=	1(géš) 41 1/4 sa_{10}.ma.na	=	2' šar 4. gín – igi.4. $<$gín$>$
sq. 10 cubits	=	2(géš) 5 sa_{10}.ma.na	=	3" šar 2. gín – 1 sa_{10}.ma.na
sq. 11 cubits	=	2(géš) 31 1/4 sa_{10}.ma.na	=	(1 šar – 10 gín) 1 sa_{10}.$<$ma.na$>$ 15 $<sa_{10}$.gín$>$
$<$sq. 12 cubits	=	sq. 2 reeds = sq. 1 ninda	=	1 šar$>$
sq. 3 reeds	=	sq. (18 cubits)	=	2 šar 15. gín

The following entries in A 681 (Fig. A1.5) contain subtractively formed numbers:

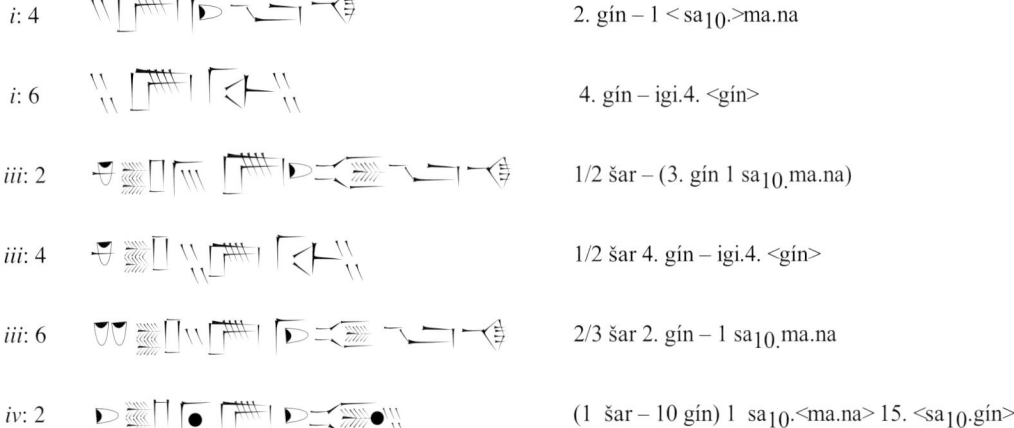

i: 4		2. gín – 1 $<sa_{10}.>$ma.na
i: 6		4. gín – igi.4. $<$gín$>$
iii: 2		1/2 šar – (3. gín 1 sa_{10}.ma.na)
iii: 4		1/2 šar 4. gín – igi.4. $<$gín$>$
iii: 6		2/3 šar 2. gín – 1 sa_{10}.ma.na
iv: 2		(1 šar – 10 gín) 1 sa_{10}.$<$ma.na$>$ 15. $<sa_{10}$.gín$>$

Fig. A1.5. Subtractively formed area numbers in the table of areas of small squares A 681.

It is interesting to note that both *curviform* (round) and *cuneiform* (sharp) numbers occur in this text. Multiples of the cubit (kùš), the šar and the sa_{10}.ma.na are written with the normal Sumerian 'ones', which are round and facing to the left, while multiples and fractions of the shekel (gín) are written with 'ones' that are slanting and cuneiform.

The text is interesting also because it is the earliest known text in which fraction of the type igi *n* (= 1/*n*) appear.[1] There are also early forms of the signs for the "basic fractions" 3' (= 1/3), 2' (= 1/2), and 3" (= 2/3). the signs for 2' and 3" are round (a rotated and crossed-over round 'one', and two rotated round 'ones', respectively), while the sign for 3' is sharp. (The corresponding round form of the early sign for 3' is the sign šú, two wedges forming a corner, followed by a rotated round 'one'.)

An additional interesting feature is that this text contains what seems to be a precursor of place value notation. Indeed, note how 1/4 sa_{10}.ma.na can expressed in three different ways in A 681:

1/4 sa_{10}.ma.na = 15. gín	(lines 1, 11)
2 1/4 sa_{10}.ma.na = 1. gín – igi.4.	(lines 3, 9)
1/4 sa_{10}.ma.na = 5.	(lines 5, 7)

In lines 1 and 11 (expressing the squares of 1 and 11 cubits, respectively), 15. gín stands for 15/60 of 1 sa_{10}.ma.na, which means that, apparently, gín here is a brief form for sa_{10}.gín. In lines 3 and 9, 2 1/4

---

1. Another text of a similar type will soon be published (D. Owen, *personal communication*). The new text is much larger than A 681 and contains several sub-tables of which one is roughly parallel to the table on A 681.

sa$_{10}$.ma.na is obviously thought of as $(3 - 1/4 \cdot 3)$ sa$_{10}$.ma.na $= 1$ gín $- 1/4$ gín. In lines 5 and 7, finally, 1/4 is thought of as 1/12 gín, written simply as 5, meaning 5/60!

Note: An ED III text of the same type as A 681, but much more extensive, is **CUNES 50-08-001**. See App. 7 below.

Appendix 2
The Old Babylonian
Combined Multiplication Table

The table on the following pages shows the *maximal* extent of the Old Babylonian combined multiplication table, with single multiplication tables for *all* attested head numbers.

2/3 of sixty	40
half of it	30
rec. 3	20
rec. 4	15
rec. 5	12
rec. 6	10
rec. 8	7 30
rec. 9	6 40
rec. 10	6
rec. 12	5
rec. 15	4
rec. 16	3 45
rec. 18	3 20
rec. 20	3
rec. 24	2 30
rec. 25	2 24
rec. 27	2 13 20
rec. 30	2
rec. 32	1 52 30
rec. 36	1 40
rec. 40	1 30
rec. 45	1 20
rec. 48	1 15
rec. 50	1 12
rec. 54	1 06 40
rec. 1	1
rec. 1 04	56 15
rec. 1 12	50
rec. 1 15	48
rec. 1 20	45
rec. 1 21	44 26 40

50 × 1	50
× 2	1 40
× 3	2 30
× 4	3 20
× 5	4 10
× 6	5
× 7	5 50
× 8	6 40
× 9	7 30
× 10	8 20
× 11	9 10
× 12	10
× 13	10 50
× 14	11 40
× 15	12 30
× 16	13 20
× 17	14 10
× 18	15
× 19	15 50
× 20	16 40
× 30	25
× 40	33 20
× 50	41 40
50 × 50	41 40

48 × 1	48
× 2	1 36
× 3	2 24
× 4	3 12
× 5	4
× 6	4 48
× 7	5 36
× 8	6 24
× 9	7 12
× 10	8
× 11	8 48
× 12	9 36
× 13	10 24
× 14	11 12
× 15	12
× 16	12 48
× 17	13 36
× 18	14 24
× 19	15 12
× 20	16
× 30	24
× 40	32
× 50	40
48 × 48	38 24

45 × 1	45
× 2	1 30
× 3	2 15
× 4	3
× 5	3 45
× 6	4 30
× 7	5 15
× 8	6
× 9	6 45
× 10	7 30
× 11	8 15
× 12	9
× 13	9 45
× 14	10 30
× 15	11 15
× 16	12
× 17	12 45
× 18	13 30
× 19	14 15
× 20	15
× 30	22 30
× 40	30
× 50	37 30
45 × 45	33 45

44 26 40 × 1	44 26 40
× 2	1 28 53 20
× 3	2 13 20
× 4	2 57 46 40
× 5	3 42 13 20
× 6	4 26 40
× 7	5 11 06 40
× 8	5 55 33 20
× 9	6 40
× 10	7 24 26 40
× 11	8 08 53 20
× 12	8 53 20
× 13	9 37 46 40
× 14	10 22 13 20
× 15	11 06 40
× 16	11 51 06 40
× 17	12 35 33 20
× 18	13 20
× 19	14 04 26 40
× 20	14 48 53 20
× 30	22 13 20
× 40	29 37 46 40
× 50	37 02 13 20
44 26 40 × 44 26 40	32 55 18 31 06 40

40 × 1	40
× 2	1 20
× 3	2
× 4	2 40
× 5	3 20
× 6	4
× 7	4 40
× 8	5 20
× 9	6
× 10	6 40
× 11	7 20
× 12	8
× 13	8 40
× 14	9 20
× 15	10
× 16	10 40
× 17	11 20
× 18	12
× 19	12 40
× 20	13 20
× 30	20
× 40	26 40
× 50	33 20
40 × 40	26 40

36 × 1	36
× 2	1 12
× 3	1 48
× 4	2 24
× 5	3
× 6	3 36
× 7	4 12
× 8	4 48
× 9	5 24
× 10	6
× 11	6 36
× 12	7 12
× 13	7 48
× 14	8 24
× 15	9
× 16	9 36
× 17	10 12
× 18	10 48
× 19	11 24
× 20	12
× 30	18
× 40	24
× 50	30
36 × 36	21 36

30 × 1	30
× 2	1
× 3	1 30
× 4	2
× 5	2 30
× 6	3
× 7	3 30
× 8	4
× 9	4 30
× 10	5
× 11	5 30
× 12	6
× 13	6 30
× 14	7
× 15	7 30
× 16	8
× 17	8 30
× 18	9
× 19	9 30
× 20	10
× 30	15
× 40	20
× 50	25
30 × 30	15

25 × 1	25
× 2	50
× 3	1 15
× 4	1 40
× 5	2 05
× 6	2 30
× 7	2 55
× 8	3 20
× 9	3 45
× 10	4 10
× 11	4 35
× 12	5
× 13	5 25
× 14	5 50
× 15	6 15
× 16	6 40
× 17	7 05
× 18	7 30
× 19	7 55
× 20	8 20
× 30	12 30
× 40	16 40
× 50	20 50
25 × 25	10 25

24 × 1	24
× 2	48
× 3	1 12
× 4	1 36
× 5	2
× 6	2 24
× 7	2 48
× 8	3 12
× 9	3 36
× 10	4
× 11	4 24
× 12	4 48
× 13	5 12
× 14	5 36
× 15	6
× 16	6 24
× 17	6 48
× 18	7 12
× 19	7 36
× 20	8
× 30	12
× 40	16
× 50	20
24 × 24	9 36

22 30 × 1	22 30
× 2	45
× 3	1 07 30
× 4	1 30
× 5	1 52 30
× 6	2 15
× 7	2 37 30
× 8	3
× 9	3 22 30
× 10	3 45
× 11	4 07 30
× 12	4 30
× 13	4 52 30
× 14	5 15
× 15	5 37 30
× 16	6
× 17	6 22 30
× 18	6 45
× 19	7 07 30
× 20	7 30
× 30	11 15
× 40	15
× 50	18 45
22 30 × 22 30	8 26 30

20 × 1	20
× 2	40
× 3	1
× 4	1 20
× 5	1 40
× 6	2
× 7	2 20
× 8	2 40

× 9	3
× 10	3 20
× 11	3 40
× 12	4
× 13	4 20
× 14	4 40
× 15	5
× 16	5 20
× 17	5 40
× 18	6
× 19	6 20
× 20	6 40
× 30	10
× 40	13 20
× 50	16 40
20 × 20	6 40
18 × 1	18
× 2	36
× 3	54
× 4	1 12
× 5	1 30
× 6	1 48
× 7	2 06
× 8	2 24
× 9	2 42
× 10	3
× 11	3 18
× 12	3 36
× 13	3 54
× 14	4 12
× 15	4 30
× 16	4 48
× 17	5 06
× 18	5 24
× 19	5 42
× 20	6
× 30	9
× 40	12
× 50	15
18 × 18	5 24
16 40 × 1	16 40
× 2	33 20
× 3	50
× 4	1 06 40
× 5	1 23 20
× 6	1 40
× 7	1 56 40
× 8	2 13 20
× 9	2 30
× 10	2 46 40
× 11	3 03 20
× 12	3 20
× 13	3 36 40
× 14	3 53 20
× 15	4 10
× 16	4 26 40
× 17	4 43 20

× 18	5
× 19	5 16 40
× 20	5 33 20
× 30	8 20
× 40	11 06 40
× 50	13 53 20
16 40 × 16 40	
	4 37 46 40
16 × 1	16
× 2	32
× 3	48
× 4	1 04
× 5	1 20
× 6	1 36
× 7	1 52
× 8	2 08
× 9	2 24
× 10	2 40
× 11	2 56
× 12	3 12
× 13	3 28
× 14	3 44
× 15	4
× 16	4 16
× 17	4 32
× 18	4 48
× 19	5 04
× 20	5 20
× 30	8
× 40	10 40
× 50	13 20
16 × 16	4 16
15 × 1	15
× 2	30
× 3	45
× 4	1
× 5	1 15
× 6	1 30
× 7	1 45
× 8	2
× 9	2 15
× 10	2 30
× 11	2 45
× 12	3
× 13	3 15
× 14	3 30
× 15	3 45
× 16	4
× 17	4 15
× 18	4 30
× 19	4 45
× 20	5
× 30	7 30
× 40	10
× 50	12 30
15 × 15	3 45

12 30 × 1	12 30
× 2	25
× 3	37 30
× 4	50
× 5	1 02 30
× 6	1 15
× 7	1 27 30
× 8	1 40
× 9	1 52 30
× 10	2 05
× 11	2 17 30
× 12	2 30
× 13	2 42 30
× 14	2 55
× 15	3 07 30
× 16	3 20
× 17	3 32 30
× 18	3 45
× 19	3 57 30
× 20	4 10
× 30	6 15
× 40	8 20
× 50	10 25
12 30 × 12 30	
	2 36 15
12 × 1	12
× 2	24
× 3	36
× 4	48
× 5	1
× 6	1 12
× 7	1 24
× 8	1 36
× 9	1 48
× 10	2
× 11	2 12
× 12	2 24
× 13	2 36
× 14	2 48
× 15	3
× 16	3 12
× 17	3 24
× 18	3 36
× 19	3 48
× 20	4
× 30	6
× 40	8
× 50	10
12 × 12	2 24
10 × 1	10
× 2	20
× 3	30
× 4	40
× 5	50
× 6	1
× 7	1 10
× 8	1 20

× 9	1 30
× 10	1 40
× 11	1 50
× 12	2
× 13	2 10
× 14	2 20
× 15	2 30
× 16	2 40
× 17	2 50
× 18	3
× 19	3 10
× 20	3 20
× 30	5
× 40	6 40
× 50	8 20
10 × 10	1 40
9 × 1	9
× 2	18
× 3	27
× 4	36
× 5	45
× 6	54
× 7	1 03
× 8	1 12
× 9	1 21
× 10	1 30
× 11	1 39
× 12	1 48
× 13	1 57
× 14	2 06
× 15	2 15
× 16	2 24
× 17	2 33
× 18	2 42
× 19	2 51
× 20	3
× 30	4 30
× 40	6
× 50	7 30
9 × 9	1 21
8 20 × 1	8 20
× 2	16 40
× 3	25
× 4	33 20
× 5	41 40
× 6	50
× 7	58 20
× 8	1 06 40
× 9	1 15
× 10	1 23 20
× 11	1 31 40
× 12	1 40
× 13	1 48 20
× 14	1 56 40
× 15	2 05
× 16	2 13 20
× 17	2 21 40

× 18	2 30
× 19	2 38 20
× 20	2 46 40
× 30	4 10
× 40	5 33 20
× 50	5 41 40
8 20 × 8 20	
	1 09 26 40
8 × 1	8
× 2	16
× 3	24
× 4	32
× 5	40
× 6	48
× 7	56
× 8	1 04
× 9	1 12
× 10	1 20
× 11	1 28
× 12	1 36
× 13	1 44
× 14	1 52
× 15	2
× 16	2 08
× 17	2 16
× 18	2 24
× 19	2 32
× 20	2 40
× 30	4
× 40	5 20
× 50	6 40
8 × 8	1 04
7 30 × 1	7 30
× 2	15
× 3	22 30
× 4	30
× 5	37 30
× 6	45
× 7	52 30
× 8	1
× 9	1 07 30
× 10	1 15
× 11	1 22 30
× 12	1 30
× 13	1 37 30
× 14	1 45
× 15	1 52 30
× 16	2
× 17	2 07 30
× 18	2 15
× 19	2 22 30
× 20	2 30
× 30	3 45
× 40	5
× 50	6 15
7 30 × 7 30	56 15

7 12 × 1	7 12
× 2	14 24
× 3	21 36
× 4	28 48
× 5	36
× 6	43 12
× 7	50 24
× 8	57 36
× 9	1 04 48
× 10	1 12
× 11	1 19 12
× 12	1 26 24
× 13	1 33 36
× 14	1 40 48
× 15	1 48
× 16	1 55 12
× 17	2 02 24
× 18	2 09 36
× 19	2 16 48
× 20	2 24
× 30	3 36
× 40	4 48
× 50	6
7 12 × 7 12	
	51 50 24
7 × 1	7
× 2	14
× 3	21
× 4	28
× 5	35
× 6	42
× 7	49
× 8	56
× 9	1 03
× 10	1 10
× 11	1 17
× 12	1 24
× 13	1 31
× 14	1 38
× 15	1 45
× 16	1 52
× 17	1 59
× 18	2 06
× 19	2 13
× 20	2 20
× 30	3 30
× 40	4 40
× 50	5 50
7 × 7	49
6 40 × 1	6 40
× 2	13 20
× 3	20
× 4	26 40
× 5	33 20
× 6	40
× 7	46 40
× 8	53 20
× 9	1
× 10	1 06 40
× 11	1 13 20
× 12	1 20
× 13	1 26 40
× 14	1 33 20
× 15	1 40
× 16	1 46 40
× 17	1 53 20
× 18	2
× 19	2 06 40
× 20	2 13 20
× 30	3 20
× 40	4 26 40
× 50	5 33 20
6 40 × 6 40	
	44 26 40
6 × 1	6
× 2	12
× 3	18
× 4	24
× 5	30
× 6	36
× 7	42
× 8	48
× 9	54
× 10	1
× 11	1 06
× 12	1 12
× 13	1 18
× 14	1 24
× 15	1 30
× 16	1 36
× 17	1 42
× 18	1 48
× 19	1 54
× 20	2
× 30	3
× 40	4
× 50	5
6 × 6	36
5 × 1	5
× 2	10
× 3	15
× 4	20
× 5	25
× 6	30
× 7	35
× 8	40
× 9	45
× 10	50
× 11	55
× 12	1
× 13	1 05
× 14	1 10
× 15	1 15
× 16	1 20
× 17	1 25
× 18	1 30
× 19	1 35
× 20	1 40
× 30	2 30
× 40	3 20
× 50	4 10
5 × 5	25
4 30 × 1	4 30
× 2	9
× 3	13 30
× 4	18
× 5	22 30
× 6	27
× 7	31 30
× 8	36
× 9	40 30
× 10	45
× 11	49 30
× 12	54
× 13	58 30
× 14	1 03
× 15	1 07 30
× 16	1 12
× 17	1 16 30
× 18	1 21
× 19	1 25 30
× 20	1 30
× 30	2 15
× 40	3
× 50	3 45
4 30 × 4 30	20 15
4 × 1	4
× 2	8
× 3	12
× 4	16
× 5	20
× 6	24
× 7	28
× 8	32
× 9	36
× 10	40
× 11	44
× 12	48
× 13	52
× 14	56
× 15	1
× 16	1 04
× 17	1 08
× 18	1 12
× 19	1 16
× 20	1 20
× 30	2
× 40	2 40
× 50	3 20
4 × 4	16
3 45 × 1	3 45
× 2	7 30
× 3	11 15
× 4	15
× 5	18 45
× 6	22 30
× 7	26 15
× 8	30
× 9	33 45
× 10	37 30
× 11	41 15
× 12	45
× 13	48 45
× 14	52 30
× 15	56 15
× 16	1
× 17	1 03 45
× 18	1 07 30
× 19	1 11 15
× 20	1 15
× 30	1 52 30
× 40	2 30
× 50	2 07 30
3 45 × 3 45	
	14 03 45
3 20 × 1	3 20
× 2	6 40
× 3	10
× 4	13 20
× 5	16 40
× 6	20
× 7	23 20
× 8	26 40
× 9	30
× 10	33 20
× 11	36 40
× 12	40
× 13	43 20
× 14	46 40
× 15	50
× 16	53 20
× 17	56 40
× 18	1
× 19	1 03 20
× 20	1 06 40
× 30	1 40
× 40	2 13 20
× 50	2 46 40
3 20 × 3 20	
	11 06 40
3 × 1	3
× 2	6
× 3	9
× 4	12
× 5	15
× 6	18
× 7	21
× 8	24
× 9	27
× 10	30
× 11	33
× 12	36
× 13	39
× 14	42
× 15	45
× 16	48
× 17	51
× 18	54
× 19	57
× 20	1
× 30	1 30
× 40	2
× 50	2 30
3 × 3	9
2 30 × 1	2 30
× 2	5
× 3	7 30
× 4	10
× 5	12 30
× 6	15
× 7	17 308
× 8	20
× 9	22 30
× 10	25
× 11	27 30
× 12	30
× 13	32 30
× 14	35
× 15	37 30
× 16	40
× 17	42 30
× 18	45
× 19	47 30
× 20	50
× 30	1 15
× 40	1 40
× 50	2 15
2 30 × 2 30	6 15
2 24 × 1	2 24
× 2	4 48
× 3	7 12
× 4	9 36
× 5	12
× 6	14 24
× 7	16 48
× 8	19 12
× 9	21 36
× 10	24
× 11	26 24
× 12	28 48
× 13	31 12
× 14	33 36
× 15	36
× 16	38 24

× 17	40 48	**1 40 × 1**	1 40	× 9	12	× 16	19 12
× 18	43 12	× 2	3 20	× 10	13 20	× 17	20 24
× 19	45 36	× 3	5	× 11	14 40	× 18	21 36
× 20	48	× 4	6 40	× 12	16	× 19	22 48
× 30	1 12	× 5	8 20	× 13	17 20	× 20	24
× 40	1 36	× 6	10	× 14	18 40	× 30	36
× 50	2	× 7	11 40	× 15	20	× 40	48
2 24 × 2 24	5 46 36	× 8	13 20	× 16	21 20	× 50	1
─────		× 9	15	× 17	22 40	1 12 × 1 12	
2 15 × 1	2 15	× 10	16 40	× 18	24		1 26 24
× 2	4 30	× 11	18 20	× 19	25 20	─────	
× 3	6 45	× 12	20	× 20	26 40		
× 4	9	× 13	21 40	× 30	40		
× 5	11 15	× 14	23 20	× 40	53 20		
× 6	13 30	× 15	25	× 50	1 06 40		
× 7	15 45	× 16	26 40	1 20 × 1 20			
× 8	18	× 17	28 20		1 46 40		
× 9	20 15	× 18	30	─────			
× 10	22 30	× 19	31 40	**1 15 × 1**	1 15		
× 11	24 45	× 20	33 20	× 2	2 30		
× 12	27	× 30	50	× 3	3 45		
× 13	29 15	× 40	1 06 40	× 4	5		
× 14	31 30	× 50	1 23 20	× 5	6 15		
× 15	33 45	1 40 × 1 40		× 6	7 30		
× 16	36		2 46 40	× 7	8 45		
× 17	38 15	─────		× 8	10		
× 18	40 30	**1 30 × 1**	1 30	× 9	11 15		
× 19	42 45	× 2	3	× 10	12 30		
× 20	45	× 3	4 30	× 11	13 45		
× 30	1 07 30	× 4	6	× 12	15		
× 40	1 30	× 5	7 30	× 13	16 15		
× 50	1 52 30	× 6	9	× 14	17 30		
2 15 × 2 15		× 7	10 30	× 15	18 45		
	5 03 45	× 8	12	× 16	20		
─────		× 9	13 30	× 17	21 15		
2 × 1	2	× 10	15	× 18	22 30		
× 2	4	× 11	16 30	× 19	23 45		
× 3	6	× 12	18	× 20	25		
× 4	8	× 13	19 30	× 30	37 30		
× 5	10	× 14	21	× 40	50		
× 6	12	× 15	22 30	× 50	1 02 30		
× 7	14	× 16	24	1 15 × 1 15			
× 8	16	× 17	25 30		1 33 45		
× 9	18	× 18	27	─────			
× 10	20	× 19	28 30	**1 12 × 1**	1 12		
× 11	22	× 20	30	× 2	2 24		
× 12	24	× 30	45	× 3	3 36		
× 13	26	× 40	1	× 4	4 48		
× 14	28	× 50	1 15	× 5	6		
× 15	30	1 30 × 1 30	2 15	× 6	7 12		
× 16	32	─────		× 7	8 24		
× 17	34	**1 20 × 1**	1 20	× 8	9 26		
× 18	36	× 2	2 40	× 9	10 48		
× 19	38	× 3	4	× 10	12		
× 20	40	× 4	5 20	× 11	13 12		
× 30	1	× 5	6 40	× 12	14 24		
× 40	1 20	× 6	8	× 13	15 36		
× 50	1 40	× 7	9 20	× 14	16 48		
2 × 2	4	× 8	10 40	× 15	18		

Appendix 3
An Old Babylonian Combined Arithmetical Algorithm

A3.1. CBS 10201. Hilprecht's Misunderstood Algorithm Text from Nippur

The first major publication of Old Babylonian mathematical and metrological texts was Hilprecht's *BE 20/1* (1906). (See Friberg, *Survey* (1982), 13-14.) In addition to the many arithmetical and metrological table texts of standard type, all from Nippur, published for the first time in that volume there were also a fragment of an interesting problem text (**CBM 12648**; Friberg, *op. cit.*, 13; Muroi, *Cent.* 31 (1989)), and a small hand tablet (**CBS 10201**; the conform transliteration in Fig. A3.1 below is based on Hilprecht's hand copy). Hilprecht, knowing nothing about the nature of Old Babylonian sexagesimal numbers in relative place value notation, mistakenly interpreted it as "Divisors of 12,960,000 and their quotients in a geometric progression". The mistake gave rise to a fanciful interpretation of 12,960,000 as the "Number of Plato" (*BE 20/1*, 29).

It was later realized (Scheil, *RA* 13 (1916), Sachs, *JCS* 1 (1947)) that the text in question was an algorithm text with a systematic computation of a series of pairs of reciprocal sexagesimal numbers.

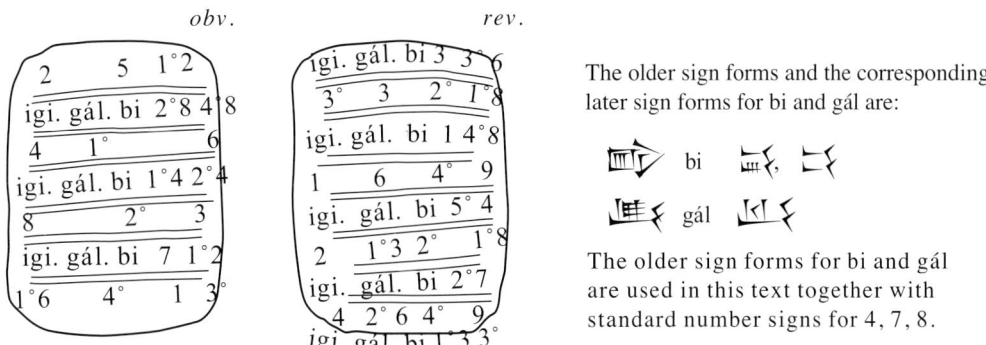

The older sign forms and the corresponding later sign forms for bi and gál are:

The older sign forms for bi and gál are used in this text together with standard number signs for 4, 7, 8.

Fig. A3.1. CBS 10201. A combined algorithm for the computation of pairs of reciprocal numbers.

Two arithmetical algorithms are combined in this brief text. The first algorithm is the Old Babylonian "doubling and halving algorithm" (Friberg, *RlA 7* (1990) Sec. 5.3b) which is based on the observation that if n, n' is a pair of reciprocals (i. e. if $n \cdot n' = $ '1' = some power of 60), then also $2 \cdot n, 1/2 \cdot n'$ is a pair of reciprocals. In CBS 10201, the first eight steps of the algorithm are recorded, with departure from the initial pair of reciprocals $n, n' = 2\ 05, 28\ 48$, where 2 05 is the cube of 5 and 28 48 the cube of 12.

The second algorithm is the "trailing part algorithm" explained in Sec. 1.4 a above. (See also Friberg, *op. cit.* Sec. 5.3b.) Thus, the numbers recorded in lines 1, 3, ..., 13, 15 can be explained as follows:

2 **05**	12 (= rec. 5)	33 **20**	18 (= rec. 3 20)
4 **10**	6 (= rec. 10)	1 06 **40**	9 (= rec. 6 40)
8 **20**	3 (= rec. 20)	2 13 **20**	18 (= rec. 3 20)
16 **40**	1 30 (= rec. 40)	4 26 **40**	9 (= rec. 6 40)

Consequently, the entries in all the odd lines in CBS 10201 are the initial lines of the application in each case of the trailing part algorithm. Hilprecht (*op. cit.*, 28) admitted his inability to explain what was going on here with the words:

> "Notwithstanding all my efforts to find a law in them or to solve the problem by the aid of competent American and European mathematicians, I have failed to get at their meaning. Suffice it to say that there seems to exist a certain relation between the first and the second number in each odd line."

A3.2. UM 29.13.21. A Fragment of a Multiple Algorithm Text from Nippur

Several related texts were published by Neugebauer and Sachs in *MCT* (1945), and by Sachs in *JCS* 1 (1947). The most extensive of those texts is the Nippur text **UM 29.13.21** (Fig. A3.2 below; Neugebauer and Sachs, *MCT* (1945), 13-15 and pl. 24). UM 29.13.21 is a small fragment of a large table text with originally five or six different examples of the application of the doubling and halving algorithm, as shown by the reconstruction of the text in Fig. A3.2 below (in a conform transliteration based on photos of the clay tablet in *MCT*). The first example in UM 29.13.21 consists of 30 steps of the doubling and halving algorithm, with the initial pair of reciprocals 2 05, 28 48. The remaining four examples consist of 10 steps each of the same algorithm, with the initial pairs (2 40, 22 30), (1 40, 36), (1 04, 56 15), and (4 03, 14 48 53 20).

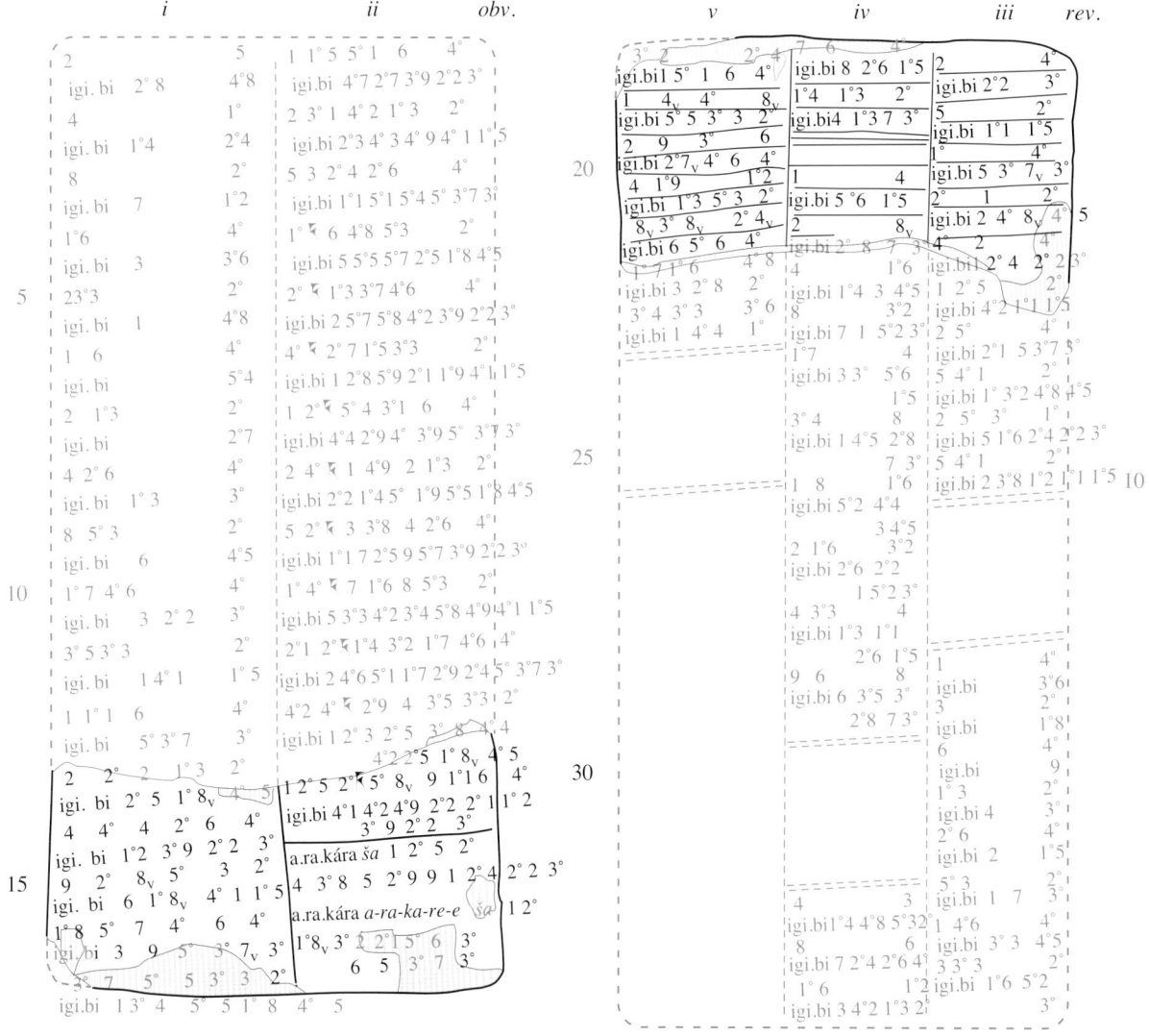

Fig. A3.2. UM 29.13.21. A text with five or six applications of the doubling and halving algorithm.

The tables in this text are applications of the doubling and halving algorithm, without references to the trailing part algorithm as in the odd lines of CBS 10201. The small text frame in the lower right corner of the obverse probably contains a couple of assignments of some kind, but it is difficult to know exactly what they are. Similar assignments seem to have been given in three small text frames on the reverse, but none of them is preserved.

The reason for the strange form of the thirtieth and last entry in the algorithm table on the obverse (1 25 20 X 58 09 11 06 40 instead of 1 26 18 09 11 06 40) was explained by Sachs (*op. cit.*) as a consequence of the appearance of a double zero in the entry 10 06 48 53 20 in the twentieth line, now lost. The entry was probably written in the form 10 X 6 48 53 20, with X separating the 10 from the 6.

A3.3. CBS 1215. An Algorithm Text with Explicit Computations

The most impressive of the Old Babylonian algorithm tables published by Sachs in *JCS* 1 (1947) is without doubt **CBS 1215** (Fig. A3.3 below). It is an elaborate combined application of the doubling and halving algorithm and the trailing part algorithm, in the case of the standard initial pair 2 05, 28 48, the same as the one in CBS 10201 and the one on the obverse of UM 29.13.21.

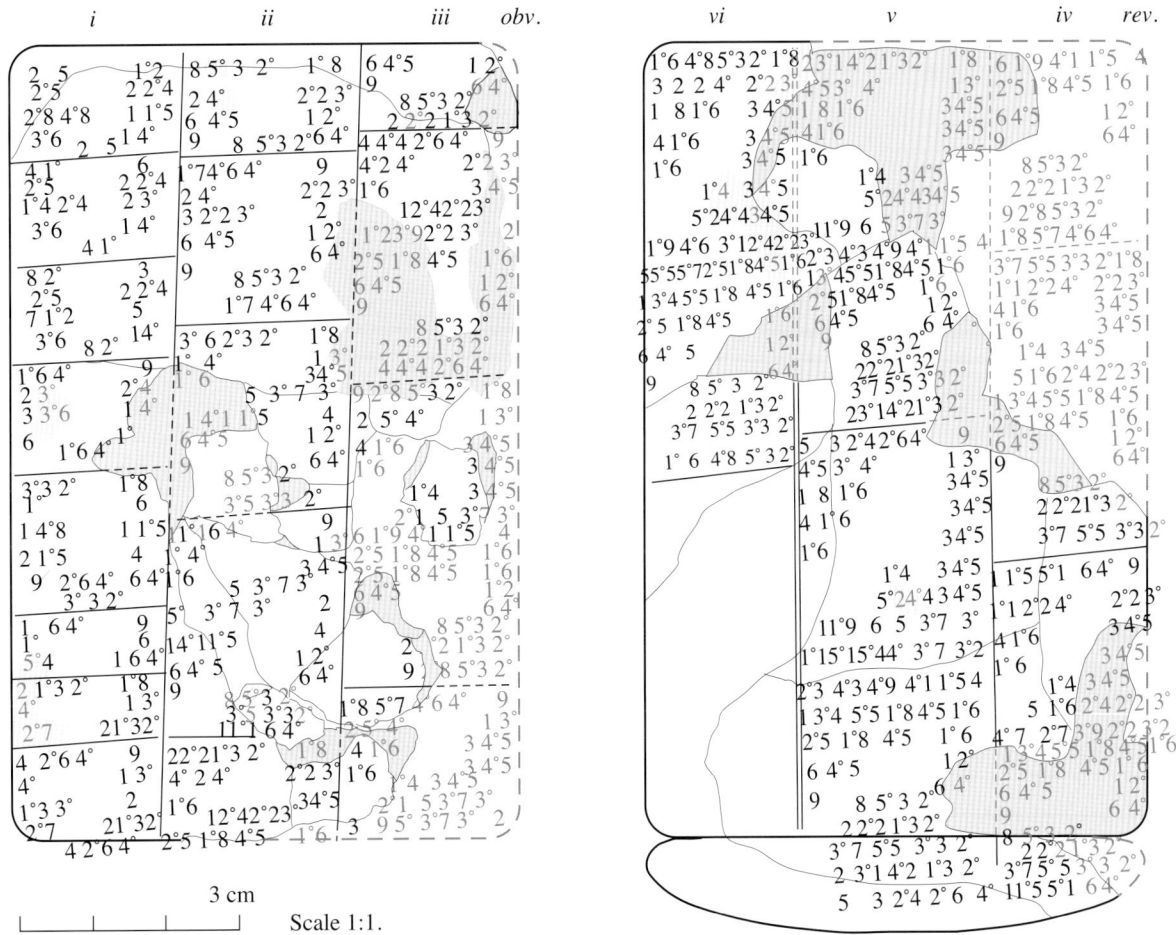

Fig. A3.3. CBS 1215. A combined algorithm. (Conform transliteration based on a photo of the clay tablet.)

There are 21 reciprocal pairs mentioned in CBS 1215:

2 05	28 48	1 11 06 40	50 37 30
4 10	14 24	2 22 13 20	25 18 45
8 20	7 12	4 44 26 40	12 39 22 30
16 40	3 36	9 28 53 20	6 19 41 15

33 20	1 48	18 57 46 40	3 09 50 37 30
1 06 40	54	37 55 33 20	1 34 55 18 45
2 13 20	27	1 15 51 06 40	47 27 39 22 30
4 26 40	13 30	2 31 42 13 20	23 43 49 41 15
8 53 20	6 45	5 03 24 26 40	11 51 54 50 37 30
17 46 40	3 22 30	10¹ 06¹ 48 53 20	5 55 57 25 18 45
35¹ 33¹ 20	1 41 15		

They are precisely the ones that can be obtained as the result of 20 steps of the doubling and halving algorithm with the initial pair 2 05, 28 48. However, for each pair it is shown how an application of the trailing part algorithm to the first number in the pair produces the second number, and, conversely, how an application of the same algorithm to the second number in the pair leads back again to the first number, which is, thus, *the reciprocal of its own reciprocal.*

The text of CBS 1215 is extensively damaged, but easy reconstruct to its original form:

Column i (## 1-8):

2 05	12
25	2 24
28 48	1 15
36	1 40
	2 05
4 10	6
25	2 24
14 24	2 30
36	1 40
	4 10
8 20	3
25	2 24
7 12	5
36	1 40
	8 20
16 40	9
2 30	24
3 36	1 40
6	10
	16 40
33 20	18
10	6
1 48	1 15
2 15	4
9	6 40
	26 40
	33 20
1 06 40	9
10	6
54	1 06 40
2 13 20	18
40	1 30
27	2 13 20
4 26 40	9
40	1 30
13 30	2
27	2 13 20
	4 26 40

1-8

Column ii (## 9-13):

8 53 20	18
2 40	22 30
6 45	1 20
9	6 40
	8 53 20
17 46 40	9
2 40	22 30
3 22 30	2
6 45	1 20
9	6 40
	8 53 20
	17 46 40
35 33 20	18
10 40	1 30
16	3 45
	5 37 30
1 41 15	4
6 45	1 20
9	6 40
	8 53 20
	35 33 20
1 11 06 40	9
10 40	1 30
16	3 45
	5 37 30
50 37 30	2
1 41 15	4
6 45	1 20
9	6 40
	8 53 20
	35 33 20
	1 11 06 40
2 22 13 20	18
42 40	22 30
16	3 45
1 24 22 30	
25 18 45	16

9-13

Column iii (## 13-16):

6 45	1 20
9	6 40
8 53 20	
2 22 13 20	
4 44 26 40	9
42 40	22 30
16	3 45
1 24 22 30	
12 39 22 30	2
25 18 45	16
6 45	1 20
9	6 40
8 53 20	
2 22 13 20	
4 44 26 40	
9 28 53 20	18
2 50 40	1 30
4 16	3 45
16	3 45
14 03 45	
21 05 37 30	
6 19 41 15	4
25 18 45	16
6 45	1 20
9	6 40
8 53 20	
2 22 13 20	
9 28 53 20	
18 57 46 40	9
2 50 40	1 30
4 16	3 45
16	3 45
14 03 45	
21 05 37 30	
3 09 50 37 30	2

13-16

Column vi (# 21):

10 06 48 53 20	18
3 02 02 40	22 30
1 08 16	3 45
4 16	3 45
16	3 45
14 03 45	
52 44 03 45	
19 46 31 24 22 30	
5 55 57 25 18 45	16
1 34 55 18 45	16
25 18 45	16
6 45	1 20
9	6 40
8 53 20	
2 22 13 20	
37 55 33 20	
10 06 48 53 20	

 # 21

Column v (## 19-20):

2 31 42 13 20	18
45 30 40	1 30
1 08 16	3 45
4 16	3 45
16	3 45
14 *03 45*	
52 44 *03 45*	
1 19 06 *05 37 30*	
23 43 49 41 15	4
1 34 55 18 45	16
25 18 45	16
6 45	1 20
9	6 40
8 53 20	
2 22 13 20	
37 55 33 20	
2 31 42 13 20	
5 03 24 26 40	9
45 30 40	1 30
1 08 16	3 45
4 16	3 45
16	3 45
14 03 45	
52 44 03 45	
1 19 06 05 37 30	
11 51 54 40 37 30	2
23 43 49 41 15	4
1 34 55 18 45	16
25 18 45	16
6 45	1 20
9	6 40
8 53 20	
2 22 13 20	
37 55 33 20	
2 31 42 13 20	
5 03 24 26 40	

19-20

Column iv (## 16-18):

6 19 41 15	4
25 18 45	16
6 45	1 20
9	6 40
8 53 20	
2 22 13 20	
9 28 53 20	
18 57 46 40	
37 55 33 20	18
11 22 40	22 30
4 16	3 45
16	3 45
14 03 45	
5 16 24 22 30	
1 34 55 18 45	16
25 18 45	16
6 45	1 20
9	6 40
8 53 20	
2 22 13 20	
37 55 33 20	
1 15 51 06 40	9
11 22 40	22 30
4 16	3 45
16	3 45
14 *03 45*	
5 16 24 22 30	
47 27 39 22 30	2
1 34 55 18 45	16
25 18 45	16
6 45	1 20
9	6 40
8 53 20	
2 22 13 20	
37 55 37 55 33 20	
1 15 51 06 40	

16-18

Trailing parts used in the computation are shaded. Lost or damaged digits are written in *italics*.

Fig. A3.4. CBS 1215. A full reconstruction of the 21 sub-algorithms of the algorithm table.

In this algorithm table, each reciprocal m is computed as the product of reciprocals of factors of the corresponding number n (and conversely). The following regular sexagesimal numbers are used in various combinations as factors in the computation of the 21 *reciprocals*:

$$24,\ \mathbf{22\ 30},\ 18,\ 12,\ 9,\ 6,\ 3\ 45,\ 3,\ 2\ 24,\ 1\ 30.$$

The following regular numbers are used in the computation of the 21 *reciprocals of reciprocals*:

$$16,\ 10,\ 6\ 40,\ 5,\ 4,\ 2\ 30,\ \mathbf{2\ 13\ 20},\ 2,\ 1\ 40,\ 1\ 20,\ 1\ 15.$$

In the table below, the first column exhibits the trailing parts of numbers appearing in the *first* halves of the sub-algorithms in CBS 1215. The second column shows the corresponding reciprocals or, rather, multipliers. In the remaining columns are given the numbers of the sub-algorithms in CBS 1215 in which these multipliers appear as factors in the computation of any one of the 21 *reciprocals*.

		1	2	3	4	5	6	7	8	9	10	11	12	13	14	15	16	17	18	19	20	21
40	1 30							7	8			11	12			15	16			19	20	
25	2 24	1	2	3																		
20	3			3																		
16	3 45											11	12	13	14	15	16	17	18	19	20	21
10	6					5	6															
6 40	9				4		6		8		10		12		14		16		18		20	
5	12	1																				
3 20	18					5		7		9		11		13		15		17		19		21
2 40	**22 30**									9	10			13	14			17	18			21
2 30	24				4																	

Note that *the fact that 22 30 appears repeatedly as a multiplier in algorithm tables like CBS 1215 may be the reason why 22 30 is one of the head numbers in the Old Babylonian combined multiplication table!* (Cf. the discussion in Sec. 2.6 f above.)

In the table below, the first column exhibits the trailing parts of numbers appearing in the *second* halves of the sub-algorithms in CBS 1215. The second column shows the corresponding reciprocals or, rather, multipliers, and in all the remaining columns are given the numbers of the sub-algorithms in CBS 1215 in which these multipliers appear as factors in the computation of any one of the 21 *reciprocals of reciprocals*.

		1	2	3	4	5	6	7	8	9	10	11	12	13	14	15	16	17	18	19	20	21
48	1 15	1				5																
45	1 20									9	10	11	12	13	14	15	16	17	18	19	20	21
36	1 40	1	2	3	4																	
30	2								8		10		12		14		16		18		20	
27	**2 13 20**								8													
18 45	16													13	14	15	16	17	18	19	20	21
24	2 30		2																			
15	4					5						11	12			15	16			19	20	
12	5			3																		
9	6 40					5					10	11	12	13	14	15	16	17	18	19	20	21
6	10				4																	

Note, that, strictly speaking, 16 is not the reciprocal of 18 45. On the other hand,

$$18\ 45 \cdot 16 = 1\ 15 \cdot 4 = 5\ (00),$$

so that 16 is a reciprocal of 18 45 in a broader sense. It is easy to see that the trailing part algorithm works just as well with this kind of generalized reciprocals as with reciprocals of the ordinary kind!

Note also that 2 13 20 is known as a head number in no single multiplication tables and in only one Old Babylonian combined multiplication table, namely MS 3974 (see Fig. 2.6.13.). On the other hand, 2 13 20 is only once involved in a multiplication in CBS 1215, and that is a simple doubling.

Appendix 4
Cuneiform Systems of Notations for Numbers and Measures

There are five major "traditional" (in addition to several exclusively "proto-literate") systems of numbers or measures in cuneiform texts from the fourth, third, and second millennia BC, namely

S	sexagesimal counting numbers	used to count people and objects
C	capacity numbers	used to measure amounts of cereals, *etc.*
M	weight (or metal) numbers	used to measure amounts of silver, *etc.*
A	area numbers	used to measure areas of fields, *etc.*
L	length numbers	used to measure lengths of straight or circular lines

For each system of numbers or measures there exists a hierarchy of progressively larger "units" and a number of "conversion rules". For each kind of unit there is a corresponding sign, and small multiples of the units are written, in most cases, *additively*, that is by repeating the signs as many times as needed. The notation is always *non-positional*, so that large multiples of a unit are converted into multiples of "higher" units by use of the conversion rules.

The simplest way of describing all kinds of systems of numbers and measures is by presenting their "factor diagrams", which exhibit in a simple and visual form both the hierarchy of units in a given system, the conversion rules, and the signs for the units.[1]

A4.1. Proto-Literate/Traditional Sexagesimal Counting Numbers

The oldest of the mentioned five major traditional systems of cuneiform numbers and measures are the sexagesimal number system and the system of area numbers. Both are attested in "proto-cuneiform" texts from the Uruk IV-III periods, around the end of the fourth millennium BC, and in slightly younger "proto-Elamite" texts, and both continued to be used in Mesopotamia, with only marginal changes and additions, throughout the whole third millennium. The three first of the factor diagrams in Fig. A4.1 below show the development in the case of the sexagesimal number system. Note the change in the orientation of the script and the transition from round "curviform" to angular cuneiform number signs. Note also the use of the word gal 'great' to increase the value of a number sign by a factor 60.

The series of factor diagrams in Fig. A4.1 demonstrate that some of the ideas behind the Babylonian sexagesimal place value system must have been clearly understood already by those who invented the original sexagesimal number system, in the proto-literate period towards the end of the fourth millennium, or even earlier.[2] Thus already the earliest form of the sexagesimal system possessed *a repetitive structure* with an alternating series of conversion factors 10, 6, 10, 6, 10, ... In addition, the sign for 60 was already from the

---

1. The use of factor diagrams was initiated by the present author in *ERBM 1-2* (1978/1979), unfortunately a work published in a very limited edition. For more accessible examples of the use of factor diagrams, see instead Friberg, *GMS 3* (1993), Friberg, *JCS* 51 (1999), and Nissen/Damerow/Englund, *ABK* (1993) Fig. 28.

beginning of the same form as the sign for 1, only slightly larger. Therefore, in a sense, the invention of the sexagesimal place value system required only one additional idea, to let also the sign for 10 · 60 be of the same form as the sign for 10. Indeed, as shown by the fourth factor diagram in Fig. A4.1, in the sexagesimal place value system the *signs* for the units 1, 10, 60, 10 · 60, 60 · 60, ... form an alternating series just like the *conversion factors* 10, 6, 10, 6, 10, ...

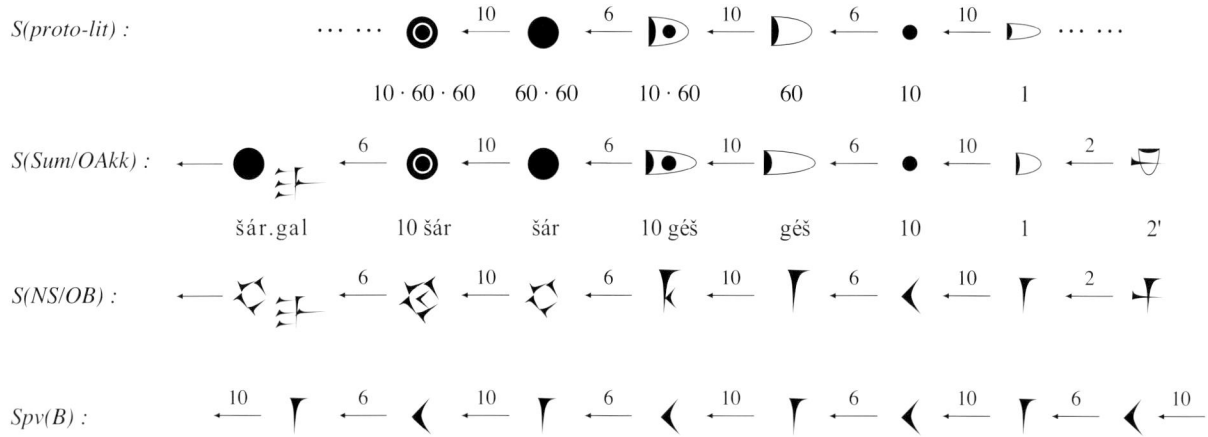

Fig. A4.1. The historical development of the factor diagram for the sexagesimal number system.

One should not forget another remarkable feature of the Old Babylonian sexagesimal place value system, namely that the place value idea was applied not only to whole numbers but also to fractions. This is indicated in the fourth factor diagram in Fig. A4.1 by letting the diagram be open in both ends.

In this connection, it is instructive to consider the following review of the various ways in which fractions could be expressed in Old Babylonian cuneiform texts (cf. Friberg, *RlA 7* (1990), Sec. 3.1).

basic fractions	3', 2', 3", 6"	(Fig. A4.2 below)
igi fractions (*n*-th parts)	igi.*n* or igi.*n*.gál (*n* = 2, 3, 4, 5, ...)	
composite fractions, such as	3" u_4 *ù* igi.5.gál 3" u_4	(YBC 7164 § 7)[3]
multiples of the shekel and the barley-corn		
metrological fractions		
sexagesimal fractions		

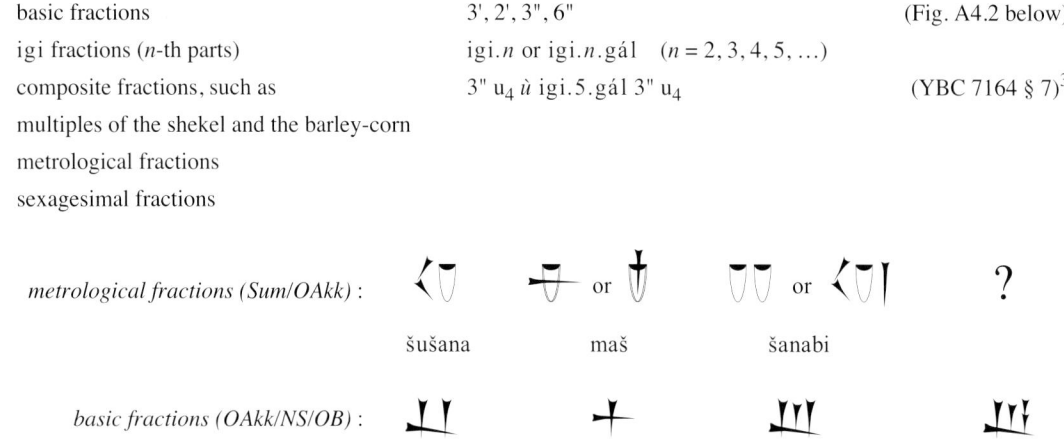

Fig. A4.2. The Neo-Sumerian/Old Babylonian basic fractions and their Old Sumerian ancestors.

Metrological fractions are discussed below, in Secs. A4 e-g. Multiples of the shekel in Sumerian adminis-

2. Actually, it is likely that sexagesimal numbers were represented by sets of "tokens" of clay in Mesopotamia and surrounding regions long before the invention of writing. See the tentative interpretation in Friberg, *OLZ* 89 (1994), 492-496, and in Sec. A4 j below, of sets of tokens inside some "spherical envelopes" of clay from Susa and Uruk.
3. See Friberg, *ChV* (2001), 127. In the cited example, 4/5 of a day is expressed as '2/3 of a day and 1/5 of 2/3 of a day'. (The explanation for this expression is that 4/5 = ;48 = ;40 + ;08 = 2/3 + 1/5 · 2/3.)

trative texts were in a way the ancestors of the Old Babylonian sexagesimal fractions, since they expressed fractions of various metrological units as multiples of a sixtieth of that unit, in the same way as 1/3 =;20 means that 1/3 = 20 sixtieths, and so on. In a similar way, the Sumerian/Old Babylonian basic fractions were descendants of Old Sumerian metrological fractions.

In Old Babylonian mathematical problem texts, all kinds of fractions from the list above can be used in the questions and the answers, but sexagesimal fractions are the only kind of fractions used in the solution procedures.

A4.2. Proto-Literate Bisexagesimal Counting Numbers

In proto-literate (that is, proto-cuneiform or proto-Elamite) texts, a special system of "bisexagesimal" counting numbers is used for specific purposes, in particular for counting food rations. The bisexagesimal system is identical with the sexagesimal system for small numbers, but counts with multiples of $2 \cdot 60$ rather than with multiples of 60 for big numbers, using special number signs for $2 \cdot 60$, $10 \cdot 2 \cdot 60$, and(?) $60 \cdot 2 \cdot 60$:

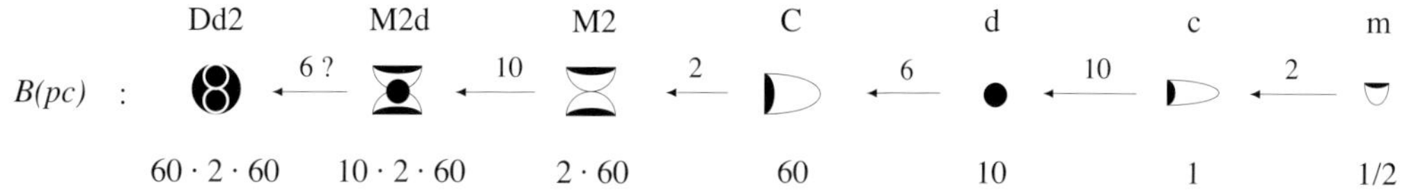

Fig. A4.3. Factor diagram for the proto-literate bisexagesimal counting umbers.

A4.3. Proto-Elamite Decimal Counting Numbers

The roots of the decimal number system appear to go as far back as the roots of the sexagesimal number system, probably to long before the invention of writing. In particular, it is likely that 3 large tetrahedrons, 2 punched tetrahedrons of medium size, and 4 small tetrahedrons in a broken spherical envelope of clay from *pre-literate* Susa (western Iran) stand for 324 workdays, counted decimally. (See Sec. A4 j below.) In *proto-literate* Susa (around the end of the fourth millennium BC) not only sexagesimal and bisexagesimal counting numbers were used in proto-Elamite inscriptions on clay tablets but also decimal counting numbers. (Although the cuneiform-like proto-Elamite script remains undeciphered, the number systems used in proto-Elamite inscriptions are now well understood. See Friberg, *ERBM 1* (1978), Damerow/Englund, *PETTY* (1989).) Here is the factor diagram for the proto-Elamite decimal number system, which appears to have been used mainly to count animals:

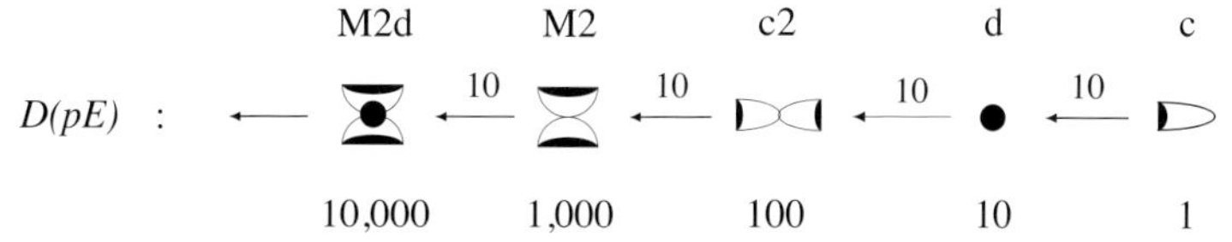

Fig. A4.4. Factor diagram for the proto-Elamite decimal system.

There never existed a special system of notations for decimal numbers in proto-cuneiform and cuneiform texts from Mesopotamia. Yet it seems to be clear that decimal numbers were used in the everyday life of the Semitic part of the population in Mesopotamia, even if almost exclusively sexagesimal numbers appear in the written records. When decimal numbers do appear in cuneiform texts, they are nearly always *counted* as multiples of 'a hundred' (me'atu), 'a thousand' (līmu), *etc.*

In cuneiform inscriptions on clay tablets from the ancient city Ebla in Syria (approximately contemporary with Old Akkadian inscriptions form Mesopotamia) only decimal numbers are used as counting numbers. (See the discussion of TM.75.G.1392 in Sec. A6 e below.) There, too, decimal numbers were *counted* in terms of 'hundreds', *etc.*, as shown by the factor diagram in Fig. A4.5 below.

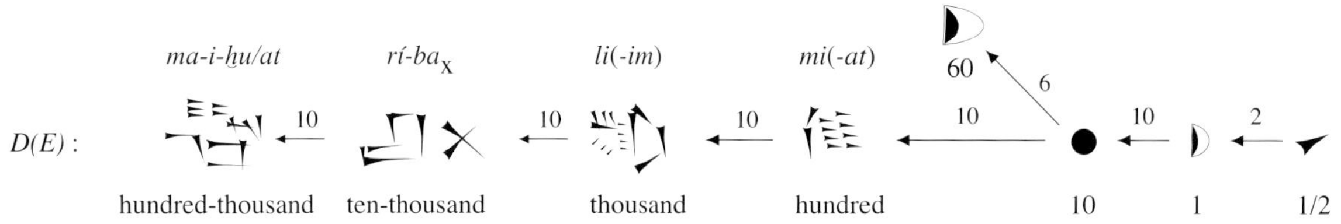

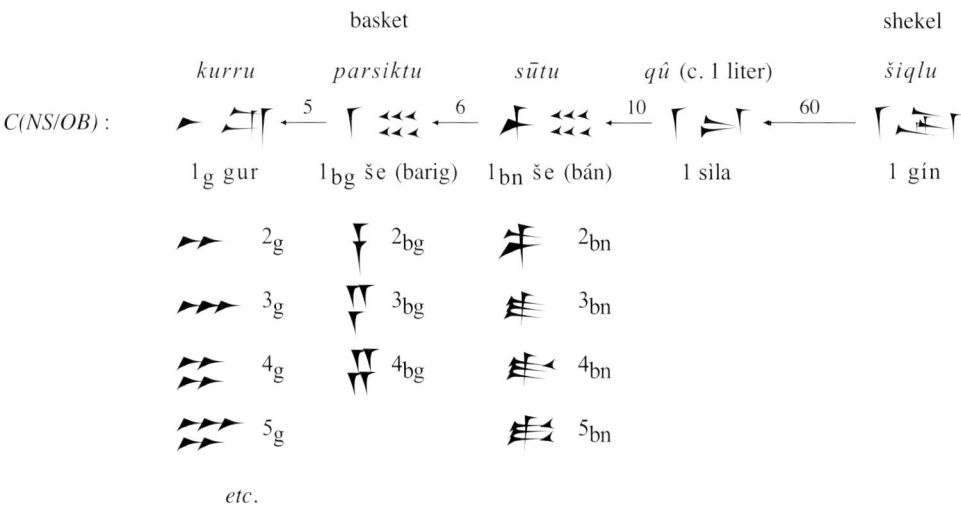

Fig. A4.5. Factor diagram for the Eblaite decimal system.

A4.4. Proto-Literate and Traditional Capacity Numbers

The system of capacity numbers used in proto-cuneiform texts has the following factor diagram:

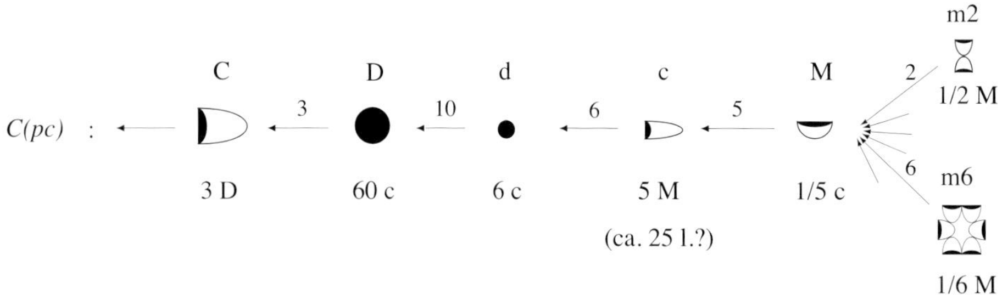

Fig. A4.6. Factor diagram for the proto-cuneiform capacity system.

(The factor diagram for the proto-Elamite capacity system is essentially the same, with only slightly different notations for the smallest fractions.)

Fig. A4.7. Factor diagram for the Neo-Sumerian/Old Babylonian capacity system.

Since the names of the units in the proto-cuneiform capacity system are not known, easily remembered surrogate names are used here instead (c and C for small and large cup-shaped signs, d and D for small and large disk-shaped signs, and M for the sign for the major fraction, a rotated large cup). Note that the values of the number signs are "context-dependent". Thus, for instance, the four signs c, d, D, and C have the values 1, 10, 60, and $60 \cdot 60$ in the sexagesimal number system, but the values 1 c, 6 c, 180 c, and 60 c in the proto-literate capacity number system (Friberg, *ERBM 1-2* (1978/1979); Friberg, *JCS* 51 (1999)).

The proto-cuneiform capacity number system ceased to be used after the end of the proto-literate period. In its place came a new, Sumerian capacity number system, which in its Neo-Sumerian/Old Babylonian form has the factor diagram shown in Fig. A4.7.

In that number system, the barig is a fraction of the gur, the bán of the barig, the sìla of the bán, and the gín (shekel) of the sìla. Multiples of the gín and of the sìla are *counted*, as so and so many gín and sìla, respectively. For multiples of the bán and the bán there are special number signs, as shown above. Multiples of the gur are again *counted*, and are written as ordinary sexagesimal non-positional counting numbers followed by the sign gur, with the exception that before gur the digits 1 through 9 are written with rotated (horizontal) wedges.

A4.5. Proto-Cuneiform and Traditional Weight Numbers

Several small clay tablets from the earliest phase of the proto-literate period are inscribed with number signs that may belong to a proto-cuneiform system of weight numbers (system *E*). The factor diagram below suggests that this system was composed of two parts, one with the basic unit c, the other with the basic unit ce (c×en), which is 16 times smaller:

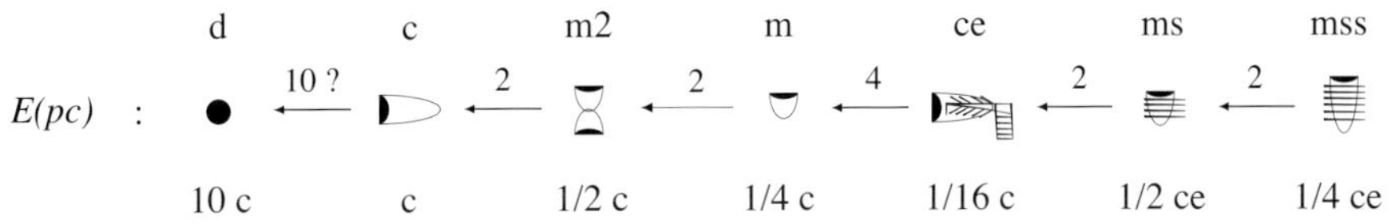

Fig. A4.8. Factor diagram for the proto-cuneiform system *E* (weight numbers?).

The fractional units are in both cases 1/2 and 1/4 of the basic unit. These features of system *E* and certain other features of the texts in which this system is used seems to suggest that system *E* was a system of weight numbers, with the upper half of the system used to measure amounts of silver and with the lower half of the system used to measure smaller amounts of the 16 times more precious gold! (See Friberg, *JCS* 51 (1999), 129-135.)

Like the proto-cuneiform system of capacity numbers, this proto-cuneiform system of weight numbers ceased to be used after the end of the proto-literate period. In its place came a Sumerian system of weight numbers, which like the proto-cuneiform system *E* seems to be composed of two parts, with the upper part suitable for measuring amounts of copper, and with the lower part suitable for measuring smaller amounts of the 180 times more precious silver (Friberg, *CDLJ* 2005:2, Sec. 2.2.8).

The form that this system of weight numbers had taken in the Neo-Sumerian/Old Babylonian period, after some minor changes, is described by the factor diagram in Fig. A4.9 below.

In that system, all weight numbers are *counted*, as multiples or fractions of the barley-corn, the shekel, the mina, and the talent. No special number signs are used, except that before gú 'talent' the digits 1 through 9 are written with rotated (horizontal) wedges.

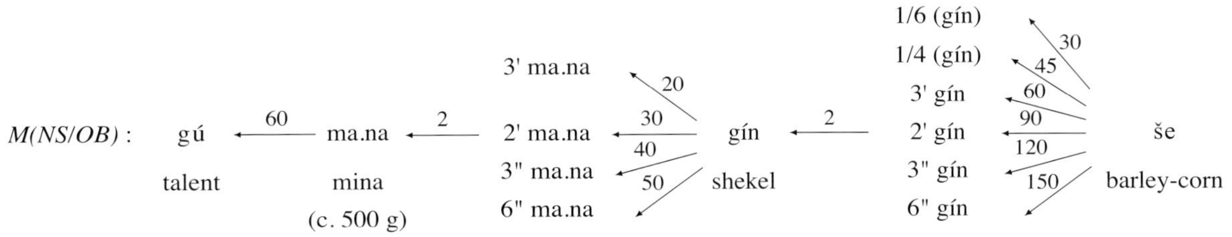

Fig. A4.9. Factor diagram for the Neo-Sumerian/Old Babylonian system *M* (weight numbers).

A4.6. Proto-Literate/Traditional Area Numbers

Like the proto-cuneiform sexagesimal number system, the proto.cuneiform area number system continued to be used, without any major changes, throughout the whole third millennium BC. The factor diagrams below show the development of the number signs in the system. Note the change in the orientation of the script and the transition from curviform (round) to cuneiform number signs.

Area numbers are often accompanied by the determinative gán (aša$_5$), meaning 'field, area'. Note that the number signs used to write the area units 60 bùr, 10 · 60 bùr, and 60 · 60 bùr in this system are the same as the number signs used to write the units 60 · 60, 10 · 60 · 60, and 60 · 60 · 60 in the sexagesimal (non-positional) number system. (The names of these high units in the area number system are not known.)

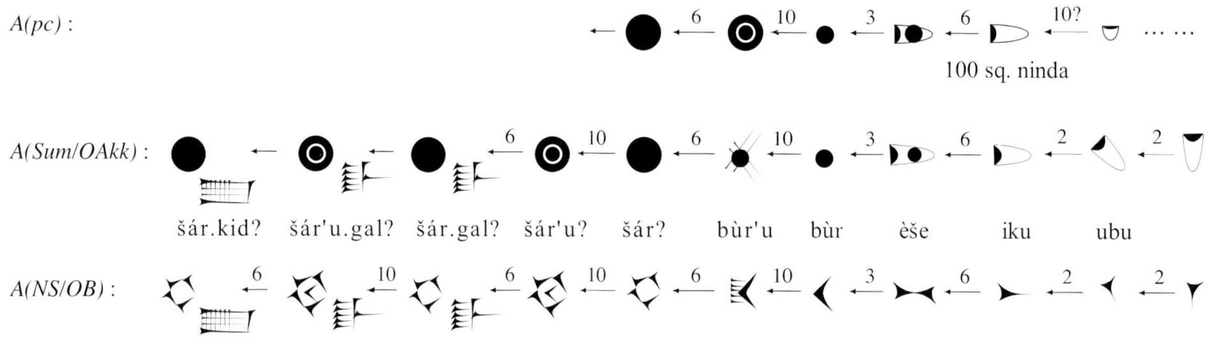

Fig. A4.10. The historical development of the factor diagram for Mesopotamian area numbers.

A4.7. Old Akkadian and Neo-Sumerian/Old Babylonian Length Numbers

In proto-cuneiform metro-mathematical field texts (Friberg, *AfO* 44/45 (1997/98)) an unnamed length unit must be the ninda, since the square of sixty times the length unit is equal to 2 bùr in field division texts like *MSVO 1*, 2 (*op. cit*., Fig. 5.1). Multiples of the ninda are not converted to higher length units in any of those texts. Thus, in the metro-mathematical exercise W 19408 (op. *cit*., Fig. 2.1), the ninda is counted in tens of sixties.

The ninda reappears in the metro-mathematical table text VAT 12593 from Shuruppak (Early Dynastic III), where the first entry mentions the large length number 10 · 60 ninda^{du} (Fig. 6.3 above). In metro-mathematical exercises from the Early Dynastic and Old Akkadian periods, such as CUNES 50-08-001 (App. 7), *DPA* 36 (Fig. A6.3 in App. 6), A 5443 (Fig. A6.9), and HS 815 (Fig. A6.10 b), various minor units are documented, as shown in the following factor diagram:

			seed-cubit		cubit								finger
L(ED III/OAkk) : ← ninda	←4—	*nikkas*	←1 1/2—	kùš.numun	←2—	giš.bad/kùš	←2—	šu.bad/*zipaḫ*	←1 1/2—	šu.dù.a	←10—		šu.si
6 k.n.		1 1/2 k.n.		(c. 1 m)		1/2 k.n.		1/4 k.n.		1/6 k.n.			1/60 k.n.

Fig. A4.11. Factor diagram for the Old Akkadian system of length numbers.

Some of the minor units ceased to be used after the Old Akkadian period. The length units appearing in Neo-Sumerian/Old Babylonian mathematical or metrological texts are shown below:

			length		rope		rod?		reed		cubit		finger	
	bēru		*?*		*eblu*		*nindanu*		*qanû*		*ammatu*		*ubānu*	
L(NS/OB) :	danna	←30—	uš	←6—	èše	←10—	ninda	←2—	gi	←6—	kùš	←30—	šu.si	←
	30 00 n.		1 00 n.		10 n.		(c. 6 m)		;30 n.		;05 n.		;00 10 n.	
							12 c.		6 c.		(c. 50 cm)		;02 c.	

Fig. A4.12. Factor diagram for the Neo-Sumerian/Old Babylonian system of length numbers.

A4.8. Proto-Cuneiform Time Numbers

A previously only vaguely suspected special proto-cuneiform system of notations for time numbers was confirmed and extensively discussed by Englund in *JESHO* 31 (1988). The system is based on a division of the year in 12 months of 30 days each. It was, apparently, used only for *administrative* purposes, such as for the computation of daily or monthly rations, or for the indication of the age of sheep and other domestic animals. The system ceased to be used after the end of the proto-literate period. The ingenious principle behind the form of the proto-cuneiform time numbers is explained by the following series of notations, where the sign U_4 has the meaning 'sun, day':

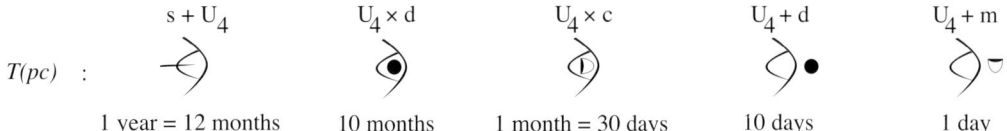

Fig. A4.13. Factor diagram for the proto-cuneiform system of time numbers.

Days were written as a sun sign with sexagesimal numbers *below* it, months as a sun sign with sexagesimal numbers *inside* it, and years as a sun sign with one, two, or several strokes (rays) *above* it.

A4.9. An Integrated Family of Numbers and Measures

A remarkable feature of all the mentioned Mesopotamian (*proto-cuneiform*, *Sumerian*, and *Old Babylonian*) systems of measures is that they are "sexagesimally adapted". What this means is that all the "conversion factors" appearing in the various factor diagrams displayed above are *small, sexagesimally regular numbers*. Indeed, all the conversion factors are equal to one of the following numbers:

1 1/2, 2, 3, 4, 5, 6, 10, 12, 30.

This fact makes it easy to count with Mesopotamian measures. Typically, in a non-trivial computation involving various numbers and measures, the need will arise to convert a relatively large sexagesimal multiple of some unit in a given system of measures into a sum of small multiples of increasingly high units of the same system. The task will be quite easy when all the conversion factors involved in the process are regular sexagesimal numbers.[4] Without sexagesimally adapted measure systems, the various Old Babylonian metrological tables would have been much messier than what they actually are!

The Old Babylonian sexagesimal number system and the systems of measures for capacity, weight, area, (volume,) and length deserve to be called an "integrated family of numbers and measures" not only because the systems of measures share the feature of being sexagesimally adapted. There are also a number of remarkable "unitary" relations between measures belonging to different systems, such as (see a) Friberg, *BagM* 28 (1997), Fig. 7.1; b) Sec. 7.3 a above; c) Friberg, *AfO* 44/45 (1997/98), Fig. 1.2):

a) 1 (cubic) sìla = the *content* of a cube with the side ;01 ninda (= 6 fingers)	systems *C* and *L*
b) 1 talent = the *weight* of a baked square brick with the side 1 cubit and the height ;01 ninda	systems *M* and *L*
b) 1 volume šar = the volume of 1 brick šar (12 00) of such bricks	systems *V* and *B*
c) 1 plane iku = the *area* of a square field with the side 1 'rope' (= 10 ninda)	systems *A* and *L*
d) 1 solid iku = the *volume* of a square prism with the side 1 'rope' and the height 1 cubit	systems *V* and *L*

---

4. Compare with the much less ingenious form of the Anglo-Saxon family of measure systems, where the various conversion factors are far from being decimally adapted. Thus, for instance, 1 mile = 5,280 feet, 1 acre = 4,840 square yards, 1 pound = 7,000 grains, and 1 gallon (US) = 231 cubic inches.

A4.10. Pre-Literate Number Tokens

In *Before Writing, Vol. 1* (1992), Schmandt-Besserat gave a detailed account of her revolutionary theory about the crucial role played by small clay-figures, so called "tokens", in the prehistory of writing.[5] According to this theory, there were seven essential steps in the early development of writing as a tool for accounting and communication: 1) the appearance in various parts of the Middle East around 8000 BC, that is at the time of the agricultural revolution, of five or six types of "plain tokens", small geometric objects in baked clay, probably used as counters and representing various agricultural products in a one-to-one correspondence; 2) in the late fifth and early fourth millennia, the gradual introduction of additional types and subtypes of tokens, so called "complex tokens"; also the occasional use of perforations, allowing groups of tokens to be strung together; 3) around the middle of the fourth millennium, at the time of the first cities and beginning state formation, an explosive proliferation of the repertory of complex tokens at a limited number of sites (mainly Susa in Iran, Uruk in Iraq, and Habuba Kabira in Syria), probably in order to represent many new kinds of products from the city workshops; particularly interesting are several "series" of tokens of fixed form but with a variable number of incised lines or dots; 4) the invention of "spherical envelopes", containing (mostly) plain tokens and often impressed with cylinder seals, sometimes for good measure also marked on the outside with more or less schematic representations of the tokens inside; 5) for a short while, around 3300 BC, the use of "impressed tablets, instead of, or together with, spherical envelopes and yielding the same kind of information, the number signs on these first clay tablets being imitations of the previously used tokens; 6) the invention of writing on clay tablets, with a large inventory of sometimes pictographic but most often abstract signs, of which the latter in many cases were two-dimensional representations of the complex tokens they replaced; 7) the complete disappearance of tokens from (almost) all excavated sites after the invention of writing.

For the history of number notations, the spherical envelopes are particularly important. Their importance derives from two hypothetical situations: Either the content of an envelope constitutes an account of a single *disbursement or delivery*, in which case the enclosed tokens record a number in a single system of number tokens. Or else, the content of an envelope constitutes a record of a single *transaction*, in which case the enclosed tokens record two numbers in two separate systems of number tokens, and there exists some simple mathematical relation between the two numbers.

There is no reason to doubt that the (mostly) plain tokens enclosed in spherical envelopes belonged to a small number of *pre-literate* systems of number tokens, very much similar to the now well known *proto-literate* systems of number notations impressed on clay tablets. It is likely that it would be possible to determine the values of many of the types of enclosed tokens if, at some point of time in the future, the contents of all known intact spherical envelopes (there are more than 80 of them) can be made known in some way, either by opening the envelopes or by looking into them by use of modern high technology. Until that has happened, one has to make do with the material that is available at present. (See, however, Amiet, *Éd. CNRS* (1986), Drilhon, et al., *Éd. CNRS* (1986), Damerow and Meinzer, *BagM* 26 (1995).)

(Extremely) tentative explanations are offered below for the contents of some selected examples of spherical envelopes.

Sb 1932 (next page) contains *1 big sphere, 6 small spheres, and 1 flat disk*. On the outside of the spherical envelope is written what looks like a number. This number can, with some imagination, be interpreted as '16 plus a fraction'. Since it is natural to assume that the tokens inside represent the same number as the markings on the outside, and that what is accounted for is an amount of barley, one is led to the conclusion that the big sphere may be a pre-literate capacity unit with the same value as the proto-literate (that is, both proto-cuneiform and proto-Elamite) unit 1 D = 60 c, that the small sphere in a similar way may correspond to 1 d = 6 c, and that the flat disk may correspond to 1 c. If that is so, then the tokens inside Sb 1932 together represent the capacity number 1 D 6 d 1 c = 16 1/6 d = 97 c.

---

5. See the constructively critical review in Friberg, *OLZ* 89 (1994), from which a part of the discussion here is borrowed. Drawings of tokens copied, with the author's kind permission, from *Before Writing*.

SB 1938 (below) contains *3 small cones, 1 very thick disk, 2 small spheres, and two flat disks of different sizes*. In view of the explanation above of the flat disk as a forerunner to the proto-literate unit 1 c, one is tempted to explain the thick disk as a forerunner to the proto-literate unit 1 C = 3 D = 30 d = 180 c. This tentative explanation has the interesting consequence that the capacity number presumably recorded in Sb 1938 may be almost precisely twice as large as the capacity number presumably recorded in Sb 1932. Indeed,

$$1 \, C \, 2 \, d \, 1 \, c \, 1/2 \, c = 193 \, 1/2 \, c = (1 + 3/40) \cdot 180 \, c, \quad \text{and} \quad 1 \, D \, 6 \, d \, 1 \, c = 97 \, c = (\text{appr.}) \, (1 + 3/40) \cdot 90 \, c.$$

This tentative explanation is further supported by the possibility that the *3 small cones* in Sb 1938 may stand for the sexagesimal number 3 · 60 = 180. Therefore, it is tempting to draw the conclusion that the tokens inside Sb 1938 may constitute a record of *a monthly disbursement of barley* to 180 persons, with each person receiving bread for 1 c of barley and beer for 3/40 of 1 c of barley. Note that

$$3/40 \cdot 1 \, c = 3/8 \cdot 1 \, M = 1/2 \, 1/8 \cdot 1 \, M = m2 \, m8.$$

The meaning of the cross-like sign impressed on the outside of Sb 1938 is not known.

W 20987, 8, finally (next page), contains *5 pyramids or tetrahedrons and 2 flat disks*. The assumption that 1 flat disk may stand for 1 c of barley can be used for the tentative interpretation of the contents of this spherical envelope, namely that what is recorded here is 2 c of barley as wages for 5 *man-days of hired labor*. The interpretation implies, in particular, that

the *wages* for 1 man-day of hired labor is 2/5 · 1 c = 2 M of barley   (12 times as much as a daily *ration*).

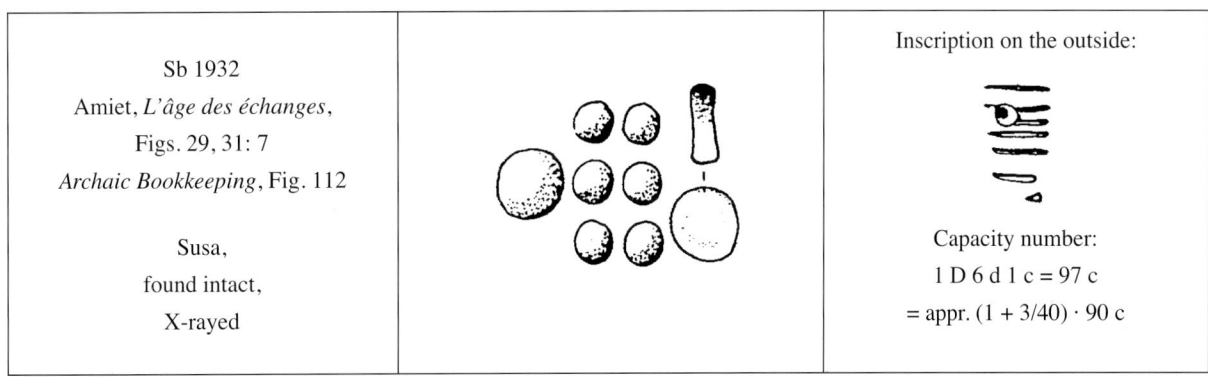

		Inscription on the outside:
Sb 1932 Amiet, *L'âge des échanges*, Figs. 29, 31: 7 *Archaic Bookkeeping*, Fig. 112 Susa, found intact, X-rayed		Capacity number: 1 D 6 d 1 c = 97 c = appr. (1 + 3/40) · 90 c

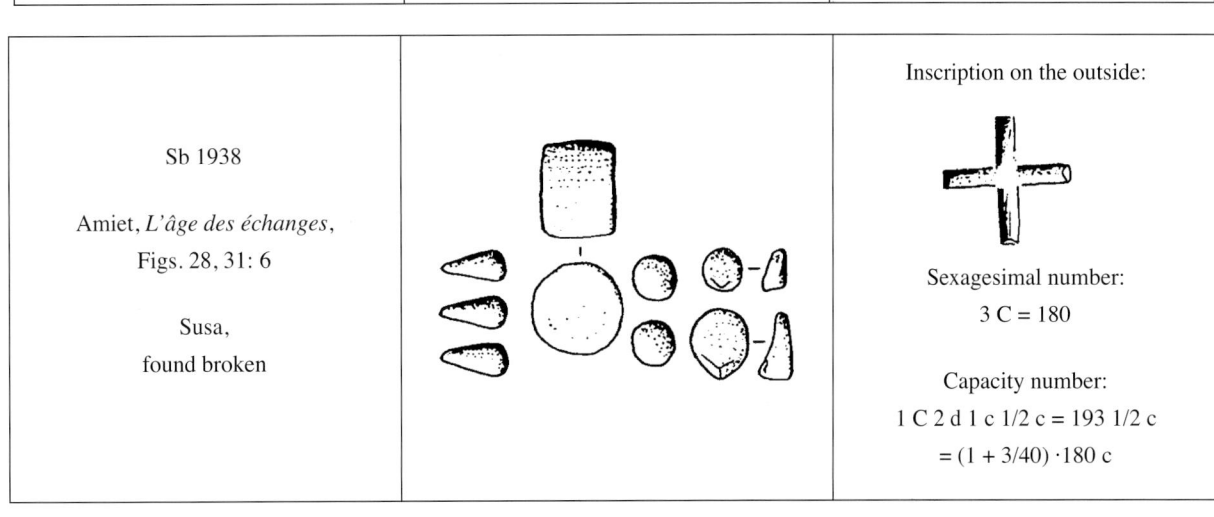

		Inscription on the outside:
Sb 1938 Amiet, *L'âge des échanges*, Figs. 28, 31: 6 Susa, found broken		Sexagesimal number: 3 C = 180 Capacity number: 1 C 2 d 1 c 1/2 c = 193 1/2 c = (1 + 3/40) · 180 c

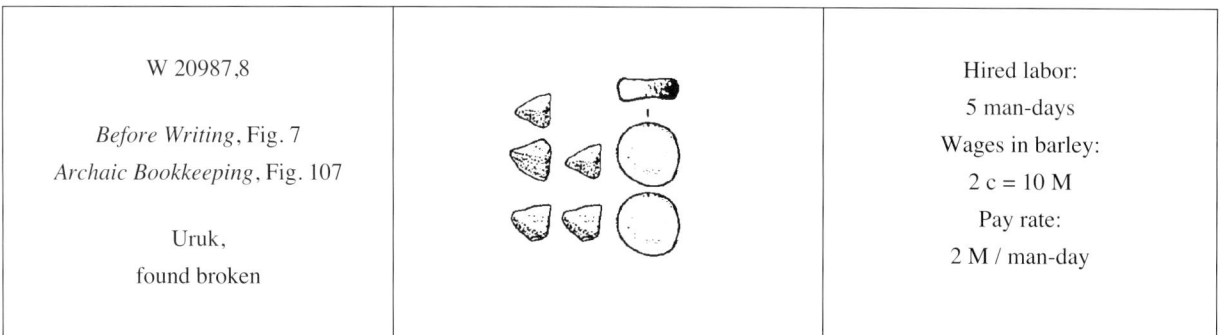

W 20987,8		Hired labor:
Before Writing, Fig. 7		5 man-days
Archaic Bookkeeping, Fig. 107		Wages in barley:
		2 c = 10 M
Uruk,		Pay rate:
found broken		2 M / man-day

If this interpretation is correct, then in **Sb 1967** (below) the *2 big spheres and 4 small spheres* can be explained as 2 D 4 d = 24 d = 360 · 2 M, the wages for something like 360 man-days of hired labor. Consequently, it seems to be reasonable to assume that the *3 big tetrahedrons, the 2 punched middle size tetrahedrons and the 4 small tetrahedrons* inside Sb 1967 stand for 3(100) 2(10) 4 man-days. The corresponding upwards adjusted pay rate in the case of Sb 1967 would then be

$$360 \cdot 2 \text{ M for } 324 \text{ man-days} = (1 + 1/9) \cdot 2 \text{ M / man-day.}$$

This line of reasoning may also help to explain the contents of the spherical envelope **Sb 1927** (next page). The tokens inside Sb 1927 are clearly different from the ones inside the spherical envelopes discussed so far. It seems reasonable, therefore, to assume that they stand for units of sexagesimal counting numbers. The markings on the outside, probably to be read from right to left, as later the proto-Elamite script, stand in a one-to-one relationship with the tokens inside. The biggest of these markings looks like the proto-literate sign for 10 · 60. Consequently, the most straightforward interpretation is that the tokens inside, *1 big punched cone, three small cones, and three lens-formed disks*, stand for the sexagesimal number 1(10 · 60) 3(60) 3(10) = 13 30 = 27 · 30, possibly representing 27 · 30 man-days = 27 man-months. It is interesting to note that if the pay rate is the same as in the case of Sb 1967, then 27 · 30 man-days correspond to an expenditure of 60 d = 6 D = 2 C of barley. Indeed,

$$27 \cdot 30 \text{ man-days} \cdot (1 + 1/9) \cdot 2 \text{ M / man-day} = 30 \cdot 30 \cdot 2 \text{ M} = 30 \cdot 2 \text{ d} = 6 \text{ D} = 2 \text{ C.}$$

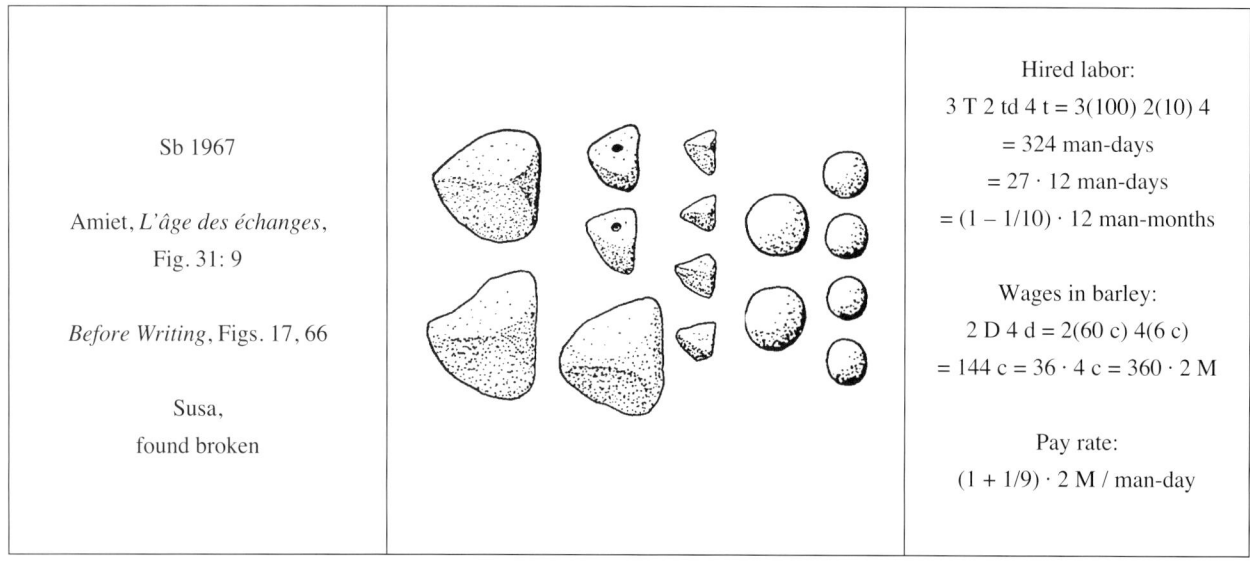

Sb 1967		Hired labor:
		3 T 2 td 4 t = 3(100) 2(10) 4
Amiet, *L'âge des échanges*,		= 324 man-days
Fig. 31: 9		= 27 · 12 man-days
		= (1 – 1/10) · 12 man-months
Before Writing, Figs. 17, 66		Wages in barley:
		2 D 4 d = 2(60 c) 4(6 c)
Susa,		= 144 c = 36 · 4 c = 360 · 2 M
found broken		Pay rate:
		(1 + 1/9) · 2 M / man-day

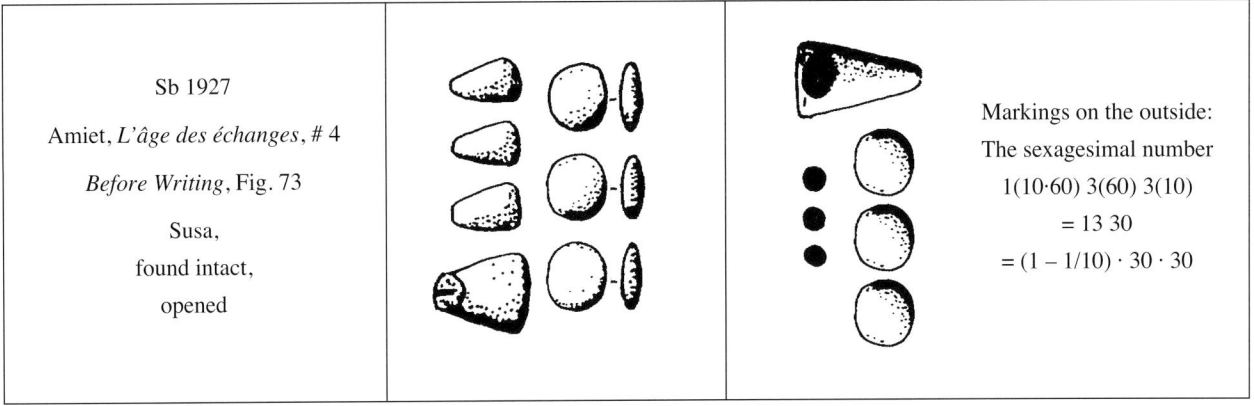

Through arguments of this kind, one is led to the following series of tentative identifications of several kinds of plain tokens inside spherical envelopes with proto-literate number signs:

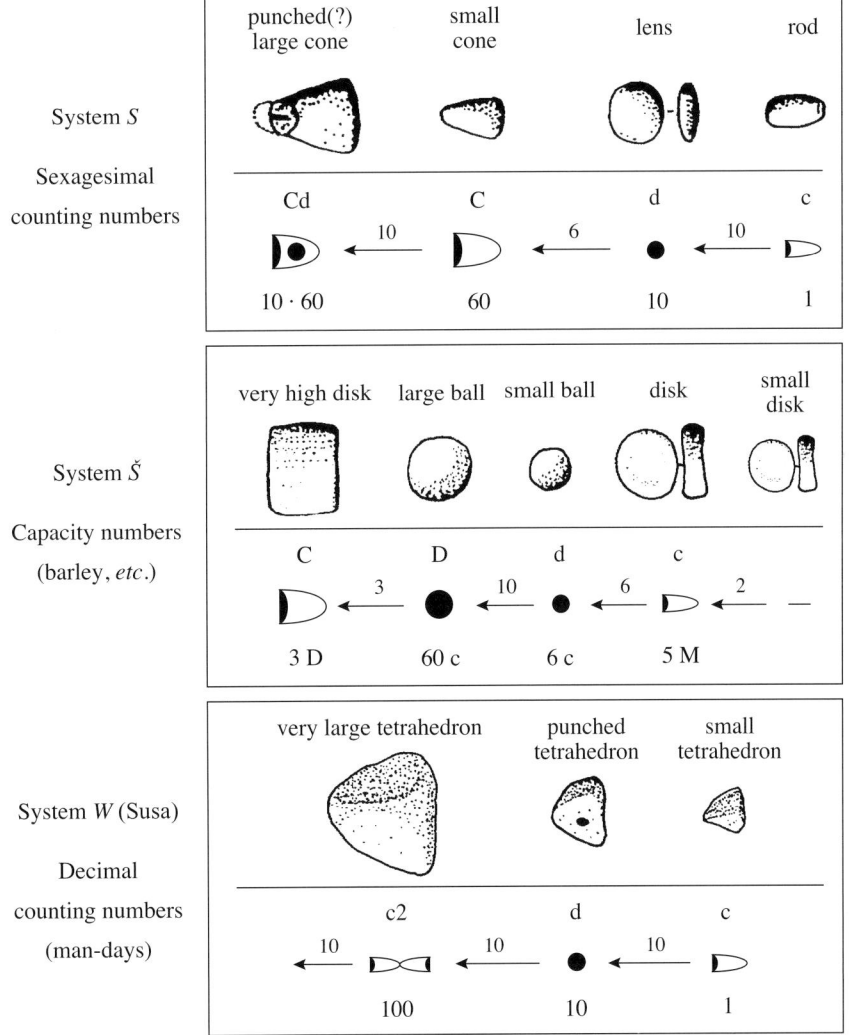

Fig. A4.14. Tentative identification of pre-literate number tokens with proto-literate number signs.

Note that while the proto-literate number signs are polyvalent, so that only the context can fix their values, the pre-literate number signs are, by necessity, monovalent. Indeed, if, for instance, tokens for sexagesimal and capacity numbers are present together inside a spherical envelope, then there is no context that can be used to separate them from each other.

Note also that even if the analysis above should turn out to be close to the truth, there are many known spherical envelopes with contents that cannot be explained in terms of only the simple factor diagrams above for the pre-literate systems S, Š, and W. An example from the Schøyen collection is **MS 4631**, a spherical envelope containing *4 tetrahedrons, 4 crescents with two incised lines, 1 sphere, and 1 rod with a grove*. (To be edited by P. Damerow and R. K. Englund in a separate publication, together with a number of proto-literate texts from the Schøyen Collection.)

More promising is the spherical envelope **MS 4632**, which contains *1 large ball and 8 small balls*, together possibly representing 1 D 8 d = 18 d = 18 · 30 M. In addition it contains *5 small cones*, possibly representing the sexagesimal number 5 · 60, and *3 small tokens of an irregular shape*, which may or may not be badly formed rods, standing for the number 3. Thus, if the three odd tokens are disregarded, the remaining tokens contained in MS 4632 can be tentatively interpreted as

18 · 30 M (of barley) for 5 · 60 (man-days), at a pay rate of (1 – 1/10) · 2 M / man-day.

This interpretation is supported by the observation that MS 4632, W 20987, 8, and Sb 1967, all seem to be concerned with pay rates of about the same size, namely

9/10 · 2 M / man-day	in MS 4632	(from an unknown site)
2 M / man-day	in W 20987, 8	(from Uruk)
10/9 · 2 M / man-day	in Sb 1967	(from Susa)

The meaning of the factors 10/9 and 9/10 remains a mystery, of course, and it is also difficult to know why the (alleged) man-days are represented by tetrahedrons in two of the cases but by cones in the third. The explanation may be that the tetrahedrons stand for time numbers, used to count *days worked*, while the cones stand for sexagesimal numbers, used to count *people working*.

A schematic representation of the contents of MS 4632 is offered below. A color photo of the spherical envelope and its tokens (courtesy P. Damerow) is shown in App. 10.

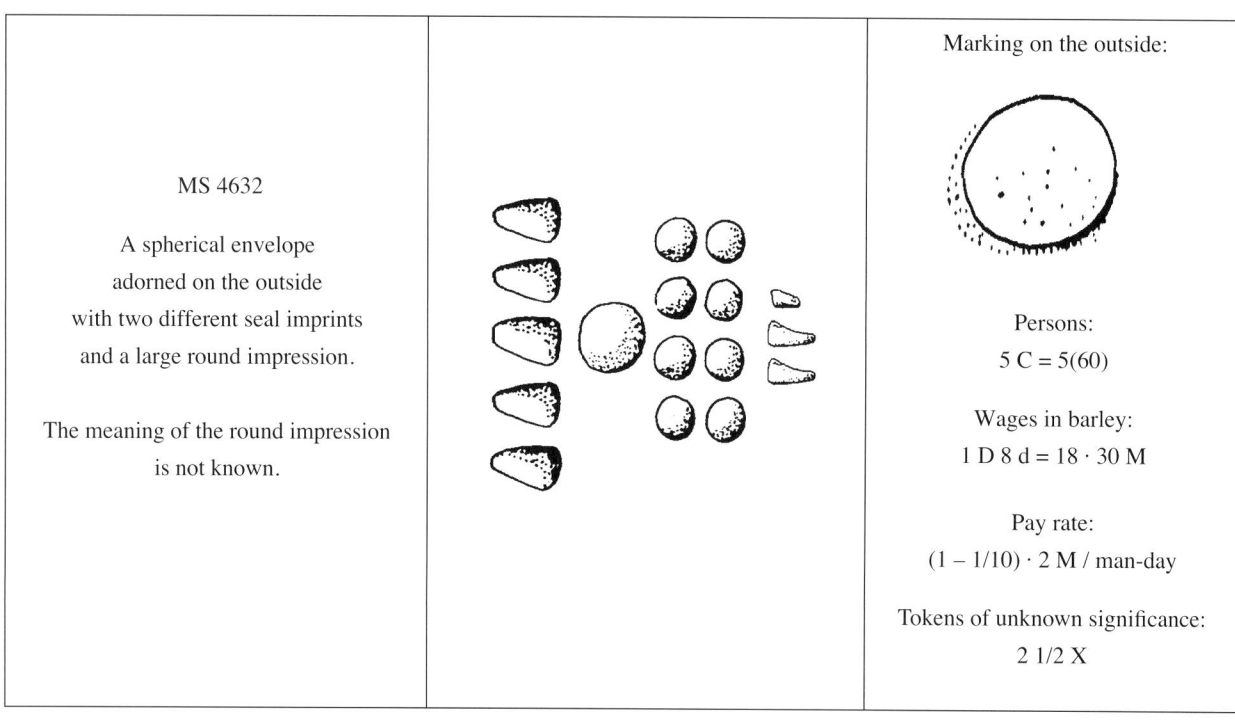

MS 4632

A spherical envelope
adorned on the outside
with two different seal imprints
and a large round impression.

The meaning of the round impression
is not known.

Marking on the outside:

Persons:

5 C = 5(60)

Wages in barley:
1 D 8 d = 18 · 30 M

Pay rate:
(1 – 1/10) · 2 M / man-day

Tokens of unknown significance:
2 1/2 X

Appendix 5
Old Babylonian Complete Metrological Tables

The proper use of the various Sumero-Akkadian systems of measures and measure notations was learned in the scribal schools of the ancient Middle East, through a careful study of texts belonging to a corpus of metro-mathematical exercises, and through the copying and committing to memory of various metrological lists or tables. A single or combined "metrological list" contains the *standard notations*, in ascending order, for a selected assortment of measures belonging to one or several of the systems C (capacity), M (weight or metal), A (area), and L (length). A single or combined "metrological table" is a single or combined metrological list which contains not only the standard notations for the selected measures but also their values as multiples of a certain "basic unit".

There are reasons to believe that all preserved Old Babylonian metrological lists and tables are more or less abridged versions of an original "canonical" combined metrological table. The basic units in this canonical table are the sìla (system C), the gú 'talent' or the ma.na (system M), the šar (system A), and for linear measures the ninda (system Ln) or the kùš 'cubit' (system Lc). In combined metrological lists or tables, the sub-tables for the various systems follow each other in the fixed order C, M, A, Ln, Lc. The attested "ranges" for the respective sub-tables are

C:	from	3' sìla	$(20 \cdot 60^{-1}$ sìla$)$	to	2 šár.gal gur	$(10 \cdot 60^4$ sìla$)$	149 lines
M:	from	2' še 'b.c.'	$(10 \cdot 60^{-2}$ ma.na$)$	to	1(géš) gú	$(1 \cdot 60^2$ ma.na$)$	128 lines
A:	from	3' šar	$(20 \cdot 60^{-1}$ šar$)$	to	2 šár.gal aša$_5$	$(1 \cdot 60^4$ šar$)$	109 lines
Ln:	from	1 šu.si 'finger'	$(10 \cdot 60^{-2}$ ninda$)$	to	1(géš) danna	$(30 \cdot 60^2$ ninda$)$	133 lines

A few known Old Babylonian large table texts, such as the prism **AO 8865** (*MKT 1* (1935), 69), contain both metrological and arithmetical sub-tables, apparently again in a fixed order. (For further examples, see Friberg, *RlA 7* (1990), Sec. 5.1, 543 b, or below, in Sec. A5 d.) It is likely that this fixed order is a clue to the order in which various kinds of mathematical or metrological table texts were studied in some of the Old Babylonian scribal schools: first multiplication tables and tables of squares, then metrological tables, and last of all tables of square sides and cube sides.

A5.1. The Complete Metrological Table for System C(NS/OB)

The 6-sided prism **MS 2723** (Fig. 3.1.3 above) is an almost perfectly preserved complete metrological *table* for system C(NS/OB), with 159 entries. Another, almost complete table for system C is the first sub-table of the combined metrological table **YBC 4633,** in the first five columns on the obverse of a less than perfectly preserved six-columns clay tablet. A photo (not very clear) was published by Nemet-Nejat in *JNES* 54 (1995), Fig. 8. The table ends with the entry [šár.gal gur] 10.

A transliteration of an almost complete metrological *list* for system C on **VAT 2596**, a clay cylinder with 134 entries in 6 columns, was published by Meissner in *BAP* (1893), 58. The list ends with the entry šàr.1.šu.kam še gur '60 šár gur of barley'.

Another well preserved metrological list for system C is the four-columns clay tablet **BM 94889** (Al-Rawi, unpublished), extending from 1/3 sìla še to 1(šár)×1.gal še gur (*obv. i - rev. i*).

A third complete metrological list for system *C*, ending with the entry 1 šár.gal gur, is the first subtable of the combined metrological list **CBM 10990+** (see Fig. A5.5 below).

The range is the same for these three metrological lists as for the metrological table YBC 4633.

The complete metrological table for system *C*, as it appears on MS 2723 (Fig. 3.1.3), is reproduced below. It can be used as a handy reference in the study of other metrological tables for system *C*.

The complete metrological table for the Old Babylonian system *C*

1	gín	=	1	($\cdot\ 60^{-1}$ sìla)		1(barig)	še	=	1	($\cdot\ 60$ sìla)
2	gín	=	2	($\cdot\ 60^{-1}$ sìla)		1(barig)	1(bán)	=	1 10	($\cdot\ 1$ sìla)
3	gín	=	3	($\cdot\ 60^{-1}$ sìla)		1(barig)	2(bán)	=	1 20	($\cdot\ 1$ sìla)
4	gín	=	4	($\cdot\ 60^{-1}$ sìla)		1(barig)	3(bán)	=	1 30	($\cdot\ 1$ sìla)
5	gín	=	5	($\cdot\ 60^{-1}$ sìla)		1(barig)	4(bán)	=	1 40	($\cdot\ 1$ sìla)
6	gín	=	6	($\cdot\ 60^{-1}$ sìla)		1(barig)	5(bán)	=	1 50	($\cdot\ 1$ sìla)
7	gín	=	7	($\cdot\ 60^{-1}$ sìla)		2(barig)	še	=	1	($\cdot\ 60$ sìla)
8	gín	=	8	($\cdot\ 60^{-1}$ sìla)		2(barig)	1(bán)	=	2 10	($\cdot\ 1$ sìla)
9	gín	=	9	($\cdot\ 60^{-1}$ sìla)		2(barig)	2(bán)	=	2 20	($\cdot\ 1$ sìla)
10	gín	=	10	($\cdot\ 60^{-1}$ sìla)		2(barig)	3(bán)	=	2 30	($\cdot\ 1$ sìla)
11	gín	=	11	($\cdot\ 60^{-1}$ sìla)		2(barig)	4(bán)	=	2 40	($\cdot\ 1$ sìla)
12	gín	=	12	($\cdot\ 60^{-1}$ sìla)		2(barig)	5(bán)	=	2 50	($\cdot\ 1$ sìla)
13	gín	=	13	($\cdot\ 60^{-1}$ sìla)		3(barig)	še	=	3	($\cdot\ 60$ sìla)
14	gín	=	14	($\cdot\ 60^{-1}$ sìla)		3(barig)	1(bán)	=	3 10	($\cdot\ 1$ sìla)
15	gín	=	15	($\cdot\ 60^{-1}$ sìla)		3(barig)	2(bán)	=	3 20	($\cdot\ 1$ sìla)
16	gín	=	16	($\cdot\ 60^{-1}$ sìla)		3(barig)	3(bán)	=	3 30	($\cdot\ 1$ sìla)
17	gín	=	17	($\cdot\ 60^{-1}$ sìla)		3(barig)	4(bán)	=	3 40	($\cdot\ 1$ sìla)
18	gín	=	18	($\cdot\ 60^{-1}$ sìla)		3(barig)	5(bán)	=	3 50	($\cdot\ 1$ sìla)
19	gín	=	19	($\cdot\ 60^{-1}$ sìla)		4(barig)	še	=	4	($\cdot\ 60$ sìla)
3'	sìla	=	20	($\cdot\ 60^{-1}$ sìla)		4(barig)	1(bán)	=	4 10	($\cdot\ 1$ sìla)
2'	sìla	=	30	($\cdot\ 60^{-1}$ sìla)		4(barig)	2(bán)	=	4 20	($\cdot\ 1$ sìla)
3"	sìla	=	40	($\cdot\ 60^{-1}$ sìla)		4(barig)	3(bán)	=	4 30	($\cdot\ 1$ sìla)
6"	sìla	=	50	($\cdot\ 60^{-1}$ sìla)		4(barig)	4(bán)	=	4 40	($\cdot\ 1$ sìla)
1	sìla	=	1	($\cdot\ 1$ sìla)		4(barig)	5(bán)	=	4 50	($\cdot\ 1$ sìla)
1 3'	sìla	=	1 20	($\cdot\ 60^{-1}$ sìla)		1	gur	=	5	($\cdot\ 60$ sìla)
1 2'	sìla	=	1 30	($\cdot\ 60^{-1}$ sìla)		1(gur) gur	1(barig)	=	6	($\cdot\ 60$ sìla)
1 3"	sìla	=	1 40	($\cdot\ 60^{-1}$ sìla)		1(gur) gur	2(barig)	=	7	($\cdot\ 60$ sìla)
1 6"	sìla	=	1 50	($\cdot\ 60^{-1}$ sìla)		1(gur) gur	3(barig)	=	8	($\cdot\ 60$ sìla)
2	sìla	=	2	($\cdot\ 1$ sìla)		1(gur) gur	4(barig)	=	9	($\cdot\ 60$ sìla)
3	sìla	=	3	($\cdot\ 1$ sìla)		2(gur)	gur	=	10	($\cdot\ 60$ sìla)
4	sìla	=	4	($\cdot\ 1$ sìla)		3(gur)	gur	=	15	($\cdot\ 60$ sìla)
5	sìla	=	5	($\cdot\ 1$ sìla)		4(gur)	gur	=	20	($\cdot\ 60$ sìla)
6	sìla	=	6	($\cdot\ 1$ sìla)		5(gur)	gur	=	25	($\cdot\ 60$ sìla)
7	sìla	=	7	($\cdot\ 1$ sìla)		6(gur)	gur	=	30	($\cdot\ 60$ sìla)
8	sìla	=	8	($\cdot\ 1$ sìla)		7(gur)	gur	=	35	($\cdot\ 60$ sìla)
9	sìla	=	9	($\cdot\ 1$ sìla)		8(gur)	gur	=	40	($\cdot\ 60$ sìla)
1(bán)	še	=	10	($\cdot\ 1$ sìla)		9(gur)	gur	=	45	($\cdot\ 60$ sìla)
1(bán)	1 sìla	=	11	($\cdot\ 1$ sìla)		10	gur	=	50	($\cdot\ 60$ sìla)
1(bán)	2 sìla	=	12	($\cdot\ 1$ sìla)		11(gur)	gur	=	55	($\cdot\ 60$ sìla)
1(bán)	3 sìla	=	13	($\cdot\ 1$ sìla)		12(gur)	gur	=	1	($\cdot\ 60^2$ sìla)
1(bán)	4 sìla	=	14	($\cdot\ 1$ sìla)		13(gur)	gur	=	1 05	($\cdot\ 60$ sìla)
1(bán)	5 sìla	=	15	($\cdot\ 1$ sìla)		14 (gur)	gur	=	1 10	($\cdot\ 60$ sìla)
1(bán)	6 sìla	=	16	($\cdot\ 1$ sìla)		15(gur)	gur	=	1 15	($\cdot\ 60$ sìla)
1(bán)	7 sìla	=	17	($\cdot\ 1$ sìla)		16(gur)	gur	=	1 20	($\cdot\ 60$ sìla)
1(bán)	8 sìla	=	18	($\cdot\ 1$ sìla)		17(gur)	gur	=	1 25	($\cdot\ 60$ sìla)
1(bán)	9 sìla	=	19	($\cdot\ 1$ sìla)		18(gur)	gur	=	1 30	($\cdot\ 60$ sìla)
2(bán)	še	=	20	($\cdot\ 1$ sìla)		19(gur)	gur	=	1 35	($\cdot\ 60$ sìla)
3(bán)	še	=	30	($\cdot\ 1$ sìla)		20	gur	=	1 40	($\cdot\ 60$ sìla)
4(bán)	še	=	40	($\cdot\ 1$ sìla)		30	gur	=	2 30	($\cdot\ 60$ sìla)
5(bán)	še	=	50	($\cdot\ 1$ sìla)		40	gur	=	3 20	($\cdot\ 60$ sìla)

50	gur	=	4 10	(· 60 sìla)
1(géš)	gur	=	5	(· 60² sìla)
1(géš) 10	gur	=	5 50	(· 60 sìla)
1(géš) 20	gur	=	6 40	(· 60 sìla)
1(géš) 30	gur	=	7 30	(· 60 sìla)
1(géš) 40	gur	=	8 20	(· 60 sìla)
1(géš) 50	gur	=	9 10	(· 60 sìla)
2(géš)	gur	=	10	(· 60² sìla)
3(géš)	gur	=	15	(· 60² sìla)
4(géš)	gur	=	20	(· 60² sìla)
5(géš)	gur	=	25	(· 60² sìla)
6(géš)	gur	=	30	(· 60² sìla)
7(géš)	gur	=	35	(· 60² sìla)
8(géš)	gur	=	40	(· 60² sìla)
9(géš)	gur	=	45	(· 60² sìla)
10(géš)	gur	=	50	(· 60² sìla)
11(géš)	gur	=	55	(· 60² sìla)
12(géš)	gur	=	1	(· 60³ sìla)
13(géš)	gur	=	1 05	(· 60² sìla)
14(géš)	gur	=	1 10	(· 60² sìla)
15(géš)	gur	=	1 15	(· 60² sìla)
16(géš)	gur	=	1 20	(· 60² sìla)
17(géš)	gur	=	1 25	(· 60² sìla)
18(géš)	gur	=	1 30	(· 60² sìla)
19(géš)	gur	=	1 35	(· 60² sìla)
20(géš)	gur	=	1 40	(· 60² sìla)
30(géš)	gur	=	2 30	(· 60² sìla)
40(géš)	gur	=	3 20	(· 60² sìla)
50(géš)	gur	=	4 10	(· 60² sìla)
1(šár)	gur	=	5	(· 60³ sìla)

1(šár) 10(géš)	gur	=	5 50	(· 60² sìla)
1(šár) 20(géš)	gur	=	6 40	(· 60² sìla)
1(šár) 30(géš)	gur	=	7 30	(· 60² sìla)
1(šár) 40(géš)	gur	=	8 20	(· 60² sìla)
1(šár) 50(géš)	gur	=	9 10	(· 60² sìla)
2(šár)	gur	=	10	(· 60³ sìla)
3(šár)	gur	=	15	(· 60³ sìla)
4(šár)	gur	=	20	(· 60³ sìla)
5(šár)	gur	=	25	(· 60³ sìla)
6(šár)	gur	=	30	(· 60³ sìla)
7(šár)	gur	=	35	(· 60³ sìla)
8(šár)	gur	=	40	(· 60³ sìla)
9(šár)	gur	=	45	(· 60³ sìla)
10(šár)	gur	=	50	(· 60³ sìla)
11(šár)	gur	=	55	(· 60³ sìla)
12(šár)	gur	=	1	(· 60⁴ sìla)
13(šár)	gur	=	1 05	(· 60³ sìla)
14(šár)	gur	=	1 10	(· 60³ sìla)
15(šár)	gur	=	1 15	(· 60³ sìla)
16(šár)	gur	=	1 20	(· 60³ sìla)
17(šár)	gur	=	1 25	(· 60³ sìla)
18(šár)	gur	=	1 30	(· 60³ sìla)
19(šár)	gur	=	1 35	(· 60³ sìla)
20(šár)	gur	=	1 40	(· 60³ sìla)
30(šár)	gur	=	2 30	(· 60³ sìla)
40(šár)	gur	=	3 20	(· 60³ sìla)
50(šár)	gur	=	4 10	(· 60³ sìla)
šár.1.gal	gur	=	5	(· 60⁴ sìla)
šár.2.gal	gur	=	10	(· 60⁴ sìla)

A5.2. The Complete Metrological Table for System M(NS/OB)

The 6-sided prism **NBC 2513** (Fig. A5.1 below) is an almost perfectly preserved complete metrological table for system M(NS/OB), with 128 entries. A hand copy of the text was published by Nies and Keiser in *BIN 2*, 36 (1920). They also published a photo of the prism (*op. cit.*, pl. 66).

The complete metrological table for system M, as it appears on NBC 2513, is reproduced below:

The complete metrological table for the Old Babylonian system M

2'	še	=	10	(· 60⁻³ ma.na)	
1	še	=	20	(· 60⁻³ ma.na)	
1 2'	še	=	30	(· 60⁻³ ma.na)	
2	še	=	40	(· 60⁻³ ma.na)	
2 2'	še	=	50	(· 60⁻³ ma.na)	
3	še	=	1	(· 60⁻² ma.na)	
4	še	=	1 20	(· 60⁻³ ma.na)	
5	še	=	1 40	(· 60⁻³ ma.na)	
6	še	=	2	(· 60⁻² ma.na)	
7	še	=	2 20	(· 60⁻³ ma.na)	<
8	še	=	2 40	(· 60⁻³ ma.na)	
9	še	=	3	(· 60⁻² ma.na)	
10	še	=	3 20	(· 60⁻³ ma.na)	
11	še	=	3 40	(· 60⁻³ ma.na)	
12	še	=	4	(· 60⁻² ma.na)	

13	še	=	4 20	(· 60⁻³ ma.na)	
14	še	=	4 40	(· 60⁻³ ma.na)	
15	še	=	5	(· 60⁻² ma.na)	
16	še	=	5 20	(· 60⁻³ ma.na)	
17	še	=	5 40	(· 60⁻³ ma.na)	<
18	še	=	6	(· 60⁻² ma.na)	
19	še	=	6 20	(· 60⁻³ ma.na)	
20	še	=	6 40	(· 60⁻³ ma.na)	
21	še	=	7	(· 60⁻² ma.na)	
22	še	=	7 20	(· 60⁻³ ma.na)	
22 2'	še	=	7 30	(· 60⁻³ ma.na)	
23	še	=	7 40	(· 60⁻³ ma.na)	
24	še	=	8	(· 60⁻² ma.na)	
25	še	=	8 20	(· 60⁻³ ma.na)	
26	še	=	8 40	(· 60⁻³ ma.na)	<

27	še	=	9	$(\cdot\ 60^{-2}$ ma.na$)$	
28	še	=	9 20	$(\cdot\ 60^{-3}$ ma.na$)$	
29	še	=	9 40	$(\cdot\ 60^{-3}$ ma.na$)$	
1/6		=	10	$(\cdot\ 60^{-2}$ ma.na$)$	
1/6 5	še	=	11 40	$(\cdot\ 60^{-3}$ ma.na$)$	
1/6 10	še	=	13 20	$(\cdot\ 60^{-3}$ ma.na$)$	
1/4		=	15	$(\cdot\ 60^{-2}$ ma.na$)$	
1/4 5	še	=	16 40	$(\cdot\ 60^{-3}$ ma.na$)$	
1/4 10	še	=	18 20	$(\cdot\ 60^{-3}$ ma.na$)$	
3'	gín	=	20	$(\cdot\ 60^{-2}$ ma.na$)$	<
2'	gín	=	30	$(\cdot\ 60^{-2}$ ma.na$)$	
3"	gín	=	40	$(\cdot\ 60^{-2}$ ma.na$)$	
6"	gín	=	50	$(\cdot\ 60^{-2}$ ma.na$)$	
1	gín	=	1	$(\cdot\ 60^{-1}$ ma.na$)$	
1 1/6		=	1 10	$(\cdot\ 60^{-2}$ ma.na$)$	
1 1/4		=	1 15	$(\cdot\ 60^{-2}$ ma.na$)$	
1 3'	gín	=	1 20	$(\cdot\ 60^{-2}$ ma.na$)$	
1 2'	gín	=	1 30	$(\cdot\ 60^{-2}$ ma.na$)$	
1 3"	gín	=	1 40	$(\cdot\ 60^{-2}$ ma.na$)$	
1 6"	gín	=	1 50	$(\cdot\ 60^{-2}$ ma.na$)$	<
2	gín	=	2	$(\cdot\ 60^{-1}$ ma.na$)$	
3	gín	=	3	$(\cdot\ 60^{-1}$ ma.na$)$	
4	gín	=	4	$(\cdot\ 60^{-1}$ ma.na$)$	
5	gín	=	5	$(\cdot\ 60^{-1}$ ma.na$)$	
6	gín	=	6	$(\cdot\ 60^{-1}$ ma.na$)$	
7	gín	=	7	$(\cdot\ 60^{-1}$ ma.na$)$	
8	gín	=	8	$(\cdot\ 60^{-1}$ ma.na$)$	
9	gín	=	9	$(\cdot\ 60^{-1}$ ma.na$)$	
10	gín	=	10	$(\cdot\ 60^{-1}$ ma.na$)$	
11	gín	=	11	$(\cdot\ 60^{-1}$ ma.na$)$	<
12	gín	=	12	$(\cdot\ 60^{-1}$ ma.na$)$	
13	gín	=	13	$(\cdot\ 60^{-1}$ ma.na$)$	
14	gín	=	14	$(\cdot\ 60^{-1}$ ma.na$)$	
15	gín	=	15	$(\cdot\ 60^{-1}$ ma.na$)$	
16	gín	=	16	$(\cdot\ 60^{-1}$ ma.na$)$	
17	gín	=	17	$(\cdot\ 60^{-1}$ ma.na$)$	
18	gín	=	18	$(\cdot\ 60^{-1}$ ma.na$)$	
19	gín	=	19	$(\cdot\ 60^{-1}$ ma.na$)$	
3'	ma.na	=	20	$(\cdot\ 60^{-1}$ ma.na$)$	
2'	ma.na	=	30	$(\cdot\ 60^{-1}$ ma.na$)$	
3"	ma.na	=	40	$(\cdot\ 60^{-1}$ ma.na$)$	<
6"	ma.na	=	50	$(\cdot\ 60^{-1}$ ma.na$)$	
1	ma.na	=	1	$(\cdot\ 1$ ma.na$)$	
1 3'	ma.na	=	1 20	$(\cdot\ 60^{-1}$ ma.na$)$	
1 2'	ma.na	=	1 30	$(\cdot\ 60^{-1}$ ma.na$)$	
1 3"	ma.na	=	1 40	$(\cdot\ 60^{-1}$ ma.na$)$	
1 6"	ma.na	=	1 50	$(\cdot\ 60^{-1}$ ma.na$)$	
2	ma.na	=	2	$(\cdot\ 1$ ma.na$)$	
3	ma.na	=	3	$(\cdot\ 1$ ma.na$)$	

4	ma.na	=	4	$(\cdot\ 1$ ma.na$)$	<
5	ma.na	=	5	$(\cdot\ 1$ ma.na$)$	
6	ma.na	=	1	$(\cdot\ 1$ ma.na$)$	
7	ma.na	=	7	$(\cdot\ 1$ ma.na$)$	
8	ma.na	=	8	$(\cdot\ 1$ ma.na$)$	
9	ma.na	=	9	$(\cdot\ 1$ ma.na$)$	
10	ma.na	=	10	$(\cdot\ 1$ ma.na$)$	
11	ma.na	=	11	$(\cdot\ 1$ ma.na$)$	
12	ma.na	=	12	$(\cdot\ 1$ ma.na$)$	
13	ma.na	=	13	$(\cdot\ 1$ ma.na$)$	
14	ma.na	=	14	$(\cdot\ 1$ ma.na$)$	<
15	ma.na	=	15	$(\cdot\ 1$ ma.na$)$	
16	ma.na	=	16	$(\cdot\ 1$ ma.na$)$	
17	ma.na	=	17	$(\cdot\ 1$ ma.na$)$	
18	ma.na	=	18	$(\cdot\ 1$ ma.na$)$	
19	ma.na	=	19	$(\cdot\ 1$ ma.na$)$	
20	ma.na	=	20	$(\cdot\ 1$ ma.na$)$	
30	ma.na	=	30	$(\cdot\ 1$ ma.na$)$	
40	ma.na	=	40	$(\cdot\ 1$ ma.na$)$	
50	ma.na	=	50	$(\cdot\ 1$ ma.na$)$	<
1	gú	=	1	$(\cdot\ 60$ ma.na$)$	
1 gú 10	ma.na	=	1 10	$(\cdot\ 1$ ma.na$)$	
1 gú 20	ma.na	=	1 20	$(\cdot\ 1$ ma.na$)$	
1 gú 30	ma.na	=	1 30	$(\cdot\ 1$ ma.na$)$	
1 gú 40	ma.na	=	1 40	$(\cdot\ 1$ ma.na$)$	
1 gú 50	ma.na	=	1 50	$(\cdot\ 1$ ma.na$)$	
2	gú	=	2	$(\cdot\ 60$ ma.na$)$	
3	gú	=	3	$(\cdot\ 60$ ma.na$)$	
4	gú	=	4	$(\cdot\ 60$ ma.na$)$	
5	gú	=	5	$(\cdot\ 60$ ma.na$)$	<
6	gú	=	6	$(\cdot\ 60$ ma.na$)$	
7	gú	=	7	$(\cdot\ 60$ ma.na$)$	
8	gú	=	8	$(\cdot\ 60$ ma.na$)$	
9	gú	=	9	$(\cdot\ 60$ ma.na$)$	
10	gú	=	10	$(\cdot\ 60$ ma.na$)$	
11	gú	=	11	$(\cdot\ 60$ ma.na$)$	
12	gú	=	12	$(\cdot\ 60$ ma.na$)$	
13	gú	=	13	$(\cdot\ 60$ ma.na$)$	
14	gú	=	14	$(\cdot\ 60$ ma.na$)$	
15	gú	=	15	$(\cdot\ 60$ ma.na$)$	<
16	gú	=	16	$(\cdot\ 60$ ma.na$)$	
17	gú	=	17	$(\cdot\ 60$ ma.na$)$	
18	gú	=	18	$(\cdot\ 60$ ma.na$)$	
19	gú	=	19	$(\cdot\ 60$ ma.na$)$	
20	gú	=	20	$(\cdot\ 60$ ma.na$)$	
30	gú	=	30	$(\cdot\ 60$ ma.na$)$	
40	gú	=	40	$(\cdot\ 60$ ma.na$)$	
50	gú	=	50	$(\cdot\ 60$ ma.na$)$	
1(géš)!	gú	=	1	$(\cdot\ 60^{2}$ ma.na$)$	

The last couple of entries in this table are 50 gú 50 '50 talents = 50 $(\cdot\ 60$ minas)' and 1 gú 1 kù.babbar '60! talents = 1 $(\cdot\ 60^{2}$ minas) of silver'. Here, the sign for 60 should have been a vertical wedge, but it was written by mistake as a horizontal wedge. (This is clear on the photo of the prism.) Horizontal wedges are normally used for 1, 2, ..., 9 talents, not for sixties of talents!

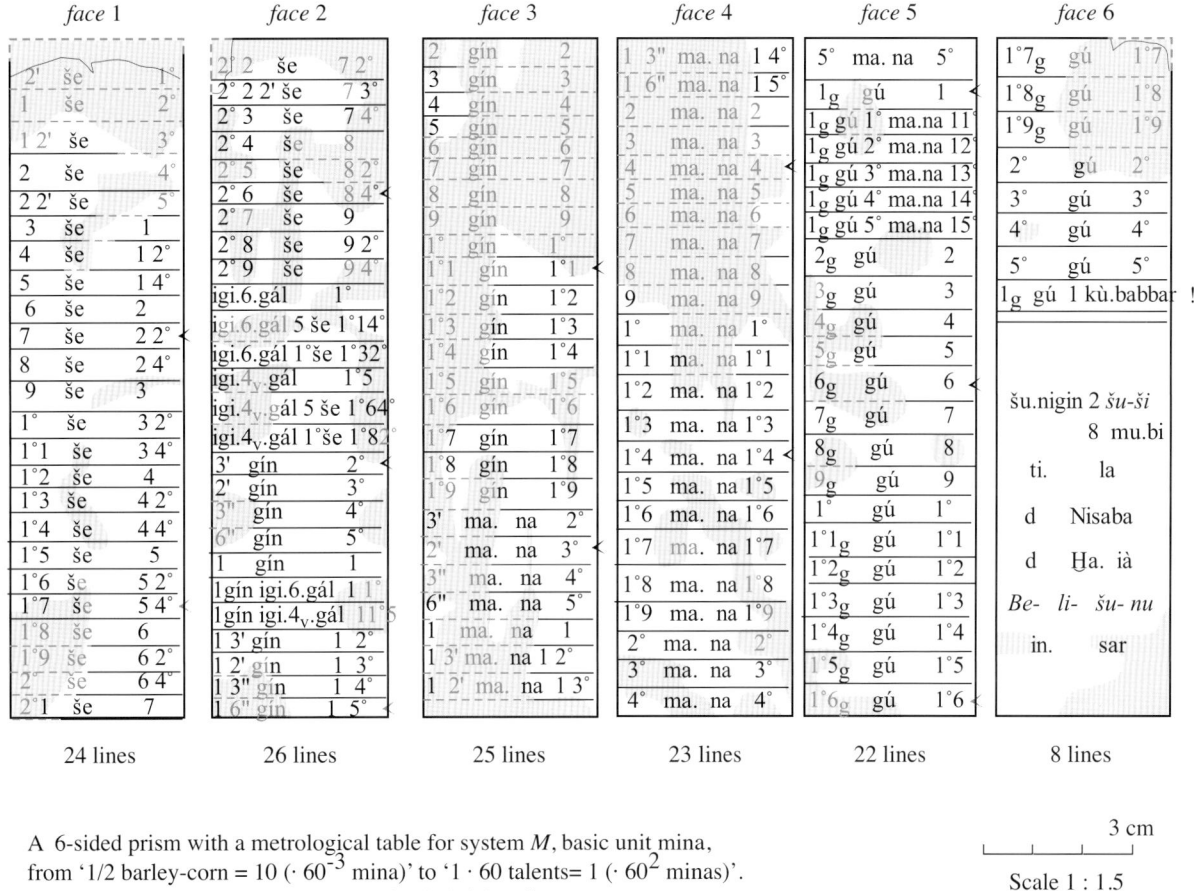

face 1 (24 lines)

2'	še	1°
1	še	2°
1 2'	še	3°
2	še	4°
2 2'	še	5°
3	še	1
4	še	1 2°
5	še	1 4°
6	še	2
7	še	2 2°
8	še	2 4°
9	še	3
1°	še	3 2°
1°1	še	3 4°
1°2	še	4
1°3	še	4 2°
1°4	še	4 4°
1°5	še	5
1°6	še	5 2°
1°7	še	5 4°
1°8	še	6
1°9	še	6 2°
2°	še	6 4°
2°1	še	7

face 2 (26 lines)

2° 2	še	7 2°
2° 2 2'	še	7 3°
2° 3	še	7 4°
2° 4	še	8
2° 5	še	8 2°
2° 6	še	8 4°
2° 7	še	9
2° 8	še	9 2°
2° 9	še	9 4°
igi.6.gál		1°
igi.6.gál 5 še		1°14°
igi.6.gál 1 še		1°32°
igi.4$_v$.gál		1°5
igi.4$_v$.gál 5 še		1°64°
igi.4$_v$.gál 1 še		1°8°
3'	gín	2°
2'	gín	3°
3"	gín	4°
6"	gín	5°
1	gín	1
1 gín igi.6.gál		1 1°
1 gín igi.4$_v$.gál		1 1 5°
1 3'	gín	1 2°
1 2'	gín	1 3°
1 3"	gín	1 4°
1 6"	gín	1 5°

face 3 (25 lines)

2	gín	2
3	gín	3
4	gín	4
5	gín	5
6	gín	6
7	gín	7
8	gín	8
9	gín	9
1°	gín	1°
1°1	gín	1°1
1°2	gín	1°2
1°3	gín	1°3
1°4	gín	1°4
1°5	gín	1°5
1°6	gín	1°6
1°7	gín	1°7
1°8	gín	1°8
1°9	gín	1°9
3'	ma. na	2°
2'	ma. na	3°
3"	ma. na	4°
6"	ma. na	5°
1	ma. na	1
1 3'	ma. na	1 2°
1 2'	ma. na	1 3°

face 4 (23 lines)

1 3"	ma. na	1 4°
1 6"	ma. na	1 5°
2	ma. na	2
3	ma. na	3
4	ma. na	4°
5	ma. na	5
6	ma. na	6
7	ma. na	7
8	ma. na	8
9	ma. na	9
1°	ma. na	1°
1°1	ma. na	1°1
1°2	ma. na	1°2
1°3	ma. na	1°3
1°4	ma. na	1°4
1°5	ma. na	1°5
1°6	ma. na	1°6
1°7	ma. na	1°7
1°8	ma. na	1°8
1°9	ma. na	1°9
2°	ma. na	2°
3°	ma. na	3°
4°	ma. na	4°

face 5 (22 lines)

5°	ma. na	5°
1$_g$	gú	1
1$_g$ gú 1°	ma.na	11°
1$_g$ gú 2°	ma.na	12°
1$_g$ gú 3°	ma.na	13°
1$_g$ gú 4°	ma.na	14°
1$_g$ gú 5°	ma.na	15°
2$_g$	gú	2
3$_g$	gú	3
4$_g$	gú	4
5$_g$	gú	5
6$_g$	gú	6
7$_g$	gú	7
8$_g$	gú	8
9$_g$	gú	9
1°	gú	1°
1°1$_g$	gú	1°1
1°2$_g$	gú	1°2
1°3$_g$	gú	1°3
1°4$_g$	gú	1°4
1°5$_g$	gú	1°5
1°6$_g$	gú	1°6

face 6 (8 lines)

1°7$_g$	gú	1°7
1°8$_g$	gú	1°8
1°9$_g$	gú	1°9
2°	gú	2°
3°	gú	3°
4°	gú	4°
5°	gú	5°
1$_g$ gú	1 kù.babbar	!

šu.nigin 2 šu-ši
8 mu.bi
ti.   la
d   Nisaba
d   Ḫa. ià
Be- li- šu- nu
in.   sar

A 6-sided prism with a metrological table for system *M*, basic unit mina, from '1/2 barley-corn = 10 (· 60^{-3} mina)' to '1 · 60 talents= 1 (· 60^2 minas)'. Altogether 128 lines. Variant number sign in igi.4$_v$.gál.

3 cm
Scale 1 : 1.5

Fig. A5.1. NBC 2513. System *M*(NS/OB), from 2' še = 10 to 1 (00) gú = 1.
(The hand copy here of the text is based on Nies' and Keiser's hand copy and photos in *BIN* 2.)

An interesting and fairly well preserved metrological *list* for system *M* is Clay, **BRM 4, 41** (1923). It has the same range as NBC 2513, proceeding from [1/2 še] kù.babbar to 1 (00) gú kù.babbar. A less well preserved metrological list for the same range is **VAT 1155** (Meissner, *Beiträge* (1893), 58). On the other hand, the sub-table for system *M* in the combined metrological list **CBM 10990+** (Figs. A5.5-6 below) has a much greater range (182 lines), from 1 še kù.babbar to 1 šár.gal gú.

A5.3. The Complete Metrological Table for System A(NS/OB)

The clay tablet **A 21984** (Fig. A5.2 below; Greengus, *OBT Ishchali* (1979), 292) is an almost perfectly preserved complete metrological table for system *A*(NS/OB), with 109 entries, from 1/3 šar = 20 to 2 šár.gal aša$_5$ = 1. (The conform transliteration of the text in Fig. A5.2 below is based on Greengus' hand copy in *OBT Ishchali*.)

The range of the sub-table for system A in the combined metrological *list* **CBM 10990+** (Fig. A5.5 below) is about the same, from 1 šar a.šà to 1 šár.gal aša$_5$ (106 lines).

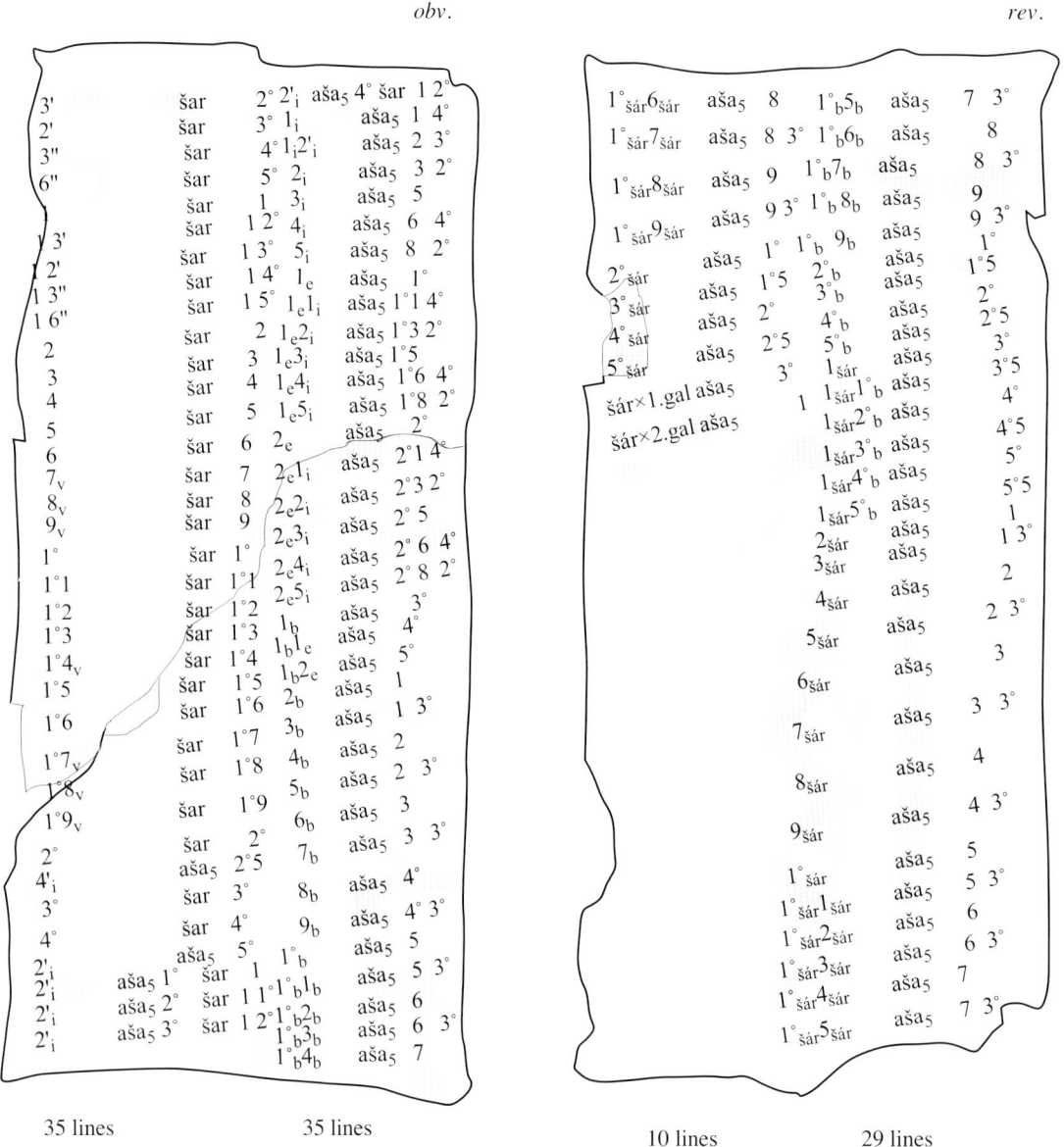

Fig. A5.2. A 21984. System *A*, from 3' šar = 20 to šár×2.gal aša₅ = 1.

The complete metrological table for system *A*, as it appears on A 21984, is reproduced below.

The complete metrological table for the Old Babylonian system *A*.

3'	šar	=	20	($\cdot\ 60^{-1}$ šar)		4	šar	=	4	($\cdot$ 1 šar)
2'	šar	=	30	($\cdot\ 60^{-1}$ šar)		5	šar	=	5	($\cdot$ 1 šar)
3"	šar	=	40	($\cdot\ 60^{-1}$ šar)		6	šar	=	6	($\cdot$ 1 šar)
6"	šar	=	50	($\cdot\ 60^{-1}$ šar)		7	šar	=	7	($\cdot$ 1 šar)
1	šar	=	1	($\cdot$ 1 šar)		8	šar	=	8	($\cdot$ 1 šar)
1 3'	šar	=	1 20	($\cdot\ 60^{-1}$ šar)		9	šar	=	9	($\cdot$ 1 šar)
1 2'	šar	=	1 30	($\cdot\ 60^{-1}$ šar)		10	šar	=	10	($\cdot$ 1 šar)
1 3"	šar	=	1 40	($\cdot\ 60^{-1}$ šar)		11	šar	=	11	($\cdot$ 1 šar)
1 6"	šar	=	1 50	($\cdot\ 60^{-1}$ šar)		12	šar	=	12	($\cdot$ 1 šar)
2	šar	=	2	($\cdot$ 1 šar)		13	šar	=	13	($\cdot$ 1 šar)
3	šar	=	3	($\cdot$ 1 šar)		14	šar	=	14	($\cdot$ 1 šar)

15	šar	=	15	(· 1 šar)
16	šar	=	16	(· 1 šar)
17	šar	=	17	(· 1 šar)
18	šar	=	18	(· 1 šar)
19	šar	=	19	(· 1 šar)
20	šar	=	20	(· 1 šar)
4'(iku)	aša$_5$	=	25	(· 1 šar)
30	šar	=	30	(· 1 šar)
40	šar	=	40	(· 1 šar)
2'(iku)	aša$_5$	=	50	(· 1 šar)
2'(iku) aša$_5$ 10	šar	=	1	(· 60 šar)
2'(iku) aša$_5$ 20	šar	=	1 10	(· 1 šar)
2'(iku) aša$_5$ 30	šar	=	1 20	(· 1 šar)
2'(iku) aša$_5$ 40	šar	=	1 30	(· 1 šar)
1(iku)	aša$_5$	=	1 40	(· 1 šar)
1 2'(iku)	aša$_5$	=	2 30	(· 1 šar)
2(iku)	aša$_5$	=	3 20	(· 1 šar)
3(iku)	aša$_5$	=	5	(· 60 šar)
4(iku)	aša$_5$	=	6 40	(· 1 šar)
5(iku)	aša$_5$	=	8 20	(· 1 šar)
1(èše)	aša$_5$	=	10	(· 60 šar)
1(èše) 1 (iku)	aša$_5$	=	11 40	(· 1 šar)
1(èše) 2 (iku)	aša$_5$	=	13 20	(· 1 šar)
1(èše) 3 (iku)	aša$_5$	=	15	(· 60 šar)
1(èše) 4 (iku)	aša$_5$	=	16 40	(· 1 šar)
1(èše) 5 (iku)	aša$_5$	=	18 20	(· 1 šar)
2(èše)	aša$_5$	=	20	(· 60 šar)
2(èše) 1 (iku)	aša$_5$	=	21 40	(· 1 šar)
2(èše) 2 (iku)	aša$_5$	=	23 20	(· 1 šar)
2(èše) 3 (iku)	aša$_5$	=	25	(· 60 šar)
2(èše) 4 (iku)	aša$_5$	=	26 40	(· 1 šar)
2(èše) 5 (iku)	aša$_5$	=	28 20	(· 1 šar)
1(bùr)	aša$_5$	=	30	(· 60 šar)
1(bùr) 1 (èše)	aša$_5$	=	40	(· 60 šar)
1(bùr) 2 (èše)	aša$_5$	=	50	(· 60 šar)
2(bùr)	aša$_5$	=	1	(· 60^2 šar)
3(bùr)	aša$_5$	=	1 30	(· 60 šar)
4(bùr)	aša$_5$	=	2	(· 60^2 šar)
5(bùr)	aša$_5$	=	2 30	(· 60 šar)
6(bùr)	aša$_5$	=	3	(· 60^2 šar)
7(bùr)	aša$_5$	=	3 30	(· 60 šar)
8(bùr)	aša$_5$	=	4	(· 60^2 šar)
9(bùr)	aša$_5$	=	4 30	(· 60 šar)
10(bùr)	aša$_5$	=	5	(· 60^2 šar)

11(bùr)	aša$_5$	=	5 30	(· 60 šar)
12(bùr)	aša$_5$	=	6	(· 60^2 šar)
13(bùr)	aša$_5$	=	6 30	(· 60 šar)
14(bùr)	aša$_5$	=	7	(· 60^2 šar)
15(bùr)	aša$_5$	=	7 30	(· 60 šar)
16(bùr)	aša$_5$	=	8	(· 60^2 šar)
17(bùr)	aša$_5$	=	8 30	(· 60 šar)
18(bùr)	aša$_5$	=	9	(· 60^2 šar)
19(bùr)	aša$_5$	=	9 30	(· 60 šar)
20(bùr)	aša$_5$	=	10	(· 60^2 šar)
30(bùr)	aša$_5$	=	15	(· 60^2 šar)
40(bùr)	aša$_5$	=	20	(· 60^2 šar)
50(bùr)	aša$_5$	=	25	(· 60^2 šar)
1(šar)	aša$_5$	=	30	(· 60^2 šar)
1(šar) 10 (bùr)	aša$_5$	=	35	(· 60^2 šar)
1(šar) 20 (bùr)	aša$_5$	=	40	(· 60^2 šar)
1(šar) 20 (bùr)	aša$_5$	=	45	(· 60^2 šar)
1(šar) 40 (bùr)	aša$_5$	=	50	(· 60^2 šar)
1(šar) 50 (bùr)	aša$_5$	=	55	(· 60^2 šar)
2(šar)	aša$_5$	=	1	(· 60^2 šar)
3(šar)	aša$_5$	=	1 30	(· 60^2 šar)
4(šar)	aša$_5$	=	2	(· 60^3 šar)
5(šar)	aša$_5$	=	2 30	(· 60^2 šar)
6(šar)	aša$_5$	=	3	(· 60^3 šar)
7(šar)	aša$_5$	=	3 30	(· 60^2 šar)
8(šar)	aša$_5$	=	4	(· 60^3 šar)
9(šar)	aša$_5$	=	4 30	(· 60^2 šar)
10(šar)	aša$_5$	=	5	(· 60^3 šar)
11(šar)	aša$_5$	=	5 30	(· 60^2 šar)
12(šar)	aša$_5$	=	6	(· 60^3 šar)
13(šar)	aša$_5$	=	6 30	(· 60^2 šar)
14(šar)	aša$_5$	=	7	(· 60^3 šar)
15(šar)	aša$_5$	=	7 30	(· 60^2 šar)
16(šar)	aša$_5$	=	8	(· 60^3 šar)
17(šar)	aša$_5$	=	8 30	(· 60^2 šar)
18(šar)	aša$_5$	=	9	(· 60^3 šar)
19(šar)	aša$_5$	=	9 30	(· 60^2 šar)
20(šar)	aša$_5$	=	10	(· 60^3 šar)
30(šar)	aša$_5$	=	15	(· 60^3 šar)
40(šar)	aša$_5$	=	20	(· 60^3 šar)
50(šar)	aša$_5$	=	25	(· 60^3 šar)
šár×1.gal	aša$_5$	=	30	(· 60^3 šar)
šár×2.gal	aša$_5$	=	1	(· 60^4 šar)

IM 43415 (Al-Rawi, unpublished) is another complete metrological table for system *M*. It differs in some details from A 21984, omitting the following 5 entries present in that text:

from 2(èše) 1 (iku) aša$_5$ 21 40 to 2(èše) 5 (iku) aša$_5$ 28 20.

On the other hand, it includes the following 5 extra lines:

igi.6.gál šar 10, igi.4.gál šar 15, 1 šar igi.6.gál šar 1 10, 1 šar igi.4.gál šar 1 15, 1(bùr) 1(èše) 3(iku) 45

A5.4. The Complete Metrological Tables for Systems *Ln* and *Lc(NS/OB)*

As mentioned above, in Sec. 3.4, the most extensive known Old Babylonian metrological table for system *Ln* (with the ninda as its basic unit) is the sub-table for system *Ln* on the 6-sided prism **Ash. 1923.366**, extending from 1 finger = 10 to 60 danna = 30 (105 lines). Another, quite extensive metrological table for system *Ln*, with some additional entries for fractions of a cubit, is ***UET 7, 114***, an early Old Babylonian text from Ur (Gurney (1974). The conform transliteration in Fig. A5.3 below is based on Gurney's hand copy).

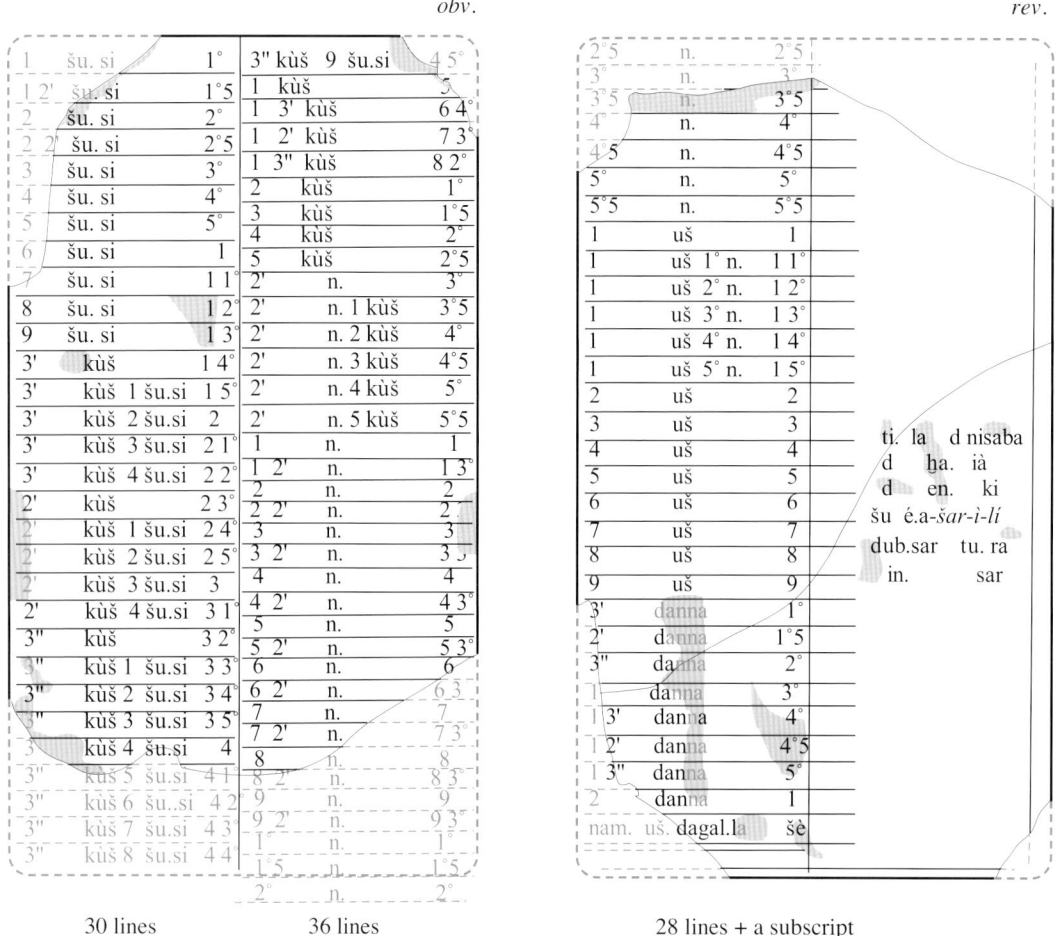

obv.   *rev.*

30 lines   36 lines   28 lines + a subscript

Fig. A5.3. *UET 7*, 114. System *Ln*, from 1 šu.si = 10 to 2 danna = 1.

Here is a transliteration of all but the last eight lines (those from 1/3 to 2 danna) on *UET 7*, 114:

The complete metrological table for the Old Babylonian system *Ln*

1	šu.si	=	10	($\cdot\, 60^{-2}$ ninda)
1 2'	šu.si	=	15	($\cdot\, 60^{-2}$ ninda)
2	šu.si	=	20	($\cdot\, 60^{-2}$ ninda)
2 2'	šu.si	=	25	($\cdot\, 60^{-2}$ ninda)
3	šu.si	=	30	($\cdot\, 60^{-2}$ ninda)
4	šu.si	=	40	($\cdot\, 60^{-2}$ ninda)
5	šu.si	=	50	($\cdot\, 60^{-2}$ ninda)
6	šu.si	=	1	($\cdot\, 60^{-1}$ ninda)
7	šu.si	=	1 10	($\cdot\, 60^{-2}$ ninda)
8	šu.si	=	1 20	($\cdot\, 60^{-2}$ ninda)
9	šu.si	=	1 30	($\cdot\, 60^{-2}$ ninda)
3'	kùš	=	1 40	($\cdot\, 60^{-2}$ ninda)
3'	kùš 1 šu.si	=	1 50	($\cdot\, 60^{-2}$ ninda)
3'	kùš 2 šu.si	=	2	($\cdot\, 60^{-1}$ ninda)
3'	kùš 3 šu.si	=	2 10	($\cdot\, 60^{-2}$ ninda)
3'	kùš 4 šu.si	=	2 20	($\cdot\, 60^{-2}$ ninda)

2'	kùš	=	2 30	($\cdot\, 60^{-2}$ ninda)
2'	kùš 1 šu.si	=	2 40	($\cdot\, 60^{-2}$ ninda)
2'	kùš 2 šu.si	=	2 50	($\cdot\, 60^{-1}$ ninda)
2'	kùš 3 šu.si	=	3	($\cdot\, 60^{-1}$ ninda)
2'	kùš 4 šu.si	=	3 10	($\cdot\, 60^{-2}$ ninda)
3"	kùš	=	3 20	($\cdot\, 60^{-2}$ ninda)
3"	kùš 1 šu.si	=	3 30	($\cdot\, 60^{-2}$ ninda)
3"	kùš 2 šu.si	=	3 40	($\cdot\, 60^{-2}$ ninda)
3"	kùš 3 šu.si	=	3 50	($\cdot\, 60^{-2}$ ninda)
3"	kùš 4 šu.si	=	4	($\cdot\, 60^{-1}$ ninda)
3"	kùš 5 šu.si	=	4 10	($\cdot\, 60^{-2}$ ninda)
3"	kùš 6 šu.si	=	4 20	($\cdot\, 60^{-2}$ ninda)
3"	kùš 7 šu.si	=	4 30	($\cdot\, 60^{-2}$ ninda)
3"	kùš 8 šu.si	=	4 40	($\cdot\, 60^{-2}$ ninda)
3"	kùš 9 šu.si	=	4 50	($\cdot\, 60^{-2}$ ninda)
1	kùš	=	5	($\cdot\, 60^{-1}$ ninda)

1 3'	kùš	=	6 40	($\cdot\ 60^{-2}$ ninda)
1 2'	kùš	=	7 30	($\cdot\ 60^{-2}$ ninda)
1 3"	kùš	=	8 20	($\cdot\ 60^{-2}$ ninda)
2	kùš	=	10	($\cdot\ 60^{-1}$ ninda)
3	kùš	=	15	($\cdot\ 60^{-1}$ ninda)
4	kùš	=	20	($\cdot\ 60^{-1}$ ninda)
5	kùš	=	25	($\cdot\ 60^{-1}$ ninda)
2'	ninda	=	30	($\cdot\ 60^{-1}$ ninda)
2'	ninda 1 kùš	=	35	($\cdot\ 60^{-1}$ ninda)
2'	ninda 2 kùš	=	40	($\cdot\ 60^{-1}$ ninda)
2'	ninda 3 kùš	=	45	($\cdot\ 60^{-1}$ ninda)
2'	ninda 4 kùš	=	50	($\cdot\ 60^{-1}$ ninda)
2'	ninda 5 kùš	=	55	($\cdot\ 60^{-1}$ ninda)
1	ninda	=	1	($\cdot$ 1 ninda)
1 2'	ninda	=	1 30	($\cdot\ 60^{-1}$ ninda)
2	ninda	=	2	($\cdot$ 1 ninda)
2 2'	ninda	=	2 30	($\cdot\ 60^{-1}$ ninda)
3	ninda	=	3	($\cdot$ 1 ninda)
3 2'	ninda	=	3 30	($\cdot\ 60^{-1}$ ninda)
4	ninda	=	4	($\cdot$ 1 ninda)
4 2'	ninda	=	4 30	($\cdot\ 60^{-1}$ ninda)
5	ninda	=	5	($\cdot$ 1 ninda)
5 2'	ninda	=	5 30	($\cdot\ 60^{-1}$ ninda)
6	ninda	=	6	($\cdot$ 1 ninda)
6 2'	ninda	=	6 30	($\cdot\ 60^{-1}$ ninda)
7	ninda	=	7	($\cdot$ 1 ninda)
7 2'	ninda	=	7 30	($\cdot\ 60^{-1}$ ninda)
8	ninda	=	8	($\cdot$ 1 ninda)
8 2'	ninda	=	8 30	($\cdot\ 60^{-1}$ ninda)
9	ninda	=	9	($\cdot$ 1 ninda)
9 2'	ninda	=	9 30	($\cdot\ 60^{-1}$ ninda)
10	ninda	=	10	($\cdot$ 1 ninda)

10 2'	ninda	=	10 30	($\cdot\ 60^{-1}$ ninda)
11	ninda	=	11	($\cdot$ 1 ninda)
11 2'	ninda	=	11 30	($\cdot\ 60^{-1}$ ninda)
12	ninda	=	12	($\cdot$ 1 ninda)
12 2'	ninda	=	12 30	($\cdot\ 60^{-1}$ ninda)
13	ninda	=	13	($\cdot$ 1 ninda)
13 2'	ninda	=	13 30	($\cdot\ 60^{-1}$ ninda)
14	ninda	=	14	($\cdot$ 1 ninda)
14 2'	ninda	=	14 30	($\cdot\ 60^{-1}$ ninda)
15	ninda	=	15	($\cdot$ 1 ninda)
20	ninda	=	20	($\cdot$ 1 ninda)
25	ninda	=	25	($\cdot$ 1 ninda)
30	ninda	=	30	($\cdot$ 1 ninda)
35	ninda	=	35	($\cdot$ 1 ninda)
40	ninda	=	40	($\cdot$ 1 ninda)
45	ninda	=	45	($\cdot$ 1 ninda)
50	ninda	=	50	($\cdot$ 1 ninda)
55	ninda	=	55	($\cdot$ 1 ninda)
1	uš	=	1	($\cdot$ 60 ninda)
1	uš 10 ninda	=	1 10	($\cdot$ 1 ninda)
1	uš 20 ninda	=	1 20	($\cdot$ 1 ninda)
1	uš 30 ninda	=	1 30	($\cdot$ 1 ninda)
1	uš 40 ninda	=	1 40	($\cdot$ 1 ninda)
1	uš 50 ninda	=	1 50	($\cdot$ 1 ninda)
2	uš	=	2	($\cdot$ 60 ninda)
3	uš	=	3	($\cdot$ 60 ninda)
4	uš	=	4	($\cdot$ 60 ninda)
5	uš	=	5	($\cdot$ 60 ninda)
6	uš	=	6	($\cdot$ 60 ninda)
7	uš	=	7	($\cdot$ 60 ninda)
8	uš	=	8	($\cdot$ 60 ninda)
9	uš	=	9	($\cdot$ 60 ninda)

The continuation in Ash. 1923-366 beyond the level 10 uš = (1/3 danna) is as follows:

10	uš	=	10	($\cdot$ 60 ninda)
11	uš	=	11	($\cdot$ 60 ninda)
12	uš	=	12	($\cdot$ 60 ninda)
13	uš	=	13	($\cdot$ 60 ninda)
14	uš	=	14	($\cdot$ 60 ninda)
2'	danna	=	15	($\cdot$ 60 ninda)
16	uš	=	16	($\cdot$ 60 ninda)
17	uš	=	17	($\cdot$ 60 ninda)
18	uš	=	18	($\cdot$ 60 ninda)
19	uš	=	19	($\cdot$ 60 ninda)
3"	danna	=	20	($\cdot$ 60 ninda)
10	danna	=	5	($\cdot\ 60^2$ ninda)
11	danna	=	5 30	($\cdot$ 60 ninda)
12	danna	=	6	($\cdot\ 60^2$ ninda)
13	danna	=	6 30	($\cdot$ 60 ninda)
14	danna	=	7	($\cdot\ 60^2$ ninda)
15	danna	=	7 30	($\cdot$ 60 ninda)
16	danna	=	8	($\cdot\ 60^2$ ninda)
17	danna	=	8 30	($\cdot$ 60 ninda)

1	danna	=	30	($\cdot$ 60 ninda)
1 2'	danna	=	45	($\cdot$ 60 ninda)
1 3"	danna	=	50	($\cdot$ 60 ninda)
2	danna	=	1	($\cdot\ 60^2$ ninda)
3	danna	=	1 30	($\cdot$ 60 ninda)
4	danna	=	2	($\cdot\ 60^2$ ninda)
5	danna	=	2 30	($\cdot$ 60 ninda)
6	danna	=	3	($\cdot\ 60^2$ ninda)
7	danna	=	3 30	($\cdot$ 60 ninda)
8	danna	=	4	($\cdot\ 60^2$ ninda)
9	danna	=	4 30	($\cdot$ 60 ninda)
18	danna	=	9	($\cdot\ 60^2$ ninda)
19	danna	=	9 30	($\cdot$ 60 ninda)
20	danna	=	10	($\cdot\ 60^2$ ninda)
30	danna	=	15	($\cdot\ 60^2$ ninda)
40	danna	=	20	($\cdot\ 60^2$ ninda)
50	danna	=	25	($\cdot\ 60^2$ ninda)
1(géš)	danna	=	30	($\cdot\ 60^2$ ninda)

The most extensive known Old Babylonian metrological table for system *Lc* (with the cubit as its basic unit) is the sub-table for system *Lc* on the fragment ***UET 7*, 115** (Gurney 1974), an early Old Babylonian text from

Ur. By a lucky accident, both the beginning and the end of this sub-table is preserved. A partial reconstruction of he the text is suggested in Fig. A5.4 below:

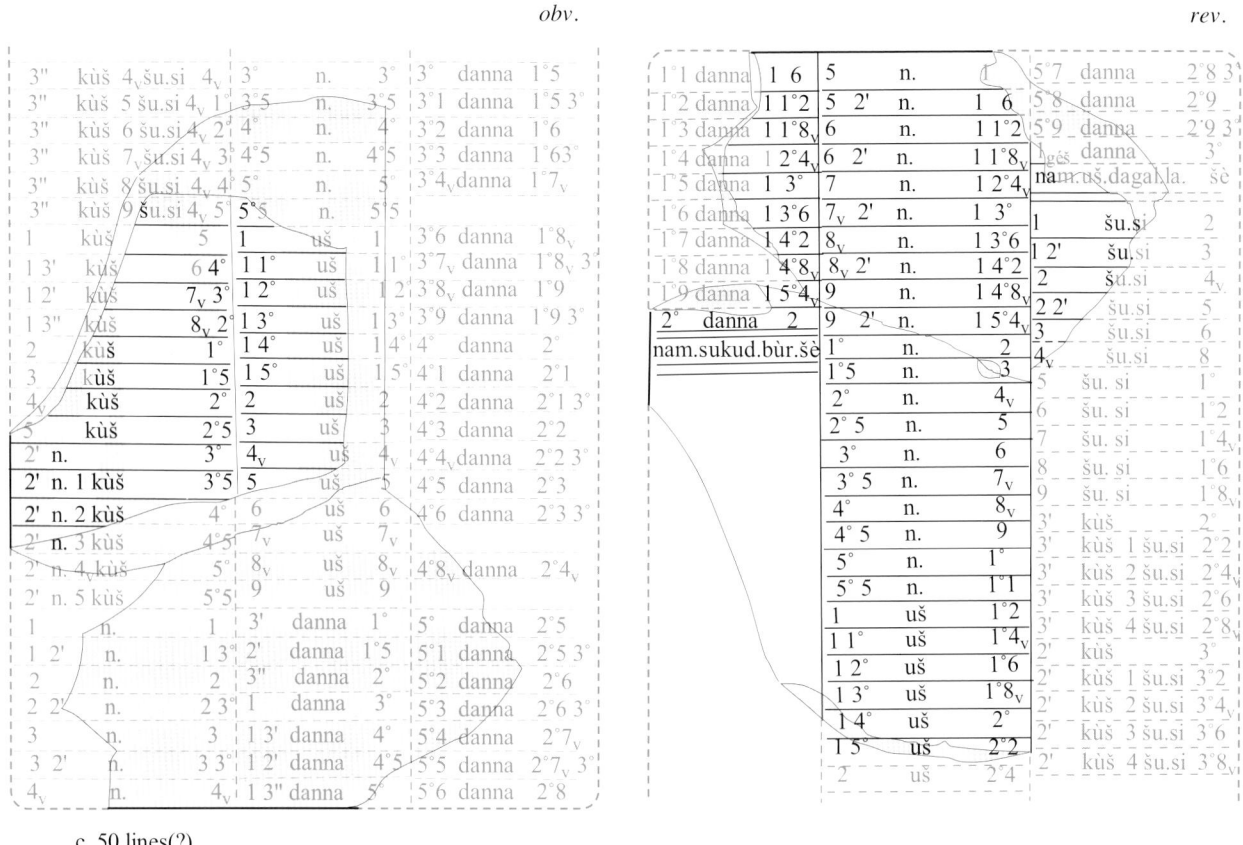

Fig. A5.4. UET 7, 115. System *Ln*, from 1 šu.si = 10 to 1 00 danna = 30, followed by system *Lc*, from 1 šu.si = 02 to 20 danna = 2.

Note: in the second column on the reverse, '1 10 uš' stands for 1;10 uš = 1 uš 10 ninda, *etc*.

On the 6-sided prism Ash. 1923-366, the sub-table for system *Lc* on face 3 extends only from 1 šu.si = 2 ($\cdot 60^{-1}$ cubit) to 10 ninda = 2 ($\cdot 60$ cubits). The sub-table for system *Lc* on the large fragment BM 92698 extends from [1 šu.si = 2 ($\cdot 60^{-1}$ cubit)] to 2 danna = 12 ($\cdot 60^2$ cubits). (See the suggested reconstruction of the table in Thureau-Dangin, *RA* 27 (1930), 116).)

Below is given a suggested reconstruction of the sub-table for *Lc* on *UET 7*, 115:

The complete metrological table for the Old Babylonian system *Lc*

1	šu.si	=	2	($\cdot 60^{-1}$ kùš)	3'	kùš	=	20	($\cdot 60^{-1}$ kùš)
1 2'	šu.si	=	3	($\cdot 60^{-1}$ kùš)	3'	kùš 1 šu.si	=	22	($\cdot 60^{-1}$ kùš)
2	šu.si	=	4	($\cdot 60^{-1}$ kùš)	3'	kùš 2 šu.si	=	24	($\cdot 60^{-1}$ kùš)
2 2'	šu.si	=	5	($\cdot 60^{-1}$ kùš)	3'	kùš 3 šu.si	=	26	($\cdot 60^{-1}$ kùš)
3	šu.si	=	6	($\cdot 60^{-1}$ kùš)	3'	kùš 4 šu.si	=	28	($\cdot 60^{-1}$ kùš)
4	šu.si	=	8	($\cdot 60^{-1}$ kùš)	2'	kùš	=	30	($\cdot 60^{-1}$ kùš)
5	šu.si	=	10	($\cdot 60^{-1}$ kùš)	2'	kùš 1 šu.si	=	32	($\cdot 60^{-1}$ kùš)
6	šu.si	=	12	($\cdot 60^{-1}$ kùš)	2'	kùš 2 šu.si	=	34	($\cdot 60^{-1}$ kùš)
7	šu.si	=	14	($\cdot 60^{-1}$ kùš)	2'	kùš 3 šu.si	=	36	($\cdot 60^{-1}$ kùš)
8	šu.si	=	16	($\cdot 60^{-1}$ kùš)	2'	kùš 4 šu.si	=	38	($\cdot 60^{-1}$ kùš)
9	šu.si	=	18	($\cdot 60^{-1}$ kùš)	3"	kùš	=	40	($\cdot 60^{-1}$ kùš)

3"	kùš 1 šu.si	=	42	(· 60⁻¹ kùš)		35	ninda	=	7	(· 60 kùš)
3"	kùš 2 šu.si	=	44	(· 60⁻¹ kùš)		40	ninda	=	40	(· 60 kùš)
3"	kùš 3 šu.si	=	46	(· 60⁻¹ kùš)		45	ninda	=	9	(· 60 kùš)
3"	kùš 4 šu.si	=	48	(· 60⁻¹ kùš)		50	ninda	=	10	(· 60 kùš)
3"	kùš 5 šu.si	=	50	(· 60⁻¹ kùš)		55	ninda	=	11	(· 60 kùš)
3"	kùš 6 šu.si	=	52	(· 60⁻¹ kùš)		1	uš 20 ninda	=	16	(· 60 kùš)
3"	kùš 7 šu.si	=	54	(· 60⁻¹ kùš)		1	uš 30 ninda	=	18	(· 60 kùš)
3"	kùš 8 šu.si	=	56	(· 60⁻¹ kùš)		1	uš 40 ninda	=	20	(· 60 kùš)
3"	kùš 9 šu.si	=	58	(· 60⁻¹ kùš)		1	uš 50 ninda	=	22	(· 60 kùš)
1	kùš	=	1	(· 1 kùš)		2	uš	=	24	(· 60 kùš)
1 3'	kùš	=	1 20	(· 60⁻¹ kùš)		3	uš	=	36	(· 60 kùš)
1 2'	kùš	=	1 30	(· 60⁻¹ kùš))		1	uš	=	12	(· 60 kùš)
1 3"	kùš	=	1 40	(· 60⁻¹ kùš)		1	uš 10 ninda	=	14	(· 60 kùš)
2	kùš	=	2	(· 1 kùš)		4	uš	=	48	(· 60 kùš)
3	kùš	=	3	(· 1 kùš)		5	uš	=	1	(· 60² kùš)
4	kùš	=	4	(· 1 kùš)		6	uš	=	1 12	(· 60 kùš)
5	kùš	=	5	(· 1 kùš)		7	uš	=	1 24	(· 60 kùš)
2'	ninda	=	6	(· 1 kùš)		8	uš	=	1 36	(· 60 kùš)
2'	ninda 1 kùš	=	7	(· 1 kùš)		9	uš	=	1 48	(· 60 kùš)
2'	ninda 2 kùš	=	8	(· 1 kùš)		3'	danna	=	2	(· 60² kùš)
2'	ninda 3 kùš	=	9	(· 1 kùš)		2'	danna	=	3	(· 60² kùš)
2'	ninda 4 kùš	=	10	(· 1 kùš)		3"	danna	=	4	(· 60² kùš)
2'	ninda 5 kùš	=	11	(· 1 kùš)		1	danna	=	6	(· 60² kùš)
1	ninda	=	12	(· 1 kùš)		1 3'	danna	=	8	(· 60² kùš)
1 2'	ninda	=	18	(· 1 kùš)		1 2'	danna	=	9	(· 60² kùš)
2	ninda	=	24	(· 1 kùš)		1 3"	danna	=	10	(· 60² kùš)
2 2'	ninda	=	30	(· 1 kùš)		2	danna	=	12	(· 60² kùš)
3	ninda	=	36	(· 1 kùš)		3	danna	=	18	(· 60² kùš)
3 2'	ninda	=	42	(· 1 kùš)		4	danna	=	24	(· 60² kùš)
4	ninda	=	48	(· 1 kùš)		5	danna	=	30	(· 60² kùš)
4 2'	ninda	=	54	(· 1 kùš)		6	danna	=	36	(· 60² kùš)
5	ninda	=	1	(· 60 kùš)		7	danna	=	42	(· 60² kùš)
5 2'	ninda	=	1 06	(· 1 kùš)		8	danna	=	48	(· 60² kùš)
6	ninda	=	1 12	(· 1 kùš)		9	danna	=	54	(· 60² kùš)
6 2'	ninda	=	1 18	(· 1 kùš)		10	danna	=	1	(· 60³ kùš)
7	ninda	=	1 24	(· 1 kùš)		11	danna	=	1 06	(· 60² kùš)
7 2'	ninda	=	1 30	(· 1 kùš)		12	danna	=	1 12	(· 60² kùš)
8	ninda	=	1 36	(· 1 kùš)		13	danna	=	1 18	(· 60² kùš)
8 2'	ninda	=	1 42	(· 1 kùš)		14	danna	=	1 24	(· 60² kùš)
9	ninda	=	1 48	(· 1 kùš)		15	danna	=	1 30	(· 60² kùš)
9 2'	ninda	=	1 54	(· 1 kùš)		16	danna	=	1 36	(· 60² kùš)
10	ninda	=	2	(· 60 kùš)		17	danna	=	1 42	(· 60² kùš)
15	ninda	=	3	(· 60 kùš)		18	danna	=	1 48	(· 60² kùš)
20	ninda	=	4	(· 60 kùš)		19	danna	=	1 54	(· 60² kùš)
25	ninda	=	5	(· 60 kùš)		20	danna	=	2	(· 60³ kùš)
30	ninda	=	6	(· 60 kùš)						

A5.5. Old Babylonian (and Other) Combined Metrological Lists

Three large fragments (**CBM 10990+19185+19757**; Figs. A5.5-6 below) of an OB *complete combined metrological list* for systems *C, M, A, L* were published in Hilprecht *BE 20/1* (1906), text 29. In Figs. A5.5-6 below is presented a conform transliteration of the text, within a suggested reconstructed outline of the clay tablet.

By a lucky accident, the final lines of the sub-tables for systems *C, M,* and *A,* and the initial lines of the sub-tables for systems *M, A,* and *L* are preserved. Only the initial line for system *C* and the final line for system *L* are missing. The ranges of the four sub-tables appear to have been as follows:

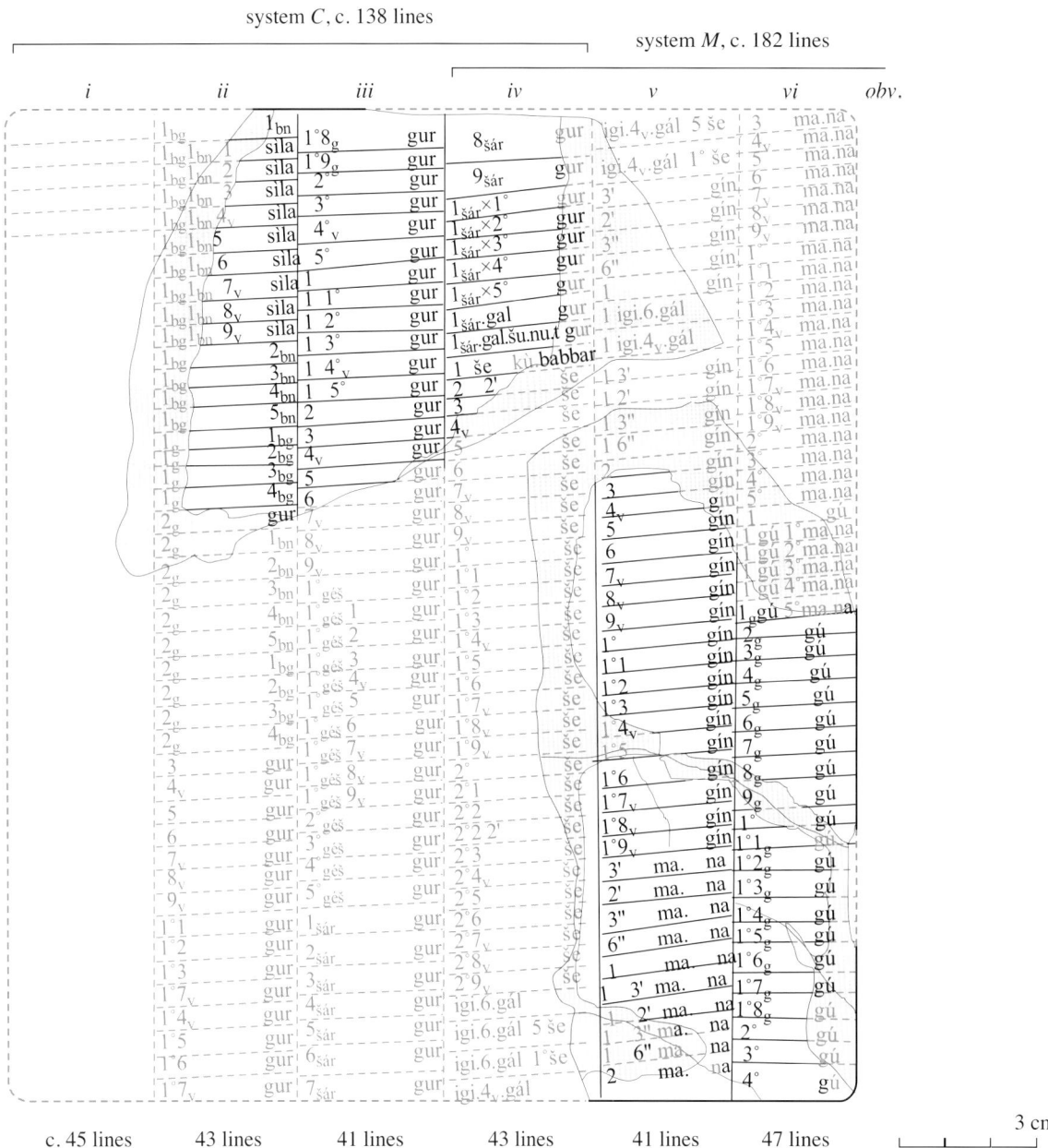

Fig. A5.5. CBM 10990+, *obv*. Three fragments of an Old Babylonian combined metrological list.
(The conform transliteration here of the text is based on Hilprecht's hand copy and photos.)

system *C*:	from [?]	to 1 šár.gal gur	and 1 šár.gal šu.nu.tag gur
system *M*:	from 1 še kù.babbar	to 1 šár.gal gú	and 1 šár.gal šu.nu.tag gú
system *A*:	from 1 šar a.šà	to 1 šár.gal aša₅	and 1 šár.gal šu.nu.tag aša₅
system *L*:	from 1 šu.si	to either [20 danna] or [1(géš) danna]	

If the sub-list for system *L* ended with 20 danna, then there was space for a subscript in *rev*. col. *xiii*. The
phrase šu.nu.tag 'the hand cannot touch (it)' in the lists for systems *C*, *M*, and *A* refers, presumably, to the
next higher unit after šár.gal, for which the scribe did not know any name.

Fig. A5.6. CBM 10990+, *rev.*

Another, quite extensive combined metrological list is **RS 21.10** from Ras Shamra or Ugarit (c. 1300 BC), published as text 144 in Nougayrol, *Ugaritica* 5 (1968). The text is written in cuneiform script on a 3-columns clay tablet, but makes use of local notations for large numbers and measures, different from the ones used in Old Babylonian texts. The ranges of its three sub-tables are:

system *C*:	from [?]	to 1-*šu* <sig₇> gur še	(= 1 šár.gal gur of barley)
system *M*:	from 1/2 še kù.babbar	to [1-*šu* gú.un kù.babbar]	(= 60 talents of silver)
system *A*:	from 1/3 šar a.šà	to 2(sig₇) aša₅ a.šà	(= 2 šár aša₅ of field)

A second example of a post-Old Babylonian and peripheral cuneiform metrological text is **VAT 9840+9889**, a large fragment of a clay tablet from Assur (Schroeder, *KAV* (1920)). It contains a metrological list for system *C*, from [1 sìla]? to (probably) [*lim lim* (a million) *imēru*], where 1 *imēru* = 10 bán = 100 sìla.

A5.6. Old Babylonian Combined Metrological Tables

There are several known examples of Old Babylonian *multiple or combined metrological tables*. Some of them have been mentioned already. Here follows a brief catalog:

UM 55.21.78 (2N-T 530): a fragment of a possibly 8-columns clay tablet with metrological tables for[1]

system *C*	from [?]	to [?]	[*obv. i - iii*]
system *M*	from [?]	to 1(géš) gú = (1) kù.babbar	*obv.* [*iv*]- *vi*
system *A*	from 1/3 šar = 20	to 1(šár) aša$_5$, [*etc.*]	*obv. vi - viii*
system *Ln*	from 1 šu.si = 10	to 1(géš) danna = 30	*rev. i - iv*
system *Lc*	from [1 šu.si] = 2	to [?]	*rev. iv - [vi]*
[square sides]?	from [?]	to [?]	*rev.*[*vi*] *- [viii]*

YBC 4633: a badly preserved 4-columns clay tablet with metrological tables for[2]

system *C*	from 1/3 sìla = [20]	to [1 šár.gal] gur = 10	*obv. i - iii*
system *M*	from 1/3 še = 10 gín	to [1(géš) gú = 1]	*obv. ii i- vi*
system *A*	from 1/3 šar = 20	to [5 šar] = 5 (the table is unfinished)	*obv. vi*

MS 3869/13: a fragment of a 4-columns clay tablet with metrological tables for[3]

system *A*	from [1/3 šar = 20]	to 1 šár.gal aša$_5$ = 30	*obv. i - iv*
system *Ln*	from 1 šu.si = 10	to 1 danna = 30	*obv. iv - rev. ii*
system *Lc*	from 1 šu.si = 2	to [1 danna = 1]	*rev. vii - iv?*

Ash. 1923.366: a 6-sided prism with 5 more or less well preserved faces, with tables for[4]

system *Ln*	from [1] šu.si = 10	to [2(géš)) danna = 1	faces 1-2
system *Lc*	from 1 šu.si = 2	to 10 ninda = 2	face 3
square sides	from 1.e 1 íb.si$_8$	to 1.[e 1 íb.si$_8$]	faces 3-5
[cube sides]?	from [1.e 1 ba.si]$^?$	to [?]	[face 6]

AO 8865: a 6-sided prism (dated to Samsuiluna 1) with 5 more or less well preserved faces, with tables for[5]

system *Ln*	from 1 šu.si = 10	to [20] danna = 10	faces 1-2
system *Lc*	from [1 šu.si = 2]	to [?]	[face 3]
square sides	from [1.e 1.àm íb.si$_8$]	to 1.e 1(00).àm íb.si$_8$	faces 4-5
cube sides	from 1.e 1.àm ba.si	to 17 46 40.e 40.àm ba.si	faces 5-6

BM 92698: a fragment of a 4-columns clay tablet with tables for [6]

system *Ln*	from [1 šu.si = 10]?	to 2 danna = 1	*obv. i - ii*
system *Lc*	from [1 šu.si = 2]?	to 2 danna = 12	*obv. iii - iv*
squares	from 1 a.rá 1 1	to [1(00) a.rá 1(00) 1]	*rev. i**
square sides	from 1.e 1 íb.si$_8$	to [1(00 00).e 1(00).àm íb.si$_8$]	*rev. ii**
cube sides	from 1.e 1 ba.si$_8$.e	to [1(00 00 00).e 1(00) ba.si$_8$.e]	*rev. iii**

A5.7. On Prisms, Cylinders, and a Family of Subscripts

The fact that MS 2723 is inscribed on a 6-sided prism makes it a member of a quite small family of known Old Babylonian mathematical or metrological texts, comprising only the following items:

MS 2723:	a 6-sided prism with a table for system *C*	(159 lines)
NBC 2513:	a 6-sided prism with a table for system *M*	(128 lines)
Ash. 1923.366:	a 6-sided prism with tables for systems *Ln*, *Lc*, square sides, [cube-sides]	(c. 300 lines)
AO 8865:	a 6-sided prism with tables for systems *Ln*, *Lc*, square sides, cube-sides	(c. 300 lines)
VAT 2576 (*BAP*, 58):	a cylinder with a metrological table for system *C*	(8 columns, 133 lines)
A 7897 (*MCT*, 25):	a cylinder with a combined multiplication table (until 1 30 ×)	(13 columns, c. 812 lines)
AO 8862 (*MKT 1-2*):	a 4-sided prism with mixed mathematical problems	(c. 812 lines)

---

1. Cf. Neugebauer and Sachs, *JCS* 36 (1984), section C.
2. Cf. Nemet.Nejat, *JNES* 54 (1995), 258-259.
3. See Figs. 3.5.1-2 above.
4. Cf. van der Meer, *Syllabaries* (1938), 156; Neugebauer and Sachs, *MCT* (1945), 34.
5. Cf. Thureau-Dangin, *RA* 27 (1930), 73-78; Neugebauer, *MKT 1* (1935), 69-72, 89-90, Proust, *TMN* (2004), Annexe 8.
6. Cf. Thureau-Dangin, *RA* 27 (1933), 116; Neugebauer, *MKT 1* (1935), 69-70, 73; *MKT 2* (1935), pl. 61.

(The prisms and cylinders are quite small, with densely written text. Thus, MS 2723 measures 12 cm × 5.5 cm, AO 8862 16.8 cm × 7.3 cm, AO 8865 20 × 8 cm, and A 7897 22.5 cm x 10.6 cm.)

AO 8862 and AO 8865 are both probably from Larsa (as is also, by the way, the combined table text BM 92698). Ash. 1923-366, AO 8865, VAT 2576, AO 8862 (as well as the combined table texts MS 3869/13 and BM 92698 and the combined list text CBM 10990+) make use of variant number signs for 4, 7, 8 (and in some cases also 9). Therefore, all these texts can be assumed to be from southern Mesopotamian sites and early Old Babylonian. Only MS 2723 makes use of standard number signs, which may or may not be significant.

In this connection, it may be of interest to recall that, as mentioned in Sec. 3.4 above, the tables for systems *Ln* and *Lc* in BM 92698 end with the subscripts nam.uš.sag aša$_5$.šè 'for lengths and fronts of fields in general' and nam.sukud.bùr.saḫar.šè 'for heights and depths of mud in general', and that the tables in *UET 7*, 114 and *UET 7*, 115 (both from Ur) end with similar subscripts.

There is an additional interesting feature common to several of the mentioned large table texts, namely invocations to Nisaba, goddess of grain, writing, and wisdom, who was worshipped in Uruk, Umma, and other southern Mesopotamian cities. In particular, AO 8862, the 4-sided prism with problem texts, is inscribed near the top of the first side of the prism with the single word ^{d}nisaba.[7][8]

The following five texts have subscripts mentioning Nisaba and one or two other divine names:

MS 2723	ti.la / ^{d}nisaba / ^{d}ḫa.ià /	By the life of Nisaba, Haia,	(Fig. 3.1.3)
	igi ^{d}nisaba / ù ^{d}ḫa.ià /	before Nisaba and Haia,	
	^{m}*li-iš-lul* / *ši-im*-dingir	Lishlul.shimīl <wrote it>,	
	ḫe.en.ša$_6$	may it please (them).	
NBC 2513	šu.nigin 2 *šu-ši* 8 mu.bi /	Together 2 sixty 8 its names (= lines).	(Fig. A5.1)
	ti.la / ^{d}nisaba / ^{d}ḫa.ià /	By the life of Nisaba, Haia,	
	be-li-šu-nu / in.sar	Belishunu wrote it.	
UET 7, 114	ti.la / ^{d}nisaba / ^{d}ḫa.ià / ^{d}en.ki /	By the life of Nisaba, Haia, Enki,	(Fig. A5.3)
	šu è.a-*šar-ì-lí* /	the hand of Easharili	
	dub.sar tur.a / in.sar	the young scribe, wrote it.	
MS 3845	ti.la ^{d}nisaba / ù ^{d}ḫa.ià /	By the life of Nisaba and Haia,	(Fig. 2.6.10)
	^{m}šu.šaḫan / in.sar	Shushahan wrote (it),	
	^{m}en.zu-*i-qí-ša-am* / ^{lú}gù.dé.a	Suen-iqisham, the Gudea-man.	
IM 73355	ti.la / ^{d}nisaba / ù ^{d}ḫa.ià /	By the life of Nisaba, and Haia,	(Fig. A5.7)
	na-wi-ir / in.sar	Nawir wrote it.	

An additional example of a text of this kind is AO 8865, a 6-sided prism with tables for systems *Ln*, *Lc*, square sides, and cube-sides, and with the following subscript:

AO 8865	4 13 mu.bi.im /	4 13 are its names (= lines).	(*MKT 1*, 72)
	iti zíz.a u$_4$ 5.kam /	The month *Šabat*, the 5th day,	
	[mu *sa*]-*am-su-i-lu-na* [lugal].e	the year Samsuiluna became king.	

In addition, there is on the upper face of the prism an eroded, but still readable inscription (see Proust, *SCIAM-VS* 6 (2005)), never noticed by the original editor of the text:

^{d}nisaba ^{d}ḫa.[ià]   Nisaba and Ha*ia*[9]

Of these texts with similar subscripts, MS 2723 is a 6-sided prism with a complete metrological table for system *C*, NBC 2513 is a 6-sided prism with a complete metrological table for system *M*, *UET 7*, 114 is a clay tablet with a complete metrological table for system *Ln*, MS 3845 is a (large fragment of) a clay tablet with a combined multiplication table (the standard table of reciprocals and the 10 first multiplication tables), while

---

7. Cf. Høyrup, *LWS* (2002), 164, fn. 194: "This invocation is the closest any Old Babylonian mathematical text comes to connecting its topic with religious or other esoteric matters." Clearly Høyrup knew of no other examples.

8. IM 31247, a large fragment of a mathematical problem text in 4 columns (Bruins, *Sumer* 9 (1953) has a single ^{d}nisaba as a subscript. (There may have been additional lines of the subscript in the missing part of the text).

9. Cf. the doxology ^{d}nisaba zà.mí 'praise be Nisaba' that was typically inserted at the end of Old Babylonian list texts from Nippur, such as the "List of trees and wooden objects". (See Veldhuis *EEN* (1997), 167, 252.) Phrases like 'By the life of Nisaba and Haia', on the other hand, appear to be known only from the prisms and other texts mentioned above.

IM **73355** is a double table of powers (Arnaud, *TL* (1994). The conform transliteration in Fig. A5.7 below is based on Arnaud's hand copy of the text). Thus all the mentioned texts with similar subscripts are large arithmetical or metrological table texts, except IM 73355, a relatively small table text

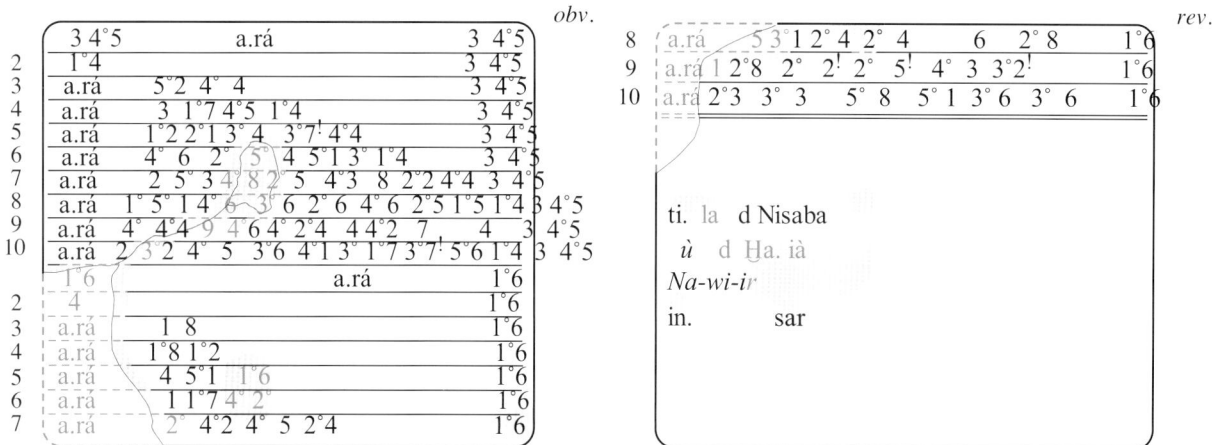

Fig. A5.7. IM 73355 (from Larsa). A double table of powers: The ten first powers of 3 45, and of 16.

Note: The two tables on IM 73355 are ascending tables of powers, in contrast to the tables on MS 2242 and MS 3879 (Fig. 1.4.1), which are descending tables of powers. As mentioned in Sec. 1.4, MS 2242 and MS 3879 are probably factorization exercises, in the former case to find a factorization of 46 20 54 51 30 14 03 5 (the 6th power of 3 45), in the latter case to find a factorization of 3 11 06 10 42 48 57 36 (the 12th power of 12).

Several interesting conjectures can be made with departure from the discussion above. To begin with, the form of the subscripts on the two mentioned MS texts, MS 2723 and MS 3845, suggests that those texts are from southern Mesopotamia, most likely from either Ur (as *UET 7*, 114) or Larsa (as IM 73355). [10]

More important, however, is what the brief catalogs above suggest about the origin of Old Babylonian mathematical and metrological table texts. It is likely that extensive combined metrological *lists*, like CBM 10990+ (Fig. A5.5-6), for instance, existed even *before* the invention of place value notation for sexagesimal numbers. Then, as an essential part of the revolution of Mesopotamian computational practices that was caused by the invention of place value notation, *a fixed series of table texts* was composed, consisting of, in this order

the combined multiplication table (the standard table of reciprocals followed by 41 selected multiplication tables),
complete metrological tables for systems *C, M, A, Ln*, and *Lc*, arithmetical table texts for squares, square sides, and cube-sides.

The *original* series of canonical table texts can be assumed to have been inscribed in one of the southern Mesopotamian cities (Ur, or Larsa, or Uruk) on a series of clay cylinders or 6-sided clay prisms, all of them being dedicated to Nisaba, the goddess of writing and wisdom, accompanied by other deities.

The reason for writing the canonical table texts on cylinders or prisms is obvious. Such cylinders and prisms always have a hole through them along the central axis. Each cylinder or prism can be assumed to have been mounted vertically on a stand, with a wooden stick through the hole. A number of scribe school students could then be placed around the object, each student being required to copy the column from the extensive table text that was directly in front of him onto an individual small clay tablet. After a while, the text could be rotated, and the students could start to copy new columns without changing places.

The suggested way of copying individual sub-tables from large table texts can possibly explain why there is so little overlapping of known examples of single multiplication tables or excerpts from metrological tables. The small table texts are, of course, the students' individual copies, while the large clay tablets or cylinders or prisms are the teachers' model texts.

10. **Ist. Ni 4908**, *from Nippur*, is very small fragment of a four-sided prism, published in Proust, *TMN* (2004), vol. 2, 23. The fragment is inscribed with a small part of a table for system *C* on what remains of the first face, and with a small part of a table for system *Ln* on what remains of the fourth face. There is no preserved subscript.

Appendix 6
Metro-Mathematical Cuneiform Texts from the Third Millennium BC[1]

There are very few known mathematical texts from the Neo-Sumerian Ur III period (c. 2100-2000 BC), only three tables of reciprocals (**HS 201, Ist. Ni. 374**, and **Ist. T. 7375**; see Sec. 2.5 above, and App. 1, Fig. A1.2), a very interesting text about the growth of a herd of cows (**AO 5499**; Nissen, Damerow, and Englund *ABK* (1993), Fig. 76), and (possibly) ***RTC* 413**, a small lenticular tablet with an exercise concerning bricks, with sexagesimal numbers in place value notation (Friberg, *RlA 7* (1990), Fig. 3; Fig. 7.3.3 above). This absence of known mathematical texts from Ur III may be due to archaeological bad luck, but there may also have been little place for mathematics (as opposed to accounting) in the super-bureaucratic environment of Ur III.

AO 5499, *RTC* 413, and all other known Mesopotamian mathematical texts from the third millennium, before the invention of place value notation, are really "metro-mathematical", in the sense that they are *simultaneously mathematical and metrological exercises*. It is instructive to try to find out how metro-mathematical problems were formulated and solved *without* the use of sexagesimal numbers in place value notation. At the same time, it is instructive to look for similarities between the well known types of mathematical problems in Old Babylonian mathematical texts and the types of problems that one finds in the small corpus of pre-Old-Babylonian mathematical texts.[2]

From the Old Akkadian period (c. 2340-2200 BC), for instance, quite a few small mathematical hand tablets are known. They will be discussed briefly below. (Cf. Friberg, *RlA 7* (1990), Sec. 4.4.)

A6.1. Two Old Akkadian Applications of the Field Expansion Procedure

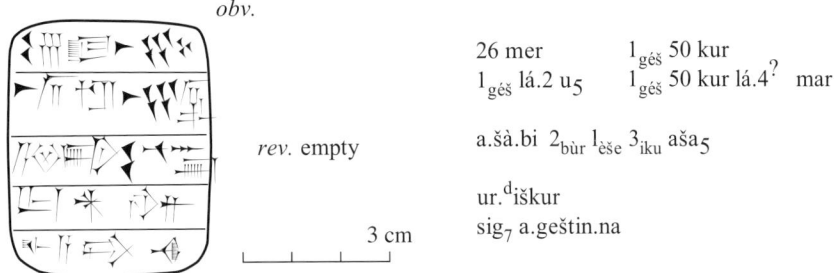

obv.

rev. empty

3 cm

26 mer $1_{géš}$ 50 kur
$1_{géš}$ lá.2 u₅ $1_{géš}$ 50 kur lá.4$^?$ mar

a.šà.bi $2_{bùr}$ $1_{èše}$ 3_{iku} aša₅

ur.^{d}iškur
sig₇ a.geštin.na

(A computer-aided reproduction of Limet's orignal hand copy of the clay tablet.)

Fig. A6.1. *DPA* 34. An Old Akkadian application of the (proto-literate) field expansion procedure.

---

1. Friberg, *CDLJ* 2005:2 is an expanded, more elaborate version of the discussion in the present appendix.
2. The study of pre-Old-Babylonian mathematical texts was initiated by Powell in his discussion of "Sumerian area measures and the alleged decimal substratum" in *ZA* 62 (1972), of "The antecedents of Old Babylonian place notation" in *HM* 3 (1976), and of "Sumerian numeration and metrology" in his dissertation (1977).

***DPA* 34** (Fig. A6.1 above), published in Limet, *DPA* (1973),[3] is a small hand tablet mentioning the sides and the area of an almost trapezoidal field, as well as the name of the writer and a geographic name. The short sides of the quadrilateral are given as '26 north' and '58 south', while the long sides are '1 géš 50 east' (géš = 60) and '1 géš 46[?] west'. The corresponding half-sums are 42 (ninda) and 1 géš 48 (ninda). The ratio of the two half-sums is 18 : 7, or approximately 5 : 2. The product of the half-sums is, on one hand, equal to an *almost round area number* (cf. Secs. 1.2 and 8.1 b), sinceš

$$1 \text{ géš } 48 \text{ n.} \cdot 42 \text{ n.} = (1 + 1/5) \cdot 1 \text{ géš } 30 \text{ n.} \cdot (1 + 1/6) \cdot 36 \text{ n.} = (1 + 1/5) \cdot (1 + 1/6) \cdot 5 \text{ géš sq. n.,}$$

where the ratio of 1 géš 30 and 36 is exactly 5 : 2. On the other hand, the product of the half-sums is approximately equal to a *round area number*, since

$$(1 + 1/5) \cdot (1 + 1/6) \cdot 54 \text{ géš sq. n.} = (1 + 1/5) \cdot 1 \text{ šár } 3 \text{ géš sq. n.} = 1 \text{ šár } 15 \text{ géš } 36 \text{ sq. n.} = 2 \text{ } 1/2 \text{ bùr (36 šar).}$$

Thus, the sides of the nearly trapezoidal quadrilateral seem to have been chosen in such a way that the resulting area, computed by means of the (proto-literate) quadrilateral area rule (see Sec. 6 a) is simultaneously a round and an almost round area number, and so that the ratio of the two half-sums is close to a simple 5 : 2. This situation is typical for an application of the (proto-literate) field expansion rule. (See the discussion of almost round area numbers and the proto-literate field expansion procedure in Sec. 8.1 b above, and, in particular, the illustration in Fig. 8.1.3.)

A similar, perhaps more convincing, example is the hand tablet **A 786** (Luckenbill (1930), *OIP 14*, 116):

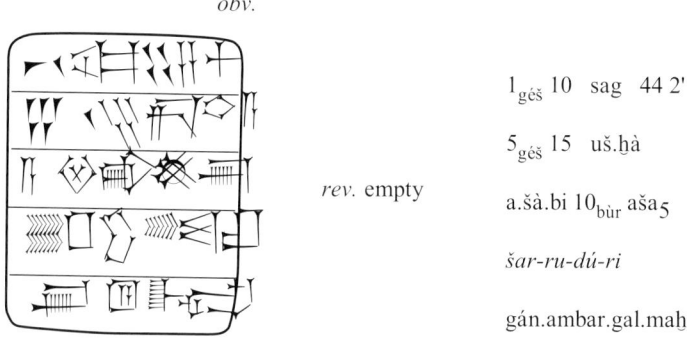

obv.

rev. empty

1 géš 10   sag 44 2'

5 géš 15   uš.ḫà

a.šà.bi 10 bùr aša5

šar-ru-dú-ri

gán.ambar.gal.maḫ

Fig. A6.2. A 786. Another Old Akkadian application of the (proto-literate) field expansion procedure.

This text begins by specifying the sides of a long and narrow trapezoidal field. The given unequal 'fronts' (sag) are 1 géš 10 and 44 1/2 ninda (half-sum 57 n. 3 c.), and two equal 'lengths' (uš.ḫi.a) are both 5 géš 15 ninda. Thus, the ratio of the length to the half-sum of the widths is again roughly 5 : 1. Next, the text states that the area of the field with the given sides is 10 bùr, *a large and round area number*. The last two lines of the text contain the name of the writer, 'The-king-is-my-wall', and the name of a field 'The-field-by-the-large-and-grand-marsh', which sounds like a fictive name.

Since the area of the field is a round number and its length is an almost round number (note that 5 15 = 7 · 45), it is likely that this is another example of an application of the field expansion procedure. As a matter of fact, the construction of the numbers in the text can be explained as follows:

The author of the text wanted to construct a trapezoid with the area 10 bùr = 5 šár sq. ninda, and with the ratio of the length to the half-sum of the widths close to 5 : 1. For that purpose, he started with a rectangle with the false initial length 4 géš 30 n., the false initial front 54 n. = 1/5 · 4 géš 30 n., and consequently the false initial area 4 šár 3 géš sq. n. (Note that 4 géš 3 = 3[5]; cf. *DPA* 39 in Fig. A5.11 below.) The area deficit was then close to 1 šár sq. n., 1/4 of the initial area. In order to more than halve the deficit, he extended the false length by 1/6, and obtained the new length (1 + 1/6) · 4 géš 30 n. = 5 géš 15 n. and the new area (1 + 1/6) · 4

---

3. The hand copies of *DPA* 34, 36-39 in Figs. A6.1, 3, 5,10a, 11 are computer-aided reproductions of Limet's original hand copies. Similarly, the hand copy of A 786 in Fig. A6.2 is a computer-aided reproduction of Luckenbill's original hand copy.

šár 3 géš sq. n. = 4 šár 43 géš 30 sq. n. The new area deficit was then 16 géš 30 sq. n., that is close to 1/17 of the new area. In order to slightly more than eliminate this new deficit, he extended the false front by 1/16, and obtained the new front $(1 + 1/16) \cdot 54$ n. = 57 n. 4 1/2 cubits, slightly more than 57 n. 3 c. Then, the twice extended area of the rectangle became

$$(1 + 1/16) \cdot (1 + 1/6) \cdot 4 \text{ šár 3 géš sq. n.} = \text{appr. 5 géš 15 n.} \cdot 57 \text{ n. 3 c.} = \text{appr. 5 šár 33 sq. n.}$$

The author of the text probably did not bother to compute the area of the twice extended rectangle, but instead immediately declared it to be 10 bùr (5 šár sq. n.). Moreover, since he wanted the final result to be a trapezoid rather than a rectangle, he chose to let the upper front be 1 10 n. The lower front must then be set equal to $2 \cdot 57$ n. 3 c. – 1 10 n. = 44 1/2 n., which is the value given in the text.

A6.2. Old Akkadian Square-Side-and-Area Exercises

A group of Old Akkadian small mathematical texts, first published in Limet, *DPA* (1973), were studied by Powell in *HM* 3 (1976). One of them is the field expansion text *DPA* 34 (Fig. A6.1 above), while four of the texts are squaring exercises or, more exactly, "square-side-and-area exercises".

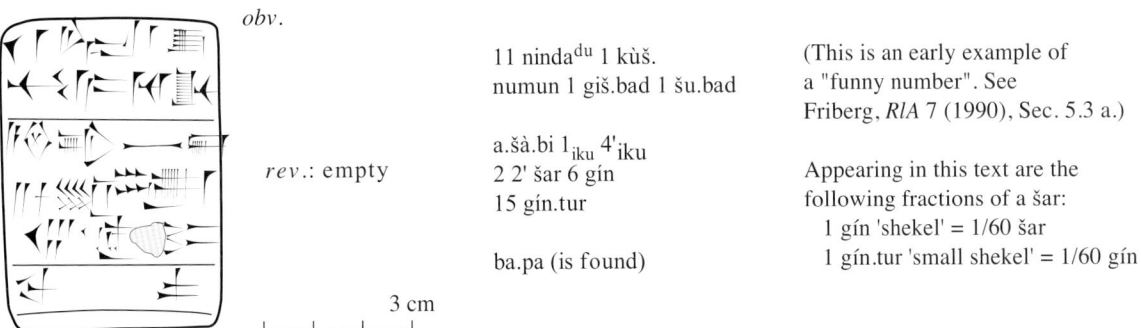

obv.

rev.: empty

3 cm

11 ninda^{du} 1 kùš.
numun 1 giš.bad 1 šu.bad

a.šà.bi 1_{iku} 4'_{iku}
2 2' šar 6 gín
15 gín.tur

ba.pa (is found)

(This is an early example of
a "funny number". See
Friberg, *RlA* 7 (1990), Sec. 5.3 a.)

Appearing in this text are the
following fractions of a šar:
  1 gín 'shekel' = 1/60 šar
  1 gín.tur 'small shekel' = 1/60 gín

Fig. A6.3. *DPA* 36. An Old Akkadian square-side-and-area exercise.

***DPA* 36** (Fig. A6.3 above) was discussed by Powell in *HM* 3 (1976). Powell discovered a mistake in the text, which he interpreted as an error introduced by the writer, probably a student, when "he apparently lost track of the correct sexagesimal place" in the course of the computation, based on "a mental construct analogous to Old Babylonian place notation".

However, just as *DPA* 34 and A 786 in Sec. A6.a above, this text, too, can be explained as an exercise based on the use of almost round length numbers, rather than on the use of sexagesimal numbers in place value notation. Indeed, the given length number in this text, silently understood to be the side of a square, is expressed in terms of the ninda and three smaller Old Akkadian units of length,

kùš.numun 'seed-cubit' = 1/6 ninda, and its fractions giš.bad = 1/2 kùš.numun, šu.bad = 1/4 kùš.numun.

Therefore the given length number can be interpreted as

$$s = 11 \text{ n. 1 1/2 1/4 k.n.} = 11 \text{ 1/4 n. 1/4 k.n.} = (1 + 1/8) \cdot 10 \text{ n.} + 1/24 \text{ n.}$$

In other words, the given square side is *an almost round length number plus a small added length number*, what may be called a "fractionally and marginally expanded length number". The square of this length number was probably computed in a few simple steps, tentatively reconstructed below.

The first step was, presumably, to compute the square of the first two terms of the length number:

$$\text{sq. } (10 \text{ n.} + 1/8 \cdot 10 \text{ n.}) = \text{sq. 10 n.} + 2 \cdot 1/8 \cdot \text{sq. 10 n.} + \text{sq. 1/8} \cdot \text{sq. 10 n.}$$
$$= 1 \text{ iku} + 2 \cdot 1/8 \text{ iku} + 1/8 \cdot 1/8 \text{ iku} = 1 \text{ 1/4 iku} + 1/8 \cdot 12 \text{ 1/2 šar}$$
$$= 1 \text{ 1/4 iku 1 1/2 šar 3 1/2 gín 15 gín.tur.}$$

Geometrically, the computation can be explained (see Fig. A6.4) as an application of the easily observed binomial rule, that *the square of a length composed of two unequal parts is a square composed of four parts, namely two unequal squares and two equal rectangles.*

A second application of the same rule would then show (see again Fig. A6.4) that

$$\text{sq. } (10 \text{ n.} + 1/8 \text{ of } 10 \text{ n.} + 1 \text{ šu.bad.})$$
$$= \text{sq. } (10 \text{ n.} + 1/8 \cdot 10 \text{ n.}) + 2 \cdot (10 \text{ n.} + 1/8 \text{ of } 10 \text{ n.}) \cdot 1 \text{ šu.bad} + \text{sq. } (1 \text{ šu.bad})$$
$$= 1\ 1/4 \text{ iku } 1\ 1/2 \text{ šar } 3\ 1/2 \text{ gín } 15 \text{ gín.tur} + 2 \cdot (25 \text{ gín} + 1/8 \cdot 25 \text{ gín}) + 6\ 1/4 \text{ gín.tur}$$
$$= 1\ 1/4 \text{ iku } 2\ 1/2 \text{ šar } 6\ 1/4 \text{ gín.tur.}$$

The computation may actually have been facilitated if the author of *DPA* 36 had recourse to a table of squares of multiples of fractions of the ninda, such as the šu.bad.[4]

The slightly incorrect answer in *DPA* 36 (with 6 1/4 gín instead of 6 1/4 gín.tur) is

$$\text{sq. } s = 1\ 1/4 \text{ iku } 2\ 1/2 \text{ šar } 6 \text{ gín } 15 \text{ gín.tur } (= 1\ 1/4 \text{ iku } 2\ 1/2 \text{ šar } 6\ 1/4 \text{ gín}).$$

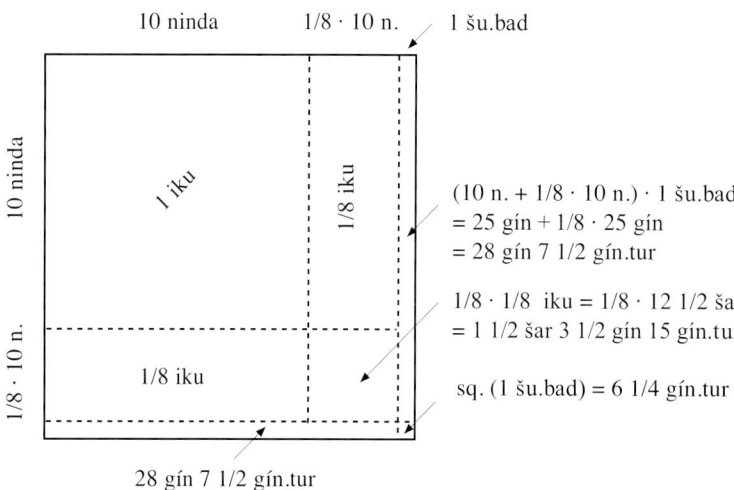

Fig. A6.4. *DPA* 36. The area of the square of a fractionally and marginally expanded length number.

Another Old Akkadian square-side-and-area exercise is **DPA 37** (Limet, *DPA* (1973)):

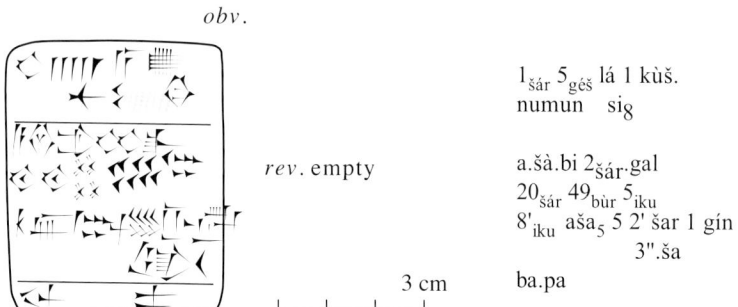

Fig. A6.5. *DPA* 37. Another Old Akkadian square-side-and-area exercise.

In this text, the notation sá (or si₈) 'it is equal' after a single given length number indicates that this number is the common length of all sides of a square. In the present case, the given number can be characterized as a "fractionally expanded and marginally contracted length number". Indeed,

$$s = 1 \text{ šár } 5 \text{ géš (ninda)} - 1 \text{ kùš.numun} = (1 + 1/12) \cdot 1 \text{ šár (ninda)} - 1/6 \text{ ninda.}$$

Therefore, the square of this length number can be computed in a few easy steps in, essentially, the same way as the square of the corresponding number in *DPA* 36. See Fig. A6.6 below. Note the use there of the rule that

---

4. Cf. CUNES 50-008-001, the Early Dynastic arithmetical-metrological table text discussed in App. 7 below.

the square of the difference of two lengths is equal to two unequal squares minus two equal rectangles. Note also that the given side of the square in *DPA* 37 is quite unrealistic. Indeed, since 1 ninda = appr. 6 meters, it follows that the side of the square is, approximately,

$$\text{1 šár 5 géš ninda} - \text{1 kùš.numun} = \text{3,900 ninda} - \text{1/6 ninda} = \text{23.4 kilometers} - \text{1 meter.}$$

There is a curious error in *DPA* 37. The area mentioned in the answer exceeds the correctly computed area of the square by 1/8 iku 5 1/2 šar = 18 šar.[5] Apparently, the student who wrote the text became confused by the 1 èše 1/2 iku resulting from the preceding step of the computation (see Fig. A6.6 below) and computed the last term of sq. *s* not as 1/36 · 1 šar, as he should have done, but as

$$\text{1/36} \cdot \text{(1 èše 1/2 iku} - \text{1 šar)} = \text{1/6} \cdot \text{(1 iku 8 1/3 šar} - \text{1/6 šar)} = \text{18 1/18 šar} - \text{1/36 šar} = \text{18 1/36 šar.}[6]$$

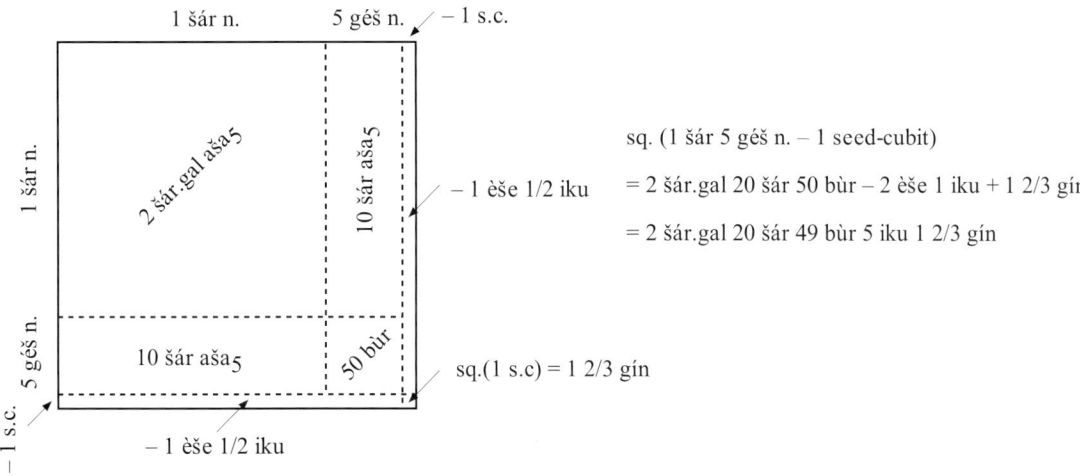

Fig. A6.6. *DPA* 37. The area of the square of a fractionally expanded and marginally contracted length number.

A 5446 (Fig. A6.7 below), published in Whiting *ZA* 74 (1984),[7] is an Old Akkadian hand tablet with notes about two mathematical assignments given to two named individuals. The second assignment, given to a student named Ur-Ištaran, is identical with the statement of the problem in the first line of *DPA* 37, so, perhaps, *DPA* 37 is Ur-Ištaran's answer to the assignment given to him according to A 5446.

Fig. A6.7. A 5446. Two assignments of field-side-and-area problems to named students.

The first of the two assignments on A 5446, given to a student named Meluḫḫa, is to find the area of a square with the side 28 šár (ninda) − 1 šu.dù.a (where 1 šu.dù.a = 1/3 cubit = 1/36 ninda). The answer to this

---

5. The sign used for 1/8 šar in *DPA* 37 is the cuneiform variant of the "curviform" sign for 1/8 šar used in some of the *BIN 8* texts. See Powell, *ZA* 62 (1972), 218: Fig. 2.

6. An alternative explanation of the error was given by Whiting in *ZA* 74 (1984), namely that the author of the text may have used sexagesimal place value notation at an intermediate stage of his solution procedure. In place value notation, the correct answer is 1 10 24 38 20;01 40 šar, while the answer given in the text is 1 10 24 38 38;01 40 šar, which Whiting explains as a dittography with the 38 instead of 20 in the units place because of the 38 in the nearby sixties place.

7. The hand copies of A 5443 and A 5446 in Figs. A6.7 and 9 are based on the hand copies in Whiting, *ZA* 74 (1984).

assignment is not given, but should have been

$$sq. (28 \text{ šár} - 1 \text{ šu.dù.a})$$
$$= 26 \text{ šár.gal } 8 \text{ šár} - 3 \text{ bùr } 2 \text{ iku} + 2 \text{ 2/3 gín.tur } 1/3 \text{ of } 1/3 \text{ gín.tur}$$
$$= 26 \text{ šár.gal } 7 \text{ šár } 56 \text{ bùr } 2 \text{ èše } 4 \text{ iku } 2 \text{ 2/3 gín.tur } 1/3 \text{ of } 1/3 \text{ gín.tur.}$$

The next text, **Ash. 1924.689** = Gelb (1970), **MAD 5, 112** is possibly a hybrid between texts like A 5446 with a brief list of assignments, and texts like *DPA* 36-37 with answers to assignments. (The hand copy in Fig. A6.8 below is based on photos of the clay tablet.)

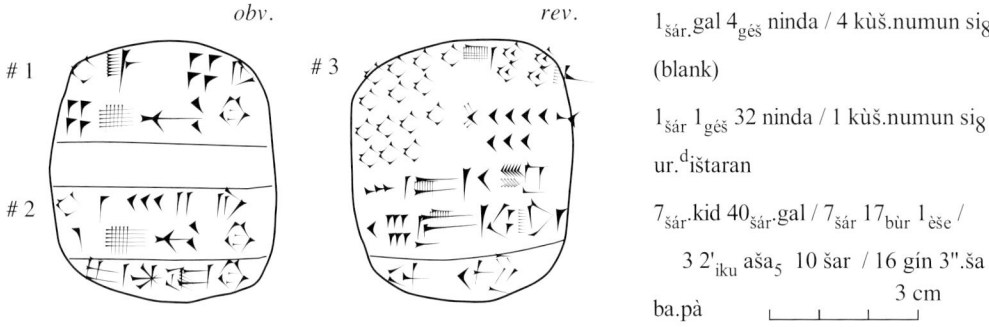

Fig. A6.8. Ash. 1924.689. Three square-side-and-area exercises: two questions and one answer.

On the obverse of this text, there are two assignments for square-side-and-area exercises, one followed by a blank, the other followed by the name Ur-Ištaran, the same name as in A 5446. The given length number in the first assignment is bigger than all the length numbers in *DPA* 36-37 and A 5446:

$$1 \text{ šár.gal } 4 \text{ géš ninda } 4 \text{ kùš.numun} = \text{appr. } 1{,}296 \text{ km} + 1{,}440 \text{ m} + 4 \text{ m.}$$

This is an unrealistically large length number. The corresponding area is also immense. Note that

$$sq. 1 \text{ šár ninda} = 2 \text{ šár.gal aša}_5 = 2 \cdot 60^2 \text{ bùr} = 2 \cdot 60^2 \cdot \text{appr. } 64{,}800 \text{ square meters} = \text{appr. } 466.56 \text{ sq. km}$$
$$sq. 10 \text{ šár ninda} = 3 \text{ šár.líl } 20 \text{ šár.gal aša}_5 = \text{appr. } 46{,}656 \text{ sq. km}$$
$$sq. 1 \text{ šár.gal ninda} = 60^6 \text{ šar} = 2 \cdot 60^4 \text{ bùr} = 2 \cdot 60^4 \cdot \text{appr. } 64{,}800 \text{ square meters} = \text{appr. } 1{,}679{,}616 \ (6^8) \text{ sq. km}$$

The name for 60^4 bùr is not known. (Cf. App. 4, Fig. A4.14.)[8]

The area number recorded in assignment # 3 on the reverse of Ash. 1924.689 is

$$A = 7 \text{ šár.kid } 40 \text{ šár.gal } 7 \text{ šár } 17 \text{ bùr } 1 \text{ èše } 3 \text{ 1/2 iku aša}_5 \text{ 10 šar } 16 \text{ gín } 3'' \text{ <gín>.}$$

This area number can be shown to be the answer to a square-side-and-area exercise. It is not a perfect square, although it is very close to one. Apparently, the one who wrote the text made a small error near the end of his computation of the square of the side

$$s = 15 \text{ šár } 10 \text{ géš } 4 \text{ ninda} - 1 \text{ kùš } 1/2 \text{ šu.si.}$$

(See Friberg, *CDLJ* 2005:2, Sec. 4.7.7 ff.) The precise cause of the error is difficult to establish.

The very small clay tablet **A 5443** (Whiting, *op. cit.* and Fig. A6.9 below) is an Old Akkadian square-side-and-area exercise in which the numbers don't make sense. (The given length number is also strangely written with one of the signs for 10 géš misplaced.)

On the other hand, both the length number and the area number in A 5443 have, to some extent, the form of "funny numbers". Thus, the given length number contains the digit 3 written 5 times, and the area number begins with the digit 1 repeated three times (1 šár.gal 1 šár×u 1 šár).

---

8. **TM.75.G.2200**, a small lexical text from Ebla (Pettinato, **MEE 3, 72** (1981)), consists of 10 entries on the theme šár. (Cf. the small Early Dynastic lexical text *TSS* 190, Fig.6.2.3 above.) The first four entries are šár.ki (a place name?), šár.gal, šár.kid, šár.li. Therefore, it is possible that šár.li is a name for 60^4 bùr. Another lexical text from Ebla with names for big numbers is Pettinato, **MEE 4, 78** (1982), where entries 25-30, the last six entries of the text, are *ma-i-at* (10^5), *ma-ḫu-at* (10^5, or 10^6?), šár'u ($10 \cdot 60$ bùr, or $10 \cdot 60^2$), šár.gal (60^2 bùr, or 60^3), šár.kid (60^3 bùr, or 60^4?), and šár.šár *bur-ḫi-da-rí-ga* (60^4 bùr?, or 60^5?),

The area of a square with the side 33 géš ninda is 36 géš 18 bùr = 36 šár 18 bùr. Therefore, a square with the side 33 géš 33 ninda 3 cubits (since 1 *nikkas* = 1/4 ninda = 3 cubits), the length number recorded in the first text box on A 5443, clearly cannot have the area recorded in the second text box, 1 šár.gal 11 šár 27 bùr 5 iku 1/2 šar 3 2/3 gín 5 gín.<tur>.

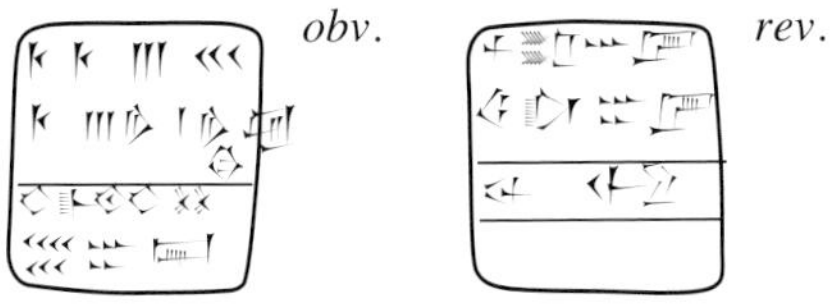

$3_{géš'u} \, 3_{géš} \, 3_u$ ninda 3 ninda 1 ník.kas$_x$ si$_8$ /

$1_{šár}$.gal $1_{šár'u} \, 1_{šár} \, 2_{bùr'u} \, 7_{bùr} \, 5_{iku}$ aša$_5$

2' šar 3 gín 3".ša 5 gín.<tur> /

ba.pà

3 cm

Fig. A6.9. A 5443. An Old Akkadian square-side-and-area exercise with numbers that don't make sense.

A partial explanation of the strange numbers in A 5443 is that they may be an example of experimentation with the traditional Sumerian number systems, in an attempt to make them decimal. (Known examples of such experiments are two Old Babylonian mathematical texts from the city Mari. See Chambon, *FlMar* 6 (2002), Proust, *FlMar* 6 (2002), and Soubeyran, *RA* 78 (1984).)

Now, suppose that in this text

$$1 \, géš \times u = 10 \, géš = 1,000, \, 1 \, géš = 100.$$

Then 33 géš 33 ninda 1 *nikkas* in this text means 3,333 ninda 3 cubits, and

$$\text{sq. } 3,333 \text{ n. } 3 \text{ c.} = (\text{sq. } 3,333 + 1/2 \cdot 3,333 + 1/16) \text{ sq. n.} = (11,108,889 + 1,666 \, 1/2 + 1/16) \text{ sq. n.}$$
$$= 11,110,555 \, 1/2 \, 1/16 \text{ sq. n.} = 111,105 \text{ iku } 55 \, 1/2 \text{ šar } 3 \, 2/3 \text{ gín } 5 \text{ gín.tur.}$$

Comparing this result with the area number recorded on A 5443, one is led to the conclusion that the decimalized area system used in this text may have had units with the following decimal values:

$$1 \, šár.gal = 100,000 \, iku, 1 \, šár \times u = 10 \, šár = 10,000 \, iku, 1 \, šár = 1,000 \, iku.$$

Under these assumptions, several of the initial and final digits in the computed area number are the same as the corresponding digits in the recorded area number. Unfortunately, however, the digits in the middle are not the same in the computed and the recorded area numbers. Presumably, the lack of agreement is due to some counting error, but an attempt to establish the precise nature of that error has not been successful.

It is interesting to note, by the way, that the given length number in A 5443 can be interpreted as

$$3,333 \, ninda \, 3 \, cubits = 1/3 \cdot 10,000 \, ninda - 1 \, cubit.$$

This is the explanation for the four initial digits 1 in sq. 3,333 n. 3 c., since

$$\text{sq. } [1/3 \cdot 10,000 \text{ n.}] = 1/9 \cdot 100,000,000 \text{ sq. n.} = 11,111,111 \, 1/9 \text{ sq. n.}$$

A6.3. Old Akkadian Metric Division Exercises

Among the Old Akkadian mathematical texts discussed by Powell in *HM 3* (1976) are three very interesting "metric division exercises" (Figs. A6.10-11 below).

This is the brief text of **DPA 38** (Limet, *op. cit.*; Fig. A6.10, left):

$2_{géš}$ 40$_v$ uš.ḫi /
sag 1 iku /
sag.bi 3 kùš.numun / 1 giš.bad 1 šu.bad

2 gèš 40 the lengths.
(What is) the front (if the field is) 1 iku?
Its front (is) 3 1/2 1/4 seed-cubits.

It is clear that the front *s* (silently understood as the front of a rectangle) can be computed as the area $A = 1$ iku divided by the length $u = 2$ géš 40 (ninda). It is also clear that 2 géš 40 = 16 · 10 is a regular sexagesimal number. However, *without place value notation the front cannot have been computed as the area times the reciprocal of the length.* Instead, the front may have been computed by use of a method based on *factorization*.

Start, for instance, with a square of area 1 iku, hence with the length and the front both equal to 10 ninda. Multiply the length of this square by one of the factors in the given length number, and divide the front by the same factor. The result will be a rectangle, still with the area 1 iku. The process can be repeated, until the final rectangle has the required length,. The front of the final rectangle will then be the answer to the problem.

The solution procedure can have taken the form of a numerical algorithm, similar to some of the operations with many-place regular sexagesimal numbers discussed in Secs. 1.4-1.5 above:

10 n.	10 n.	(10 n. · 10 n. = 1 iku)
40 n.	2 n. 3 k.n.	(4 · 10 n. = 40 n., and 1/4 · 10 n. = 2 1/2 n = 2 n. 3 k.n.)
2 géš 40 n	3 1/2 1/4 k.n.	(4 · 40 n. = 2 géš 40 n., and 1/4 · 2 n. 3 k.n. = 3 1/2 1/4 k.n.)

(Here k.n. is an abbreviation for kùš.numun, 1/2 k.n. = 1 giš.bad, and 1/4 k.n. = 1 šu.bad.)

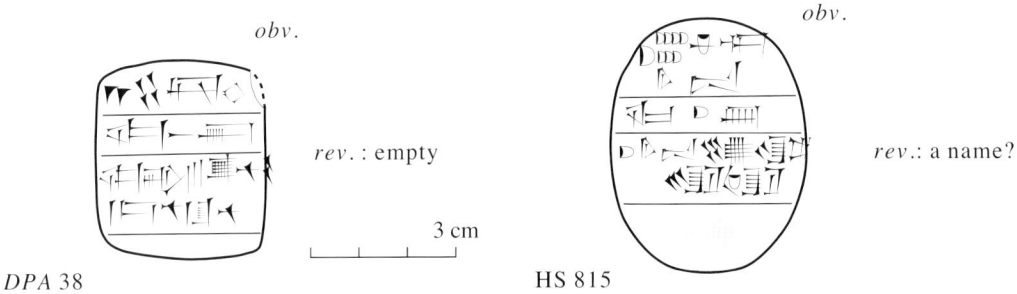

Fig. A6.10. *DPA* 38 and HS 815, two Old Akkadian metric division exercises.

HS 815 (Pohl, *TMH 5*, text 65; Westenholz, *ECTJ* (1975), text 65; Fig. A6.10 above) is a similar text, probably from Old Akkadian Nippur. The length u is given as 1 géš 7 1/2 ninda^{du}, the area A is again 1 iku, and the answer is given in the form

$$s = 1 \text{ ninda}^{du} 5 \text{ kùš } 2 \text{ šu.dù.a } 3 \text{ šu.si } 3' \text{ šu.si } (= 1 \text{ ninda } 5 \text{ 2/3 cubits } 3 \text{ 1/3 fingers}).$$

(Note that in this text the system of units of length measure is not the same as in *DPA* 38.) However, the given length is here, too, a regular sexagesimal number (times 1 ninda), since 1 géš 7 1/2 = 9 · 7 1/2 = 1/2 · 27 · 5. The solution may again have been obtained by use of a numerical algorithm:

2 1/2 n.	40 n.	(2 1/2 n. · 40 n. = 1 iku)
7 1/2 n.	13 n. 4 c.	(3 · 2 1/2 n. = 7 1/2 n., and 1/3 · 40 n. = 13 n. 4 c.)
22 1/2 n.	4 n. 5 c. 10 f.	(3 · 7 1/2 n. = 22 1/2 n., and 1/3 · 13 n. 4 c. = 4 n. 5 c. 10 f.)
1 géš 7 1/2 n.	1 n. 5 2/3 c. 3 1/3 f.	(3 · 22 1/2 n. = 1 géš 7 1/2 n., and 1/3 · 4 n. 5 c. 10 f. = 1 n. 5 2/3 c. 3 1/3 f.)

DPA **39** (Limet, *op. cit.*; Fig. A6.11 below) is a third text of the same type, except that no answer is given in this text, where instead a blank and the phrase 'to be found' follow the phrase 'its front (is)'. The given length is 4 géš 3 ninda, where

4 géš 3 (the 5th power of 3) is a regular sexagesimal number.

The solution may once again have been obtained by use of a factorization method. However, an added difficulty in this case is that the answer cannot be expressed in *presently known* Old Akkadian units of length. Assume, for the sake of the argument, that the small fractional length unit 1 še 'barley-corn' = 1/6 finger was in use already in the Old Akkadian period, although there is no textual evidence to support this assumption.[9]

If this assumption is admitted, then the solution algorithm may have been as follows:

1 n.	100 n.	(1 n. · 100 n. = 1 iku)
3 n.	33 n. 4 c.	
9 n.	11 n. 1 1/3 c.	
27 n.	3 n. 8 1/3 c. 3 1/3 f.	
1 géš 21 n.	1 n. 2 2/3 c. 4 1/3 f. 2/3 b.c.	
4 géš 3 n.	4 2/3 c. 8 f. 2/3 b.c. and 1/3 of 2/3 b.c.	

---

9. See Powell, *RIA 7* (1990), "Masse und Gewichte", Sec. I.2, and Friberg, *GMS 3* (1993), texts 4, 5, 9, and 12.

Thus the answer in this case would be

$$s = 4\ 2/3 \text{ cubits } 8 \text{ fingers } 2/3 \text{ barley-corn and } 1/3 \text{ of } 2/3 \text{ barley-corn.}$$

However, this answer would no fit into the small space available on the clay tablet!

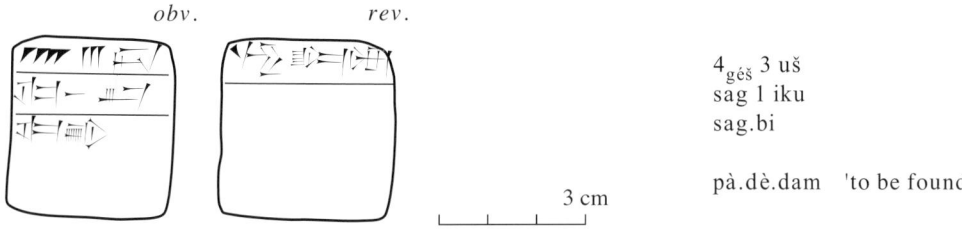

4$_{g\acute{e}\check{s}}$ 3 uš
sag 1 iku
sag.bi

pà.dè.dam   'to be found'

Fig. A6.11. *DPA* 39, an Old Akkadian mathematical assignment with a metric division problem.

A6.4. IM 58045. An Old Akkadian Trapezoid Partition Problem

The Old Akkadian hand tablet **IM 58045** (2N-T 600), a text from Nippur, is by its find site in a collapsed house firmly dated to the reign of Šarkallišarri. There is drawn on it a trapezoid with a transversal line parallel to the fronts of the trapezoid. The lengths of all four sides of the trapezoid, but not the length of the transversal, are indicated in the drawing. See Fig. A6.12 below.

The length of the trapezoid is given as 2 gi 'reeds' = 12 cubits (1 ninda). The lengths of the two fronts are 3 reeds – [1 cubit] = 17 cubits and 1 reed 1 cubit = 7 cubits, respectively. It is known from a number of Old Babylonian mathematical that 17, 13, 7 is the basic example of a "trapezoid triple", which means that the area of a trapezoid with the fronts 17 and 7 is bisected or "equipartitioned", that is divided in two equal parts by a transversal of length 13. (See Friberg, *RlA 7* (1990), Sec. 5.4 k.) The triple therefore satisfies the "(trapezoid) transversal equation"

$$\text{sq. } 17 + \text{sq. } 7 = 2 : \text{sq. } 13 \quad (\text{sq. } m + \text{sq. } n = 2 \text{ sq. } d, \text{ where } d \text{ is the transversal, } m \text{ and } n \text{ the parallel fronts}).$$

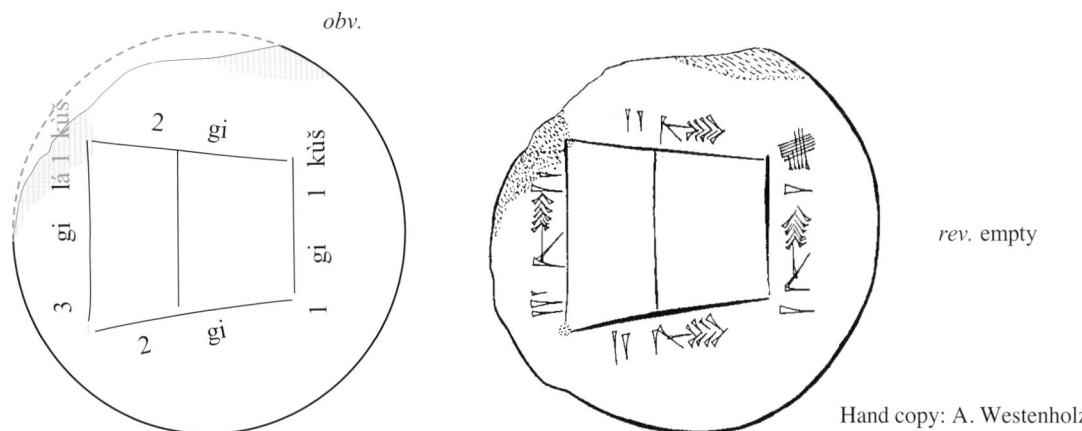

Fig. A6.12. IM 58045. An Old Akkadian trapezoid equipartition exercise.

The background to this equation is the following: Generally, according to the Babylonian trapezoid area rule, the area of a trapezoid with the fronts m and n is divided in two equal halves by a transversal d if the triple m, d, n satisfies the equation

$$\text{sq. } m - \text{sq. } d = \text{sq. } d - \text{sq. } n, \quad \text{or, equivalently,} \quad m + \text{sq. } n = 2 \text{ sq. } d.$$

Indeed (cf. Fig. 10.3.4), if two concentric squares have the sides m and n, respectively, and if the area of a third concentric square with the side d is the half-sum of the areas of the two given squares, then a trapezoid with the fronts *m* and *n* is "equipartitioned" by a transversal of length *d*, in the sense that the area of the trapezoid is divided in two equal parts by the transversal. See Fig. A6. 13 below:

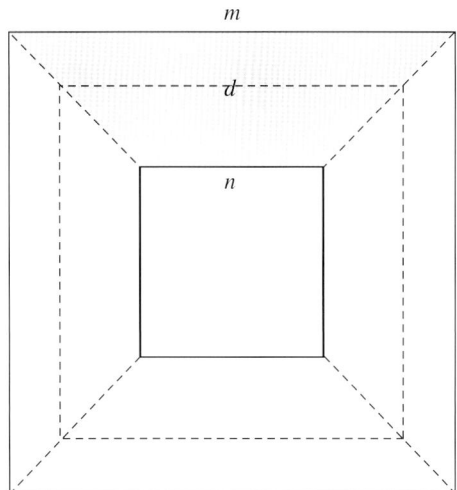

In an "equipartitioned trapezoid" with the fronts *m* and *n*, the transversal *d* divides the trapezoid in two parts of equal area.

This happens exactly when
sq. *m* − sq. *d* = sq. *d* − sq. *n*
or, equivalently, when
sq. *m* + sq. *n* = 2 sq. *d*.

Example: *m, d, n* = 17, 13, 7,
sq. 17 = 4 49,
sq. 13 = 2 49,
sq. 7 = 49.

Fig. A6.13. A geometric demonstration of the equipartitioned trapezoid equation.

The examples of Old Akkadian mathematical texts discussed above, in Secs. A6.a-d, show that the Old Babylonians inherited their way of doing mathematics from their Old Akkadian predecessors. This is shown, in particular, by the central role played by squares, rectangles, and trapezoids in both Old Babylonian and Old Akkadian mathematics, as well as by the use of regular sexagesimal numbers in the Old Akkadian metric division problems.

A6.5. TM.75.G.1392 (Ebla). A Division Algorithm in Decimal Numbers

There are only two known Old Babylonian division exercises. One of them is MS 3871 (Sec. 1.3 above), where the problem apparently is to find by which factor the number 4 37 46 40 has to be multiplied in order to get 11 34 26 40. The text is in the form of an assignment, without solution procedure. Since both numbers are *regular* sexagesimal numbers, it is likely that the Old Babylonian teacher's intention was that the division problem should be solved by use of the "trailing part algorithm", that is by systematically eliminating common regular factors in the two numbers.

The other known example is MS 2317 (Sec. 7.1 above), and the related exercise in *UET 5*, 121 § 2, where the funny number 1 01 01 01 is to be divided by 13. Since, in this case, both numbers are *non-regular* sexagesimal numbers, the trailing part algorithm cannot be used. It was suggested in Sec. 7.1 that it was the Old Babylonian teacher's intention that the problem should be solved by finding approximate solutions to the successive equations

$$? \cdot 13 = 1\ 00, ? \cdot 13 = 10\ 00, ? \cdot 13 = 1\ 00\ 00, ? \cdot 13 = 10\ 00\ 00, ? \cdot 13 = 1\ 00\ 00\ 00,$$

and then combining the results in a proper way.

A division algorithm of this kind was actually used in **TM.75.G.1392** (Friberg, *VO* 6 (1986); Fig. A6.14 below), a mathematical text from the city Ebla in Syria, dating from c. 2250 BC, thus almost 500 years older than the majority of the Old Babylonian mathematical texts. (The hand copy of the text in Fig. A6.14 below is based on photos of the clay tablet, courteously provided by A. Archi.)

The number notations in TM.75.G.1392 are the ones described by the factor diagram in App. 4, Fig. A4.5. The measures mentioned in the text belong to the system of capacity measures used at Ebla, with

1 gú.bar = 20 šú×níg.sag, and 1 šú×níg.sag = 6 an.zam$_x$.

There is no explicit statement of the question in this text, which begins as follows:

3 gú.bar 4._r an.zam_x / 1 mi gú.bar   3 gú.bar 4 an.zam_x — 1 hundred gú.bar.

It is not directly clear what this means. A likely interpretation seems to be that the text gives the answer to a question of the following type:

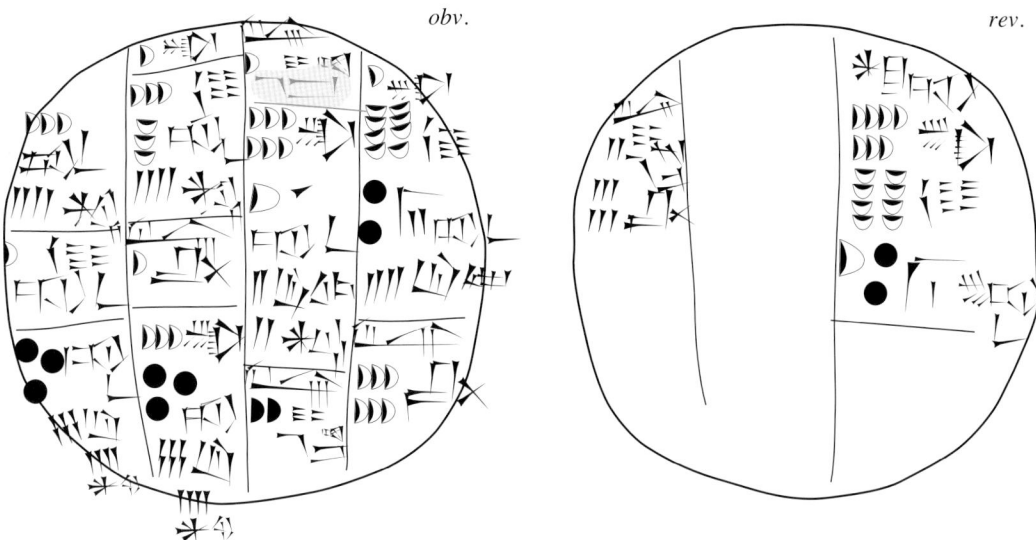

obv.   rev.

Fig. A6.14. TM.75.G.1392. A text from Ebla with a systematic solution algorithm for a division exercise.

The market rate for a certain commodity is 33 gú.bar of the commodity for 1 gú.bar (of barley).
What is the price in barley for 260,000 gú.bar of the commodity?

To us, this appears to be a simple numerical division problem, to divide 260,000 by 33. To an Eblaite scribe, it was a difficult "metrological" division problem, to divide 260,000 gú.bar by 33 gú.bar. The successive steps of the division algorithm in the text are (in a simplified form):[10]

3 4/120 gú.bar	is the (approximate) price of	1 <i>mi</i> (100) gú.bar
30 6/20 4/120 gú.bar	"	1 <i>li</i> (1,000) (gú.bar)
303 4/120 gú.bar	"	1 <i>rí-ba</i>_x (10,000) (gú.bar)
3,030 6/20 4/120 gú.bar	"	1 <i>ma-i-at</i> (100,000) (gú.bar)
6,060 1/2 2/20 2/120 gú.bar	"	2 <i>ma-i-at</i> (200,000) (gú.bar)
1,820 – 2 4/20 gú.bar	"	6 <i>rí-ba</i>_x (60,000) (gú.bar)
together: 7,880 - 1 še gú.bar	"	2 <i>ma-i-ḫu</i> 6 <i>rí-ba</i>_x (260,000) (gú.bar)

Note that the answer, in two text boxes on the reverse, explicitly mentions še 'barley':

an.šè.gú 7 <i>li</i> 8_r <i>mi</i> 1_{géš} 20 lá 1._r še gú.bar /   total: 7 thousand 8 hundred 1(60) 20 – 1 gú.bar of barley,
lú 2._r <i>ma-i-ḫu</i> 6._r <i>rí-ba</i>_x   that of 2 hundredthousand 6 tenthousand.

(Here, the subscripts 'r' and '.' denote <i>rotated</i> and <i>cuneiform</i> number signs, respectively.)

The division algorithm can be explained as follows (cf. again the discussion of MS 2317 in Sec. 7.1 a). Given that 33 gú.bar of the unnamed commodity is worth 1 gú.bar of barley, then each unit of barley corresponds to 33 times as many units of the commodity. Therefore the question is

Which capacity measure times 33 equals 260,000 gú.bar?

The solution was found by first calculating approximate solutions to the successive equations

---

10. According to the new, collated version of the text in Archi, <i>RA</i> 83 (1989), 3-4.

$$? \cdot 33 = 1 \text{ gú.bar}, \quad ? \cdot 33 = 10 \text{ gú.bar}, \quad ? \cdot 33 = 100 \text{ gú.bar}, \quad ? \cdot 33 = 1,000 \text{ gú.bar},$$
$$? \cdot 33 = 10,000 \text{ gú.bar}, \quad ? \cdot 33 = 100,000 \text{ gú.bar},$$

and then combining the results in the proper way. This was done, apparently, in the following way:

$4 \text{ an.zam}_x \cdot 33$	$= 1 \text{ gú.bar}$	$(+ 2 \text{ šú} \times \text{níg.sag})$
$6 \text{ šú} \times \text{níg.sag } 4 \text{ an.zam}_x \cdot 33$	$= 10 \text{ gú.bar}$	$(+ 1 \text{ gú.bar})$
$3 \text{ gú.bar } 4 \text{ an.zam}_x \cdot 33$	$= 100 \text{ gú.bar}$	$(+ 2 \text{ šú} \times \text{níg.sag})$
$30 \text{ gú.bar } 6 \text{ šú} \times \text{níg.sag } 4 \text{ an.zam}_x \cdot 33$	$= 1,000 \text{ gú.bar}$	$(+ 1 \text{ gú.bar})$
$303 \text{ gú.bar } 4 \text{ an.zam}_x \cdot 33$	$= 10,000 \text{ gú.bar}$	$(+ 2 \text{ šú} \times \text{níg.sag})$
$3,030 \text{ gú.bar } 6 \text{ šú} \times \text{níg.sag } 4 \text{ an.zam}_x \cdot 33$	$= 100,000 \text{ gú.bar}$	$(+ 1 \text{ gú.bar})$
$6,060 \ 1/2 \text{ gú.bar } 2 \text{ šú} \times \text{níg.sag } 2 \text{ an.zam}_x \cdot 33$	$= 200,000 \text{ gú.bar}$	$(+ 7 \text{ šú} \times \text{níg.sag})$
$1,818 \text{ gú.bar } 4 \text{ šú} \times \text{níg.sag} \cdot 33$	$= 60,000 \text{ gú.bar}$	$(+ 12 \text{ šú} \times \text{níg.sag})$
$7,879 \text{ gú.bar} \cdot 33$	$= 260,000 \text{ gú.bar}$	$(+ 7 \text{ gú.bar})$

The two first of these computations are so easy that they are not mentioned in the text. Note the round-off to the nearest whole gú.bar in the final answer.

The idea behind an algorithm of the indicated kind was to proceed in a series of steps, using in each step, in order to save time and labor, the results obtained in the preceding steps. Thus, the barley for 10 units of the unnamed commodity was *precisely* ten times as much as the barley for 1 unit of the commodity, the barley for 100 units *approximately* 10 times as much as for 10 units, *etc.*

It is possible to give a modern interpretation of the division algorithm in TM.75.G.1392, highlighting the importance of the text from the point of view of history of mathematics. Indeed, one can describe the algorithm as follows, using modern mathematical terminology: It starts by computing successively better approximations to 1/33 (or rather to $10^n \cdot 1/33$, for $n = 2, 3, \ldots$):

$100/33$	$=$	$3 \ 1/33$	$=$ appr.	$3 \ 1/30,$
$1,000/33$	$=$	$30 \ 10/33$	$=$ appr.	$30 \ 1/3,$
$10,000/33$	$=$	$303 \ 1/33$	$=$ appr.	$303 \ 1/30,$
$100,000/33$	$=$	$3,030 \ 10/33$	$=$ appr.	$3,030 \ 1/3.$

Then these preliminary approximations are used to find approximations also to 200,000/33, to 60,000/33, and, finally, to 260,000/33.

What is done in the first phase of this computation is not very far removed from showing that the *common fraction* 1/33 can be represented by the *periodic infinite decimal fraction* 0.030303… !

A6.6. TM.75.G.2346 (Ebla). Another Decimal Division Algorithm

Another mathematical text from Ebla is **TM.75.G.2346** (Archi, *RA* 83 (1989)). (The hand copy of the text in Fig. A6.15 below is based on photos of the clay tablet, courteously provided by A. Archi.)

There is no explicit statement of the question in this text, either. It begins as follows:

4 še gú.bar 4.ᵣ níg.sagšu / 1 *mi-at* níg.sagšu   4 gú.bar 4 níg.sagšu of barley — 1 hundred níg.sagšu.

(Here níg.sagšu is the same capacity unit as the one that was called šú×níg.sag in TM.75.G 1392.) A likely interpretation seems to be that the text gives the answer to a question of the following type:

The market rate for a certain commodity is 24 níg.sagšu of the commodity for 1 gú.bar of barley. What is the price in barley for 10,000 níg.sagšu of the commodity?

The successive steps of the division algorithm in the text are (in a simplified form):

4 še gú.bar 4 níg.sagšu	is the (approximate) price of	1 *mi-at* (100) níg.sagšu
20 1/2 še gú.bar 8 níg.sagšu	"	5 *mi* (500) níg.sagšu
41 1/2 še gú.bar 4 níg.sagšu	"	1 *li-im* (1,000) níg.sagšu
210 − 2 še gú.bar 8 níg.sagšu	"	5 *li-im* (5,000) níg.sagšu
420 − 2 še gú.bar 4 níg.sagšu	"	1 *rí-ba*$_x$ (10,000) níg.sagšu

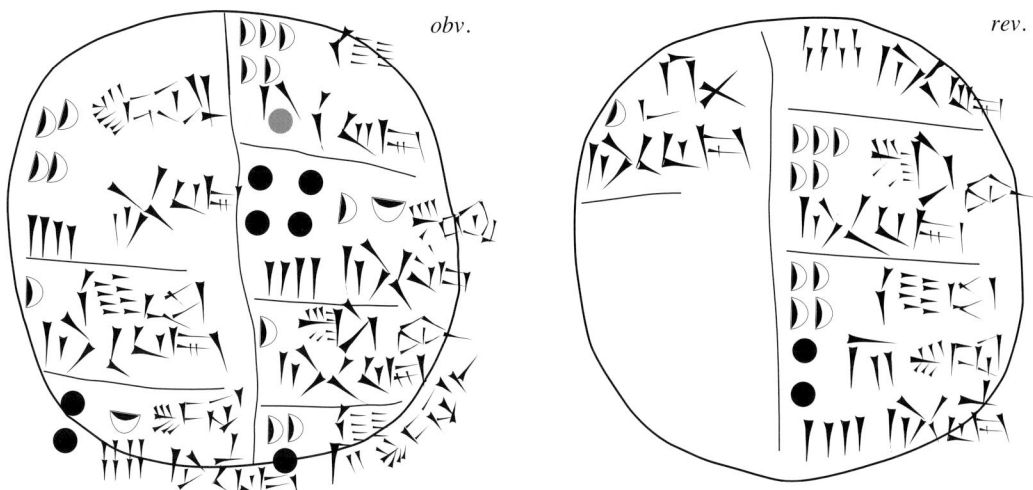

obv.   *rev.*

Fig. A6.15. TM.75.G.2346. A text from Ebla with another division exercise.

The division algorithm can be explained as follows. Given that 24 níg.sagšu of the unnamed commodity is worth as much as 1 gú.bar of barley, the question is, essentially,

What (decimal) number times 24 n.s. equals 10,000 n.s.?

The solution was found by first calculating approximate solutions to the successive equations

$? \cdot 24$ n.s. = 100 n.s., $? \cdot 24$ n.s. = 500 n.s., $? \cdot 24$ n.s. = 1,000 n.s., $? \cdot 24$ n.s. = 5,000 n.s., $? \cdot 24$ n.s. = 10,000 n.s

This seems to have been done, essentially, in the following way:

$$
\begin{array}{rcrcr}
4 \cdot 24 + 4 & = & 96 + 4 & = & 100 \\
20\ 1/2 \cdot 24 + 8 & = & 492 + 8 & = & 500 \\
41\ 1/2 \cdot 24 + 4 & = & 996 + 4 & = & 1{,}000 \\
210 - 2 \cdot 24 + 8 & = & 4{,}992 + 8 & = & 5{,}000 \\
\mathbf{420 - 3\ 1/2 \cdot 24 + 4} & = & 9{,}996 + 4 & = & 10{,}000 \\
\end{array}
$$

There is a curious error in the last line of the text, 420 − 2 instead of 420 − 3 1/2. (Cf. the discussion of subtractive notations for numbers in App. 1.) The error can be explained if the text is a copy, presumably made by a student, of a correct original, presumably made by a teacher. Then the student can have been confused by the subtraction of 2 in the previous line, which made him write − 2 instead of − 3 1/2 also in the last line.

Just as in the case of TM.75.G 1392, it is possible to give an anachronistic modern interpretation of the division algorithm in TM.75.G.2346, highlighting the importance of that text, too, from the point of view of history of mathematics. The idea seems to have been to compute successively better approximations to $10^n \cdot 1/24$ and $10^n \cdot 5/24$ as follows:

$$
\begin{array}{rclclcl}
100/24 & = & 4\ 4/24 & = & \text{appr.} & 4\ 4/20, \\
500/24 & = & 20\ 1/2\ 8/24 & = & \text{appr.} & 20\ 1/2\ 8/20, \\
1{,}000/24 & = & 41\ 1/2\ 4/24 & = & \text{appr.} & 41\ 1/2\ 4/20, \\
5{,}000/24 & = & 208\ 8/24 & = & \text{appr.} & 208\ 8/20, \\
10{,}000/24 & = & 416\ 1/2\ 4/24 & = & \text{appr.} & 416\ 1/2\ 4/20. \\
\end{array}
$$

Thus, what the author of this text did is close to showing that the *common fractions* 1/24 and 5/24 can be represented by the *infinite decimal fractions* 0.041666... and 0.208333...!

TM.75.G 2346 can also be understood as *a table of data for interest problems* of the following kind (cf. the Old Babylonian artificial interest problem VAT 8521 discussed in Sec. 2.4 a above):

The interest is 4 níg.sagšu per gú.bar (= 1/5). When the initial barley plus interest is given, find the initial barley.

The table on TM.75.G 2346 gives the answer to this question in five simple cases, when the initial barley plus interest is 100, 500, 1,000, 5,000, or 10,000 níg.sagšu:

$$
\begin{aligned}
4 \text{ še gú.bar} \cdot (1 + 1/5) \ \ + 4 \text{ níg.sagšu} &= \ \ \ \ 100 \text{ níg.sagšu,} \\
20 \ 1/2 \text{ še gú.bar} \cdot (1 + 1/5) \ \ + 8 \text{ níg.sagšu} &= \ \ \ \ 500 \text{ níg.sagšu,} \\
41 \ 1/2 \text{ še gú.bar} \cdot (1 + 1/5) \ \ + 4 \text{ níg.sagšu} &= \ \ 1,000 \text{ níg.sagšu,} \\
208 \text{ še gú.bar} \cdot (1 + 1/5) \ \ + 8 \text{ níg.sagšu} &= \ \ 5,000 \text{ níg.sagšu,} \\
416 \ 1/2 \text{ še gú.bar} \cdot (1 + 1/5) \ \ + 4 \text{ níg.sagšu} &= 10,000 \text{ níg.sagšu.}
\end{aligned}
$$

A6.7. TSS 50, 671 (Shuruppak). Sexagesimal Metric Division Exercises (ED IIIa)

There are two known Old Sumerian division exercises from Shuruppak in the Early Dynastic IIIa period, about the middle of the 3rd millennium BC. One of them is **TSS 50**, where the barley in a granary is to be divided into rations of 7 sìla each. The answer is given, correctly, in non-positional sexagesimal numbers (App. 4, Fig. A4.1, system $S(OS)$). Details of the solution procedure are not provided.

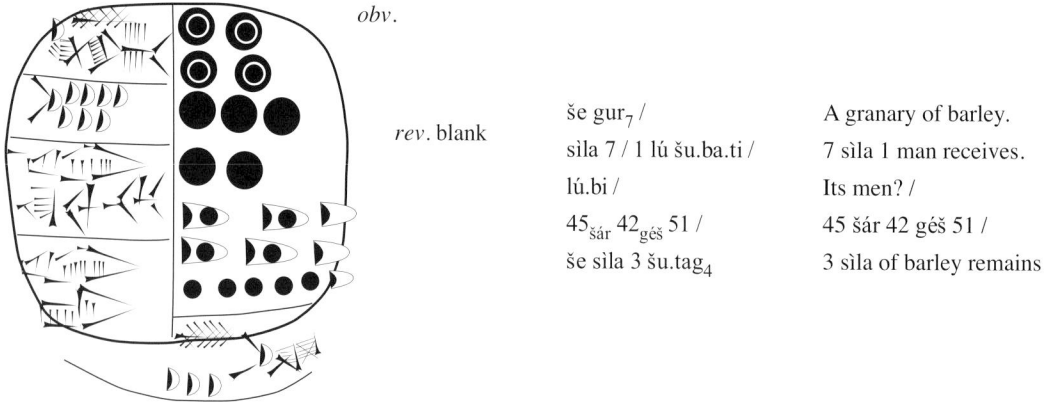

obv.

rev. blank

še gur₇ /   A granary of barley.

sìla 7 / 1 lú šu.ba.ti /   7 sìla 1 man receives.

lú.bi /   Its men? /

45$_{šár}$ 42$_{géš}$ 51 /   45 šár 42 géš 51 /

še sìla 3 šu.tag₄   3 sìla of barley remains.

Fig. A6.16. *TSS* 50. A division exercise from Shuruppak, mentioning a large capacity number, a 'granary'.

A related text, also from Shuruppak, is **TSS 671**. Both *TSS* 50 and *TSS* 671 were first published in Jestin, *TSS* (1937). Photos of the clay tablets can be found in Høyrup, *HM* 9 (1982). The hand copies of *TSS* 50 and *TSS* 671 in Figs. A6.16-17 are based on Høyrup's photos.

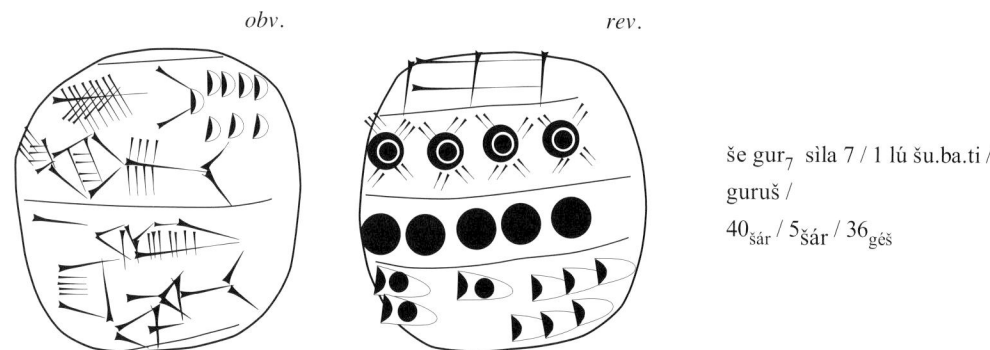

obv.   *rev.*

še gur₇ sìla 7 / 1 lú šu.ba.ti /

guruš /

40$_{šár}$ / 5$_{šár}$ / 36$_{géš}$

Fig. A6.17. *TSS* 671. A related division exercise from Shuruppak.

The two texts are concerned with the same division exercise, in both cases with a question of the form:

All the barley contained in 1 'granary' is to be divided into rations,
so that each man (lú) or worker (guruš) receives 7 sìla. How many men get their rations?

The correct answer is:

45 šár 42 géš 51 rations, with 3 sìla left over (šu.tag$_4$).

Counting backwards, one finds that the barley contained in 1 'granary' must have been

7 sìla · 45 šár 42 géš 51 + 3 sìla = 5 šár.gal 20 šár sìla.

This result agrees fairly well with the known fact that in later cuneiform texts, both Sumerian and Babylonian, gur$_7$ 'granary' was the name of a very large capacity unit, equal to 1 šár gur. The gur, in its turn, was another large capacity unit, equal 5 00 (300) sìla in Old Babylonian mathematical texts, but of varying size in Sumerian administrative texts, depending on from what site and from which period the texts originate. Hence, what *TSS* 50 seems to tell us is that in Shuruppak, in the middle of the third century, 1 gur$_7$ was equal to 1 šár gur, with each gur equal to 5 20 (320) sìla.

Actually, however, two kinds of gur are documented in texts from Shuruppak, in some texts the gur.mah 'mighty gur', equal to 8 géš (480) sìla, in other texts the líd.ga, equal to 4 géš (240) sìla. Hence, either the gur$_7$ was equal to 40 géš gur.mah, alternatively 1 šár 20 géš líd.ga, in Shuruppak, or else the number of gur in a gur$_7$ was reduced in *TSS* 50, for some unknown reason, to 2/3 of its ordinary value in the former case, or increased to 4/3 of its ordinary value in the latter case.

Melville (*UOS* (2002), 237-252) gives a convincing explanation of how the answers in *TSS* 50 and *TSS* 671 can have been computed by an ED III school boy *without recourse to sexagesimal numbers in place value notation*. Although Melville fails to realize it, the solution algorithm he proposes is closely related to the solution algorithm used in the Ebla text TM.75.G 1392 (see above). Somewhat simplified and refined, Melville's explanation goes as follows in the case of *TSS* 50:

barley	men receiving rations	remainder	
1 bán = 10 sìla	1	3 sìla	10 sìla = 1 ration + 3 sìla
1 barig = 6 bán	8	4 sìla	6 · 3 sìla = 2 rations + 4 sìla
1 gur.mah = 8 barig	1 géš 8	4 sìla	8 · 4 sìla = 4 rations + 4 sìla
10 gur.mah	11 géš 25	5 sìla	10 · 4 sìla = 5 rations + 5 sìla
1 géš gur.mah	1 šár 8 géš 34	2 sìla	6 · 5 sìla = 4 rations + 2 sìla
10 géš gur.mah	11 šár 25 géš 42	6 sìla	10 · 2 sìla = 2 rations + 6 sìla
40 géš gur.mah	45 šár 42 géš 51	3 sìla	4 · 6 sìla = 3 rations + 3 sìla

In the case of *TSS* 671, the author of the text apparently made a fatal mistake halfway through the solution algorithm, which was then propagated to the remaining steps of the solution procedure:[11]

barley	men receiving rations	remainder	
1 bán = 10 sìla	1	3 sìla	10 sìla = 1 ration + 3 sìla
1 barig = 6 bán	8	4 sìla	6 · 3 sìla = 2 rations + 4 sìla
1 gur.mah = 8 barig	1 géš 8	4 sìla	8 · 4 sìla = 4 rations + 4 sìla
10 gur.mah	11 géš **24**	––	10 · 4 sìla = 4 bán = **4 rations!**
1 géš gur.mah	1 šár 8 géš 24	—	
10 géš gur.mah	11 šár 24 géš	—	
40 géš gur.mah	45 šár 36 géš	—	

A6.8. Examples of Complicated Designs

The idea of designs with rotational symmetry like the mazes exhibited on MS 4516 (Sec. 8.3.c above) had ancient roots in Mesopotamia. Two interesting examples are provided by a detail from *UE 3*, **518** (Legrain (1936)), the Early Dynastic seal of Mesannipadda, king of Kish, shown in Fig. A6.18 below, and by a detail from an *UE 3*, **393**, an even older seal, shown in Fig. A6.19 below.

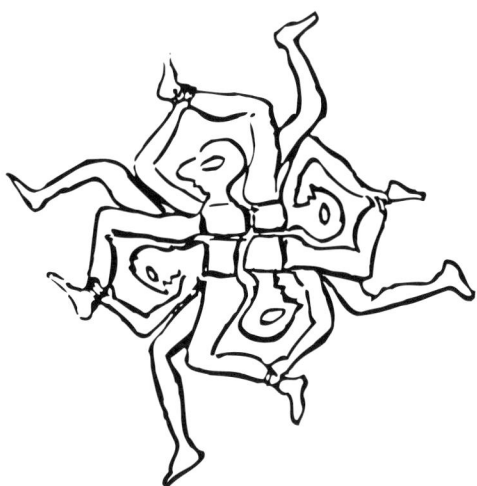

Fig. A6.18. Enlarged detail of *UE 3* (1936), 518, a seal imprint from the Early Dynastic period (c. 2500 BC).

This figure exhibits a fourfold rotational symmetry: the design remains the same if it is rotated by a quarter of a full rotation. It is not unlikely that the one who constructed this design got his inspiration from some geometric figure

Fig. A6.19. Enlarged detail of *UE 3* (1936), 393, a seal imprint from the Jemdet Nasr period (c. 2900 BC).

This design, too, exhibits a fourfold rotational symmetry. Also in this case, it is not unlikely that the design is a paraphrase of some geometric figure, such as, for instance, Fig. 8.2.3, left, in Sec. 8.2.

---

11. In *HM* 3 (1976), 432, Powell wrote that *TSS* 671 was "written by a bungler who did not know the front from the back of his tablet, did not know the difference between standard numerical notation and area notation, and succeeded in making half a dozen writing errors in as many lines, but nevertheless was not without a modicum of ability and probably finished school with a low passing grade, took a post with the government and became a bureaucrat".

An even more interesting design with a *fivefold* rotational symmetry can be found on **VA 5953** (Andrae, *BPK* 58 (1937); Moortgat, *Tammuz* (1949), Pl. 31 b), a small marble plaque (apparently a mold) from Old Babylonian Babylon. The figure on the plaque is a pentagram formed of five intricately entangled naked strong men, enclosing a central regular pentagon. (The hand copy in Fig. A6.20 below, based on the photo of the plaque in *BPK* 58, was made for the author by Tomas Wahlberg.)

Fig. A6.20. VA 5953. A marble plaque with a pentagram of five entangled strong men.

This plaque with its design shows that the idea of pentagrams and pentagons was not unknown in Mesopotamia. This is interesting, because the Susa tablet *TMS* 2 (Fig. 8.2.15 above) shows a regular hexagon (6-side) and a regular heptagon (7-side), but there is no similar clay tablet with a drawing of a regular pentagon (5-side). On the other hand, the entry in line 26 of the table of constants BR mentions the constant '1 40' for a 5-side. (See Sec. 8.2 e above.)

Another example of an early interest in complicated designs of a mathematical nature is a doodle on the reverse of **TSS 973**, a text from Early Dynastic Shuruppak, showing three examples of intricate knots drawn with one continuous line. The reproduction of the doodle in Fig. A6.21 below is a detail borrowed from the hand copy of *TSS* 973 in Jestin (1973).

Knotted curves with 3, 4, and 5 nodes.

An early example of the inclination to look for variations on a mathematical theme.

These drawings are inserted in a vacant place on the reverse of a clay tablet with an undeciphered Sumerian literary text.

Fig. A6.21. *TSS* 973. Geometric doodles on the reverse of an Early Dynastic clay tablet.

Note, in particular, the three- and fourfold rotational symmetries of the two figures in this doodle with three and four nodes, respectively. (The third figure, with five nodes, seems to be a failed attempt to construct a similar figure with a fivefold rotational symmetry.) The figure with three nodes is essentially unchanged after being rotated a third or two thirds of a full revolution, and the figure with four nodes is essentially unchanged after being rotated one, two, or three quarters of a full revolution.

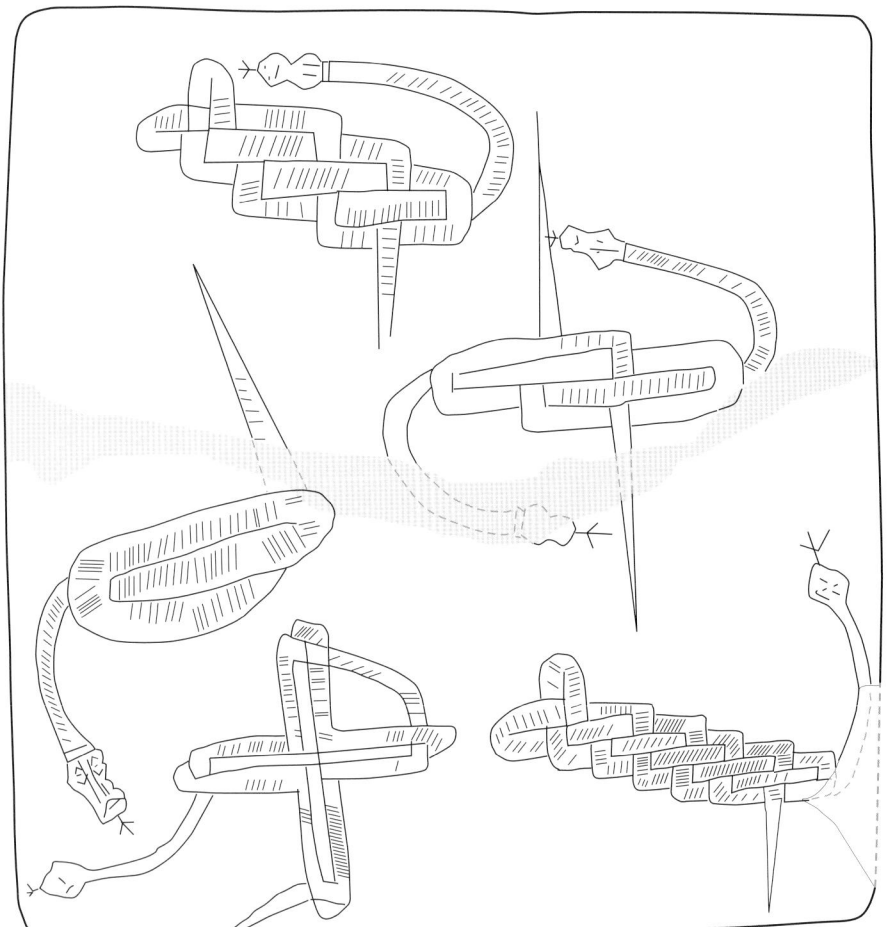

rev.

Doodles on the reverse of a clay tablet from ED IIIa Shuruppak.
These drawings can be understood as another early example of a mathematical theme text.

The obverse is inscribed with a lexical list, the so called Titles and Professions List.

Fig. A6.22. VAT 9130, *rev*. Geometric doodles on the reverse of an Early Dynastic clay tablet (c. 2600 BC).

Knots of another kind are the knotted snakes inscribed as doodles on the reverse of **VAT 9130** (= Deimel, *SF* (1923), 75). The hand copy in Fig. A6.22 is based on a photo of the reverse of the clay tablet in Nissen, Damerow, and Englund, *ABK* (1993), Fig. 89.

Appendix 7
A Combined Metro-Mathematical Table Text with Areas of Large and Small Squares (ED IIIb)

A7.1. CUNES 50-08-001. An Early Dynastic Metro-Mathematical Table Text

CUNES 50-08-001[1] (Figs. A7.1-2 below) is a combined metro-mathematical table text, closely related to A 681, the table from Adab of areas of small squares (Fig. A1.4 above). Although it is unprovenanced, there are reasons to believe that it comes from Zabalam, and it appears to be Early Dynastic III, not quite as early as the texts from Shuruppak, but earlier than (most) of the texts from Adab. It is a large square tablet, probably fired in antiquity, meticulously inscribed with 5 sub-tables in 7 columns on the obverse and 3 on the reverse (see the hand copy below). The text ends with the following subscript in col. *x*:

sanga [x] / sanga kù si / lú dub.sar / lugal.ḫé.gál sud / ir da / sanga kù si / sanga zíz x mu zu x

The subscript gives the name of the scribe (dub.sar), a priest (sanga), *etc.*

The 5 sub-tables on CUNES 50-08-001 are all "tables of areas of squares", organized as follows:

A.	Areas of squares with sides given in	multiples of the **ninda**, from sq. (1 n.) to sq. (10(šár) n.)	cols. *i - v*
B.		multiples of the ***nikkas***, from sq. (1 *n.*) to sq. (10 *n.*)	cols. *v - vi*
C.		multiples of the **kùš.numun**, from sq. (1 k.n.) to sq. 10 k.n.)	cols. *vi - vii*
D.		multiples of the **giš.bad**, from sq. (1 g.b.) to sq. (10 g.b.)	cols. *vii - viii*
E.		multiples of the **šu.bad**, from sq. (1 š.b.) to sq. (10 š.b.)	cols. *viii - ix*

The mentioned fractions of the ninda are previously well known, in particular since they appear in the five Old Akkadian mathematical exercises ***DPA* 36** (Fig. A6.3 above), ***DPA* 37** (Fig. A6.5), **A 5446** (Fig. A6.7), **Ash. 1924.689** (Fig. A6.8), and **A 5443** (Fig. A6.9). The present text shows that the fractions of the ninda known from the mentioned Old Akkadian exercises were in use already before the Old Akkadian period. Conversely, it is likely that the authors of the mentioned Old Akkadian exercises had recourse to a metro-mathematical table text of the same kind as CUNES 50-08-001!

In Table A, the name of the length unit is mentioned only in the first entry, in accordance with the convention, observed from the earliest proto-cuneiform metro-mathematical texts (see, for instance, Nissen/Damerow/Englund, *ABK* (1993), Fig. 50) to the latest Seleucid mathematical texts, that multiples of the ninda usually are written as sexagesimal numbers without explicit mention of the ninda.

Here is the simple factor diagram for the non-positional sexagesimal numbers appearing in Table A:

_____

1. This text is published here in close cooperation with Rudi Mayr, who supplied the excellent hand copy along with numerous comments and suggestions, and with the kind permission of David I. Owen, Curator of Tablet Collections, Dept. of Near Eastern Studies, Cornell University, Ithaca, New York.

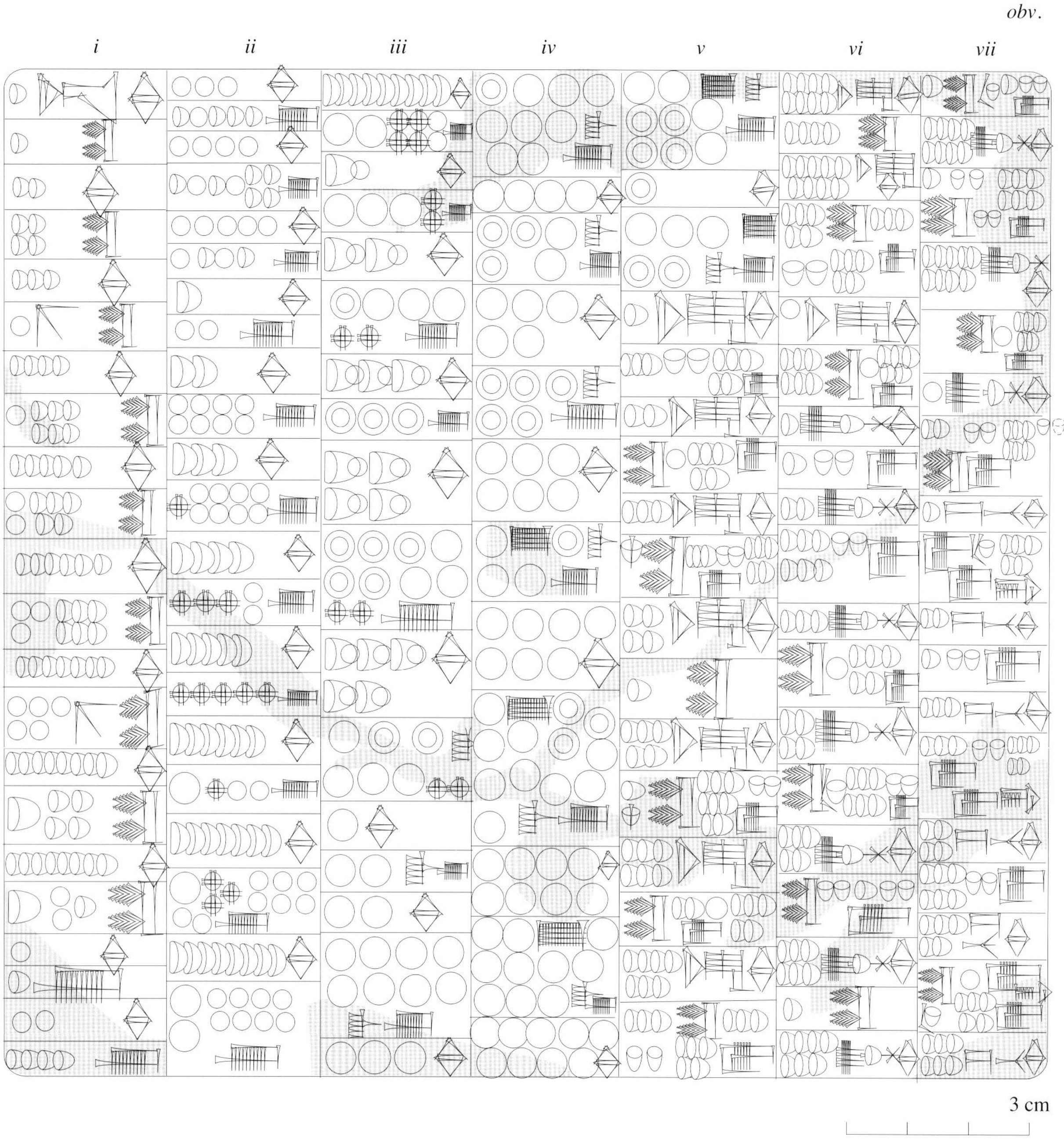

obv.

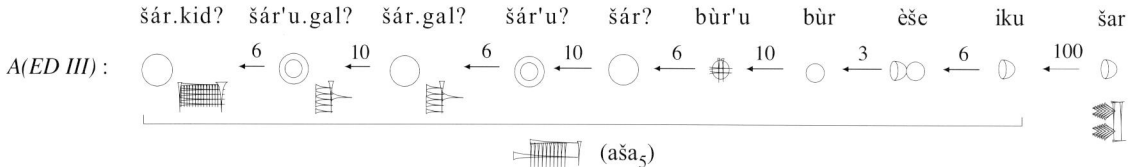

3 cm

Fig. A7.1. CUNES 50-08-001, *obv.* A combined table of areas of large and small squares (ED IIIb).

The factor diagram for the area numbers appering in Table A is more complicated:

A(ED III):

From the iku upwards all area numbers are followed by the qualifier aša₅ 'field, area'. Therefore, there is no risk of confusion, although 1 and iku, 10 and bùr, *etc.,*are written with the same number signs.

Fig. A7.2. CUNES 50-08-001, *rev.*

The transliteration below of **Table A** shows that the table is organized as follows:

a) sides from 1 ninda to 9 ninda	
b) sides from 1(u) ninda to 5(u) ninda	u = 10
c) sides from 1(géš) ninda to 9(géš) ninda	géš= 60
d) sides from 1(géš'u) ninda to 5(géš'u) ninda	géš'u = 10 · 60
e) sides from 1(šár) ninda to 9(šár) ninda	šár = 60 · 60
f) side 1(šár'u) ninda	šár'u = 10 · 60 · 60

The remaining sub-tables on CUNES 50-08-001 (**Tables B-E**) are tables of areas of squares with sides measured in multiples of certain fractions of the ninda. Here is the factor diagram for the Early Dynastic III / Old Akkadian fractions of the ninda:

$$L(ED/OAkk): \;\; \leftarrow \text{ninda} \xleftarrow{4} \textit{nikkas} \xleftarrow{1\,1/2} \underset{\text{seed-cubit}}{\text{kùš.numun}} \xleftarrow{2} \underset{\text{cubit}}{\text{giš.bad/kùš}} \xleftarrow{2} \text{šu.bad/}\textit{zipaḫ} \xleftarrow{1\,1/2} \text{šu.dù.a} \xleftarrow{10} \underset{\text{finger}}{\text{šu.si}}$$

The fractions of the šar appearing in Tables B-E are the following ones:

	3″	2′	3′		gín	gín.bi	gín.ba.gín
basic fractions:				shekel fractions:			
	2/3	1/2	1/3		1/60	1/60 · 1/60	1/60 · 1/60 · 1/60

The basic fractions appear frequently in administrative or mathematical cuneiform texts, with beginning in the Early Dynastic III period. The shekel fractions in the form shown above are attested exclusively in CUNES 50-08-001. Their Sumerian names can be translated as follows:

gín 'shekel'   gín.bi 'its shekel'   gín.ba.gín 'shekel of its shekel'

The computation of the entries in **Table A** is relatively simple, in view of the factor diagram for system *A* (see above), from which follows, in particular, that

1 iku = 1(géš) 40 ninda,   1 èše = 10(géš) ninda,   1 bùr = 30(géš) ninda.

A. ninda (cols. *i - v*)

sq. (1 n.)	=	1	šar	sq. (6(géš) n.)	=	1(šár) 12(bùr)	aša₅
sq. (2 n.)	=	4	šar	sq. (7(géš) n.)	=	1(šár) 38(bùr)	aša₅
sq. (3 n.)	=	10 − 1	šar	sq. (8(géš) n.)	=	2(šár) 8(bùr)	aša₅
sq. (4 n.)	=	16	šar	sq. (9(géš) n.)	=	2(šár) 42(bùr)	aša₅
sq. (5 n.)	=	25	šar	sq. (10(géš) n.)	=	3(šár) 20(bùr)	aša₅
sq. (6 n.)	=	[3]6	šar	sq. (20(géš) n.)	=	13(šár) [20(bùr)]	aša₅
sq. (7 n.)	=	50 − 1	šar	sq. (30(géš) n.)	=	30(šár)	aša₅
sq. (8 n.)	=	1(géš) 4	šar	sq. (40(géš) n.)	=	53(šár) 20(bùr)	aša₅
sq. (9 n.)	=	1(géš) 21	šar	sq. (50(géš) n.)	=	[1(šár)].gal 23(šár) 20(bùr)	aša₅
sq. (10 n.)	=	[1(iku)]	aša₅	sq. (1(šár) n.)	=	2(šár).gal	aša₅
sq. (20 n.)	=	[4(iku)	aša₅]	sq. (2(šár) n.)	=	8(šár).[gal]	aša₅
sq. (30 n.)	=	1(èše) 3(iku)	aša₅	sq. (3(šár) n.)	=	1[8(šár).gal]	aša₅
sq. (40 n.)	=	2(èše) 4(iku)	aša₅	sq. (4(šár) n.)	=	32(šár).gal	aša₅
sq. (50 n.)	=	1(bùr) 1(èše) 1(iku)	aša₅	sq. (5(šár) n.)	=	50(šár).gal	aša₅
sq. (1(géš) n.)	=	2(bùr)	aša₅	sq. (6(šár) n.)	=	1(šár).[kid] 12(šár).gal	aša₅
sq. (2(géš) n.)	=	8(bùr)	aša₅	sq. (7(šár) n.)	=	1(šár).kid 38(šár).gal	[aša₅]
sq. (3(géš) n.)	=	18(bùr)	aša₅	sq. (8(šár) n.)	=	2(šár).kid 8(šár).gal	aša₅
sq. (4(géš) n.)	=	[3]2(bùr)	aša₅	sq. (9(šár) n.)	=	2(šár).kid [4]2(šár).gal	aša₅
sq. (5(géš) n.)	=	50(bùr)	aša₅	sq. (10(šár) n.)	=	3(šár).kid 20(šár).gal	aša₅

The computation of the entries in the *nikkas* table (**Table B**) is complicated. It must have begun with the computation of the first entry, perhaps as follows, with 1 gín = 1/60 šar, 1 gín.bi = 1/60 gín:

1 *nikkas* = 1/4 ninda
1 ninda · 1 *nikkas* = 1/4 sq. ninda (šar) = 15 gín
1 sq. *nikkas* = 1/16 sq. ninda = 1/4 · 15 gín = 3 gín + 1/4 · 3 gín = 3 gín 45 gín.bi = 3 2/3 gín 5 gín.bi

In sexagesimal place value notation, the computation would have been considerably simpler:

1 *nikkas* = ;15 ninda, sq. (1 *nikkas*) = sq. ;15 n. = ;03 45 sq. n.

With departure from this result, the entries in the *nikkas* sub-table can have been computed as follows:

B. *nikkas* = 1/4 ninda (cols. *v - vi*)   explanation

sq. (1 *n.*)	= 3 3" 5 gín	1 sq. *nikkas*	= 1/16 šar	= 3 2/3 gín 5 gín.bi
sq. (2 *n.*)	= šar 15 gín	sq. (1/2 ninda)	= 1/4 šar	= 15 gín
sq. (3 *n.*)	= 2' šar 3 3" 5 gín	9 sq. *nikkas*	= 33 gín + 45 gín.bi	= 1/2 šar 3 2/3 gín 5 gín.bi
sq. (4 *n.*)	= 1 šar	sq. (1 ninda)		= 1 šar
sq. (5 *n.*)	= [1 2' šar] 3 3" 5 gín	25 sq. *nikkas*	= 1_g 31 2/3 gín + 2_g 5 gín.bi	= 1 1/2 šar 3 2/3 gín 5 gín.bi
sq. (6 *n.*)	= šar 2 1[5] gín	sq. (1 1/2 ninda)	2 1/4 šar	= 2 šar 15 gín
sq. (7 *n.*)	= 3 šar 3 3" 5 gín	49 sq. *nikkas*	= 2_g 59 2/3 gín + 4_g 5 gín.bi	= 3 šar 3 2/3 gín 5 gín.bi
sq. (8 *n.*)	= 4 šar	sq. (2 ninda)		= 4 šar
sq. (9 *n.*)	= 5 šar 3 3" 5 gín	1_g 21 sq. *nikkas*	= 4_g 57 gín + 6_g 45 gín.bi	= 5 šar 3 2/3 gín 5 gín.bi
sq. (10 *n.*)	= 6 šar 15 gín	sq. (2 1/2 ninda)	6 1/4 šar	= 6 šar 15 gín

Note: In actual practice, 2 *nikkas* would never appear in this form, it would be replaced by 2' ninda. Similarly, 3 *nikkas* would be replaced by 2' ninda 1 *nikkas*, *etc*. For that reason, a rather plausible alternative explanation of the construction of the squares of odd multiples of the *nikkas* in Table B is that they were computed by use of the geometrically obvious *binomial rule*, according to which sq. ($a + b$) = sq. $a + 2 a \cdot b$ + sq. b. By use of this rule, it could be shown, for instance, that

sq. (3 *n.*) = sq. (1/2 n. 1 *n.*) = sq. (1/2 n.) + 1 n. · 1 n. + sq. (1 *n.*) = 1/4 šar + 1/4 šar + 3 2/3 gín 5 gín.bi
sq. (5 *n.*) = sq. (1 n. 1 *n.*) = sq. (1 n.) + 2 n. · 1 n. + sq. (1 *n.*) = 1 šar + 1/2 šar + 3 2/3 gín 5 gín.bi
sq. (7 *n.*) = sq. (1 1/2 n. 1 *n.*) = sq. (1 1/2 n.) + 3 n. · 1 n. + sq. (1 *n.*) = 2 1/4 šar + 3/4 šar + 3 2/3 gín 5 gín.bi
sq. (9 *n.*) = sq. (2 n. 1 *n.*) = sq. (2 n.) + 4 n. · 1 n. + sq. (1 *n.*) = 4 šar + 1 šar + 3 2/3 gín 5 gín.bi

The computation of the entries in the kùš.numun table (**Table C**) must have begun with the computation of the first entry, perhaps as follows:

1 kùš.numun = 1/6 ninda
1 ninda · 1 kùš.numun = 1/6 sq. ninda (šar) = 10 gín
1 sq. kùš.numun = 1/36 sq. ninda = 1/6 · 10 gín = 1 2/3 gín

Here, too, the computation is simpler in sexagesimal place value notation:

1 kùš.numun = ;10 ninda, sq. (1 kùš.numun) = sq. ;10 n. = ;01 40 sq. n.

All the entries in the kùš.numun table can have been computed as follows:

C. kùš.numun = 1/6 n. (cols. *vi - vii*)   explanation

sq. (1 k.n.)	= 1 3" gín	1 sq. kùš.numun	= 1/36 šar	= 1 2/3 gín
sq. (2 k.n.)	= 6 3" gín	4 sq. kùš.numun	= 4 gín + 2 2/3 gín	= 6 2/3 gín
sq. (3 k.n.)	= šar 15 gín	sq. (1/2 ninda)	= 1/4 šar	= 15 gín
sq. (4 k.n.)	= šar 3' 6 3" gín	16 sq. kùš.numun	= 16 gín + 10 2/3 gín	= 1/3 šar 6 2/3 gín
sq. (5 k.n.)	= [šar 3" 1 3" gín]	25 sq. kùš.numun	= 25 gín + 16 2/3 gín	= 2/3 šar 1 2/3 gín
sq. (6 k.n.)	= 1 šar	sq. (1 ninda)		= 1 šar
sq. (7 k.n.)	= [1 šar 3' 1 3" gín]	49 sq. kùš.numun	= 49 gín + 32 2/3 gín	= 1 1/3 šar 1 2/3 gín
sq. (8 k.n.)	= 1 3" šar [6 3" gín]	1_g 4 sq. kùš.numun	= 1_g 4 gín + 42 2/3 gín	= 1 2/3 šar 6 2/3 gín
sq. (9 k.n.)	= 2 šar 1[5 g]ín	sq. (1 1/2 ninda)	2 1/4 šar	= 2 šar 15 gín
sq. (10 k.n.)	= [2 3" šar] 6 3" gín	1_g 40 sq. kùš.numun	= 1_g 40 gín + 1_g 6 2/3 gín	= 2 2/3 šar 6 2/3 gín

As in the case of Table B, an alternative explanation of the computation of the squares of some multiples of the kùš.numun is that they were computed as follows, by use of the binomial rule:

sq. (4 k.n.) = sq. (1/2 n. 1 k.n.) = sq. (1/2 n.) + 1 n. · 1 k.n. + sq. (1 k.n.) = 15 gín + 10 gín + 1 2/3 gín
sq. (5 k.n.) = sq. (1/2 n. 2 k.n.) = sq. (1/2 n.) + 1 n. · 2 k.n. + sq. (2 k.n.) = 15 gín + 20 gín + 6 2/3 gín
sq. (7 k.n.) = sq. (1 n. 1 k.n.) = sq. (1 n.) + 2 n. · 1 k.n. + sq. (1 k.n.) = 1 šar + 1/3 šar + 1 2/3 gín
sq. (8 k.n.) = sq. (1 n. 2 k.n.) = sq. (1 n.) + 2 n. · 2 k.n. + sq. (2 k.n.) = 1 šar + 2/3 šar + 6 2/3 gín
sq. (10 k.n.) = sq. (1 1/2 n. 1 k.n.) = sq. (1 1/2 n.) + 3 n. · 1 k.n. + sq. (1 k.n.) = 2 šar 15 gín + 1/2 šar + 1 2/3 gín

The entries in the giš.bad table (**Table D**) can have been computed as follows:

1 giš.bad = 1/2 kùš.numun = 1/12 ninda
1 ninda · 1 giš.bad = 1/12 šar = 5 gín
1 sq. giš.bad = 1/12 · 5 gín = 25 gín.bi = 1/3 gín 5 gín.bi

In sexagesimal place value notation:

1 giš.bad = ;05 n., sq. (1 giš.bad) = sq. ;05 n. = ;00 25 sq. n.

All the entries in the giš.bad table can have been computed as follows:

D. giš.bad = 1/12 ninda (cols. *vii - viii*)explanation

sq. (1 g.b.)	= gín 3' 5 gín.bi	1 sq. giš.bad	= 1/3 gín 5 gín.bi
sq. (2 g.b.)	= 1 3" gín	sq. (1 kùš.numun)	= 1 2/3 gín
sq. (3 g.b.)	= 3 [3" gín] 5 [gín.bi]	sq. (1 *nikkas*)	= 3 2/3 gín 5 gín.bi
sq. (4 g.b.)	= 6 3" gín	sq. (2 kùš.numun)	= 6 2/3 gín
sq. (5 g.b.)	= šar 10 gín 3' 5 gí[n.bi]	25 sq. giš.bad = 8 1/3 gín + 2$_g$ 5 gín.bi	= 10 1/3 gín 5 gín.bi
sq. (6 g.b.)	= šar 15 gín	sq. (3 kùš.numun)	= 15 gín
sq. (7 g.b.)	= šar 3' gín 3' 5 gí[n.bi]	49 sq. giš.bad = 16 1/3 gín + 4$_g$ 5 gín.bi	= 1/3 šar 1/3 gín 5 gín.bi
sq. (8 g.b.)	= [šar 3' 6 3" gín]	sq. (4 kùš.numun)	= 1/3 šar 6 2/3 gín
sq. (9 g.b.)	= [šar 2' 3 3" gín 5 gín.bi]	sq. (3 *nikkas*)	= 1/2 šar 3 2/3 gín 5 gín.bi
sq. (10 g.b.)	= [šar 3"] 1 3" gín	sq. (5 kùš.numun)	= 2/3 šar 1 2/3 gín

It is, of course, also possible that the squares of odd multiples of the giš.bad were computed by use of the binomial rule, in much the same way as in the case of Table B.

Finally, the entries in the šu.bad table (**Table E**) can have been computed as follows:

1 šu.bad = 1/2 giš.bad = 1/4 kùš.numun = 1/24 ninda
1 sq. šu.bad = 1/4 sq. giš.bad = 1/4 · 25 gín.bi = 6 1/4 gín.bi = 6 gín.bi 15 gín.ba.gín
In sexagesimal place value notation: 1 šu.bad = ;02 30 n., sq. (1 šu.bad) = sq. ;02 30 n. = ;00 06 15 sq. n.

E. šu.bad = 1/24 ninda (cols. *viii - x*)explanation

sq. (1 š.b.)	= [gín].bi.ta 6 [15] gín.[ba.gín]	1 sq. šu.bad	= 6 gín.bi 15 gín.ba.gín
sq. (2 š.b.)	= gín 3' 5 gín.bi	sq. (1 giš.bad)	= 1/3 gín 5 gín.bi
sq. (3 š.b.)	= 56 gín$^!$ 15 [gín.ba.gín]	9 sq. šu.bad	= 2/3 gín 16 gín.bi 15 gín.ba.gín
sq. (4 š.b.)	= 1 3" gín	sq. (2 giš.bad)	= 1 2/3 gín
sq. (5 š.b.)	= 2 [2' gín].ta gín.[bi] 6 1[5] gín.[ba.gín]	25 sq. šu.bad	= 2 1/2 gín 6 gín.bi 15 gín.ba.gín
sq. (6 š.b.)	= 3 3" gín 5 [gín.bi]	sq. (3 giš.bad)	= 3 2/3 gín 5 gín.bi
sq. (7 š.b.)	= 5 g[ín gín.bi] 6 [15] gín.[ba.gín]	49 sq. šu.bad	= 5 gín 6 gín.bi 15 gín.ba.gín
sq. (8 š.b.)	= 6 3" gín	sq. (4 giš.bad)	= 6 2/3 gín
sq. (9 š.b.)	= 8 3' gín gín.bi.ta 6 15 gín.ba.gín.ta	1$_g$ 21 sq. šu.bad	= 8 1/3 gín 6 gín.bi 15 gín.ba.gín
sq. (10 š.b.)	= [šar 10] gín [3' 5 gín].bi	sq. (5 giš.bad)	= 10 1/3 gín 5 gín.bi

(In entries 5 and 9 of Table E, the suffix .ta is a Sumerian postposition with the meaning 'of'.)

It is, of course, also possible that the squares of odd multiples of the šu.bad were computed by use of the binomial rule, in much the same way as in the case of Table B.

Note: The factor diagram above for the fractions of the ninda shows that Tables B-E on CUNES 50-08-001 are tables for the fractions of the ninda in decreasing order, from the *nikkas* to the šu.bad. The question then arises why there are no similar tables for the two remaining fractions of the ninda, namely the šu.dù.a and the šu.si. A possible answer to this question may be that the author of the combined table text originally planned to include separate tables (Tables F-G) also for those fractions. Since the space required for two more tables would have been approximately three columns, it appears that the author of the text had made space on the clay tablet for those two tables, plus the subscript. However, while constructing Table E he must have realized that the entries were getting increasingly complex, and increasingly non-realistic, with ever smaller shekel fractions, so he simply abandoned the continuation of his project and stopped after Table E.

A7.2. A Parallel Text from Adab (ED IIIb)

The Adab text **A 681** (Fig. A1.4), an ED IIIb table of areas of squares with sides measured in multiples of the kúš 'cubit' is an obvious parallel to Table D on CUNES 50-08-001. On the other hand, there are many differences between the two table texts, as shown below:

CUNES 50-08-001 Table D		A 681		place value	
side	area	side	area	side	area
1 giš.bad	gín 3' 5 gín.bi	1 cubit	1 sa$_{10}$ ma.na 15 gín	;05	;00 25
2 giš.bad	1 3" gín	2 cubits	2. gín – 1 ma.na	;10	;01 40
3 giš.bad	3 [3" gín] 5 [gín.bi]	3 cubits	4. gín – igi.4	;15	;03 45
4 giš.bad	6 3" gín	4 cubits	6. gín 3" ma.na	;20	;06 40
5 giš.bad	šar 10 gín 3' 5 gí[n.bi]	5 cubits	10 gín 3'. 5.	;25	;10 25
6 giš.bad	šar 15 gín	6 cubits	15. gín	;30	;15
7 giš.bad	šar 3' gín 3' 5 gí[n.bi]	7 cubits	3'. šar 3'. 5.	;35	;20 25
8 giš.bad	[šar 3' 6 3" gín]	8 cubits	2' šar – (3 gín 1 sa$_{10}$.ma.na)	;40	;26 40
9 giš.bad	[šar 2' 3 3" gín 5 gín.bi]	10 – 1 c.	2' šar 4. gín – igi.4.	;45	;33 45
10 giš.bad	[šar 3"] 1 3" gín	10 cubits	3" šar 2. gín – 1 sa$_{10}$.ma.na	;50	;41 40
		11 cubits	(1 šar – 10 gín) 1 sa$_{10}$ 15	;55	;50 25
		3 reeds	2 šar 15. gín	1;30	2;15

The comparison shows that although the two texts are roughly contemporary, there are so many differences between them that they can hardly be copies of one and the same original table text.

A7.3. The Historical Importance of the Combined Table Text CUNES 50-08-001

All previously known mathematical cuneiform texts from the Old Akkadian or ED III periods, such as ***TSS 188***, ***TSS 926***, **VAT 12593**, **MS 3047** in Ch. 6 above, **A 681** in App. 1, or ***TSS 50***, ***TSS 671***, and ***DPA 34-38***, *etc.* in App. 6, are probably students' exercises, or excerpts from teachers' master texts. CUNES 50-08-001 is the first known example of an original, superbly organized and well executed mathematical text of great complexity, dating from around the middle of the 3rd millennium BC. It demonstrates much more clearly than any other known text that serious education in mathematics must have started very early in Mesopotamia.

The circumstance that CUNES 50-08-001 and A 681 are parallel texts yet with many differences in details shows that already in the ED IIIb period, which ended c. 2350 BC, various "schools" in mathematics, each with its own mathematical terminology, had developed in Mesopotamia. A similar conclusion, by the way, can be drawn about education in mathematics in the ED IIIa period, c. 2500 BC, based on a comparison of the two more or less parallel table texts MS 3047, *obv.* (Fig. 6.4, top) and VAT 12593 (Fig. 6.3). This is the same as the situation in the Old Babylonian period, when each major city seems to have had its own "school" of mathematics.

CUNES 50-08-001 is, of course, a forerunner to both Old Babylonian tables of squares and Old Babylonian metrological tables. What is just as interesting is that CUNES 50-08-001 with its long initial Table A (38 entries) and the shorter Tables B-E (10 entries each) is organized just like the Old Babylonian combined algorithm text UM 29.13.21 (Fig. A3.2), which begins with a long table with 30 entries, only to continue with 4 or 5 shorter tables with 10 entries each.

Like Old Babylonian metrological tables for system *A*, CUNES 50-08-001 is concerned with very large area numbers (in Table A) and very small area numbers (in Table E). Both the very large and the very small area numbers are obviously of quite unrealistic sizes and demonstrate a universal tendency of mathematical texts to be concerned more with completeness than with usefulness.

A completely unexpected feature of CUNES 50-08-001 is that it is a clear forerunner of the invention of sexagesimal numbers in place value notation towards the end of the 3rd millennium. It counts multiples of the area measure bùr in sixties (šár), sixties of sixties (šár.gal) and sixties of sixties of sixties (šár.kid). More surprisingly, it counts fractions of the square ninda (šar) in sixtieths (gín), sixtieths of sixtieths (gín.bi), and sixtieths of sixtieths of sixtieths (gín.ba.gín). If it had not been for the existence of a competing, and ultimately victorious, series of small fractions in the system of weight measures (see below), this way of counting with small sexagesimal fractions could easily have led to the invention of sexagesimal place value notation already in the middle of the 3rd millennium!

A7.4. CUNES 47-12-176. An Old Akkadian Lexical Text with Fractions of the Mina[2]

CUNES 47-12-176 (Fig. A7.3 below) is a fairly well preserved clay tablet inscribed with a long series of weight numbers steadily decreasing from [10 ma.na] '10 minas' (c. 500 g) to 1/2 sa_{10} gín 'exchange shekel' (c. 1/40 g). On one hand, it appears to be a metrological list for weight measures. On the other hand, it may be a forerunner to the weight section in the encyclopedic lexical series ur_5.ra = *ḫubullu* (Hh XVI 417-452 = Landsberger *et al., MSL 10* (1970), 15-16, 49-50, 60-61), which goes from na_4.1.gú.un, na_4.50.ma.na, *etc.*, to na_4.1.še, na_4.2'.še, na_4.3'.še. (See Sec. 4 e above.)

The fractions of the mina in CUNES 47-12-176 are the same as the fractions of the šar in A 681 (see Fig. A1.5 above), with 1 sa_{10}.ma.na = 1/3 gín and 1 sa_{10}.gín = 1/60 sa_{10}.ma.na.

The list of weight measures in CUNES 47-12-176 is structured as follows:

1.	From [10 ma.na] to [1] ma.na, [3".š]a ma.na, [2'] ma.na, [2' ma.na]$^?$ –10 gín, [3'.š]a [ma.na]	col. *i*
2.	From [20 gín]$^?$, [15 gín] to 4 gín, 3 gín sa_{10}×15, 2 gín sa_{10}×10, 1 gín sa_{10}×5	cols. *ii - iii*
3.	sa_{10}×2 ma.na, 2' gín, sa_{10}×1 ma.na sa_{10}×15	col. *iii*
4.	igi.3.gál, igi.4.gál, igi.6.gál	col. *iii*
5.	sa_{10}×3".ša, sa_{10}×3'.ša 5 gín.tur (with 1 gín.tur 'small shekel' = 1/60 gín; cf. *DPA* 36, Fig. A6.3)	col. *iii*
6.	From sa_{10}×15 gín to [sa_{10}×1] gín and [sa_{10}×2']$^?$ gín.	cols. *iii - iv*

The list is peculiar in several ways, since it is not everywhere straightforward and regularly decreasing. Thus, for instance, the end of the section 2 of the list is curious, since

3 gín sa_{10}×15, 2 gín sa_{10}×10, 1 gín sa_{10}×5 has to be interpreted as
= 3 gín 15 sa_{10}<.gín>, 2 gín 10 sa_{10}<.gín>, 1 gín 5 sa_{10}<.gín>
= 3 gín + 1/36 of 3 gín, 2 gín + 1/36 of 2 gín, 2 gín + 1/36 of 2 gín.

The whole section 3 is fairly straightforward. It can be interpreted as

sa_{10}×2 ma.na, 2' gín, sa_{10}×1 ma.na sa_{10}×15
= 2 sa_{10} ma.na, 1/2 gín, 1 sa_{10} ma.na 15 sa_{10}<.gín>
= 40 gín.tur, 30 gín.tur, 25 gín.tur.

---

2. This text, too, is published here in close cooperation with Rudi Mayr, and with the kind permission of David I. Owen, Curator of Tablet Collections, Dept. of Near Eastern Studies, Cornell University, Ithaca, New York.

Section 4 is quite straightforward:

igi.3.gál, igi.4.gál, igi.6.gál = 1/3, 1/4, 1/6 <gín> = 20 gín.tur, 15 gín.tur, 10 gín.tur.

It is followed by section 5, which again is somewhat curious, since it can be interpreted as

$sa_{10}\times3''$.ša, $sa_{10}\times3'$.ša 5 gín.tur
= 2/3 sa_{10}.ma.na, 1/3 sa_{10}.ma.na 5 gín.tur
= 40 sa_{10}.gín, 35 sa_{10}.gín = 13 1/3 gín.tur, 11 2/3 gín.tur.

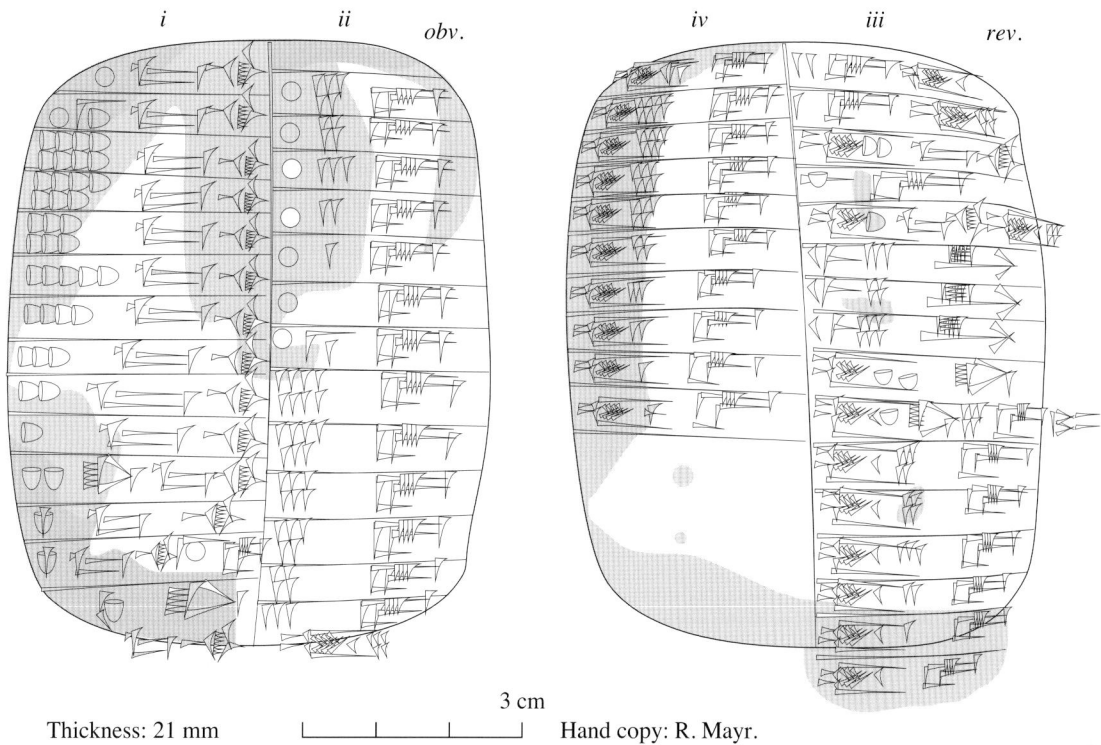

i　　　　*ii*　　*obv.*　　　　　*iv*　　　*iii*　　*rev.*

3 cm

Thickness: 21 mm　　　Hand copy: R. Mayr.

Fig. A7.3. CUNES 47-12-176. A decreasing list of weight measures (Old Akkadian or ED IIIb).

The complicated system of weight measures (ED IIIa-b) figuring in CUNES 47-12-176 is composed of two parts, with the upper part (talent, mina, shekel) suitable for measuring amounts of copper, and with the lower part (sa_{10}.ma.na 'exchange mina' = 1/180 mina and sa_{10}.gín 'exchange shekel' = 1/180 shekel) suitable for measuring smaller amounts of the 180 times more precious silver (Friberg, *CDLJ* 2005:2, Sec. 2.2.8). It was soon to be abandoned in favor of the simplified Neo-Sumerian/Old Babylonian system of weight measures, where the exchange mina had disappeared, and where the exchange shekel had been replaced by the še 'barley-corn'.

Note that just as the ED III system of fractions of the mina was used in the Adab table of areas of squares A 681 to denote fractions of the šar (square ninda), so in Old Babylonian mathematical texts the shekel and the barley-corn were used to denote fractions of the šar. See, for instance, YBC 4669, *rev.*, # 6 (Sec. 11. 3 above), where the area of a rectangle with the sides 3 cubits 1 finger and 2 cubits 6 fingers is given as 2 2/3 shekels 20 1/2 barley-corns. See also Scheil (1938), *RA* 35, texts 1, 2, and Proust, *TMN* (2004), Sec. 7.4.1, all mentioned below.

The historical importance of CUNES 47-12-176 is that it seems to support a conjecture that the origin of the Old Babylonian metrological lists and tables may have been metrological sections of Old Akkadian or Early Dynastic lexical texts.

A7.5. *RA* 35, Texts 1-2 and IM 96183. OB Table Texts Related to CUNES 50-08-001

It is an easily observed rule that all Mesopotamian mathematical texts from before the invention of place value notation for sexagesimal numbers are metro-mathematical, while Old Babylonian mathematical texts, in particular Old Babylonian arithmetical table texts, with very few exceptions are predominantly numerical. A notable exception is a pair of Old Babylonian metro-mathematical table texts from Susa (Scheil (1938), ***RA* 35, texts 1, 2**; Neugebauer/Sachs, *MCT* (1945), 6-10; Figs. A7 4-5 below). Together, they make up a combined table of areas of rectangles, squares, and circles, organized as follows:

A. Areas of rectangles,	from	1/2 cubit · 1/3 cubit = 12 1/2 barley-corn	to	25 ninda · 20 ninda = 16(šár) 40(bùr)
B. Areas of squares,	from	sq. 1 finger = 1 1/2 barley-corn	to	sq. (4 ninda) = 16 šar
C. Areas of circles,	from	;05 · sq. 1/2 ninda = 1 1/4 shekel	to	;05 · sq. (5(géš) ninda) = 4(bùr) 3(iku)

The detailed structure of Table A is quite interesting. It begins with three rectangles with the sides

$$u_1, s_1 = 1/2 \text{ c.}, 1/3 \text{ c.}, \quad u_2, s_2 = 2 \cdot (u_1, s_1), \quad u_3, s_3 = 3 \cdot (u_1, s_1).$$

With few exceptions, all the remaining rectangles are such that

$$u_{k+1}, s_{k+1} = (1 + 1/n) \cdot u_k, u_k \quad \text{where } 1 + 1/n = (n+1)/n \text{ is a regular sexagesimal number.}$$

For this to happen, it is necessary that the pair $(n, n + 1)$ is a pair of "regular sexagesimal twins", in the actual cases always $(2, 3), (3, 4), (4, 5), (5, 6), (8, 9), (9, 10),$ or $(15, 16)$.

Note that, due to the condition $s_{k+1} = u_k$, most of the successive rectangles in Table A are linked together in the same way as the successive rectangles in **MS 2729** and **MS 2728** (Fig. 1.1.2 above).

There is also a group of 7 known hand tablets of a common format, all from Nippur, discussed together by Proust in ***TMN* (2004), Sec. 7.4.1**. In each one of the 7 texts, within a frame in the lower right corner, the side of a square is given, it is asked "What is its area?", and the answer is given in traditional area numbers. The computation of the area by use of sexagesimal numbers is explicitly given outside the frame. The contents of the 7 exercises are described in the following brief catalog:

1.	sq. (2 fingers)	= 3' b.c.	sq. ;00 20	= ;00 00 06 40	**UM 29.15.192**
2.	sq. (1/3 cubit 1/2 finger)	= c. 9 1/5 b.c.	sq. ;01 45	= ;00 03 03 45	**IM 57828**
3.	sq. (1/3 cubit 3 fingers)	= 13 1/4 (error for 14 1/12) b.c.	sq. ;02 10	= ;00 04 41 40	**Ist. Ni 18**
4.	sq. (2/3 cubit 9 fingers)	= 1/4 sh. [25 1/12 b.c.]	sq. ;04 50	= ;00 23 21 40	**IM 57846**
5.	sq. (1 cubit)	= 3' sh. 15 b.c.	sq. ;05	= ;00 25	**CBS 11318**
6.	sq. (1 cubit 2 fingers)	= 3' sh. 25 3' b.c.	sq. ;05 20	= ;00 28 26 40	**UM 55.21.76**
7.	sq. (1 ninda 4 cubits)	= 1 3" šar 6 3" sh.	sq. 1;20	= 1;46 40	**NBC 8082**

This group of exercises, all of precisely the same kind but with different numerical data, nicely illustrates the   observation (made above in connection with the discussion in Secs. 2.4 b and 2.6 f of hand tablets with excerpts from tables of squares, square sides, cube sides, quasi-cube sides, or combined multiplication tables) that Old Babylonian mathematics teachers liked to hand out series of similar, yet different problems to the students in their classes (or, maybe, to single students).

It is interesting that all the computations with sexagesimal numbers in the 7 Nippur texts above (except possibly the first one) can be regarded as excerpts from the big table text **IM 96183**, a large Old Babylonian special table of squares (Fig. A7. 6 below).[3] That table text is of type b_v, like the much shorter text MS 3906 (Fig. 2.1.2, top). It is likely that it has a metrological background, since if it is assumed that all the numbers being squared in IM 96183 are multiples of the ninda, then it is easy to convert those numbers into so and so many ninda, cubits, and fingers (sometimes in more ways than one). That is precisely what is done in the interpretation below of the structure of IM 96183:

1. From 8 50 to 10. Step: 5. (15 lines.)	col. *i*
Interpretation: **From 8 ninda 10 cubits to 10 ninda. Step: 1 cubit**.	
2. From 10 30 to 20. Step: 30. (20 lines.)	cols. *i-ii*

---

3. This is a previously unpublished text. I want to thank F. Al-Rawi for kindly giving me permission to discuss it here.

> Interpretation: **From 10 ninda to 20 ninda. Step: 1/2 ninda.**
>
> 3. From 22 30 to 1. Step: 2 30.   (16 lines.)   cols. *ii-iii*
> > Interpretation: **From 4 1/2 cubits to 1 ninda. Step: 1/2 cubit.**
>
> 4. From 1 05 to 4. Step: 5   (33 lines; 1 05, 2 05, 3 05 missing).   cols. *iii-iv*
> > Interpretation: a) **From 6 1/2 fingers to 2/3 cubits 4 fingers. Step: 1/2 finger**.
> >
> > b) **From 1 ninda to 4 ninda. Step: 1 cubit.**
>
> 5. 4 30, 4 45, 4 50.   (3 lines.)   col. *iv*
> > Interpretation: **2/3 cubits 7 fingers, 2/3 cubit 8 1/2 fingers, 2/3 cubits 9 fingers**
>
> 6. From 5 to 6. Step. 10.   (7 lines.)   col. *v*
> > Interpretation: **From 1 cubit to 1 cubit 6 fingers. Step: 1 finger.**
>
> 7. From 6 40 to 10. Step: 50.   (5 lines.)   col. *v*
> > Interpretation: **From 1 cubit 10 fingers to 2 cubits. Step: 5 fingers.**
>
> 8. From 10 10 to 11. Step: 10.   (6 lines.)   col. *v*
> > Interpretation: **From 2 cubits 1 finger to 2 cubits 6 fingers. Step: 1 finger.**
>
> 9. 12, 12 30, 13 20.   (3 lines.)   col. *v*
> > Interpretation: **2 1/3 cubits 2 fingers, 2 1/2 cubits, 2 2/3 cubits.**
>
> 10. From 15, 16 40, 17 30, 18 20, 20, to 58 20, and 1. (32 lines; 5 lines missing).   cols. *v-vi*
> > Interpretation: **3 cubits, 3 1/3 cubits, 3 1/2 cubits, 3 2/3 cubits, 4 cubits, *etc*., to 11 2/3 cubits, 1 ninda.**

In view of the proposed explanation of IM 96183, it is clear that also the much smaller special tables of squares discussed in Sec. 2.1 b above can be explained in a similar way. Thus **MS 3937** (Fig. 2.1.2, top), where the numbers being squared proceed from 10 to 2 05 in steps of 5, can be interpreted as a table for the squares of ;00 10 n. = 1 finger, ;00 15 n. = 1 1/2 finger, *etc*., all the way to ;02 05 n. = 13 fingers. Similarly, **MS 3906** (Fig. 2.1.2, bottom), where the numbers being squared proceed from 10 to 4 in steps of 10, can be interpreted as a table for the squares of ;00 10 n. = 1 finger, ;00 20 n. = 2 fingers, *etc*., all the way to ;04 n. = 24 fingers.

The proposed explanation of IM 96183 is important also for the reason that it shows that IM 96183 is in a sense halfway between an ED III metro-mathematical table text for areas of squares such as CUNES 50-08-001 and an Old Babylonian metrological list or table of standard type or, for that matter, an Old Babylonian standard table of squares!

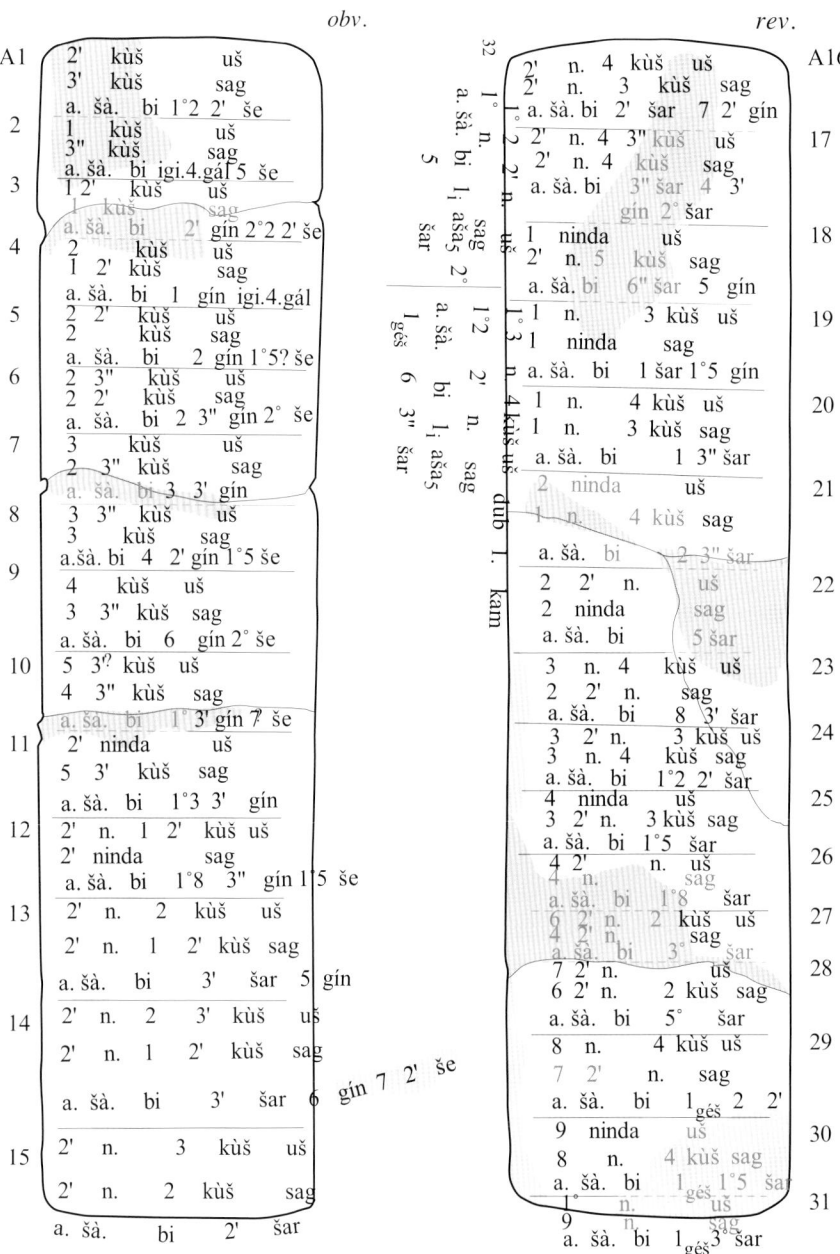

Fig. A7.4. *RA* 35, text 1. An Old Babylonian metro-mathematical table of areas of rectangles.
(The conform transliteration here is based on V. Scheil's original hand copy.)

obv.

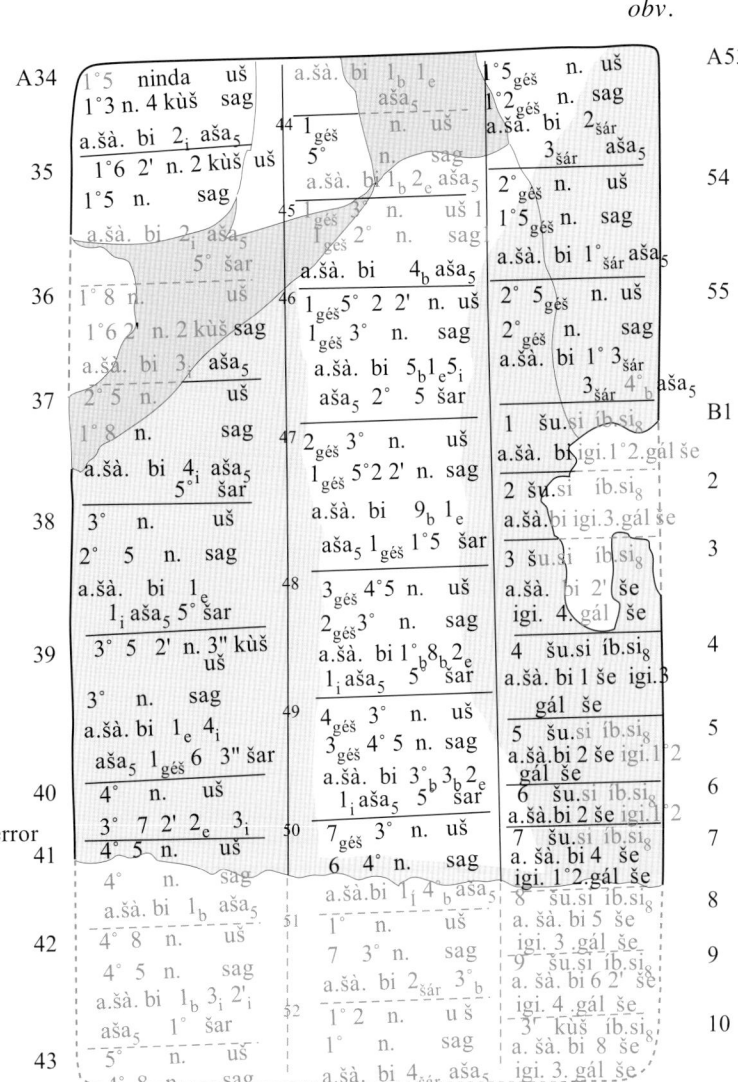

Fig. A7.5. *RA* 35, text 2. Continuation of text 1. Areas of rectangles and squares.
The table text continues on the reverse, where there is also a table of areas of circles.
(The conform transliteration here is based on V. Scheil's original hand copy.)

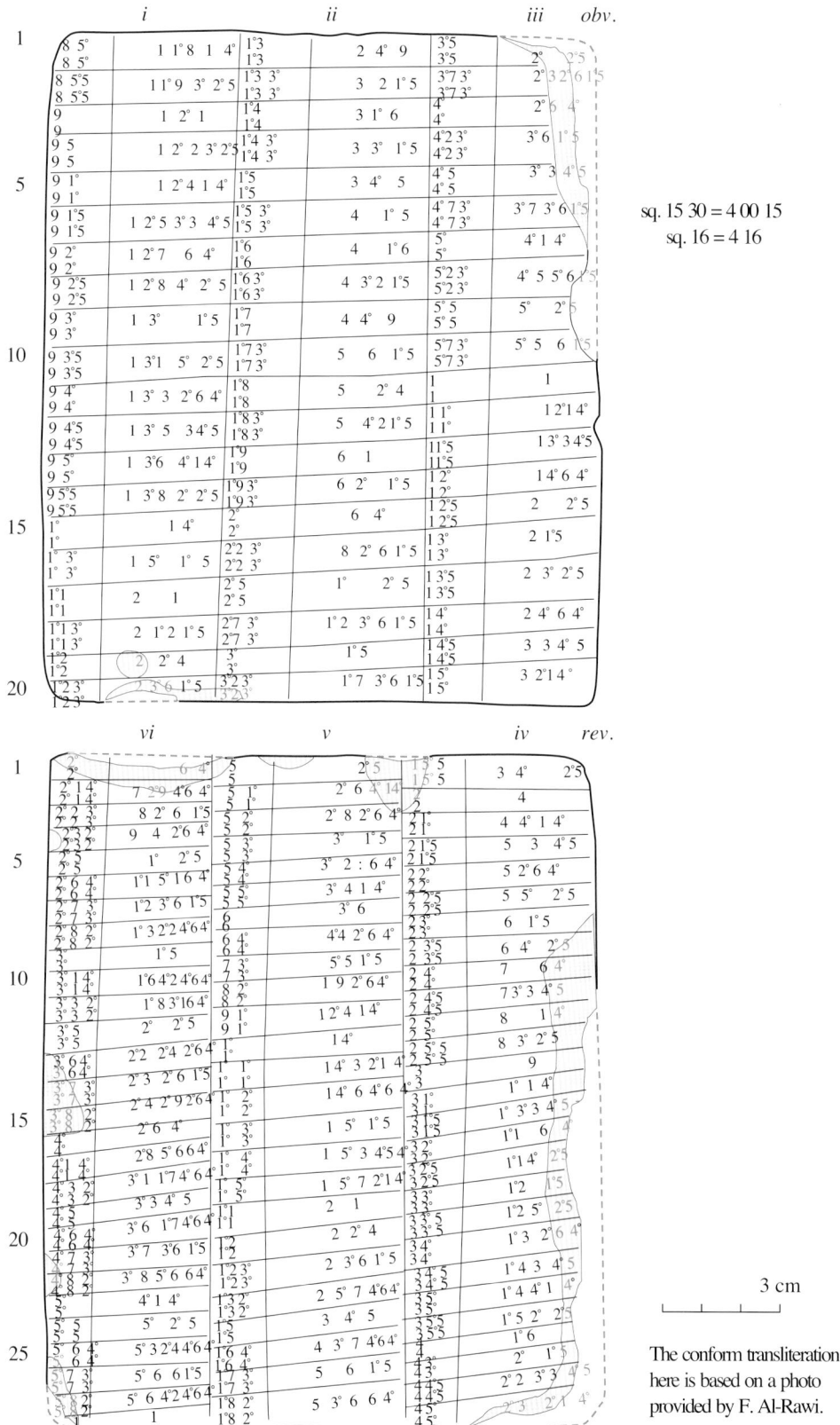

Fig. A7.6. IM 96183. An Old Babylonian special table of squares, with a metrological background.

Appendix 8
Plimpton 322, a Table of Parameters for igi–igi.bi Problems

Plimpton 322 (Fig. A8.1 below) is a large fragment of an Old Babylonian table text from Larsa. It was published by Neugebauer and Sachs in *MCT* (1945) as text A, with a clear photo on pl. 25. Neugebauer and Sachs immediately realized that this was an extremely important text for the history of mathematics, showing that Old Babylonian mathematics was much more sophisticated than had been known before.

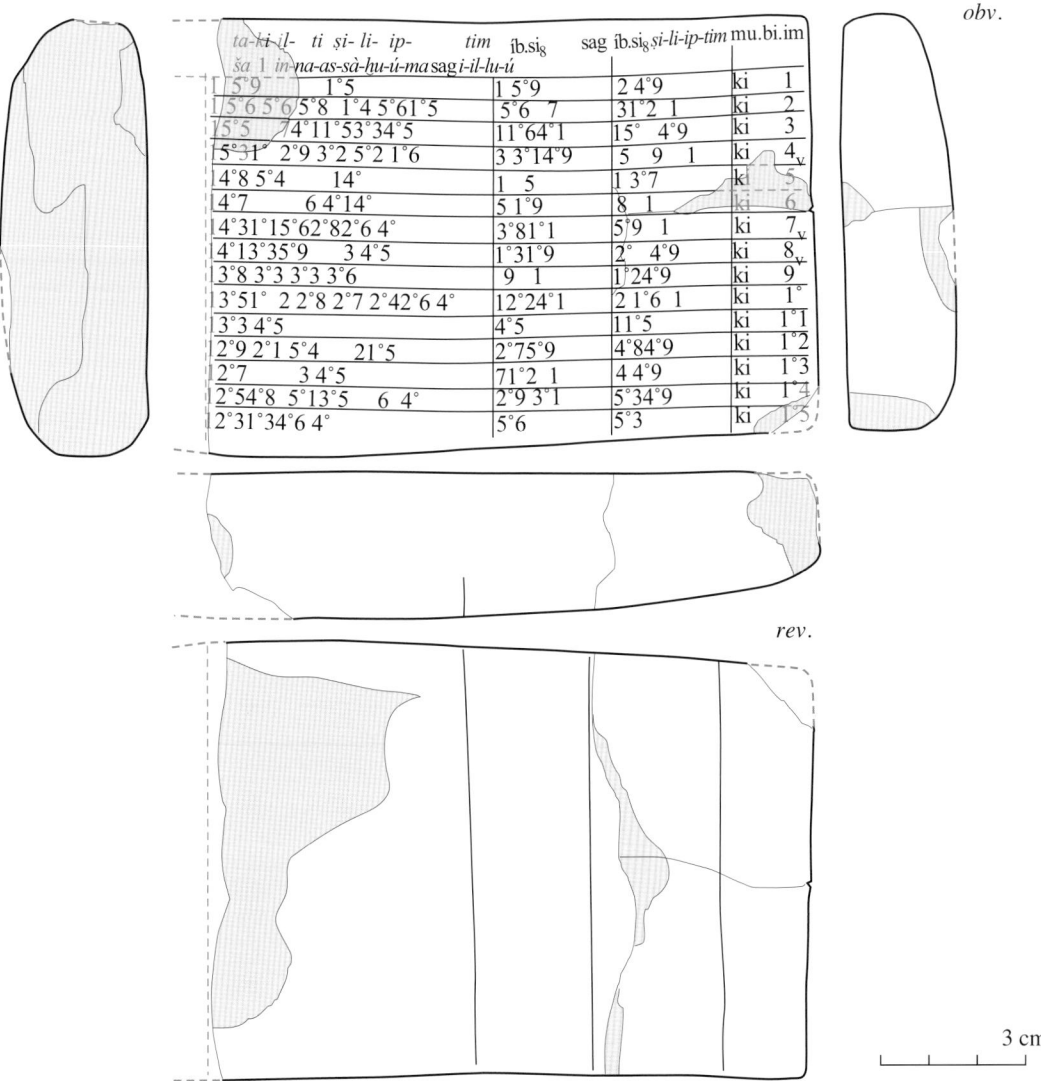

obv.

ta-ki-il- ti ṣi- li- ip- tim ša 1 in-na-as-sà-ḫu-ú-ma	íb.si₈ sag i-il-lu-ú	sag íb.si₈	si-li-ip-tim	mu.bi.im
1°59° 1°5	1 5°9	2 4°9	ki	1
1°5°6 5°6/5°8 1°4 5°61°5	5°6 7	31°2 1	ki	2
1°5°5 74°11°53°34°5	11°64°1	15° 4°9	ki	3
1°5°3 1° 2°9 3°2 5°2 1°6	3 3°14°9	5 9 1	ki	4ᵥ
1°4°8 5°4 14°	1 5	1 3°7	ki	5
1°4°7 6 4°14°	5 1°9	8 1	ki	6
1°4°31°15°62°82°6 4°	3°81°1	5°9 1	ki	7ᵥ
1°4°13°35°9 3 4°5	1°31°9	2° 4°9	ki	8ᵥ
1°3°8 3°3 3°3 3°6	9 1	1°24°9	ki	9°
1°3°51° 2 2°8 2°7 2°42°6 4°	12°24°1	2 1°6 1	ki	1°
1°3°3 4°5	4°5	11°5	ki	1°1
1°2°9 2°1 5°4 21°5	2°75°9	4°84°9	ki	1°2
1°2°7 3 4°5	71°2 1	4 4°9	ki	1°3
1°2°54°8 5°13°5 6 4°	2°9 3°1	5°34°9	ki	1°4
1°2°31°34°6 4°	5°6	5°3	ki	1°5

rev.

3 cm

Fig. A8.1. Plimpton 322. A table of parameters for 15 igi-igi.bi equations.

433

An attempt made in *MCT* to explain how the table of numbers in Plimpton 322 had been constructed was only partly successful. Moreover, some fateful mistakes were made in the analysis of the meaning of the table, in particular the claims that Plimpton 322 was "a text of purely number theoretical character", and that the purpose of the table in the text was the construction of a series of rational right triangles with the angles between the diagonal and the long side of the triangle decreasing "almost linearly". As a result of these mistakes in the first publication of the text, several attempts were subsequently made by a series of authors to find new or improved interpretations.

A complete and definitive explanation of the method used for the construction of the table of numbers in Plimpton 322 was given in Friberg, *HM* 8 (1981), 277-318, together with detailed discussions of interesting numerical errors in the table and of the apparent purpose of the text.[1] The explanation proposed in *HM 8* is strongly supported by the new evidence unexpectedly provided by MS 3052 § 2 and MS 3971 § 3 (see Secs. 10.2 b and 10.1 c above). This circumstance makes is motivated to repeat and refine here some of the arguments first put forward by the present author in 1981.

A8.1. Plimpton 322. A Description of the Preserved Part of the Table Text

Plimpton 322 is a large fragment of an Old Babylonian clay tablet with 15 lines of text, arranged in 4 columns, all with their individual headings. The curvature of the fragment suggests that a third, or more, of the tablet has been lost. A reasonable estimate is that the missing third, or so, can have contained about four narrow columns.

In the transliteration below of the text of Plimpton 322, errors are indicated with bold style:

íb.si$_8$ sag	íb.si$_8$ *şiliptim*	mu.bi.im	
1 59	2 49	ki.1	
56 07	**3 12 01**	ki.2	col. 2: 56 = 50 + 6 instead of 50 06
1 16 41	1 50 49	ki.3	col. 3: 12 01 instead of 13 = 12 + 1
3 31 49	5 09 01	ki.4	
1 05	1 37	k[i.5]	
5 19	8 01	[ki.6]	
38 11	59 01	ki.7	
13 19	20 49	ki.8	col. 1: 59 = 45+14 instead of 45 14
9 01	12 49	ki.9	col. 2: 9 instead of 8
1 22 41	2 16 01	ki.10	
45	1 15	ki.11	
27 59	48 49	ki.12	
7 12 01	4 49	ki.13	col. 2: 7 12 01 instead of 2 41
29 31	53 59	ki.14	
56	53	ki.15	

The headings above the four preserved columns are as follows:

[*x-k*]*i*-[*i*]*l-ti şi-li-ip-tim* /	the xx of the cross-over (diagonal),
[*ša* 1 *in*]*-na-as-sà-ḫu-ú-ma* sag *i-il-lu-ú*	that *1 is* torn off of and the front comes up
íb.si$_8$ sag	the square side of the front

---

1. The pretentious and polemical attempt by Robson in *HM* 28 (2001) to find an alternative explanation of the table on Plimpton 322 is so confused and misleading that it should be completely disregarded, with the exception of the improved reading of the word *i-il-lu-ú* in the second line of the heading over the first preserved column, and the dating of the text, *op. cit.*, 172, "to the 60 years or so before the siege and capture of Larsa by Hammurabi of Babylon in 1762 BCE". Cf. the verdict of Muroi, *HSJ* 12 (2003), note 4: "The reader should carefully read this paper written in a non-scientific style, because there are some inaccurate descriptions of Babylonian mathematics and several mistakes in Figure 1, Tables, and transliterations." A briefer and less polemical, but still pointless, version of the same story can be found in Robson, *AMM* 109 (2002).

íb.si₈ *si-li-ip-tim*   the square side of the cross-over
mu.bi.im   its name (= line)   (Cf. the subscript of MS 3049.)

What this means is far from clear at first sight. Nevertheless, the mention of diagonals, fronts, and square sides suggests that in some way the sides and diagonals of a series of rectangles (or right triangles) are involved, and that some of the numbers are computed as square sides. In other words, the vocabulary suggests that the numbers in the table in some way emanate from applications of the Old Babylonian diagonal rule (cf. Fig. 8.2.3 above).

All the numbers in the first column on Plimpton 322 have trailing parts that are square numbers:

$$15 \, (00) = \text{sq.} \, 30, \quad 3 \, 45 = \text{sq.} \, 15, \quad 16 = \text{sq.} \, 4, \quad 1 \, 40 = \text{sq.} \, 10, \quad 6 \, 40 = \text{sq.} \, 20, \quad 36 = \text{sq.} \, 6.$$

The heading states vaguely that if […] is subtracted from the diagonal, then the result will be the front. Obviously, what this means is that if […] is subtracted from the *squares* of the diagonals, then the result will be the *squares* of the fronts. It is not difficult to find out that what is subtracted must be the square of 1.

Clearly, then, the numbers in the first preserved column of Plimpton, and also those numbers with 1 subtracted, can be assumed to be squares of regular, or at least semi-regular, sexagesimal numbers. In Sec. 1.5 a above is described the Old Babylonian algorithm for computing the square sides of semi-regular squares by use of a variant of the trailing part algorithm. (See the discussion in that section of the three hand tablets MS 2318/2, *UET 6/2* 222, and A 30279.)

As a test of the proposed explanation of the numbers in the first preserved column on Plimpton 322, the mentioned algorithm will be used below on the numbers 1 59 00 15 and 1 56 56 58 14 50 06 15 (corrected here) in lines 1-2 of that column, and on those two numbers with 1 subtracted.

For the number in line 1, dual applications of the trailing part algorithm give the following result:

30	**1 59 00 15** (00)	4		30	**59 00 15** (00)	4
2 49	7 56 01			**1 59**	3 56 01	
	1 24 30				**59 30**	

Thus,

$$\text{sqs.} \, 1 \, 59 \, 00 \, 15 = \text{sqs.} \, 7 \, 56 \, 01 \cdot 30 = 2 \, 49 \cdot 30 = 1 \, 24 \, 30, \text{ and}$$
$$\text{sqs.} \, 59 \, 00 \, 15 = \text{sqs.} \, 3 \, 56 \, 01 \cdot 30 = 1 \, 59 \cdot 30 = 59 \, 30.$$

Consequently, in Babylonian *relative* sexagesimal numbers,

$$\text{sq.} \, 1 \, 24 \, 30 - \text{sq.} \, 1 = \text{sq.} \, 59 \, 30.$$

This means that the number triple **1 24 30, 1, 59 30** is an example of a "diagonal triple", satisfying the Old Babylonian "diagonal rule" for the sides and the diagonal of a rectangle.

Similarly, for the number in line 2 of the first preserved column, the algorithms proceed as follows:

30	**1 56 56 58 14 50 06 15 (00)**	4		30	**56 56 58 14 50 06 15 (00)**	4
5	7 47 47 52 59 20 25	2 24		5	3 47 47 52 59 20 25	2 24
5	18 42 42 55 10 25	2 24		5	9 06 42 55 10 25	2 24
5	44 54 31 00 25	2 24		5	21 52 07 00 25	2 24
5	1 47 46 50 25	2 24		**56 07**	52 29 04 49	
5	4 18 40 25	2 24			4 40 35	
3 13	10 20 49				23 22 55	
	16 05				1 56 54 35	
	1 20 25				**58 27 17 30**	
	6 42 05					
	33 30 25					
	2 47 32 05					
	1 23 46 02 30					

The computation to the left shows that

$$1 \, 56 \, 56 \, 58 \, 14 \, 50 \, 06 \, 15 = 10 \, 20 \, 49 \cdot \text{sq.} \, (5^5 \cdot 30),$$

and that, consequently,

$$\text{sqs. } 1\ 56\ 56\ 58\ 14\ 50\ 06\ 15 = 3\ 13 \cdot 5^5 \cdot 30 = 1\ 23\ 46\ 02\ 30.$$

In the same way, the computation to the right shows that

$$56\ 56\ 58\ 14\ 50\ 06\ 15 = 52\ 29\ 04\ 49 \cdot \text{sq. } (5^3 \cdot 30),$$

and that

$$\text{sqs. } 56\ 56\ 58\ 14\ 50\ 06\ 15 = 56\ 07 \cdot 5^3 \cdot 30 = 58\ 27\ 17\ 30.$$

Consequently, in Babylonian *relative* sexagesimal numbers,

$$\text{sq. } 1\ 23\ 46\ 02\ 30 - \text{sq. } 1 = \text{sq. } 58\ 27\ 17\ 30.$$

Thus, **1 23 46 02 30, 1, 58 27 17 30** is another example of a diagonal triple.

The two couples of examples above were worked through in all details for the following reason. The first couple of examples showed that 2 49 and 1 59 are what may be called the "factor-reduced cores" of the square sides of 1 59 00 15 and 59 00 15. This result must be significant, since the numbers 1 59 and 2 49 are recorded in the *first* line of the columns with the headings 'the square side of the front' and 'the square side of the diagonal', respectively.

Similarly, the second couple of examples shows that 3 13 and 56 07 are the factor-reduced cores of the square sides of 1 56 56 58 14 56 15 and 56 56 58 14 56 15. This result, too, must be significant, since the numbers 56 07 and 3 12 01 are recorded in the second line of the columns with the headings 'the square side of the front' and 'the square side of the diagonal'. (Here 3 12 01 in the text is, of course, a trivial, and easily explained, mistake for 3 12 + 1 = 3 13.)

The inevitable conclusion is that what is recorded in *all* lines of columns 2 and 3 on the preserved part of the clay tablet are *the factor-reduced cores of the square sides of the numbers recorded in column 1, and of the same numbers with 1 subtracted.*[2]

It is easy, although laborious, to check that the mentioned conclusion is essentially correct for all lines of the table on Plimpton 322, with one exception. In line 13, the number 7 12 01 in column 3 is an obvious mistake, since the first step of the trailing part algorithm in this case is the computation 27 00 03 45 · 16 = 7 12 01. Thus, the author of Plimpton 322 may have forgotten to compute the square side of 7 12 01. The correct entry in line 13 of column 2 would be 2 41 (= sqs. 7 12 01).

A8.2. Related Texts: Texts with igi–igi.bi Problems

The purpose of the tables on Plimpton 322 was discussed in Friberg, *HM* 8 (1981), § 5. The explanation suggested there is now confirmed by MS 3052 § 2 and by the equations for 'five cross-overs' (diagonals or rectangles)[3] in MS 3971, Secs. 3 a-b (Sec. 10.1 c above). Consider, for instance,

---

2. The circumstance that for all lines (but one) of the table the pair of numbers in columns 2 and 3 are coprime, that is that they are without common prime factors, is probably accidental. (The only exception is the pair 45, 1 15 in line 11; cf. MS 3052 § 2.) Thus, it is not the result of some number-theoretical consideration. Instead it is simply the result of the desired elimination of regular square factors from the numbers in column 1 and from same numbers with 1 subtracted. Moreover, it is not correct, as claimed by Neugebauer and Sachs in *MCT*, by Robson in *HM* 28, *etc.*, that the numbers recorded in columns 2 and 3 are the diagonals and fronts, respectively, of a series of (rational) right triangles! In particular, 56 07 and 3 13 in line 2 are not the diagonal and the front of such a triangle, neither are 56 and 53 in line 15. This fact is explained by the observation that in the application of the trailing part algorithm six regular square factors are removed from 1 56 56 58 14 56 15 while only four such factors are removed from 56 56 58 14 56 15, and similarly in the case of line 15.

3. The term *ṣiliptum* can refer not only to the diagonal of a rectangle, but also to the rectangle itself. Thus, in MS 3971, the best translation of the phrase 5 ṣilpātum may be '5 rectangles'. Cf. the tables of constants *MCT* Ud and Ue where 1 is the constant for a *ṣiliptum*, probably meaning that the area A of a rectangle with length u and front s is equal to $1 \cdot u \cdot s$.

MS 3971 § 3 e

1	ki.5	5th.
	1 12 igi 50 igi.bi	1 12 the igi, 50 the igi.bi.
2	1 12 [ù] 50 [gar.gar] / 2 02	1 12 and 50 *heap*, 2 02.
	2' 2 02 gaz 1 01 /	1/2 of 2 02 break, 1 01.
3	1 01 du₇.du₇ 1 02 01 /	1 01 (make) butt (itself), 1 02 01.
4	1 *i-na* 1 02 01 zi 2 01 in.sì /	1 from 1 02 01 tear off, 2 01 it gives.
5	2 01.e 11 íb.si₈	2 01 makes 11 equalsided.
	11 sag ki.5	11, the 5th front.

This is the solution procedure for a quite peculiar special case of a rectangular-linear system of equations of the type illustrated in Fig. 11.2.10. The rectangular-linear system of equations itself is only vaguely indicated by the words '1 12 is the igi, 50 the igi.bi'. This means that the length and the front of a rectangle are given as a pair of reciprocal sexagesimal numbers, $u = 1\ 12$ and $s = 50$. Apparently, these numbers are used as data to set up the following system of equations:

> The half-sum of the length and the front of a rectangle is $p/2 = 1\ 01$.
> The area of the rectangle is $A = 1\ (00\ 00)$.
> What is the half-difference $q/2$?

A geometric solution to this metric algebra problem is illustrated in Fig. A8.2 below:

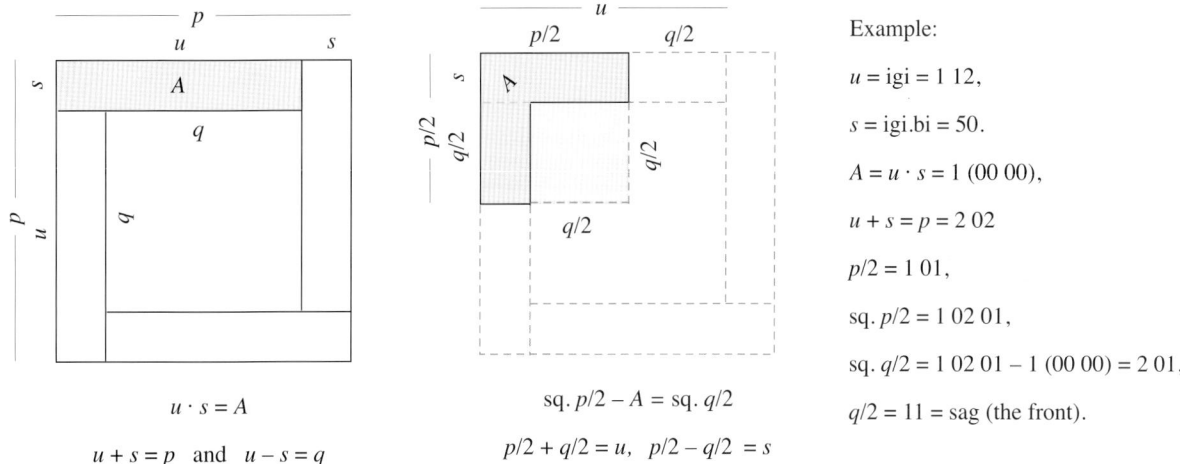

Example:

$u = \text{igi} = 1\ 12$,

$s = \text{igi.bi} = 50$.

$A = u \cdot s = 1\ (00\ 00)$,

$u + s = p = 2\ 02$

$p/2 = 1\ 01$,

sq. $p/2 = 1\ 02\ 01$,

sq. $q/2 = 1\ 02\ 01 - 1\ (00\ 00) = 2\ 01$,

$q/2 = 11 = \text{sag (the front)}$.

$u \cdot s = A$

$u + s = p$ and $u - s = q$

sq. $p/2 - A = $ sq. $q/2$

$p/2 + q/2 = u, \quad p/2 - q/2 = s$

Fig. A8.2. MS 3971 § 3 e. Finding the half-difference $q/2 = (\text{igi} - \text{igi.bi})/2$ in a roundabout way.

The fact that the five exercises in § 3 are referred to in the text as 5 *ṣilpātum* '5 cross-overs' and that in each exercise the half-differences $q/2$ are referred to as s a g 'the front' suggests that a more correct interpretation of these exercises would be as metric algebra problems of the following type:

> In a rectangle, the length is $b = 1\ (00)$ and the diagonal is $c = (\text{igi} + \text{igi.bi})/2$,
> where igi, igi.bi is a given pair of reciprocals with igi· igi.bi = 1 (00 00).
> What is the front a?

The solution procedure in each one of the five examples can be interpreted as a simple application of the Old Babylonian diagonal rule:

> sq. front = sq. diagonal − sq. length.

As shown by Fig. A8.2, an alternative way of answering the problem would be to say that

> If in a rectangle the length is 1 (00) and the diagonal (igi + igi.bi)/2, then the front is (igi− igi.bi)/2.

In other words, in a situation like the one depicted in the right half of Fig. A8.2, with $A = 1\ (00\ 00)$,

The triple $p/2$, sqs. A, $q/2$ = (igi + igi.bi)/2, 1 (00), (igi − igi.bi)/2 is a diagonal triple.

The following table shows the values of all numerical parameters appearing in MS 3971 § 3 a-e:

MS 3971 § 3 a-e

§ 3	igi	igi.bi	$p/2$	sq. $p/2$	sq. $q/2$	$q/2 = sag$
a	1 04	56;153	[1 00;07 30]	[1 00 15;00 56 15]	[15;00 56 15]	[3;52 30]
b	1 40	6	1 08	1 17 04	17 04	32
c	1 30	40	1 05	1 10 25	10 25	25
d	1 20	45	1 02;30	1 05 06;15	5 06;15	17;30
e	1 12	50	1 01	1 02 01	2 01	11

The relatively well preserved exercises MS 3971 § 3 b-e are all formulated in the same way, and are ordered with decreasing values in the first column. What remains of the badly preserved first exercise does not quite fit into this pattern. Its igi-value is smaller than the others and should have been the last one, not the first. In addition, the presentation of the given values in § 3 a has the form '1 04 the igi, and the igi.bi 56 15' while the presentation of the given values in the other exercises have the form '1 30 the igi, 40 the igi.bi', *etc*. Therefore, it is possible that § 3 a and § 3 b-e are borrowed from two different sources. (In MS 3052 § 2 (Sec. 10.2 b), the phrase is just 'the igi (is) 2'.)

It is unfortunate that the solution procedure of § 3 a is not preserved, in particular as it would have been interesting to see if the square of the number 1 00;07 30 was correctly computed. The square can be computed in various ways, for instance by use of the binomial rule, as follows:

sq. 1 00;07 30 = sq. 1 00 + 2 · 1 00 · ;07 30 + sq. ;07 30 = 1 00 00 + 15 + ;00 56 15 = 1 00 15;00 56 15.

Note that in MS 3971 § 3 the computations could have been carried one or two steps further. When both $p/2$ = (igi + igi.bi)/2 and $q/2$ = (igi − igi.bi)/2 had been computed, it would have been simple to check the result by computing also the corresponding values of igi and igi.bi, which should be equal to the initially given values. Such computations of the initially given values are commonplace in Old Babylonian mathematics, and probably served to check simultaneously the correctness of the method and the correctness of the computations.

The computations in MS 3971 § 3 can be compared with the computations in another Old Babylonian "igi-igi.bi-text", BM 85200 + VAT 6599 ## 14-19 (Neugebauer, *MKT 2* (1935), pl. 39; Høyrup, *LWS* (2002), 142-144). BM 85200 + VAT 6599 is a large fragment of a mathematical recombination text from Sippar, with 30 exercises for a túl.sag, an excavated room in the form of a rectangular prism. Of the six exercises ## 14-19, four are more or less corrupt and of no interest here. The remaining two exercises, ## 16 and 18, are closely related. Here is the text of # 18:

BM 85200 + VAT 6599 # 18

1	túl.sag	An excavated room.
	ma-la igi uš	As much as the igi is the length,
	ma-la igi.bi sag	as much as the igi.bi is the front,
	ma-la nigin igi igi.b[i gam-*m*]a	as much as the sum of the igi (and) the igi.b*i is the depth.*
	30 saḫar.ḫi.a ba.zi /	30 the mud I tore out.
	igi igi.bi *ù* gam en.nam /	The igi, the igi.bi, and the depth are what?
2	za.e	You:
	igi 12 du₈.a 5 *t*[*a-ma*]*r*	The opposite of 12 resolve, 5 you *see*.
3	5 *a-na* 30 saḫar.ḫi.a *i-ši* / 2 30 *ta-mar*	5 to 30, the mud, raise, 2 30 you see.
	2′ 2 30 *ḫe-pé šu-*[*tam-ḫir*]	1/2 of 2 30 break (and) make eq*ualsided*,
	[1 33 4]5 *ta-*[*mar*] /	*1 33 45 you see.*
4	1 *i-na* 1 33 45 ba.zi 3[3 4]5 *ta-mar*	1 from 1 33 45 tear out, *33 45* you see.
5	45 íb.si₈ / *a-na* 1 15 daḫ.ḫa *ù* ba.zi	45, the square side to 1 15 join and tear out,
	2 *ù* 30 *ta-mar*	2 and 30 you see.
	ne-pé-šum	The doing.

The rectangular prism in this exercise has the length u, the front s, and the depth h, where

$$u = \text{igi}, s = \text{igi.bi}, \text{hence } u \cdot s = 1$$
$$u + s \text{ "}= h\text{"}$$
$$V = u \cdot s \cdot h = 30, \text{hence } h = 30.$$

Here u and s are measured in ninda, while h is measured in cubits. Therefore, when the text says that the depth is equal to the sum of the length (the igi) and the front (the igi.bi), it really means that the depth is as many ninda as the sum of the length and the front. Measured in cubits, however, as it normally is, the depth is 12 times the sum of the length and the front. Conversely, the sum of the length and the front is equal to ;05 times the height. Therefore, the problem in # 18 can be reduced to the following system of equations

$$A = \text{igi} \cdot \text{igi.bi} = 1,$$
$$m = \text{igi} + \text{igi.bi} = ;05 \cdot 30 = 2;30.$$

The remaining computations in # 18 (and the counterparts in # 16) are shown in tabular form below:

BM 85200 + VAT 6599 # # 18, 16

	$p/2$	sq. $p/2$	sq. $q/2$	$q/2$	igi	igi.bi
# 18	1;15	1;33 45	;33 45	;45	2	;30
# 16	1;05	1;10 25	;10 25	;25	1;30	;40

For the readers' convenience, sexagesimal colons were inserted in the two tables above. (It is necessary to keep track of the *absolute* values of sexagesimal numbers whenever square roots or sums of the numbers are computed.) Note that in the table above for the values of the parameters in MS 3971 § 3 the semicolons have been arbitrarily placed so that in all cases igi · igi.bi = 1 00 00, while in the table for the values of the parameters in BM 85200 + VAT 6599, # # 18, 16, the semicolons have been placed, by necessity, so that in all cases igi · igi.bi = 1. (How the semicolons are placed in each case depends, of course, on how big the considered rectangles are assumed to be.)

Note that he values of the diagonal triple $p/2$, 1, $q/2$ in the igi-igi.bi-texts mentioned above are

1;15, 1, ;45	(= ;15 times the diagonal triple 5, 4, 3)	in BM 85200+ # 18
1;05, 1, ;25	(= ;05 times the diagonal triple 13, 12, 5)	in BM 85200+ # 16
1 00;07 30, 1 00, 3;52 30	(= ;07 30 times the diagonal triple 8 01, 8 00, 31)	in MS 3971 § 3 a
1 08, 1 00, 32	(= 4 times the diagonal triple 17, 15, 8)	in MS 3971 § 3 b
1 05, 1 00, 25	(= 5 times the diagonal triple 13, 12, 5)	in MS 3971 § 3 c
1 02;30, 1 00, 17;30	(= 2;30 times the diagonal triple 25, 24, 7)	in MS 3971 § 3 d
1 01, 1 00, 11		in MS 3971 § 3e

A third Old Babylonian igi-igi.bi-text is YBC 6967, discussed in Sec. 11.2 n above. The problem in that text is the following rectangular-linear system of equations:

$$\text{igi} \cdot \text{igi.bi} = 1\ (00), \quad \text{igi} - \text{igi.bi} = 7.$$

The first equation is not explicitly stated in the text but must have been in the mind of the author of the problem. The computations in YBC 6967 are shown in tabular form below:

YBC 6967

$q/2$	sq. $q/2$	sq. $p/2$	$p/2$	igi	igi.bi
3;30	12;15	1 12;15	8;30	12	5

Note that in this text,

$$\text{sq. } p/2 - \text{sq. } q/2 = \text{sq. } 8;30 - \text{sq. } 3;30 = 1\ 00\ (= \text{igi} \cdot \text{igi.bi}) \text{ is } not \text{ a square number.}$$

Therefore, as correctly observed by Robson in *HM* 28 (2001), 184, no construction of a (rational) diagonal triple is part of the solution procedure in the case of YBC 6967.

A8.3. A Suggested Reconstruction of the Lost Columns on Plimpton 322

The table above showing the values of numerical parameters appearing in MS 3971 § 3 can be expected to be a partial parallel to the *original* table on Plimpton 322. The table for MS 3971 § 3 has separate columns for the following parameters:

igi   igi.bi $p/2$ sq. $p/2$ sq. $q/2$ $q/2 = $ sag

On Plimpton 322, the columns for sq. $p/2$ and sq. $q/2$ are combined into a single column. The heading of that column can now be understood as saying:

> [a.šà *ta-k*]*i-il-ti și-li-ip-tim* /
> [*ša* 1 *in*]*-na-as-sà-ḫu-ú-ma*
> <a.šà *ta-ki-il-ti*> sag *i-il-lu-ú*

The field (= square) *of the ho*lder of the cross-over,
(from) which 1 is torn out, then
<the field of the holder of> the front comes up.

Here it is assumed that *takīlti șiliptim* and *takīlti* sag were the terms used for $p/2$ and $q/2$, respectively. (See Sec. 11.2 n above for a discussion of the meaning of the term *takīltum*.) The term used in the heading for the 'square' of a number must be very short, because there is very little space left for it in the lost beginning of the heading. The use of a.šà with the meaning 'square (of a number)' is known from several of the series texts in *MKT 1-2*, for instance in the following phrases in the series text VAT 7537: a.šà uš 'the field of the length' (= sq. *u*), a.šà sag 'the field of the front' (= sq. *s*), and a.šà a.na uš ugu sag diri 'the field of as much as the length is beyond the front' (= sq. $(u - s)$).[4]

The table below is a suggested reconstruction of the whole original text of Plimpton 322, with columns for the values of the following parameters:

[igi]   [igi.bi]   [$q/2$]   [$p/2$]   sq. $p/2$ & sq. $q/2$ $((q/2))$ $((p/2))$ #

Here $((q/2))$, in the text called íb.si$_8$ sag, and $((p/2))$, íb.si$_8$ șiliptim, are the "factor-reduced cores" of $q/2$ and $p/2$, obtained through elimination of regular sexagesimal factors by use of the trailing part algorithm.

Plimpton 322 reconstructed, with the errors corrected

igi	igi.bi	*takīlti* sag	*takīlti* șiliptim	a.šà *takīlti* șiliptim *ša* 1 *inassaḫuma* sag *illû*	íb.si$_8$ sag	íb.si$_8$ șiliptim	mu. bi.im
2 24	25	59 30	1 24 30	1 59 00 15	1 59	2 49	ki.1
2 22 13 20	25 18 45	58 27 17 30	1 23 46 02 30	1 56 56 58 14 50 06 15	56 07	3 13	ki.2
2 20 37 30	25 36	57 30 45	1 23 06 45	1 55 07 41 15 33 45	1 16 41	1 50 49	ki.3
2 18 53 20	25 55 12	56 29 04	1 22 24 16	1 53 10 29 32 52 16	3 31 49	5 09 01	ki.4
2 15	26 40	54 10	1 20 50	1 48 54 01 40	1 05	1 37	ki.5
2 13 20	27	53 10	1 20 10	1 47 06 41 40	5 19	8 01	ki.6
2 09 36	27 46 40	50 54 40	1 18 41 20	1 43 11 56 28 26 40	38 11	59 01	ki.7
2 08	28 07 30	49 56 15	1 18 03 45	1 41 33 45 14 03 45	13 19	20 49	ki.8
2 05	28 48	48 06	1 16 54	1 38 33 36 36	8 01	12 49	ki.9
2 01 30	29 37 46 40	45 56 06 40	1 15 33 53 20	1 35 10 02 28 27 24 26 40	1 22 41	2 16 01	ki.10
2	30	45	1 15	1 33 45	45	1 15	ki.11
1 55 12	31 15	41 58 30	1 13 13 30	1 29 21 54 02 15	27 59	48 49	ki.12
1 52 30	32	40 15	1 12 15	1 27 00 03 45	2 41	4 49	ki 13
1 51 06 40	32 24	39 21 30	1 11 45 20	1 25 48 51 35 06 40	29 31	53 59	ki.14
1 48	33 20	37 20	1 10 40	1 23 13 46 40	56	53	ki.15

---

4. Another brief notation for 'square' in some mathematical cuneiform texts is nígin, probably used in this capacity because the cuneiform sign for this word has the form of a square. Cf., for instance, Robson's reconstruction, in *MMTC* (1999), 49, of the beginning of an inscription on *TMS* 2, *obv*. as [nígin sag] sag.7 '*the square of the front* of the 7-side'.

According to this suggested reconstruction, there were originally eight columns on Plimpton 322. Apparently, nearly half the clay tablet is missing. The curvature of the fragment seems to confirm this conjecture. (See Fig. A8.3 below.) The unusual format is, of course, dictated by the large number of columns. (There are several known astronomical cuneiform texts from the Seleucid period which, for similar reasons, are written on tablets of similar formats. See, for instance, Neugebauer, *ACT* (1955), text 5 (photo on pl. 255 in *ACT 3*), a table for "New moons for S.E. 146-148", with originally 12 columns. That clay tablet is broken, in about the same way as Plimpton 322, along a vertical line near the middle of the tablet.[5] Also the extensive 6-place table of reciprocals AO 6456 (*MKT 1*, 14-22) is of a similar format.)

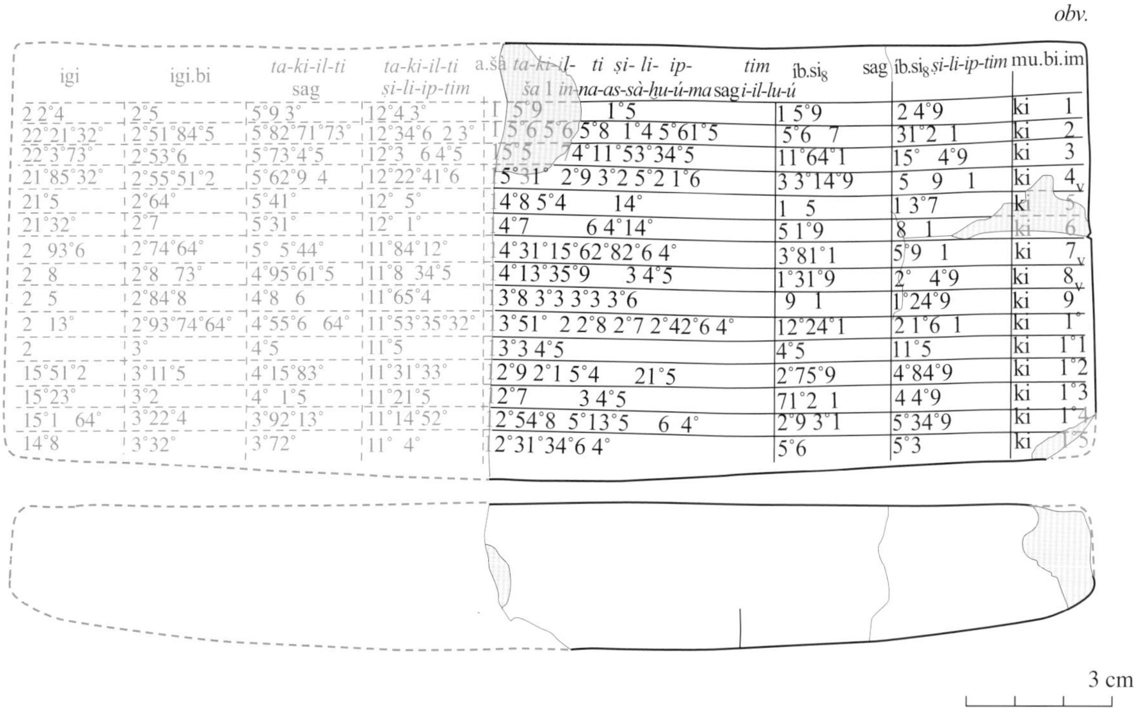

obv.

igi	igi.bi	ta-ki-il-ti sag	ta-ki-il-ti si-li-ip-tim	a.ša ta-ki-il- ti ši-li- ip- tim ša 1 in na-as-sà-hu-ú-ma sag i-il-lu-ú	ib.siģ	sag ib.siģ si-li-ip-tim	mu.bi.im
2 2°4	2°5	5°9 3°	12°4 3°	[1] 5°9 1°5	1 5°9	2 4°9	ki 1
22°21°32°	2°51°84°5	5°82°71°73	12°34°6 2 3°	[1] 5°6 5°6 5°8 1°4 5°61°5	5°6 7	31°2 1	ki 2
22°3°73°	2°53°6	5°73°4°5	12°3 6 4°5	15°5 74°11°53°34°5	11°64°1	15° 4°9	ki 3
21°85°32°	2°55°51°2	5°62°9 4	12°22°41°6	5°31° 2°9 3°2 5°2 1°6	3 3°14°9	5 9 1	ki 4ᵥ
21°5	2°64°	5°41°	12° 5°	4°8 5°4 14°	1 5	1 3°7	ki 5
21°32°	2°7	5°31°	12° 1°	4°7 6 4°14°	5 1°9	8 1	ki 6
2 93°6	2°74°64°	5° 5°44°	11°84°12°	4°31°15°62°82°6 4°	3°81°1	5°9 1	ki 7ᵥ
2 8	2°8 73°	4°95°61°5	11°8 34°5	4°13°35°9 3 4°5	1°31°9	2° 4°9	ki 8ᵥ
2 5	2°84°8	4°8 6	11°65°4	3°8 3°3 3°3 3°6	9 1	1°24°9	ki 1⁻
2 13°	2°93°74°64°	4°55°6 64°	11°53°35°32°	3°51° 2 2°8 2°7 2°42°6 4°	12°24°1	2 1°6 1	ki 1⁻
2	3°	4°5	11°5	3°3 4°5	4°5	11°5	ki 1°1
15°51°2	3°11°5	4°15°83°	11°31°33°	2°9 2°1 5°4 21°5	2°75°9	4°84°9	ki 1°2
15°23°	3°2	4° 1°5	11°21°5	2°7 3 4°5	71°2 1	4 4°9	ki 1°3
15°1 64°	3°22°4	3°92°13°	11°14°52°	2°54°8 5°13°5 6 4°	2°9 3°1	5°34°9	ki 1°4
14°8	3°32°	3°72°	11° 4°	2°31°34°6 4°	5°6	5°3	ki 1°5

3 cm

Fig. A8.3. A suggested reconstruction of Plimpton 322.

A8.4. The OB Rectangle Parameter Equations. Restrictions on the Parameters

The explanation of what is going on in Plimpton 322 and in the igi-igi.bi texts discussed above is as follows: It is easy to demonstrate, for instance by use of Fig. A8.4 below, that

> If c, b, a is any given diagonal triple of sexagesimal numbers with $b = 1$ and b greater than a, then $c + a, c - a$ = igi, igi.bi is a pair of reciprocal sexagesimal numbers with igi greater than igi.bi, and c, b, a = (igi + igi.bi)/2, 1, (igi – igi.bi)/2.

Conversely

> If igi, igi.bi is any given pair of reciprocal sexagesimal numbers with igi greater than igi.bi, then c, b, a = (igi + igi.bi)/2, 1, (igi – igi.bi)/2 is a diagonal triple with b greater than a and satisfying the Old Babylonian diagonal equation sq. c – sq. a = sq. b.

This converse assertion, too, can easily be demonstrated by use of a geometric model. (See, again, Fig. A8.4

---

5. Cf. the following statement in Vincente, *ZA* 85 (1995), 235: "Long tablets tend to break along a horizontal line, at the point of maximum thickness …". If this observation is correct, then in a similar way very wide tablets, like Plimpton 322, would tend to break along a *vertical* line at the point of maximum thickness. In the case of Plimpton 322, the exact place of the vertical break was determined, of course, by the many deeply impressed initial digits 1 in the first preserved column.

below). In the following, the set of equations $c, b, a = (\text{igi} + \text{igi.bi})/2, 1, (\text{igi} - \text{igi.bi})/2$ will be referred to as "the Old Babylonian rectangle parameter equations".

According to an Old Babylonian convention, for a diagonal triple c, b, a to be equal to the *diagonal, the length, and the front of a rectangle,* it is required that $a < b$. It is easy to see that

If $c, b, a = (\text{igi} + \text{igi.bi})/2, 1, (\text{igi} - \text{igi.bi})/2$,
where igi, igi.bi is a pair of reciprocal sexagesimal numbers, then
$a > 0$ precisely when igi > 1, igi.bi < 1, and $a < b$ precisely when igi $<$ sqs. $2 + 1 = $ appr. 2;24 51 10.

To verify the second statement, note that if $a < b = 1$, then sq. $c = $ sq. $a + 1 < 2$. Consequently, $c < $ sqs. 2 and therefore igi $= c + a < $ sqs. $2 + 1$. (See Fig. A8.4 b.) If $a \geq 1$, on the other hand, then sq. $c = $ sq. $a + 1 \geq 2$, so that $c \geq $ sqs. 2 and therefore igi $= c + a \geq $ sqs. $2 + 1$.

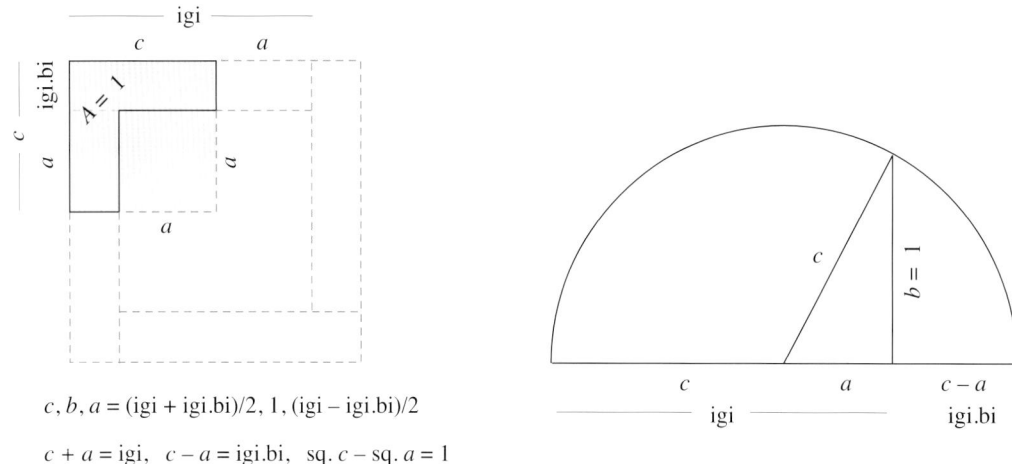

$c, b, a = (\text{igi} + \text{igi.bi})/2, 1, (\text{igi} - \text{igi.bi})/2$

$c + a = \text{igi}, \quad c - a = \text{igi.bi}, \quad \text{sq.}\, c - \text{sq.}\, a = 1$

Fig. A8.4. Two alternative geometric derivations of the Old Babylonian rectangle parameter equations.

Consider a given pair igi, igi.bi of regular sexagesimal reciprocal numbers satisfying the condition that $1 < $ igi $< $ sqs. $2 + 1 = $ appr. 2;24 51 10. The trailing part algorithm can be used to successively remove the fractional part of the number called igi. After a finite number of steps, what remains is a regular sexagesimal integer. Consider, for instance, 2;18 53 20 in line 4 of the first column of the reconstructed tables on Plimpton 322. An application of the trailing part algorithm shows that

2;18 53 20 · 3 = 6;56 40,   6;56 40 · 3 = 20;50,   20;50 · 6 = 2 05. Consequently,
2;18 53 20 = 2 05 · igi 6 · igi 3 · igi 3 = 2 05 · igi 54.

The argument shows that

The number igi can always be written in the form igi $= m \cdot$ igi n,
where m and n are regular sexagesimal integers.

Moreover, if m and n have any common factors, those can be removed. Therefore:

The condition $1 < $ igi $< $ sqs. $2 + 1 = $ appr. 2;24 51 10 can be replaced by the simpler condition that
igi $= m \cdot$ igi n, where m and n are regular sexagesimal integers, without common factors,
with m between n and $n \cdot$ 2;25.

It is clear that the Old Babylonian rectangle parameter equations, even with the restrictions on the parameters described above, can be used to generate infinitely many different pairs of numbers a, c with sq. $c - $ sq. $a = 1$. Therefore, the question remains which method the author of the tables on Plimpton 322 may conceivably have used to select the 15 pairs a, c that he computed and (presumably) recorded in the columns for *takīlti* sag and *takīlti ṣiliptim.*

In order to try to answer this question, a sensible first step is to make a survey of the values of igi $= m \cdot$ igi n that were actually used in the computation of the fifteen pairs a, c. Those values are:

12 · igi 5	2 05 · igi 54	54 · igi 25	1 21 · igi 40	15 · igi 8
1 04 · igi 27	9 · igi 4	32 · igi 15	2 · igi 1	50 · igi 27
1 15 · igi 32	20 · igi 9	25 · igi 12	48 · igi 25	9 · igi 5

Here n varies between 1 and 54, m between 2 and 2 05. The obvious conclusion is that the author of Plimpton 322 decided to use only values for n with $1 \leq n < 1\ 00\ (= 60)$. The table below shows which values for igi and igi.bi he could then obtain with the mentioned restrictions on m and n.

n	$n \cdot 2;25$	m	igi $= m \cdot$ igi n	igi.bi $= n \cdot$ igi m	#
1	2;25	2	2	;30	11
2	4;50	3	1;30	;40	22
3	7;15	4	1;20	;45	29
		5	1;40	;36	18
4	9;40	5	1;15	;48	31
		9	2;15	;26 40	5
5	12;05	6	1;12	;50	32
		8	1;36	;37 30	20
		9	1;48	;33 20	15
		12	2;24	;25	1
6	14;30	–	–	–	–
8	19;20	9	1;07 30	;53 20	34
		15	1;52 30	;32	13
9	21;45	10	1;06 40	;54	35
		16	1;46 40	;33 45	16
		20	2;13 20	;27	6
10	24;10	–	–	–	–
12	29	25	2;05	;28 48	9
15	36;15	16	1;04	;56 15	37
		32	2;08	;28 07 30	8
16	38;40	25	1;33 45	;38 24	21
		27	1;41 15	;35 33 20	17
18	43;30	25	1;23 20	;43 12	27
20	48;20	27	1;21	;44 26 40	28
24	58	25	1;02 30	;57 36	38
25	1 00;25	27	1;04 48	;55 33 20	36
		32	1;16 48	;46 52 30	30
		36	1;26 24	;41 40	24
		48	1;55 12	;31 15	12
		54	2;09 36	;27 46 40	7
27	1 05;15	32	1;11 06 40	;50 37 30	33
		40	1;28 53 20	;40 30	23
		50	1;51 06 40	;32 24	14
		1 04	2;22 13 20	;25 18 45	2
30	1 12;30	–	–	–	–
32	1 17;20	45	1;24 22 30	;42 40	26
		1 15	2;20 37 30	;25 36	3
36	1 27	–	–	–	–
40	1 36;40	1 21	2;01 30	;29 37 46 40	10
45	1 48;45	1 04	1;25 20	;42 11 15	25
48	1 56	–	–	–	–
50	2 00;50	1 21	1;37 12	;37 02 13 20	19
54	2 10;30	2 05	2;18 53 20	;25 55 12	4

All values of igi $= m \cdot$ igi n and igi.bi $= n \cdot$ igi m, when $1 \leq n < 1\ 00$ and $n < m < n \cdot 2;25$.

In the first column of this table, *n* varies over *all* regular sexagesimal integers between 1 and 1 00, ordered by size. In the second column, those *n* values are multiplied by 2;25, and in the third column, *m* varies over *all* regular integers between *n* in the first column and *n* · 2;25 in the second column. Only *m* values with no factors in common with *n* are allowed, in order to avoid duplications. The numbers in the last column can be used to sort the igi values by size. Numbers from 1 to 15 in that last column are written in bold style. They correspond to the 15 lines in Plimpton 322.

There are 38 igi values in this table, varying from 2;24 (# 1) to 1;02 30 (# 38). The first 15 igi values are the ones used for the computation of lines 1 through 15 on Plimpton 322. This means that the table on Plimpton 322 was *exhaustive*; there are no igi values missing in the interval considered, igi between 2;24 (# 1) and 1;48 (# 15)![6] Presumably it was the author's intention to record all the 38 lines of the complete table on Plimpton 322, but his work was interrupted when he had only had time to finish the inscription on the obverse and to draw the lines separating the columns on the reverse.

The "normalized" rectangles with the length 1 00 and with the fronts and diagonals computed by use of the rectangle parameter equations and the igi values listed in the table above vary from a nearly square rectangle (# 1) to a very thin rectangle (# 38). (See the examples in Fig. A8.5 below.)

It is interesting that the igi values 1;40, 1;30, 1;20, 1;12, and 1;04 appearing in the five diagonal problems MS 3971 §§ 3 a-e correspond to ## 18, 22, 29, 32, and 37 in the table above, all *beyond* the range of the table on Plimpton 322. (See the middle register of Fig. A8.5.) In particular, the rectangle in the case of MS 3971 § 3 a is very thin. The igi values in the two excavation problems BM 85200 ## 18 and 16 are 2 (# 11) and 1;30 (# 22), the first two of the 38 igi values in the table above!

The rectangle parameter equations for a *normalized* rectangle, one with the length 1, can be expressed in terms of either the *sexagesimal fractions* igi and igi.bi or the *integers m* and *n*:

$$\text{If} \quad c, b, a = (\text{igi} + \text{igi.bi})/2, \ 1, \ (\text{igi} - \text{igi.bi})/2, \text{ where igi} = m \cdot \text{igi } n,$$
$$\text{then} \quad c, b, a = (\text{sq. } m + \text{sq. } n)/(2 \, m \cdot n), \ 1, \ (\text{sq. } m - \text{sq. } n)/(2 \, m \cdot n).$$

Enlarging the rectangle by the factor $2 \, m \cdot n$, one gets the following variant of the rectangle parameter equations for a "primitive" rectangle:

$$c, b, a = (\text{sq. } m + \text{sq. } n), \ 2 \, m \cdot n, \ (\text{sq. } m - \text{sq. } n), \text{ with } m \text{ and } n \text{ regular and } n < m < n \cdot (\text{sqs. } 2 + 1)(*)$$

If *m* and *n* have no common factor, then also the diagonal, the length, and the front of such a primitive rectangle have no common factor (except for the common factor 2 if both *m* and *n* are odd integers).

As a consequence of the mentioned restrictions on the parameters *m* and *n*, it was not possible to find by use of the Old Babylonian rectangle parameter equations *all* primitive rectangles with integers for the diagonal, the length, and the front. Indeed, because the Babylonians operated only with regular sexagesimal numbers and sexagesimal fractions, the equations above marked with an asterisk are able to produce only primitive diagonal triples *c*, *b*, *a* with *b* equal to a *regular* sexagesimal integer, and with *a* < *b*. This means that even simple diagonal triples like 29, 21, 20 (*m*, *n* = 7, 3), or 29, 20, 21 (*m*, *n* = 5, 2, igi = 2;30 > sqs. 2 + 1), and 33, 56, 65 (*m*, *n* = 7, 4) would be missed by the Old Babylonian method. (For a further discussion of this question, see Friberg, *HM* 8 (1981), 301-302.)

The Old Babylonian mathematical tradition was still alive in the Late Babylonian and Seleucid periods in Mesopotamia, 1500 years after the Old Babylonian period. (For examples, see Friberg, Hunger, and Al-Rawi, *BagM* 21 (1990) §§ 1-2; Friberg, *GMS 3* (1993); Friberg, *BagM* 28 (1997) §§ 1, 4; Friberg, *BagM* 30 (1999); and Friberg, *CTMMA 2* (2005).) In the present connection, it is interesting that in the Seleucid mathematical recombination text **AO 6484** (see below) § **7 a-d** are four igi-igi.bi problems closely related to the diagonal problems in MS 3971 § 3 and to the tables on Plimpton 322. This is even more interesting in view of the circumstance that the vocabulary of this Seleucid text bears little resemblance to the vocabulary of the Old Babylonian igi-igi.bi problems in MS 3971 and BM 85200 + VAT 6599.

---

6. Robson's six "missing lines" in *HM* 28 (2001), Table 8, all violate the restriction that *n* should be smaller than 1 00.

AO 6484 § 7 a-d (Neugebauer, *MKT 1*, 98-99; Høyrup, *LWS* (2002), 390-391)

a	1	igi *u* igi-*bu-ú* 2 : : 33 20	igi and igi.bi 2 00 00 33 20.
		igi *u* igi-*bu-ú* [*ki-ma-a ma-și* … … … … …]? /	igi and igi.bi *how much* … … … …
	2	• 30 rá-*ma* 1 : : 16 40 :	ḫ 30 go, then 1 00 00 16 40.
		1 : : 16 40 [• 1 : : 16 40 rá-*ma*]	1 00 00 16 40 *ḫ 1 00 00 16 40 go, then*
		[1 : : 33 20 04 37 46 40 :] /	1 00 00 33 20 04 57 46 40.
	3	1 ta *lìb-bi* lá-*ma*	1 from inside remove, then
		re-ḫe 33 <20> 04 37 46 40	remains 33 <20> 04 37 46 40.
		mi [• *mi lu-*rá-*ma*]	What *ḫ what may I go, then*
		[33 <20> 04 37 46 40 :] /	33 <20> 04 37 46 40?
	4	44 43 20 • 44 43 20 rá-*ma*	44 43 20 ḫ 44 43 20 go, then
		33 <20> 04 37 4[6 40 :]	33 <20> 04 37 46 *40.*
		[44 43 20 *a-na* 1 : : 16 40 tab-*ma*] /	44 43 20 to 1 00 00 16 40 repeat, then
	5	1 : 45 igi-*ú*	1 00 45, the igi.
		44 04 43 20 ta 1 : : 16 40 lá[-*ma*]	44 04 43 20 from 1 00 00 16 40 rem*ove, then*
		[59 15 33 20 igi-*bu-ú*]	59 15 33 20, the igi.bi.
b	1	igi *u* igi-*bu-ú* 2 03	igi and igi.bi 2 03
		• 30 rá-*ma* 1 01 30 [:]	ḫ 30 go, then 1 01 30.
		[1 01 30 • 1 01 30 rá-*ma* 1 03 02 15 :] /	1 01 30 ḫ 1 01 30 go, then 1 03 02 15.
	2	1 ta *lìb-bi* lá-*ma re-ḫe* 3 02 15	1 from inside remove, then remains 3 02 15.
		mi • *mi* [*lu-*rá-*ma* 3 02 15 : … … … …] /	What ḫ *what may I go, 3 02 15* … … … …?
	3	13 30 • 13 30 {•} r[á-*m*]*a* 3 02 15 :]	13 30 ḫ 13 30 {ḫ} go, *the*n 3 02 15.
		13 30 [*a-na* 1 01 30 tab-*ma* 1 15 igi-*ú*] /	13 30 *to 1 01 30 repeat, then 1 15, the igi.*
	4	13 30 ta 1 01 30 lá-*ma* 48 igi-*b*[*u-ú*]	13 30 from 1 01 30 remove, then 48, the igi.*bi.*
c	1	igi *u* igi-*bu-ú* 2 05 26 40	igi and igi.bi 2 05 26 40
		• 30 rá-*ma* 1 02 43 [20 :]	ḫ 30 go, then 1 02 43 *20.*
	2	[1 02 43 20 • 1 02 43 20] / rá-*ma*	*1 02 43 20 ḫ 1 02 43 20 go, then*
		1 05 34 04 37 46 40 :	1 05 34 04 37 46 40.
		1 ta *lìb-bi* lá-*m*[*a*]	1 from inside remove, then
		[*re-ḫe* 5 34 04 37 46 40]	*remains 5 34 04 37 46 40.*
	3	[*mi* • *mi*] / *lu-*rá-*ma* […]	*What ḫ what* may I go, then …
		[5 34] 04 37 46 40 :	*5 34* 04 37 46 40?
		18 16 40 [• 18 16 40 rá-*ma*]	18 16 40 ḫ *18 16 40 go, then*
		[5 34 04 37 46 40 :] /	*5 34 04 37 46 40.*
	4	18 16 40 [*a-na* 1 02 43 20 tab-*ma*]	18 16 40 *to 1 02 43 20 repeat, then*
		[1 21 igi-*ú*]	1 21, the igi.
		18 16 [40 ta 1 02 43 20 lá-*ma*] /	18 16 40 *from 1 02 43 20 remove,* then
	5	*re-ḫe* 44 26 40 igi-*bu-ú*	44 26 40, the igi.bi.
d	1	igi *u* igi-*bu-ú* 2 : 15	igi and igi.bi 2 00 15
		• 30 rá-*ma* 1 : 07 30 :	ḫ 30 go, then 1 00 07 30.
		1 [: 07 30 • 1 : 07 30 rá-*ma*]	*1 00 07 30 ḫ 1 00 07 30 go, then*
		[1 : 15 : 56 15 :] /	*1 00 15 00 56 15.*
	2	1 ta *lìb-bi* lá-*ma*	1 from inside remove, then
		re-ḫe 15 : 56 15	remains 15 00 56 15.
		mi • *mi lu-*rá-[*ma* 15 : 56 15 :] /	What ḫ what may I go, *then 15 00 56 15?*
	3	3 52 30 • 3 52 30 rá-*ma* 15 : 56 15 :	3 52 30 · 3 52 30 go, then 15 00 56 15.
		3 5[2 30 *a-na* 1 : 07 30 tab-*ma*] /	*3 52 30 to 1 00 07 30 repeat, then*
	4	1 04 igi-*ú*	1 04, the igi.
		3 52 30 ta 1 : 07 30 lá-*ma*	3 52 30 from 1 00 07 30 remove, then
		5[6 15 igi-*bu-ú*]	*56 15, the igi.bi.*

Plimpton 322

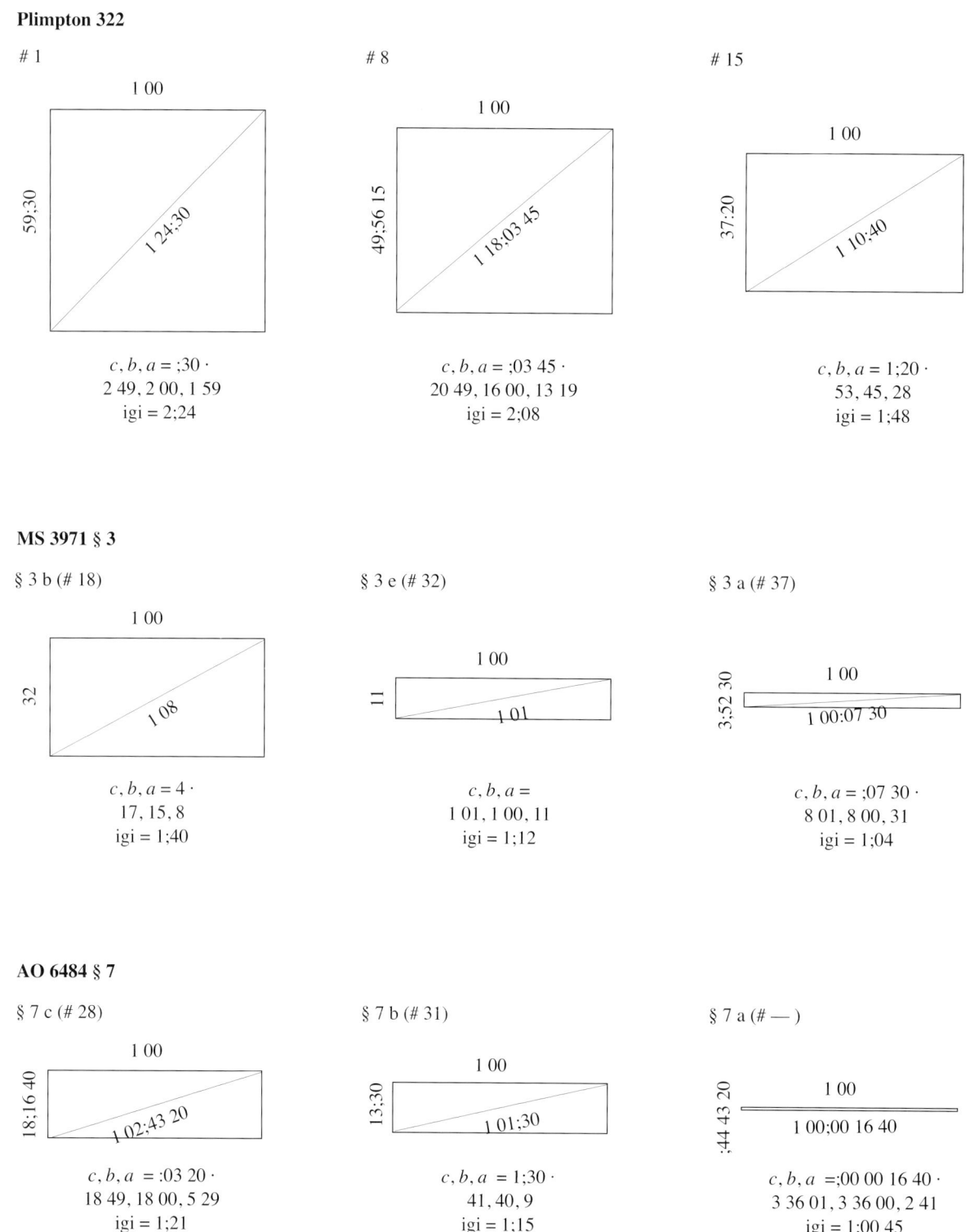

 # 1

1 00

59;30

1 24;30

c, b, a = ;30 ·
2 49, 2 00, 1 59
igi = 2;24

 # 8

1 00

49;56 15

1 18;03 45

c, b, a = ;03 45 ·
20 49, 16 00, 13 19
igi = 2;08

 # 15

1 00

37;20

1 10;40

c, b, a = 1;20 ·
53, 45, 28
igi = 1;48

MS 3971 § 3

§ 3 b (# 18)

1 00

32

1 08

c, b, a = 4 ·
17, 15, 8
igi = 1;40

§ 3 e (# 32)

1 00

11

1 01

c, b, a =
1 01, 1 00, 11
igi = 1;12

§ 3 a (# 37)

1 00

3;52 30

1 00;07 30

c, b, a = ;07 30 ·
8 01, 8 00, 31
igi = 1;04

AO 6484 § 7

§ 7 c (# 28)

1 00

18;16 40

1 02;43 20

c, b, a = :03 20 ·
18 49, 18 00, 5 29
igi = 1;21

§ 7 b (# 31)

1 00

13;30

1 01;30

c, b, a = 1;30 ·
41, 40, 9
igi = 1;15

§ 7 a (# —)

1 00

;44 43 20

1 00;00 16 40

c, b, a =;00 00 16 40 ·
3 36 01, 3 36 00, 2 41
igi = 1;00 45

Fig. A8.5. Examples of rectangles constructed by use of the Old Babylonian rectangle parameter equations.

Fig. A8.5 above is a partial survey of the shapes of rectangles associated with the diagonal triples appearing in Plimpton 322, MS 3971 § 3, and AO 6484 § 7 (arbitrarily scaled so that b = 1 00).

The following table shows the numerical parameters appearing in AO 6484 § 7:

AO 6484 § 7 a-d

	$p/2$	sq. $p/2$	sq. $q/2$	$q/2$	igi-ú	igi-*bu-ú*
c	1 02;43 20	1 06 34;04 37 46 40	6 34;04 37 46 40	18;16 40	1 21	44;26 40
b	1 01;30	1 03 02;15	3 02;15	13;30	1 15	48
d	1 00;07 30	1 00 15;00 56 15	15;00 56 15	3;52 30	1 04	56;15
a	1 00;00 16 40	1 00 00;33 20 04 57 46 40	;33 20 04 57 46 40	;44 43 20	1 00;45	59;15 33 20

The first three the igi values in this table, 1;21 = 27 · igi 20, 1;15 = 5 · igi 4, and 1;04 = 16 · igi 15, correspond to ## 28, 31, and 37 in the completed Plimpton 322 table. The fourth igi value, 1;00 45 = 1 21 · igi 1 20 in § 7 a does not satisfy the condition of the Plimpton 322 table that $n < 1\ 00$. The rectangle constructed with departure from this igi value is extremely thin, as shown in the lower register of Fig. A8.5. The author of AO 6484 may have included an example with this extreme igi value for just that reason. Another possibility is that he devised the exercise with this extreme igi value in order to show his students how to count with sexagesimal numbers with several "internal double zeros". [7]

There is a curious *pair of errors* in the same exercise, AO 6484 § 7 a. In line 3, sq. $p/2$ – 1 = sq. $q/2$ is incorrectly computed as 33 04 37 46 40 instead of 33 **20** 04 37 46 40. Then, in the lost right half of line 3, and in the left half of line 4, it is claimed that sq. 44 43 20 = 33 04 37 46 40 (so that $q/2$ = 44 33 20). However, this *correct* value for $q/2$ cannot have been computed as the square side of the *incorrect* value 33 04 37 46 40, and the square of this correct value is not equal to that incorrect value. Since the incorrect value 33 04 37 46 40 appears twice in lines 3-4, it cannot be the result of a simple copying error. Instead, the double error can be explained in the following interesting way.

Assume that the author of AO 6484 had at his disposal *a table text like Plimpton 322*, where one of the lines was like the line for AO 6484 § 7 a in the table above, but with the incorrect value 33 20 57 46 40 in the sq. $q/2$ column. He could then fill out the details of the problem, using the numerical parameters from this line of his table text, like an Old Babylonian school boy doing his assignment while copying the numerical parameters from his hand tablet, and produce the text of an exercise like AO 6484 § 7 a with the double error in it.

An even more interesting possibility is that the author of AO 6484 had at his disposal an Old Babylonian igi-igi.bi text, from which he wanted to copy four igi-igi.bi exercises. He would then, in some way, have to *translate* the text of the exercises from the Old Babylonian mathematical jargon to his own very much different Seleucid mathematical jargon. A simple way of doing that would be to first extract just the numerical parameters from the original text and then fill in the details again using his own mathematical vocabulary. If he had copied one of the numerical parameters incorrectly, namely the value of sq. $q/2$, the result would be the same as if he had used a line from a table text of the Plimpton 322 type with an error in it.

A8.5. The Purpose of the Tables on Plimpton 322

A number of Old Babylonian round "hand tablets" from Ur were discussed in Friberg, *RA* 94 (2000), § 2. Most of those hand tablets are inscribed with literary exercises on the obverse, and mathematical "notes" on the reverse. Those notes can be interpreted as tabular presentations of the data for various mathematical exercises, much like one line of the table above for the numerical parameters appearing in AO 6484 § 7.

---

7. In Seleucid mathematical and astronomical texts a special cuneiform "separation sign" was sometimes used to indicate both the end of a sentence and missing sexagesimal places ("double zeros"). In AO 6484, for instance, the separation sign sign takes the form of a pair of oblique wedges, one on top of the other. In the transliteration above of AO 6484 § 7, the separation sign is represented by a colon. It is often difficult to distinguish this Seleucid cuneiform separation sign from the Seleucid cuneiform multiplication sign, written as GAM (a pair of oblique wedges, one placed above and slightly to the left of the other). In the transliteration above of AO 6484 § 7, this multiplication sign is represented by a fat dot •.
In the Late Babylonian multiplication table BM 141493 (Fig. 2.6.15 above), the multiplication sign is written as three oblique wedges in an oblique line.

Thus, for instance, the hand tablet ***UET 6/2 274*** (*op. cit.*, § 2 e) contains the numerical parameters occurring in the solution procedure for a quadratic-linear system of equations (sq. u + sq. s = 12 30, $s = u \cdot$ igi 7). In other cases, *several* hand tablets contain the numerical parameters for a number of related mathematical exercises. Examples are the hand tablets ***UET 6/2 233, 254, 293, 298*** (*op. cit.*, § 2 h) with numerical parameters for computations of the cost in man-days and silver for the excavation of canals of various sizes, and ***UET 6/2 290, 218, 452***, and ***374*** (*op. cit.*, § 2 i) with numerical parameters for computations of the capacity measures of cylindrical containers of various sizes.

The purpose of such hand tablets is not known, but a reasonable conjecture seems to be that they were notes taken by students when a teacher demonstrated problems in class, and that at the end of the school day the students took the hand tablets home, where they were supposed to write their own mathematical clay tablets, filling in the details of the statements of the problems and of the solution procedures. (All the hand tablets discussed in *RA* 94 were found in a rich man's house at the main street of Ur, probably left there by the rich man's son.)

It has been noted above that there are surprisingly *few duplicates* of small arithmetical or metrological table texts. In the same way, there are few, if any, duplicates of single mathematical problem texts or of mathematical hand tablets. The reason for this phenomenon can be either that teachers wanted to hand out *similar but different* assignments to their students, if they had several of them, or that they wanted to hand out several similar but different assignments to each student, if they had only one or a few. The big combined multiplication tables or metrological tables served, of course, as sources for the small arithmetical or metrological table texts. In a similar way, mathematical *theme texts* or *series texts*, large clay tablets with long series of related mathematical problems, could serve as sources for many related single problem texts. *Plimpton 322, with its extensive table of parameters for a long series of related mathematical problems, seems to be something halfway between a combined arithmetical table text and a mathematical theme text*! It could obviously serve as a rich source of data for mathematical hand tablets of the *UET 6/2* type. Note, by the way, that (according to the dealer, E. J. Banks, who sold it) Plimpton 322 is a tablet from Larsa, which means both that it is from a site close to Ur (see the map in Fig. 9.2) and that it is early Old Babylonian just like the *UET 6/2* hand tablets from Ur.

Each line of numbers in the (reconstructed) Plimpton 322 table

| [igi] | [igi.bi] | [$q/2$] | [$p/2$] | sq. $p/2$ & sq. $q/2$ | $((q/2))$ | $((p/2))$ | # |

can be understood as the numerical parameters for a single mathematical exercise like the ones in MS 3971 § 3. In each one of the exercises in MS 3971 § 3, the half-sum $p/2$ (*takīlti ṣiliptim*) is computed, for a given choice of the parameters igi and igi.bi. It is, apparently, silently understood that this half-sum is the diagonal of a normalized rectangle, that is a rectangle with the length 1 (or 1 00). The objective of the exercise is to compute the front of the rectangle, in MS 3971 § 3 explicitly called sag. This is done by squaring $p/2$, subtracting 1 (or 1 00 00), and computing the square side. A catch with this explanation is that the sequence of computed parameters in MS 3971 § 3 is

| igi | igi.bi | $p/2$ | sq. $p/2$ | sq. $q/2$ | $q/2 = $ sag |

which is not the same as the sequence of parameters in Plimpton 322. The reason is, probably, that Plimpton 322 is a combined table; it can be used as well for the case when the diagonal is known and the front is computed as for the case when the front is known and the diagonal is computed, and therefore none of the parameters $q/2$ and $p/2$ is given priority over the other.

The Plimpton 322 table can also be used for a related but slightly different kind of mathematical problems, exemplified by AO 6484 § 7, namely rectangular-linear systems of equations where either the sum p or the difference q of igi and igi.bi is known and the task is to compute the values of igi and igi.bi. (The equation igi $\cdot$ igi.bi = 1 (or 1 (00 00)) is, of course, silently understood.)

It remains to explain the role of the two columns $((q/2))$ and $((p/2))$ for íb.si$_8$ sag and íb.si$_8$ *ṣiliptim* in Plimpton 322. As mentioned, it is likely that this couple of rather strange headings is to be understood as abbreviations for the more comprehensible pair of phrases

íb.si$_8$ <à.ša *takīlti*> sag	the square side of the <square of the holder for the> front
íb.si$_8$ <à.ša *takīlti*> ṣiliptim	the square side of the <square of the holder for the> diagonal.

Even after this clarification, the meaning of the pair of headings remains somewhat obscure. As explained above, the numbers in this pair of columns are the *factor-reduced cores* of the square sides of the squares of the front and the diagonal. They were included in the table for the reason that the simplest way to compute the square sides of the squares of the front and the diagonal was to use the trailing part algorithm and to compute the square sides of those squares as the products of the square sides of the factor-reduced cores and the square sides of all the removed regular factors.

An additional complication of the issue is the circumstance that there are two errors in the numbers recorded in the column for sq. $p/2$ in Plimpton 322, lines 2 and 8. Those errors would have made it difficult to compute correctly the square sides of those two numbers. For that reason, a reasonably modified explanation of the numbers in the columns called íb.si$_8$ sag and íb.si$_8$ *ṣiliptim* is that they were computed directly as the factor-reduced cores of the front and the diagonal in each line of the table. It is easy to understand why, since it was obviously easier to compute the factor-reduced cores of the front and the diagonal than to compute the factor-reduced cores of the square sides of the squares of the front and the diagonal. In addition, if the factor-reduced cores of the diagonal and the front were known, this would simplify the computation of the squares of the diagonal and the front.

Note that the explanation proposed above of the columns íb.si$_8$ sag and íb.si$_8$ *ṣiliptim* has the important implication that it was never the intention of the author of Plimpton 322 to reduce his series of *normalized* diagonal triples (with the length equal to 1 in each triple) to a corresponding series of *primitive* diagonal triples (with the front, the length, and the diagonal equal to integers without common factors). Therefore, the interpretation of Plimpton 322 as a witness of an Old Babylonian occupation with number-theoretical questions must be abandoned.

Note also, by the way, that it has been assumed throughout this discussion that the topic of the tables on Plimpton 322 was rectangles with diagonals, not right triangles! Rectangles are more important in Old Babylonian mathematical texts than right triangles. (Cf. the remark in Høyrup, *LWS* (2002), directly under Fig. 77.) Furthermore, the only known instances of Old Babylonian mathematical texts explicitly mentioning the diagonal of a right triangle are **IM 55357** (Friberg, *HM* 8 (1981), Fig. 6.2; Høyrup, *LWS* (2002), 231-234) where it is called uš gíd 'the long length', not *ṣiliptum*, and Bruins and Rutten, *TMS* 1 (Fig. 8.2.6 above), where it is called uš 'the length'.

A8.6. The Diagonal Rule in the Corpus of Mathematical Cuneiform Texts

In an essay in *ChV* (2001), Damerow discussed the extent to which the "theorem of Pythagoras" (in this book called the Babylonian diagonal rule) was known and applied in mathematical cuneiform texts. In particular, in footnote 92 on p. 286 of the essay, Damerow listed all mathematical cuneiform texts known to him, in which the theorem is used either directly or indirectly.

Without mentioning Damerow's essay, Høyrup wrote a brief note in *LWS* (2002), 385-387, with a discussion of nine examples known to him of applications in Old Babylonian mathematical texts of the "Pythagorean rule". Høyrup believes he has made a relevant observation concerning the geographical origin of the texts with applications of the Pythagorean rule, namely that they are from the Mesopotamian "periphery", since BM 85194, BM 85196, and VAT 6598 belong to Group 6A from Sippar, Db$_2$-146 to Group 7B from Eshnunna, and *TMS* 1, 3 and 19 to Group 8 from Susa, with only YBC 8633, with its *incorrect* application of the Pythagorean rule, being a southern text, belonging to Group 4 from Uruk. (See the map, Fig. 9.2.) The remaining example, Plimpton 322, has previously been assumed to be from the southern city Larsa (Group 1), but Høyrup makes the (unwarranted) conjecture that it may just as well "if not better" be a northern text belonging to Group 6.

The Høyrup/Damerow list is updated below, and made considerably more complete and explicit.

Old Babylonian texts

BM 85194 ## 21-22	two problems for a chord in a circle	Fig. 11.1.3 above
BM 85196 # 9	a pole-against-a-wall problem	Friberg, *UL* (2005), Sec. 3.1 b
BM 96957+ §§ 5-7	problems for the diagonal of a gate	Sec. 11.1 d above
IM 55357	a geometric algorithm for a right triangle	Høyrup, *LWS* (2002), 231
IM 67118 (= Db$_2$-146)	a rectangle with given diagonal and area	Fig. 11.2.6 above
MS 3049 § 1	problems for a chord in a circle	Fig. 11.1.2 above
———— § 5	a problem for the inner diagonal of a gate	Sec. 11.1.5 above
MS 3052 § 1 c-d	problems for a mud wall with a drilled hole	Sec. 10.2 a above
———— § 2	an igi–igi.bi problem	Sec. 10.2 b above
MS 3876	problems for 20 equilateral triangles	Sec. 11.3 above
MS 3971 § 2	a rectangle with given diagonal and area	Sec. 10.1 b above
———— § 3	five igi-igi.bi problems	Sec. 10.1 c above
———— § 4	a scaling problem for a rectangle with diagonal	Sec. 10.1 d above
Plimpton 322	a table of parameters for igi–igi.bi problems	App. 8
TMS 1	a symmetric triangle inscribed in a circle	Fig. 8.2.6 above
TMS 2	a regular 6-front and a regular 7-front	Fig. 8.2.15 above
TMS 19 § 1	a simple problem for the diagonal of a rectangle	
———— § 2	an extremely complicated problem for a diagonal	Høyrup, *LWS* (2002), 194-200
VAT 7531	two examples of "Heronic" triangles	Friberg, *UL* (2005), Sec. 3.7 c
YBC 7289	a square with diagonals	Fig. 8.2.8 above
YBC 8633	one symmetric and two right triangles	Høyrup, *LWS* (2002), 254-257

Late Babylonian/Seleucid texts

AO 6484 # 6 (§ 5 a)	a system of quadratic equations for a rectangle	Neugebauer, *MKT 1* (1935), 104
———— # 7 (§ 4 b)	a problem for a symmetric triangle	Friberg, *BagM* 28 (1997), Fig. 6.3
———— # 8 (§ 5 b)	the side of a square with given diagonal	Neugebauer, *MKT 1* (1935), 104
———— ## 14-17 (§ 7)	four igi–igi.bi problems	App. 8, Sec. A8 d above
BM 34568	seventeen problems for rectangles with diagonals	Neugebauer, *MKT 3* (1937), 20
———— # 12	a pole-against-a-wall problem	Friberg, *UL* (2005), Sec. 3.1 b
VAT 7848 § 1	the height of an equilateral triangle	Friberg, *BagM* 28 (1997), § 6 d
———— § 2	the diagonal of a right triangle	Friberg, *BagM* 28 (1997), § 6 d
———— § 3	the height of a symmetric trapezoid	Friberg, *BagM* 28 (1997), § 6 d
W 23291 § 4 a-c	problems for symmetric or equilateral triangles	Friberg, *BagM* 28 (1997), § 4

Tables of constants

TMS 3 (BR) 31 1 25 igi.gub *šà* bar.dá *šà* nigin	1 25, the constant of the cross-bar of a square
———————— 32 1 15 igi.gub *šà* bar.dá *šà* uš *ù* sag	1 15, the constant of the cross-bar of a length-and-front
YBC 7243 (NSe) 101 24 51 10 *ṣi-li-ip-tum* íb.si$_8$	1 24 51 10, the cross-over of an equalside

(BR 31 and NSe 10 give two different approximations to the length of the diagonal of a normalized square with the side 1, while BR 32 gives the length of the diagonal of the Old Babylonian favorite normalized solution to the diagonal equation, the rectangle with $c, b, a = 1\ 15, 1, 45$.)

The updated list contradicts Høyrup's mentioned observation, as well as his conjecture: Plimpton 322 is obviously closely related to MS 3052 § 2 and MS 3971 § 3, and both texts belong to Group 3 from Uruk, a central southern site. Moreover, the Uruk text MS 3971 § 2 is a close parallel to the Eshnunna text Db$_2$-146, with the same numerical data, and as a Larsa text of a particular tabular format, Plimpton 322 is of roughly the same age as Db$_2$-146 (Robson, *AMM* 109 (2002), 110-111).

Besides, the discussion of Plimpton 322 in the present appendix forcefully contradicts Høyrup's opinion (*op. cit.*, fn. 475), according to which

> "Friberg's own proposal – that the (Plimpton) table was meant to provide parameters from which second-degree equations could be constructed – though not impossible does not fit the Old Babylonian habit of constructing problems from known very simple solutions. A new perspicacious (sic!)[8] analysis is [Robson 2001]."

---

8. See footnote 1 of the present appendix for a totally different opinion about the alleged perspicacity of Robson's analysis.

The discussion above also strongly contradicts Robson's unfairly patronizing and disdainful attitude towards the mathematical abilities of the (actually quite skillful and insightful) original authors of Old Babylonian mathematical texts, expressed in Robson, *HM* 28 (2001) in words such as

"I certainly do not feel justified in referring to authors and copyists of OB mathematics as 'mathematicians,' with the connotations of creativity and professionalism this word carries; I prefer the more neutral 'scribes.'"   (*op. cit.*, 171)

"Most earlier analyses of the (Plimpton) tablet [e.g. Friberg 1981] have started from the assumption that the tablet is exhaustive and have attempted to determine the criteria by which the scribe chose his starting points, whether p, q generators or reciprocal pairs. All of these attempts seek a single rule for choosing those starting points (e.g. Friberg's "restrictions on the parameters" [1981, 284]). But are we justified in assuming that the concept of mathematical completeness would have meant anything at all in the early second millennium BC? Or that the scribe must have generated his starting numbers (p, q or reciprocals) using a single algorithm? And does it really matter? ...... We are (then) guilty of acting like Pingree's 'treasure hunters seeking pearls in the dung heap,' privileging the apparently modern at the expense of the obviously ancient.
If on the other hand we are interested in what Plimpton 322 might have been for, then its degree of completeness is an issue. Did it matter to its ancient compiler? Would it even have been a meaningful issue for him?"   (*op. cit.*, 195)

"Is this (Robson's own) explanation for the scribe's choices any less historically plausible than previous scholars' ...? We have to remember that we are trying to (re)construct what a real human being, nearly 4000 years ago, might have thought and done to produce the figures on Plimpton 322, not an idealised mathematical automaton ...." (*op. cit.*, 199)

"... we need to grasp the challenge of glimpsing what is often called the Big Picture: that is, to look beyond our currently favourite 'texts' to begin exploring the mathematical environment and mindset of the ancient world and accept that it is disturbingly alien in character".   (*op. cit.*, 202)

Appendix 9
Many-Place Squares of Squares in Late Babylonian Mathematical Texts

The Old Babylonian tradition of operating with many-place sexagesimal numbers (see the examples in §§ 1.4-5 above) survived intact into the Late Babylonian and Seleucid periods in the second half of the first millennium BC and was, of course, a necessary prerequisite for the well known stunning development of Late Babylonian/Seleucid mathematical astronomy.

A survey of Late Babylonian/Seleucid texts exhibiting many-place regular sexagesimal numbers can be found in Friberg, *CTMMA 2* (2005), 295-296. In that survey are enumerated all known representatives of the following distinct categories of such texts:

1. A standard table of reciprocals, followed by a table of squares of 3-place regular sexagesimal numbers.
2. Ten fragments of many-place tables of pairs of reciprocals n, igi. n, with n between '1' and '2'.
3. A fragment of a many-place table of regular reciprocals n, igi. n, with n between '4' and '8'(?).
4. Two many-place tables of regular reciprocals n, igi. n, with n between '1' and '3'.
5. A many-place table of regular reciprocals n, igi. n, with n between '1' and '4'.
6. Several fragments of many-place lists of regular squares from sq. '1' to sq. '2'.
7. Two fragments of many-place lists of regular squares of squares from sq. sq. '1' to sq. sq. '2'.
8. An explicit computation of sq. sq. 3^{23}, the fourth power of a 7-place regular number.
9. Four examples of the use of a factorization algorithm for many-place regular sexagesimal numbers.

See, in particular, the presentation in Friberg, *BagM* 30 (1999), *BagM* 31 (2000) of **W 23021**, a round clay tablet with a quite interesting series of applications of a factorization algorithm for the computation of reciprocals of many-place regular sexagesimal numbers. A simpler text of the same kind is von Weiher, *SpTU 5*, **316** (1998). It is a Late Babylonian parallel to MS 2242 and MS 3037 (see Fig. 1.4.1 in Ch. 1 above). (In Fig. A9.1 below, some apparent mistakes in the hand copy have been corrected.)

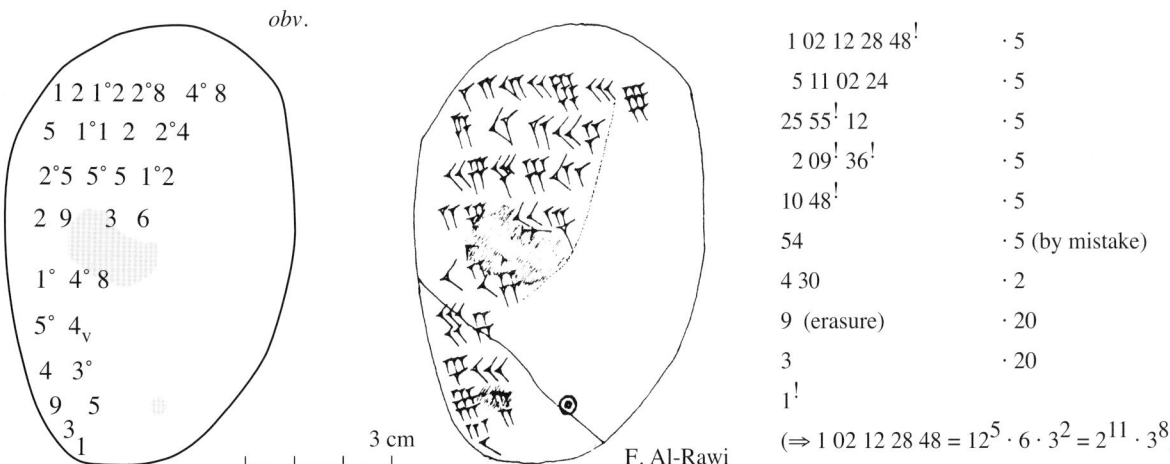

Fig. A9.1. *SpTU 5*, 316. A Late Babylonian hand tablet with a factorization algorithm for a 5-place number.

453

Some of the texts mentioned in the survey in Friberg, *CTMMA 2* (2005) have not yet been published. Two of them are interesting enough to be discussed in some detail below.

A9.1. Squares of Squares of Many-Place Regular Sexagesimal Numbers

BM 55557, a Late Babylonian/Seleucid fragment of a list of squares of squares of many-place regular sexagesimal numbers ("a table of 4th powers") was published in Britton, *JCS* 43-45 (1991-93). See the new hand copy in Fig. A9.2 below. Britton showed that his table of squares of squares was intimately related with the many previously known Late Babylonian/Seleucid fragments of many-place tables of reciprocals and fragments of lists of squares of many-place regular sexagesimal numbers. Apparently, most such fragments of many-place tables of reciprocals are parts of clay tablets with excerpts from a large combined table with a long list of many-place regular sexagesimal numbers, and their reciprocals. The beginning of such a list is exhibited in Friberg, *op. cit.*, Table 2, where it is called "The reconstructed First Tablet of a total twelve-place table of reciprocals". Here are the first few lines of that table (with the addition of headings over the columns and of a first column listing the indices r, s for the numbers $n = 2^r \cdot 3^s$):

r	s	$n = 2^r \cdot 3^s$	igi n
0	0	1	1
-18	-11	1 00 16 53 53 20	59 43 10 50 52 48
15	-2	1 00 40 53 20	59 19 34 13 07 30
-2	5	1 00 45	59 15 33 20
-20	-6	1 01 02 06 33 45	58 58 56 38 24
13	3	1 01 26 24	58 35 37 30
-4	10	1 01 30 33 45	58 31 39 35 18 31 06 40
	etc.	*etc.*	*etc.*

If there ever was such a First Tablet, then it was probably accompanied by another large tablet with the squares of *the same list* of many-place regular sexagesimal numbers, and possibly even by a third large tablet with all the squares of squares of *the same list* of numbers. Reconstructions of the content of the First Tablet and the two associated lists are presented at the end of this paragraph.

BM 55557 is clearly the upper right corner of a relatively large clay tablet. The original width of the clay tablet can be estimated from a reconstruction of the first few lines of the table text:

[1 04 14 56 27 48 28 07 55 21 34 22 30 13] 23 22 30 52 44 03 45	(21-place)
[1 05 58 14 30 41 45 31 01 24 05 45] 36	(13-place)
[1 06 1 6 08 59 52 16 43 11 54 24 1]8 45 52 44 03 45	(17-place)

How could Britton find out the nature of the numbers inscribed on this fragment and reconstruct them to their original lengths? Normally, it is relatively easy to explain a many-place number found on a clay tablet or a fragment. If the whole number is there, a factorization of the number by use of the trailing part algorithm laboriously but surely leads to the goal. If the beginning of the number is preserved, and if the number is at most 11-place or has a reciprocal that is at most 11-place, then a look in Gingerich's systematically arranged table of 11-place regular sexagesimal numbers and their reciprocals (*TAPS 55* (8) (1965)) quickly leads to the answer. However, in the case of BM 55557, none of these conditions is satisfied. Large parts of the beginnings of the numbers originally recorded on BM 55557 are lost, and only the ends of more than 13-place numbers are preserved on the fragment. (The first of the reconstructed numbers on the fragment is a 20-place number.)

On the other hand, if the end of a many-place regular sexagesimal number is preserved, it can be enough to get a hold on the situation. Indeed, look at the trailing parts of the numbers in the first four lines of the fragment: 3 45, 36, 3 45, 6 40. Since all these trailing parts are squares of regular sexagesimal numbers, it is immediately clear that BM 55557 must be a table of squares of regular sexagesimal numbers. On the other hand, it is easy to check that BM 55557 is not an excerpt of the usual kind from the Late Babylonian/Seleucid table of squares of many-place regular sexagesimal numbers.

A detailed and fascinating account of how to proceed from there is given by Britton, *op. cit.*, 73-74.

Remark: The reconstruction in Fig. A9.2 of the lost left half of the text of BM 55557 was made under the assumption that the author of the text made no mistakes, and that he carefully maintained a vertical alignment of ones and tens in the numbers he recorded. It is clear from the uneven positioning of the starting points of the reconstructed numbers that this assumption was too optimistic.

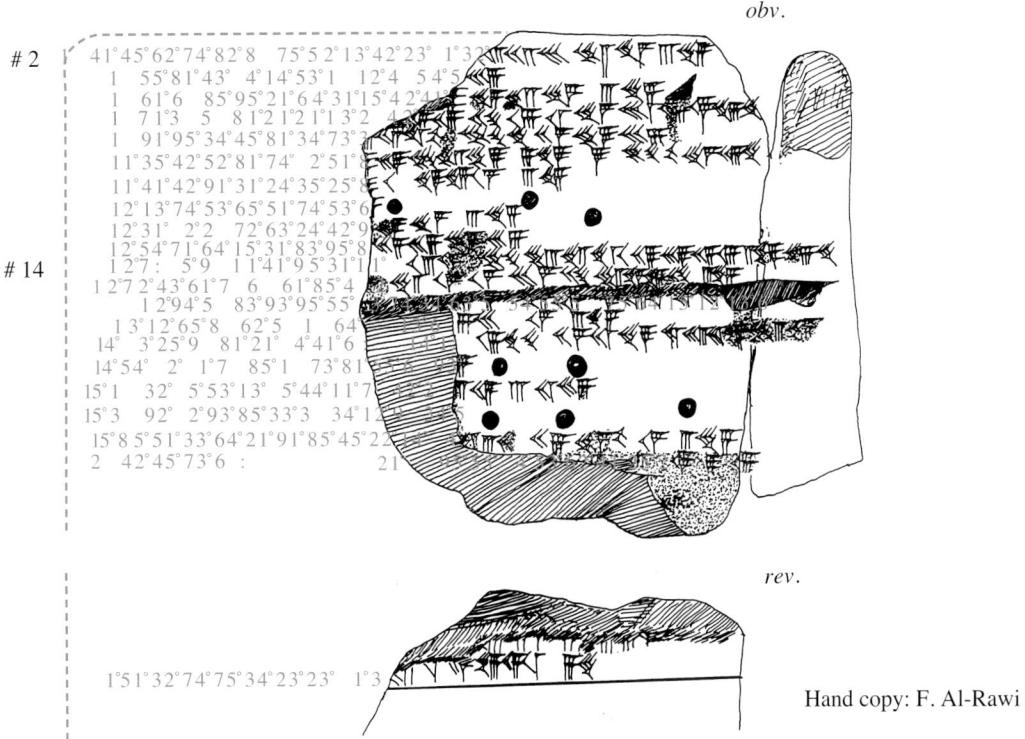

Fig. A9.2. BM 55557. A fragment of an excerpt from the Late Babylonian table of squares of squares.

Now take a look at **BM 32584**, found by Farouk Al-Rawi in the archives of the British Museum. The reconstruction in Fig. A9.3 below shows that BM 32584 is a fragment from the middle of a clay tablet, so that both the beginnings and the ends of the recorded numbers are missing. In addition, the numbers are often exceedingly long. Therefore, it could have been very difficult to find an interpretation of the numbers on BM 32584 if not, by a lucky accident, BM 55557 had just recently been published when BM 32584 was found. Under the circumstances, it was relatively easy to match the longest preserved part of a number on BM 32584, namely

[1] **27 <00> 59 01 14 19 53 11 10 13 14 59 53 42 41 10 28 3**[7 25 16 02 57 46 40]   (25-place)

with the corresponding preserved part of a number on BM 55557, namely

[1 27 00 59 01 14 19 53 11 10] **13 14 59 53 42 41 10 28 37 25 16 02 57 46 40**.

After that, it was easy to find other matches. Consequently, it became clear that BM 32584, too, is a table of squares of squares, and the mentioned matching pair corresponds to number 14 in the complete table of squares of squares (see below). Moreover, as shown in Fig. A9.3, it is likely that originally nearly the whole table of 100 squares of squares was recorded on BM 325854, with about 45 lines on the obverse, 46 on the reverse, and 6 on the edge. (There are some gaps in the sequence.)

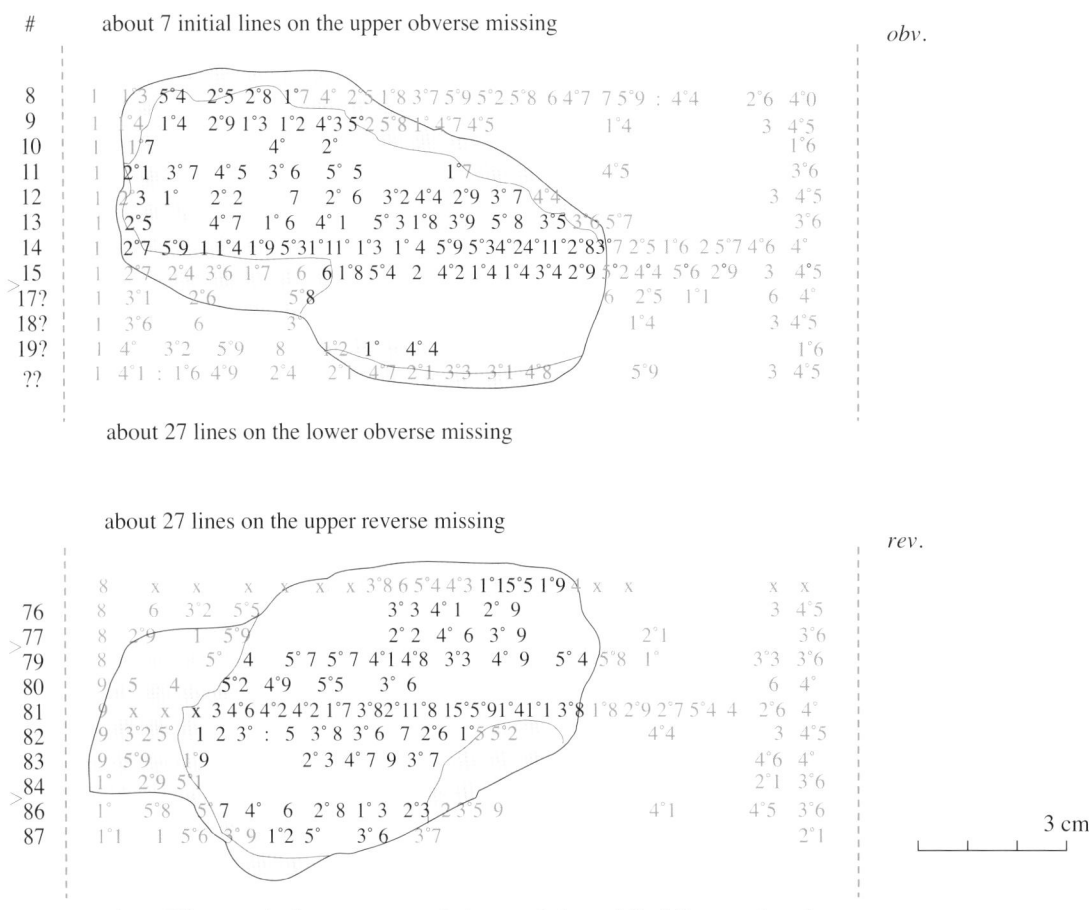

Fig. A9.3. BM 32584. Another fragment of an excerpt from the Late Babylonian table of squares of squares.

A9.2. An Explicit Late Babylonian Multiplication Algorithm

The Late Babylonian/Seleucid joined fragment **BM 34601** = Sp 2, 76 + 759 (see Fig. A9.4 below) was published as no. 1644 in Sachs, *LBAT* (1955). It is clear that the fragment looks like what remains of a clay tablet with an explicit computation of the product of two many-place sexagesimal numbers, with the result of the computation recorded underneath a slanting column of "partial products". Since there are traces remaining of 9 partial products, it is also clear that the multiplier must have been at least a 9-place number, and since one of the partial products has 13 preserved places, the multiplicand must have been at least a 13-place number.

Many futile attempts were made by the present author over several years to find out the true nature of the multiplication on BM 34601. Then a letter arrived from John Britton who reported that he had ben able to identify one of the broken lines on BM 34601 as the number

$$[3]\ 03\ 13\ 15\ 33\ 54\ 58\ 1[9\ 24\ 11\ 01\ 39]\ 06\ 45 = 5 \cdot 3^{25} \qquad \text{(s14-place)}$$

He did not tell how he had managed to make this identification, but the reconstruction of the original text given in Fig. A9.4, top, reveals that the number that Britton had identified is the only number originally recorded on the clay tablet of which the trailing part is preserved on the fragment. So, maybe he reasoned as follows: Since 45 is a regular sexagesimal number and is in a position which suggests that it may be the final place of a many-place number, it may be a good start to assume that it really *is* the last place of some regular many-place number. If that is so, then it should be possible to simplify that number through multiplication with a factor 4. The result of such a multiplication is:

$$4 \cdot [\ldots] \, 03 \, 13 \, 15 \, 33 \, 54 \, 58 \, 1[\ldots \ldots \ldots] \, 06 \, 45 = [\ldots] \, 12 \, 53 \, 02 \, 15 \, 39 \, 5[\ldots \ldots \ldots \ldots] \, 27.$$

The next obvious step is to multiply by 20 (the reciprocal of 3):

$$20 \cdot [\ldots] \, 12 \, 53 \, 02 \, 15 \, 39 \, 5[\ldots \ldots \ldots \ldots] \, 27 = [\ldots]4 \, 17 \, 40 \, 45 \, 13 \, [\ldots \ldots \ldots \ldots \ldots] \, 09.$$

Since 9 is a square, one can then suspect that [...]4 17 40 45 13 [... ... ...] 09, too, is a square. After some more work(!), one can then arrive at the conclusion that it really is, and that

$$[4 \, 0]4 \, 17 \, 40 \, 45 \, 13 \, [17 \, 45 \, 52 \, 14 \, 42 \, 12] \, 09 = sq. \, 20 \, 10 \, 40 \, 80 \, 30 \, 02 \, 7 = sq. \, 3^{23}. \qquad \text{(13-place)}$$

Therefore, the mentioned broken line with the last place 45 can be reconstructed as

$$3 \cdot 15 \cdot [4 \, 0]4 \, 17 \, 40 \, 45 \, 13 \, [17 \, 45 \, 52 \, 14 \, 42 \, 12] \, 09 = [3] \, 03 \, 13 \, 15 \, 33 \, 54 \, 58 \, 1[9 \, 24 \, 11 \, 01 \, 39] \, 06 \, 45. \qquad \text{(14-place)}$$

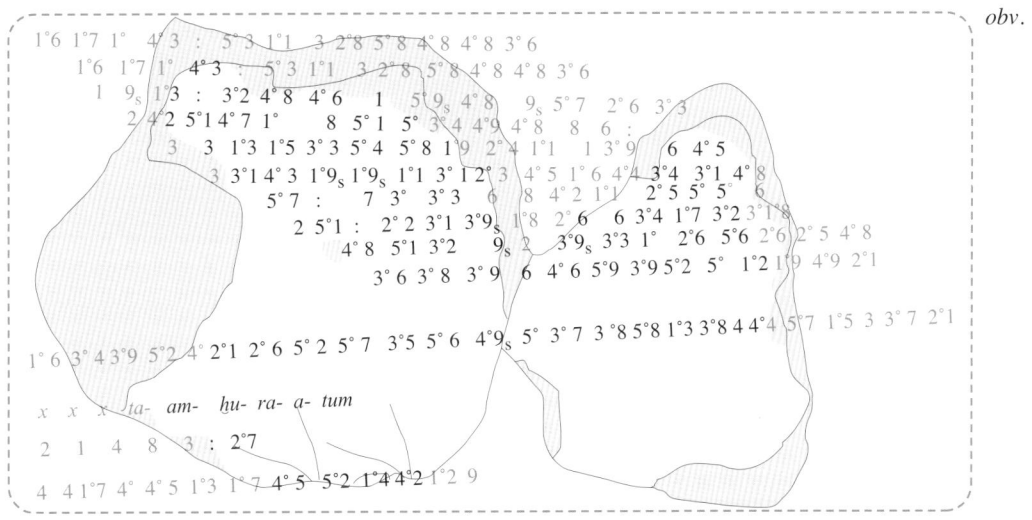

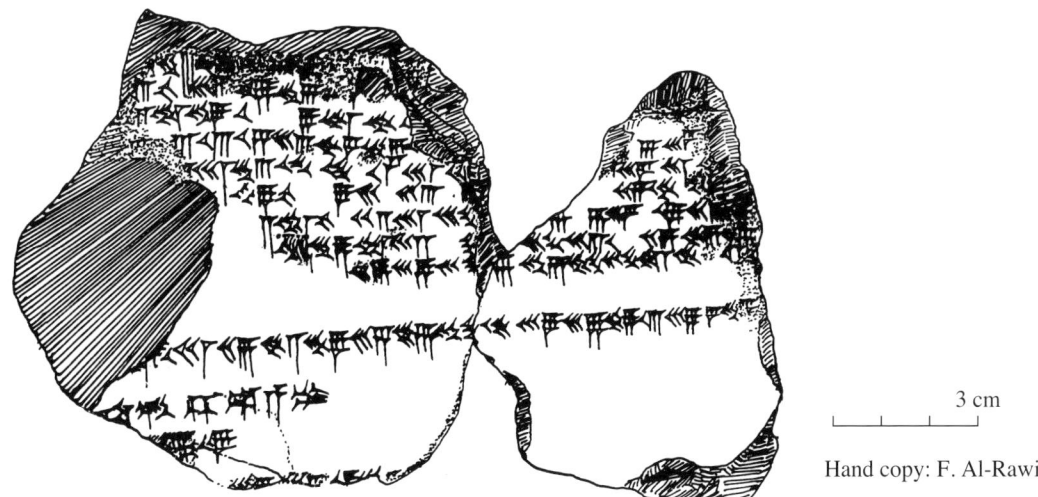

Fig. A9.4. BM 34601. An algorithm for the computation of the square of the square of a 7-place number.

With departure from this first identification, the present author then managed to reconstruct all the damaged lines on BM 34601 as the products of 4 04 17 40 45 13 17 45 52 14 42 12 09 with the following multipliers:

$$\ldots, \ldots, 17, 40, 45, 52, 14, 42, 12, \text{ and } 9.$$

Thus, the original text on BM 34601 appeared to have been the following, with the underlined digits and letters representing what is preserved on the fragment:

```
··· ··· ··· ··· ··· ··· ··· ··· ··· ··· ··· ··· ··· ··· ··· ··· ··· ···
   ··· ··· ··· ··· ··· 43 ··· ··· ··· ··· ··· ··· ··· ··· ··· ··· ··· ···
      1 09 13 00 32 48 46 01 59 48 09 57 26 33
         2 42 51 47 10 08 51 50 34 49 48 08 06 00
            3 03 13 15 33 54 58 19 24 11 01 39 06 45
               3 31 43 19 19 11 31 23 45 16 44 34 31 48
                  57 00 07 30 33 06 08 42 11 25 50 50 06
                     2 51 00 22 31 39 18 26 06 34 17 32 30 18
                        48 51 32 09 02 39 33 10 26 56 26 25 48
                           36 38 39 06 46 59 39 52 50 12 19 49 21


   ··· ··· ··· ··· ··· ··· 21 26 52 57 35 56 49 50 37 38 58 13 38 04 4 ··· ··· ··· ··· ··· ···


   ··· ··· ··· ta-am-ḫu-ra-a-tum
   2 01 04 08 03 00 27
   4 04 17 40 45 13 17 45 52 14 42 12 09
```

So far, so good, but something is still amiss here, since the number beneath all the partial products *cannot*, as expected, be identified with the sum of all those partial products.

On the other hand, the meaning of the last three (reconstructed) lines of the text is clear enough. The word *tamḫurātum* is almost certainly the plural of *tamḫartum*, a word which has the meaning 'square number' in the Seleucid mathematical text **AO 6484 § 2** (*MKT 1* (1935), 97). The number below this word, [2 01 04 08 03] 00 27 is the sexagesimal representation of the 23rd power of 3, and the last number, [4 04 17 40 45 13 17] 45 52 14 42 [12 09] has already been identified as the square of the 23rd power of 3, and as the multiplicand in the recorded multiplication algorithm.

A reasonable conclusion is that MS 36401 is what remains of an explicit multiplication algorithm for the computation of the square of 4 04 17 40 45 13 17 45 52 14 42 12 09, in its turn the square of 2 01 04 08 03 00 27. If this conclusion is correct, then the original text of MS 36401 can have been of the form shown in Fig. A9.5 below, with or without the explicit mention of the multiplicand 4 04 17 40 45 13 17 45 52 14 42 12 09 above the computation, and the multiplier 4 04 17 40 45 13 17 45 52 14 42 12 09 to the left of the computation. Thus, the correct result of the computation should be that the square of the square of 2 01 04 08 03 00 27 is the 25-place regular sexagesimal number

$$16\ 34\ 39\ 52\ 40\ 21\ 26\ 52\ 57\ 35\ 56\ 49\ 50\ 37\ 38\ 58\ 13\ 38\ 04\ 44\ 57\ 15\ 03\ 37\ 21 = 3^{92}.$$

However, there is still a catch. A comparison of the suggested reconstruction in Fig. A9.5 with what is preserved of the computation on BM 46301 reveals that BM 46301 is *an imperfect copy* of a clay tablet with the correct computation. Indeed, the student who wrote BM 46301 copied the first five partial products correctly, but after he had copied the *first* partial product 45 · 2 01 04 08 03 00 27 = 3 03 13 15 33 54 58 19 24 11 01 39 06 45, he inadvertently skipped three partial products and continued with the partial product that followed after the *second* instance of the number 3 03 13 15 33 54 58 19 24 11 01 39 06 45. The omitted lines of the computation are written in bold style in Fig. A9.5. Note that the sum of *all* the partial products was copied correctly!

An interesting observation is that the number computed on BM 46301 is *just beyond the range* of the 100 entries in the Late Babylonian First Table of squares of squares, if such a table ever existed! Indeed, the last entry in that table is the square of the square of 1 58 31 06 40, while the computed number on BM 46301 is the square of the square of 2 01 04 08 03 00 27.

```
 4  04 17 40 45 13 17 45 52 14 42 12 09

 ×

 4   16 17 10 43 00 53 11 03 28 58 48 48 36
04       16 17 10 43 00 53 11 03 28 58 48 48 36
17        1 09 13 00 32 48 46 01 59 48 09 57 26 33
40          2 42 51 47 10 08 51 50 34 49 48 08 06 00
45            3 03 13 15 33 54 58 19 24 11 01 39 06 45
13              52 55 49 49 47 52 50 56 19 11 08 37 57
17               1 09 13 00 32 48 46 01 59 48 09 57 26 33
45                 3 03 13 15 33 54 58 19 24 11 01 39 06 45
52             3 31 43 19 19 11 31 23 45 16 44 34 31 48
14              57 00 07 30 33 06 08 42 11 25 50 50 06
42             2 51 00 22 31 39 18 26 06 34 17 32 30 18
12                48 51 32 09 02 39 33 10 26 56 26 25 48
09 _____ 36 38 39 06 46 59 39 52 50 12 19 49 21
   16 34 39 52 40 21 26 52 57 35 56 49 50 37 38 58 13 38 04 44 57 15 03 37 21


··· ··· ···  ta-am-ḫu-ra-a-tum
 2 01 04 08 03 00 27
 4 04 17 40 45 13 17 45 52 14 42 12 09
```

Fig. A9.5. The computation of the square of the square of the 23rd power of 3 (corrected).

BM 46301 is not the only example of a Late Babylonian fragment of a clay tablet with an explicit multiplication algorithm. Indeed, the two fragments von Weiher, *SpTU 5*, **317** (1998) (Fig. A9.6) appear to be what remains of one or two clay tablets of that kind.

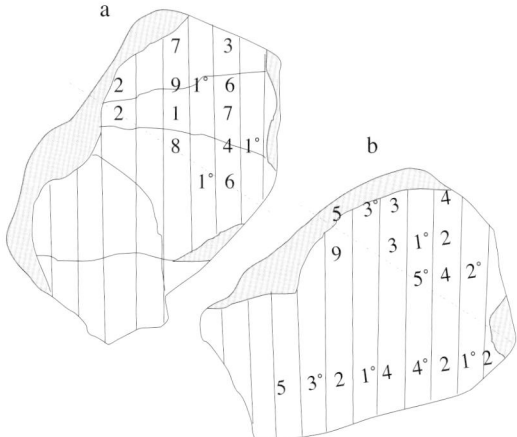

Fig. A9.6. *SpTU 5*, **317**. Two fragments of clay tablets with explicit multiplication algorithms.

In Fig. A9.6, the copies of the two fragments, here called fragments a and b, are placed as if they once were part of the same clay tablet (as claimed by von Weiher). However, the numbers on fragment a are written in horizontal lines, while the numbers on fragment b are written in slanting lines, so it is more likely that the two fragments are not that closely related. Note that fragment b appears to contain not only parts of a few partial products, but also part of the sum of all the partial products. Note also that in both fragments *all the ones and all the tens are inscribed neatly in separate columns*!

In fragment a, the preserved numbers are

```
··· ··· ··· · 07 03 ··· ··· ··· ··· ···
··· x2 09 16 ··· ··· ··· ··· ······
 2 01 07 ··· ··· ··· ··· ···
 8 04 1x ··· ··· ··· ··· ···
16 ··· ··· ··· ···· ···· ···
```

If these numbers are partial products, they should all be equal to some integer between 1 and 59 times the multiplicand, which means that they should all be simple multiples of each other. It is, indeed, easy to check that $4 \cdot \underline{2\ 01\ 07}$ [...] $= \underline{8\ 04}\ 28$ [...], and that $8 \cdot \underline{2\ 01\ 07}$ [...] $= \underline{16}\ 08\ 56$ [...]. Therefore, the multiplicand in the case of fragment a ought to be 2 01 07 [...], or this number divided by some integer between 2 and 7.

In fragment b, the preserved numbers are

$$
\begin{array}{c}
\cdots \cdots \cdots \cdot 07\ 03\ \cdots \cdots \cdots \cdots \cdots \cdots \\
\cdots \text{x5}\ 33\ 04\ \cdots \cdots \cdots \cdots \cdots \cdots \\
9\ 03\ 12\ \cdots \cdots \cdots \cdots \cdots \cdots \\
54\ 20 \cdots \cdots \cdots \cdots \cdots \cdots \\
5\ 32\ 14\ 42\ 12\ \cdots \cdots \cdots \cdots \cdots \cdots
\end{array}
$$

Here $10 \cdot 54\ 20$ [...] $= 9\ 03\ 20$ [...], which is a little bit too much, compared with the partial product 9 03 12 [...], which indicates that the hand copy is not quite correct. Whatever the case may be, it seems to be impossible, to get any further with the reconstruction of the original texts of the two explicit multiplication algorithms.

On the following four pages are presented attempted reconstructions of

a) the Late Babylonian "first tablet" of many-place reciprocals and squares

b) the Late Babylonian "first tablet" of many-place squares of squares.

The Late Babylonian First Tablet of Many-Place Reciprocals and Squares (Reconstructed)

r	s	n	igi n	sq. n	line
-2	5	1 00 45	59 15 33 20	1 01 30 33 45	1
-20	-6	1 01 02 06 33 45	58 58 56 38 24	1 02 05 17 25 04 28 03 59 03 45	2
13	3	1 01 26 24	58 35 37 30	1 02 54 52 24 57 36	3
-4	10	1 01 30 33 45	58 31 39 35 18 31 06 40	1 03 03 24 11 33 59 03 45	4
-5	-8	1 01 43 42 13 20	58 19 12	1 03 30 23 41 07 29 22 57 46 40	5
11	8	1 02 12 28 48	57 52 13 20	1 04 29 50 06 57 01 26 24	6
-7	-3	1 02 30	57 36	1 05 06 15	7
8	-5	1 03 12 35 33 20	56 57 11 15	1 06 35 29 18 34 24 11 51 06 40	8
-9	2	1 03 16 52 30	56 53 20	1 06 44 30 59 45 56 15	9
6	0	1 04	56 15	1 08 16	10
4	5	1 04 48	55 33 20	1 09 59 02 24	11
-14	-6	1 05 06 15	55 17 45 36	1 10 38 33 09 03 45	12
2	10	1 05 36 36	54 52 10 51 51 06 40	1 11 44 40 19 33 36	13
1	-8	1 05 50 37 02 13 20	54 40 30	1 12 15 22 56 55 27 17 51 36 17 46 40	14
-16	-1	1 05 55 04 41 15	54 36 48	1 12 25 10 42 58 29 28 21 33 45	15
17	8	1 06 21 18 43 12	54 15 12 30	1 13 23 00 45 14 28 50 18 14 24	16
-1	-3	1 06 40	54	1 14 04 26 40	17
-3	2	1 07 30	53 20	1 15 56 15	18
12	0	1 08 16	52 44 03 45	1 17 40 20 16	19
-5	7	1 08 20 37 30	52 40 29 37 46 40	1 17 50 52 05 23 26 15	20
10	5	1 09 07 12	52 05	1 19 37 34 27 50 24	21
-8	-6	1 09 26 40	51 50 24	1 20 22 31 51 06 40	22
8	10	1 09 59 02 24	51 26 25 11 06 40	1 21 37 45 36 55 17 45 36	23
-10	-1	1 10 18 45	51 12	1 22 23 50 51 33 45	24
5	-3	1 11 06 40	50 37 30	1 24 16 47 24 26 40	25
-12	4	1 11 11 29 03 45	50 34 04 26 40	1 24 28 12 58 45 57 07 44 03 45	26
3	2	1 12	50	1 26 24	27
18	0	1 12 49 04	49 26 18 30 56 15	1 28 22 25 43 32 16	28
1	7	1 12 54	49 22 57 46 40	1 28 34 24 36	29
0	-11	1 13 09 34 29 08 08 53 20	49 12 27	1 29 12 19 26 34 23 19 49 38 08 36 52 20 44 26 40	30
-17	-4	1 13 14 31 52 30	49 09 07 12	1 29 24 25 04 54 26 00 56 15	31
16	5	1 13 43 40 48	48 49 41 15	1 30 35 49 04 44 32 38 24	32
-2	-6	1 14 04 26 40	48 36	1 31 26 58 06 25 11 06 40	33
14	10	1 14 38 58 33 36	48 13 31 06 40	1 32 52 33 46 00 30 52 24 57 36	34
-4	-1	1 15	48	1 33 45	35
11	-3	1 15 51 06 40	47 27 39 22 30	1 35 53 30 12 20 44 26 40	36
-6	4	1 15 56 15	47 24 46 40	1 36 06 30 14 03 45	37
9	2	1 16 48	46 52 30	1 38 18 14 24	38
-9	-9	1 17 09 37 46 40	46 39 21 36	1 39 13 44 30 27 09 37 46 40	39
7	7	1 17 45 36	46 17 46 40	1 40 46 37 03 21 36	40
-11	-4	1 18 07 30	46 04 48	1 41 43 30 56 15	41
4	-6	1 19 00 44 26 40	45 33 45	1 44 02 57 02 46 15 18 31 06 40	42
2	-1	1 20	45	1 46 40	43
0	4	1 21	44 26 40	1 49 21	44
15	2	1 21 55 12	43 56 43 07 30	1 51 50 53 11 02 24	45
-2	9	1 22 00 45	43 53 44 41 28 53 20	1 52 06 03 00 33 45	46
-3	-9	1 22 18 16 17 46 40	43 44 24	1 52 54 02 06 26 38 54 09 22 57 46 40	47
13	7	1 22 56 38 24	43 24 10	1 54 39 42 25 41 22 33 36	48
-5	-4	1 23 20	43 12	1 55 44 26 40	49
-7	1	1 24 22 30	42 40	1 58 39 08 26 15	50

r	s	n	igi n	sq. n	line
8	-1	1 25 20	42 11 15	2 01 21 46 40	51
6	4	1 26 24	41 40	2 04 24 57 36	52
-12	-7	1 26 48 20	41 28 19 12	2 05 35 12 16 06 40	53
4	9	1 27 28 48	41 09 08 08 53 20	2 07 32 45 01 26 24	54
3	-9	1 27 47 29 22 57 46 40	41 00 22 30	2 08 27 20 47 51 55 11 45 04 31 36 17 46 40	55
-14	-2	1 27 53 26 15	40 57 36	2 08 44 45 43 03 59 03 45	56
1	-4	1 28 53 20	40 30	2 11 41 14 04 26 40	57
-1	1	1 30	40	2 15	58
-3	6	1 31 07 30	39 30 22 13 20	2 18 23 45 56 15	59
12	4	1 32 09 36	39 03 45	2 21 33 27 56 09 36	60
-6	-7	1 32 35 33 20	38 52 48	2 22 53 23 27 31 51 06 40	61
10	9	1 33 18 43 12	38 34 48 53 20	2 25 07 07 45 38 18 14 24	62
-8	-2	1 33 45	38 24	2 26 29 03 45	63
7	-4	1 34 48 53 20	37 58 07 30	2 29 49 50 56 47 24 26 40	64
-10	3	1 34 55 18 45	37 55 33 20	2 30 10 09 44 28 21 33 45	65
5	1	1 36	37 30	2 33 36	66
-13	-10	1 36 27 02 13 20	37 19 29 16 48	2 35 02 43 17 40 04 56 17 46 40	67
3	6	1 37 12	37 02 13 20	2 37 27 50 24	68
-15	-5	1 37 39 22 30	36 51 50 24	2 38 56 44 35 23 26 15	69
18	4	1 38 18 14 24	36 37 15 56 15	2 41 03 40 35 05 51 21 36	70
1	11	1 38 24 54	36 34 47 14 34 04 26 40	2 41 25 30 44 00 36	71
0	-7	1 38 45 55 33 20	36 27	2 42 34 36 38 04 46 25 11 06 40	72
-17	0	1 38 52 37 01 52 30	36 24 32	2 42 56 39 06 41 36 18 48 30 56 15	73
-2	-2	1 40	36	2 46 40	74
13	-4	1 41 08 08 53 20	35 35 44 31 52 30	2 50 28 27 01 56 52 20 44 26 40	75
-4	3	1 41 15	35 33 20	2 50 51 33 45	76
11	1	1 42 24	35 09 22 30	2 54 45 45 36	77
-7	-10	1 42 52 50 22 13 20	34 59 31 12	2 56 24 25 47 34 08 17 07 09 37 46 40	78
9	6	1 43 40 48	34 43 20	2 59 09 32 32 38 24	79
-9	-5	1 44 10	34 33 36	3 00 50 41 40	80
6	-7	1 45 20 59 15 33 20	34 10 18 45	3 04 58 34 44 55 33 52 55 18 31 06 40	81
-11	0	1 45 28 07 30	34 08	3 05 23 39 26 00 56 15	82
4	-2	1 46 40	33 45	3 09 37 46 40	83
2	3	1 48	33 20	3 14 24	84
-16	-8	1 48 30 25	33 10 39 21 36	3 16 13 45 25 10 25	85
17	1	1 49 13 36	32 57 32 20 37 30	3 18 50 27 52 57 36	86
0	8	1 49 21	32 55 18 31 06 40	3 19 17 25 21	87
-1	-10	1 49 44 21 43 42 13 20	32 48 18	3 20 42 43 44 47 22 29 36 40 49 22 57 46 40	88
15	6	1 50 35 31 12	32 33 07 30	3 23 50 35 25 40 13 26 24	89
-3	-5	1 51 06 40	32 24	3 25 45 40 44 26 40	90
-5	0	1 52 30	32	3 30 56 15	91
10	-2	1 53 46 40	31 38 26 15	3 35 45 22 57 46 40	92
-7	5	1 53 54 22 30	31 36 17 46 40	3 36 14 38 01 38 26 15	93
8	3	1 55 12	31 15	3 41 11 02 24	94
-10	-8	1 55 44 26 40	31 06 14 24	3 43 15 55 08 38 31 06 40	95
6	8	1 56 38 24	30 51 51 06 40	3 46 44 53 22 33 36	96
5	-10	1 57 03 19 10 37 02 13 20	30 45 16 52 30	3 48 21 56 58 25 38 07 33 28 02 51 11 36 17 46 40	97
-12	-3	1 57 11 15	30 43 12	3 48 52 54 36 33 45	98
21	6	1 57 57 53 16 48	30 31 03 16 52 30	3 51 55 41 38 32 25 57 30 14 24	99
3	-5	1 58 31 06 40	30 22 30	3 54 06 38 21 14 04 26 40	100

The Late Babylonian First Tablet of Many-Place Squares of Squares (Reconstructed)

sq. sq. n		
1 03 03 24 11 33 59 03 45	1	
1 04 14 56 27 48 28 07 55 21 34 22 30 13 23 22 30 52 44 03 45	2	
1 05 58 14 30 41 45 31 01 24 05 45 36	3	
1 06 16 08 59 52 16 43 11 54 24 18 45 52 44 03 45	4	
1 07 13 05 08 12 12 11 32 04 28 26 29 16 11 44 31 36 17 46 40	5	
1 09 19 53 44 58 13 47 33 25 41 00 59 40 24 57 36	6	
1 10 38 33 09 03 45	7	
1 13 54 25 28 17 40 25 18 37 59 52 58 06 57 07 59 00 44 26 40	8	
1 14 14 29 13 12 43 52 58 10 47 45 14 03 45	9	
1 17 40 20 16	10	
1 21 37 45 36 55 17 45 36	11	
1 23 10 22 07 26 32 44 29 37 44 03 45	12	
1 25 47 16 41 53 18 39 58 35 36 57 36	13	
1 27 00 59 01 14 19 53 11 10 13 14 59 53 42 41 10 28 37 25 16 02 57 46 40	14	25-place
1 27 24 36 17 06 06 18 54 02 42 14 14 34 29 52 44 56 29 03 45	15	
1 29 45 08 39 39 55 50 37 43 46 35 05 45 55 07 56 41 51 21 36	16	
1 31 26 58 06 25 11 06 40	17	
1 36 06 30 14 03 45	18	
1 40 32 59 08 12 10 44 16	19	
1 41 00 16 49 24 21 47 21 33 31 48 59 03 45	20	
1 45 40 20 17 08 51 07 38 15 56 09 36	21	
1 47 40 13 23 56 10 05 45 40 44 26 40	22	
1 51 03 20 55 31 30 54 41 17 04 22 03 25 03 21 36	23	
1 53 09 20 29 38 53 33 03 41 29 03 45	24	
1 58 23 05 26 21 09 41 04 11 51 06 40	25	
1 58 55 13 36 42 19 18 54 52 21 40 41 03 20 26 44 00 14 03 45	26	
2 04 24 57 36	27	
2 10 09 55 50 45 36 24 12 52 17 08 16	28	
2 10 45 15 53 38 53 09 36	29	
2 12 37 36 12 42 56 59 42 05 30 01 33 07 07 23 00 12 43 01 32 19 02 17 31 10 52 24 01 58 31 06 40	30	33-place
2 13 13 36 20 47 53 49 21 23 08 00 58 03 45 52 44 03 45	31	
2 16 47 48 37 09 18 38 24 36 08 29 16 34 10 33 36	32	
2 19 23 00 22 46 13 57 13 48 20 57 36 47 24 26 40	33	
2 23 45 57 35 56 19 46 06 08 54 19 55 05 57 02 13 24 05 45 36	34	
2 26 29 03 45	35	
2 33 15 13 21 42 41 46 14 38 16 50 41 58 31 06 40	36	
2 33 56 49 27 18 02 52 02 45 14 03 45	37	
2 41 03 40 35 05 51 21 36	38	
2 44 06 24 01 30 43 49 43 23 06 37 28 48 12 43 47 09 37 46 40	39	
2 49 15 59 44 24 02 16 01 41 22 33 36	40	
2 52 28 01 42 59 04 37 44 03 45	41	
3 00 26 13 54 18 46 25 13 51 02 59 42 30 09 36 07 54 04 04 26 40	42	
3 09 37 46 40	43	
3 19 17 25 21	44	
3 28 29 59 56 16 25 20 16 16 53 45 36	45	
3 29 26 35 50 42 15 48 22 48 59 03 45	46	
3 32 26 32 31 51 09 38 40 37 50 37 47 19 20 04 25 48 47 17 51 36 17 46 40	47	25-place
3 39 07 19 30 16 37 39 56 01 10 22 24 09 12 57 36	48	
3 43 15 55 08 38 31 06 40	49	
3 54 38 22 43 14 18 41 29 03 45	50	

sq. sq. *n*		
4 05 28 58 07 36 17 46 40	51	
4 17 59 20 27 48 05 45 36	52	
4 22 52 01 46 29 04 57 39 34 04 26 40	53	
4 31 07 56 28 41 04 55 14 04 24 57 36	54	
4 35 00 53 12 03 48 46 36 47 07 03 41 53 27 30 07 55 39 00 35 54 32 58 36 02 57 46 40	55	29-place
4 36 15 32 27 22 59 57 30 48 32 45 00 52 44 03 45	56	
4 49 01 31 47 41 49 11 26 25 11 06 40	57	
5 03 45	58	
5 19 13 28 43 33 17 45 14 03 45	59	
5 33 58 35 57 53 54 10 48 20 44 09 36	60	
5 40 17 29 45 16 46 43 23 37 38 59 05 40 44 26 40	61	
5 50 59 28 21 24 47 19 30 28 46 23 47 05 51 21 36	62	
5 57 37 40 19 37 44 03 45	63	
6 14 09 16 26 59 28 23 08 04 51 54 24 11 51 06 40	64	
6 15 50 50 25 38 12 09 24 32 09 15 14 56 29 03 45	65	
6 33 12 57 36	66	
6 40 39 03 48 41 30 35 50 50 33 45 05 52 04 49 42 42 57 46 40	67	
6 53 14 54 40 39 56 09 36	68	
7 01 03 44 30 10 38 15 14 59 46 48 59 03 45	69	
7 12 20 44 01 52 23 21 23 40 53 29 33 50 58 33 36	70	
7 14 18 05 47 03 38 14 52 52 48 21 36	71	
7 20 31 13 47 31 18 10 30 17 59 34 40 43 09 50 56 47 24 26 40	72	
7 22 30 48 41 34 39 28 11 06 11 19 36 17 08 45 47 30 57 07 44 03 45	73	
7 42 57 46 40	74	
8 04 21 26 40 28 16 27 38 06 54 43 11 55 19 04 01 58 31 06 40	75	
8 06 32 55 33 41 29 03 45	76	
8 29 01 59 22 46 39 21 36	77	
8 38 39 29 15 53 25 11 13 24 53 31 47 19 46 53 56 27 28 17 07 09 37 46 40	78	25-place
8 54 57 57 41 48 33 49 54 58 10 33 36	79	
9 05 04 52 49 55 36 06 40	80	
9 30 16 14 19 03 46 42 42 17 38 21 18 01 59 14 11 38 18 29 27 54 04 26 40	81	
9 32 51 02 30 05 38 36 07 26 15 52 44 03 45	82	
9 59 19 23 47 09 37 46 40	83	
10 29 51 21 36	84	
10 41 45 55 53 43 20 47 01 12 38 30 25	85	
10 58 57 46 28 13 23 02 35 09 41 45 36	86	
11 01 56 39 12 50 36 37 21	87	
11 11 25 22 04 22 26 02 14 20 20 45 21 24 48 37 42 19 22 49 02 21 22 51 11 36 17 46 40	88	
11 32 32 02 23 05 53 06 57 02 57 58 12 38 00 57 36	89	
11 45 37 43 10 16 33 08 28 38 31 06 40	90	
12 21 34 37 44 03 45	91	
12 55 50 48 53 39 53 57 51 36 17 46 40	92	
12 59 21 2521 57 37 00 58 56 29 56 29 03 45	93	
13 35 22 21 42 40 53 45 36	94	
13 50 47 24 07 39 19 22 58 23 14 47 14 34 04 26 40	95	
14 16 54 43 41 01 41 28 38 33 12 57 36	96	
14 29 10 57 01 49 49 42 52 33 21 34 39 03 46 10 47 16 37 21 53 29 26 26 41 20 28 31 56 02 57 46 40	97	33-place
14 33 06 53 41 21 34 41 01 48 59 03 45	98	
14 56 30 42 20 35 17 25 27 07 19 14 33 08 04 18 36 03 27 21 36	99	
15 13 27 47 53 42 32 30 13 52 11 24 46 25 11 06 40	100	

Appendix 10
Color Photos of 70 Selected Texts

1.1.1 b. MS 2728, one of two linked triples of consecutive OB multiplication exercises.

1.2.1 a. MS 2831. OB computations of the squares of 5 round or almost round numbers.

1.1.3. MS 3955 *obv*. Four OB multiplication exercises with funny numbers.

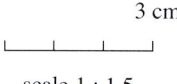

3 cm

scale 1 : 1.5

1.3.1. MS 3871. An OB division exercise: 11 34 26 40 / 4 37 46 40 = 2 30.

1.4 .1 a. MS 2242. An OB factorization of 46 20 54 51 30 14 03 45 (the sixth power of 3 45) by use of the trailing part algorithm.

1.4.5. MS 3264. (The upper right corner is a late addition to a damaged clay tablet.)
1) 1 01 30 33! 45 (the square of 5, tripled 12 times).
   Its reciprocal (not correct): 1 27 47 29 23 57 46 40 (the 9th power of 5, doubled 21 times).
2) 1 30 48 06 02 15 25, a 7-place regular sexagesimal number (5, tripled 25 times).
   Its reciprocal: 39 38 48 38 28 37 02 08 43 37! 09 43 15 53 05 11 06 40, an 18-place regular
   sexagesimal number written in one line on the obverse and two on the reverse.

3 cm

scale 1 : 1.5

1.5.1 a. MS 2318/2. The first line of an OB square
side algorithm. The number 6 in the left margin is
the square side of 36, the last place of the regular
sexagesimal number 1 59 34 27 12 36.

1.5.1 b. MS 2731. An OB computation of the square of
 a number with a funny number (7 07) as a factor.

1.5.5. MS 2351. The regular many-place number 13 22 50 54 59 09 29 58 26 43 17 31 51 06 40,
written in two lines on the obverse and continued onto the reverse. The square of the square of
14 48 53 20 (the 5th power of 20).

2.1.1 a. MS 2794. An OB abridged standard table of squares.

3 cm

2.1.2 a. MS 3937. An OB special table of squares. (Squares of successive multiples of 5.)

2.2.1 b. MS 2185. An excerpt from an OB table of square sides (lines 21-40). Type a'. Colophon (subscript): ^dNisaba.

2.3.2 a. MS 3973/1. An excerpt from an OB table of cube sides (lines 31-100). Type b.

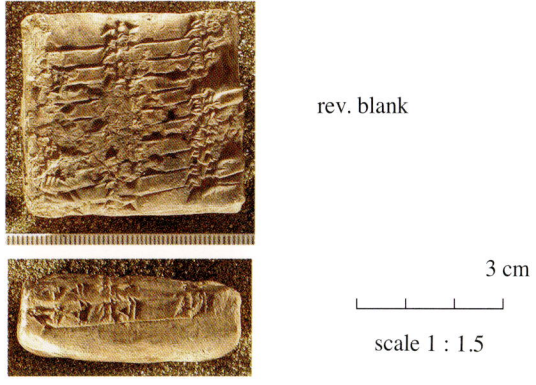

rev. blank

3 cm

scale 1 : 1.5

2.4.1 a. MS 3899. An excerpt from an OB table of $n \cdot n \cdot (n + 1)$ sides (lines 1-12.)

2.4.1 b. MS 3048. An excerpt from an OB table of $n \cdot (n + 1) \cdot (n + 2)$ sides (lines 1-30.)

2.5.1. MS 3874. A large fragment of an OB standard table of reciprocals (28 lines), with a colophon. Variant number signs for 4, 6, 7, 8, and 40, and an early OB form of the sign for .bi.

3 cm

2.5.2 b. MS 3890. A standard table of reciprocals of type a (27 lines), mutilated in antiquity by an irate teacher. Standard OB number signs.

2.6.1 b. MS 2184/3. A single multiplication table (12 ×) of type a'.

2.6.1 c. MS 3044/3. A single multiplication table (45 ×) of the concise type b* (23 lines).

3 cm

scale 1 : 1.5

2.6.10. MS 3845. A large fragment of an Old Babylonian multiple multiplication table of type a (11 sub-tables).

2.6.14. MS 3849. An atypical single multiplication table (50 ×) of type c.

scale 1 : 1.5

3 cm

2.6.12. MS 3974. A combined multiplication table of type c with 40 sub-tables.

2.6.13. MS 3974, *rev*.

face 1   *face 2*   *face 3*   *face 4*

face 5   *face 6*

3 cm

scale 1 : 1.5

3.1.3. MS 2723. A 6-sided prism with a metrological table for system *C*, basic unit sìla, from '1 shekel of barley = 1 ($\cdot 60^{-1}$ sìla)' to '2 $\cdot 60^{3}$ gur = 10 ($\cdot 60^{4}$ sìla)', altogether 159 lines.

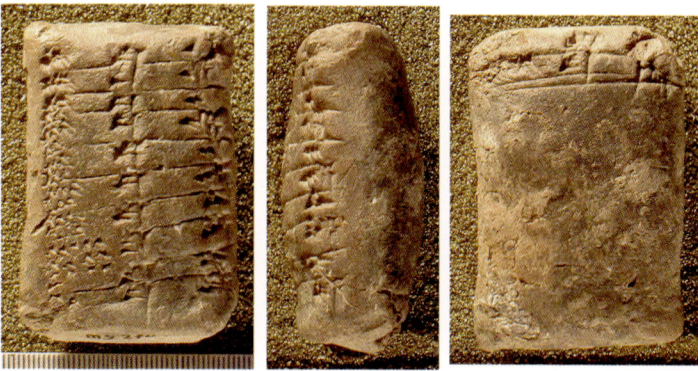

3.1.2 c. MS 2704 . A metrological table for system C, basic unit sìla (12 lines), from '2 · 60^2 gur = 10 (· 60^3 sìla)' to '13 · 60^2 gur = 1 05 (· 60^3 sìla)'.

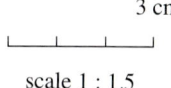

3 cm

scale 1 : 1.5

3.2.4. MS 3925 . A brief metrological table for system M, basic unit shekel Exceeds the usual range for such tables. The 4 lines on the obverse are repeated on the reverse, in a different hand.

3 cm

scale 1 : 1.5

3.2.1 a. MS 2186. Metrological table
for system M, from '1/2 barley-corn =
10 ($\cdot\, 60^{-3}$ mina)' to '1 shekel = 1 ($\cdot\, 60^{-1}$ mina)'.

3 cm

scale 1 : 1.5

3.3.1 a. MS 2735. A metrological table for system *A*, from '1/2 iku = 50 (sq. ninda)' to '1 èše = 10 (· 60 sq. ninda)'.

3.3.1 b. MS 2768. A metrological table for system *A*, from '1 šar = 30 (· 60^2 sq. ninda)' to '10 šar = 5 (· 60^3 sq. ninda)'.

3.4.1 a. MS 3869/11. A metrological table for system *Ln*, from '1 finger = 10 (· 60^{-2} ninda)' to '5 ninda = 5 (· 1 ninda)'.

3.4.1 b. MS 2705. A metrological table for system *Ln*, from '1 uš = 1 (· 60 ninda)' to '1 danna = 30 (· 60 ninda)'.

3.5.1. MS 3869/13. A fragment of an OB combined metrological table for systems *A* (107 lines), *Ln* (92 lines), and *Lc* (72 lines).

3 cm

scale 1 : 1.5

3.5.3. MS 3869/14, *obv*. A multiple metrological table for systems *M* and *A*.

3.5.4. MS 3869/14, *rev*. A multiple metrological table for systems *M* and *A* (continued).

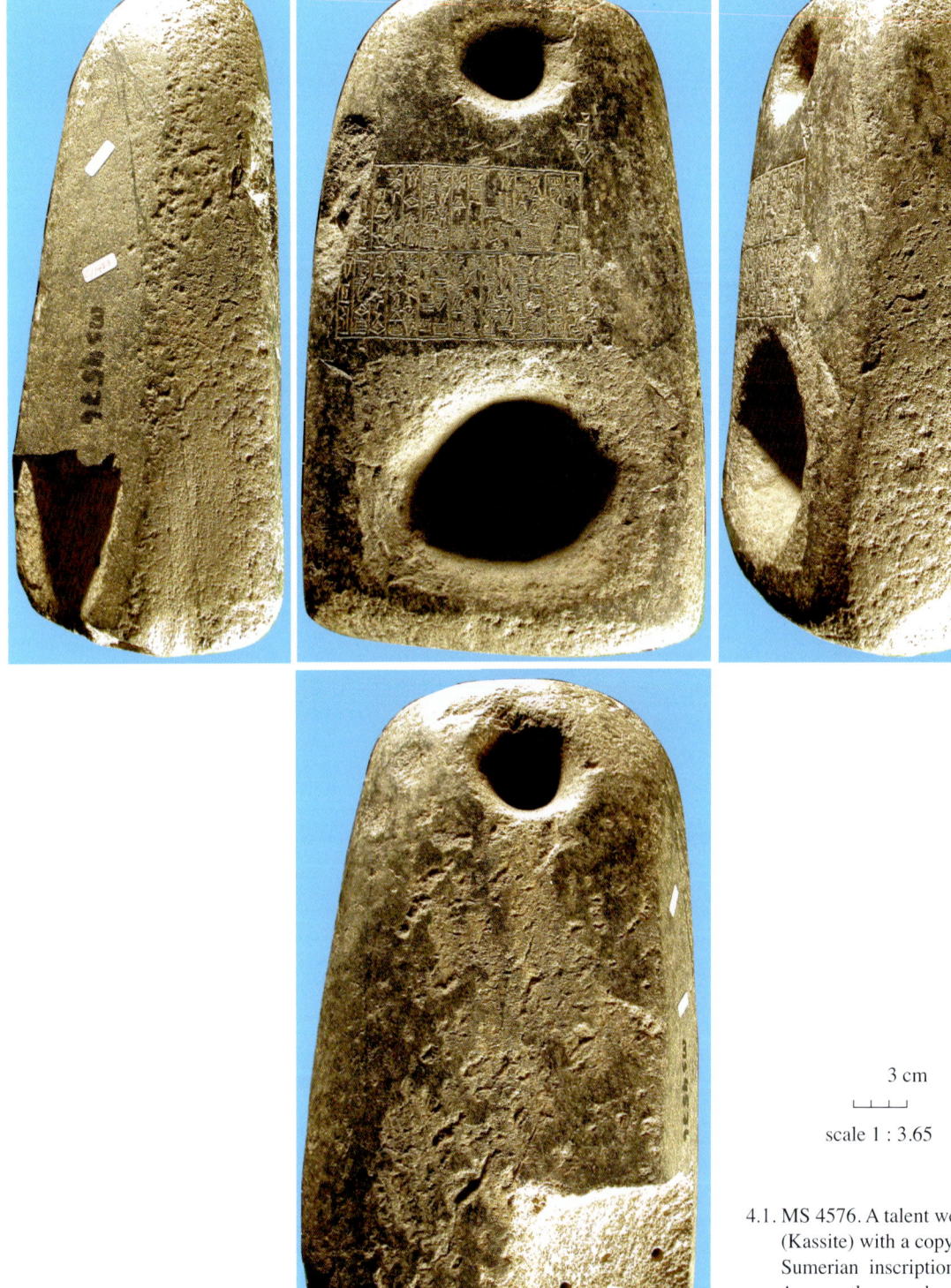

3 cm

scale 1 : 3.65

4.1. MS 4576. A talent weight (Kassite) with a copy of a Sumerian inscription. Apparently reused as a door socket.

4.3. MS 2481. Diorite weight, 1 mina. 478.2 g, 5.3 × 12 cm.
Shaped like a stretched barrel. With an inscription in 3 lines.

3 cm

scale 1 : 1.5

Hand copies: F. Al-Rawi

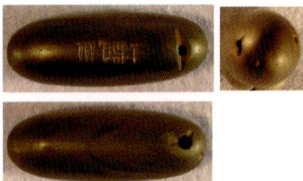 

4.4. MS 2837. Hematite weight, 3 shekels. 24 g, 1.5 × 4.1 cm.
Ellipsoidal. Pierced for a string. Inscription: 3 gín.

4.5. MS 2836. Agate weight, 1/3 shekel. 3g, 0.9 × 1.8 × 0.9 cm.
A duck with its head turned back. Pierced for a string. Inscription: 3' gín.

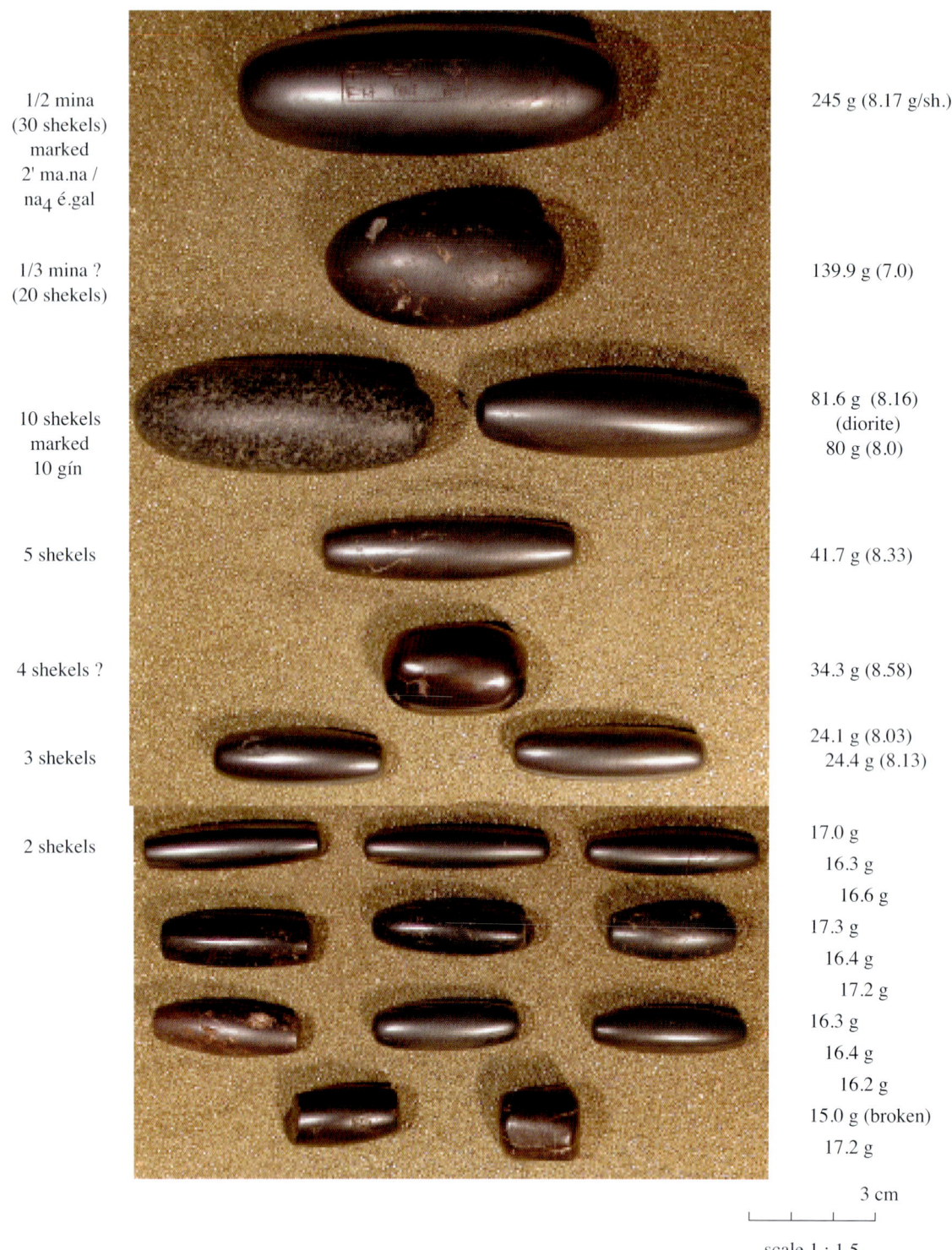

1/2 mina (30 shekels) marked 2' ma.na / na$_4$ é.gal		245 g (8.17 g/sh.)
1/3 mina ? (20 shekels)		139.9 g (7.0)
10 shekels marked 10 gín		81.6 g (8.16) (diorite) 80 g (8.0)
5 shekels		41.7 g (8.33)
4 shekels ?		34.3 g (8.58)
3 shekels		24.1 g (8.03) 24.4 g (8.13)
2 shekels		17.0 g 16.3 g 16.6 g 17.3 g 16.4 g 17.2 g 16.3 g 16.4 g 16.2 g 15.0 g (broken) 17.2 g

3 cm

scale 1 : 1.5

4.6 a. MS 5088/1-19. Weight stones from a bronze pot, ranging from 1/2 mina to 2 shekels.

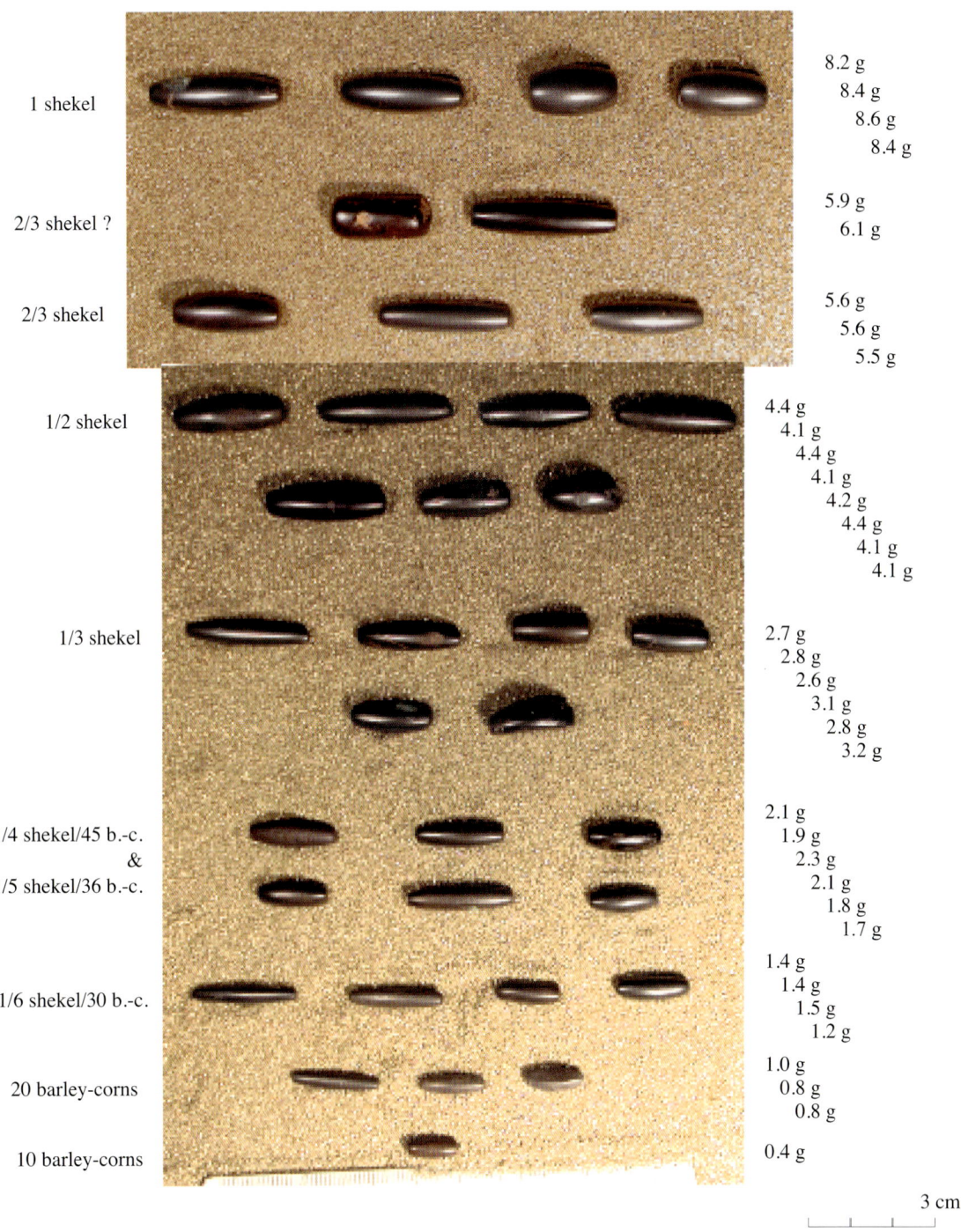

1 shekel	8.2 g 8.4 g 8.6 g 8.4 g
2/3 shekel ?	5.9 g 6.1 g
2/3 shekel	5.6 g 5.6 g 5.5 g
1/2 shekel	4.4 g 4.1 g 4.4 g 4.1 g 4.2 g 4.4 g 4.1 g 4.1 g
1/3 shekel	2.7 g 2.8 g 2.6 g 3.1 g 2.8 g 3.2 g
1/4 shekel/45 b.-c. & 1/5 shekel/36 b.-c.	2.1 g 1.9 g 2.3 g 2.1 g 1.8 g 1.7 g
1/6 shekel/30 b.-c.	1.4 g 1.4 g 1.5 g 1.2 g
20 barley-corns	1.0 g 0.8 g 0.8 g
10 barley-corns	0.4 g

3 cm

4.6 b. MS 5088/20-55. Weight stones from the same bronze pot, ranging from 1 shekel to 10 barley-corns.

5.1. MS 1984. A Neo-Sumerian field plan with a central field and five peripheral regions.

3 cm

scale 1 : 1.5

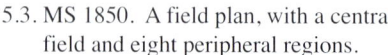

6.2.1. MS 3047 *obv*. Before cleaning.

5.3. MS 1850. A field plan, with a central
field and eight peripheral regions.

6.2.1. MS 3047 *obv./rev.* Two metro-mathematical tabular exercises from the early Sumerian ED III period.

7.2.1 b. MS 2832. A market rate exercise
with regular data.

7.2.2 a. MS 2268/19. A market rate exercise
with non-regular data.

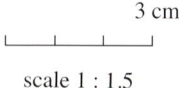

3 cm

scale 1 : 1.5

7.1.1. MS 2317. A division
exercise for the funny number
1 01 01 01.

7.3.1. MS 2221. *Obv.*: Computations of four carrying numbers for bricks and mud.
      Rev.: A computation of a combined work norm for carrying bricks.

7.4.2. MS 1844. A numerical
solution algorithm for an
inheritance problem. The
shares form a geometric
progression.

3 cm

scale 1 : 1.5

8.1.2 a. MS 2107. A trapezoid with an almost round area number.

3 cm

scale 1 : 1.5

8.1.2 b. MS 3908. A trapezoid divided into three stripes,
and a complete set of associated numerical parameters.

8.1.12. MS 1938/2, *obv*. A trapezoid divided into six stripes, the areas forming an arithmetical progression.
8.2.14. MS 1938/2, *rev*. A circle inscribed in a regular hexagon, a certain distance away from the sides of the hexagon.

8.1.1 c. MS 3051. An equilateral triangle inscribed in a circle.

3 cm

scale 1 : 1.5

8.2.2 a. MS 2192. An equilateral triangular band
divided into a chain of trapezoids.

8.1.1 b. MS 2985. A circle inscribed in a square,
a certain distance away from the sides of the square.

8.2.2 b. MS 3050. A square with diagonals, inscribed in a circle.

8.3.1. MS 4515. An Old Babylonian square labyrinth with two paths, one good and one bad.

8.3.13. MS 4516. A geometric theme text with 8 assorted mazes.

3 cm

scale 1 : 1.5

8.3.14 a. MS 3940. A geometric pattern
superimposed on a dense grid of guide lines.

3 cm

scale 1 : 1.5

8.3.9. MS 3194. An Old Babylonian rectangular labyrinth with two paths, one good and one bad.

8.3.14 b. MS 3031. A drawing of
the house plan for Nur Adad's
palace at Larsa.

9.1. MS 1686. An Ur-Isin king list.

3 cm

scale 1 : 1.5

9.3. MS 2855. An antediluvian
      Sumerian king list.

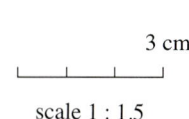

10.1.1. MS 3971, an OB double-
 column mathematical
 recombination text
 from Uruk.

3 cm

scale 1 : 1

3 cm

scale 1 : 1.5

10.2.2, 10.2.13. MS 3052. An Old Babylonian mathematical recombination text with 8 problems belonging to 4 separate themes.
Obv. and upper part of *rev.*: Five problems for mud walls with breaches and/or with holes drilled through them,
all leading to interesting metric algebra problems.
Rev., lower part: Three badly preserved problems for a 'diagonal', an 'excavation', and a 'square'.

10.3.1, 10.3.5. MS 2792. Two Old Babylonian problems for top layers added to a ramp built by four officers and their men.

3 cm

scale 1 : 1.5

3 cm

scale 1 : 1

11.1.1, 11.1.4. MS 3049. A Kassite(?) fragment of a mathematical recombination text, with the Old Babylonian diagonal rule as its
general topic, but with five separate themes, according to a summary in the subscript.
The first theme is circles with chords (on the obverse). The last theme is the "interior diagonal" of a gate in a wall (on the reverse).

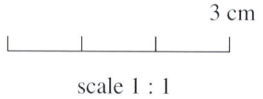

3 cm

scale 1 : 1

11.2.1, 11.2.9. MS 5112.
A large fragment of a Kassite
mathematical recombination text
with originally 23 problems.

Obv.: Metric algebra problems for
one or more squares.
Rev.: Metric algebra problems for
the sides of a rectangle.

3 cm

scale 1 : 1

11.3.1. MS 3876. A clay
tablet of an unusual format,
Late Kassite.

Three related problems for
20 equilateral triangles and
a 'horn-figure' made in copper.

App. 4 i. MS 4632.
A spherical envelope and
all the tokens it contained,
except for 1 small ball which
still sticks to the inside of the
spherical envelope.

Photo: P. Damerow

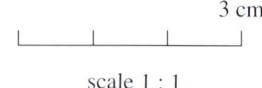

3 cm

scale 1 : 1

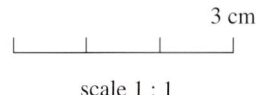

A7.1. CUNES 50-08-001, *obv*. A combined metro-mathematical table text with areas of squares (ED IIIb).

Photos copyright, Cornell University, Dept. of Near Eastern Studies.
I would like to thank Lisa Bajwa for making the extraordinary photos,
and David I. Owen, curator of tablet collections, for providing them.
(The clay tablet is dusted with ammonium chloride, for better contrast.)

3 cm

scale 1 : 1

A7.2. CUNES 50-08-001, *rev.*

3 cm

scale 1 : 1

A7.3. CUNES 47-12-176. A decreasing list of weight measures, possibly an extract from a lexical text
(ED IIIb/Old Akkadian).

Photos copyright, Cornell University, Dept. of Near Eastern Studies. I would like to thank Lisa Bajwa for
making the extraordinary photos, and David I. Owen, curator of tablet collections, for providing them.

A7.6. IM 96183.
Photo: F. Al-Rawi.

3 cm

Vocabulary for the MS Texts

A. Akkadian Terms (Cf. Black/George/Postgate, *A Concise Dictionary of Akkadian* (1999).)

adi	until	
a-di		MS 3052 § 1
		MS 5112 § 5
agurru(m)	square brick	
a-gú-ru		MS 2221
aḫāru(m)	to be behind(?)	
a-ḫa-ri-im		MS 3876 # 3
akālu(m)	to eat	
šu-ta-ki-il (multiply)		MS 3049
amāru(m)	to see	
a-ma-ri-k[a]		MS 3971 § 3
		MS 3052 § 1
		MS 2792
ta-mar, ša ta-mu-r[u]		MS 3049
		MS 3876
ana	to, for, *etc.*	
a-na		*passim*
annû(m)	this	
an-ni-im		MS 3876 # 3
appu(m)	nose, tip, rim	
ap-pi		MS 3876 # 3
arammu(m)	ramp	
a-ra-mu		MS 2792
arḫu(m)	cow; half brick	
ar-ḫu		MS 2221
aššum	in order to	
aš-šum		MS 3971 § 3
		MS 3052 § 1
		MS 2792
		MS 5112
bāmtu(m)	halfpart	
ba-ma-at, ba-am-ta		MS 5112 § 5
		MS 3876 # 3
banû(m)	to build	
ab-ni		MS 5112 § 6
ib-ni-šu-nu-ti		MS 3876 # 2
ebēru(m)	to stretch across	
e-bi-ir (cross-section)		MS 3052 § 1
erēbu(m)	to enter	
e-ru-ub		MS 3049 § 5
elēnu(m)	above, over	
e-le-nu		MS 3052 § 1
elû(m)	to come up	
šu-li		MS 3049
i-il-a-kum		MS 5112 § 6
uš-te-li		MS 3876 # 2

eperu(m)	mud, earth	
e-pe-rum		MS 2221
epēšu(m)	to do, to make	
li-pé-eš, te-pé-eš		MS 3049 § 5
te-pu-uš, te-pé-eš		MS 3876
erēšu(m)	to request	
er-ši-šu		MS 3052 § 1 d
eṣēpu(m)	to double	
e-ṣí-ip		MS 3049 § 1
te-ṣi-im-ma, e-ṣi-im-ma		MS 3876
gašāru(m)	to become strong	(be repaired)
ag-šu-ur, ug-šu-ur,		
ag-šu-ra-am-ma		MS 3052 § 1
ḫarāmu(m)	to cut off	
uḫ-ra-am, ta-aḫ-ra-am-[mu]		MS 3052 § 1
ḫarāṣu(m)	to break off	
kud-aṣ		MS 3052 § 1
ḫu-ru-uṣ		MS 5112 § 3
ḫepû(m)	to break	
2' (a) ḫe-ep-pe-e	take 1/2 of *a*	MS 3971
(a) ḫe-pé-e		MS 3049
2' (a) ḫe-pe, ba-ma-at (a) ḫe-pe		MS 3876
igāru(m)	wall	
i-ga-ri-im, i-ga-ri		MS 3049 § 5
igigubbû(m)	constant	
igi-gub-e		MS 3876
ina	from	
i-na		*passim*
indanu(m)	ninda(?) (rate of change)	
in-da-nim, in-da-nu		MS 3052 § 1
		MS 2792
išdu(m)	base	
iš-di-im		MS 3052 § 1
išten	one	
iš-te-en		MS 2792
		MS 3876 # 3
ištu	from	
iš-tu		MS 3052 § 1
		MS 2792 # 2
itti	with	
it-ti		MS 5112 §§ 6, 12
-ka	your	
		MS 3971 § 3
		MS 3052 § 1
		MS 3876 # 3

kamāru(m)	to heap (join together)	
ku-mu-ur		MS 3049 § 5
kapāpu(m)	to curve	
[ak-pu-up]		MS 3049 § 1
kabāru(m)	to be thick	
ik-bi-ir		MS 3052 § 1
kī maṣi	how much?	
		MS 3052 § 1
		MS 3876 # 3
kīma	like	
ki-ma		MS 3049 § 1
		MS 5112 § 8
kīa (or *kiyā*)	how much?	
ki-ia-a		MS 3971 § 1
kīam	so	
ki-a-am		MS 3049
		MS 3876 # 3
kubru(m)	thickness	
ku-ub-ru-ša		MS 3052 § 1 c
kuburrû(m)	thickness	
ku-bu-ri		MS 3049 § 5
kullu(m)	to hold	
re-eš-ka li-ki-il		MS 3052 §§ 1, 4
		MS 5112 *obv.*
ša re-eš-ka ú-ka-a[l-lu]		MS 5112 § 7
-kum	to, for you	
qà-bu-kum		MS 5112 *obv.*
lapātu(m)	to inscribe	
lu-pu-ut, ta-la-pa-at		MS 3052 § 1
		MS 5112
leqû(m)	to take	
le-qé-e		MS 3049 § 4
		MS 5112 § 3
te-le-qé		MS 3876 # 3
li-, lu-	let	
		MS 3052 § 1
libittu(m)	mudbrick	
li-bi-tum		MS 2221
lū	let	
lu		MS 3052 § 1
-ma	and, then	
		passim
mahāru(m)	to be equal	
im-ta-aḫ-ru		MS 3876
matnu(m)	string	
ma-at-nim		MS 3049 § 1
meḫru(m)	copy	
me-eḫ-ra-am		MS 3049 § 1
mimma	anything	
mi-im-ma ú-la	zero	MS 3052 § 1 d
minu(m)	what?	
mi-na, mi-[nu]		MS 3971 § 1
		MS 3049 § 1
		MS 3876 # 3
mi-nam		MS 5112
mišlu(m)	half	
mi-ši-il		MS 3876 # 3

muḫḫu(m)	top	
mu-uḫ-ḫi, mu-uḫ-ḫi-im		MS 3052 § 1
muttarrittu(m)	descent (perpendicular)	
mu-ut-ta-ri-tum, mu-ut-ta-ri-tim		MS 3049 § 1
nadanu(m)	to give	
at-ta-di, at-ta-an-din,		
at-ta-an-din-nu		MS 3971 § 1
nadû(m)	to lay down	
i-di		MS 3049 § 1
nalbattu(m)	brick mold (rectangle?)	
na-al-ba-tum		MS 3049 § 4
nasāḫu(m)	to tear off	
ta-na-sa-aḫ		MS 3876
našû(m)	to lift	
i-ši		MS 3049
		MS 3876
nazbalu(m)	carrying	
na-az-ba-lam		MS 2792
nēpešu(m)	doing, procedure	
né-pé-[šum]		MS 3049 § 1
palāku(m)	to mark out	
p[u-lu-u]k-šu-nu-ši-im		MS 3052 § 1
palāšu(m)	to drill through	
ap-lu-ší		MS 3052 § 1
pānû(m)	earlier	
pa-nu, pa-ni, [pa]-ni-tum		MS 5112 § 9
parāṣu(m)	to breach	
pa-ri-iṣ		MS 3052 § 1
pašāru(m)	to release, solve	
pa-šar		MS 3049 § 1
perṣu(m)	breach	
pe-er-ṣum, pe-er-ṣi, pe-er-ṣa-am		MS 3052 § 1
pilku(m)	marked off sector	
pi-il-kum		MS 2792
qabû(m)	to say	
qà-bu		MS 5112
qaqqaru(m)	ground (area)	
qá-qá-ar-ka, qá-qá-ra		MS 3876 # 3
rebû(m)	fourth	
re-bu-ú		MS 2792
rēšu(m)	head	
re-eš-ka li-ki-il		see *kullu(m)*
saḫāru(m)	to turn back (become narrow)	
su-ḫu-ur, suḫ?-ur?		MS 3052 § 1
samnu(m)	eighth (1/8)	
sa-am-na-tim, sa-am-na-ti-šu		MS 3876
ṣiliptu(m)	crossed over (diagonal)	
ṣi-li-ip-tum		MS 3971 §§ 2, 4
		MS 3052 § 2
ṣi-l[i-i]p-ti, ṣi-il-pa-tum		MS 3971 § 3
sittu(m)	remainder	
si-tum		MS 5112
si-i-tam(?) *or* si-ni-tam(?)		MS 3876
ša	that which	
		passim
-ša	its	
		MS 3052 § 1

šakānu(m)	to set	
ša-ak-na-at, ša-ak-nu		MS 3049 § 1
šalšu(m)	third	
ša-al-šu		MS 2792
šalšatu(m)	third (MB)	
3'-*ti*		MS 3049 § 4
šanû(m)	second	
ša-nu-ú		MS 2792
ša-ni-tim		MS 3876
šapiltu(m)	remainder	
ša-pi-il-ti		MS 5112 § 6
šâti	that	
ša-ti		MS 3876 # 3
šâtu(m)	the same(?)	
ša-a-tam		MS 5112 § 5
šina	two	
ši-na		MS 3049 § 1
		MS 3876 # 1
šinipât	two-thirds	
ši-ni-pa-at		MS 3049 § 4
šīpu(m)	metal-covered(?)	
ši-pi-e		MS 3876 # 3
-*šu*	its, -th	
		MS 3971 § 1
		MS 2792
		MS 3049
		MS 5112
šū	that	
šu-ú		MS 3971 § 1 d
šumma	if	
šum-ma		MS 3049 § 5
		MS 3876 # 3
šuāti	that, the same	
šu-a-ti		MS 3052 § 1 d
-*šunūti*	them	
-*šu-nu-ti*		MS 3049 § 5
		MS 3876 # 2
târu(m)	to turn back	
tu-úr-ma		MS 3049 § 1
ù	and, plus	
		passim
ubānu(m)	finger	
ú-ba-nim		MS 3049 § 4
-*ša*	its	
		MS 3052 § 1
		MS 3049 subscr.
ula	not	
ú-la		MS 3052 § 1 d
uššu(m)	foundation	
uš-šu		MS 3052 § 1 d
(w)arādu(m)	to go down	
ur-dam		MS 3049 § 1
(w)āṣītu(m)	extension	
wa-ṣi-tam		MS 5112 §§ 1, 10
zabālu(m)	to carry	
i-za-bi-lu]		MS 2792

B. Sumerian Terms / Sumerograms

(Cf. *Pennylvania Sumerian Dictionary* at
<http://psd.museum.upenn.edu/epsd/index.html>.)

ak	to do	
		MS 3052 § 1
		MS 5112
.àm	suffix for numbers and measures	
		passim
a.na	whatever, as much as(?)	
		MS 3052 § 1 d
an.na	upper	
		MS 3971 § 1
		MS 3052 § 1
		MS 2792
an.ta	upper	
		MS 3049 § 1
a.rá	steps(?) (times)	
		MS 5112
		MS 3876
a.šà	field (area)	
		MS 3971
		MS 3052 § 2
		MS 5112
aša₅ (or gán)	field (figure)	
		MS 3052 § 1 d
		MS 5112 §§ 7, 11
		MS 3876
bàd	city wall (hexagon?)	
		MS 3876 # 3
ba.si₈	"likeside" (square root)	
		MS 3049
.bi	its	
		MS 3049 subscr.
		MS 3876 # 3
bar	cross-over(?) (diagonal)	
		MS 3049 § 5
bùr	30 00 sq. ninda	(c. 64,800 m²)
		MS 3971 § 1
dagal	width	
		MS 3052 § 1
dagal.la		MS 3049 § 5
daḫ	to join (add to sth.)	
		MS 3971
		MS 3052
		MS 5112
dal	transversal	
		MS 2792
		MS 3049 § 1
dar	to crush	
2' (*a*) dar	take 1/2 of *a*	MS 5112
diri	to be beyond	
		MS 3052 § 1
		MS 2792
		MS 2830
		MS 5112 § 11

du$_8$	to release (compute a reciprocal)	
		MS 3971
		MS 3052
		MS 2792
		MS 5112
ba.du$_8$		MS 5112 § 5
du$_7$.du$_7$	to butt (multiply)	
		MS 3971
		MS 3052
		MS 2792
.e	this(?)	
		MS 3049 § 5
.e	ergative suffix(?)	
		MS 3971
		MS 3052 § 1
		MS 2792
egir	later	
		MS 5112 §§ 9, 13
en.nam	what?	
		passim
èše	10 00 sq. ninda (c. 21,600 m^2)	
		MS 3971 § 1
érin	soldier	
		MS 2792
gán (or aša$_5$)	field (figure)	
		MS 3876
gán.si	horn-figure (icosahedron?)	
		MS 3876 # 3
gán.za.na	gaming-piece-field (equilateral triangle)	
		MS 3876
gar.gar	to heap (join together)	
		passim
gar.ra	set	
		MS 2792
		MS 3876
gaz	to break	
2′ (a) gaz	take 1/2 of *a*	MS 3971 § 2, 3
		MS 3052
		MS 2792
gi	reed (c. 3 m)	
		MS 3971 § 1
gín	shekel (1/60)	
		MS 3971 § 1
		MS 2830
		MS 3876 # 3
gi.na	true, correct	
		MS 3971 § 1
gi$_4$.na		MS 3876 # 1
gú I	talent	
		MS 3876 # 3
gú II	edge	
		MS 3052 § 1
gu$_7$.gu$_7$	to eat (to square)	
		MS 5112
gúr	arc (circle)	
		MS 3052 § 4

gur	capacity measure	
		MS 3052 § 4
		MS 3876 # 3
ḫa.la	share	
		MS 2830
ḫé.gar	shall set	
		MS 3971 § 1
		MS 3052 § 2
ḫi.a	plural ending	
		MS 2792
		MS 3049 § 4
íb.si$_8$	"equalsided" (square, square root)	
		MS 3971
		MS 3052
		MS 5112
igi I	opposite (reciprocal)	
igi *(a)* du$_8$		*passim*
igi-*a-šu*		MS 5112 § 5
igi *ù* igi.bi (or *igi-bé-e*)		MS 3052 § 2
		MS 3971 § 3
igi II	earlier	
		MS 5112 § 13
igi.(*a*).gál	"the opposite of *a*" (the *a*-th part)	
		MS 2830
		MS 1844
		MS 5112
		MS 3876 # 3
igi.bi	"its opposite" (its reciprocal)	
		MS 3971 § 3
(or *igi-bé-e*)		MS 3052 § 2
igi.gub	constant	
		MS 3876
iku	1 40 sq. ninda (c. 3,600 m^2)	
		MS 3971 § 2
íl	to raise (multiply)	
		MS 3052 § 1
im.dù.a	mud wall	
		MS 3052
im.gíd.da	long tablet (table text)	
		MS 3049 § 5
im.šu	hand tablet (assignment)	
		MS 3052 subscr.
		MS 5112 subscr.
in.sì	it gives (is the result)	
		MS 3971
		MS 3052
		MS 2792
ká	gate	
		MS 3049 § 5
ki	with	
		MS 2221 § 2
ki.1, ki.2, *etc.*	1st, 2nd, *etc.*	
		MS 3971
		MS 2792
		MS 5112
ki.lá	excavation(?)	
		MS 3052 subscr.
		MS 2792

ki.ta	lower	
		MS 3971
		MS 3052 § 1
kù.babbar	silver	
		MS 2830
		MS 3925
kud	to cut	
		MS 3052 § 1 c
kùš	cubit (c. 1/2 m)	
		passim
ma.na	mina (c. 1/2 kg)	
		MS 2830
.meš	plural ending	
		passim
mu	name	
		MS 3049 § 4
múr	middle	
		MS 3052 § 1
.ne.ne	plural ending	
		MS 2792
nigin(?)	square	
		MS 3049 subscr.
nim	to lift (multiply with a factor)	
		MS 3971
		MS 3052 § 1
		MS 2792
ninda	rod(?) (c. 6 m)	
		passim
nu	not	
nu.zu	I don't know	MS 3052 § 1 d
		MS 5112 § 5
rá (or du)	to go (multiply)	
		MS 5221
ba.rá		MS 5221 § 5
sag I	front, head (short side)	
		passim
sag II	head (initial value)	
		MS 3971 § 1
sukud	height	
		MS 3052 § 1
		MS 2792
		MS 3049 § 5
sag.kak	peghead (triangle)	
		MS 3049 subscr.
		MS 3876 # 2
saḫar	mud (volume)	
		MS 3052
		MS 2792
		MS 3876 # 3
sig$_4$	brick	
		MS 3049 § 4
suḫuš	base	
		MS 3052 § 1 d

sukud	height	
		MS 3052
		MS 3049
šà	inside	
		MS 3052 § 1
		MS 3049 § 5
šagina	officer	
		MS 2792
šar (or sar)	1 sq. ninda (c. 36 m^2)	
		MS 3971 § 1
		MS 3876 # 3
še	barley	
		MS 3052 § 2
		MS 3876 # 3
.šè	for	
		MS 1844
šeš	brother	
		MS 2830
šu.si	finger (c. 1.6 cm)	
		MS 3971 § 1
		MS 3052 § 1
		MS 2792
		MS 3049
šu.nigin	together	
		MS 5112 subscr.
.ta.ám	each	
		MS 5112
		MS 3876
.ta		MS 3876 # 3
tab	to repeat	
		MS 5112
e.tab		MS 3971 § 2
		MS 3052 § 1
		MS 2792
tab.ba		MS 3876 # 3
téš.a.si	"sameside" (square, square root)	
		MS 5112
tur	small	
		MS 1844
ugu	over	
		MS 3052 § 1
		MS 2792
		MS 2830
		MS 5112 § 11
urudu	copper	
		MS 3876 # 3
uš I	length (long side)	
		passim
uš II	1 00 ninda (c. 360 m)	
		MS 3052
za.e	you	
		MS 3052
		MS 2792
		MS 3049 § 1
		MS 5112

zi	to tear off (subtract)		za.na (see gán.za.na)		
		MS 3971			MS 3876 # 3
		MS 3052			
		MS 2792	zu	to know	
		MS 5112	nu.zu		MS 3052 § 1 d
		MS 3876			MS 5112 § 5
zi.zi		MS 3049 § 1	-zu	your	
					MS 3052 § 1

Index of Subjects

Systematic catalog of discussed mathematical cuneiform texts in the Schøyen Collection

Index of Texts

(Bold type indicates pages with hand copies of the texts, **C** indicates a color photo in App. 10.)

References

0. How to Get a Better Understanding of Mathematical Cuneiform Texts

Friberg, Jöran (1985) "Babylonian Mathematics" in *The History of Mathematics from Antiquity to the Present. A Selective Bibliography* (J. W. Dauben, ed.), New York.

—— (1997) "'Seed and Reeds Continued'. Another metro-mathematical topic text from Late Babylonian Uruk" *BagM* 28, 251-365, pl. 45-46.

—— (2000) "Mesopotamian mathematics" in *The History of Mathematics from Antiquity to Present. A Selective Annotated Bibliography. Revised edition on CD-ROM* (J. W. Dauben, ed.), American Mathematical Society, 128-152.

—— (2001) "La matematica" (in English) in *Storia della scienza, I. La scienza antica*. Rome, 388-408.

Høyrup, Jens (1990) "Algebra and naive geometry. An investigation of some basic aspects of Old Babylonian mathematical thought" *AfO* 17, 27-69, 262-354.

—— (1996) "Changing trends in the historiography of Mesopotamian mathematics: An insider's view" *HS* 34, 1-32.

—— (2002) *Lengths, Widths, Surfaces. A portrait of Old Babylonian algebra and its kin*. New York, Berlin, *etc.*: Springer.

Neugebauer, Otto (1935-37 (reprinted 1973)) *Mathematische Keilschrift-Texte I-III*, Berlin: Springer.

—— and A. Sachs (1945) *Mathematical Cuneiform Texts* (AOS 29) New Haven: American Oriental Society.

Proust, Christine (2004) *Tablettes mathématiques de Nippur (Mésopotamie, début du deuxième millénaire avant notre ère). 1. Reconstitution du cursus scolaire. 2. Édition des tablettes d'Istanbul*. Dissertation, Université Paris 7: Histoire des sciences.

Robson, Eleanor (1999) *Mesopotamian mathematics 2100-1600 BC: Technical Constants in Bureaucracy and Education*. Oxford: Clarendon Press.

Thureau-Dangin, François (1938) *Textes mathématiques babyloniens*, Leiden: E. J. Brill.

1. Old Babylonian Arithmetical Hand Tablets

Arnaud, Daniel (1994) *Texte aus Larsa. Die epigraphischen Funde der 1. Kampagne In Senkereh-Larsa 1933,* Berlin.

Britton, John P. (1991-93) "A table of 4th powers and related texts from Seleucid Babylon" *JCS* 43-45, 71-87.

Friberg, Jöran (1990) "Mathematik" (in English) in *Reallexikon der Assyriologie und Vorderasiatischen Archäologie 7* (Dietz Otto Edzard, ed.), Berlin, New York, 531-585.

—— (1997) "'Seed and Reeds Continued'. Another metro-mathematical topic text from Late Babylonian Uruk" *BagM* 28, 251-365, pl. 45-46.

—— (1997/98) "Round and almost round numbers in proto-literate metro-mathematical field texts" . *AfO* 44/45, 1-58.

—— (1999) "A Late Babylonian factorization algorithm for the computation of reciprocals of many-place sexagesimal numbers"

BagM 30, 139-161, 2 pl.

—— (2000) "Nachtrag. Korrigendum zum Friberg in *BagM* 30, 1999" *BagM* 31.

—— (2000) "Mathematics at Ur in the Old Babylonian period" *RA* 94, 97-188.

—— (2005) "Nos. 72-77. The learned tradition: Mathematical texts." In *CTMMA 2 (*Ira Spar and Wilfred George Lambert, *eds.*), New York, 288-314.

Gingerich, Owen (1965) "Eleven-digit regular sexagesimals and their reciprocals" *TAPS* 55(8), 3-38.

Høyrup, J. (2002) *Lengths, Widths, Surfaces. A portrait of Old Babylonian algebra and its kin*. New York, Berlin, *etc.*: Springer.

Nemet-Nejat, Karen R. (2002) "Square tablets in the Yale Babylonian Collection" In *Under One Sky. Astronomy and Mathematics in the Ancient Near East* (John M. Steele and Annette Imhausen, eds.), Münster, 253-281.

Neugebauer, Otto (1935 (reprinted 1973)) *Mathematische Keilschrift-Texte I*, Berlin: Springer.

—— and A. Sachs (1945) *Mathematical Cuneiform Texts* (AOS 29) New Haven: American Oriental Society.

Nissen, Hans J., Peter Damerow, and Robert K. Englund (1993) *Archaic Bookkeeping, Early Writing and Techniques of Economic Administration in the Ancient Near East*, Chicago, London: The University of Chicago Press.

Robson, Eleanor (2002) "More than metrology: Mathematics education in an Old Babylonian scribal school" In *Under One Sky. Astronomy and Mathematics in the Ancient Near East* (John M. Steele and Annette Imhausen, eds.), Münster, 325-365.

Sachs, Abraham J. (1947) "Babylonian mathematical texts I: Reciprocals of regular sexagesimal numbers" *JCS* 1, 219-240.

von Weiher, Egbert (1998) *Uruk, Spätbabylonische Texte aus dem Planquadrat U 18* (SpTU 5), Mainz.: Deutsches Archäologisches Institut Abteilung Baghdad.

2.   Old Babylonian Arithmetical Table Texts

Aaboe, Asger (1968-1969) "Two atypical multiplication tables from Uruk" *JCS* 22, 88-91.

Edzard, Dietz Otto, (1969) "Eine altsumerische Rechentafel" *Festschrift von Soden* (AOAT 1), Kevelaer and Neukirchen-Vluyn.

Friberg, Jöran (1990) "Mathematik" (in English) in *RlA 7* (Dietz Otto Edzard, ed.), Berlin, New York, 531-585.

—— (1993) "On the structure of cuneiform metrological table texts from the -1st millennium" *GMS* 3, 383-405.

—— (1996) "Pyramids and cones in ancient mathematical texts. New hints of a common tradition" *PCHM* 6, 80-95

—— (1997) "'Seed and Reeds Continued'. Another metro-mathematical topic text from Late Babylonian Uruk" *BagM* 28, 251-365, pl. 45-46.

—— (2000) "Mathematics at Ur in the Old Babylonian period" *RA* 94, 97-188.

—— (2005) *Unexpected Links Between Egyptian and Babylonian Mathematics*, Singapore, *etc.*: World Scientific.

Hilprecht, Hermann Vollrat (1906) *Mathematical, Metrological and Chronological Tablets from the Temple Library in Nippur* (BE 20/1), Philadelphia.

Høyrup, Jens (1993) "'Remarkable numbers' in Old Babylonian mathematical texts: A note on the psychology of numbers" *JNES* 52, 281-286.

—— (2002) *Lengths, Widths, Surfaces. A portrait of Old Babylonian algebra and its kin*. New York, Berlin, *etc.*: Springer.

Neugebauer, Otto (1935-37 (reprinted 1973)) *Mathematische Keilschrift-Texte I-III*, Berlin: Springer.

—— and A. Sachs (1945) *Mathematical Cuneiform Texts* (AOS 29) New Haven: American Oriental Society.

Oelsner, Joachim (1999) "Zu den mathematischen Keilschrifttexten aus Mesopotamien" *OLZ* 94, 5-16.

―――― (2001) "HS 201 – Eine Reziprokentabelle der Ur III-Zeit" in *Changing Views on Ancient Near Eastern Mathematics* (Jens Høyrup and Peter Damerow, eds.) Berlin, 53-59

Parker, Richard A. (1972) *Demotic Mathematical Papyri*, Providence, R. I., London.

Proust, Christine (2004) *Tablettes mathématiques de Nippur (Mésopotamie, début du deuxième millénaire avant notre ère). 1. Reconstitution du cursus scolaire. 2. Édition des tablettes d'Istanbul*. Dissertation, Université Paris 7: Histoire des sciences.

Robson, Eleanor (1999) *Mesopotamian mathematics 2100-1600 BC: Technical Constants in Bureaucracy and Education*. Oxford: Clarendon Press.

―――― (2003/2004). Review of J. Høyrup and P. Damerow (eds.), Changing Views on Ancient Near Eastern Mathematics (BBVO 19). *AfO* 50, 356-362.

Scheil, Vincent (1915) "Les tables 1 igi x gal-bi, *etc.*" *RA* 12, 195-198.

3.   Old Babylonian Metrological Table Texts

Friberg, Jöran (1990) "Mathematik" (in English) in *RlA 7* (Dietz Otto Edzard, ed.), Berlin, New York, 531-585.

―――― (1993) "On the structure of cuneiform metrological table texts from the -1st millennium" *GMS* 3, 383-405.

―――― (1997) "'Seed and Reeds Continued'. Another metro-mathematical topic text from Late Babylonian Uruk" *BagM* 28, 251-365, pl. 45-46.

―――― (2000) "Mathematics at Ur in the Old Babylonian period" *RA* 94, 97-188.

Greengus, Samuel (1979) *Old Babylonian Tablets from Ishchali and Vicinity* (PIHANS 44), Istanbul.

Gurney, O. R. (1974) *Middle Babylonian Legal Documents and Other Texts* (UET 7), London.

Hilprecht, Hermann Vollrat (1906) *Mathematical, Metrological and Chronological Tablets from the Temple Library in Nippur* (BE 20/1), Philadelphia.

Meissner, Bruno (1893) *Beiträge zum altbabylonischen Privatrecht* (AB 11), Leipzig.

Nemet-Nejat, Karen R. (1995) "Systems for learning mathematics in Mesopotamian scribal schools" *JNES* 54, 241-260.

Neugebauer, Otto and A. Sachs (1984) "Mathematical and metrological texts" *JCS* 36, 243-251.

Powell, M. A., Jr. (1990) "Masse und Gewichte" in *RlA 7* (Dietz Otto Edzard, ed.), 457-530.

Robson, Eleanor (2004) "Mathematical cuneiform texts in the Ashmolean Museum, Oxford." *SCIAMVS* 5, 3-66.

van der Meer, Petrus E. (1938) *Syllabaries A, B[1] and B with miscellaneous lexicographical texts from the Herbert Weld Collection* (OECT 4), Oxford/London.

4.   Mesopotamian Weight Stones

Belaiew, N. T. (1929) "Au sujet de la valeur probable de la mine sumérienne" *RA* 26, 115-127.

Brinkman, John (1968) *A Political History of Post-Kassite Babylonia 1158-722 B.C.* (AnOr 43), Rome.

Friberg, Jöran (1997) "'Seed and Reeds Continued'. Another metro-mathematical topic text from Late Babylonian Uruk" *BagM* 28, 251-365, pl. 45-46.

—— (2000) "Mathematics at Ur in the Old Babylonian period" *RA* 94, 97-188.

Landsberger, Benno, Erica Reiner, Miguel Civil (1970) *The series ḪAR-ra = ḫubullu. Tablets XVI, XVII, XIX, and Related Texts* (MSL 10) Rome: Biblical Institute Press.

Melville, Duncan (2002) "Weighing stones in ancient Mesopotamia" *HM* 29, 1-12.

Neugebauer, Otto and A. Sachs (1945) *Mathematical Cuneiform Texts* (AOS 29) New Haven: American Oriental Society.

Picchioni, Sergio A. (1980) "La direzione della scrittura cuneiforme e gli archivi di Tell Mardikh Ebla" *OrNS* 49, 225-251.

Powell, M. A., Jr. (1977) *Sumerian Numeration and Metrology* (Dissertation, University of Minnesota 1971, on microfilm) Ann Arbor, MI.

—— (1979) "Ancient Mesopotamian weight metrology: Methods, problems and perspectives" In *AOAT 203* (Powell, Marvin and R. Sack, eds.) Neukirchen-Vluyn, 71-109.

Veldhuis, Niek (1997) *Elementary Education at Nippur. The Lists of Trees and Wooden Objects*, Groningen.

5.   Neo-Sumerian Field Plan Texts (Ur III)

Allotte de la Fuÿe, François-Maurice (1915) "Un cadastre de Djokha" *RA* 12, 47-54.

Deimel, Anton (1923) *Schultexte aus Fara* (WVDOG 43), Leipzig.

—— (1930) "Miszellen, 26" *Or* 5 ed. 2, 60-62.

Dunham, Sally (1986) "Sumerian words for foundation. Part I: Temen" *RA* 80, 31-64.

Nissen, Hans J., Peter Damerow, and Robert K. Englund (1993) *Archaic Bookkeeping, Early Writing and Techniques of Economic Administration in the Ancient Near East*, Chicago, London: The University of Chicago Press.

Quillien, Jacques (2003) "Deux cadastres de l'époque d'Ur III" *RHM* 9, 9-31.

Schneider, N. (1930) "Die Geschäftsurkunden aus Drehem und Djoha in den Staatlichen Museen (VAT) zu Berlin" *Or* 47-49, pl. 127.

Stephens, Ferris J. (1953) "A surveyor's map of a field" *JCS* 7, 1-4.

Thureau-Dangin, François (1897) "Un cadastre chaldéen" *RA* 4, 13-27.

—— (1903) *Recueil de tablettes chaldéennes*, Paris.

6. An Old Sumerian Metro-Mathematical Table Text (Early Dynastic III)

Deimel, Anton (1923) *Schultexte aus Fara* (WVDOG 43), Leipzig.

Friberg, Jöran (1997) "'Seed and Reeds Continued'. Another metro-mathematical topic text from Late Babylonian Uruk" *BagM* 28, 251-365, pl. 45-46.

—— (1997/98) "Round and almost round numbers in proto-literate metro-mathematical field texts" . *Archiv für Orientforschung* 44/45, 1-58.

Jestin, Raymond R. (1937) *Tablettes sumériennes de Shuruppak*, Paris.

Nissen, Hans J., Peter Damerow, and Robert K. Englund (1993) *Archaic Bookkeeping, Early Writing and Techniques of Economic Administration in the Ancient Near East*, Chicago, London: The University of Chicago Press.

Pettinato, G. (1981) *MEE 3: Testi lessicali monolingui della biblioteca L. 2769*. Naples.

Picchioni, Sergio A. (1980) "La direzione della scrittura cuneiforme e gli archivi di Tell Mardikh Ebla" *OrNS* 49, 225-251.

7.   Old Babylonian Hand Tablets with Practical Mathematics

Al-Rawi, F. N. H. and M. Roaf (1984) "Ten Old Babylonian mathematical problems from Tell Haddad, Himrin." *Sumer* 43, 195-218.

Friberg, Jöran (1990) "Mathematik" (in English) in *RlA 7* (Dietz Otto Edzard, ed.), Berlin, New York, 531-585.

—— (1999/2000) "Review of E. Robson, Mesopotamian Mathematics 2100-1600 BC." *AfO* 46/47, 309-317.

—— (2000) "Mathematics at Ur in the Old Babylonian period" *RA* 94, 97-188.

—— (2001) "Bricks and mud in metro-mathematical cuneiform texts" in *Changing Views on Ancient Near Eastern Mathematics* (Jens Høyrup and Peter Damerow, eds.) Berlin, 61-154.

Nemet-Nejat, Karen R. (1995) "Systems for learning mathematics in Mesopotamian scribal schools" *JNES* 54, 241-260.

—— (2002) Square tablets in the Yale Babylonian Collection. In *Under One Sky. Astronomy and Mathematics in the Ancient Near East* (John M. Steele and Annette Imhausen, eds.), Münster, 253-281.

Neugebauer, Otto and A. Sachs (1945) *Mathematical Cuneiform Texts* (AOS 29) New Haven: American Oriental Society.

Robson, Eleanor (1999) *Mesopotamian mathematics 2100-1600 BC: Technical Constants in Bureaucracy and Education*. Oxford: Clarendon Press.

—— (2000) "Mathematical cuneiform texts in Philadelphia, Part I: Problems and calculations." *SCIAMVS* 1: 11-48.

Sachs, Abraham J. (1952) "Babylonian mathematical texts II: Approximations of reciprocals of irregular numbers in an Old Babylonian text. III: The problem of finding the cube root of a number" *JCS* 6, 151-156.

8.   Old Babylonian Hand Tablets with Geometric Exercises

Baqir, Taha (1962) "Tell Dhiba'i: New mathematical texts" *Sumer* 18, 11-14, pl. 1-3.

Bruins, Evert M. (1953) "Revision of the mathematical texts from Tell Harmal" *Sumer* 9, 241-253.

—— (1962) "Interpretation of cuneiform mathematics" *Physis* 4, 277-340.

—— (1984) "Requisites for the interpretation of ancient mathematics." *Janus* 71, 107-34.

—— and Marguerite Rutten (1961) *Textes mathématiques de Suse* (MDP 34), Paris.

Friberg, Jöran (1990) "Mathematik" (in English) in *RlA 7* (Dietz Otto Edzard, ed.), Berlin, 531-585.

—— , Hermann Hunger, and Farouk N. Al-Rawi (1990) "'Seed and Reeds', a metro-mathematical topic text from Late Babylonian Uruk" *BagM* 21, 483-557, pl. 46-48.

—— (1997) "'Seed and Reeds Continued'. Another metro-mathematical topic text from Late Babylonian Uruk" *BagM* 28, 251-365, pl. 45-46.

—— (1997/98) "Round and almost round numbers in proto-literate metro-mathematical field texts" . *AfO* 44/45, 1-58.

—— (2000) "Mathematics at Ur in the Old Babylonian period" *RA* 94, 97-188.

—— (2004) "Nos. 72-77. The learned tradition: Mathematical texts." In *Cuneiform Texts from the Metropolitan Museum of Art, 2* (I. Spar and W. G. Lambert, *eds.*) New York, NY: 288-314.

—— (2005) *Unexpected Links Between Egyptian and Babylonian Mathematics*, Singapore, *etc.*

Høyrup, Jens (2002) *Lengths, Widths, Surfaces. A portrait of Old Babylonian algebra and its kin*. New York, Berlin, *etc*.: Springer.

Kern, Hermann (1982) *Labyrinthe: Erscheinungsformen und Deutungen: 5000 Jahre Gegenwart eines Urbilds*, München.

—— (2000) *Through the Labyrinth: Designs and Meaning Over 5,000 Years*, Prestel, USA.

Muroi, Kazuo (1992) "Reexamination of Susa mathematical text no. 3: Alleged value π ≈ 3 1/8" *HSJ* 2, 45-49.

—— (2000) "Quadratic equations in the Susa mathematical text no. 21" *Sciamvs* 1, 3-10.

Nemet-Nejat, Karen R. (2002) "Square tablets in the Yale Babylonian Collection" In *Under One Sky. Astronomy and Mathematics in the Ancient Near East* (John M. Steele and Annette Imhausen, eds.), Münster, 253-281.

Neugebauer, Otto and A. Sachs (1945) *Mathematical Cuneiform Texts* (AOS 29) New Haven: American Oriental Society.

Parker, Richard A. (1972) *Demotic Mathematical Papyri*, Providence, R. I., London.

Robson, Eleanor (1999) *Mesopotamian mathematics 2100-1600 BC: Technical Constants in Bureaucracy and Education*. Oxford: Clarendon Press.

Vaiman, Aisak A. (1963) "Istolkovanie geometriceskih postoyannyh iz suzskogo klinopisnogo spiska I (Suzy) (Interpretation of geometric constants in the cuneiform tablet I from Susa)" *VDI* 1963, 75-86.

9.   The Beginning and the End of the *Sumerian King List*

Black, Jeremy A., Graham G. Cunningham, Esther Flückiger-Hawker, Eleanor Robson, and Gabor Zólyomi (1998-) *The Electronic Text Corpus of Sumerian Literature*. <http://www-etcsl.orient.ox.ac.uk>

Finkelstein, J. J. (1963) "The antediluvian kings: A University of California tablet" *JCS* 17, 39-51.

Friberg, Jöran (1992) "Numbers and counting in the Ancient Near East" in *The Anchor Bible Dictionary* (D. N. Freedman, ed.) New York, *etc*., 1139-1146.

—— (1999) "Proto-literate counting and accounting in the Middle East. Examples from two new volumes of proto-cuneiform texts." *JCS* 51, 107-137.

—— (2000) "Mathematics at Ur in the Old Babylonian period" *RA* 94, 97-188.

Gurzadyan, V. G. (2000) "Astronomy and the fall of Babylon" *Sky & Telescope* 100(1), 40-45. http://arxiv.org/abs./physics/0311114

Hilprecht, Hermann Vollrat (1906) *Mathematical, Metrological and Chronological Tablets from the Temple Library in Nippur* (BE 20/1), Philadelphia.

Jacobsen, Thorkild (1939) *The Sumerian King List* (AS 11) Chicago.

Lambert, W. G. (1973) "A new fragment from a list of antediluvian kings and Marduk's chariot" in M. A. Beek, et al. (eds) *Symbolae biblicae et mesopotamicae Francisco Mario Theodoro De Liaghre Böhl dedicatae*. Leiden, Brill.

Langdon, S. H. (1923) *The Weld-Blundell collection, 2. Historical Inscriptions, Containing Principally the chronological Prism W-B. 444* (OECT 2), Oxford.

Manning, W., B. Kromer, et al. (2001) "Anatolian tree rings and a new chronology for the East Mediterranean bronze-iron ages" *Science* 294, 2532-2535. <http://www.sciencemag.org/cgi/content/full/294/5551/2532>

Oelsner, Joachim (2001) "HS 201 – Eine Reziprokentabelle der Ur III-Zeit" in *Changing Views on Ancient Near Eastern Mathematics* (Jens Høyrup and Peter Damerow, eds.) Berlin, 53-59

Poebel, Arno (1914) *Historical and Grammatical Texts* (PBS 5), Philadelphia.

Schnabel, Paul (1923) *Berossos und die babylonisch-hellenistische Literatur*, Leipzig/Berlin.

Sigrist, Marcel and Peter Damerow (2001) *Mesopotamian Year Names. Neo-Sumerian and Old Babylonian Date Formulae.* <http://www.cdli.ucla.edu/dl/yearnames/yn_index.htm>

Sollberger, Edmond (1954) "New lists of the kings of Ur and Isin" *JCS* 8, 135-136.

Steiner, Gerd (1988) "Der 'reale Kern' in den 'legendären' Zahlen von Regierungsjahren der ältesten Herrscher Mesopotamiens." *ASJ* 10, 129-152.

Steinkeller, Piotr (2003) "An Ur III manuscript of the Sumerian King List" *in* Walther Sallaberger, K. Volk and A. Zgoll (eds.) *Literatur, Politik und Recht in Mesopotamien* (FS Wilcke), Wiesbaden, 267-292.

Young, Dwight W. (1988) "A mathematical approach to certain dynastic spans in the Sumerian King List" *JNES* 47, 123-129.

10.   Three Old Babylonian Mathematical Problem Texts from Uruk

Baqir, Taha (1951) "Some more mathematical texts" *Sumer* 7, 28-45 & plates.

Civil, Miguel, M. W. Green, W. G. Lambert (1979) *Ea A = nâqu, Aa A = nâqu, with their Forerunners and Related Texts* (MSL XIV), Rome.

Clay, Albert Tobias (1923) *Epics, hymns, omens, and other texts* (BRM 4), New Haven.

Friberg, Jöran (1990) "Mathematik" (in English) in *RlA 7* (Dietz Otto Edzard, ed.), Berlin, 531-585.

—— (1997) "'Seed and Reeds Continued'. Another metro-mathematical topic text from Late Babylonian Uruk" *BaM* 28, 251-365, pl. 45-46.

—— (2000) "Mathematics at Ur in the Old Babylonian period" *RA* 94, 97-188.

—— (2001) "Bricks and mud in metro-mathematical cuneiform texts" in *Changing Views on Ancient Near Eastern Mathematics* (Jens Høyrup and Peter Damerow, eds.) Berlin, 61-154.

Gandz, Solomon (1948) "Studies in Babylonian mathematics 1: Indeterminate analysis in Babylonian mathematics." *Osiris* 8, 12-40.

Goetze, Albrecht (1951) "A mathematical compendium from Tell Harmal" *Sumer* 7, 126-155.

—— (2002) *Lengths, Widths, Surfaces. A portrait of Old Babylonian algebra and its kin.* New York, Berlin, *etc.*: Springer.

Heimpel, Wolfgang (2004) "AO 7667 and the meaning of ba-an-gi4" *CDLJ* 2004:1. <http://cdli.ucla.edu/pubs/cdlj/2004/cdlj2004_001.html>

Huber, Peter (1955) "Zu einem mathematischen Keilschrifttext." *Isis* 46, 104-106.

Muroi, K. (1992) "Wall and dike construction problems in the Babylonian mathematical text YBC 4673" *SBM* 2, 1-12.

Neugebauer, Otto (1935-37 (reprinted 1973)) *Mathematische Keilschrift-Texte I-III*, Berlin: Springer.

—— and A. Sachs (1945) *Mathematical Cuneiform Texts* (AOS 29) New Haven: American Oriental Society.

Robson, Eleanor (1999) *Mesopotamian mathematics 2100-1600 BC: Technical Constants in Bureaucracy and Education.* Oxford: Clarendon Press.

—— (2001) "Neither Sherlock Holmes nor Babylon: A reassessment of Plimpton 322" *HM* 28, 167-206.

Thureau-Dangin, François (1938) *Textes mathématiques babyloniens*, Leiden: E. J. Brill.

11.   Three Problem Texts Not Belonging to Any Known Group of Texts

Arnaud, Daniel (1994) *Texte aus Larsa. Die epigraphischen Funde der 1. Kampagne In Senkereh-Larsa 1933*, Berlin.

Baqir, Taha (1951) "Some more mathematical texts" *Sumer* 7, 28-45 & plates.

Bruins, Evert M. and Marguerite Rutten (1961) *Textes mathématiques de Suse* (MDP 34), Paris.

——— (1981) "Methods and traditions of Babylonian mathematics, 2. An Old Babylonian catalogue text with equations for squares and circles" *JCS* 33, 57-64.

Friberg, Jöran (1982) *A Survey of Publications on Sumero-Akkadian Mathematics, Metrology and Related Matters (1854–1982)*. Dep-Math CTH–GU (Gothenburg) 1982-17.

——— (1990) "Mathematik" (in English) in *Reallexikon der Assyriologie und Vorderasiatischen Archäologie 7* (Dietz Otto Edzard, ed.), Berlin, New York, 531-585.

———, Hermann Hunger, and Farouk N. Al-Rawi (1990) "'Seed and Reeds', a metro-mathematical topic text from Late Babylonian Uruk" *BagM* 21, 483-557, pl. 46-48.

——— (1993) "On the structure of cuneiform metrological table texts from the -1st millennium" *GMS* 3, 383-405.

——— (1996) "Pyramids and cones in ancient mathematical texts. New hints of a common tradition" *PCHM* 6, 80-95

——— (1997) "'Seed and Reeds Continued'. Another metro-mathematical topic text from Late Babylonian Uruk" *BagM* 28, 251-365, pl. 45-46.

——— (1999) "Proto-literate counting and accounting in the Middle East. Examples from two new volumes of proto-cuneiform texts." *JCS* 51, 107-137.

——— (2000) "Mathematics at Ur in the Old Babylonian period" *RA* 94, 97-188.

——— (2005) *Unexpected Links Between Egyptian and Babylonian Mathematics*, Singapore, *etc.*

Goetze, Albrecht (1951) "A mathematical compendium from Tell Harmal" *Sumer* 7, 126-155.

Greengus, S (1979) *Old Babylonian Tablets from Ishchali and vicinity* (PIHANS 44) Istanbul.

Gundlach, Karl-Bernhard and Wolfram von Soden (1963) "Einige altbabylonische Texte zur Lösung "quadratischer Gleichungen"" *AmSUH* 26, 248-263.

Heath, T. L. (1926; Dover edition 1956) *Euclid. The Thirteen Books of The Elements, I-III*. New York: Dover Publications.

Høyrup, Jens (2002) *Lengths, Widths, Surfaces. A portrait of Old Babylonian algebra and its kin*. New York, Berlin, *etc.*: Springer.

Kilmer, Anne Draffkorn (1960) "Two new lists of key numbers for mathematical operations." *OrNS* 29, 273-308.

King, Leonard William (1900) *CT 9*, London.

Leemans, W. F. and Evert M. Bruins (1951) "Un texte vieux-babylonien concernant des cercles concentriques" *RAI* 2, Paris, 31-35.

Muroi, Kazuo (1989) "Extraction of cube sides in Babylonian mathematics" *Centaurus* 31, 181-188.

——— (1998) "Expressions of a unit in Babylonian mathematics" *ASJ* 20, 121-125.

——— (2000) "Quadratic equations in the Susa mathematical text no. 21" *SCIAMVS* 1, 3-10.

——— (2001) "Inheritance problems in the Susa mathematical text no. 26" *HSJ* 10, 226-234.

——— (2003) "Mathematical term *takīltum* and completing the square in Babylonian mathematics" *HSJ* 12, 254-263.

Neugebauer, Otto (1935-37 (reprinted 1973)) *Mathematische Keilschrift-Texte I-III*, Berlin: Springer.

——— and A. Sachs (1945) *Mathematical Cuneiform Texts* (AOS 29) New Haven: American Oriental Society.

Proust, Christine (2004) *Tablettes mathématiques de Nippur (Mésopotamie, début du deuxième millénaire avant notre ère). 1. Reconstitution du cursus scolaire. 2. Édition des tablettes d'Istanbul*. Dissertation, Université Paris 7: Histoire des sciences.

Robson, Eleanor (1997) "Three Old Babylonian methods for dealing with 'Pythagorean' triangles" *JCS* 49, 51-72.

⸻ (1999) *Mesopotamian mathematics 2100-1600 BC: Technical Constants in Bureaucracy and Education*. Oxford: Clarendon Press.

⸻ (2001) "Neither Sherlock Holmes nor Babylon: A reassessment of Plimpton 322" *HM* 28, 167-206.

Sachs, Abraham J. (1952) "Babylonian mathematical texts II: Approximations of reciprocals of irregular numbers in an Old Babylonian text. III: The problem of finding the cube root of a number" *JCS* 6, 151-156.

App. 1. Subtractive Notations for Numbers in Mathematical Cuneiform Texts

Delaporte, Louis (1911) "Document mathématique de l'époque des rois d'Our" *RA* 8, 131-133.

Friberg, Jöran (1999) "Proto-literate counting and accounting in the Middle East. Examples from two new volumes of proto-cuneiform texts." *JCS* 51, 107-137.

Hilprecht, Hermann Vollrat (1906) *Mathematical, Metrological and Chronological Tablets from the Temple Library in Nippur* (BE 20/1), Philadelphia.

Oelsner, Joachim (2001) "HS 201 – Eine Reziprokentabelle der Ur III-Zeit" in *Changing Views on Ancient Near Eastern Mathematics* (Jens Høyrup and Peter Damerow, eds.) Berlin, 53-59.

App. 3. An Old Babylonian Combined Arithmetical Algorithm

Friberg, Jöran (1982) *A Survey of Publications on Sumero-Akkadian Mathematics, Metrology and Related Matters (1854–1982)* (Department of Mathematics CTH–GU, (Gothenburg) 1982-17).

⸻ (1990) "Mathematik" (in English) in *Reallexikon der Assyriologie und Vorderasiatischen Archäologie 7* (Dietz Otto Edzard, ed.), Berlin, New York, 531-585.

Hilprecht, Hermann Vollrat (1906) *Mathematical, Metrological and Chronological Tablets from the Temple Library in Nippur* (BE 20/1), Philadelphia.

Muroi, Kazuo (1989) "Extraction of cube sides in Babylonian mathematics" *Centaurus* 31, 181-188.

Neugebauer, Otto and A. Sachs (1945) *Mathematical Cuneiform Texts* (AOS 29) New Haven: American Oriental Society.

Sachs, Abraham J. (1947) "Babylonian mathematical texts I: Reciprocals of regular sexagesimal numbers" *JCS* 1, 219-240.

Scheil, Vincent (1916) "Le texte mathématique 10201 du Musée de Philadelphie" *RA* 13, 138-142.

App. 4. Cuneiform Systems of Notations for Numbers and Measures

Amiet, Pierre (1986) *L'age des échanges inter-iraniens*, Paris.

⸻ (1986) "Approche physique de la comptabilité à l'époque d'Uruk. Les bulles-enveloppes de Suse" *Éditions du CNRS*, Paris, 331-334.

Damerow, Peter and Robert K. Englund (1989) *The Proto-Elamite Texts from Tepe Yahya*, Cambridge, MA.

Damerow, Peter and Hans-Peter Meinzer (1995) "Computertomografische Untersuchung ungeöffneter archaischer Tonkugeln aus

Uruk: W 20987,9, W 20987,11 und W 20987,12" *BagM* 26, 7-33, pl.1-4.

Drilhon, F., M. Laval-Jeantet, and A. Lahmi (1986) "Étude en laboratoire de seize bulles Mésopotamiennes appartenant au Département des Antiquités Orientales" *Éditions du CNRS*, Paris, 335-344.

Englund, Robert K. (1988) "Administrative timekeeping in Ancient Mesopotamia" *JESHO* 31, 121-185.

Friberg, Jöran (1978) "Early Roots of Babylonian Mathematics 1. A method for the decipherment ... of proto-Sumerian and proto-Elamite semi-pictographic inscriptions" *Department of Mathematics CTH–GU 1978-9*, Gothenburg.

—— (1979) "Early Roots of Babylonian Mathematics 2, Metrological relations in a group of semi-pictographic tablets of the Jemdet-Nasr type, probably from Uruk-Warka. *Department of Mathematics CTH–GU 1979-15*, Gothenburg.

—— (1990) "Mathematik" (in English) in *RlA 7* (Dietz Otto Edzard, ed.), Berlin, New York, 531-585.

—— (1993) "On the structure of cuneiform metrological table texts from the -1st millennium" *GMS* 3, 383-405.

—— (1994) "Preliterate counting and accounting in the Middle East. A constructively critical review of Schmandt-Besserat's Before Writing" *OLZ* 89, 477-502.

—— (1997/98) "Round and almost round numbers in proto-literate metro-mathematical field texts" . *AfO* 44/45, 1-58.

—— (1999) "Proto-literate counting and accounting in the Middle East. Examples from two new volumes of proto-cuneiform texts." *JCS* 51, 107-137.

—— (2001) "Bricks and mud in metro-mathematical cuneiform texts" in *Changing Views on Ancient Near Eastern Mathematics* (Jens Høyrup and Peter Damerow, eds.) Berlin, 61-154.

—— (2001) "I primi sistemi di notazione numerica" (in English) In *Storia della scienza, I. La scienza antica*. Rome, 320-326.

—— (2005) "On the alleged counting with sexagesimal place value numbers." *CDLJ* 2005:2. <http://cdli.ucla.edu/pubs/cdlj/2005/cdlj2005_002.html>

Nissen, Hans J., Peter Damerow, and Robert K. Englund (1993) *Archaic Bookkeeping, Early Writing and Techniques of Economic Administration in the Ancient Near East*, Chicago, London: The University of Chicago Press.

Schmandt-Besserat, Denise (1992) *Before Writing, Vol. 1: From Counting to Cuneiform*, Austin, TX.

App. 5.   Old Babylonian Complete Metrological Tables

Arnaud, Daniel (1994) *Texte aus Larsa. Die epigraphischen Funde der 1. Kampagne In Senkereh-Larsa 1933*, Berlin.

Bruins, Evert M. (1953) "Revision of the mathematical texts from Tell Harmal" *Sumer* 9, 241-253.

Clay, A. T. (1923) *Epics, hymns, omens. and other texts* (BRM 4) New Haven.

Friberg, Jöran (1990) "Mathematik" (in English) in *RlA 7* (Dietz Otto Edzard, ed.), Berlin, New York, 531-585.

Greengus, Samuel (1979) *Old Babylonian Tablets from Ishchali and Vicinity* (PIHANS 44), Istanbul.

Gurney, O. R. (1974) *Middle Babylonian Legal Documents and Other Texts* (UET 7), London.

Hilprecht, Hermann Vollrat (1906) *Mathematical, Metrological and Chronological Tablets from the Temple Library in Nippur* (BE 20/1), Philadelphia.

Høyrup, Jens (2002) *Lengths, Widths, Surfaces. A portrait of Old Babylonian algebra and its kin*. New York, Berlin, *etc.*: Springer.

Nemet-Nejat, Karen R. (1995) "Systems for learning mathematics in Mesopotamian scribal schools" *JNES* 54, 241-260.

Neugebauer, Otto (1935-37 (reprinted 1973)) *Mathematische Keilschrift-Texte I-III*, Berlin: Springer.

—— and A. Sachs (1945) *Mathematical Cuneiform Texts* (AOS 29) New Haven: American Oriental Society.

Nies, James B. and Clarence E. Keiser (1920) *Historical, Religious and Economic Texts and Antiquities* (BIN 2) New Haven, CT.

Nougayrol, Jean, et al. (1968) *Ugaritica V*, Paris.

Proust, Christine (2004) *Tablettes mathématiques de Nippur (Mésopotamie, début du deuxième millénaire avant notre ère). 1. Reconstitution du cursus scolaire. 2. Édition des tablettes d'Istanbul.* Dissertation, Université Paris 7: Histoire des sciences.

—— (2005) A propos d'un prisme du Louvre: aspects de l'enseignement des mathématiques en Mésopotamie. *SCIAMVS* 6, 3-32.

Schroeder, Otto (1920) *Keilschrifttexte aus Assur verschiedenen Inhalts*, Leipzig.

Thureau-Dangin, François (1930) "La graphie du système sexagésimal" *RA* 27, 73-78.

van der Meer, Petrus E. (1938) *Syllabaries A, B¹ and B with miscellaneous lexicographical texts from the Herbert Weld Collection* (OECT 4), Oxford/London.

Veldhuis, Niek (1997) *Elementary Education at Nippur. The Lists of Trees and Wooden Objects*, Groningen.

App. 6.   Metro-Mathematical Cuneiform Texts from the Third Millennium BC

Andrae, W. (1937) *Berichte aus den Preußischen Kunstsammlungen* 58, 34-35.

Archi, Alfonso (1989) "Tables de comptes eblaïtes" *RA* 83, 1–6.

Chambon, Grégory (2002) "Trois documents pédagogiques de Mari" *FlMar* 6, 497-503.

Deimel, Anton (1923) *Schultexte aus Fara* (WVDOG 43), Leipzig.

Englund, Robert K. and Jean-Pierre Grégoire (1991) *The Proto-Cuneiform Texts from Jemdet Nasr* (MSVO 1), Berlin.

Friberg, Jöran (1986) "The early roots of Babylonian mathematics 3: Three remarkable texts from ancient Ebla" *VO* 6, 3–25.

—— (1990) "Mathematik" (in English) in *Reallexikon der Assyriologie und Vorderasiatischen Archäologie 7* (Dietz Otto Edzard, ed.), Berlin, New York, 531-585.

—— (1993) "On the structure of cuneiform metrological table texts from the -1st millennium" *GMS* 3, 383-405.

—— (2005) "On the alleged counting with sexagesimal place value numbers." *CDLJ* 2005:2. <http://cdli.ucla.edu/pubs/cdlj/2005/cdlj2005_002.html>

Gelb, Ignace Jay (1970) *Sargonic Texts in the Ashmolean Museum* (MAD 5), Oxford.

Høyrup, Jens (1982) "Investigations of an Early Sumerian division problem" *HM* 9, 19–36.

Jestin, Raymond R. (1937) *Tablettes sumériennes de Shuruppak*, Paris.

Legrain, Léon (1936) *Archaic Seal-Impressions* (UE 3), London/Philadelphia.

Limet, Henri (1973) *Étude de documents de la période d'Agadé appartenant à l'Université de Liège*, Paris.

Luckenbill, Daniel D. (1930) *Inscriptions from Adab* (OIP 14), Chicago.

Melville, Duncan (2002) "Ration computations at Fara: Multiplication or repeated addition?" In *Under One Sky. Astronomy and Mathematics in the Ancient Near East* (John M. Steele and Annette Imhausen, eds.), Münster, 237-252.

Moortgat, A. (1949) *Tammuz. Der Unsterblichkeitsglaube in der Altorientalischen Bildkunst.* Berlin.

Nissen, Hans J., Peter Damerow, and Robert K. Englund (1993) *Archaic Bookkeeping, Early Writing and Techniques of Economic Administration in the Ancient Near East*, Chicago, London: The University of Chicago Press.

Pettinato, Giovanni (1981) *Testi lessicali monolingui della biblioteca L. 2769* (MEE 3), Naples.

—— (1982) *Testi lessicali monolingui della biblioteca L. 2769* (MEE 4), Naples.

Pohl, Alfred (1935) *Vorsargonische und sargonische Wirtschaftstexte* (TMH 5), Leipzig.

Powell, M. A., Jr. (1972) "Sumerian area measures and the alleged decimal substratum." *ZA* 62, 165-221.

—— (1975) Review of Limet, Étude de documents de la période d'Agadé (1973) *JCS* 27, 180-188.

—— (1976) "The antecedents of Old Babylonian place notation and the early history of Babylonian mathematics" *HM* 3, 417–439.

—— (1976) "Two notes on metrological mathematics in the Sargonic period." *RA* 70, 97-102.

—— (1977) *Sumerian Numeration and Metrology* (Dissertation, University of Minnesota 1971, on microfilm) Ann Arbor, MI.

—— (1990) "Masse und Gewichte" in *RlA 7* (Dietz Otto Edzard, ed.), Berlin, New York, 457-530.

Proust, C. (2002) "Numération centésimale de position à Mari." *FlMar* 6, 513-516.

Soubeyran, Denis (1984) "Textes mathématiques de Mari" *RA* 78, 19-48.

Westenholz, Aage (1975) *Early Cuneiform Texts in Jena. PreSargonic and Sargonic Documents from Nippur and Fara*, København.

Whiting, R. M. (1984) "More evidence for sexagesimal calculations in the Third Millennium B. C." *ZA* 74: 59-66.

App. 7.   A Combined Metro-Mathematical Table Text with Areas of Large and Small Squares (ED IIIb)

Nissen, H. J., P. Damerow, and R. K. Englund (1993) *Archaic Bookkeping, Early Writing and Techniques of Economic Administration in the Ancient Near East*. Chicago/London: The University of Chicago Press.

Proust, Christine (2004) *Tablettes mathématiques de Nippur (Mésopotamie, début du deuxième millénaire avant notre ère). 1. Reconstitution du cursus scolaire. 2. Édition des tablettes d'Istanbul*. Dissertation, Université Paris 7: Histoire des sciences.

Scheil, V. (1938) "Tablettes susiennes. Exercises scolaires. Calcul des surfaces." *RA* 35, 92-103.

App. 8.   Plimpton 322, a Table of Parameters for igi–igi.bi Problems

Damerow, Peter (2001) "Kannten die Babylonier den Satz des Pythagoras? Epistemologische Anmerkungen zur Natur der babylonischen Mathematik" in *Changing Views on Ancient Near Eastern Mathematics* (Jens Høyrup and Peter Damerow, eds.), Berlin, 219-310.

Friberg, Jöran (1981) "Plimpton 322, Pythagorean triples, and the Babylonian triangle parameter equations" *HM* 8, 277-318.

—— , Hermann Hunger, and Farouk N. Al-Rawi (1990) "'Seed and Reeds', a metro-mathematical topic text from Late Babylonian Uruk" *BagM* 21, 483-557, pl. 46-48.

—— (1990) "Mathematik" (in English) in *RlA 7* (Dietz Otto Edzard, ed.), Berlin, New York, 531-585.

—— (1993) "On the structure of cuneiform metrological table texts from the -1st millennium" *GMS* 3, 383-405.

—— (1997) "'Seed and Reeds Continued'. Another metro-mathematical topic text from Late Babylonian Uruk" *BagM* 28, 251-365, pl. 45-46.

—— (1999) "A Late Babylonian factorization algorithm for the computation of reciprocals of many-place sexagesimal numbers" *BagM* 30, 139-161, 2 pl.

Høyrup, Jens (1998) "Pythagorean 'rule' and 'theorem'– mirror of the relation between Babylonian and Greek mathematics" *Roskilde University, Section for Philosophy and Science Studies* 1998(3), 1-15.

Neugebauer, Otto (1935-37 (reprinted 1973)) *Mathematische Keilschrift-Texte I-III*, Berlin: Springer.

———— and A. Sachs (1945) *Mathematical Cuneiform Texts* (AOS 29) New Haven: American Oriental Society.

———— (1955) *Astronomical Cuneiform Texts. Babylonian Epheremides of the Seleucid Period for the Motion of the Sun, the Moon, and the Planets.* New York, Heidelberg, Berlin, Springer-Verlag.

Robson, Eleanor (1999) *Mesopotamian mathematics 2100-1600 BC: Technical Constants in Bureaucracy and Education.* Oxford: Clarendon Press.

———— (2001) "Neither Sherlock Holmes nor Babylon: A reassessment of Plimpton 322" *HM* 28, 167-206.

———— (2002) "Words and pictures: New light on Plimpton 322." *AMM* 109, 105-120.

Vincente, Claudine-Adrienne (1995) "The Tell Leilān recension of the Sumerian King List" *ZA* 85, 234-270.

App. 9.   Many-Place Squares of Squares in Late Babylonian Mathematical Texts

Britton, John P. (1991-93) "A table of 4th powers and related texts from Seleucid Babylon" *JCS* 43-45, 71-87.

Friberg, Jöran (1999) "A Late Babylonian factorization algorithm for the computation of reciprocals of many-place sexagesimal numbers" *BagM* 30, 139-161, 2 pl.

———— (2000) "Nachtrag. Korrigendum zum Friberg in *BaM* 30, 1999" *BagM* 31.

Gingerich, Owen (1965) "Eleven-digit regular sexagesimals and their reciprocals" *Transactions of the American Philosophical Society* 55(8), 3-38.

Neugebauer, Otto (1935-37 (reprinted 1973)) *Mathematische Keilschrift-Texte I-III*, Berlin: Springer.

Sachs, Abraham J. (1955) *Late Babylonian Astronomical and Related Texts. Copied by T. G. Pinches and J. N. Strassmaier,* Providence.

von Weiher, Egbert (1998) *Uruk, Spätbabylonische Texte aus dem Planquadrat U 18* (SpTU 5), Mainz.: Deutsches Archäologisches Institut Abteilung Baghdad.

Sources and Studies in the

History of Mathematics and Physical Sciences

Continued from page ii

E. Kheirandish
The Arabic Version of Euclid's *Optics*, Volumes I and II

J. Lützen
Joseph Liouville 1809–1882: Master of Pure and Applied Mathematics

J. Lützen
The Prehistory of the Theory of Distributions

G.H. Moore
Zermelo's Axiom of Choice

O. Neugebauer
A History of Ancient Mathematical Astronomy

O. Neugebauer
Astronomical Cuneiform Texts

F.J. Ragep
Nasīr al-Dīn al-Tūsī's *Memoir on Astronomy*
(al-Tadhkira fi cilm al-hay'a)

B.A. Rosenfeld
A History of Non-Euclidean Astronomy

G. Schubring
Conflicts between Generalization, Rigor, and Intuition: Number Concepts Underlying the Development of Analysis in 17–19th Century France and Germany

J. Sesiano
Books IV to VII of Diophantus' *Arithemetica*: In the Aribic Translation Attributed to Qustä ibn Lüqä

L. Sigler
Fibonacci's Liber Abaci: A Translation into Modern English of Leonardo Pisano's Book of Calculation

B. Stephenson
Kepler's Physical Astronomy

N.M. Swerdlow/O. Neugebauer
Mathematical Astronomy In Copernicus' De Revolutionibus

G.J. Toomer (Ed.)
Apolonius Conics Books V to VII: The Arabic Translation of the Lost Greek Original in the Version of Banū Mūsā, Edited, with English Translation and Commentary by G.J. Toomer

G.J. Toomer (Ed.)
Diocles on Burning Mirrors: The Arabic Translation of the Lost Greek Original, Edited, with English Translation and Commentary by G.J. Toomer

C. Truesdell
The Tragicomical History of Thermodynamics, 1822–1854

I. Tweddle
James Stirling's Methodus Differentialis: An Annotated Translation of Stirling's Text

K. von Meyenn/A. Hermann/V.F. Weiskopf (Eds.)
Wolfgang Pauli: Scientific Correspondence II: 1930–1939

K. von Meyenn (Ed.)
Wolfgang Pauli: Scientific Correspondence 111: 1940–1949

K. von Meyenn (Ed.)
Wolfgang Pauli: Scientific Correspondence IV, Part 1: 1950–1952

K. von Meyenn (Ed.)
Wolfgang Pauli: Scientific Correspondence IV, Part 2: 1953–1954

J. Stedall
The Arithmetic of Infinitesimals: John Wallis 1656

Printed in Singapore